Nanotechnology Research Directions for Societal Needs in 2020

Science Policy Reports

The series Science Policy Reports presents the endorsed results of important studies in basic and applied areas of science and technology. They include, to give just a few examples: panel reports exploring the practical and economic feasibility of a new technology; R & D studies of development opportunities for particular materials, devices or other inventions; reports by responsible bodies on technology standardization in developing branches of industry.

Sponsored typically by large organizations – government agencies, watchdogs, funding bodies, standards institutes, international consortia – the studies selected for Science Policy Reports will disseminate carefully compiled information, detailed data and in-depth analysis to a wide audience. They will bring out implications of scientific discoveries and technologies in societal, cultural, environmental, political and/or commercial contexts and will enable interested parties to take advantage of new opportunities and exploit on-going development processes to the full.

For other titles published in this series, go to
http://www.springer.com/series/8882

Mihail C. Roco • Chad A. Mirkin
Mark C. Hersam

Nanotechnology Research Directions for Societal Needs in 2020

Retrospective and Outlook

Mihail C. Roco
National Science Foundation
4201 Wilson Boulevard
Arlington, VA 22230
USA

Chad A. Mirkin
Department of Chemistry
Northwestern University
2145 Sheridan Road
Evaston, IL 60208
USA

Mark C. Hersam
Department of Engineering
and Materials Science
Northwestern University
2220 Campus Drive
Evaston, IL 60208
USA

The cover picture suggests the integration of various nanotechnology-based solutions in the design of a blended hybrid-wing-body concept for future subsonic commercial aircraft. It represents a radical departure from conventional subsonic aircraft design. Mechanical and thermal insulation properties of the nanocomposite will allow for "morphing" airframe and propulsion structures that can change their shape or properties on demand to improve aerodynamic efficiency and respond to damage. Composite materials derived from low density, high strength carbon nanotube-based fibers and durable nanoporous matrixes will enable the production of ultra-lightweight multifunctional airframes and with embedded lightning strike protection. Nanotexturing will create surfaces that are naturally resistant to ice accretion thereby eliminating the need for runway deicing and in-flight ice mitigation. Replacement of heavy copper wiring cables with carbon nanotube wires will enable significant reductions in aircraft weight. Distributed autonomous networks of nanotechnology based state sensors powered by high efficiency energy harvesting (thermoelectric, piezoelectric, or photovoltaics) will enable real-time monitoring of the overall health and performance of the aircraft leading to reduced emissions and noise and improved safety. The design, developed by NASA and Massachusetts Institute of Technology, is for a 354 passenger aircraft that would be available for commercial use in 2030–2035 and would enable a reduction in aircraft fuel consumption by 54% over a Boeing 777 baseline aircraft. (Courtesy of NASA and MIT).

Copyright 2011 by WTEC. The U.S. Government retains a nonexclusive and nontransferable license to exercise all exclusive rights provided by copyright. This document is sponsored by the National Science Foundation (NSF) under a cooperative agreement from NSF (ENG-0844639) to the World Technology Evaluation Center, Inc. The first co-editor was supported by the NSF Directorate for Engineering. The Government has certain rights in this material. Any writings, opinions, findings, and conclusions expressed in this material are those of the authors and do not necessarily reflect the views of the United States Government, the authors' parent institutions, or WTEC. A selected list of available WTEC reports and information on obtaining them appears on the inside back cover of this report. This document is available at http://www.wtec.org/nano2.

ISBN 978-94-007-1167-9   e-ISBN 978-94-007-1168-6
DOI 10.1007/978-94-007-1168-6
Springer Dordrecht Heidelberg London New York

Library of Congress Control Number: 2011928076

© WTEC, 2011
No part of this work may be reproduced, stored in a retrieval system, or transmitted in any form or by any means, electronic, mechanical, photocopying, microfilming, recording or otherwise, without written permission from the Publisher, with the exception of any material supplied specifically for the purpose of being entered and executed on a computer system, for exclusive use by the purchaser of the work.

Cover design: estudio Calamar S.L.

Printed on acid-free paper

Springer is part of Springer Science+Business Media (www.springer.com)

Acknowledgements

We at WTEC wish to thank all the participants for their valuable insights and their dedicated work in conducting this international study of nanoscience and nanotechnology. Appendix 2 has a complete list of the "cast of thousands" of experts around the world who shared their valuable time with us. The expert panelists who wrote chapters of this report need specific mention: Dawn Bonnell, Jeff Brinker, Mark Hersam, Evelyn Hu, Mark Lundstrom, Chad Mirkin, André Nel, and Jeff Welser. They were assisted in the writing by Mamadou Diallo, James Murray, Mark Tuominen, and Stuart A. Wolf.

For making this study possible, our sincere thanks go to the many who contributed some of their research funds: Michael Reischman NSF/ENG, Clark Cooper and Shaochen Chen NSF/CMMI, Robert Wellek NSF/CBET, Robert Trew and Pradeep Fulay NSF/ECCS, Christine Rohlfing NSF/MPS, Shankar Basu NSF/CISE, and Hongda Chen USDA/NIFA. Our international hosts supported the costs of the workshops abroad, and provided unparalleled hospitality. Among others, they included in Hamburg: Christos Tokamanis, EU; Helmut Dosch, DESY—in Tsukuba: Kazunobu Tanaka, JST, Japan; Jo-Won Lee, MEST, Korea; Maw-Kuen Wu, NSC, Taiwan—and in Singa—re: Freddy Boey, Nanyang Technological University, Singapore; Chennupati Jagadish, Australian National University; Chen Wang, National Center for Nanoscience and Technology (NCNST), China; Jayesh R. Bellare, IIT Bombay, India; Salman Alrokayan, King Abdullah Institute for Nanotechnology, Saudi Arabia.

And, of course, Mike Roco provided the guiding light for the whole effort, very much engaged in coordination, writing and editing, and in pushing everyone along to get the best possible results.

R.D. Shelton
President, WTEC

WTEC Mission

WTEC provides assessments of international research and development in selected technologies under awards from the National Science Foundation (NSF), the Office of Naval Research (ONR), and other agencies. Formerly part of Loyola College, WTEC is now a separate nonprofit research institute. Michael Reischman, Deputy Assistant Director for Engineering, is NSF Program Director for WTEC. WTEC's mission is to inform U.S. scientists, engineers, and policymakers of global trends in science and technology. WTEC assessments cover basic research, advanced development, and applications. Panelists are leading authorities in their field, technically active, and knowledgeable about U.S. and foreign research programs. As part of the assessment process, panels visit and carry out extensive discussions with foreign scientists and engineers abroad. The WTEC staff helps select topics, recruits expert panelists, arranges study visits to foreign laboratories, organizes workshop presentations, and finally, edits and publishes the final reports. See http://wtec.org for more information or contact Dr. R. D. Shelton at Shelton@ScienceUS.org.

WTEC Study on Nanotechnology Research Directions for Societal Needs in 2020

Study Coordinator
Mihail C. Roco, Ph.D.
Senior Advisor for Nanotechnology,
National Science Foundation,
4201 Wilson Boulevard,
Arlington, VA 22230

WTEC Panel and Other Contributors

Chad A. Mirkin, Ph.D.
(Co-chair)
Northwestern University
Department of Chemistry
2145 Sheridan Road
Evanston, IL 60208

Mark Hersam, Ph.D.
(Co-chair)
Northwestern University
Department of Materials
Science & Engineering
2220 Campus Drive
Evanston, IL 60208

Dawn Bonnell, Ph.D.
University of Pennsylvania
Department of Materials
Science and Engineering
3231 Walnut Street
Rm 112-A
Philadelphia, PA 19104

C. Jeffrey Brinker, Ph.D.
University of New Mexico
Department of Chemical and
Nuclear Engineering
1001 University Blvd. SE
Albuquerque, NM 87131
and
Sandia National Laboratories
Self-Assembled Materials
Dept 1002 Albuquerque
NM 87131

Mamadou Diallo, Ph.D.
California Institute of
Technology
Division of Engineering and
Applied Science
1200 East California Blvd.
Mail Stop 139-74 Pasadena
CA 91125

Evelyn L. Hu, Ph.D.
Harvard School of
Engineering and Applied
Sciences, 29 Oxford Street
Cambridge, MA 02138

Mark Lundstrom, Ph.D.
Purdue University
School of Electrical &
Computer Engineering
DLR Building, NCN Suite,
207 S. Martin Jischke Dr.
West Lafayette, IN 47907

James Murday, Ph.D.
USC Office of Research
Advancement
701 Pennsylvania Avenue
NW Suite 540
Washington, DC 20004

André Nel, M.D., Ph.D.
UCLA Department of
Medicine and California
NanoSystems Institute
10833 Le Conte Ave.
52-175 CHS
Los Angeles, CA 90095

Mark Tuominen, Ph.D.
University of Massachusetts
Center for Hierarchical
Manufacturing and
MassNanoTech,
411 Hasbrouck Laboratory
Amherst, MA 01003

Jeffrey Welser, Ph.D.
Semiconductor Research
Corp.
1101 Slater Road, Suite 120
Durham, NC 27703
and
IBM Almaden Research
Center
650 Harry Road
San Jose, CA 95120

Stuart A. Wolf, Ph.D.
University of Virginia
NanoStar
395 McCormick Road
Charlottesville, VA 22904

External Reviewers to the Study

Eric Isaacs, Argonne National Laboratory
Martin Fritts, Nanotechnology Characterization Laboratory
Naomi Halas, Rice University
Robert Langer, MIT
Emilio Mendez, Brookhaven National Laboratory
Gunter Oberdörster, URMC
Gernot Pomrenke, AFOSR
David Shaw, SUNY Buffalo
Richard Siegel, RPI
Sandip Tiwari, Cornell University
George Whitesides, Harvard University

World Technology Evaluation Center, INC. (WTEC)

R.D. Shelton, President
V.J. Benokraitis, Vice President for Operations, Project Manager
Geoffrey M. Holdridge, Vice President for Government Services
Patricia Foland, Director of Information Systems
Grant Lewison (Evaluametrics, Ltd.), Advance Contractor, Europe
David Kahaner (Asian Technology Information Program), Advance Contractor, Asia
Patricia M.H. Johnson, Director of Publications
Haydon Rochester, Jr., Lead Editor

Foreword

Impacts, Lessons Learned, and International Perspectives for Nanotechnology to 2020

The accelerating pace of discovery and innovation and its increasingly interdisciplinary nature leads, at times, to the emergence of converging areas of knowledge, capability, and investment; nanotechnology is a prime example. It arose from the confluence of discoveries in physics, chemistry, biology, and engineering around the year 2000. At that time, a global scientific and societal endeavor was initiated, focused by two key factors: (1) an integrative definition of nanotechnology based on distinctive behaviors of matter at the nanoscale and the ability to systematically control and engineer those behaviors,[1] and (2) articulation of a long-term vision and goals for the transformative potential of nanotechnology R&D to benefit society[2] that included a 20-year vision for the successive introduction of four generations of nanotechnology products.[3] The definition and long-term vision for nanotechnology paved the way for the U.S. National Nanotechnology Initiative (NNI), launched in 2000, and also inspired sustained R&D programs in the field by Japan, Korea, the European Community, Germany, China, and Taiwan. In fact, over 60 countries established nanotechnology R&D programs at a national level between 2001 and 2004. A new wave of R&D investments by Russia, Brazil, India, and several Middle East countries began after the second generation of nanotechnology products came to market about 2006. The U.S. nanotechnology commitment is significant: cumulative NNI funding since 2000 amounts to more than $12 billion, including about $1.8 billion in 2010, placing the NNI second only to the space program in terms of civilian science and technology investment.[4]

This report outlines the foundational knowledge and infrastructure development achieved by nanotechnology in the last decade and explores the potentials of the U.S. and global nanotechnology enterprise to 2020 and beyond. It aims to redefine the R&D goals for nanoscale science and engineering integration, and to establish nanotechnology as a general-purpose technology in the next decade. The vision for the future of nanotechnology presented here draws on scientific insights from U.S. experts in the field, examinations of lessons learned, and international perspectives shared by participants from 35 countries in five international brainstorming meetings

---

[1] Proposed in an international benchmarking study: Siegel, R., E. Hu, and M.C. Roco, eds. 1999. *Nanostructure science and technology*. Washington, DC: National Science and Technology Council. Also published in 1999 by Springer.

[2] Roco M.C., R.S. Williams, and P. Alivisatos, eds. 1999. *Nanotechnology research directions: Vision for the next decade*. IWGN Workshop Report 1999. Washington, DC: National Science and Technology Council. Also published in 2000 by Springer. Available online: http://www.wtec.org/loyola/nano/IWGN.Research.Directions/.

[3] Roco, M.C. 2004. Nanoscale science and engineering: Unifying and transforming tools. *AIChE J.* 50(5):890–897.

[4] Lok, C. 2010. Small Wonders. *Nature* 467:18–21. (2 September).

hosted or co-hosted by the principal authors of this report.[5] The report was peer reviewed and received input from various stakeholders' public comments at the website http://wtec.org/nano2/. It aims to provide decision makers in academia, industry, and government with a nanotechnology community perspective of productive and responsible paths forward for nanotechnology R&D.

Only 10 years after adopting the definition and long-term vision for nanotechnology, the NNI and other programs around the world have achieved remarkable results in terms of scientific discoveries that span better understanding of the smallest living structures, uncovering the behaviors and functions of matter at the nanoscale, and creating a library of nanostructured building blocks for devices and systems. Myriad R&D results include technological breakthroughs in such diverse fields as advanced materials, biomedicine, catalysis, electronics, and pharmaceuticals; and expansion into new fields such as energy resources and water filtration, agriculture and forestry, and integration of nanotechnology with other emerging areas such as quantum information systems, neuromorphic engineering, and synthetic and system nanobiology. New fields have emerged such as spintronics, plasmonics, metamaterials, and molecular nanosystems. "Nanomanufacturing" is already under way and is a growing economic focus. Nanotechnology has come to encompass a rich infrastructure of multidisciplinary professional communities, advanced instrumentation, user facilities, computing resources, formal and informal education assets, and advocacy for nanotechnology-related societal benefit. Communication, coordination, research, and regulation efforts have gained momentum in addressing ethical, legal, and social implications (ELSI) and environmental, health, and safety (EHS) aspects of nanotechnology.

Many nanotechnology breakthroughs have begun to impact the marketplace: 2009 values for nanotechnology-enabled products are estimated at about $91 billion in the United States and $254 billion worldwide. Current developments presage a burgeoning economic impact: trends suggest that the number of nanotechnology products and workers worldwide will double every 3 years, achieving a $3 trillion market with six million workers by 2020. The governance mandate has broadened steadily so that in addition to promoting scientific discovery and technological innovation, it increasingly advances social innovation by proactively addressing many complex issues of responsible development of a new technology. Nanotechnology R&D has become a socio-economic target in all developed countries and in many developing countries—an area of intense international collaboration and competition.

And yet, nanoscale science, engineering, and technology are still in a formative stage, with most of their growth potential ahead and in still-emerging directions. Ambitious goals for several key scientific achievements over the next decade include an increase of about 5,000 times in X-ray source brilliancy for direct

---

[5] For more information, see Appendix A. U.S. and International Workshops.

measurement of nanostructures and of about 10,000 times in computational capabilities of nanostructures. Key areas of emphasis over the next decade are:

- Integration of knowledge at the nanoscale and of nanocomponents in nanosystems with deterministic and complex behavior, aiming toward creating fundamentally new products
- Better control of molecular self-assembly, quantum behavior, creation of new molecules, and interaction of nanostructures with external fields in order to build materials, devices, and systems by modeling and computational design
- Understanding of biological processes and of nano-bio interfaces with abiotic materials, and their biomedical and health/safety applications, and nanotechnology solutions for sustainable natural resources and nanomanufacturing
- Governance to increase innovation and public-private partnerships; oversight of nanotechnology safety and equity building on nascent models for addressing EHS, ELSI, multi-stakeholder and public participation; and increasing international collaborations in the process of transitioning to new generations of nanotechnology products. Sustained support for education, workforce preparation, and infrastructure all remain pressing needs

Overall, it is predicted that continuing research into the systematic control of matter and a focus on innovation at the nanoscale will accelerate in the first part of the next decade, especially in the next 5 years, underpinning a growing revolution in technology and society. Nanotechnology already is having a major impact in the development of many sectors, ranging from electronics to textiles; by 2020, it will be a broad-based technology, seamlessly integrated with most technologies and applications used by the masses, driven by economics and by the strong potential for achieving previously unavailable solutions in medicine, productivity, sustainable development, and human quality of life.

Contents

The Long View of Nanotechnology Development: The National Nanotechnology Initiative at 10 Years.......... 1
Mihail C. Roco

Investigative Tools: Theory, Modeling, and Simulation 29
Mark Lundstrom, P. Cummings, and M. Alam

Enabling and Investigative Tools: Measuring Methods, Instruments, and Metrology.......... 71
Dawn A. Bonnell, Vinayak P. Dravid, Paul S. Weiss, David Ginger, Keith Jackson, Don Eigler, Harold Craighead, and Eric Isaacs

Synthesis, Processing, and Manufacturing of Components, Devices, and Systems 109
Chad A. Mirkin and Mark Tuominen

Nanotechnology Environmental, Health, and Safety Issues.......... 159
André Nel, David Grainger, Pedro J. Alvarez, Santokh Badesha, Vincent Castranova, Mauro Ferrari, Hilary Godwin, Piotr Grodzinski, Jeff Morris, Nora Savage, Norman Scott, and Mark Wiesner

Nanotechnology for Sustainability: Environment, Water, Food, Minerals, and Climate 221
Mamadou Diallo and C. Jeffrey Brinker

Nanotechnology for Sustainability: Energy Conversion, Storage, and Conservation 261
C. Jeffrey Brinker and David Ginger

Applications: Nanobiosystems, Medicine, and Health 305
Chad A. Mirkin, André Nel, and C. Shad Thaxton

Applications: Nanoelectronics and Nanomagnetics.......... 375
Jeffrey Welser, Stuart A. Wolf, Phaedon Avouris, and Tom Theis

Applications: Nanophotonics and Plasmonics.......... 417
Evelyn L. Hu, Mark Brongersma, and Adra Baca

Applications: Catalysis by Nanostructured Materials 445
Evelyn L. Hu, S. Mark Davis, Robert Davis, and Erik Scher

Applications: High-Performance Materials and Emerging Areas 467
Mark Hersam and Paul S. Weiss

**Developing the Human and Physical Infrastructure
for Nanoscale Science and Engineering** 501
James Murday, Mark Hersam, Robert Chang, Steve Fonash,
and Larry Bell

**Innovative and Responsible Governance of Nanotechnology
for Societal Development** 561
Mihail C. Roco, Barbara Harthorn, David Guston, and Philip Shapira

Selected Bibliography (2000–2009) 619

Appendices 631

Author Index 681

Subject Index 685

Preface

Executive summary

Nanotechnology is the control and restructuring of matter at the nanoscale, at the atomic and molecular levels in the size range of about 1–100 nm, in order to create materials, devices, and systems with fundamentally new properties and functions due to their small structure. The 1999 "Nano1" report *Nanotechnology Research Directions: Vision for Nanotechnology in the Next Decade* described nanotechnology as a broad-based, multidisciplinary field projected to reach mass use by 2020 and offering a new approach to education, innovation, learning, and governance—a field expected to revolutionize many aspects of human life.[6] Nanotechnology can profoundly affect the ways we live, how healthy we are, what we produce, how we interact and communicate with others, how we produce and utilize new forms of energy, and how we maintain our environment.

Ten years have passed since that first "Nano1" U.S. National Science and Technology Council report on the prospects for nanotechnology. During this past decade, research and development in nanotechnology has made astonishing progress and has now provided a clearer indication of its potential. This new report ("Nano2") examines the last decade's progress in the field and uncovers the opportunities for nanotechnology development in the United States and around the world in the next decade. It summarizes what has been achieved with the investments made since 2000, but more importantly, it describes the expected targets for nanotechnology R&D in the next decade and beyond and how to achieve them in the context of societal needs and other emerging technologies.

The Nano2 report incorporates views of leading experts from academia, industry, and government shared among U.S. representatives and those from over 35 other economies in four forums held between March and July 2010. These began with a brainstorming meeting in Chicago (United States) and included U.S.-multinational workshops in Hamburg, Germany (involving European Union and U.S. representatives); Tokyo, Japan (involving Japan, South Korea, Taiwan, and U.S. representatives); and Singapore (involving Singapore, Australia, China, India, Saudi Arabia, and U.S. representatives). Participants came from a wide range of disciplines, including the physical and biological sciences, engineering, medicine, social sciences, economics, and philosophy.

Keywords Nanoscale, science, and engineering • Research, education, innovation • Forecast • Governance • Societal implications • International perspective

---

[6] Roco, M.C., R.S. Williams, and P. Alivisatos, eds. 1999. *Nanotechnology research directions: IWGN workshop report. Vision for nanotechnology R&D in the next decade*. Washington, DC: National Science and Technology Council. Also published in 2000 by Springer.

Outline of the Report

This international report documents the progress made in nanotechnology from 2000 to 2010 and lays out a vision for progress in nanotechnology from 2010 to 2020, in four broad categories of interest:

1. *Methods and tools* of nanotechnology for investigation, synthesis, and manufacturing
2. *Safe and sustainable development* of nanotechnology for responsible and effective management of its potential; this includes environmental, health, and safety (EHS) aspects and support for a sustainable environment in terms of energy, water, food, raw materials, and climate
3. *Nanotechnology applications* for advances in biosystems and medicine; information technology; photonics and plasmonics; catalysis; and high-performance materials, devices, and systems
4. *Societal dimensions*, including education, investing in physical infrastructure, and governance of nanotechnology for societal benefit

This report is addressed to the academic community, private sector, government agencies, and generally to nanotechnology stakeholders. It aims specifically to provide input for planning of nanotechnology R&D programs to those producing, using, and governing this emerging field. Significant examples of nanotechnology discoveries and achievements since 2000 and the goals to 2020 are listed in Table 1 near the end of the chapter, arranged according to the aforementioned four categories. In addition, four sidebars II-V graphically illustrate several high-impact applications of nanotechnology (in electronics, biomedical and catalysts) and U.S. infrastructure investments to support progress in nanotechnology as of 2010.

Progress Since 2000

The broad consensus of forum participants is of strong progress since 2000 in the following areas.

Overall

- The viability and societal importance of nanoscale science, engineering, and technology applications have been confirmed, while extreme predictions, both pro and con, have receded. Advancements in scientific foundation and physical infrastructure were inspired by the 1999 unifying definition and vision of "Nano1."
- Nanotechnology has been recognized as a revolutionary field of science and technology, comparable to the introduction of electricity, biotechnology, and digital information revolutions. Between 2001 and 2008, the numbers of discoveries, inventions, nanotechnology workers, R&D funding programs, and markets all increased by an average annual rate of 25%. The worldwide market for products incorporating

Preface   xvii

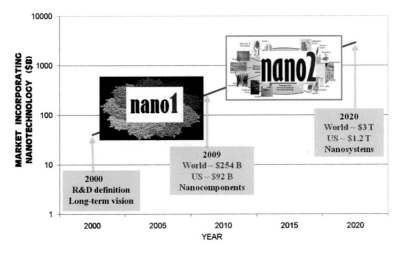

Fig. 1 Market of final products incorporating nanotechnology: the long-term vision for 2000–2020 (solid line, see Chap. 1) and outcomes in 2009 (survey by Lux Research, Chapter "Innovative and Responsible Governance of Nanotechnology for Societal Development"). The R&D focus evolves from fundamental discoveries in 2000–2010 (Nano1 in the figure) to applications-driven fundamental and nanosystem research in 2010–2020 (Nano2)

nanotechnology reached about $254 billion in 2009 (Fig. 1, Chapter "Innovative and Responsible Governance of Nanotechnology for Societal Development").

Methods and Tools

- New instrumentation has allowed femtosecond measurements with atomic precision in domains of engineering relevance. Single-phonon spectroscopy and sub-nanometer measurements of molecular electron densities have been performed. Single-atom and single-molecule characterization methods have emerged that allow researchers to probe the complex and dynamic nature of nanostructures in previously impossible ways Chapter "Measuring, Methods, Instruments, and Metodology".
- Simulation from basic principles has expanded to assemblies of atoms 100 times larger than in 2000, and "materials by design" can now be done for a few polymeric and other nanostructures Chapter "Theory, Modeling, and Simulation".
- Fundamental structure–function studies for nanomaterials have led to the discovery and development of important new phenomena such as plasmonics, negative index of refraction in IR/visible wavelength radiation, Casimir forces, nanofluidics, nanopatterning, teleportation of information between atoms, and bio-interactions at the nanoscale. Other nanoscale phenomena are better understood and quantified, such as quantum confinement, polyvalency, and shape anisotropy. Each has become the foundation for new domains in science and engineering.
- An illustration is the discovery of spin torque transfer (the ability to switch the magnetization of nanomagnet using a spin polarized current), which has

significant implications for memory, logic, sensors, and nano-oscillators. A new class of devices has been enabled, as exemplified by the worldwide competition to develop spin torque transfer random access memory (STT-RAM), which will be fully commercialized in the next decade.
- Scanning probe tools for printing one molecule or nanostructure high on surfaces over large areas with sub-50 nm resolution have become reality in research and commercial settings. This has set the stage for developing true "desktop fab" capabilities that allow researchers and companies to rapidly prototype and evaluate nanostructured materials or devices at point of use.

Safe and Sustainable Development

- There is greater recognition of the importance of nanotechnology-related environmental, health, and safety (EHS) issues for the first generation of nanotechnology products, and of ethical, legal, and social implications (ELSI) issues. Considerable attention is now being paid to building physico-chemical-biological understanding, regulatory challenges for specific nanomaterials, governance methods under conditions of uncertainty and knowledge gaps, risk assessment frameworks, and life cycle analysis based on expert judgment, use of voluntary codes, and incorporation of safety considerations into the design and production stages of new nano-enabled products. Increased attention includes modes of public participation in decision making and overall anticipatory governance with respect to nanotechnology.
- Nanotechnology has provided solutions for about half of the new projects on energy conversion, energy storage, and carbon encapsulation in the last decade.
- Entirely new families have been discovered of nanostructured and porous materials with very high surface areas, including metal organic frameworks, covalent organic frameworks, and zeolite imidazolate frameworks, for improved hydrogen storage and CO_2 separations.
- A broad range of polymeric and inorganic nanofibers and their composites for environmental separations (membrane for water and air filtration) and catalytic treatment have been synthesized. Nanocomposite membranes, nanosorbents, and redox-active nanoparticles have been developed for water purification, oil spill cleanup, and environmental remediation.

Towards Nanotechnology Applications

- Many current applications are based upon relatively simple "passive" (steady function) nanostructures used as components to enable or improve products (e.g., nanoparticle-reinforced polymers). However, since 2005, more sophisticated products with "active" nanostructures and devices have been introduced to meet needs not addressed by current technologies (e.g., point-of-care molecular diagnostic tools and life-saving targeted drug therapeutics).

nano2

II. Nanoelectronic and nanomagnetic components incorporated into common computing and communication devices, in production in 2010

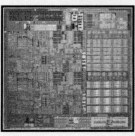

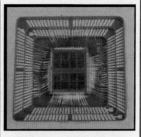

32 nm complementary metal oxide semiconductor (CMOS) processor technology by Intel (2009), (gate length of 30 nm) with high-K / metal gate. This technology is used to make integrated circuit (IC) chips that will be available in a wide variety of laptop, desktop, and server computer systems, giving higher speed, higher density, and lower power.

90 nm thin-film storage (TFS) flash flexmemory by Freescale (2010) for next-generation microcontrollers, utilizing silicon nanocrystals as the charge storage layer. The nanocrystal layer enables higher-density arrays, lower-power operation, faster erase times, and improved reliability. Micro-controllers are the "brains" of a wide variety of industrial and consumer products.

16 megabit magnetic random access memory (MRAM) by Everspin (2010) is based on nanometer-scale magnetic tunnel junctions. These memories have many industrial and commercial applications, such as saving data during a system crash, enabling resume-play features, quick storage and retention of data encryption during shutdown, and retention of vehicle data in an accident for later analysis.

- Entirely new classes of materials have been discovered and developed, both scientifically and technologically. These include one-dimensional nanowires and quantum dots of various compositions, polyvalent noble metal nanostructures, graphene, metamaterials, nanowire superlattices, and a wide variety of other particle compositions. A periodic table of nanostructures is emerging, with entries defined by particle composition, size, shape, and surface functionality.

- Entirely new concepts have been proven: first quantum device was built and tested in 2010, first artificial cell with synthetic genome was completed, and first hierarchical nanostructures have been calculated.
- A versatile library has been invented of new nanostructures and surface patterning methods that are fueling the development of the field. These include commercialized systems such as a large variety of nanoparticles, nanolayers, nanostructured polymers, metals, ceramics and composites, optical and "dip-pen" nanolithography, nanoimprint lithography, and roll-to-roll processes for manufacturing graphene and other nanosheets. This said, nanotechnology is still in a formative phase from the standpoints of characterization methods, the level of empiricism in synthesis and manufacturing, and the development of complex nanosystems. More fundamental R&D is needed to address these limitations.
- New processes and nanostructures have been formulated using basic principles from quantum and surface sciences to molecular bottom-up assembly, and have been combined with semi-empirical, top-down miniaturization methods for integration into products. Nanotechnology has enabled or facilitated novel research in areas such as quantum computing, computer and communications devices (see sidebar II), nanomedicine, energy conversion and storage, water purification, agriculture and food systems, aspects of synthetic biology, aerospace, geoengineering, and neuromorphic engineering.
- Nanoscale medicine has made significant breakthroughs in the laboratory, advanced rapidly in clinical trials, and made inroads in applications of biocompatible materials, diagnostics, and treatments, see side bar III. Advanced therapeutics such as Abraxane are now commercialized and making a significant impact in treating different forms of cancer. The first point-of-care nano-enabled medical diagnostic tools such as the Verigene System are now being used around the world to rapidly diagnose disease. In addition, over 50 cancer-targeting drugs based on nanotechnology are in clinical trial in the United States alone. Nanotechnology solutions are enabling companies such as Pacific Biosciences and Illumina to offer products that are on track to meet the $1000 genome challenge.
- There has been extensive penetration of nanotechnology into several critical industries. Catalysis by engineered nanostructured materials impacts 30–40% of the U.S. oil and chemical industries see sidebar IV Chapter "Applications: Catalysis by Nanostructured Materials"; semiconductors with features under 100 nm constitute over 30% of that market worldwide and 60% of the U.S. market Chapter "The Long view of Nanotechnology Development: The NNI at Ten Years"; molecular medicine is a growing field. The state of the art in nanoelectronics has progressed rapidly from microscale devices to the realm of 30 nm and is continuing this trajectory to even smaller feature sizes. These and many other examples show nanotechnology is well on its way to reaching the goal set in 2000 for it to become a "general-purpose technology" with considerable economic impact.
- In the United States, the financial investment in nanotechnology R&D has been considerable over the last 10 years. The cumulative U.S. Government funding of nanotechnology now exceeds US$12 billion, placing it among the largest U.S. civilian technology investments since the Apollo Moon-landing program (*Nature*, Sept. 2010, p. 18). Industry has recognized the importance of nanotechnology and the central role of government in the NNI R&D. The estimated

Preface xxi

market for products incorporating nanotechnology is about $91 billion in 2009 in the United States Chapter "Innovative and responsible Governance of Nanotechnology for Societal Development". Finally, approximately 60 countries have adopted nanotechnology research programs, making nanotechnology one of the largest and most competitive research fields globally.

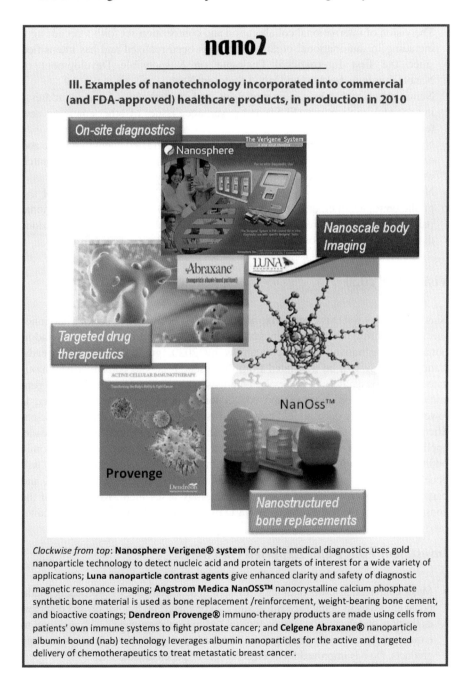

nano2

III. Examples of nanotechnology incorporated into commercial (and FDA-approved) healthcare products, in production in 2010

Clockwise from top: **Nanosphere Verigene® system** for onsite medical diagnostics uses gold nanoparticle technology to detect nucleic acid and protein targets of interest for a wide variety of applications; **Luna nanoparticle contrast agents** give enhanced clarity and safety of diagnostic magnetic resonance imaging; **Angstrom Medica NanOSS™** nanocrystalline calcium phosphate synthetic bone material is used as bone replacement /reinforcement, weight-bearing bone cement, and bioactive coatings; **Dendreon Provenge®** immuno-therapy products are made using cells from patients' own immune systems to fight prostate cancer; and **Celgene Abraxane®** nanoparticle albumin bound (nab) technology leverages albumin nanoparticles for the active and targeted delivery of chemotherapeutics to treat metastatic breast cancer.

Societal Dimensions

- Various activities have led to establishment of an international community of nanotechnology professionals, a sophisticated R&D infrastructure, and diverse manufacturing capabilities spanning the chemical, electronics, advanced materials, and pharmaceutical industries.
- The vision of international collaboration and competition set forth a decade ago, including in multinational organizations, has been realized and has intensified since the first International Dialogue on Responsible Development of Nanotechnology, held in the United States in 2004.
- Nanotechnology has become a model for, and an intellectual focus in, addressing societal implications (ELSI) and governance issues of other emerging new technologies.
- Nanotechnology has catalyzed overall efforts in and attracted talent to science and engineering in the last decade worldwide. Key education networks and research user facilities in the United States in 2010 are illustrated in sidebar IV.
- Nanotechnology has become a model for informal science education of the public on topics of emerging technologies and for building strategic educational partnerships between researcher institutions and public education institutions that benefit the educational goals of both.

Vision For 2020

Nanotechnology R&D is expected to accelerate the succession of science and innovation breakthroughs towards nanosystems by design, and to lead to many additional and qualitatively new applications by 2020, guided by societal needs. Nanotechnology will be translated from the research labs to consumer use, motivated by responsiveness to societal challenges such as sustainability; energy generation, conservation, storage, and conversion; and improved healthcare that is lower-cost and more accessible. During the first decade, the main driver was scientific discovery accruing from curiosity-driven research. During the next decade, application-driven research will produce new scientific discoveries and economic optimization leading to new technologies and industries. Such translation will benefit society but will require new approaches in accountable, anticipatory, and participatory governance, and real-time technology assessment. Key points of the consensus vision for nanotechnology R&D over the next decade are noted below.

Investment Policy and Expected Outcomes

- Major continued investment in basic research in nanotechnology is needed, but additional emphasis in going forward should also be placed on innovation and commercialization, on job creation, and on societal "returns on investment," with measures to ensure safety and public participation. With each new generation of nanotechnology products, there is improved focus on economic and societal outcomes.

nano2

IV. Examples of nanotechnology in commercial catalysis products for applications in oil refining, 2010

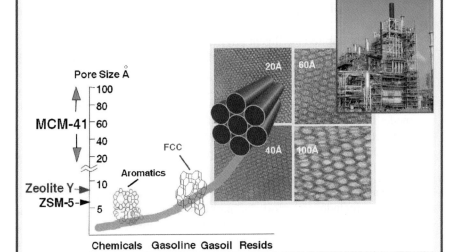

ExxonMobil, Chevron, Dow Chemical, and other oil companies use nanostructured catalysts developed since 2000 for more efficient upgrading of crude oil into transportation fuels and petrochemicals. Redesigned mesoporous silica materials, like MCM-41, along with improved zeolites, are used in a variety of processes such as fluid catalytic cracking (FCC) for producing gasoline from heavy gas oils, and transalkylation for producing para-xylene and related building blocks for the manufacture of polyesters. The global industry upgrades 80+ million barrels per day (MBD) of petroleum to fuels and chemicals; many of the streams (cuts) are catalytically processed, some multiple times.

Shape-selective catalysis for managing molecular size and shape is a key application area for modern nanotechnology. The image at left represents many of the nano-engineered catalyst agglomerates of nanoparticles applied to different specialized purposes. Better control of the size, shape, and surface orientation of supported catalyst materials, and better methods to control particulate porosity over multiple length scales, lead to such benefits as increased activity, selectivity, and energy efficiency. In addition, replacement of precious metals by base metal catalysts tailored at the nanoscale is being used to improve productivity and reduce processing costs.

Nano-engineered materials now constitute 30–40% of the huge global catalyst market, which has total annual sales of $18–20 billion. The broader, value-added impact of catalytic processing on the U.S. economy alone is estimated at several hundred billion dollars per year.

- The frontiers of nanotechnology research will be transformed in areas such as:
  - understanding nanoscale phenomena and processes using direct measurements and simulations
  - the classical/quantum physics transition in nanostructures and devices
  - multiscale self-assembly of materials from the molecular or nanostructure level upwards

- interaction of nanostructures with external fields
- complex behavior of large nanosystems
- efficient energy harvesting, conversion, and storage with low-cost, benign materials
- understanding of biological processes and of bio-physicochemical interactions at the nano-bio interface with abiotic materials
- creation of molecules, materials, and complex systems by design from the nanoscale
- biologically inspired intelligent physical systems for computing
- artificial organs, including the use of fluid networks and nanoscale architectures for tissue regeneration
- personalized instruction for K–12 students on nanotechnology in the form of affordable electronic books that incorporate 3D visual imagery/audio/tactile modes of communication to permit self-paced individualized learning
- direct control and feedback of prosthetics by external sensing of brain activity and/or by direct coupling into the peripheral nervous system associated with the artificial limb

- An innovation ecosystem will be further developed for applications of nanotechnology, including support for multidisciplinary participation, multiple sectors of application, entrepreneurial training, multi-stakeholder-focused research, regional hubs, private–public partnerships, gap funding, and legal and tax incentives.
- Nanotechnology will continue its widespread penetration into the economy as a general-purpose technology, which—as with prior technologies such as electricity or computing—is likely to have widespread and far-reaching applications across many sectors. For example, nanoelectronics including nanomagnetics has a pathway to devices (including logic transistors and memory devices) with feature sizes below 10 nm and is opening doors to a whole host of innovations, including replacing electron charge as the sole information carrier. Many other vital industries will experience evolutionary, incremental nanotechnology-based improvements in combination with revolutionary, breakthrough solutions that drive new product innovations.
- Nanotechnology is expected to be in widespread use by 2020. There is potential to incorporate nanotechnology-enabled products and services into almost all industrial sectors and medical fields. Resulting benefits will include increased productivity, more sustainable development, and new jobs.
- Nanotechnology governance in research, education, manufacturing, and medicine programs will be institutionalized for optimum societal benefits.

Research Methods and Tools

- New theories on nanoscale complexity, tools for direct measurements, simulations from basic principles, and system integration at the nanoscale will accelerate discovery.

Preface

V. U.S. Nanotechnology infrastructure in 2010: (*top*) **Key education networks**; (*bottom*) **Research user facilities** (see also Research Centers in Appendix D)

- Tools for simulation of and physical intervention in cellular processes at the molecular scale will establish scientific bases for health/medical interventions, largely completing the conversion of biology into a quantitative "physico-chemical science" rather than an empirical science.

- *In situ* characterization tools for operating nanodevices will accelerate innovation in electronics and energy sectors, while *in situ* probes in realistic environments will enable environmental monitoring, food safety, and civil defense.
- In-depth understanding of principles and methods of nanotechnology will be a condition of competitiveness in sectors such as advanced materials, information technology devices, catalysts, and pharmaceuticals. Development of precompetitive nanoscale science and engineering platforms will provide the foundation for innovation in diverse industry sectors.

Education and Physical Infrastructure

- Multidisciplinary, horizontal, research-to-application-continuum, and system-application education and training will be integrated by the unifying scientific and engineering goals and through new education and training organizations.
- A network of regional hub sites— "Nanotechnology Education Hub Network"— should be established as a sustainable national infrastructure for accelerating nanotechnology education and to implement horizontal, vertical, and integration in educational systems.
- Nanotechnology will enable portable devices that will allow individualized learning anywhere and anytime, as well other modalities of learning using techniques such as brain–machine interaction.
- It will be important to continue to create and maintain centers and hubs as research and user facilities, as well as test beds for development and maturation of innovative nano-enabled device and system concepts. Remote access capabilities will be significantly expanded.

Safe and Sustainable Development

- A focus on nanotechnology EHS hazards and ELSI concerns must be routinely integrated into mainstream nanotechnology research and production activities to support safer and more equitable progress of existing and future nanotechnology generations.
- Simulations of nanoparticle exposure, bio-distribution, and interaction with biological systems will be integrated in risk assessment frameworks, together with life cycle analysis, use of control standards nanomaterials and expert judgments.
- Application of nanotechnology will significantly lower costs and make economic solar energy conversion costs by about 2015 in the United States, and water desalinization by 2020–2025, depending on the region. Nanotechnology will continue to provide breakthrough solutions for over 50% of new projects on energy conversion, energy storage, and carbon encapsulation.
- By 2020 nanotechnology will have extended the limit of sustainability in water resources by 10 years. The nanostructured membranes and materials with large surface areas discovered in the last decade will be optimized and scaled-up for

a variety of applications, including water filtration and desalination, hydrogen storage, and carbon capture.

Nanotechnology Applications

- A library of nanostructures (particles, wire, tubes, sheets, modular assemblies) of various compositions will be developed in industrial-scale quantities.
- New applications expected to emerge in the next decade range from very low-cost, long-life, and high-efficiency photovoltaic devices, to affordable high-performance batteries enabling electric cars, to novel computing systems, cognitive technologies, and radical new approaches to diagnosis and treatment of diseases like cancer.
- As nanotechnology grows in a broader context, it will enable creation or advancements in new areas of research such as synthetic biology, cost-effective carbon capture, quantum information systems, neuromorphic engineering, geoengineering using nanoparticles, and other emerging and converging technologies.
- Nanotechnology developments in the next decade will allow systematic design and manufacturing of nanotechnology products from basic principles, through a move towards simulation-based design strategies that use an increasing amount of fundamental science in applications-driven R&D, as defined in the Pasteur quadrant (Stokes 1997, *Pasteur's Quadrant: Basic Science and Technological Innovation*, Brookings Institution Press).
- Nanotechnology developments will allow increasing the power of computers by about 100,000 times since 2010 and building billion sensor networks by 2020.
- Nanomedicine will revolutionize the way we diagnose and treat people, and in most cases, substantially lower the cost of health care. Personalized and point-of-use diagnostic systems will be used extensively to quickly determine the health of a patient and his or her ability to be treated with specific therapeutics. On the therapeutic side, nanomaterials will be the key to enabling gene therapies for widespread use and an effective means of dealing with antibiotic resistance and the so-called "superbugs."

Strategic Priorities

- *Continue support for fundamental research, education, and physical infrastructure* to change the nanoscale science and engineering frontiers using individual, group, center, and network approaches, with particular focus on direct investigative methods, complex behavior at the nanoscale and how nanoscale behavior controls the microscale/macroscale performance of materials and devices.
- *Promote focused R&D programs*, such as "signature initiatives," "grand challenges," and other kinds of dedicated funding programs, to support the development of priority tools, manufacturing capabilities in critical R&D areas,

and a nanotechnology-adapted innovation ecosystem. Due to its pervasiveness, nanotechnology will progressively be integrated with the developments in other emerging and converging technologies.
- *Advance partnerships* between industry, academia, NGOs, multiple agencies, and international organizations. Give priority to creation of additional regional "nano-hubs" for R&D, system-oriented academic centers, earlier nanotechnology education, nanomanufacturing, and nanotechnology EHS.
- *Support precompetitive R&D platforms*, system application platforms, private–public consortia, and networks in areas such as health, energy, manufacturing, commercialization, sustainability, and nanotechnology EHS and ELSI. The platforms will ensure a "continuing" link between nanoscale fundamental research and applications, across disciplines and sectors.
- *Promote global coordination* to develop and maintain viable international standards, nomenclatures, databases, and patents and other intellectual property protections. Explore international co-funding mechanisms for these activities. Seek international coordination for nanotechnology EHS activities (such as safety testing and risk assessment and mitigation) and nanotechnology ELSI activities (such as broadening public participation and addressing the gaps between developing and developed countries). An international co-funding mechanism is envisioned for maintaining databases, nomenclature, standards, and patents.
- *Develop experimental and predictive methods* for exposure and toxicity to multiple nanostructured compounds.
- *Support horizontal, vertical, and system integration in nanotechnology education*, to create or expand regional centers for learning and research, and to institutionalize nanoscience and nanoengineering educational concepts for K-16 students. Use incentives and competitive methods to harness the energy generated by the students and professors themselves to discover nanotechnolo gy.
- *Use nanoinformatics and computational science prediction tools* to develop a cross-disciplinary, cross-sector information system for nanotechnology materials, devices, tools, and processes.
- *Explore new strategies for mass dissemination, public awareness, and participation* related to nanotechnology R&D, breaking through gender, income, and ethnicity barriers. This is a great challenge in the next 10 years.
- *Institutionalize—create standing organizations and programs to fund and guide nanotechnology activities*—in R&D, education, manufacturing, medicine, EHS, ELSI, and international programs. Important components are the incentive-based, bottom-up programs for research, education, and public participation.

Conclusions

Several strategic lessons have been learned in the last 10 years:
- There is a need for continued, focused investment in theory, investigation methods, and innovation at the nanoscale to realize nanomaterials and nanosystems

- by design for revolutionary new products, because nanotechnology is still in a formative phase. Modeling and simulation methods are essential for nanoscale design and manufacturing processes.
- The potential of nanotechnology to support sustainable development in water, energy, minerals, and other resources is higher than realized in the last 10 years; increased R&D focus is needed.
- Nanotechnology EHS needs to be addressed on an accelerated path as an integral part of the general physico-chemical-biological research program and as a condition of application of the new technology. Knowledge is needed not only for the first generation, but also for the new generations of active nanostructures and nanosystems.
- Besides new emerging areas, more traditional industries may provide opportunities for large-scale application of nanotechnology in areas such as mineral processing, plastics, wood and paper, textiles, agriculture, and food systems.
- Multi-stakeholder and public participation in nanotechnology development is essential in order to better address societal dimensions; efforts in this area need to increase.
- Public-private partnerships need to be extended in research and education.

Nanotechnology is still in an early phase of development, and fundamental understanding and tools are still in the pipeline of new ideas and innovations. Key research themes have been driven by open discovery in the last decade. In the next decade, nanotechnology R&D is likely to shift its focus to socioeconomic needs–driven governance, with significant consequences for science, investment, and regulatory policies. Likewise, R&D investment will increasingly focus on science and engineering systems—some with complex and large architectures – that have societal relevance.

It will be vital over the next decade to focus on four distinct aspects of progress in nanotechnology: (1) how nanoscale science and engineering can improve understanding of nature, protect life, generate breakthrough discoveries and innovation, predict matter behavior, and build materials and systems by nanoscale design – *knowledge progress*; (2) how nanotechnology can generate medical and economic value – *material progress*; (3) how nanotechnology can promote safety in society, sustainable development, and international collaboration – *global progress*; (4) how responsible governance of nanotechnology can enhance quality of life and social equity – *moral progress*.

September 30, 2010

Mihail C. Roco
Chad A. Mirkin
Mark Hersam[7]

---

[7] For the institutional affiliations of authors, please see Appendix 2. List of Participants and Contributors.

Table 1 Key nanotechnology achievements since 2000 and goals to 2020. The story of the past decade of nanotechnology R&D can be described by the developments in fundamental knowledge (e.g., plasmonics), evolutionary or integrative approaches (e.g., integration of nanoelectronics and optoelectronics), and revolutionary approaches (e.g., nano-enabled drug delivery using Abraxane) and how these impact each other. Table 1 helps to illustrate ways that future achievements can build on past ones

Main Achievements/Discoveries since 2000	Fundamental goals/"Holy Grails" to Attain and Barriers to Overcome by 2020
Theory, Modeling, and Simulation	
Discovery of fundamental mechanical, optical, electronic, magnetic, and biological phenomena, properties, and processes at the nanoscale	New theories on complexity for concurrent phenomena, and system integration at the nanoscale that will accelerate discovery
Quantum effects were identified and measured in a series of nanostructures, such as quantum dots, nanotubes, and nanowires	Fundamental understanding of the transition from quantum to classical physics behavior in nanoscale devices and systems Control and use of quantum phenomena in nanomaterials and systems
Emergence of non-equilibrium Green Function (NEGF) as a useful theory of electronic devices	Excited-state electronic structure frameworks including electronic correlation effects to address 10,000 atoms (and realistic coupled electron-ion dynamics for 1000). This 100X advance would open doors to new electron-scale understanding and potentially to high-throughput evaluation of new nanomaterials for artificial photosynthesis and other applications for which energy transport is central
Identification of teleportation of information between two atoms, which would enable quantum computing	Controlled teleportation of information between two atoms in nanoscale systems
Advances in atom and nanoparticle level simulations: • *ab initio*, excited-state electronic structure frameworks with realistic treatment of electronic correlation effects • Molecular Dynamic (MD) simulations with chemical bonding • self-assembly of functionalized nanoparticles • advances in multiscale simulation by coupling of electronic structure theory with MD methods • modeling of some nanoparticle-reinforced polymeric composites	10,000 times increase in computational capacity to enable: • full Hartree-Foch *ab initio* simulations of quantum dots • simulation of self-assembly of programmed materials. • automatic generation of force field and reactive force fields for molecular dynamic simulations of materials • multiscale whole-cell modeling of tandem solar cells and light-emitting devices
Theory of plasmons in metallic nanoparticles and plasmon enhancement of optical processes in molecular and semiconductor systems	Control and use of plasmonics in nanoscale systems

(continued)

Table 1 (continued)

Main Achievements/Discoveries since 2000	Fundamental goals/"Holy Grails" to Attain and Barriers to Overcome by 2020
Developing understanding of interfaces between biotic and abiotic, natural and manmade, materials, and of nanosystems from the nanoscale	Predictive approach for compatibility and assembly of biotic and abiotic materials
Statistical theories of complex nanostructured materials and devices	General approaches to multiscale/multi-phenomena simulation for computational design of nanoscale materials, devices, and integrated nanosystems from basic principles using new models and theories. Simulations will address processes such as self-assembly, catalysis, and dynamics of complex systems
Software packages allowing rapid simulation of optical properties of nanostructures and atomistic-level simulations of nanotransistors	Simulation of quantum transport/current flow at the molecular scale in active nanodevices such as a nanotransistor
Measurement, Instrumentation, and Standards	
Femtosecond observation of nanoscale interactions (displacement of atoms) in chemical processes	Simultaneous atomic resolution, 3D imaging with chemical specificity and temporal resolution of the nanoscale phenomena. Tools for measuring and restructuring matter with atomic precision, for time resolution of chemical reactions, and for domains of engineering and biological relevance
Single-charge, single-spin, spin excitation and bond vibrations probed at the atomic scale	
3D tracking at the single-molecule level of protein motors, enzymes, liposomes, and other bionanostructures	3D internal structure imaging with chemical specificity at atomic resolution of an individual protein
Routine patterning on surfaces becoming scaled up to be useful in applications	Generalized use of reference standard materials and measurement methods in nanoelectronics, biomedical field, nanomanufacturing, and other areas
Measuring probes of continuum properties with atomic resolution, e.g., dielectric function, work function	Develop *in situ* instrumentation for nanomanufacturing process control
	Develop easy-to-use instrumentation for non-specialists and educational applications
Synthesis, Assembly, and Manufacturing	
Creation in the laboratory of a library of nano-components such as particle, tubes, sheets, and 3D structures	Develop a library of nanostructures (particles, wire, tubes, sheets, modular assemblies) of various compositions in industrial-scale quantities
Creation of relatively simple self-assembled nanostructures	Fundamental understanding of the pathways for self assembly or controllable assembly of atoms or molecules into larger, hierarchical, and stable nanostructures and nanosystems. Better understand the role often played by the presence of a catalytic material or directing structure

(continued)

Table 1 (continued)

Main Achievements/Discoveries since 2000	Fundamental goals/"Holy Grails" to Attain and Barriers to Overcome by 2020
New concepts of 3D programmable assembly (using electrostatic, chemical, and biological interactions) have been tested in laboratory. New polymeric molecules for self-assembling purposes and hierarchical polymeric materials have been obtained by design	Systematic approach for design and manufacturing of scalable, hierarchical, directed assembling to three-dimensional structures and devices; programmable assembly
Directed assembly using block copolymers, e.g., using grapheo epitaxy for data storage	Use both equilibrium and non-equilibrium processes in devices and systems
Creating bio-inspired nanostructures in laboratory conditions (see Chap. 4)	Expand environmentally benign nanoscale processes
First molecular machines by design were built	Manufacture nanoproducts that replace hazardous or insufficient materials and existing products
	Use existing infrastructure such as lithography and roll-to-roll facilities to create new nanomanufacturing methods
Discovery of graphene (in 2004), its unique properties, and its rapid movement toward large-scale production	New physics (photons and electron behavior, assembly tools) leading to new applications (faster graphene transistors, 98% transparent, linking of nanosystems, sensors, composite materials) (Chap. 4)
Introduction of graphene as a viable material for transparent electrodes and large-area roll-to-roll processes for its manufacture	Sustainably produce carbon nanomaterials integrated into a broad range of electronics products
Introduction of graphene as an electronic material, and introduction of processes for its large-area production, to replace indium	Use graphene as a foundation for a new generation of planar devices complementing or replacing silicon
Production of chirality-separated carbon nanotubes	Manufacture pure samples of carbon nanotubes to eliminate the need for post-manufacturing sorting
Discovery of metamaterials, their unique properties, and their rapid movement toward large-scale production	Develop manufacturing methods for nanomaterials
Commercialization of new approaches to molecular and materials-based printing on the nanoscale based upon scanning probe systems (e.g., dip pen, polymer pen, and block copolymer lithographies, etc.)	Develop desktop fabs, like desktop printers, that allow researchers to rapidly prototype devices at the point-of-use at low cost without the requirement of a clean room
Contact printing techniques based upon soft elastomers that have become extensively used research tools	Molecular printing techniques that enable single-protein molecule positioning on a surface and exquisite control over important surface-stimulated processes such as stem cell differentiation (at large scale)
Commercialization of high-resolution light-based techniques such as nanoimprint lithography, which are beginning to be used in the semiconductor industry	Establish full manufacturing tool set and design rules for integration using nanoimprint lithography
	Nano-imprinting as a replacement for lithography in some high-resolution applications

(continued)

Table 1 (continued)

Main Achievements/Discoveries since 2000	Fundamental goals/"Holy Grails" to Attain and Barriers to Overcome by 2020
Realization of video displays using nanotechnology	Mass use of economic, large, and flexible displays
New measurement principles and devices for sensors using nanoscale structures and phenomena	Develop nanoscale sensor capacity for process monitoring, health monitoring, and environmental benchmarking and monitoring
Nanotechnology Environmental, Health, and Safety Issues	
Development of the concept that the unique properties of engineered nanomaterials allow a wide range of interactions with biomolecules and biological processes that can constitute the basis of nanomaterial hazard as well as be the cornerstone for new diagnosis and treatment options	Further understanding of the nano-bio interface through development of improved instrumentation, rapid-throughput and *in silico* methodologies that lead to a deeper insight into the biophysicochemical interactions that are required for hazard screening, risk assessment, and safety design of nanomaterials and for improved diagnostics and therapeutics
Demonstration that the use of toxicological injury pathways at the cellular level constitute a robust scientific basis for knowledge generation about potentially hazardous properties of nanomaterials. The demonstration that oxygen radical production is an important toxicological injury mechanism has resulted in the development of a hierarchical oxidative stress response pathway as the basis for performing hazard ranking of engineered nanomaterials that produce reactive oxygen species biotically and abiotically	Use of toxicological injury pathways as the basis for high-throughput screening platforms, which would enable large-volume screening, hazard ranking, and prioritization of this information for limited and focused animal experimentation. Although animal experimentation is still necessary to validate the predictiveness of *in vitro* screening approaches, incremental knowledge generation by smart *in vitro* procedures could ultimately reduce to a minimum animal experimentation and accompanying costs
Understanding of the importance of developing validated and widely accepted methods for *in vitro* and *in vivo* screening of nanomaterial hazards to allow scientists to develop a risk assessment platform commensurate with the growth of nanotechnology	Develop predictive toxicological screening methods that allow the correct balance between *in vitro* and *in vivo* screening that can be executed by high-throughput technology and *in silico* decision-making tools that speed up the rate of knowledge generation
	Develop experimental and predictive methods for exposure and toxicity to multiple nanostructured compounds and on multiple routes
Establishment of public–private partnerships that have been effective in promoting nano-EHS awareness as well as risk reduction strategies, e.g., the DuPont and Environmental Defense Nano Risk Framework (2007)	Active industry participation in nano-EHS, including in hazard and risk assessment, lifecycle analysis, non-confidential product information disclosure, and implementation of safe-by-design strategies

(continued)

Table 1 (continued)

Main Achievements/Discoveries since 2000	Fundamental goals/"Holy Grails" to Attain and Barriers to Overcome by 2020
Voluntary efforts by leading international scientists to develop harmonized protocols that can be validated by round-robin testing, e.g., the International Alliance of Nano Harmonization	Internationally accepted standards for nanomaterial hazard assessment and risk assessment strategies for the product life cycle
Implementation of the first examples of high-throughput screening assays for nanoparticle hazard assessment	Develop as an integral part of new program development the nano-informatics and *in silico* decision-making tools that can help to model and predict nanomaterial hazard, risk assessment, and safe design of nanomaterials
Realization that potential nanoscale hazards must be evaluated as a function of particle size, and that nanomaterials are not necessarily dangerous	
Nanotechnology for Sustainability: Environment, Climate, and Natural Resources	
Awareness of the interdependent, ecosystem-wide implications of human activity on the Earth and the potential of nanotechnology to provide some solutions	Develop a coordinated approach to use nanotechnology innovation for breakthrough solutions in sustainable development
Synthesis of a broad range of polymeric and inorganic nanofibers for environmental separations (filtration, membranes) and catalysis	Solar-powered photocatalytic systems and separation systems (e.g., nanoporous membranes with ion-channel mimics) that extract clean water, energy, and valuable elements (e.g., nutrients and minerals) from impaired water, including wastewater, brackish water, and seawater, with ~99% water recovery
Emergence of electrospinning as a versatile technique for the synthesis of polymeric, inorganic, and hybrid organic-inorganic nanofibers	
Development of nanocomposite membranes (e.g., zeolite nanocomposite reverse osmosis membranes and superhydrophobic nanowire membranes), nanosorbents (e.g. magnetic iron oxide nanoparticles) and redox-active nanoparticles (e.g., zero valent iron nanoparticles) for water purification, oil spill clean-up, and environmental remediation	Integrate functionalized nanofibers and nanoparticles into systems to develop improved separation and catalytic systems for • pollution abatement • environmental remediation • green manufacturing
Proposals of carbon capture methods using nanotechnology	Capture carbon and nitrogen using nanostructures and reuse at industrial scale
Discovery of high-porosity nanostructured materials such as metal organic frameworks (MOFs), covalent organic frameworks (COFs), and zeolite imidazolate frameworks (ZIFs) for hydrogen storage and carbon sequestration	Apply multifunctional sorbent/membrane systems with embedded MOFs, COFs and ZIFs than can selectively extract CO_2 from flue gases and convert it to useful by-products
Proposals of geoengineering concepts and experiments using sulfate or magnetic nanoparticles in upper atmosphere for sunlight reflection	Develop international projects on geoengineering with control of the Earth cooling effect and with respect for biodiversity and environmental safety

(continued)

Table 1 (continued)

Main Achievements/Discoveries since 2000	Fundamental goals/"Holy Grails" to Attain and Barriers to Overcome by 2020
Methods for efficient use of raw materials using nanotechnology have been researched	Develop more efficient and environmentally acceptable separation systems for recovering critical minerals such as rare earth elements (REE) from mine tailings and wastewater from mineral/metallurgical extraction and processing plants. Develop nontoxic, cost-effective REE substitutes and reduce and (eventually) eliminate the release of toxic pollutants into soil, water, and air.
Nanotechnology for Sustainability: Energy	
Rapid improvement in efficiency and scalability of using nanotechnology for solar energy conversion	Mass and economic use of nanotechnology for solar energy conversion by 2005–2006 in United States
New solution-processable formulations of inorganic semiconductors for large-area, low-cost photovoltaics	Increase module efficiency and decrease production and installation costs to achieve a path towards solar electricity at $1/Wp installed
Increased power conversion efficiency of nanostructured organic solar cells by nearly 800% since 2000	Use of nanoparticles and quantum dots in carrier multiplication and hot-carrier collection strategies to overcome the Shockley-Queisser 31% efficiency limit in thin-film photovoltaic devices
	Improve efficiency of organic photovoltaics from less than 1% to greater than 8% through nanoscale phase separation and nanoengineering of device architecture
	Improve lifetime of organic photovoltaics through nanotechnology
	Replace crystalline silicon with cheap and earth-abundant alternative materials, such as iron disulfide, for photovoltaics
Increased the power density of Li-ion batteries by over 50%, enabling practical hybrid electric vehicles	Nanotechnology-enabled batteries for electric vehicles with large distance range of action
First industrial-scale metal organic frameworks (MOFs) synthesized by BASF in 2010 for hydrogen storage (scale up and deployment of MOFs in gas storage [H_2 and CH_4])	Increase efficiency of single junction photovoltaic devices to over 31% through multi-exciton generation, hot carrier harvesting, or other novel phenomena to beat the thermodynamic limit
Green light emission and other basic discoveries enabling solid-state lighting	Increase efficiency of solid state lighting to greater than 50% through nano-enhanced field intensity (plasmonics) and emission rate and field coupling capabilities

(continued)

Table 1 (continued)

Main Achievements/Discoveries since 2000	Fundamental goals/"Holy Grails" to Attain and Barriers to Overcome by 2020
Nanobiosystems and Nanomedicine	
Development of diagnostic methods that are sensitive down to picomole and attomole levels and allow for multiple analytes to be assessed simultaneously by lab-on-a-chip approaches	Point of care (POC) medical diagnostics: Many-order-of-magnitude increased sensitivity, selectivity, and multiplexing capabilities at low cost to enable point-of-care diagnosis and treatment; these capabilities will allow clinicians to track and treat disease – in some cases, years earlier than with conventional tools. Nanodiagnostic tools will become the backbone of clinical medicine by 2020, making the transition from remote labs to hospitals and then eventually to homes.
	Biological diagnostics: Routine live cell imaging with the ability to identify and quantify the key components of a cell (nucleic acids, small molecules, and metal ions) that enable a new way of studying, diagnosing, and treating some of the most debilitating diseases (cancer, cardiovascular disease, and Alzheimer's disease)
	Nonintrusive diagnostics based on breath and saliva nanoscale detection
Abraxane®, the first nanotherapeutic proven to be effective for breast cancer is FDA-approved and a multibillion dollar pharmaceutical; it consists of nanoparticle drug delivery systems, including liposomal, polymer, and albumin nanospheres	Nanotherapeutics: Overcome many challenges such as pharmacokinetics, biodistribution, targeting, tissue penetration, etc., to support widespread adoption by industry of nanotherapeutics
	At least 50% of all drugs used in 2020 will be enabled by nanotechnology; many of these will be for diseases like glioblastoma, pancreatic cancer, and ovarian cancer, where patient prognosis is grim with current therapies
	Widespread adoption of nanomaterials by the pharmaceutical community to increase the effectiveness of chemotherapeutics while eliminating toxic side effects
More than 50 of U.S. pharmaceutical companies have nanotechnology-based solutions for treating cancer in clinical testing (*Science*, Oct. 2010)	50% of drugs for pancreatic cancer and ovarian cancer will be nanotechnology-enabled
Gene therapy enabled by nanomaterials in experimental laboratories; first human trials of siRNA involving a nanomaterial delivery system	Clinical approval of gene therapy for treating a wide range of diseases, including many forms of cancer
	Inexpensive gene sequencing enabled by nanotechnology

(continued)

Preface   xxxvii

Table 1 (continued)

Main Achievements/Discoveries since 2000	Fundamental goals/"Holy Grails" to Attain and Barriers to Overcome by 2020
Use of temperature-sensitive polymer fibers to coat cell culture dishes for the purposes of cell sheet engineering and demonstrating that the technology can be used for repair of a damaged myocardium, cornea, or esophageal lining (Japan)	Use nano-architectures and synthetic pro-morphogens for tissue engineering, including stem cell therapy, construction of new organs (e.g., the whole heart or bladder), and spinal cord regeneration
	Widespread use by 2020 of nano-enabled tissue constructs for repair of cardiac damage (in heart attack victims)
Controlled development of molecules to promote tissue repair and regeneration *in situ*	Stem cells: Use nanobiology and nanomedicine to aid in understanding and control of stem cell differentiation and the transition of stem cells to widespread medicinal application; these advances will be fueled by advances in diagnostics, intracellular gene regulation, and high-resolution patterning tools
	Multifunctional nanoparticle delivery systems that can be used for drug and siRNA delivery, as well as a combination of both; the multifunctional platform can be further endowed with controllable nano-valves, attachment of surface ligands for cancer tissue, and inclusion or attachment of imaging modalities
	Widespread use by 2020 of nanotechnology-enabled stem cell-based therapies for spinal cord regeneration
Achievement of nanoscale control in synthetic biology	Use synthetic biology in regenerative medicine, biotechnology, pharmaceuticals, and energy applications
	Economic impact: translation of many bionanomaterials to the medical arena, with the market size for these nanomedicine advances growing to $200 billion by 2020, by varying estimates, and in the process dramatically lowering health care costs
Nanoelectronics and Nanomagnetics	
Discovery of the quantum spin Hall effect and demonstration of spin transfer torque, which enable direct control of electron spin and magnetic domains by electrical current (see Chap. 9)	Discover multiferroic/magneto-electric materials that will enable spin and magnetic domain control with voltage instead of current
	Discover room-temperature collective behavior of carriers in novel materials such as graphene or topological insulators to enable lower-energy nanoelectronic devices

(continued)

Table 1 (continued)

Main Achievements/Discoveries since 2000	Fundamental goals/"Holy Grails" to Attain and Barriers to Overcome by 2020
First fundamental experiments on quantum computing using small numbers of quantum bits	Realize quantum computers for specific uses
Continuation of Moore's Law	Achieve 3D, near-atomic-level control of reduced-dimensional materials to enable novel nanoelectronic and nanomagnetic behavior
Scaling of CMOS to 30 nm dimensions, including an approximate 1 nm gate insulator, with monolayer accuracy across a 300 nm wafer	
	Combine lithography and self-assembly to pattern semi-arbitrary structures down to 1 nm precision
Research, design, and first manufacturing of MRAM nonvolatile memory device	Lower switching current densities and decrease error rate due to thermal fluctuations in the magnetization reversal process in MRAM
	Achieve cost-effective architectures of integrated memory and logic using MRAM
Elucidation of electronic, optical, and thermal properties of carbon nanotubes and graphene, and establishment of a new class of electronic materials: carbon electronics	Discover a new logic device capable of switching with energy on the order of a few kT, potentially utilizing an alternative state variable for representing information
Discovery of very long spin lifetimes in nitrogen-vacancy (NV) centers in diamond; the quantum state of these centers can be initialized, manipulated, and measured with high fidelity at room temperature	Develop a quantum repeater for long-range quantum communication involving teleportation of quantum bits
Nanophotonics and Plasmonics	
Achievement of slowed light in solid state nanophotonic structures (Chap. 11); which enables applications and information systems never before available to photonic systems, such as delaying and storing optical signals	Store light with millisecond storage times or longer: although "slowed" light has been demonstrated, truly stored light has not; this could come about through extremely high-Q, low-loss resonant structures
Rapid advances in the field of plasmonics, and the innovations that accompany plasmonics, such as ultra-high resolution optical imaging	Use plasmonic enhanced-emission and detection to achieve controlled absorption and emission of light from single molecules
First demonstrations of metamaterials (materials with reverse diffraction index) at visible and near-infrared wavelengths	Create "superlenses" for ultra-high-resolution imaging and "cloaking" at multiple wavelengths (a change of paradigm in optics)
Realization of ultralow-threshold lasers, with thresholds of tens of nanoWatts	Achieve "thresholdless" lasers, where efficiency of energy transfer is so great that lasing can be initiated with minuscule power input to achieve exceptionally high power gains

(continued)

Table 1 (continued)

Main Achievements/Discoveries since 2000	Fundamental goals/"Holy Grails" to Attain and Barriers to Overcome by 2020
Nanostructured Catalysts	
Initial ability to characterize some catalytic processes in "the working state"	Achieve a complete snapshot of a multistep catalytic process
Control over the size, structure, and crystalline face of nano-sized catalysts	Ensure robustness and stability of nano-sized catalysts
Demonstrated of ability to monitor "single turnover events" (single catalytic events)	Overall goal: precise control of composition and structure of catalysts over length scales spanning 1 nm to 1 µm, allowing the efficient control of reaction pathways
Nanostructured catalysts introduced in production after 2000 represent 30–40% worldwide of all catalysts	New nanostructured catalysts will cover at least 50% of the market worldwide by 2020
Emerging Uses of Nanomaterials	
Establishment of synthesis and separation strategies for producing monodisperse nanomaterials, e.g., chirality-separated carbon nanotubes	Develop a complete library of monodisperse nanomaterials at industrial-scale quantities
Realization of bulk nanocomposites/coatings with predictable and unique properties based on monodisperse nanoscale building blocks (e.g., transparent conductors based on carbon nanotubes and graphene)	Realize hierarchical nanostructured materials with independent tunability of previously coupled properties, (e.g., decoupling optical and electrical properties for photovoltaics, decoupling electrical and thermal properties for thermoelectrics)
Evolution from microcrystalline to nanocrystalline metals, polymer micro-composites to nano-composites, and microscale to nanoscale particle coatings	Nanocomposite coatings with improved mechanical, thermal, chemical, electrical, magnetic, and optical properties compared to current state of the art
Realization of lighter and higher conductivity materials for airplane, satellite, and spacecraft wiring	Realize nanocomposites for structural components, thus enabling 40% weight reduction in airplane designs with better overall performance
Realization of nanofluidic devices and systems	Achieve scalable nanofluidic systems for processing in biotechnology, pharmaceuticals, and chemical engineering
Introduction of cellulose wood fibers in nanocomposite materials	Mass use in nanotechnology of renewable and earth-abundant raw materials
Research Facility Infrastructure	
Rapid expansion of interdisciplinary nanoscience user centers, including large-scale facilities, as engines for interdisciplinary science and engineering discovery	Expand the breadth of interdisciplinary center capabilities and extend the geographical distribution for more widespread access
Establishment of over 150 interdisciplinary research centers and user facilities in the United States and many others worldwide, providing broad access to fabrication and characterization facilities	Create open access centers and hubs as test beds for development and maturation of innovative nano-enabled device and system concepts

(continued)

Table 1 (continued)

Main Achievements/Discoveries since 2000	Fundamental goals/"Holy Grails" to Attain and Barriers to Overcome by 2020
Creation of the Nanoscale Computation Network in 2002, redesign of the National Nanotechnology Infrastructure Network in 2003, and establishment of the Network for Informal Science Education in 2004, providing more democratic and global access to nanoscale science and engineering knowledge and tools	Achieve widespread use of net-based remote control of instrumentation and local technician support at research facilities to enable reduced travel requirements and increased access by students
Education Infrastructure	
Nanotechnology has begun to foster interdisciplinary perspectives on science and engineering at all levels of STEM education	Embed nanoscale science and engineering education in internationally benchmarked standards and curriculum at all levels of education, but especially in the K–12 developmental progression
NanoHub, NISE, and the NACK Nano Education Portal of the NCLT are providing web-based access to nanoscale science and engineering resources	Establish a network of regional hub sites – the "Nanotechnology Education Hub Network" – with a sustainable infrastructure
Publication of more than 50 textbooks for college-level nanoscale science and engineering courses leading to degrees with nanoscale science and engineering minors (and some majors) and/or certificates	Migrate education for the nanoscale from a supplement to traditional disciplines into its own specialties, e.g., nano-education organizations, degrees, and professional disciplines
Nanotechnology has emerged as a topic of interest on websites, in exhibits, and in educational programs at science museums around the world and across the country, including at Walt Disney World's Epcot Center	Incorporate nanoscale science and engineering into all levels of STEM education
Governance	
Establishment of specific methods for governance of nanotechnology: bottom-up multi-agency governance approach, multi stakeholder assessment, scenario development	Emplace new principles and organizations for risk governance of new generations of nanotechnology products and processes with increased complexity, dynamics, biology contents, and uncertainty
Establishment of an international community of professionals and organizations, including for nanotechnology EHS and ELSI	Prepare the knowledge, people, and regulatory capacity to address mass use of nanotechnology by 2020
Developments in nomenclature, patents, standards, and standard materials	Create internationally recognized reference materials, terminology, materials certification, and measurement standards for nanomaterials
Creation of investment programs across disciplines, areas of application, and funding agencies	Institutionalize funding programs for nanotechnology research, education, manufacturing, and nanotechnology EHS and ELSI

(continued)

Table 1 (continued)

Main Achievements/Discoveries since 2000	Fundamental goals/"Holy Grails" to Attain and Barriers to Overcome by 2020
Establishment of a "Nanotechnology in Society" network with funding comparable to that of hard science projects	Institutionalize earlier integration of societal implications programs with hard-science programs, and of research, production and regulation organizations
Establishment of nanoinformatics as a new field for communication, design, manufacturing, and medicine in nanotechnology	Develop a national and international network for nanoinformatics
Development of a "scenarios approach" for nanotechnology foresight and governance	Increase international funding mechanisms for nanotechnology areas of common use and benefit as nanotechnology becomes a progressively more socio-economic undertaking
Initiation of funding programs for industry-inspired fundamental research with industry sectors using Collaborative Boards for Advancing Nanotechnology (CBAN)	Integrate discovery and innovation programs into public-private-partnership platforms where academics, industry, economists, and regulators are involved in all stages of the innovation process

The Long View of Nanotechnology Development: The National Nanotechnology Initiative at 10 Years*, **

Mihail C. Roco**

Abstract A global scientific and societal endeavor was set in motion by the nanotechnology vision formulated in 1999 that inspired the National Nanotechnology Initiative (NNI) and other national and international R&D programs. Establishing foundational knowledge at the nanoscale has been the main focus of the nanotechnology research community in the first decade. As of 2009, this new knowledge underpinned about a quarter of a trillion dollars worldwide market, of which about $91 billion was in U.S. products that incorporate nanoscale components. Nanotechnology is already evolving towards becoming a general-purpose technology by 2020, encompassing four generations of products with increasing structural and dynamic complexity: (1) passive nanostructures, (2) active nanostructures, (3) nanosystems, and (4) molecular nanosystems. By 2020, the increasing integration of nanoscale science and engineering knowledge and of nanosystems promises mass applications of nanotechnology in industry, medicine, and computing, and in better comprehension and conservation of nature. Nanotechnology's

---

* "If I were asked for an area of science and engineering that will most likely produce the breakthroughs of tomorrow, I would point to nanoscale science and engineering often called simply 'nanotechnology.'" (Neil Lane, April 1, 1998, NSF Testimony in U.S. Congress)

"Some of these [nanotechnology] research goals will take 20 or more years to achieve. But that is why there is such a critical role for the federal government." (President Bill Clinton, Speech announcing NNI at Caltech, January 20, 2000)

**This chapter is based on the author's experience in the nanotechnology field, as founding chair of the NSET Subcommittee coordinating the NNI and as a result of interactions in international nanotechnology policy arenas. The opinions expressed here are those of the author and do not necessarily reflect the position of NSTC/NSET or NSF.

M.C. Roco (✉)
National Science Foundation, 4201 Wilson Boulevard, Arlington, VA 22230, USA
e-mail: mroco@nsf.gov

rapid development worldwide is a testimony to the transformative power of identifying a concept or trend and laying out a vision at the synergistic confluence of diverse scientific research areas.

This chapter provides a brief perspective on the development of the NNI since 2000 in the international context, the main outcomes of the R&D programs after 10 years, the governance aspects specific to this emerging field, lessons learned, and most importantly, how the nanotechnology community should prepare for the future.

Keywords Nanoscale science and engineering • Research opportunities • Research outcomes • Public-private partnerships • Governance • Return on investment • International perspective

1 The Import of a Research-Oriented Definition of Nanotechnology

The National Science Foundation (NSF) established its first program dedicated to nanoparticles in 1991 and from 1997 to 1998 funded a cross-disciplinary program entitled "Partnerships in Nanotechnology" [1]. However, only in 1998–2000 were fragmented fields of nanoscale science and engineering brought together under a unified science-based definition and a 10-year R&D vision for nanotechnology. These were laid out in the 1999 National Science Foundation workshop report, *Nanotechnology Research Directions* [2], which was adopted in 2000 as an official document by National Science and Technology Council (NSTC). These were significant steps toward establishing nanotechnology as a defining technology of the twenty-firt century. The definition of nanotechnology (see sidebar) was agreed to in 1998–1999 after consultation with experts in over 20 countries [3] and achieved some degree of international acceptance. This definition is based on novel behavior of matter and ability of scientists to restructure that matter at an intermediate length scale. This is conceptually different from the previous definitions used before 1999 that were focused on either small features under a given size, ultra-precision engineering, ultra-dispersions, or creating patterns of atoms and molecules on surfaces. The internationally generated vision published in 1999 provides guidance for nanotechnology discovery and innovation in this cross-disciplinary, cross-sector domain. It has become clear after about 60 countries developed nanotechnology activities by 2004 that without this definition and corresponding long-term vision, nanotechnology would have not been developed on the same accelerated, conceptually unifying and transforming path [4].

> **DEFINITION OF NANOTECHNOLOGY (SET OUT IN *Nanotechnology Research Directions*, 1999)[1]**
>
> Nanotechnology is the ability to control and restructure the matter at the atomic and molecular levels in the range of approximately 1–100 nm, and exploiting the distinct properties and phenomena at that scale as compared to those associated with single atoms or molecules or bulk behavior. The aim is to create materials, devices, and systems with fundamentally new properties and functions by engineering their small structure. This is the ultimate frontier to economically change materials properties, and the most efficient length scale for manufacturing and molecular medicine. The same principles and tools are applicable to different areas of relevance and may help establish a unifying platform for science, engineering, and technology at the nanoscale. The transition from single atoms or molecules behavior to collective behavior of atomic and molecular assemblies is encountered in nature, and nanotechnology exploits this natural threshold.

In 2010, the International Standardization Organization (ISO) Technical Committee 229 on nanotechnologies [5] issued a definition of nanotechnology that essentially has the same elements as those of the 1999 definition: the application of scientific knowledge to manipulate and control matter in the nanoscale range to make use of size- and structure-dependent properties and phenomena distinct from those at smaller or larger scales. Full acceptance and use of the ISO definition in the environmental, health, and safety (EHS) community has not yet been resolved [6]. Nevertheless, in 1999 as now, there has been a shared acceptance of the value of clearly defining nanotechnology to support common language and purpose for scientific discourse, engineering, education, manufacturing, commerce, regulation, and tracking of investments. Defining the long-term vision for nanotechnology development is especially critical because of nanotechnology's rapid emergence as a fundamentally new scientific and engineering paradigm and because of its broad implications for societal wellbeing.

The 1999 unifying definition of and long-term vision for nanotechnology paved the way for the U.S. National Nanotechnology Initiative, announced in January 2000. The main reasons for beginning the NNI were to fill major gaps in fundamental knowledge of matter and to pursue the novel and economic applications anticipated for nanotechnology. Coherent and sustained R&D programs in the field were soon announced by other nations: Japan (April 2001), Korea (July 2001), the European Community (March 2002), Germany (May 2002), China (2002), and Taiwan (September 2002). Over 60 countries established programs at a national level between 2001 and 2004, partially inspired or motivated by the NNI. However,

---

[1] Roco et al. [2].

the first and largest such program was the NNI itself. Its cumulative funding since 2000 of more than $12 billion, including about $1.8 billion in 2010, places the NNI second only to the space program in the U.S. civilian science and technology investments. This 2010 international study, involving experts from over 35 countries, aims to redefine the goals for nanotechnology development for the next decade.

2 Indicators of Nanotechnology Development Globally, 2000–2020

Six key indicators, described below and in Table 1, help portray the value of investments in nanotechnology development and associated science breakthroughs and technological applications. These indicators show average annual growth rates worldwide of approximately 25% between 2000 and 2008. The average growth rates of all indicators fell by more than half worldwide during the financial crisis of 2009. They appear to be returning to higher rates in 2010 compared to 2009, but with significant differences between countries and domains of relevance.

1. *The number of researchers and workers involved in one domain or another of nanotechnology* was estimated at about 400,000 in 2008, of which about 150,000 were in the United States. The estimate made in 2000 that there would be two million nanotechnology workers worldwide by about 2015 (800,000 in the United States) would have been realized if the 25% rate growth had continued. The initial 2000 estimation for quasi-exponential growth in the nanotechnology workforce [9] held

Table 1 Six key indicators of nanotechnology development in the world and the United States[a]

World/US/	People Primary workforce	SCI papers	Patent applications	Final products market	R&D funding public + private	Venture capital
2000 (actual)	~60,000 /25,000/	18,085 /5,342/	1,197 /405/	~$30 B /$13 B/	~$1.2 B /$0.37 B/	~$0.21 B /$0.17 B/
2008 (actual)	~400,00 /150,000/	70,287 /15,000/	12,776 /3,729/	~$200 B /$80 B/	~$15 B /$3.7 B/	~$1.4 B /$1.17 B/
2000–2008 average growth	~25%	~18%	~35%	~25%	~35%	~30%
2015 (2000 estimate[b])	~2,000,000 /800,000/			~1,000 B /$400 B/		
2020 (extrapolation)	~6,000,000 /2,000,000/			~$3,000 B /$1,000 B/		

[a] Global figures are indicated in bold text; U.S. figures are indicated in gray. Science Citation Index (SCI) papers and patent applications were searched by title-abstract keywords, using the method described in Chen and Roco [7]. Venture capital estimations were made by Lux Research; see Sect. 8.11 in chapter "Innovative and Responsible Governance of Nanotechnology for Societal Development"
[b] Roco and Bainbridge [8]

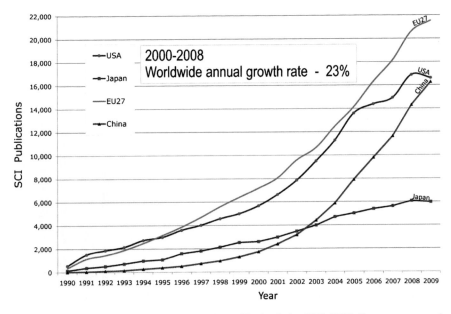

Fig. 1 Nanotechnology publications in the Science Citation Index 1990–2009. Data was generated from an online search in the Web of Science using a "title-abstract" search in SCI database for nanotechnology by keywords (Courtesy of H. Chen, Y. Dang, and M. Roco 2010)

up to 2008, and because of new generations of nanotechnology products envisioned to enter the market within the next few years, it is expected to continue.

2. *The number of Science Citation Index (SCI) papers* reflecting discoveries in the area of nanotechnology reached about 65,000 in 2008 as compared to 18,085 in 2000, based on a title-abstract keyword search [7]. The increase is rapid and uneven around the world, as suggested by Fig. 1. About 4.5% of SCI papers published in 2008 in all areas included nanoscale science and engineering aspects.

3. *Inventions reflected by the number of patent applications filed* in the top 50 depositories was about 13,000 in 2008 (of which 3,729 were filed at the U.S. Patent and Trade Office, USPTO), as compared to about 1,200 in 2000 (of which 405 were filed at USPTO) [10, 11], with an annual growth rate of about 35%, as shown in Fig. 2. The patent applications in over 50 national or international patent depositories were searched by using the title-abstract keyword search. About 0.9% of patent applications published worldwide, and about 1.1% at USPTO in 2008 in all areas, included nanoscale science and engineering aspects.

4. *The value of products incorporating nanotechnology as the key component* reached about $200 billion in value worldwide in 2008, of which about $80 billion was in the United States (these products relied on relatively simple nanostructures). The estimation made in 2000 [8] for a product value of $1 trillion by 2015, of which $800 billion would be in the United States, still appears to hold (see Fig. 3). The market is doubling every 3 years as a result of successive introduction of new products. The Lux Research estimate for the 2009 market worldwide was about $254 billion (See Sect. 8.11 in chapter "Innovative and Responsible

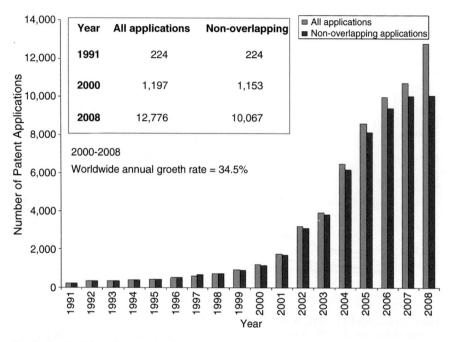

Fig. 2 Total number of nanotechnology patent applications in 15 leading patent depositories in the world from 1991–2008. Two sets of data are reported based on the number of all nanotechnology patent applications and the number of non-overlapping nanotechnology patent applications (by considering one patent application per family of similar patents submitted at more than one depository) [12]

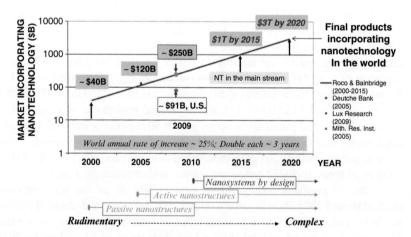

Fig. 3 Market timeline: projection for the worldwide market of finite products that incorporate nanotechnology (estimation made in 2000 at NSF; [8]). These estimations were based on direct contacts with leading experts in large companies with related R&D programs in the United States, Japan, and Europe, as part of the international study completed between 1997 and 1999 [3]

Table 2 Examples of penetration of nanotechnology in several industrial sectors. The market percentage and its absolute value affected by nanotechnology are shown for 2010

U.S.	2000	2010	Est. in 2020
Semiconductor industry	0 (with features <100 nm) 0 (new nanoscale behavior)	60% (~$90B) 30% (~$45B)	100% 100%
New nanostructured catalysts	0	~35% (~35B impact)	~50%
Pharmaceutics (therapeutics and diagnostics)	0	~15% (~$70B)	~50%
Wood	0	0	~20%

Governance of Nanotechnology for Societal Development"), about on the 2000 estimated curve, although the Lux estimate for the value of U.S. nanotechnology products in 2009 of about $91 billion was about 10% under the 2000 estimated growth curve.
5. *Global nanotechnology R&D annual investment* from private and public sources reached about $15 billion in 2008, of which about $3.7 billion was in the United States, including the Federal Government contribution of about $1.55 billion.
6. *Global venture capital investment in nanotechnology* reached about $1.4 billion in 2008, of which about $1.17 billion was in the United States (courtesy of Lux Research 2010). Venture capital funds decreased about 40% during the 2009 financial crisis (see Sect. 8.11 in chapter "Innovative and Responsible Governance of Nanotechnology for Societal Development").

Because of the technological and economic promise, nanotechnology has penetrated the emerging and classical industries especially after 2002–2003. The increase in nanocomponent complexity and the proportion of nanotechnology penetration is faster in emerging areas such as nanoelectronics and slower in more classical industry sectors such as wood and paper industry as illustrated in Table 2. Penetration of nanotechnology in key industries is related to the percentage industry spends on R&D. Penetration of nanotechnology in two biomedical areas is exemplified in chapter "Innovative and Responsible Governance of Nanotechnology for Societal Development" (Sect. 8.10).

Figure 4 shows the balance of Federal nanotechnology investments and return on investments (outputs) in the United States in 2009. Other specific indicators of the national investment in nanotechnology have increased significantly in the United States since 2000:

- The specific annual Federal R&D nanotechnology expenditure per capita has grown from about $1 in fiscal year 2000 to about $5.7 in 2010.
- The fraction of the Federal R&D nanotechnology investment as compared to all actual Federal R&D expenditures grew from 0.39% to about 1.5% in 2008.

Qualitative changes also are important to evaluating the impact of the NNI, even if there is no single indicator to characterize them. These include (1) the creation of a vibrant multidisciplinary, cross-sector, international community of professionals and

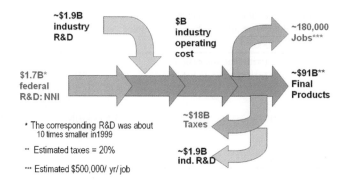

Fig. 4 Estimation of the outcomes of U.S. Federal investment in nanotechnology R&D in 2009. The figure shows an annual balance between investments and outputs. (*) The corresponding R&D was about ten times smaller in 1999 when the fundamental research for 2009 products may have started, (**) The estimated market of products where nanoscale components are essential; taxes are estimated based on Council of Chemical Research average estimation for chemical industry, (***) Estimated number of nanotechnology-related jobs, assuming about $500,000/year/job

organizations engaged in various dimensions of the nanotechnology enterprise; (2) changes in the scientific research culture that are coming about through energizing interdisciplinary academic research collaborations with industry and the medical field; and (3) increasingly unified concepts for engineering complex nanostructures "from the bottom up" for new materials, biology and healthcare technologies, digital information technologies, assistive cognition technologies, and multicomponent systems.

3 Two Foundational Steps in Nanotechnology Development

In 2000, it was estimated that nanotechnology would grow in two foundational phases from passive nanostructures to complex nanosystems by design (illustrated in Figs. 5 and 6):

1. The first foundational phase (2001–2010), which was focused as anticipated on inter-disciplinary research at the nanoscale, took place in the first decade after defining the long-term vision. Its main results are discovery of new phenomena, properties, and functions at the nanoscale; synthesis of a library of components as building blocks for potential future applications; tool advancement; and improvement of existing products by incorporating relatively simple nanoscale components. This phase, dominated by a science-centric ecosystem, might be called "Nano1."
2. The second foundational phase (2011–2020), will be focused on nanoscale science and engineering integration, is projected to transition towards direct measurements with good time resolution, science-based design of fundamentally new products, and general-purpose and mass use of nanotechnology. The focus of R&D and applications is expected to shift towards more complex nanosystems, new areas of relevance, and fundamentally new products. This phase is expected to be

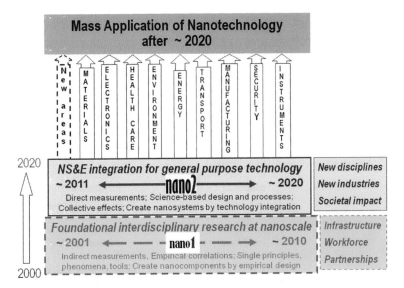

Fig. 5 Creating a new field and community in two foundational phases ("NS&E" is nanoscale science and engineering)

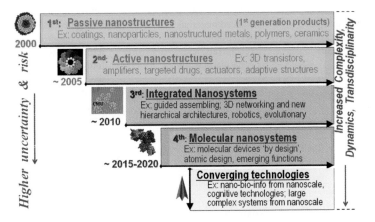

Fig. 6 Timeline for the beginning of industrial prototyping and nanotechnology commercialization: introduction of a new genarations of products and productive processes in 2000–2020 [4, 13]

dominated by an R&D ecosystem driven by socio-economic considerations; it might be called "Nano2."

The transition from the Nano1 phase to the Nano2 phase is focused on achieving direct measurements at the nanoscale, science-based design of nanomaterials and nanosystems, and general-purpose technology integration (Table 3). Several R&D targets for achievement by 2020 are presented in this chapter and detailed in this volume.

After 2020, nanotechnology R&D is projected to develop closely with other emerging and converging technologies, creating new science and engineering

Table 3 Transition between the two predominant phases in nanotechnology development, 2000–2020

Interval	2001–2010 ("Nano1")	2011–2020 ("Nano2")
Measurements	Indirect, using time and volume averaging approaches	Direct, with atomic precision in the biological or engineering domains, and femtosecond resolution
Phenomena	Discovery of individual phenomena and nanostructures	Complex simultaneous phenomena; nanoscale integration
New R&D paradigms	Multidisciplinary discovery from the nanoscale	Focus on new performance; new domains of application; an increased focus on innovation
Synthesis and manufacturing processes	Empirical/semi-empirical; dominant: top-down miniaturization; nanoscale components; polymers and hard materials	Science based design; increasing molecular bottom-up assembly; nanoscale systems; increasingly bio-based processes
Products	Improved existing products by using nanocomponents	Revolutionary new products enabled by creation of new systems; increasing bio-medical focus
Technology	From fragmented domains to cross-sector clusters	Towards emerging and converging technologies
Nanoscience and engineering penetration into new technologies	Advanced materials, electronics, chemicals, and pharmaceuticals	Increasing to: nanobiotechnology, energy resources, water resources, food and agriculture, forestry, simulation-based design methods; cognitive technologies
Education	From micro- to nanoscale–based	Reversing the pyramid of learning by earlier learning of general nanotechnology concepts [14]
Societal impact	Ethical and EHS issues	Mass application; expanding sustainability, productivity, and health; socio-economic effects
Governance	Establish new methods; science- centric ecosystem	User-centric ecosystem; increasingly participatory; techno-socio-economic approach
International	Form an S&T community; establish nomenclature, patent, and standards organizations	Global implications for economy, balance of forces, environment, sustainability

domains and manufacturing paradigms [15, 16]. In 1999/2000, a convergence was reached in defining the nanoscale world because typical phenomena in material nanostructures were better measured and understood with a new set of tools, and new nanostructures had been identified at the foundations of biological systems, nanomanufacturing, and communications. The new challenge for the next decade is

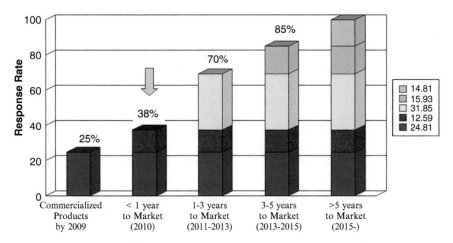

Fig. 7 An accelerating transition to new commercial products is estimated after 2011 (after Fig. 4–31, NCMS [18])

building systems from the nanoscale that will require the combined use of nanoscale laws, biological principles, information technology, and system integration. After 2020, one may expect divergent trends as a function of system architectures. Several possible divergent trends are system architectures based on guided molecular and macromolecular assembling, robotics, biomimetics, and evolutionary approaches.

A shift in research towards "active nanostructures" that change their composition or state during use has been noted in the rapid increase of related publications since 2005 [17]. The percent of papers on active nanostructures more than doubled to 11% of total nanotechnology papers in 2006. An observed transition to introduction of nanosystems appears to be correlated to commercial interests (Fig. 7) [18]; more than 50% of 270 surveyed manufacturing companies expressed interest in production or design using nanoscale science and engineering by about 2011.

4 Genesis and Structure of the National Nanotechnology Initiative

As the then Chair of the NSTC's Interagency Working Group on Nanoscale Science, Engineering, and Technology (IWGN),[2] the author had the opportunity to propose the National Nanotechnology Initiative with annual budget of about 1/2 billion U. S. dollars on March 11, 1999, at a meeting of the White House Economic Council (EC) and the Office of Science and Technology Policy (OSTP) as part of

---

[2] The IWGN was superseded in August 2000 by the Nanoscale Science, Engineering and Technology (NSET) Subcommittee of the NSTC Committee on Technology. In 1999 Neil Lane was the Director of OSTP, and Tom Kalil was Deputy Director of the White House National Economic Council and the White House co-chair of the IWGN. Jim Murday was the secretary of IWGN.

a competition for a national research priority to be funded in fiscal year 2001. The approval process moved to the Office of Management and Budget (OMB) in November 1999, the Presidential Council of Advisors in Science and Technology (PCAST) in December 1999, and the Executive Office of the President in January 2000. Hearings were held in the House and Senate of the United States Congress in the Spring of 2000. In November 1999, the OMB recommended nanotechnology as the only new R&D initiative for fiscal year (FY) 2001. On December 14, 1999, PCAST highly recommended that the President fund nanotechnology R&D. Thereafter, it was a quiet month: the Executive Office of the President advised the working group to restrain from speaking to the media because a White House announcement would be made.

President Clinton announced the National Nanotechnology Initiative at a speech at the California Institute of Technology (Caltech) in January 2000, asking listeners to imagine the new world that nanotechnology could make possible. After his speech, the IWGN moved firmly to prepare the Federal plan for R&D investment in nanotechnology and to identify key opportunities and potential participation of various agencies in the proposed initiative. House and Senate hearings brought the needed recognition and feedback from Congress. Representing the working group, the author spoke to major professional societies (the American Chemical Society, the Institute for Electric and Electronics Engineering, the American Society of Mechanical Engineering, and the American Institute of Chemical Engineering), and attended national meetings in about 20 countries to introduce the new U.S. nanotechnology initiative. The NNI has been implemented since FY 2001, with unbroken support from the Clinton, Bush, and Obama Administrations.

A challenge in the early years of the initiative, with so many new developments, was maintaining consistency, coherence, and original thinking. The definition of nanotechnology, the initiative's name, and the name of the National Nanotechnology Coordination Office (NNCO) were decided in 1999–2000. The NNI's name was proposed on March 11, 1999, but it was held under "further consideration" until the Presidential announcement, due to concerns from several professional societies and committees that it did not explicitly include the word "science." The simple name "National Nanotechnology Initiative" was selected to better show its relevance to society.

The NNI is a long-term R&D program that began in FY 2001 with participation from eight Federal agencies: the Departments of Defense, Energy, and Transportation, the Environmental Protection Agency, the National Aeronautics and Space Administration, the National Institutes of Health, the National Institute of Standards and Technology, and the National Science Foundation. As of 2010, the NNI coordinates the nanotechnology-related activities of 25 Federal departments and independent agencies. Table 4 lists the full membership in 2010.

The NSTC coordinates the initiative through the efforts of the agency members of the Nanoscale Science and Engineering (NSET) Subcommittee of the NSTC Committee on Technology. Assisting the NSET Subcommittee is the NNCO, which provides technical and administrative support. The NSET Subcommittee has chartered four working groups: the Global Issues in Nanotechnology (GIN) Working Group; the Nanomanufacturing, Industry Liaison, and Innovation (NILI) Working

Table 4 NNI members (25 Federal departments and agencies) in September 2010

Federal agencies with budgets dedicated to Nanotechnology Research and Development
Consumer Product Safety Commission (CPSC)
Department of Defense (DOD)
Department of Energy (DOE)
Department of Homeland Security (DHS)
Department of Justice (DOJ)
Department of Transportation (DOT, including the Federal Highway Administration, FHWA)
Environmental Protection Agency (EPA)

Food and Drug Administration (FDA, Department of Health and Human Services)
Forest Service (FS, Department of Agriculture)
National Aeronautics and Space Administration (NASA)
National Institute for Occupational Safety and Health (NIOSH, Department of Health and Human Services)
National Institute of Food and Agriculture (NIFA, Department of Agriculture)
National Institutes of Health (NIH, Department of Health and Human Services)
National Institute of Standards and Technology (NIST, Department of Commerce)
National Science Foundation (NSF)

Other participating agencies
Bureau of Industry and Security (BIS, Department of Commerce)
Department of Education (DOEd)
Department of Labor (DOL)
Department of State (DOS)
Department of the Treasury (DOTreas)
Intelligence Community (IC)
Nuclear Regulatory Commission (NRC)
U.S. Geological Survey (USGS, Department of the Interior)
U.S. International Trade Commission (USITC, a non-voting member)
U.S. Patent and Trademark Office (USPTO, Department of Commerce)

Group; the Nanotechnology Environmental & Health Implications Working Group (NEHI); and the Nanotechnology Public Engagement & Communications Working Group (NPEC).

4.1 The NNI Organizing Principles

The NNI's long-term view of nanotechnology development aims to enable exploration of a new domain of scientific knowledge and incorporation of a transformational general-purpose technology into the national technological infrastructure, with a 20-year view to reach some degree of systematic control of matter at the nanoscale and mass use [19]. The vision that "systematic control of matter at the nanoscale will lead to a revolution in technology and economy for societal benefit" is still the guiding principle of the initiative.

During the 10-year time span of fiscal years 2001–2010, a thriving interdisciplinary nanotechnology community of about 150,000 contributors has emerged in the

United States, along with a flexible R&D infrastructure consisting of about 100 large nanotechnology-oriented R&D centers, networks, and user facilities, and an expanding industrial base of about 3,000 companies producing nanotechnology-enabled products. Considering the complexity and rapid expansion of the U.S. nanotechnology infrastructure, the participation of a coalition of academic, industry, business, civic, governmental, and nongovernmental organizations in nanotechnology development is becoming essential and complementary to the centralized approach of the NNI. The leadership role of the Federal government through the NNI must continue in support of basic research, restructuring the education pipeline, and guiding responsible development of nanotechnology as a transformative scientific schema, as envisioned in 2000. At the same time, however, the emphasis of government leadership in nanotechnology development is changing toward increasing support of R&D for innovation, nanomanufacturing, and societal benefit, while the private sector's responsibility is growing for funding R&D in nanotechnology applications. Since 2006, private nanotechnology R&D funding in the U.S. has excluded public funding.

Several means to ensure accountability are built into the twenty-first Century Nanotechnology Research and Development Act that governs the NNI (P.L. 108–153, 15 USC 7501, of the U.S. Congress, December 3, 2003). With extensive input from NSET Subcommittee agency members, the NNI organizations submit to Congress every February an annual report on the NNI and a combined nanotechnology budget request. OMB manages and evaluates the NNI budget crosscut. Following the *Nanotechnology Research Directions* report published in 2000, NNI leadership prepares a Strategic Plan every 3 years (2004, 2007, and 2010). The NNI is evaluated every 3 years by the National Research Council of the National Academies and periodically by PCAST in its role as the National Nanotechnology Advisory Panel. Ad hoc evaluations by the Government Accountability Office and other organizations help ensure best use of taxpayer funds and respect for the public interest.

The organizing principles of the NNI have undergone two main stages between 2001 and 2010, and a third stage is projected to begin in FY 2011:

1. *Between FY 2001 and FY 2005*, nanotechnology research under the NNI was focused on five modes of investment: (1) fundamental research, (2) priority research areas, (3) centers of excellence, (4) infrastructure, and (5) societal implications and education. The second mode, collectively known as "grand challenges," focused on nine specific R&D areas directly related to applications of nanotechnology; they also were identified as having the potential to realize significant economic, governmental, and societal impact in about a decade. These priority research grand challenge areas were:

    – Nanostructured materials by design
    – Manufacturing at the nanoscale
    – Chemical-biological-radiological-explosive detection and protection
    – Nanoscale instrumentation and metrology
    – Nano-electronics, -photonics, and -magnetics
    – Healthcare, therapeutics, and diagnostics
    – Efficient energy conversion and storage

- Microcraft and robotics
- Nanoscale processes for environmental improvement
- Focused research programs and major infrastructure initiatives in the first 5 years led to the formation of the U.S. nanoscale community, a strong R&D infrastructure, and new nanotechnology education programs.

2. *Between FY 2006 and FY 2010*, nanotechnology research under the NNI was focused on four goals and seven or eight investment categories [20, 21]. The goals are to (1) advance a world-class research and development program; (2) foster the transfer of new technologies into products for commercial and public benefit; (3) develop and sustain educational resources, a skilled workforce, and the supporting infrastructure and tools to advance nanotechnology; and (4) support responsible development of nanotechnology. The NNI investment categories (originally seven, amended in 2007 to eight categories), called program component areas (PCAs), are:

- Fundamental nanoscale phenomena and processes
- Nanomaterials
- Nanoscale devices and systems
- Instrumentation research, metrology, and standards for nanotechnology
- Nanomanufacturing
- Major research facilities and instrumentation acquisition
- Environment, health, and safety
- Education and societal dimensions

3. *Beginning in fiscal year 2011*, the NNI will introduce three research and development "signature initiatives" for important long- and short-term application opportunities[3]: (1) Nanotechnology Applications for Solar Energy, (2) Sustainable Nanomanufacturing, and (3) Nanoelectronics for 2020 and Beyond. Other research "signature initiatives," and plans to enhance the innovation ecology and societal outcomes of nanotechnology, are under consideration [22].

4.2 NNI Investment in Nanotechnology R&D

The NNI's total R&D investment for nanotechnology has increased about 6.6-fold in the past decade, from $270 million in FY 2000 to about $1.8 billion in FY 2010, as shown in Fig. 8. All numbers shown in the figure are actual spending, except for FY 2010, which shows estimated spending for the current year, and FY 2011, which shows the requested budget for next year. The FY 2009 spending shown does not include $511 million in additional funding under the American Recovery and Reinvestment Act (ARRA). The 2011 budget request shown here does not include Department of Defense (DOD) earmarks included in previous years ($117 million in 2009).

---

[3] Details are available at http://www.nano.gov/html/research/signature_initiatives.html

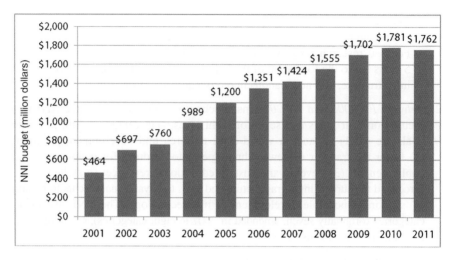

Fig. 8 NNI budgets for fiscal years 2001–2011 in millions U.S. dollars. The 2009 budget does not include the one-time supplemental ARRA funding in 2009 of $511 million

Table 5 estimates various individual government budgets and the European Union (EU) budgets for nanotechnology globally using the NNI definition and direct contacts with program managers in other countries. The 2009 government investments around the word totaled about $7.8 billion, of which $1.7 billion was in the United States (through the NNI), without including the one-time ARRA funding in 2009 of $511 million. Although the figures in Table 5 for other countries' nanotechnology investments are just a general gauge of activity, it appears in very broad terms that whereas U.S. nanotechnology investment is rising, it is rising more slowly than the investment of other nations (Fig. 9).

The government nanotechnology R&D investments are plotted in Fig. 9 for EU, Japan, US and "others" as defined in Table 5. One notes the change of global investment rates about 2000 after the announcement of NNI and about 2005–2006 corresponding to the introduction of the second generation of nanotechnology products (first industry prototypes based on active nanostructures). In 2006, industry nanotechnology R&D investment exceeded respective public investment in both the US and worldwide.

5 Governance of Nanotechnology

Governing any emerging technology requires specific approaches [24], and for nanotechnology in particular, consideration of its potential to fundamentally transform science, industry, and commerce, and of its broad societal implications. It should be stressed that the technology governance approach needs to be focused on many facets, not only on risk governance [25]. Properly taking into account the roles and views of

Table 5 Estimated government nanotechnology R&D expenditures, 2000–2010 ($ millions/year)

Region	2000	2001	2002	2003	2004	2005	2006	2007	2008	2009	2010
EU+	200	~225	~400	~650	~950	~1,050	~1,150	~1,450	1,700	1,900	
Japan	245	~465	~720	~800	~900	~950	950	~950	~950	~950	
USA[a]	270	464	697	862	989	1,200	1,351	1,425	1,554	1,702+511[a]	~1,762
Others	110	~380	~550	~800	~900	~1,100	~1,200	~2,300	~2,700	~2,700	
Total	825	1,534	2,367	3,112	3,739	4,200	4,651	6,125	6,904	7,252; 7,763[b]	
est. U.S. % of EU	135	206	174	133	104	114	117	98	91	90; 116[b]	
est. U.S. % of Total	33	30	29	28	26	29	29	24	23	22; 28[b]	

Explanatory notes: for the EU+ figures, both national and EU funding is included; EU+ includes EU member countries and Switzerland as a function of year. The category "Others" includes Australia, Canada, China, Russia, Israel, Korea, Singapore, Taiwan, and other countries with nanotechnology R&D. Budget estimates use the nanotechnology definition as defined by the NNI (this definition does not include MEMS, microelectronics, or general research on materials) (see [2, 23]; and http://nano.gov). A fiscal year begins in the United States on October 1 and in most other countries 6 months later, around April 1. In the table above, [a] denotes the one-time supplemental ARRA nanotechnology related funding and [b] (the higher figure) includes the one-time supplemental ARRA funding

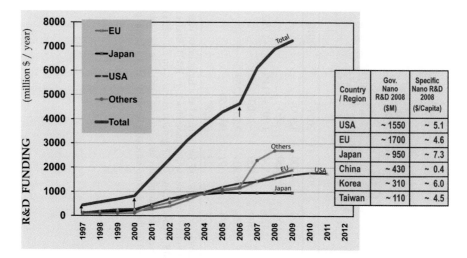

Fig. 9 2000–2009 Federal/national government R&D funding. (Budget estimates use the NNI definition of nanotechnology.). The *arrows* on the graph represent *2000*: announcement of NNI and introduction of commercial prototypes of "passive nanostructures" (first generation of nanotechnology products shown in Fig. 6, about 2000), and *2006*: introduction of commercial prototypes of "active nanostructures" (second generation of nanotechnology products, about 2005–2006); also, in 2006, industry nanotechnology R&D investment exceeded respective public investment in the U.S. and worldwide. Specific nanotechnology R&D per capita (inset table) uses the national nanotechnology expenditures and effective expenditure for all other R&D programs

the various stakeholders in the society—including their perceptions of science and technology, human behavior factors, and the varying social impacts of the technology—is an increasingly important factor in the development of any emerging, breakthrough technology. Optimizing societal interactions, R&D policies, and risk governance for nanotechnology development can enhance economic competitiveness and democratization, but all stakeholders must be equally invested. Chapter "Innovative and Responsible Governance of Nanotechnology for Societal Development" on innovative and responsible governance discusses four basic functions of the governance and four basic levels (or generations). Below are illustrated the application of the four basic functions or characteristics of effective governance of nanotechnology, that it should be (1) transformative, (2) responsible, (3) inclusive, and (4) visionary.

5.1 Transformative and Responsible Development of Nanotechnology

The goal of achieving transformative and responsible development of nanotechnology has guided many NNI decisions, with a recognition that investments must have a good return, the benefit-to-risk ratio must be justifiable, and societal concerns must be addressed. The goal of being transformative is being addressed not only through fundamental and applications-focused R&D and investment policy but also through implementing new modes of advocacy for innovation, resource-sharing, and cross-sector communication. NNI agencies introduced manufacturing at the nanoscale as a grand challenge in 2002, and at about the same time, NSF established its first research program on this topic, "Nanomanufacturing." In the next 4 years, NSF made awards to four Nanoscale Science and Engineering Centers (NSECs) on nanomanufacturing and the National Nano-manufacturing Network (NNN). Since 2006, the NNN has developed partnerships with industry and academic units, programs of the National Institute for Standards and Technology (NIST), National Institutes of Health, Department of Defense (DOD), and Department of Energy (DOE). The NNI agencies also established a new approach for interaction with various industry sectors, to augment previous models: the Consultative Boards for Advancing Nanotechnology (CBAN). DOE, NIST, DOD, and other agencies likewise have established individual programs to support advanced nanotechnology R&D. Several outcomes, such as science and technology platforms inspired or directly supported by NNI investment have been noted in various areas such as in instrumentation (Sandia National Laboratory), nanoparticles (DuPont), nano-components (General Electric) and carbon nanotube cables and sheets (National Reconnaissance Office [NRO], see Fig. 10).

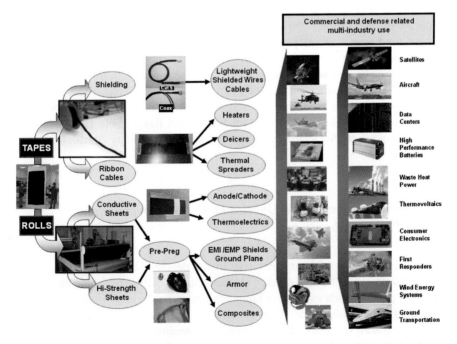

Fig. 10 Platform for carbon nanotube based cables and sheets in (Courtesy of Peter L. Antoinette, Nanocomp Technologies, Inc., and R. Ridgley, NRO, 2010)

In examples of programs focused on studying and advancing societal aspects of nanotechnology R&D, in 2004–2005, NSF began establishing new kinds of networks with national goals and outreach, focused on high school and undergraduate nanotechnology education (the National Centers for Learning and Teaching in Nanoscale Science and Engineering), nanotechnology in society (the Centers for Nanotechnology in Society), and informal nanotechnology science education (the Nanoscale Informal Science Education Network). Other aspects of pursuing responsible development of nanotechnology include the NNI's considerable and growing focus on nanotechnology environmental, health, and safety (nanotechnology EHS, or nanoEHS) research its interagency and international consultations on standards and regulation.

To support the NNI agency focus on the issues of transformative and responsible development of nanotechnology, the NSET Subcommittee established the NILI Nanomanufacturing, Industry Liaison, and Innovation NEHI (Nanotechnology Environmental and Health Issues), and GIN (Global Issues in Nanotechnology) Working Groups. The *transformative function* for nanotechnology development is addressed in chapters "Investigative Tools: Theory, Modeling, and Simulation" and "Enabling and Investigative Tools: Measuring Methods, Instruments, and Metrology" for tools, chapter "Synthesis, Processing, and manufacturing of Components, Devices, and Systems" for manufacturing, chapters "Applications: Nanobiosystems, Medicine, and Health," "Applications: Nanoelectronics and Nanomagnetics," "Applications: Nanophotonics and Plasmonics," "Applications: Catalysis by Nanostructured Materials," "Applications: High-Performance Materials and Emerging Areas" for applications, and chapter "Innovative and Responsible Governance of Nanotechnology for Societal Development" for innovation. The *responsible function* for nanotechnology development is discussed in chapter "Nanotechnology Environmental, Health, and Safety Issues" for nanotechnology environmental, health, and safety, in chapters "Nanotechnology for Sustainability: Environment, Water, Food, Minerals, and Climate" and "Nanotechnology for Sustainability: Energy Conversion, Storage, and Conservation", and in chapter "Innovative and Responsible Governance of Nanotechnology for Societal Development" for nanotechnology ethical, legal, and social issues (ELSI).

5.2 Inclusiveness in the Development and Governance of Nanotechnology

Addressing the goal of being inclusive in the development and governance of nanotechnology may be illustrated by (1) the inclusion of diverse stakeholders in the planning process (such as industry input, requesting public comments for planning documents, and holding workshops and dialogs with multiple partners in the process of producing the nanotechnology research directions reports of 1999 and 2010) and in preparation of various reports on societal implications (beginning with [8]);

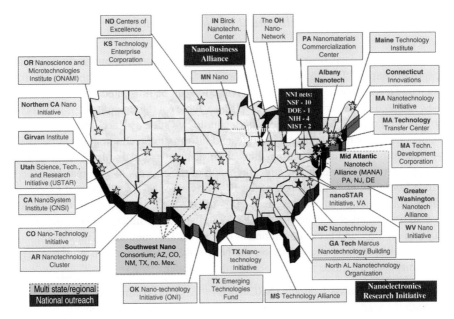

Fig. 11 As of 2009, 34 nanotechnology regional, state, and local initiatives existed in the United States (including one in Hawaii)

(2) partnering of all interested Federal agencies through the NSET Subcommittee (Table 3); (3) opening NNI strategy development process to the public in meetings and online (e.g., see http://strategy.nano.gov/); (4) R&D programs requiring all relevant disciplines and sectors of activity to work together; (5) supporting a network of 34 regional, state, and local nanotechnology alliances in the United States (http://nano.gov/html/meetings/nanoregional-update/ and Fig. 11); and (6) supporting international dialogs on nanotechnology (the first in 2004 with 25 nations and the European Union, the third in 2008 with 49 nations and the EU), and the United States also participates actively and regularly in a number of other international fora for nanotechnology (ISO, OECD, International Risk Governance Council, etc.) that are focused on development of appropriate international standards, terminology, regulations, etc. To help advance the progress towards responsibility and inclusiveness in nanotechnology development, NSF has established two centers for nanotechnology in society.

Regarding international aspects of nanotechnology governance, a multidisciplinary, international forum is needed in order to better address the nanotechnology scientific, technological, and infrastructure development challenges. Optimizing societal interactions, R&D policies, and risk governance for the converging new technologies can enhance economical competitiveness and democratization. The International Risk Governance Council [26] has provided an independent international perspective for a framework for identification, assessment, and mitigation of risk.

5.3 Vision in the Development of Nanotechnology

The goal of being visionary in the development of nanotechnology is discussed at length in chapter "Innovative and Responsible Governance of Nanotechnology for Societal Development." Support for this function can be illustrated by the long-term view adopted since the beginning of NNI (see Table 6); the integration of nanotechnology with other long-term emerging technologies such as R&D programs at the intersection of nanotechnology, biology, and information technology; development of long-term government partnerships with academia and industry, such as the Nanoelectronics Research Initiative; inclusion of the concept of anticipatory governance from the beginning as part of the 10-year vision; NSF support for the Centers for Nanotechnology in Society since 2004 to provide a foundation in this regard; and setting grand challenges (2001–2005) and signature initiatives (see below) in 2010 to identify and focus development on key R&D issues for future years.

Several observers' comments on the NNI governance approach illustrate their recognition of the value and uniqueness of the model:

- National Research Council (NRC 2002): "…[T]he committee was impressed with the leadership and level of multiagency involvement in the NNI."
- The Presidential Council of Advisors in Science and Technology [29] endorsed the governing approach adopted by NNI: "[The Council] supports the NNI's high-level vision and goals and the investment strategy by which those are to be achieved."

Table 6 A long-term view (2000–2020) drives the NNI

The NNI was designed as a science project after extensive planning, 1997–2000
Long-term view (*Nanotechnology Research Directions* 1999)
Definitions and international benchmarking (Nanostructure Science &Technology, Siegel et al. [3])
Science and Engineering Priorities and Grand Challenges (NSTC/NSET, 2000)
Societal implications [27]
Plans for government agencies (National plans and budgets, 2001-)
Public engagement brochure ("Nanotechnology: Shaping the word atom by atom," [28])

Combined four time scales in planning ("grand-challenge" approach 2001–2005; "program component area" approach 2006–2010 "signature initiatives" approach after 2011)

Four time scales:
Vision: 10–20 years (Nano1 in 2000 and Nano2 in 2010 studies)
Strategic plans: 3 years (2000, 2004, 2007, 2010)
Annual budget: 1 year (2000, 2003, 2005, 2006-)
Management decisions: 1 month (meetings of the NSET subcommittee)

Four management levels:
Agency research programs
Agencies' principals
National executive (NSTC/OSTP)
Legislative (United States Congress)

- PCAST [30]: "NNI ... has had 'catalytic and substantial impact' on the growth of the U.S. nanotechnology industry"; "...[I]n large part as a result of the NNI the United States is today, by a wide range of measures, the global leader in this exciting and economically promising field."
- "NNI is a new way to run a national priority," Charles Vest, president of the National Academy of Engineering, at the March 23, 2005, PCAST meeting reviewing the NNI for Congress.
- "The NNI story could provide a useful case study for newer research efforts into fields such as synthetic biology, renewable energy or adaptation to climate change. These are the kind of areas in which science, applications, governance and public perception will have to be coordinated across several agencies. ..[F]or emerging areas like this, the concept of NNI is a good one," David Rejeski, director of the S&T Innovation program at the Woodrow Wilson International Center for Scholars in a *Nature* interview on September 2, 2010 [31].
- "Nanotechnology has become a model and an intellectual focus in addressing societal implications and governance methods of emerging new technologies," David Guston [32].

6 Lessons Learned

6.1 Objectives That Have Not Been Fully Realized After 10 Years

- General methods for achieving nanoscale "materials by design" and composite materials: the delay is because the direct theory, modeling, and simulation tools and measurement techniques with sufficient resolution were not ready.
- Sustainable development projects: Nanotechnology for energy solutions received momentum only after 5 years, nanotechnology for water filtration and desalination and climate research still has only limited funding; it is not clear if the delay in funding nanotechnology R&D for these topics is because of insufficient pull and collaboration from respective stakeholders that are less organized than in other sectors.
- Widespread public awareness of nanotechnology; the awareness figure remains low, at about 30%; this is a challenge for increasing public participation in governance.

6.2 On Target in 2010, Even if Doubted in 2000

- A steep growth rate in scientific papers and inventions: the rate for nanotechnology has been quasi-exponential (18–35% annually), at rates at least two times higher than the average for all scientific fields.

- Significant advancement in interdisciplinary research and education: nanotechnology R&D has led to creation of many multidisciplinary projects, organizations, and communities [33].
- Estimation that U.S. nanotechnology R&D investment will grow by about 30% annual growth rate (government and private sector, in-depth vertical development and new areas of horizontal development): the rate (see earlier in this chapter) held at 25–30% from 2000 to 2008.

6.3 Better than Expected After 10 Years

- Major industry involvement after 2002–2003: as examples, more than 5,400 U.S. companies had papers, patents, and/or products in 2008 (see chapter "Innovative and Responsible Governance of Nanotechnology for Societal Development"); and Moore's law has continued for the past 10 years, despite serious doubts raised in 2000 about the trend being able to continue into the nanoscale regime.
- Unanticipated discoveries and advances in several science and engineering fields, including plasmonics, metamaterials, spintronics, graphene, cancer detection and treatment, drug delivery, synthetic biology, neuromorphic engineering, and quantum information systems.
- The formation and growing strength of the international nanotechnology community, including in nanotechnology EHS and ELSI: these developments have surpassed expectations, and the debut of governance studies was unanticipated.

6.4 Main Lessons Learned After 10 Years

- There is a need for continued, focused investment in theory, direct measurement, and simulation at the nanoscale; nanotechnology is still in the formative phase.
- Besides R&D in new nanostructured metals, polymers, and ceramics, excellent opportunities for nanotechnology R&D exist in classical industries, such as textiles, wood and paper, plastics, and agricultural and food systems. Improved mechanisms are needed for public–private partnerships to establish consortia or platforms for targeted development programs.
- There is a need to better connect science and engineering to translational research and creation of jobs.
- There is a need to continue to increase multistakeholder and public participation in nanotechnology governance.

7 Closing Remarks

The combined vision of the contributors to this report for the future of nanotechnology research and development to 2020—as a whole and in each domain (tools, manufacturing, applications, infrastructure, governance, etc.)—is presented in chapters

"Enabling and Investigative Tools: Measuring Methods, Instruments, and Metrology," "Synthesis, Processing, and manufacturing of Components, Devices, and Systems," "Nanotechnology Environmental, Health, and Safety Issues," "Nanotechnology for Sustainability: Environment, Water, Food, Minerals, and Climate," "Nanotechnology for Sustainability: Energy Conversion, Storage, and Conservation," "Applications: Nanobiosystems, Medicine, and Health," "Applications: Nanoelectronics and Nanomagnetics," "Applications: Nanophotonics and Plasmonics," "Applications: Catalysis by Nanostructured Materials," "Applications: High-Performance Materials and Emerging Areas," "Developing the Human and Physical Infrastructure for Nanoscale Science and Engineering," "Innovative and Responsible Governance of Nanotechnology for Societal Development" and the Executive Summary. The goal has been to provide a long-term and timely vision for nanotechnology R&D, with input from leading experts in the nanotechnology community.

Overall, it appears that the NNI has been the major driver for nanoscience and nanotechnology developments and applications in the United States and in the World for close to a decade, but that many nations besides the United States are continuing to rapidly expand their nanotechnology-related R&D programs in recognition of the fundamental scientific, economic, and social value of doing so.

Besides impacting products, tools, and healthcare, it is inevitable that nanotechnology R&D will also impact learning, imagination, infrastructure, inventions, public acceptance, culture, laws, and the architecture of various other socio-economic factors. From 1997–2000, the U.S. scientific establishment developed a vision for nanotechnology R&D, and in the first 10 years of the National Nanotechnology Initiative, 2001–2010, that vision has become a reality. This volume is intended to extend that vision into the next 10 years, to 2020 (and beyond).

A main impetus for the original development of the NNI was the long-term view, based on an intellectual drive towards exploiting new phenomena and processes, developing a unified science and engineering platform from the nanoscale, and using molecular and nanoscale interactions to radically improve the efficiency of manufacturing. Complementary to these goals has been the promise of broad societal benefit from pursuing nanotechnology R&D, including an anticipation of $1 trillion per year by 2015 of products where nanotechnology plays a key role, which would require two million workers with nanotechnology-related skills. Because the rate of market increase is expected to follow the trends in papers and patents of about a 25% increase per year in the previous 10 years, one may estimate that by 2020, there will be about $3 trillion in products that incorporate nanotechnology as a key performance component. The nanotechnology markets and related jobs are expected to double each 3 years.

Nanotechnology is evolving toward new scientific and engineering challenges in areas such as assembly of nanosystems, nanobiotechnology and nanobiomedicine, development of advanced tools, environmental preservation and protection, and pursuit of societal implication studies. All trends for papers, patents, and worldwide investments are still expected to have quasi-exponential growth, with potential inflexion points occurring within several years. There is a need for continuing long-term planning, interdisciplinary activities, and anticipatory measures involving interested stakeholders.

In the next 10 years, the challenges of nanotechnology will likely take new directions, because there is a transition occurring within several dominant development trends:

- From a focus on creating single nanoscale components to a focus on creating active, complex nanosystems
- From specialized or prototype research and development to mass use in advanced materials, nanostructured chemicals, electronics, and pharmaceuticals
- From applications in advanced materials, nanoelectronics, and the chemical industry, expanding into new areas of relevance such as energy, food and agriculture, nanomedicine, and engineering simulations from the nanoscale where competitive solutions are expected
- From starting at rudimentary first-principles understanding of the nanoscale to accelerating development of knowledge, where the rate of discovery remains high and significant changes continually occur in application areas
- From almost no specialized infrastructure to well-institutionalized programs and facilities for nanotechnology research, education, processes, manufacturing, tools, and standards

While expectations from nanotechnology may have been overestimated in the short term, the long-term implications for the impact of nanotechnology on healthcare, productivity, and environmental protection appear now to be underestimated, provided that proper consideration is given in coming years to educational and social issues.

It will be imperative over the next decade to focus on four distinct aspects of nanotechnology development that are discussed in this volume: (1) better comprehension of nature, leading to knowledge progress; (2) economic and societal solutions, leading to material progress; (3) international collaboration on sustainable development, leading to global progress; and (4) people working together for equitable governance, leading to moral progress.

References

1. National Science Foundation (NSF), Partnership in nanotechnology" program announcement (NSF, Arlington, 1997), Available online: http://www.nsf.gov/nano
2. M.C. Roco, R.S. Williams, P. Alivisatos (eds.), *Nanotechnology Research Directions: Vision for the Next Decade. IWGN Workshop Report 1999* (National Science and Technology Council, Washington, DC, 1999), Available online: http://www.wtec.org/loyola/nano/IWGN.Research.Directions/. Published by Kluwer, currently Springer, 2000
3. R. Siegel, E. Hu, M.C. Roco (eds.), *Nanostructure Science and Technology* (National Science and Technology Council, Washington, DC, 1999). Published by Kluwer, currently Springer, 1999
4. M.C. Roco, Nanoscale science and engineering: unifying and transforming tools. AICHE J. **50**(5), 890–897 (2004)
5. International Standards Organization (ISO), TC 229: Nanotechnologies (2010), http://www.iso.org/iso/iso_technical_committee.html?commid=381983

6. G. Lövestam, H. Rauscher, G. Roebben, B. Sokull Klüttgen, N. Gibson, J.-P. Putaud, H. Stamm, *Considerations on a Definition of Nanomaterial for Regulatory Purposes* (Joint Research Center of the European Union, Luxembourg, 2010), Available online: http://www.jrc.ec.europa.eu/
7. H. Chen, M. Roco, *Mapping Nanotechnology Innovations and Knowledge*. Global and Longitudinal Patent and Literature Analysis Series (Springer, Berlin, 2009)
8. M.C. Roco, W. Bainbridge (eds.), *Societal Implications of Nanoscience and Nanotechnology* (Springer, Boston, 2001)
9. M.C. Roco, Broader societal issues of nanotechnology. J. Nanopart. Res. **5**(3–4), 181–189 (2003a)
10. Z. Huang, H. Chen, Z.K. Che, M.C. Roco, Longitudinal patent analysis for nanoscale science and engineering: country, institution and technology field analysis based on USPTO patent database. J. Nanopart. Res. **6**, 325–354 (2004)
11. Z. Huang, H.C. Chen, L. Yan, M.C. Roco, Longitudinal nanotechnology development (1991–2002): National Science Foundation funding and its impact on patents. J. Nanopart. Res. **7**(4–5), 343–376 (2005)
12. Y. Dang, Y. Zhang, L. Fan, H. Chen, M.C. Roco, Trends in worldwide nanotechnology patent applications: 1991 to 2008. J. Nanopart. Res. **12**(3), 687–706 (2010)
13. M.C. Roco, Nanotechnology's future. Sci. Am. **295**(2), 21 (2006)
14. M.C. Roco, Converging science and technology at the nanoscale: opportunities for education and training. Nat. Biotechnol. **21**(10), 1247–1249 (2003b)
15. M.C. Roco, Coherence and divergence in science and engineering megatrends. J. Nanopart. Res. **4**(1–2), 9–19 (2002)
16. M.C. Roco, W. Bainbridge (eds.), *Converging Technologies for Improving Human Performance* (Springer, Boston, 2003)
17. V. Subramanian, J. Youtie, A.L. Porter, P. Shapira, Is there a shift to "active nanostructures"? J. Nanopart. Res. **12**(1), 1–10 (2009). doi:10.1007/s11051-009-9729-4
18. National Center for Manufacturing Science (NCMS), 2009 NCMS survey of the U.S. nanomanufacturing industry. Prepared under NSF Award Number DMI-0802026 (NCMS, Ann Arbor, 2010), Available online: http://www.ncms.org/blog/post/10-nsfnanosurvey.aspx
19. M.C. Roco, *National Nanotechnology Initiative – Past, Present, Future*. Handbook on Nanoscience, Engineering and Technology, 2nd edn. (Taylor and Francis, Oxford, 2007), pp. 3.1–3.26
20. Nanoscale Science, Engineering, and Technology Subcommittee of the National Science and Technology Council Committee on Technology (NSTC/NSET), *The National Nanotechnology Initiative Strategic Plan* (NSTC/NSET, Washington, DC, 2004), Available online: http://www.nano.gov/html/res/pubs.html
21. Nanoscale Science, Engineering, and Technology Subcommittee of the National Science and Technology Council Committee on Technology (NSTC/NSET), *The National Nanotechnology Initiative Strategic Plan* (NSTC/NSET, Washington, DC, 2007), Available online: http://www.nano.gov/html/res/pubs.html
22. Nanoscale Science, Engineering, and Technology Subcommittee of the National Science and Technology Council Committee on Technology (NSTC/NSET), *The National Nanotechnology Initiative Strategic Plan* (NSTC/NSET, Washington, DC, 2010), Available online: http://www.nano.gov/html/res/pubs.html
23. M.C. Roco, International perspective on government nanotechnology funding in 2005. J. Nanopart. Res. **7**, 707–712 (2005)
24. M.C. Roco, Possibilities for global governance of converging technologies. J. Nanopart. Res. **10**, 11–29 (2008). doi:10.1007/s11051-007-9269-8
25. M.C. Roco, O. Renn, Nanotechnology risk governance, in *Global Risk Governance: Applying and Testing the IRGC Framework*, ed. by O. Renn, K. Walker (Springer, Berlin, 2008), pp. 301–325
26. O. Renn, M.C. Roco (eds.), *White Paper on Nanotechnology Risk Governance* (International Risk Governance Council, Geneva, 2006)

27. National Science Foundation (NSF), *Societal Implications of Nanoscience and Nanotechnology* (NSF, Arlington, 2001), (Also published by Kluwer Academic Publishing, 2001), Available online: http://www.nsf.gov/crssprgm/nano/reports/nsfnnireports.jsp
28. Interagency Working Group on Nanoscience (NSTC/IWGN), Engineering and Technology of the National Science and Technology Council Committee on Technology, *Nanotechnology: Shaping the World Atom by Atom* (NSTC/IGWN, Washington, DC, 1999), Available online: http://www.wtec.org/loyola/nano/IWGN.Public.Brochure/
29. Presidential Council of Advisors on Science and Technology (PCAST), *The National Nanotechnology Initiative at Five Years: Assessment and Recommendations of the National Nanotechnology Advisory Panel* (Office of Science and Technology Policy, Washington, DC, 2005), Available online: http://www.whitehouse.gov/administration/eop/ostp/pcast/docsreports/archives
30. Presidential Council of Advisors on Science and Technology (PCAST), *Report to the President and Congress on the Third Assessment of the National Nanotechnology Initiative, Assessment and Recommendations of the National Nanotechnology Advisory Panel* (Office of Science and Technology Policy, Washington, DC, 2010)
31. C. Lok, Nanotechnology: small wonders. Nature **467**, 18–21 (2010)
32. D. Guston, *Encyclopedia of Nano-science and Society* (Sage Publications, Thousand Oaks, 2010)
33. Committee for the Review of the National Nanotechnology Initiative, *Small Wonders, Endless Frontiers. A Review of the National Nanotechnology Initiative* (National Academy Press, Washington, DC, 2002), Available online: http://www.nano.gov/html/res/small_wonders_pdf/smallwonder.pdf

Investigative Tools: Theory, Modeling, and Simulation*

Mark Lundstrom, P. Cummings, and M. Alam

Keywords Theory • multiscale modeling • Computer simulations • Ab initio • Density functional theory • Molecular dynamics • High performance computing • Cyber-infrastructure • Nanomaterials and nanosystems by design • International perspective

As subsequent chapters in this report will describe, theory, modeling, and simulation (TM&S) play a significant role in almost every branch of nanotechnology. TM&S consists of three distinct components. A *theory* can be defined as a set of scientific principles that explains phenomena—a succinct description of a class of problems. *Modeling* is the analytical/numerical applications of theory to solve specific problems. *Simulation* aims to faithfully render the physical problem in the greatest possible detail, so that the critical features emerge organically—not as a consequence of the ingenuity, insights, high level abstractions, and simplifications that characterize modeling. Each of the three TM&S components plays an important role, but the opportunities of the next decade will require a stronger emphasis on the modeling component. Multiscale modeling, in particular, will be essential in addressing the next decade's challenges in technology exploration and nanomanufacturing. Finally, it should be understood that each subdiscipline of nanotechnology has its own TM&S community; these communities share many commonalities in underlying theoretical foundations, numerical and computational methods, and modeling approaches. This chapter focuses on issues, challenges, and opportunities common to TM&S across the broad spectrum of nanotechnology.

---

*With contributions from: M. Ratner, W. Goddard, S. Glotzer, M. Stopa, B. Baird, R. Davis.

M. Lundstrom (✉) and M. Alam
School of Electrical and Computer Engineering, Purdue University, DLR Building, NCN Suite, 207 S. Martin Jischke Dr., West Lafayette, IN 47907, USA
e-mail: lundstro@ecn.purdue.edu

P. Cummings
Vanderbilt University, 303 Olin Hall, VU Station B 351604, Nashville, TN 37235, USA

1 Vision for the Next Decade

1.1 Changes in the Vision Over the Last 10 Years

As noted when the U.S. national nanotechnology research agenda was first conceived, "Fundamental understanding and highly accurate predictive methods are critical to successful manufacturing of nanostructured materials, devices and systems" ([1], p. 25). Over the past 10 years, the focus in TM&S research has been on elucidating fundamental concepts that relate the structure of matter at the nanoscale to the properties of materials and devices. As a result, theory, modeling, and simulation have played an important role in developing a fundamental understanding of nanoscale building blocks. Computational capability has increased by more than a factor of 1,000, leading to more ambitious simulations and wider use of simulation. For example, the number of atoms simulated by classical molecular dynamics for 10 ns time durations has increased from fewer than ten million in 2000 to nearly one billion in 2010. As discussed in chapter "Enabling and Investigative Tools: Measuring Methods, Instruments, and Metrology", first-principles theory is increasingly coupled with characterization and metrology in the atomic-resolution regime. Over the past decade, new theoretical approaches and computational methods have also been developed and are maturing, but our understanding of self-assembly, programmed materials, and complex nanosystems and their corresponding architectures is still primitive, as is our ability to support design and nanomanufacturing. New challenges must now be addressed.

1.2 Vision for the Next 10 Years

The promise of nanotechnology lies in the possibility of helping address the great challenges faced by human society by engineering matter on the nanoscale—i.e., in building human-scale systems with nanoscale building blocks. Nanotechnology is inherently a multiscale, multiphenomena challenge. Building on the substantial progress over the past decade, the focus for the next decade must be on making multiscale modeling and simulation pervasive. Multiscale modeling will be essential to support technology exploration, design, and nanomanufacturing. Faster computers and improved computational methods will be an important part of the solution, but a unified, conceptual framework for multiscale thinking and simulation must be developed as well. The right approach must be used for the scale of interest [2], and approaches at one scale must be related to those at adjacent scales. To support design, appropriate, high level abstractions for designing complex nanosystems must be developed. As nanotechnology moves from science to applications, access to user-friendly software design tools will become more and more important.

Looking to the future, TM&S scientists must continue to clarify underlying concepts, develop nanoscale building blocks, improve computational methods and devise new ones, but the focus must shift to applications, which entails addressing multiscale/multi-phenomena problems. Addressing the multiscale challenge will require connecting "bottom-up" and "top-down" thinking (i.e., atomistic, first-principles simulations with phenomenological macroscopic simulations). The goal is to make multiscale/multiphenomena modeling and simulation of practical problems a reality across the broad spectrum of nanotechnology. By doing so, TM&S will support the development of nanotechnologies that address society's grand challenges.

2 Advances in the Last 10 Years and Current Status

2.1 Advances in Nanotechnology: The Role of TM&S

The last decade (2000–2010) witnessed many advances in nanoscience and nanotechnology. Theory, modeling, and simulation have played a supportive, sometimes even critical, role in these advances. A few examples follow.

- *Understanding of current flow at the molecular scale.* Theoretical and computational studies were critical in understanding measurements of electron current flow in molecules [3–7].
- *Evolution of microelectronics into nanoelectronics.* TM&S helped to identify limits, issues, and possibilities for transistor scaling (e.g., [8]).
- *Discovery of graphene and development of carbon-based electronics.* Closely coupled experimental/theoretical/computational studies shed light on the physics of carbon-based electronics [9–12].
- *Emerging applications of spin torque.* Theoretical predictions [13, 14] were experimentally demonstrated [15, 16].
- *Discovery of iron-based high-Tc superconductors* [17]. Very large-scale simulations showed how the electron pairing arises that is the key to high-Tc superconductivity [18].
- *Ten-orders-of-magnitude improvement in the sensitivity of biosensors.* The performance of a wide range of biosensors [19–21] was interpreted within a scaling theory of diffusion-limited transport of nanoparticles [22, 23].

2.2 Advances in Theory: Enabling Modeling and Simulation

Nanoscience requires new theoretical ideas and new modeling approaches as well as more powerful computational capabilities. New methods and advances in computation permit larger and more complex problems to be addressed, but analytical and coarse-grained models are central to multiscale modeling. These sorts of models (e.g., the Marcus model for electron transfer and transport, the Langmuir-Hinshelwood model for adsorption, the Shockley diode equation, etc.) are also the way scientists and engineers think about, envision, and imagine the world. TM&S

is especially effective when results the results of complex simulations are expressed in the language used by experimentalists and designers.

The last decade has witnessed many advances in the underlying theory of nanoscience. Examples of accomplishments include:

- Advances in *ab initio* theory beyond density functional theory (DFT), including improved GW algorithms; DFT + dynamical mean field theory for strongly correlated systems; the DFT + Σ approach for cheap, approximate, parameter-free treatments of many-body effects [3, 4]; and increased use of hybrid functionals and post-Hartree-Fock methods (e.g., [24, 25]).
- Advances in linear-scaling quantum mechanics. Building on foundations established in the 1990s, advances continued in areas such as purification theory [26] and new boundary conditions for divide and conquer [27].
- Reactive force fields for reactive dynamics simulation of materials processes [28]. Reactive force fields allow one to simulate structure with chemistry, enabling potential impact on drug discovery and catalysis, both of which require precise differentiation between the energy costs of various reaction pathways.
- Enhanced sampling techniques such as metadynamics [29] for free energy surfaces to find important configurations for specific systems and improved methods for determining rate constants [30].
- Development of the non-equilibrium Green's function (NEGF) approach as a conceptual [31] and computational [5] framework for describing quantum transport at the quantum and atomistic scales (Fig. 1). Applications range from molecular electronics to practical applications in the semiconductor industry, where the approach has been used to connect the underlying material properties to improving the performance of the transistor.
- New statistical theories for conduction in nanostructured materials such as carbon nanotubes and semiconductor nanowire networks (Fig. 2) and phase-segregated organic solar cells, leading to quantitative predictions for the performance of devices [9, 10].

2.3 Advances in Computing: Powering TM&S

While theory provides the foundation for modeling and simulation, computing makes it possible. Over the past decade significant computing advances have been achieved. As illustrated in the examples of Sect. 8, these advances have contributed to the increasing role of simulation in nanoscience and have set the stage for an even more ambitious agenda for TM&S in the next decade.

Examples of advances in computing that have been important for TM&S include the following:

- High-performance computing (HPC) power (as measured by the most powerful machine in the biannual top-500 supercomputers list, http://top500.org) has increased by three orders of magnitude since 2000, which enables larger simulations

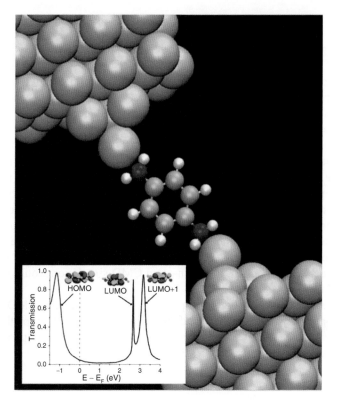

Fig. 1 First-principles density functional theory (NEGF-DFT) simulations of the conductance of molecular junctions have led to improved understanding and quantitative agreement with experiments. The figure shows the calculated atomic geometry of benzene-diamine suspended between Au electrodes; the inset shows the computed electron transmission vs. energy [5]

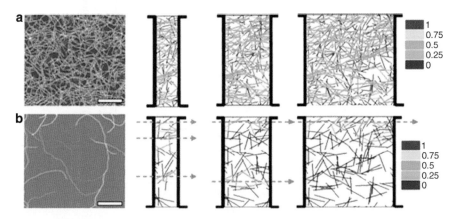

Fig. 2 SEM images (*left*) and simulated normalized current distribution for carbon nanotube networks with high (**a**) and low (**b**) coverage. Measurements were reported on SWNTs grown by chemical vapor deposition with systematically varying degrees of alignment and coverage in transistors with a range of channel lengths and orientations perpendicular and parallel to the direction of alignment. A stick-percolation-based transport model provides a simple yet quantitative framework to interpret the sometimes-counterintuitive transport measured in the devices that cannot be reproduced by classical transport models (After Kocabas et al. [32])

and brings ambitious simulations into the realm of the "every-day" computing world of theorists and experimentalists. Leadership-class computing is necessary to address grand challenges in computational nanoscience; it is also needed to perform the atomistic level simulations that provide the fundamental data for the coarse-grained methods used in multiscale simulation.

- The impact of the exponential growth in computing power has had great impact on classical molecular dynamics (CMD) over the past decade, and this will continue into the next decade. For systems with short-range forces, the compute time is linear in system size, N, as well as linear in the number of time steps, Tsim (usually ~1 fs). The computational complexity (CC) is approximately Tsim N. What can be done in 1 day of computing? Consider the recent simulation by Schulz et al. [33] of a 5.4 M-atom system on 30,720 cores of Jaguar, the current Number 1 machine on the top-500 list. The computational complexity possible on Jaguar is TstepN = 6×1,014, from which we can project that on Jaguar, a billion-atom CMD simulation would execute at 0.6 ns/day.
- Parallel computing has had great impact. Fig. 3 shows estimated CMD computational complexity for the highest- and lowest-performing machines on the top-500 list (based on assuming CMD performance scales with benchmark performance), as well as a single central processing unit (CPU) and a graphical processing

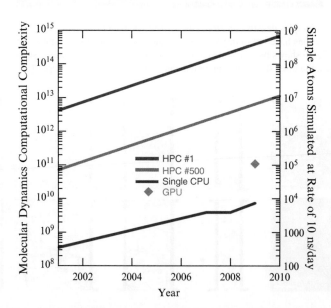

Fig. 3 *Left axis*: estimate of CMD computational complexity (equal to the product of the number of time steps and the number of atoms simulated in 1 day) on the #1 supercomputer in the top 500 (*top line*), #500 in the top 500 list (*middle line*), a single CPU (*bottom line*) and a GPU (*diamonds*). *Right axis*: For simple monatomic fluid, the number of atoms that can be simulated for 10 ns in 1 day (Data from P.T. Cummings)

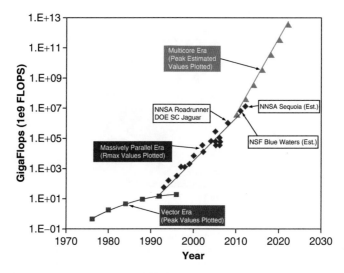

Fig. 4 Gigaflops in CPU performance vs. time (From an NRC study on modeling, simulation and games, compiled from data from http://top500.org) [34]

(or "general-purpose") unit (GPU).[1] This graph is, at best, qualitative and indicative of trends; it does, however, try to capture the impact of multicores and the impact of GPUs. As shown in Fig. 4, the introduction of multicore CPUs has considerably accelerated the rate of progress in CPU performance.

- The introduction of GPU chips, originally driven by the gaming industry, is a disruptive technology that is leading to dramatic increases in computing performance for algorithms like many-particle dynamics that are easily parallelized. For example, NVIDIA's Fermi chip announced in 2010 provides 120 times the performance of a single CPU. To achieve this performance, science codes must be rewritten for GPU chips, and new mathematical libraries must be developed. The recent development of high-level programming languages for GPUs (e.g., CUDA) is rapidly accelerating the application of GPUs to scientific computation, including computational nanoscience [35–37].

2.4 Advances in Simulation and Design of Nanomaterials

The increasingly close connection between simulation and metrology has been a significant development over the past decade. Fig. 5 shows some of the experimental

---

[1] CPUs process information serially; GPUs process information in parallel. In the past, CPUs generally have had one, two, or four cores, whereas GPUs have had a larger number of cores, up to 100 or more.

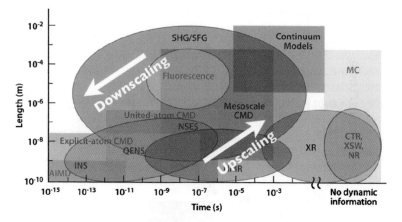

Fig. 5 Hierarchy of TM&S methods relevant to nanoscale science and technology, along with some corresponding experimental methods, and the time and length scales over which each is applicable (Courtesy of P.T. Cummings and D.J. Wesolowski), to be published). *Acronyms (and corresponding methods) indicated in the figure:* TM&S methods (*in rectangles*) *AIMD* ab initio molecular dynamics, *CMD* classical molecular dynamics, *MC* Monte Carlo experimental techniques (*in ovals*), *INS* inelastic neutron scattering, *QENS* quasi-elastic neutron scattering, *NSES* neutron spin echo spectroscopy, *NMR* nuclear magnetic resonance, *XR* X-ray reflectivity, *SHG* second harmonic generation, *SFG* sum frequency generation, CTR crystal truncation rod (an X-ray method), *XSW* X-ray standing wave, *NR* neutron reflectivity

probes that cover spatio-temporal scales that overlap with the TM&S methods. The coincidence of experiment and TM&S methods at the nanoscale offers many opportunities for strongly integrated theoretical/experimental studies (e.g., [38]).

Large investments have been made in nanotechnology research, and the resulting progress has been substantial. Most research and development in nanotechnology, however, is still done by expensive and time-consuming experiments. Industry demands faster and less expensive solutions. Modeling and simulation will be essential to addressing this challenge, and advances over the past decade have set the stage for the predictive design of materials by simulation.

Predictive materials design requires improvements in simulation methods for maximum accuracy. Over the past decade, significant progress has occurred in the individual components of a predictive materials simulation framework:

- Quantum mechanics (challenge: increased accuracy)
- Force fields (challenge: chemical reactions)
- Molecular dynamics (challenge: extract properties for materials design)
- Biological predictions (challenge: simulation in liquids and on biological timescales)

- Mesoscale dynamics (challenge: simulation on relevant time and size scales)
- Integration: (challenge: multiscale)

A key challenge for the next decade will be to develop methods that address the role of defects and disorder in nanomaterials.

Examples of the predictive simulation capabilities currently being developed include simulations of etching for damage-free semiconductor fabrication, ReaxFF reaction dynamics simulations of catalytic processes for high temperatures and pressures, reactive dynamics of hydrocarbon reactions, silicon nanowire growth, dendrimer-enhanced nanoparticle filtration for water purification, and self-assembly.

Self-assembly is increasingly a new frontier for materials research. "Assembly engineering" seeks to create building blocks that are (nearly) fully tailorable and reconfigurable, with changeable shape and/or interactions. The goal is to realistically simulate hierarchical assembly to achieve novel functions.

2.5 Advances in Multiscale Simulation and Modeling

Fig. 5 shows the range of computational methods currently employed in nanoscale TM&S, from *ab initio* methods, to atomistic and united atom molecular dynamics, to particle-based coarse-grained simulations, and finally to continuum methods. For integrated nanosystems, additional layers of high-level abstractions are required in order to capture functional and systems aspects and to enable design. Nanoscale TM&S studies increasingly traverse more than one of the scales shown in Fig. 5.

Different methods are used at different scales, but true multiscale simulation would use automatic upscaling (moving information from a more fundamental, lower scale up to a higher scale, such as force field data from *ab initio* to atomistic CMD) and downscaling (such as a finite element calculation invoking molecular simulation to obtain a diffusivity at a new state condition). Although true multiscale simulations are rare, it is common in TM&S today to use multiple methods to solve a single problem. For example, *ab initio* methods are commonly used to calibrate atomistic force fields for CMD simulations. Today, however, atomistic force fields are handcrafted in a type of "nano cottage industry," which makes it difficult to address a broad range of multiphenomena problems. Improvements in coupling different computational techniques resulted in the successful simulation of the oxidation of nanometer-sized clusters [39].

Significant progress has occurred in methods development for multiscale simulation. Upscaling methods now exist for bridging many of the scales (e.g., [40, 41]). Downscaling methods are less well developed, and true multiscale methods even less so, but specific methods tailored for narrow problem domains have been developed (e.g., [42]), and general methods are beginning to be proposed (e.g., [43]).

3 Goals, Barriers, and Solutions for the Next 5–10 Years

Theory, modeling, and simulation provide investigative tools that support nanotechnology. To support nanotechnology in the next decade, these tools must be enhanced to solve larger problems, and to do so more accurately and efficiently. With the increasing focus on applications of nanoscience, TM&S must address the larger challenge of multiscale modeling and simulation. Successfully doing to will transform research, technology development, design, and manufacturing [84].

3.1 Goals for Multiscale Theory, Modeling, and Simulation

Multiscale modeling is essential for exploring the numerous, new technology possibilities that research in nanoscience is creating. It is essential for identifying promising ideas, for developing them into real technologies, and it will be essential for nanomanufacturing. Faster computers and better computational methods will be important, but the real challenge is to develop new conceptual and modeling frameworks. New conceptual models must be developed to relate atomistic physics and chemistry to nanoscale structure and then nanoscale structure to the performance of complex materials and integrated nanosystems, finally ending up in the language that experimentalists and product designers use. These new theories will provide a computational framework for multiscale modeling, and they also will provide an intellectual framework to link those who work at the bottom with those who work at the top. Examples of multiscale modeling exist, but the goal is to make "atoms to applications modeling" pervasive across the field of nanotechnology.

Multiscale modeling should support the development of:

- Programmed materials that make use of self-assembly
- Process and bottom-up assembly models
- System-level design models that allow for billions of components and that comprehend defects and disorder
- Design methods that combine TM&S, nanoinformatics, and expert systems
- Nanomanufacturing—including nanomaterials, nanodevices, and nanosystems as well as modeling process variations and repeatability.

A longer-term goal is to develop true multiscale simulation, which involves the use of different methods at different scales with automatic upscaling (moving information from a more fundamental, lower scale up to a higher scale) and downscaling. Different methods scale differently with the number, N, of atoms in the system. Thus, an atomistic simulation could faithfully represent, for example, catalytic reactions at surfaces by treating via first-principles methods only the region in the vicinity of the reaction, and only when a reaction is probable. Addressing these problems fundamentally will require development of general, numerical methods to automatically up-scale and downscale and traverse the relevant spatio-temporal scales.

3.2 Goals for the Predictive Design of Nanomaterials

Advances over the past decade have raised the possibility of truly predictive design by simulation. The goal is to achieve predictive design for real-world applications using an overlapping hierarchy of methods to connect the fine scale to the coarse scale. The goal of realizing predictive design of materials with the accuracy to replace experiments to a large degree is within reach over the next 10 years—if the problems are carefully selected, if a sustained commitment is made to developing the new methods that will be needed, and if strong industry input is obtained to direct the research to critical needs in nanomanufacturing. To address this challenge, TM&S will need: (1) increased computing capability, (2) improved computational infrastructure, (3) critical input from experiments, and (4) to learn how to address defects, disorder, and variability in materials.

3.3 Goals for Computing and Building Block Simulations

Computational capability must address larger problems more accurately. Consider the treatment of electronic structure with Hartree-Fock methods, which require 100 times more computing power than a DFT simulation. Assuming a linear-scaling approach is employed, the calculation of the electronic structure of a 10 nm quantum dot would need an increase in computing power by approximately a factor of 10,000. Some of this increase will come from advances in computing hardware, but some must come from improved numerical methods.

The past decade witnessed a factor of 1,000 increase in computational power; the goal is to achieve another factor of 1,000 in the next decade. GPUs will bring high-performance computing to the desktop and modest supercomputer-class computing to clusters where they will impact computational nanoscience and provide increased capabilities to theorists and to experimentalists. For applications that parallelize efficiently on multicore CPUs, the benefits will be significant. Also, given the relatively low cost of GPUs, leadership-class computers will feature multi-core compute nodes with attached GPUs for acceleration of on-node computation, thus making them integral to the fabric of high-performance computing. To achieve the performance improvements beyond the factor of 1,000 thought to be possible from new computing hardware, new theoretical approaches and numerical methods will be needed.

Computation can accelerate progress in both theoretical and experimental research, so it is important to also strive to make computing and simulation more ubiquitous and put these new capabilities in the hands of people whose focus is on problem-solving rather than on computation. For example, to support the increasingly tight connection between metrology and computation, high-performance computing must become ubiquitous and user-friendly.

As investigative tools that support nanotechnology and as components in multiscale simulation, simulations must improve. For example the accuracy and speed of *ab initio* electronic structure calculations and molecular dynamics simulations must advance. Many existing, well-developed, and extensively used codes will need to be rewritten to exploit the capabilities of new hardware.

3.4 Barriers

A significant challenge for nanotechnology in general and for the development of multiscale/multiphenomena modeling in particular is the need to develop a long-term dialogue between the different disciplines spanning nanoscience and engineering. Experts must acquire knowledge from experts in other fields, which can take considerable time. Funding models often emphasize short-term achievement, making it difficult to engage in such long-term objectives. The dispersion of expertise across the globe represents another barrier, both within the TM&S community and for connecting with experimentalists seeking solutions to multiscale problems.

Scientific and engineering research are often separated, and this represents another barrier to the development of nanotechnologies. Opportunities for new technologies can be missed, and impractical outcomes and poor return on investment can also occur. To realize the promise of nanotechnology, the science and applications must be brought together in application-driven research that addresses well-defined grand challenges.

The education system's traditional departmental boundaries can also be a barrier. New pedagogical materials are being developed to convey the principles of nanotechnology to students and practicing engineers and scientists, but often they are simply extensions of traditional educational approaches. To realize the transformative potential of nanotechnology, educators must rethink the intellectual foundations of disciplinary fields from a "nano" perspective to give students the broad conceptual scientific foundation to be creative contributors to the development of nanotechnology. Programs should be encouraged that truly integrate disciplines while still giving students the depth in a single discipline that makes them useful contributors to multidisciplinary teams. Computing is increasingly sophisticated, so there is a need to train a generation of computational nanoscientists with expertise in advanced, numerical algorithms, parallel computing, and in exploiting the unique characteristics new generations of computing technology, such as GPU chips.

3.5 Solutions

To achieve the ambitious goals outlined above, a 10-year focus on four key areas is needed.

3.5.1 Multiscale Modeling and Simulation of Complex Materials and Integrated Nanosystems

Why? Shaping the properties of matter at the nanoscale to realize devices and systems at the human scale is the vision of nanotechnology. To realize this vision, nanoscience theorists and analysts must develop new conceptual frameworks for multiscale problems, new numerical approaches that span spatio-temporal scales, and new multiscale simulations that are predictive. TM&S can help identify ideas with technology potential and set the stage for experimental work.

Why now? The first decade of the NNI has given us a much better understanding of nanoscale building blocks. There is a growing appreciation that understanding integrated nanosystems involves much more than understanding the atomistic and nanoscale components. TM&S is moving up in scale to larger systems. Nanoscience has produced so many ideas that it is not possible to thoroughly explore each one experimentally, and increasingly predictive simulations have the potential to actually replace experiments to a significant extent.

Strategy: Combine "up-scaling" and "down-scaling" (top-down vs. bottom-up) approaches and people. Both viewpoints are necessary to develop unified conceptual and computational frameworks for both multiscale modeling and simulation. Support teams are needed that couple application developers with computational experts and couple industry with academia.

Drivers: Application drivers include drug discovery, catalysis design, organic photovoltaics, batteries, programmed materials, patchy particles, searches for a new logic switch, transformative thermoelectric technologies, etc.

3.5.2 Nanotechnology Fundamentals and Numerical Methods

Why? While the focus of the next decade must shift to larger, more complex systems, our fundamental understanding of the underlying processes and components is still incomplete and must not be neglected. Research on new computational methods is also needed because increases in hardware performance alone will not be sufficient.

Why now? Much has been learned over the past decade, but there is still much that is not understood about conduction in molecules, the quantum to classical transitions, thermal transport across van der Waals bonded interfaces, etc. Much more effective numerical techniques are also needed (e.g., to enable automatic up- and down-scaling in multiscale simulation.)

Strategy: Target funding for individuals and small groups that simultaneously engages experimentalists, theorists, and computational experts to address critical areas where improved understanding is needed.

Drivers: Fundamental studies should be directed at acquiring knowledge that ultimately contributes to nanotechnology effectively addressing grand challenges in energy, the environment, and health. Examples of TM&S problems to be addressed include improved understanding of many-body effects, weak force interactions, thermal dissipation in molecular systems and at van der Waals bonded interfaces, and working principles of self-assembly as well as mathematical theories that increase the speed and accuracy of simulations and permit computer-driven up- and down-scaling.

3.5.3 Pervasive High-Performance Computing

Why? Reducing CPU execution time can make high-end computing ubiquitous, thus enhancing the capabilities of theorists, experimentalists, designers, and educators.

Why now? Advances in CPUs (e.g., multicore configurations and graphical processing units) increasingly make high-end computing part of mainstream research practices. Sophisticated instrumentation requires embedded simulation. Nanomanufacturing will require a new generation of design and manufacturing tools. Powerful scripting platforms have blurred the lines between theory and simulation.

Strategy. Increase the power of scripting platform computing, create and support open-source science codes, exploit cyberinfrastructure and cloud computing to broaden access to computing, create supercomputer/software facilities that provide "end-station" service to high-end users.

Drivers. The computational end-station concept described in Sect. 4 below should be driven by national lab–university partnerships to address scientific and computational grand challenges. Cyberinfrastructure and science gateways should be exploited to make the use of simulation tools ubiquitous and to cooperatively engage both experimentalists and computational experts.

3.5.4 Education for Nanoscience and Engineering and Computational Nanoscience

Why? The promise of nanotechnology will be realized as practicing engineers and scientists and the new engineers and scientists who are being educated now learn how to transform technology by understanding and applying the new capabilities that nanoscale science provides.

Why now? The first decade of the NNI has greatly improved our understanding of nanoscale phenomena. Educational materials are beginning to appear, but the textbooks being used are still traditional, with nanoscience "add-ons." What is needed is an ambitious initiative to rethink the intellectual foundations of the separate fields of science and to develop a new class of textbooks that unites scientific knowledge, breaks down disciplinary boundaries, inspires students, and prepares them to contribute to the development of nanotechnology.

Strategy: Support teams that include experts in TM&S with experimentalists to reinvent the curricula in fields such as materials science, electronics, photonics, biomedical engineering, chemical engineering, and chemistry, etc. The educational materials produced by these teams should be made available as "open courseware" resources to serve as resources for self-study and as models for instructors worldwide. Programs to train students in the art of scientific computation using leading edge algorithms and new computing architectures should be developed and presented on a continuing basis.

Drivers: The new knowledge emerging from research should drive the new intellectual constructs and innovative educational materials developed to support nanotechnology.

4 Scientific and Technological Infrastructure Needs

Traditionally, infrastructure for TM&S has meant leadership-class supercomputing facilities, but the infrastructure needs for TM&S are much broader. The emergence of so-called cyberinfrastructure over the past 10 years has created new technologies to address several important needs. Leadership-class supercomputers are increasingly important, but access models need to evolve. Often overlooked is the fact that numerical methods, mathematical libraries, community science codes, etc., are critical parts of the infrastructure for TM&S. Finally, new types of institutes that can provide the focus and concentration of expertise needed to turn nanoscience into nanotechnologies should be considered.

4.1 Cyberinfrastructure

Cyberinfrastructure has demonstrated its effectiveness in facilitating nanotechnology R&D through initiatives such as the TeraGrid, Open Science Grid, and nanoHUB.org. In the next decade, cyberinfrastructure should play an increasingly important role in:

- Testing and validating databases (e.g., for interatomic potentials and pseudopotentials)
- Broadly disseminating new research methods and software
- Providing broad access to online simulation services for experimentalists and educators
- Disseminating cross-disciplinary and novel educational resources
- Workforce development and in-service training (e.g., parallel programming for GPU chips)
- Connecting theorists and experimentalists in collaborative research
- Technology transfer of computational tools from academia to industry

The Network for Computational Nanotechnology's nanoHUB (http://www.nanoHUB.org) is an example of the power of cyberinfrastructure. The marriage of a software development platform with cloud computing technology (e.g., the Rappture toolkit: https://nanohub.org/infrastructure/rappture/) enables computational experts to share tools with collaborators or with the nanotechnology research and educational community broadly. Live simulation services, not just software codes, are provided through a web browser. By coupling simulation services with resources for training and education in nanotechnology, a major international resource has been developed that serves more than 150,000 users annually. Broader and more creative applications of cyberinfrastructure in the next decade can enhance the impact of TM&S on the development of nanoscience and nanotechnology. Sustained funding in support of computational science and engineering cyberinfrastructure will be necessary to realize this potential.

4.2 Supercomputing Infrastructure

The NSF TeraGrid and Open Science Grid provide researchers with access to high-performance computing services, and national laboratories make leadership-class supercomputing available through competitive processes. Computational nanoscientists have benefitted from these facilities. In the future, supercomputing facilities must evolve to address the needs of computational nanoscientists.

At the leadership-class end of computing—currently petascale, and before the end of the decade, exascale—costs are very high, making such computational facilities affordable by only a few large organizations. A number of user facilities for experimentalists exist. These user facilities operate in the mode that a sponsoring Federal agency builds and manages a core facility that supplies a unique high-end resource, and then user communities (including researchers from other Federal agencies, academia, and industry) build and maintain "end-stations" consisting of specialized instrumentation to make use of the resource. As an example, the Department of Energy (DOE) manages the Spallation Neutron Source, consisting of a proton accelerator, storage ring, target building, and several core instruments, which provide researchers with neutrons via end stations that are partially or completely funded by NSF, NIH, and industry.[2] Given the cost and difficulty of operating leadership-class computing facilities, the user facility/end station model is one that is increasingly being applied. This concept is already being embraced at DOE's Oak Ridge National Laboratory in the National Center for Computational Sciences, where an end station for nanoscience is being developed by the Center for Nanophase Materials Sciences (CNMS) in collaboration with computer scientists in the Computer Science and Mathematics Division at Oak Ridge. Broad access to

---

[2] See http://www.er.doe.gov/bes/brochures/files/BES_Facilities.pdf for an overview of DOE user facilities.

Investigative Tools: Theory, Modeling, and Simulation

these kinds of facilities will be required to address computational grand challenges in nanotechnology.

4.3 Methods Development

Significant investments have been made in supercomputer facilities, but less attention has been directed at the systematic development of new computational methods. For example, to exploit the capabilities of GPU chips, much of the TM&S community's existing simulation code will need to be rewritten. Optimized math libraries are needed. In the U.S. funding model for physical, materials, and chemical sciences, researchers are funded to solve scientific problems, not to develop computational software and numerical methods. Predictive simulation of materials and multiscale modeling are areas where methods development is critical and must be funded long term. A great deal of funding already goes into theory and simulation, but no good mechanism exists for funding research on the new methods the community codes that can lead to real breakthroughs in computation. Numerical methods and algorithms and open-source science codes should be considered part of the critical infrastructure for TM&S. The education and training to make effective use of these resources should also be considered part of the TMS infrastructure.

4.4 Problem-Specific Institutes

To effectively address grand challenge problems in nanotechnology, problem-specific institutes that engage industry and academia, experimentalists and experts in TM&S, scientists, and technologists could play an important role. Participants would spend time in residency at these institutes. The synergy of different people with different expertise and experience focusing collectively on a common problem would accelerate progress in turning nanoscience into nanotechnologies.

4.5 Virtual Institutes for Simulation-Driven Research

Problem-driven institutes would focus on specific problem s, and TM&S expects would play a supporting role in teams of experimentalists and application designers. A second kind of institute, one led by experts in TM&S, should also be considered. In these institutes, teams of theorists and modeling and simulation experts, guided by experimentalists and application designers would be charged to computationally assess and prototype potential new technologies. The goal would be to set the stage for subsequent experimental work, but this kind of work would also spur the development of multiscale modeling and simulation capabilities.

5 R&D Investment and Implementation Strategies

Over the course of the next decade, significant sustained (or renewable) investments are needed in:

- Principal investigator (PI)–driven research to address nanoscience fundamentals.
- Small interdisciplinary research groups. Support should be renewable to provide the long-term continuity necessary for successful multi-disciplinary teams.
- Large-scale and problem-focused research centers and institutes that specifically address the grand challenges of complex materials and integrated, functional nanosystems.
- Large-scale, problem-driven and simulation-driven institutes that also address grand challenge problems and that set the stage for subsequent experimental programs.
- Research on computational methods, including open-source science codes, algorithms for new computing architectures, etc. Special emphasis is needed on methods specifically directed at exploiting the capabilities of new multi-core and graphical processor architectures and on revising widely-used community codes to take advantage of these new capabilities.
- Pervasive simulation (fast, user-friendly, accurate, reliable, transitioning from laboratory codes to community codes for research and education used by computational experts as well as by others). The need for user-friendly software designed for non-experts will dramatically increase to support design and nano-manufacturing.
- Strategies to promote university-industry and international partnerships.
- Cyberinfrastructure to support pervasive computing, collaboration, and dissemination of simulation services, research methods, community-wide databases, educational resources, etc.

6 Conclusions and Priorities

The past 10 years have witnessed numerous advances in nanoscience as well as in theory, modeling, and simulation for nanoscience. In particular, the understanding of and ability to simulate nanoscale building blocks has advanced substantially. The stage has been set, and the priority for theory, modeling, and simulation now is to address the challenge of turning advances in nanoscience into nanotechnologies. To address this challenge, the TMS community must focus on atoms to applications multiscale/multi-phenomena simulation. Such multiscale frameworks will make use of scale-specific simulations (e.g., quantum chemistry, molecular dynamics, electron transport, etc., codes,) so support on the underlying fundamentals of nanoscience building blocks and on numerical and computational methods cannot be neglected.

Three broad priorities for TMS during the next decade can be identified:

- Application-driven, team-based TMS research directed at identifying solutions to specific, high priority challenges in energy, the environment, health, and security.

Fig. 6 New developments in the theory of molecular conduction have resulted in quantitative simulations of molecular junctions. A joint theory and experiment collaboration showed that electrical conductance of a bipyridine-Au molecular junction can be turned "on" or "off" simply by pushing or pulling on the junction (From Quek et al. [4])

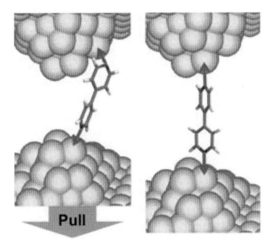

A closely-related objective should to develop broadly applicable conceptual and computational frameworks for multiscale modeling and simulation rather than problem-specific, ad hoc solutions. Priority applications include predictive design of materials and programmed materials, and design and manufacturing of large, complex, engineered nanosystems.

- Nanoscience fundamentals research addressed at proposals for new experiments and at exploring and elucidating new fields of nanoscience.
- Computational and numerical methods research directed at achieving a 10^3–10^4 increase in the size of problems that can be simulated by exploiting the capabilities of new computing architectures and at devising improved numerical algorithms. Such developments would allow simulation, computer testing, and design of complex nanosystems leading to fundamentally new products in the third and fourth generation of nanotechnology development (see preliminary chapter "The Long View of Nanotechnology Development: The National Nanotechnology Initiative at 10 Years", Fig. 6).

7 Broader Implications for Society

The promise of nanoscience lies in its potential to develop new technologies to address the great challenges that human society faces in energy, the environment, health, and security. As discussed in subsequent chapters, TM&S often plays a supporting role in the creation of nanoscience understanding and in the development of new technologies. The use of TM&S leads to shorter design cycles, which increases the impact of nanotechnology on society. In the next decade, TM&S also has an opportunity to play a leading role in the development of nanotechnology. Predictive simulation has the potential to greatly accelerate R&D in fields such as catalysis design and drug discovery. If this promise of predictive design can be realized, the impact on society would be substantial. Broader use of multiscale/multiphenomena simulation could also influence technology development directions. Nanoscience generates possibilities for new technologies, but it generates many more potential technology paths than can be experimentally explored. TM&S has the potential to rapidly explore and assess potential technologies and to then guide subsequent experimental work. In this way, TM&S can amplify the impact of nanotechnology on society.

It has been said that "The purpose of computing is insight, not numbers" [44]. Simulation coupled with powerful human interfaces and visualization can promote public understanding of nanoscience and its potential benefits to society. Moreover, the insights and understanding that emerge in the course of developing computational tools can be extremely important pedagogically. New ways of understanding nanoscience and technology facilitated by TM&S research have the potential to reshape educational paradigms and impact students at all levels, including college, pre-college, and post-college.

8 Examples of Achievements and Paradigm Shifts

8.1 Quantitative Simulation and Prediction of Molecular Conductance

Contact persons: J. B. Neaton, Lawrence Berkeley National Laboratory; L. Venkataraman, Columbia University; S. Hybertsen, Brookhaven National Laboratory

Understanding and control of electron flow through matter at the molecular level presents fundamental challenges to both experiment and theory. For more than a decade, researchers have sought to trap and "wire up" individual molecules into circuits and to probe the electron transport properties of the smallest, single electrical components imaginable. The experimental challenge lies in assembling these devices in a reproducible manner, which requires exquisite control of electrical contacts, or "alligator clips," between molecules and leads at length scales currently beyond resolution limits of experimental characterization techniques. The theoretical challenges lie in modeling the possible distribution of junction structures at the atomic scale and computing the electron transport through such junctions, a fundamentally non-equilibrium problem. Ultimate goals for both experiment and theory are elucidation of different fundamental regimes of electron transport, and the discovery of molecular structures that support a significant nonlinear electrical response, such as switching.

Since the junction conductance fundamentally derives from electron tunneling processes at the single molecule length scale, predictive theory requires the average positions of all atoms within the junction—as well as a self-consistent description of their bonding and electronic energy level alignment—when the junction is under bias.

A decade ago, atomistic first-principles methods for computing electrical conductance, based on density functional theory (DFT), were in their infancy. Although significant challenges remain, recent advances in the field have resulted in theories that provide reliable predictions for nanoscale conduction. These theories have been compared to accurate, well-characterized experiments for a broad class of molecules where amine-links provide reliable single molecule circuit formation with gold electrodes [5, 6]. In particular, the use of static DFT within standard approximations for exchange and correlation is insufficient for computing conductance and current, even in the linear response case. Direct incorporation of correlation effects in the junction conductance is essential, e.g., with a new, practical first-principles approach, DFT+Σ [3].

In an example probing novel functionality in a single molecule circuit, recent experiments showed that gold-bipyridine-gold junctions could be mechanically switched between two states, simply by pushing and pulling the junction (Fig. 6; [4, 45]). Using the DFT+Σ approach, the conductance was accurately calculated, and the conductance switching was understood to arise from controllably altering the pyridine-gold link structure. Owing to these and related advances, experiments probing molecular conductance are much more reproducible and simulations are much more predictive than they were 10 years ago.

8.2 Controlling Individual Electrons in Quantum Dots

Contact person: Michael Stopa, Harvard University

Quantum dots can be used as artificial atoms fabricated within a semiconductor heterostructure whose state can be controlled by metallic gates, voltage biases applied to source and drain contacts, and externally applied magnetic fields. These dots, which represent isolated puddles of electrons *inside* the solid, can be fabricated in groups, producing artificial molecules typically, though not exclusively, in a two-dimensional arrangement. Theory and modeling of quantum dots has taken major strides in the past decade, leading to predictions of phenomena such as: (1) single electron devices like diodes [46], (2) artificial molecular electronic structure [47], and (3) the possible realization of quantum computation [48]. The primary objectives of all of these applications have been to prepare, manipulate, and measure the state of one or several electrons in one or several dots.

In the early 2000s, semiconductor quantum dots had evolved to a stage where the electron number on a single dot could be tuned down to one and then zero reproducibly. Experiments involved measuring a current *through* the dot and varying it with source-drain biases, external voltages on gates and a magnetic field. Principally, control of the electron state in artificial atoms was achieved through the Coulomb blockade phenomenon, whereby addition of an electron to a dot required a single electron charging energy cost to be paid either by varying a gate voltage or a source-drain voltage.

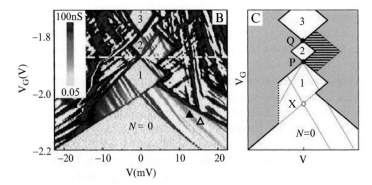

Fig. 7 "Coulomb diamond" stability diagram showing, as a function of a gate voltage and a source-drain voltage, the current through the *double-dot*. Where the current is small, particular charge states are stable (*diamonds* labeled 1, 2, and 3). The extended region on the right of the two-electron dot is a region where a particular spin state (the two-electron singlet) is the stable ground state (From Ono et al. [50])

Two major advances have occurred (both elucidated by close experiment-modeling synergy) that have enhanced the control over these systems. First, the use of "listening" devices—essentially tiny variable resistors that can sense the presence of one electronic charge on a nearby dot—were perfected [49]. Second, the phenomenon of the "Pauli blockade"—whereby electrons in the same spin could not occupy the same spatial state—was discovered and elucidated (Fig. 7; [50]). These two phenomena, together with gate control for shuttling electrons back and forth between members of complexes of dots, allowed for exquisite probing of the single or many-electron state of dots, including spin information, and for the exploration of coherent evolution of those states. Various sequences of prepare-evolve-measure the quantum state, which are the building blocks of quantum computation, have been perfected.

The ultimate goal of scalable quantum computation with its exponential speedup of the solution of several problems is still far off, though it has been made closer. Also, considerable light has been thrown on the many-body problem, the interaction of electrons and nuclei and coherent time-dependent behavior of the many-body system through these advances.

8.3 Electronic Recognition of DNA Sequence Using a Nanopore

Contact persons: A. Aksimentiev, University of Illinois Urbana-Champaign; G. Timp, University of Notre Dame

High-throughput technology for sequencing DNA has already provided invaluable information about the organization of the human genome and common variations of the genome sequence among groups of individuals. To date, however, the high cost of whole-genome sequencing limits widespread use of this method in basic research and personal medicine. Using nanopores for sequencing DNA may dramatically reduce the costs of sequencing, as they enable, in principle, direct read-out of the nucleotide sequence from the DNA strand via electrical measurement [51].

The first experiments demonstrating the possibility of detecting DNA sequences using nanopores employed a biological channel (alpha-hemolysin) suspended in a fragile lipid bilayer membrane [52]. Recent advances in nanofabrication technology have allowed nanometer-size pores to be manufactured in thin synthetic membranes, offering superior mechanical stability and the potential for integration with conventional electronics. However, the question remains on how to use such pores to read out the sequence of a DNA molecule.

Using molecular dynamics simulations as a kind of a computational microscope, several plausible strategies for sequencing DNA have been devised [53–55]. One such method is illustrated in Fig. 8. The key component is a nanometer-diameter pore in a thin capacitor membrane that consists of two conducting layers (doped silicon) separated by a layer of insulator (silicon dioxide). An external electric bias is applied across the membrane to drive a single DNA strand back and forth through the pore, while the electric potentials induced by the DNA motion are independently recorded at the top and bottom layers (electrodes) of the capacitor membrane. Another possible sequencing method exploits the unique mechanical properties

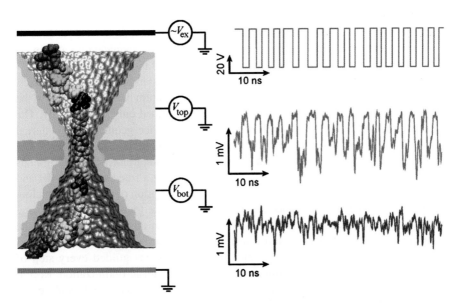

Fig. 8 Driven by an external electric field, a DNA molecule permeates a nanometer-size pore in a multilayer semiconductor membrane. The nucleotide sequence of the DNA molecule is recorded as the change of the electrostatic potential in the capacitor (From Sigalov et al. [54])

of double-stranded DNA, which permit a DNA molecule to be threaded through a nanopore smaller in diameter than the canonical DNA double helix [55]. The ability of atomic-scale simulations to provide realistic images of DNA conformations in synthetic nanopores and to relate such conformations to experimentally measured quantities has been of critical importance to the development of these sequencing methods. When implemented technologically, nanopore sequencing of DNA molecules will offer an affordable research tool to benchtop scientists across a broad range of disciplines and enable the use of whole-genome sequencing as a routine screening and diagnostics method, opening the door to personalized medicine.

8.4 Modeling and Simulation of Spin-Transfer Torque (STT) Devices

Contact persons: Sayeef Salahuddin, University of California, Berkeley; Stuart Parkin, IBM Almaden Research Center

A decade ago, the only way to switch the magnetization of a magnet was with an applied magnetic field. In 1996, Slonczewski and Berger independently predicted that switching might be possible in a nanoscale magnet by means of a spin-polarized current [13, 14]. The essential idea is that when spin polarized electrons flow through a nanoscale magnet, they transfer their spin angular momentum and thereby align the magnet in the direction of the polarized spins. Subsequently, this prediction was experimentally demonstrated by the Ralph and Buhrman groups [15] at Cornell University, and the effect is currently being considered as a candidate for a memory that is fast as well as nonvolatile, often called a universal memory.

Nonetheless, we are still at a very early stage of understanding the device physics. For example, direct measurement of spin angular momentum being transferred to the magnet (referred to as the spin torque) has been performed only very recently. The behavior of spin torque shows rich dependence on device geometry and external stimuli like voltage and magnetic field. Recently, a simulation methodology that explains the underlying physics in a simple manner suitable for device design has been developed [56]. The ferromagnet is modeled using a free electron model and the insulating material as a tunneling barrier. This description of magnetic tunneling junctions is not new; the advance lies in the fact that the transverse modes arising from a large cross section in an experimental device can now be included, and the measured torque dependence can be quantitatively explained.

STT devices are a unique example where TM&S has guided every step of development starting from the initial conception to material optimization. Recent advances provide us with a quantitative tool to analyze experiments and design devices. STT devices are currently also fueling exciting research for a new class of devices where magnetization may be switched just by applying an electric field (Fig. 9). Using current to switch the state of a magnet is a new paradigm for electronic devices.

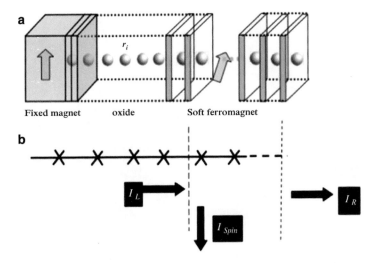

Fig. 9 (a) Schematic of an STT device. An insulator is sandwiched between two ferromagnets. The magnetization of the fixed magnet is kept unchanged. On the other hand, the magnetization of the soft ferromagnet can be changed by applying a spin polarized current through it. (b) A simple depiction of underlying physical mechanism. The difference between the spin angular momentum at the *left* and *right* interfaces of the soft ferromagnet is absorbed by the magnet itself. This subsequently tries to realign the magnetic orientation to the direction of polarized spins (From Salahuddin et al. [56])

8.5 New Materials Based on DNA-Linked Gold Nanoparticles and Polymers: A Theoretical Perspective

Contact person: George C. Schatz, Northwestern University

Gel-like materials derived from DNA-linked nanoparticles, polymers, or molecules are a new class of materials that have emerged in the last 15 years as a result of advances in synthetic methods, including the ability to routinely synthesize oligonucleotides with specified base-pair sequences, and the ability to chemically tether oligonucleotides to metal (typically gold or silver) particles or to organic molecules. The development of these materials has also been stimulated by theory and computation. Shown below (Fig. 10) is a schematic of these materials, with a DNA-functionalized gold nanoparticle shown on the left and the corresponding DNA-linked nanoparticles on the right.

These materials have quickly demonstrated their usefulness for DNA and protein detection using optical and electrical methods, and there is recent interest in using the functionalized nanoparticles for therapeutic applications, such as for siRNA delivery. Moreover a new generation of DNA-linked materials chemistry is starting to appear with the recent discovery that crystalline rather than amorphous structures can be produced when DNA links the particles together, and that the properties of these crystalline materials can be varied over a wide range by varying DNA length, base-pair sequence, and nanoparticle size and shape [58].

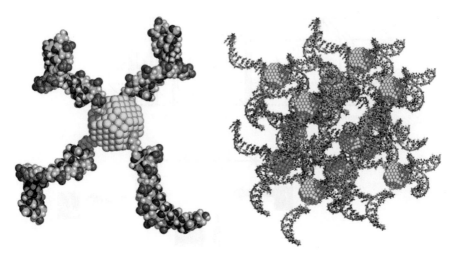

Fig. 10 *Left panel*: theoretical model of DNA-functionalized gold nanoparticle. *Right panel*: model of DNA-linked gold nanoparticle crystal (Adapted from Lee and Schatz [57])

The properties of these materials have been of great interest to theorists, but predicting these properties is a huge challenge due to the complexity of the structures. When the first DNA-linked gold nanoparticle aggregates were synthesized in the mid-1990s, it was quickly discovered that the thermal melting (dehybridization) transitions in these materials were quite different from the corresponding transitions for the same DNA structures in solution. This behavior also occurs for DNA-linked polymers, but not all DNA-linked structures show this. This work led to the development of a number of theoretical models that can roughly be grouped according to whether they treat the problem as an aggregated material that undergoes a bulk-like phase transition or as the cooperative melting of pairs of particles within an aggregate that are linked by several DNA structures.

Interest in separately observing these two mechanisms has stimulated additional experiments and the development of new classes of materials, including (1) DNA structures that can partially melt without unlinking, (2) PNA/DNA linked materials (PNA=peptide nucleic acid), and most recently, (3) DNA-linked molecular dimers, in which two organic molecules are linked by three DNA nanoparticles. The dimer experiments cleanly show that cooperative melting can occur, and a molecular dynamics study based on a newly developed coarse-grained DNA model [57] has provided an interpretation of these experiments, showing the importance of rigidity of the structure on entropy release in producing cooperativity. Theory has also provided much of the motivation for the development of DNA-linked nanoparticle crystals.

The structural properties of the DNA-functionalized and DNA-linked particles are also surprising. For example the maximum density of ss-DNA that can be attached to gold particles using thiol linkers is amazingly high (compared to DNA-functionalized silica particles, or compared to DNA-functionalized flat gold surfaces), and has been found to depend on particle diameter, being largest for the smallest-diameter particles. In addition, the length of DNA, which is either attached to just one gold particle, or through which hybridization chemistry links two particles together, is apparently much shorter than in conventional Watson-Crick DNA, and it has been proposed that perhaps the DNA structure is A-form rather than B-form in these structures. Although there has been progress with modeling these properties using atomistic molecular dynamics, there are unresolved issues that are at the limit of current simulation technology.

8.6 Performance Limit and Design Modeling for Nanoscale Biosensors

Contact person: Muhammad Alam, Purdue University

Biosensors based on silicon nanowires (Si-NW), silicon nano-cantilevers, and composites of carbon nanotubes (CNT) promise highly sensitive, dynamic, label-free, electrical detection of biomolecules with potential applications in proteomics and genomics [21, 59, 60]. In a parallel array configuration, each individual pixel (sensor) can first be functionalized with specific capture molecules. Once the analyte molecules are introduced to the sensor-array, they are exclusively captured by pixels with complementary probes. The conjugations are reflected in changes of the conductivity (DG) or that of the resonant-frequency (dw) (for cantilever sensors) of the corresponding pixel. The differential map of electrical conductivity before and after conjugation identifies the molecules present in the analyte.

The basic function of the nano-biosensor, as described above, is intuitively obvious and has been extensively verified by many research groups [19, 61–64]. Key challenges now include (1) the minimum detectable limit of analyte concentration by nano-biosensors (current estimates vary from nanomolar to femtomolar), (2) the proper design of sensitive biosensors (as a function of diameter, length, and doping, and fluidic environment, etc.), and (3) the limits of selectivity (i.e., ability to avoid false positives) of a sensor technology. During the past few years, a new theoretical framework to systematically address the above-mentioned questions and consistently interpret apparently contradictory data has been developed. In this regard, theoretical concepts of "geometry of diffusion," "screening-limited response," and "random sequential absorption" have made fundamental contributions to the understanding of biosensors.

To address the first question of minimum detection limits of nano-biosensors (for a given measurement window), a simple analytical model based on reaction diffusion theory can predict the tradeoff between average response (settling) time (t_s) and minimum detectable concentration (r_0) for nano-bio and nano-chemical sensors. The model [22] predicts a scaling relationship $\rho_0 t_s^{M_D} \sim k_D$ where M_D and k_D are dimensionality dependent constants for 1D (planar), 2D (nanowire/nanotube: NW/NT), and 3D (nano-dot) biosensors (see Fig. 11). Surprisingly, this model suggests that the improvement of the sensitivity of NW/NT sensors over planar sensors arises not because the geometry of the electrostatics, but rather due to the geometry of diffusion. The model also predicts that "femtomolar" detection, a theoretical

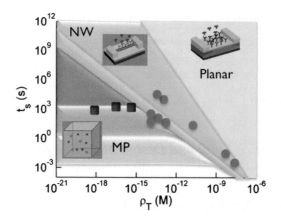

Fig. 11 Phase space of biosensor response. The different regions show the analyte concentration (ρ_T) that can be measured in a given sampling time (t_s). Classical planar sensors require more time to detect a given analyte concentration. Planar, nanowire (*NW*), and magnetic particle (*MP*) sensors behave differently due to their different diffusion characteristics. The enhanced sensitivity of MP bio-barcode sensors is due to a reduction of the diffusion bottleneck by dispersing the "sensors" with the volume of the analyte. (Symbols denote sources of experimental data: *NW* (*dots*) from Zheng et al. [64], Gao et al. [61], Stern et al. [63], Kusnezow et al. [62], Li et al. [19]; MP (*squares*) from Goluch et al. [65], Nam et al. [66], and Nam et al. [67])

possibility advanced by some groups, may actually be difficult for any practical measurement time.

Regarding the second question of electrostatic design of highly sensitive nanobiosensors, one finds that the role of screening is critical in dictating the response of biosensors, and the sensitivity of the biosensor is a nontrivial function of doping, device geometry, and fluidic environment [68]. For example, simple electrostatics calculations suggest that reduced doping would improve modulation of conductivity (for a given adsorbed charge from a biomolecule), yet device-to-device random dopant fluctuation may actually make the operation of such low-doped sensors difficult and expensive.

Finally, the problem of selectivity can be understood within the framework of random sequential absorption of capture molecules [69]. Since the theoretical maximum of RSA surface coverage is 54%, one can show that long incubation time and surface passivation by molecules (e.g., PEG) much smaller than the target molecule is the only viable approach to high selectivity. Each of these predictions has now been confirmed by numerous experiments.

8.7 First-Order Phase Transition to an Ordered Solid in Fluids Confined to the Nanoscale

Contact person: Peter T. Cummings, Vanderbilt University/Oak Ridge National Laboratory

As we shrink the size of mechanical devices to nanoscopic dimensions—such as hard disk drives, where the distance between the read head and the spinning platter is already less than 10 nm and going down exponentially—the lubrication of these systems becomes an increasingly pressing issue. Some experiments have suggested that a typical lubricant fluid becomes a solid when the distance between the surfaces confining the fluid become nanoscopic, thus rendering the lubricant useless, while other experiments contradict this finding. For over two decades, intense debate about whether or not nanoscale confinement of a fluid can induce a first order phase transition to an ordered solid structure has been an enduring controversy in nanoscience.

Clearly at the heart of the debate over the effect of nano-confinement is the inability to observe directly, via experiment, the structural changes that occur when a fluid is nanoconfined. Molecular simulations can be performed that mimic the experimental system, in which two molecularly smooth mica surfaces are gradually brought to a separation on the order of several nanometers while immersed in the fluid to be studied, and can be used to gain insight into the atomic level structural changes underlying this complex behavior. Such simulations were first attempted in the early 2000s for n-dodecane ($C_{12}H_{26}$) nanoconfined between mica-like surfaces using a united atom (UA) approach and showed a first-order phase transition to a layered and herringbone-ordered structure. Though these simulations provided valuable insight into the effects of nanoconfinement, they were based on several simplifying assumptions (driven by the available computational resources), particularly in the description of the mica surfaces. Due to major advances in both computational hardware and algorithms, these shortcomings have been eliminated and fully detailed atomistically, molecular dynamics simulations have now been performed [70, 71] that are faithful to the corresponding experimental systems in a way not previously possible and provide compelling evidence supporting the existence of a phase transition.

The key results are typified by Fig. 12; specifically, the simulations show that upon sufficient confinement, a range of fluids (from linear to cyclic alkanes) undergo a first-order phase transition to a layered and ordered solid-like structure.

The remarkable finding of this work is that even though the fluids are nonpolar (or very weakly polar), it is nevertheless the *electrostatic* interactions between the fluid and the ions in the mica that drive the fluid-solid transition—an insight that was missing from earlier simulations because the simplified models used did not involve electrostatic interactions.

8.8 Single-Impurity Metrology Through Multi-Million-Atom Simulations

Contact persons: Gerhard Klimeck, Purdue University; Sven Rogge, Delft University of Technology, The Netherlands

Improvements in electronic devices have marched along Moore's law (the doubling of the number of transistors on a chip each technology generation) for the past

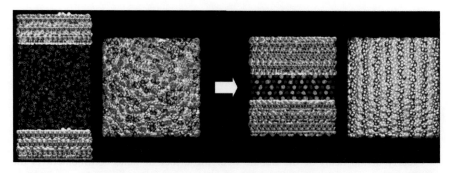

Fig. 12 Upon sufficient nanoconfinement, cyclohexane undergoes a rapid and abrupt transition to a layered and hexagonal ordered, solid-like structure. In the configurations, the mica sheets are shown and the cyclohexane rendered in stick form on the left, so that it is easy to see through the confined fluid; the confined fluid, rendered in full size without mica, is shown on the *right*, looking down from above. Cyclohexane molecules (H atoms *white*, C atoms shades of *blue*) are colored in two different shades of *gray*, making the resulting order apparent (From Docherty and Cummings [71])

40 years, reaching the ultimate limit of atomic-scale devices, high power density, and unscalable economics. To enable further computing capability advancements a replacement concept is needed. Just over 10 years ago Kane proposed a quantum computer based on hydrogen-like quantum states of individual phosphorus impurities in silicon [72]. Quantum Bits (qbits) are to be encoded in the quantum states of the individual impurities and the associated electrons and manipulated through proximal electronic gates. The concept was based on known bulk concepts, but no theory or experiment existed on gated single donors.

Early hydrogenic models were built at the University of Melbourne (UM), followed by effective mass models by various authors. Even though a single impurity serves as qbit host, the confined electron experiences the presence of around a million atoms around it (Fig. 13). A detailed understanding of the complex silicon band structure became necessary leading to a band minimum model (BMB) at UM and multi-million-atom simulations powered by NEMO3D at Purdue. The first gated single impurity spectra (see coulomb diamond streaks in Fig. 13) were measured at Delft University of Technology in experimental devices built at IMEC. Experimental evidence combined with multi-million atom simulations demonstrated the emergence of a new type of hybrid molecule system [73].

Ultra-scaled FinFET transistors bear unique fingerprint-like device-to-device differences attributed to random single impurities [74]. Through correlation of experimental data with multimillion atom simulations [75], one can now identify the impurity's chemical species and determine their concentration, local electric field and depth below the Si/SiO_2 interface. The ability to model the excited states enabled a new approach to impurity metrology critical for implementations of impurity-based quantum computers.

Fig. 13 Conductance spectroscopy through a nano-scaled transistor dominated by a single impurity. One of several measured Coulomb diamonds is depicted in the background with three different date-field-dependent single-electron wave functions bound to a single impurity (After Lansbergen et al. [73])

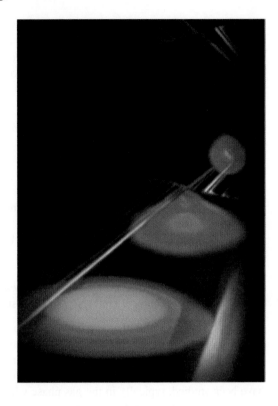

9 International Perspectives from Site Visits Abroad

Four workshops, the first in the United States followed by workshops in Europe, Japan, and Singapore, provided global perspectives on TM&S. A broad consensus on past successes and priorities for the future emerged, but each workshop also highlighted specific challenges and opportunities. In Europe, the need for a unified computational framework to support multiscale, multi-phenomena simulation was identified. Also identified were the strong opportunities for nanotechnology in biology and medicine and the need for user-friendly software designed for non-experts to support design and manufacturing in small and medium sized enterprises. The workshop in Japan stressed the need for improved human interfaces, including visualization, to support the broader application of simulation beyond the TM&S community. The power of simulation and visualization to promote public understanding of nanotechnology was also highlighted. The workshop in Singapore pointed out that multiscale modeling and simulation has not yet realized it potential and that it should be the focus in the next decade. The need for programs that develop long-term dialog between TM&S experts and experimentalists and application engineers was discussed. Finally, the importance of a considerable investment to re-think simulation algorithms and methods and to revise existing software in order to take advantage of the power of new computing architectures (e.g., GPU chips) was stressed.

A common theme that emerged from the workshops is the need for stronger collaborations, between TM&S experts and experimentalists, academia, and universities—and international collaboration and cooperative programs as well. Brief reports of the three overseas workshops follow.

9.1 United States-European Union Workshop (Hamburg, Germany)

Panel members/discussants

Lars Pastewka (co-chair), Fraunhofer Institute for Mechanics of Materials (IWM), Germany
Mark Lundstrom (co-chair), Purdue University, United States
Wolfgang Wenzel, Karlsruhe Institute of Technology (KIT) Institute of Nanotechnology, Germany
Costas Kiparissifdes, Aristotle University of Thessaloniki, Greece
Alfred Nordman, Technical University of Darmstadt, Germany

TM&S provides insights that are impossible to obtain with current experimental means. Even more important is its use for design and engineering.

Methods for nanoscale TM&S, which require an explicit incorporation of the atomic structure, have evolved over the last 20 years. A wide range of nanosystems have been studied, typically in the gas phase without interaction with the environment. For catalytic activity of nanoparticles on surfaces or nanoscale building blocks for electronic systems, for example, recent calculations now go beyond a gas-phase approximation and routinely include the crucial influence of the support. During the next 10 years, such methods need to be advanced in order to model nanosystems in liquid (e.g., during their synthesis or for biological applications) and solid environments (e.g., composite materials). Simulations will go from a basic understanding of clean systems towards real-world applications.

TM&S has advanced significantly over the past 10 years. Besides the approximately 1,000-fold increase in computing capabilities, advanced theoretical methods have been developed. A partial list includes: (1) advanced sampling techniques for free-energy calculations, (2) methods for electronic transport in realistic systems, and (3) higher accuracy exchange-correlation functionals for density functional theory. The first linear scaling O(N) quantum chemistry methods have been demonstrated, quantum/atomistic and atomistic/classical coupling has been advanced further, and reactive interatomic potentials have been developed for a large number of chemical elements. TM&S has been an enabling technology for the downscaling of MOSFETs, tailoring of multi-layer optical structures, metamaterials, and spectroscopy, among other examples.

TM&S will progress by expanding existing techniques to broader length and time scales. Methods will become more fail-safe so that a deployment to non-experts and hence a pervasive use of TM&S where appropriate can happen.

Exploration of hundreds of design alternatives will be enabled. Simulation tools will be deployed to industry, in particular to small- and medium-sized enterprises where TM&S currently finds limited use.

Further progress in TM&S requires the wide deployment of multiscale, multi-phenomena approaches. Different length scales effectively decouple, so application of theories with validity on different scales becomes possible. The interfaces between these scales need to be defined. Here, individual efforts need to occur on different physical regimes, in particular in the electronic regime, the morphological regime, and the electromagnetic regime.

Multiscale/multi-phenomena solutions to some specific "grand-challenge" problems should be funded. Such a funding strategy combines methodological advancements with specific applications. To minimize the overhead spent in developing multiscale methods and software, coordinated international action for the development of a unified computational multiscale framework should be explored. These new computational TM&S tools should make use of recently emerged and emerging computing architectures, such as computing on graphic processors and cloud computing. Specific funding to support this work needs to be available.

The interaction of nanomaterials with biological systems and in medical applications is an area of opportunity for TM&S. TM&S also has huge potential to help in the design of catalysts, composite materials, lithium ion batteries, and solar cells. TM&S will enable the design of novel hybrid nanomaterials e.g., by self-assembly of binary supercrystals, combining magnetic, optical, and electronic properties in novel ways. It will also aid the search for new high Tc superconductors, help to tailor the interface between "man-made" and biological materials, give insights into the working of tribological contacts and other buried interfaces, and advance the field of quantum computing. Smart use of multiscale methods coupled with advances in computing will greatly expand the size of problems that can be addressed.

The enabling character of TM&S and its pervasive application will lead to significantly shorter design cycles, with the obvious impact on society. For example, the 1,000-fold increase of computational capacity over the last 10 years was catalyzed itself by the use of TM&S as a design tool. In return, it catalyzed TM&S by providing ever-increasing computational capacity, and additionally transformed our society radically due to the ubiquitous presence of computers.

9.2 United States-Japan-Korea-Taiwan Workshop (Tokyo/Tsukuba, Japan)

Panel members/discussants

Akira Miyamoto (co-chair), Tohoku University, Japan
Stuart A. Wolf (co-chair), University of Virginia, United States
Nobuyuki Sano, University of Tsukuba, Japan
Satoshi Watanabe, University of Tokyo, Japan
Ching-Min Wei, Academia Sinica, China

Long-term impacts and future opportunities for nanotechnology in the field of TM&S are increasing rapidly. This is possible because of significant progress in multiscale, multiphysics simulations with sufficient informatics, human interface (visualization), and integration of experiments/measurements. The main scientific/engineering advancements in the last 10 years in the field of TM&S are summarized as follows:

- Development of highly efficient density functional theory (DFT) codes, improved DFT theory, and highly accurate QM electronic structure calculations using quantum Monte Carlo and other hybrid functionals
- Large scale simulations that include many body effects and that model the interplay between electron-phonon-atom dynamics in non-equilibrium systems
- Reactive force fields for simulations of materials growth and assembly processes
- Modeling of synthesis and assembly, and programmed materials that make use of self-assembly
- Effective screening medium approach for interfaces (e.g., Pt/water)
- Conceptual and computational tools for quantum transport including NEGF (non equilibrium Green's function) technique as well as extensions to NEGF such as combining *ab initio* molecular dynamics (MD) with NEGF, and merging NEGF with the *top-down*, MC approach for transport, and modeling transient transport/AC transport
- Progress on mesoscale simulations using multiscale, multiphysics approaches
- Simulating integrated functional nanoscale systems
- Real-time processing technologies for huge amounts of experimental data (e.g., data from light sources)

The goal of TM&S for the next 5–10 years is to realize multiscale, multiphysics simulations with sufficient informatics, human interface (visualization), and integration of experiments and/or measurements that can contribute to the progress of a variety technologies with very high social value such as a high-performance Li-ion batteries, solar cells, a polymer electrolyte fuel cell (PEFC), a solid electrolyte fuel cell (SOFC), advanced tribological technologies, and next-generation automotive catalysts.

To realize the computational capacity needed for multiscale simulation, traditional CPUs must be augmented with the new capabilities of GPUs. TM&S infrastructure should also be created to realize more dynamic collaborations between experiment and simulation for solving realistic problems in technology and society. R&D investments should be directed to promote strong coupling to industry.

During the past 10 years, nanotechnology was mainly for experts in academia or industry. The steady progress in the past 10 years summarized above, especially computer simulation coupled with a human interface (e.g., 3D visualization), will in the next 10 years have a big impact on industry and society in general. This progress also promotes societal understanding that leads to political understanding of the breadth of this technology, which is essential for the further progress and funding for nanotechnology. Although the societal understanding of the value of various technologies is considered to be a barrier for these technologies, advanced multiscale, multiphysics simulation with a sufficient human interface can promote the understanding of these technologies.

9.3 United States-Australia-China-India-Saudi Arabia-Singapore Workshop (Singapore)

Panel members/discussants

Julian Gale (co-chair), Curtin University, Australia
Mark Lundstrom (co-chair), Purdue University, United States
Michelle Simmons, University of New South Wales, Australia
Jan Ma, Nanyang Technological University, Singapore

TM&S provide insight and rationale for observations and increasingly serve as predictive tools for design and manufacturing. As recognized in the first IWGN report [76], modeling and simulation face challenges involving length scale, time scale, and accuracy. The past 10 years has seen significant progress, not just through increased computing power, but also through the development of the underlying techniques. Key developments include:

- *Linear-scaling quantum mechanics*: Although many of the foundations were laid during the 1990s, advances continued in areas such as purification theory [26] and new boundary conditions for divide and conquer [27]. Density functional calculations for thousands of atoms are now routine.
- *Reactive force-fields*: The ReaxFF method [77] saw the advent of an integrated reactive force-field that describes both multiple bond-order contributions as well as variable charge electrostatics and van der Waals terms. This was an important advance beyond the REBO model [78] that had been widely used for nanostructured carbon materials.
- *Simulation of nanostructured materials*: It has become possible to create polycrystalline materials [79] with complex interfacial structure, as well as to study the properties of extended defects [80]. Coupling different computational techniques has made it feasible to simulate the reactions of nanoparticles, as demonstrated by the oxidation of an aluminum nanometer-sized cluster [39]. Approaches to extending length and time scales in simulations of crystal growth from solvents have been developed [81].
- *Enhanced sampling of free energy surfaces*: One of the greatest challenges in simulation is to reliably search the conformational space to find the configurations that are important for a given system. Advances in this field include metadynamics [29], which increases the rate of exploration, though at the expense of time-correlation information. Similarly there have been improvements in methods for determining rate constants [30].

- *Quantum mechanical methods beyond density functional theory (DFT) for condensed phases*: While standard DFT (LDA/GGA) is computationally efficient, it suffers from several limitations. Although methods for obtaining improved reliability have long been known, the application to the solid-state, particularly with systematic basis sets, has only recently begun to become more widespread [24, 25].

Although there has been progress in multiscale/multiphysics TM&S, genuine multiscale modeling and simulation (spanning more than two scales or techniques) has yet to realize its full potential. While high-level coupling of models is routinely possible (for example via passing continuum properties or rate constants), more dynamic transfer of information remains challenging. There are emerging protocols for representation of physical metadata (e.g., [82]), but most software currently will only output, but not read and utilize, this information. Furthermore, the structure of most codes is not designed to facilitate direct coupling of different methods within the same software.

A significant barrier to the wider development of multiscale/multiphysics modeling is the need to develop a long-term dialogue between different disciplines spanning science and engineering. To effectively couple techniques requires experts from both sides to acquire knowledge of the other side, which can take considerable time. Current funding models require short-term achievement, which makes it difficult to engage in such long-term objectives. The geographic dispersity of expertise represents another barrier, both to dialogue between theorists, modelers and simulators, and to experimentalists seeking solutions to multiscale problems.

Proposed goals and strategies for TM&S are as follows:

- *Establish a virtual international centre for grand challenges in TM&S*: A virtual network should be established to focus on the solution of defined grand challenge problems. It would create a focal point for establishing strong links with leading experimental groups. Examples of the challenges that would benefit from such an approach including (a) determination of whether a sub-5 nm silicon transistor is operable, (b) design of a thermoelectric material with a figure of merit greater than three, and (c) design of an organic photovoltaic cell with an efficiency of greater than 20%.
- *Achieve automated, dynamic tools for multiscale and multiphysics TM&S*: Current software typically requires considerable human effort to establish a predetermined set of couplings. However, more effective achievement of technological outcomes would be possible by artificially intelligent linking of physics theories, models, and simulations, such that the level of the algorithm can be dynamically adjusted "on the fly" in order to achieve the solution with a target accuracy in the minimum computational time. A first step towards this goal would be real-time interactive simulation, with immersive visualization to permit human manipulation of the model.
- *Improve methods for excited states in nanotechnology*: While linear-scaling DFT allows us to describe the electronic ground state of nanodevices, many technological properties depend on the excited states of the system. Often these are

more delocalized and not as readily amenable to linear-scaling theory. Efficient methods are therefore required for going beyond ground state DFT, including determining energy levels accurately for use in coupled molecular dynamics/ electron transport calculations.

To achieve the above goals a significant increase in computing power will be required. For example, the increased use of quantum mechanical methods will require the evaluation of Hartree-Fock exchange to improve accuracy. Assuming a linear-scaling approach is employed, the calculation of the electronic structure of a 10 nm quantum dot would using HF exchange will need an increase in computing power by approximately a factor of 10,000 – well beyond the factor of 100 anticipated from advances in computing hardware.

Given the likely change in computing architectures to more complex mixtures of energy efficient, limited instruction co-processors (e.g., GPUs) with conventional cores, and the need for massively parallel applications to exploit increased number of processors (order 10^6 per machine for top 500 in 10 years), the challenge will be to rethink algorithms to best exploit these heterogeneous systems [83]. Without considerable investment in scientific tools (e.g., optimized math libraries, etc.), our ability to harness such machines may limit progress.

References

1. M.C. Roco, R.S. Williams, P. Alivisatos (eds.), *Nanotechnology Research Directions: IWGN [NSTC] Workshop Report: Vision for Nanotechnology R&D in the Next Decade* (International Technology Research Institute at Loyola College, Baltimore, 1999). Available online: http://www.nano.gov/html/res/pubs.html
2. N. Goldenfeld, L.P. Kadanoff, Simple lessons from complexity. Science **284**, 87–89 (1999)
3. S.Y. Quek, H.J. Choi, S.G. Louie, J.B. Neaton, Length dependence of conductance in aromatic single-molecule junctions. Nano Lett. **9**, 3949 (2009)
4. S.Y. Quek, M. Kamenetska, M.L. Steigerwald, H.J. Choi, S.G. Louie, M.S. Hybertsen, J.B. Neaton, L. Venkataraman, Dependence of single-molecule junction conductance on molecular conformation. Nat. Nanotechnol. **4**, 230 (2009)
5. S.Y. Quek, L. Venkataraman, H.J. Choi, S.G. Louie, M.S. Hybertsen, J.B. Neaton, Amine-gold linked single-molecule circuits: experiment and theory. Nano Lett. **7**, 3477–3482 (2007)
6. L. Venkataraman, J.E. Klare, C. Nuckrolls, M.S. Hybertsen, M.L. Steigerwald, Dependence of single-molecule junction conductance on molecular conformation. Nature **442**, 904–907 (2006a)
7. L. Venkataraman, J.E. Klare, C. Nuckrolls, M.S. Hybertsen, M.L. Steigerwald, Single-molecule circuits with well-defined molecular conductance. Nano Lett. **6**, 458–462 (2006b)
8. M. Lundstrom, Z. Ren, Essential physics of carrier transport in nanoscale MOSFETs. IEEE Trans. Electron Dev. **49**, 133–141 (2002)
9. Q. Cao, N. Kim, N. Pimparkar, J.P. Kulkarni, C. Wang, M. Shim, K. Roy, M.A. Alam, J. Rogers, H.-S. Kim, Medium scale carbon nanotube thin film integrated circuits on flexible plastic substrates. Nature **454**, 495–500 (2008)
10. Q. Cao, J. Rogers, M.A. Alam, N. Pimparkar, Theory and practice of 'striping' for improved on/off ratio in carbon nanotube thin film transistors. Nano Res. **2**(2), 167–175 (2009)

11. S. Heinze, J. Tersoff, R. Martel, V. Dercycke, J. Appenzeller, P. Avouris, Carbon nanotubes as Schottky barrier transistors. Phys. Rev. Lett. **89**, 106801 (2002). doi:10.1103/PhysRevLett.89.106801
12. A. Javey, J. Guo, Q. Wang, M. Lundstrom, H. Dai, Ballistic carbon nanotube field-effect transistors. Nature **424**, 654–657 (2003)
13. L. Berger, Emission of spin waves by a magnetic multilayer traversed by a current. Phys. Rev. B **54**, 9353–9358 (1996)
14. J.C. Slonczewski, Current-driven excitation of magnetic multilayers. J. Magn. Magn. Mater. **159**, L1–L7 (1996)
15. J.A. Katine, F.J. Albert, R.A. Buhrman, E.B. Myers, D.C. Ralph, Current-driven magnetization reversal and spin-wave excitations in Co/Cu/Co pillars. Phys. Rev. Lett. **84**, 3149–3152 (2000). doi:10.1103/PhysRevLett.84.3149
16. M. Tsoi, A.G.M. Jansen, J. Bass, W.-C. Chiang, M. Seck, V. Tsoi, P. Wyder, Excitation of a magnetic multilayer by an electric current. Phys. Rev. Lett. **80**, 4281–4284 (1998). doi:10.1103/PhysRevLett.80.4281
17. Y. Kamihara, T. Watanabe, M. Hirano, H. Hosono, Iron-based layered superconductor La[O1-xFx]FeAs (x=0.05-0.12) with TC=26 K. J. Am. Chem. Soc. **130**, 3296–3297 (2008). doi:10.1021/ja800073m
18. T.A. Maier, D. Poilblanc, D.J. Scalapino, Dynamics of the pairing interaction in the Hubbard and t-J models of high-temperature superconductors. Phys. Rev. Lett. **100**, 237001 (2008). doi:10.1103/PhysRevLett. 100.237001
19. Z. Li, Y. Chen, X. Li, T.I. Kamins, K. Nauka, R.S. Williams, Sequence-specific label-free DNA sensors based on silicon nanowires. Nano Lett. **4**, 245–247 (2004). doi:10.1021/nl034958e
20. Z. Li, B. Rajendran, T.I. Kamins, X. Li, Y. Chen, R.S. Williams, Silicon nanowires for sequence-specific DNA sensing: device fabrication and simulation. Appl. Phys. Mater. **80**, 1257 (2005). doi:10.1007/s00339-004-3157-1
21. A. Star, E. Tu, J. Niemann, J.-C.P. Gabriel, C.S. Joiner, C. Valcke, Label-free detection of DNA hybridization using carbon nanotube network field-effect transistors. Proc. Natl. Acad. Sci. U.S.A. **103**, 921–926 (2006). doi:10.1073/pnas.0504146103
22. P.R. Nair, M.A. Alam, Performance limits of nano-biosensors. Appl. Phys. Lett. **88**, 233120 (2006)
23. P.R. Nair, M.A. Alam, Dimensionally frustrated diffusion towards fractal absorbers. Phys. Rev. Lett. **99**, 256101 (2007). doi:10.1103/PhysRevLett.99.256101
24. J. Hafner, *Ab-initio* simulations of materials using VASP: density-functional theory and beyond. J. Comput. Chem. **29**, 2044 (2008). doi:10.1002/jcc.21057
25. C. Pisani, L. Maschio, S. Casassa, M. Halo, M. Schutz, D. Usvyat, Periodic local MP2 method for the study of electronic correlation in crystals: theory and preliminary applications. J. Comput. Chem. **29**, 2113 (2008). doi:10.1002/jcc.20975
26. A.M.N. Niklasson, Expansion algorithm for the density matrix. Phys. Rev. B **66**, 155115 (2002). doi:10.1103/PhysRevB.66.155115
27. L.-W. Wang, Z. Zhao, J. Meza, Linear-scaling three-dimensional fragment method for large-scale electronic structure calculations. Phys. Rev. B **77**, 165113 (2008). doi:10.1103/PhysRevB.77.165113
28. W.A. Goddard, A. van Duin, K. Chenoweth, M.-J. Cheng, S. Pudar, J. Oxgaard, B. Merinov, Y.H. Jang, P. Persson, Development of the ReaxFF reactive force field for mechanistic studies of catalytic selective oxidation processes on $BiMoO_x$. Top. Catal. **38**, 93–103 (2006). doi:10.1007/s11244-006-0074-x
29. A. Laio, F.L. Gervasio, Metadynamics: a method to simulate rare events and reconstruct the free energy in biophysics, chemistry and material science. Rep. Prog. Phys. **71**, 126601 (2008). doi:10.1088/0034-4885/71/12/126601
30. T.S. Van Erp, D. Moroni, P.G. Bolhuis, A novel path sampling method for the calculation of rate constants. J. Chem. Phys. **118**, 7762–7775 (2003). doi:10.1063/1.1562614
31. S. Datta, *Quantum Transport: Atom to Transistor* (Cambridge University Press, Cambridge, 2005)

32. C. Kocabas, N. Pimparkar, O. Yesilyurt, S.J. Kang, M.A. Alam, J.A. Rogers, Experimental and theoretical studies of transport through large scale, partially aligned arrays of single-walled carbon nanotubes in thin film type transistors. Nano Lett. **7**, 1195–1204 (2007)
33. R. Schulz, B. Lindner, L. Petridis, J.C. Smith, Scaling of multimillion-atom biological molecular dynamics simulation on a petascale supercomputer. J. Chem. Theory Comput. **5**, 2798–2808 (2009)
34. National Research Council Committee on Modeling, Simulation, and Games, *The Rise of Games and High Performance Computing for Modeling and Simulation* (The National Academies Press, Washington, DC, 2010). ISBN 978-0-309-14777-4
35. J. Anderson, S.C. Glotzer, *Applications of Graphics Processors to Molecular and Nanoscale Simulations,* Preprint (2010)
36. M. Garland, S. Le Grand, J. Nickolls, J. Anderson, J. Hardwick, S. Morton, E. Phillips, Y. Zhang, V. Volkov, Parallel computing experiences with CUDA. IEEE Micro **28**(4), 13–27, July–August (2008).
37. J.D. Owens, H. Houston, D. Lubeke, S. Green, J.E. Stone, J.C. Phillips, GPU computing. Proc. IEEE **96**, 879–899 (2008). doi:10.1109/JPROC.2008.917757
38. Z. Zhang, P. Fenter, L. Cheng, N.C. Sturchio, M.J. Bedzyk, M. Predota, A. Bandura, J. Kubicki, S.N. Lvov, P.T. Cummings, A.A. Chialvo, M.K. Ridley, P. Bénézeth, L. Anovitz, D.A. Palmer, M.L. Machesky, D.J. Wesolowski, Ion adsorption at the rutile-water interface: linking molecular and macroscopic properties. Langmuir **20**, 4954–4969 (2004)
39. P. Vashishta, R.K. Kalia, A. Nakano, Multimillion atom simulations of dynamics of oxidation of an aluminum nanoparticle and nanoindentation on ceramics. J. Phys. Chem. B **110**, 3727–3733 (2006). doi:10.1021/jp0556153
40. S. Izvekov, M. Parrinello, C.J. Burnham, G.A. Voth, Effective force fields for condensed phase systems from ab initio molecular dynamics simulation: a new method for force matching. J. Chem. Phys. **120**, 10896 (2004). doi:10.1063/1.1739396
41. D. Reith, M. Pütz, F. Müller-Plathe, Deriving effective mesoscale potentials from atomistic simulations. J. Comput. Chem. **24**, 1624–1636 (2003). doi:10.1002/jcc.10307
42. G. Csányi, G. Albaret, G. Moras, M.C. Payne, A. De Vita, Multiscale hybrid simulation methods for material systems. J. Phys. Condens. Matter **17**, R691 (2005). doi:10.1088/0953-8984/17/27/R02
43. A. Papavasiliou, I.G. Kevrekidis, Variance reduction for the equation-free simulation of multiscale stochastic systems. Multiscale Model. Simul. **6**, 70–89 (2007)
44. R.W. Hamming, *Introduction to Applied Numerical Analysis (from the Introduction)* (McGraw-Hill, New York, 1971)
45. M. Kamenetska, S.Y. Quek, A.C. Whalley, M.L. Steigerwald, H.J. Choi, S.G. Louie, C. Nuckolls, M.S. Hybertsen, J.B. Neaton, L. Venkataraman, Conductance and geometry of pyridine-linked single molecule junctions. J. Am. Chem. Soc. **132**, 6817 (2010)
46. M. Stopa, Rectifying behavior in coulomb blockades: charging rectifiers. Phys. Rev. Lett. **88**, 146802 (2002)
47. M. Rontani, F. Troiani, U. Hohenester, E. Molinari, Quantum phases in artificial molecules. Solid State Commun. **119**, 309 (2001)
48. D. Loss, D. DiVincenzo, Quantum computation with quantum dots. Phys. Rev. A **57**, 120 (1998). doi:10.1103/PhysRevA.57.120
49. C. Barthel, J. Medford, C.M. Marcus, M.P. Hanson, A.C. Gossard, Interlaced dynamical decoupling and coherent operation of a singlet-triplet Qubit. Phys. Rev. B **81**, 161308 (2010) (R)
50. K. Ono, D.G. Austing, Y. Tokura, S. Tarucha, Current rectification by Pauli exclusion in a weakly coupled double quantum dot system. Science **297**, 1313–1317 (2002). doi:10.1126/science.1070958
51. D. Branton, D.W. Deamer, A. Marziali, H. Bayley, S.A. Benner, T. Butler, M. Di Ventra, S. Gara, A. Hibbs, X. Huang, S.B. Jovanovich, P.S. Krstic, S. Lindsay, X.S. Ling, C.H. Mastrangelo, A. Meller, J.S. Oliver, Y.V. Pershin, J.M. Ramsey, R. Riehn, G.V. Soni, V. Tabard-Cossa, M. Wanunu, M. Wiggin, J.A. Schloss, The potential and challenges of nanopore sequencing. Nat. Biotechnol. **26**, 1146–1153 (2008)

52. J.J. Kasianowicz, E. Brandin, D. Branton, D.W. Deamer, Characterization of individual polynucleotide molecules using a membrane channel. Proc. Natl. Acad. Sci. U.S.A. **93**, 13770–13773 (1996)
53. C.M. Payne, X.C. Zhao, L. Vlcek, P.T. Cummings, Molecular dynamics simulation of ss-DNA translocation between copper nanoelectrodes incorporating electrode charge dynamics. J. Phys. Chem. B **112**, 1712–1717 (2008)
54. G. Sigalov, J. Comer, G. Timp, A. Aksimentiev, Detection of DNA sequences using an alternating electric field in a nanopore capacitor. Nano Lett. **8**, 56–63 (2008)
55. W. Timp, U.M. Mirsaidov, D. Wang, J. Comer, A. Aksimentiev, G. Timp, Nanopore sequencing: electrical measurements of the code of life. IEEE Trans. Nanotechnol. **9**, 281–294 (2010)
56. S. Salahuddin, D. Datta, P. Srivastava, S. Datta, Quantum transport simulation of tunneling based spin torque transfer (stt) devices: design trade-offs and torque efficiency. IEEE Electron Dev. Meet. **2007**, 121–124 (2007). doi:10.1109/IEDM.2007.4418879
57. O.-S. Lee, G.C. Schatz, Molecular dynamics simulation of DNA-functionalized gold nanoparticles. J. Phys. Chem. C **113**, 2316 (2009). doi:10.1021/jp8094165
58. S.Y. Park, A.K.R. Lytton-Jean, B. Lee, S. Weigand, G.C. Schatz, C.A. Mirkin, DNA-programmable nanoparticle crystallization. Nature **451**, 553–556 (2008)
59. A. Gupta, P.R. Nair, D. Akin, M.R. Ladisch, S. Broyles, M.A. Alam, R. Bashir, Anomalous resonance in a nanomechanical biosensor. Proc. Natl. Acad. Sci. U.S.A. **103**(36), 13362–13367 (2006)
60. J. Hahm, C.M. Lieber, Direct ultrasensitive electrical detection of DNA and DNA sequence variations using nanowire nanosensors. Nano Lett. **4**(1), 51–54 (2004)
61. Z.Q. Gao, A. Agarwal, A.D. Trigg, N. Singh, C. Fang, C.-H. Tung, Y. Fan, K.D. Buddharaju, J. Kong, Silicon nanowire arrays for label-free detection of DNA. Anal. Chem. **79**(9), 3291–3297 (2007). doi:10.1021/ac061808q
62. W. Kusnezow, Y.V. Syagailo, S. Rüffer, K. Klenin, W. Sebald, J.D. Hoheisel, C. Gauer, I. Goychuk, Kinetics of antigen binding to antibody microspots: strong limitation by mass transport to the surface. Proteomics **6**(3), 794–803 (2006)
63. E. Stern, J.F. Klemic, D.A. Routenberg, P.N. Wyrembak, D.B. Turner-Evans, A.D. Hamilton, D.A. LaVan, T.M. Fahmy, M.A. Reed, Label-free immunodetection with CMOS-compatible semiconducting nanowires. Nature **445**(7127), 519–522 (2007)
64. G.F. Zheng, F. Patolsky, Y. Cui, W.U. Wang, C.M. Lieber, Multiplexed electrical detection of cancer markers with nanowire sensor arrays. Nat. Biotechnol. **23**(10), 1294–1301 (2005)
65. E.D. Goluch, J.-M. Nam, D.G. Georganopoulou, T.N. Chiesl, K.A. Shaikh, K.S. Ryu, A.E. Barron, C.A. Mirkin, C. Liu, A bio-barcode assay for on-chip attomolar-sensitivity protein detection. Lab Chip **6**(10), 1293–1299 (2006)
66. J.-M. Nam, S.I. Stoeva, C.A. Mirkin, Bio-bar-code-based DNA detection with PCR-like sensitivity. J. Am. Chem. Soc. **126**(19), 5932–5933 (2004)
67. J.-M. Nam, C.S. Thaxton, C.A. Mirkin, Nanoparticle-based bio-bar codes for the ultrasensitive detection of proteins. Science **301**(5641), 1884–1886 (2003). doi:10.1126/science.1088755
68. P.R. Nair, M.A. Alam, Screening-limited response of nanobiosensors. Nano Lett. **8**(5), 1281–1285 (2008)
69. P.R. Nair, M.A. Alam, A theory of "Selectivity" of label-free nanobiosensors: a geometro-physical perspective. J. Appl. Phys. **107**, 064701 (2010). doi:10.1063/1.3310531
70. P.T. Cummings, H. Docherty, C.R. Iacovella, J.K. Singh, Phase transitions in nanoconfined fluids: the evidence from simulation and theory. AIChE J. **56**, 842–848 (2010). doi:10.1002/aic.12226
71. H. Docherty, P.T. Cummings, Direct evidence for fluid-solid transition of nanoconfined fluids. Soft Matter **6**, 1640–1643 (2010)
72. B.E. Kane, A silicon-based nuclear spin quantum computer. Nature **393**, 133 (1998)
73. G.P. Lansbergen, R. Rahman, C.J. Wellard, P.E. Rutten, J. Caro, N. Collaert, S. Biesemans, I. Woo, G. Klimeck, L.C.L. Hollenberg, S. Rogge, Gate induced quantum confinement transition of a single dopant atom in a Si FinFET. Nat. Phys. **4**, 656 (2008)

74. G.P. Lansbergen, C.J. Wellard, J. Caro, N. Collaert, S. Biesemans, G. Klimeck, L.C.L. Hollenberg, S. Rogge, Transport-based dopant mapping in advanced FinFETs. *in IEEE IEDM,* San Francisco, 15–17 Dec 2008. doi: 10.1109/IEDM.2008.4796794
75. G. Klimeck, S. Ahmed, H. Bae, N. Kharche, R. Rahman, S. Clark, B. Haley, S. Lee, M. Naumov, H. Ryu, F. Saied, M. Prada, M. Korkusinski, T.B. Boykin, Atomistic simulation of realistically sized nanodevices using NEMO 3-D: part I – models and benchmarks. IEEE Trans. Electron Devices **54**, 2079–2089 (2007). doi:10.1109/TED.2007.902879
76. D.A. Dixon, P.T. Cummings, K. Hess, Investigative tools: theory, modeling, and simulation (Chap. 2.7.1), in *Nanotechnology Research Directions: IWGN Workshop Report. Vision for Nanotechnology in the Next Decade,* ed. by M.C. Roco, S. Williams, P. Alivisatos (Kluwer, Dordrecht, 1999)
77. A.C.T. Van Duin, S. Dasgupta, F. Lorant, W.A. Goddard III, ReaxFF: a reactive force field for hydrocarbons. J. Phys. Chem. A **105**, 9396–9409 (2001)
78. D.W. Brenner, O.A. Shenderova, J.A. Harrison, S.J. Stuart, B. Ni, S.B. Sinnott, A second-generation reactive empirical bond order (REBO) potential energy expression for hydrocarbons. J. Phys. Condens. Matter **14**, 783 (2002)
79. T.X.T. Sayle, C.R.A. Catlow, R.R. Maphanga, P.E. Ngoepe, D.C. Sayle, Generating MnO_2 nanoparticles using simulated amorphization and recrystallization. J. Am. Chem. Soc. **127**, 12828–12837 (2005)
80. A.M. Walker, B. Slater, J.D. Gale, v Wright, Predicting the structure of screw dislocations in nanoporous materials. Nat. Mater. **3**, 715–720 (2004). doi:10.1038/nmat1213
81. S. Piana, M. Reyhani, J.D. Gale, Simulating micrometer-scale crystal growth from solution. Nature **438**, 70 (2005). doi:10.1038/nature04173
82. P. Murray-Rust, H.S. Rzepa, Chemical markup, XML, and the worldwide web. 1. Basic principles. J. Chem. Inf. Comput. Sci. **39**, 928 (1999). doi:10.1021/ci990052B
83. W.A. De Jong, Utilizing high performance computing for chemistry: parallel computational chemistry. Phys. Chem. Chem. Phys. **12**, 6896 (2010). doi:10.1039/c002859b
84. P.T. Cummings, S.C. Glotzer, *Inventing a New America Through Discovery and Innovation in Science, Engineering and Medicine: A Vision for Research and Development in Simulation-Based Engineering and Science in the Next Decade* (World Technology Evaluation Center, Baltimore, 2010)

Enabling and Investigative Tools: Measuring Methods, Instruments, and Metrology

Dawn A. Bonnell, Vinayak P. Dravid, Paul S. Weiss, David Ginger, Keith Jackson, Don Eigler, Harold Craighead, and Eric Isaacs

Keywords Scanning probe microscopy • Scanning tunneling microscopy, atomic force microscopy • Electron microscopy • Synchrotron radiation, X-ray scattering • Spatial resolution • Time resolution • Nanoscale properties • Subsurface measurements • Nanobio instrumentation • Nanolithography • Nanofabrication, nanoscale metrology • International perspective

1 Vision for the Next Decade

1.1 Changes in the Vision Over the Past 10 Years

Advances in nanotechnology investigative tools have enabled fundamentally new approaches to the research carried out during the last decade. The crucial role of tools for manipulation and characterization of matter at the nanoscale was articulated by Nobel Laureate Horst Störmer in 1999 at the first U.S. Nanotechnology Research

---

D.A. Bonnell (✉)
Department of Materials Science and Engineering, University of Pennsylvania,
3231 Walnut Street, Room 112-A, Philadelphia, PA 19104, USA
e-mail: bonnell@lrsm.upenn.edu

V.P. Dravid
NUANCE Center, Northwestern University, 2220 Campus Drive #2036, Evanston,
IL 60208-3108, USA

P.S. Weiss
California NanoSystems Institute, University of California, 570 Westwood Plaza, Building 114,
Los Angeles, CA 90095, USA

D. Ginger
Department of Chemistry, University of Washington, Box 351700,
Seattle, WA 98195-1700, USA

K. Jackson
National High Magnetic Field Laboratory, 142 Centennial Building, 205 Jones Hall,
1530 S Martin Luther King Jr. Boulevard, Tallahassee, FL 32307, USA

Directions workshop, as follows: "Nanotechnology has given us the tools... to play with the ultimate toy box of nature—atoms and molecules... [This scale] provides an impressive array of novel opportunities to mix-and-match hunks of chemistry and biology with artificially defined, person-made structures. The possibilities to create new things appear endless" ([1], p. viii). The workshop vision at that time was that the promises of nanotechnology could be realized only through "the development of new experimental tools to broaden the capability to measure and to control nanostructured matter, including developing new standards of measurement." A particular point was made to extend this recommendation to biomolecules (p. xvi).

The investments in the science of characterization tools in the last decade have resulted in exciting discoveries, enabling metrologies, and windows of opportunity for revolutionary changes in the future. It is now possible to detect the charge and spin of a single electron, image catalytic reactions in real time, track some dynamic processes with 100 femtosecond (fs)[1] time resolution, map real and imaginary contributions to dielectric properties of molecules, and control chemo-mechanical interaction of individual essential biomolecules. Electron microscopy has achieved unprecedented spatial resolution with aberration correction, demonstrated 3D tomography, and developed *in situ* capacity, even for liquid-based systems. X-ray brilliance at beam lines has increased five orders of magnitude in the last 10 years, enabling detailed observations of dynamic processes and 3D structure from x-ray scattering.

The consequence of the advances in instrumentation science of the last decade is that researchers can now envision a new generation of tools for the next decade that will:

- Revolutionize the fundamental concepts of solids and biomolecules
- Allow new views of complex systems at small scales
- Probe dynamic processes and unexplored time scales

1.2 The Vision for the Next 10 Years

The next decade will see unanticipated new discoveries of nanoscale phenomena along with the implementation of early nanoscience into novel applications. New challenges will arise in direct measurement of dynamic processes in nanoscale systems, nanomanufacturing, integration of biosystems, and in higher levels of device and system complexity. At the dawn of this decade the scope of our ability

---

[1] 10^{-15} of a second.

D. Eigler
IBM Almaden Research Center, 650 Harry Road, San Jose, CA 95120-6099, USA

H. Craighead
School of Applied and Engineering Physics, Cornell University,
212 Clark Hall, Ithaca, NY 14853, USA

E. Isaacs
Argonne National Laboratory, 9700 S Cass Avenue, Argonne, IL 60439, USA

to probe local phenomena is vastly increased. Researchers now envision capabilities that were unimaginable 10 years ago. Intriguing challenges on the horizon include:

- Atomic resolution of the three dimensional structure of a single protein with chemical specificity
- Mapping (so called) continuum properties of individual atoms in a solid
- Discovery of stable new compounds by the manipulation of atoms at room temperature
- Tracking electrons with sufficient speed to observe intermediate steps in chemical reactions
- Concurrent imaging of processes throughout and entire cell

While ambitious, achieving these goals is plausible and doing so would be the driver for the next generation of discovery and innovation.

2 Advances in the Last 10 Years and Current Status

The advances in and current status of experimental methods for characterization of structure, properties, and processes at the nanoscale are summarized here, with an emphasis on notable examples in scanning probe–based measurement approaches, electron beam–based microscopies, optical probes, scattering tools at beam lines, and nanolithography platforms.

In the last decade, new cheap and accessible methods of nanopatterning such as microcontact printing and imprint lithography allowed research and product development laboratories all over the world to easily fabricate specialized complex devices with which to explore new phenomena. The commercial availability of off-the-shelf nanoparticles in sizes ranging from one to hundreds of nanometers enabled rapid advances in the fundamental science and application of particles in products ranging from medical therapeutics to hybrid electronics. The new field of plasmonics evolved dramatically.

Instruments that access structure and properties at nanometer and atomic length scales have been, and continue to be, essential in advancing our understanding of the physics and chemistry of nanostructures. The ability to quantify nanoscale behavior is a limiting factor in understanding new physics and is a prerequisite to manufacturing. It is a basic tenet of science that, "If it can't be measured, it can't be understood." In manufacturing, if specifications can't be characterized, reliable manufacturing is not possible.

2.1 Scanning Probe-Based Approaches to Measure Structure, Properties, and Processes

At the beginning of the National Nanotechnology Initiative (NNI) in 2000, scanning tunneling microscopy and atomic force microscopy already had become routine tools for investigating surface structure and adsorption reactions. At low temperature, quantum phenomena were beginning to be explored. In addition, first-order properties such as surface potential and magnetic field variations were characterized regularly. The decade 2000–2010 witnessed a geometric expansion of the variety of nanomaterial properties that can be measured locally, some aimed at advancing fundamental science and some developed in support of technology commercialization.

2.1.1 Observing Increased Complexity: From Chemical Identification to Mapping of Vector Properties

The wide range of phenomena that can now be accessed with high-spatial-resolution scanning-probe microscopy is illustrated in Fig. 1. Davis and colleagues are among those researchers who image electronic properties with atomic resolution [3]; they have observed, for example, that in strongly correlated electron systems, the density of states may not be symmetric with respect to the Fermi energy. Figure 1a shows a conductance ratio map of a superconductor that maps such asymmetry.

Charge density waves are another manifestation of electron interactions; the phase transitions responsible for charge density waves have recently been observed in real space on oxide surfaces ([4, 5]; Fig. 1b). The last 5 years have seen particularly exciting advances in the understanding of atomic force microscopy (AFM) at atomic resolution [6, 7]. Sugimoto et al. [8] exploit local forces, taking this approach to the next level, using force differences between atomic sites for chemical identification (Fig. 1c), thus enabling mapping of chemical properties at atomic resolution.

The nanometer-scale spatial resolution achieved recently in probing continuum properties such as resistance, capacitance, dielectric function, electromechanical coupling, etc., was unanticipated. Spatial resolution approaching 5 nm in magnetic fields is achieved by improving probe tip technology. Conductance and resistance are detected with sub-nanometer spatial resolution due to stress field focusing effects under the tip. Surface potential of Ge (105)-(1 × 2) [9] (Fig. 1f) and dielectric polarization of Si (111) [10] (Fig. 1h) have been imaged at atomic resolution. A significant advance of the last decade has been the increase in the complexity of properties able to be probed at local scales. The superposition of time-varying electric fields has led to new techniques such as scanning impedance spectroscopy and dielectric constant imaging. Figure 1g shows individual defects in a carbon nanotube through use of scanning impedance methodologies. And electromechanical coupling, a vector property, can now be mapped in liquids with ~3 nm resolution, as illustrated in the characterization of protein [11] (Fig. 1i).

2.1.2 Advancing the Frontiers of Physics

Significant advances have been made in fundamental understanding of the physics of nanoscale material properties through investigations carried out at low temperature with scanning tunneling microscopy (STM). Magnetism at surfaces has been

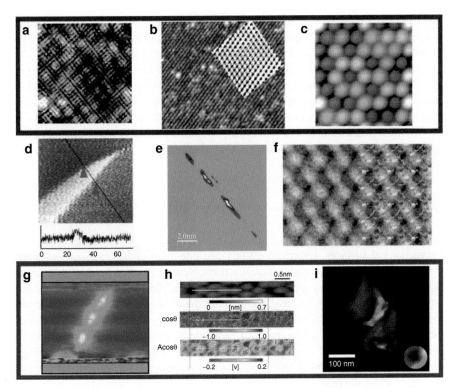

Fig. 1 Spatially resolved properties of nanoscale materials and phenomena "mapped" and/or imaged using a variety of atomic-scale characterization tools: (**a**) Conductance asymmetry in a super-conductor, a variant of tunneling spectroscopy. (**b**) STM image of K0.3MoO3 containing the two periodicities of the atomic structure and charge density waves. (**c**) Non-contact atomic force microscopy image of Si(111) with atoms colored differently for chemical force in force spectroscopy. (**d**) Magnetic force microscope image of the magnetic domain structure of a thin film taken with a focused, ion beam-milled tip (**e**) Conductance map of a HfO thin film containing electronic defects that result in leakage current. (**f**) Electrostatic potential image of the Ge(105)-(1×2) surface superimposed with the atomic model. (**g**) Scanning impedance image of a single-wall carbon nanotube conducting current. (**h**) Scanning nonlinear dielectric spectroscopy of the Si(111)-(7×7) surface in which polarization of the adatoms is obtained from the second harmonic of capacitance. (**i**) Vector piezoelectric force microscope map of local electromechanical of protein microfibrils on human tooth enamel; color indicates the orientation of the electromechanical response vector [2]

explored by spin-polarized STM. As illustrated in Fig. 1a, local details of local electronic structure—elucidated by mapping derivatives, Mott gaps, and Fourier transforms of local density of states—provide information about the interactions of electrons in solids. A long sought-after goal of the probe community was realized when Ho and coworkers [12] first demonstrated vibrational spectroscopy of organic molecules on surfaces with inelastic tunneling in a scanning tunneling microscope. Pascual [13], and others have shown the great promise of vibrational spectroscopy applied to chemical properties of adsorbed molecules. The technique is being extended to examine excitation and de-excitation channels of molecules in order to determine reaction pathways and coordinates [14].

In recent years, it has become possible to probe the low-energy spin excitation spectra of magnetic nanostructures at the atomic scale using inelastic spin excitation spectroscopy with a scanning tunneling microscope. Spin excitation spectroscopy engenders the ability to measure the energetics, dynamics, and spin configuration of single atoms and assemblies of atoms [15, 16] and was used to show that magnetic coupling of a Kondo atom to another unscreened magnetic atom can split the Kondo resonance into two peaks.

2.1.3 Advancing the Frontiers of Chemistry and Catalysis

Scanning probe microscopy has been the primary method used to observe chemical interactions at surfaces. Improvement in video rate imaging has enabled real-space studies of atomic diffusion and surface reactions. New and further-developed scanning probe tools have enabled greater control in self- and directed assembly, molecular devices, supramolecular assembly, and other areas [17] (Fig. 2). As researchers came to understand that these systems exist at the nanoscale and can be maintained far from equilibrium (a direct result of SPM measurements), they learned how to control defect types and density and ultimately to exploit those defects [18]. From there, they gained the ability to pattern all the way from the sub-nanometer scale to the wafer scale; the next accomplishment was to measure function of molecules and devices with scanning probes [19].

2.1.4 In Situ Characterization of Devices

SPM-based tools can easily be configured for *in situ* characterization. Quantifying electrical properties of nanowire and nanotube devices, for example as illustrated in Fig. 1g, has been essential to developing the present understanding of nanoelectronic and chemical sensor devices. Recent advances have extended the concept to more complex devices. Coffey and Ginger [20] simultaneously imaged the topographic structure and photogeneration rate in organic solar cells, using STM, as shown in Fig. 2.

Similarly, Bonnell and Kalinin [21] have imaged topographic structure along with the magnitude and direction of current flowing though an operating varistor (Fig. 3). The ability to examine variations in properties and processes under conditions relevant to device performance provides new information about the fundamental principles involved, as well as providing critical feedback that can facilitate product development. The photogeneration results in Fig. 3 were shown to predict complete solar cell performance. Applications that will benefit from *in situ* device characterization at the nanoscale include photovoltaics, electrochemical/ photochemical fuel generation, nanostructured batteries, thermoelectrics, supercapacitors, ferroelectric memory, and sensors, among others.

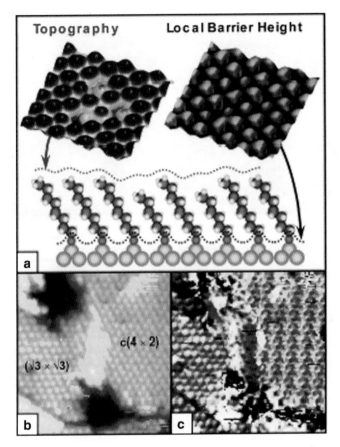

Fig. 2 Simultaneous STM measurements yield absolute tilt angles of individual molecules in self-assembled monolayers [17]

2.2 Electron Beam-Based Microscopies

Development of electron-beam instrumentation, techniques, and associated accessories has advanced at a breathtaking pace since 1999. The atomic-scale imaging (sub-0.2 nm nominal, so-called point-to-point Scherzer resolution) of beam-stable crystalline nanostructures is now readily and routinely possible given consistent and reproducible performance of both transmission electron microscopy (TEM) and scanning transmission electron microscopy (STEM) [22].

Further, most modern S/TEM instruments are capable of achieving nominal point-to-point spatial resolution approaching 0.13 nm and an information limit approaching 0.1 nm [23, 24] (Fig. 4).

The vast improvement in performance has been made possible due to a combination of high-performance field-emission sources [25] for electrons (which extend the envelope function of the contrast transfer function of the important objective lens),

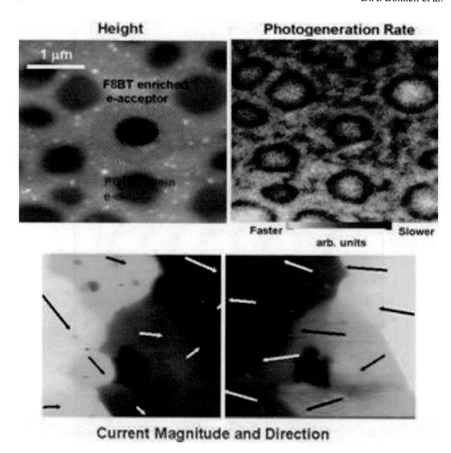

Fig. 3 (*Top*) Simultaneous imaging of structure and photo current generation in an organic solar cell (Coffey and Ginger [20]), and (*bottom*) imaging of current magnitude (size of vector) and direction across an operating oxide device [21]

improved aberration coefficients of lenses, overall stability of the microscope column, improvement in specimen stages, and adequate attention to control of the local environment at the sample. The practical realization of overall S/TEM performance has been greatly aided by innovative specimen preparation capabilities—including for biological/soft materials—coupled with image processing, simulation/modeling, and related computational developments.

2.2.1 Aberration Correction

Undoubtedly the most significant advance towards atomic-scale imaging (and many other attributes detailed later) has been the rapid emergence of aberration correctors [23, 24] for S/TEM (and for scanning electron microscopy, SEM). Until about a

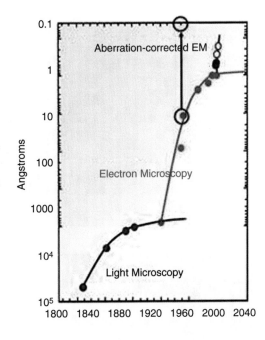

Fig. 4 Schematic illustration of the evolution of the spatial resolution of microscopy, culminating in the advent of aberration-corrected electron microscopy (*EM*): (*blue line*) light microscopy; (*green and red lines*) electron microscopy

decade ago, the spatial resolution of modern S/TEM was limited by the ubiquitous presence of spherical aberration and chromatic aberration of the principal imaging lens (objective lens) of the electron optics column. The commercial development of aberration correctors [23, 24, 26] that can be housed in commercial S/TEM columns has been the *transformative* development [26–33] in the field of electron microscopy (Fig. 5). It is interesting to note that the concept of using asymmetric lenses to improve lens performance was articulated by Richard Feynman [35] in his famous speech at the California Institute of Technology, "There is plenty of room at the bottom," which defined the concepts in modern nanotechnology.

Although modern aberration correctors employ the general concepts of quadrupole and octupole lenses to control aberrations in lenses, engineering realization of these concepts took decades to perfect and master. These remarkable new developments have also been incorporated into atomic-scale imaging with aberration-corrected SEM (Fig. 6).

In a major ongoing initiative called the Transmission Electron Aberration-Corrected Microscopy (TEAM) project [33], sub-1 Å spatial resolution was successful demonstrated in 2009 [33, 36, 37]. Several TEAM microscopes are now being developed [38]. Many commercial manufacturers have rapidly translated aberration corrections into their newer models of S/TEMs.

More recently, atomic-scale imaging with low atomic number sensitivity has also been demonstrated [39]. Concurrently, there has been remarkable progress in ancillary areas, ranging from specimen preparation tools and techniques to data analysis, processing, and mining. The combination of focused ion beam (FIB) and

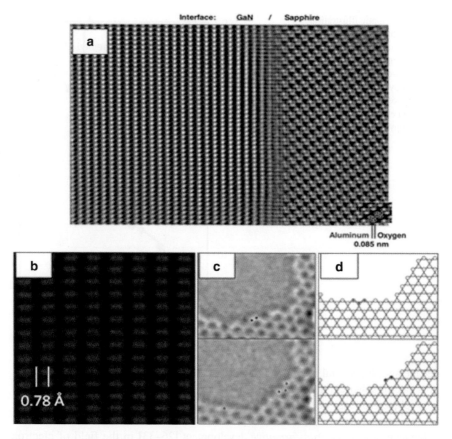

Fig. 5 Montage of atomic-scale images and phenomena enabled by aberration-corrected S/TEM. (**a**) HREM image of an interface between GaN and sapphire (Courtesy of Lawrence Berkeley National Lab.); (**b**) Silicon "dumbbells" at sub-Ångstrom resolution via HAADF/STEM (Courtesy of O. Krivanek, NION Corp. and S. Pennycook, Oak Ridge National Lab.); (**c**) Dynamics of graphene with atom-by-atom motion in successive frames. Right side illustrations show schematics of the process. (Courtesy of A. Zettl and C. Kisielowski, from Girit et al. [34])

scanning electron microscopy is one example that has had a high impact in numerous fields, including routine commercial applications in defect metrology in microelectronics [40–43].

2.3 Beam-Line Based Nanocharacterization

The last decade has witnessed dramatic advances in the capabilities of beam-Line facilities. A comparison of improvements in x-ray brilliance with those in computer processing speed provides some insight. Computer processors have increased performance at a rate that is generally considered phenomenal, 12 orders of magnitude

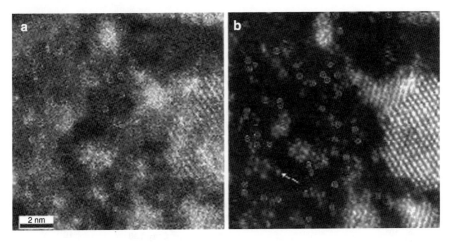

Fig. 6 SEM image (*left*) and corresponding annular dark field image (*right*) with aberration-corrected STEM instrument, demonstrating atomic-scale resolution, even in scanning electron mode [33]

in 60 years; yet, x-ray brilliance has improved ten orders of magnitude in under 30 years (see Fig. 7). The transition from third-generation to fourth-generation capability in the last decade alone represents five orders of magnitude improvement. The current levels of beam intensity and coherence allow parallel signal acquisition and time resolution at unprecedented levels. Recent progress in high pressure and *in situ* capability lead toward analysis of materials in realistic environments. At the beginning of the last decade, femtosecond control had just been demonstrated on synchrotron radiation.

The time evolution of structure is illustrated in the analysis of charge density waves on a Cr (111) surface [44]. Antiferromagnets carry no net external magnetic dipole moment, yet they have a periodic arrangement of the electron spins extending over macroscopic distances. The magnetic "noise" must be sampled at spatial wavelengths of the order of several interatomic spacings. Figure 8 shows a direct measurement of the fluctuations in the nanometer-scale superstructure of spin- and charge-density waves associated with antiferromagnetism in elemental chromium. The technique used is x-ray photon correlation spectroscopy, where coherent x-ray diffraction produces a speckle pattern that serves as a "fingerprint" of a particular magnetic domain configuration.

Materials needed for next-generation device solutions are typically complex and difficult to fabricate. For example, novel thermoelectric material systems such as cobaltates for efficient exchange of heat and electrical energy could revolutionize waste heat power generation and refrigeration, but this goal requires synthesis of new materials with tailored nanostructures to create the desired separation and optimization of the phononic and electronic transport channels. The high brightness at penetrating x-ray energies allow unprecedented nanoscale imaging of materials as they are being synthesized in complex environments.

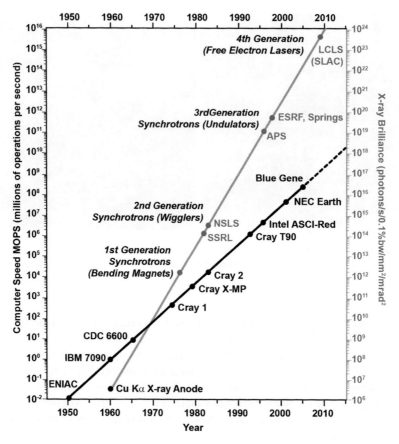

Fig. 7 Comparison of increases in computer processor speed and x-ray brilliance over the last 50 years (Adapted from information provided by Eric Isaacs)

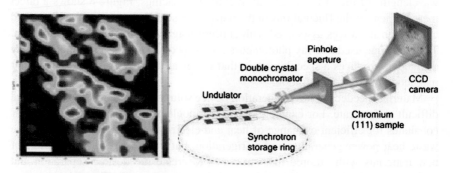

Fig. 8 Diagram of the experimental set-up and charge-coupled device (*CCD*) image of the x-ray speckle observed for the [200] lattice Bragg reflection

Researchers at next-generation synchrotron light sources and spallation neutron accelerators can watch self-assembly and directed self-assembly of nanomaterials, make tribological measurements on nanosystems, create nanomagnetic materials with novel properties, and fabricate superhard materials under pressure. Scientists are using x-rays and neutrons as nanoscale probes to:

- Solve the crystal structures of proteins
- Understand catalytic processes in toxicology measurements and atmospheric research
- Investigate plant genomics
- Image nanostructured materials, ceramics, and polymers
- Image cracks and atomic defects in structures with high resolution in real time

2.4 Instruments and Metrology in Support of Nanomanufacturing: Nanolithography

Many of the advances described above have been critical in providing the characterization necessary in the development of devices and products. Some scanning probe and electron beam-based tools have been developed for high-throughput analyses for quality control and, in some cases, for inline manufacturing. However, many of the challenges raised in earlier reports regarding standards, reference materials, and metrology tools for nanomanufacturing process control remain.

Fair progress has been made in the development of commercial nanofabrication tools that can be scaled for manufacturing (Fig. 9). Nanoimprint lithography and

Fig. 9 Lithography tools for nanoscale manufacturing. (*Left*) Molecular Imprints, Inc. (*left*, Imprio® 300, http://www.molecularimprints.com/products.php) (*Right*) Nanonex Corp. (*right*, NX-2000, http://www.nanonex.com/machines.htm)

Jet and Flash™ imprint lithography produce sub-30 nm resolution nanomaterials with <10 nm alignment over hundreds of millimeters.

Probe based techniques for positioning atoms and molecules, inducing localized chemical reactions and patterning electric, ferroelectric and magnetic fields have made great strides towards enabling nanofabrication. One of the most dramatic results of the early days of STM was from the IBM group that positioned individual atoms into quantum corrals [45]. This beautiful physics that demonstrated the ability to build structure at the atomic level was achieved at 4K or lower and not practical for manufacturing. In the last decade Morita and colleagues demonstrated atomic positioning of semiconductors at room temperature, a huge step towards the vision of building with atomic precision [46]. Lyding and colleagues [47] demonstrated tip based patterning and chemical reaction of hydrogen terminated silicon surfaces. This advance illustrated an atomic-scale lithography in the context of the computer chip industry and scale up is being explored by Zyvex. (http://www.zyvex.com/).

Innovative approaches to nanopatterning include exploiting tip-induced fields and gradients. One obvious application of nanopatterning that has motivated much research is information storage. One concept developed at IBM Zurich is the "millipede" in which a tip heats a substrate locally to write and to erase information bits. Arrays of tips geometrically expand the information density to 1 TB/sq. in. (http://www.zurich.ibm.com/st/storage/concept.html). Ferroelectric materials offer possibly the highest density information storage due to the crystallographic nature of domain boundaries, e.g., bit boundaries. Local probes induce electric fields that switch ferroelectric domains into up and down (one and zero) bits. Significant progress has been made in understanding and controlling domain patterning and bit sizes as small as 2.8 nm have been demonstrated [48].

Probe-based patterning of ferroelectric domains has also been used to control nanoparticle deposition and local chemical reaction [49, 50]. Patterned hybrid nanostructures consisting of proteins and plasmonic particles have been shown to increase efficiency in optoelectronic and energy harvesting devices [51]. The ultimate spatial resolution of <5 nm in patterning multi component nanostructures with device function opens a window of opportunity for future nanofabrication (Fig. 10).

3 Goals, Barriers, and Solutions for the Next 5–10 Years

The advances in scanning probe microscopy of the last decade have expanded our ability to relate structure and a wide range of properties/responses in real space at the nanoscale. The advances can be considered in the context of a pseudo-phase-space relating properties, length scale, and time scale (Fig. 11). Early scanning probes were predominantly in the property/length scale plane and involved single-valued functions. Advances such as those described above have accessed

Fig. 10 Surface potential image ferroelectric domains patterned on a lead zirconate titanate with an electric field from an SPM tip. The domains control chemical reactivity such that 5 nm metallic particles are deposited and peptides or optically active molecules selectively attached to produce a photo active switch (Courtesy of the Bonnell group)

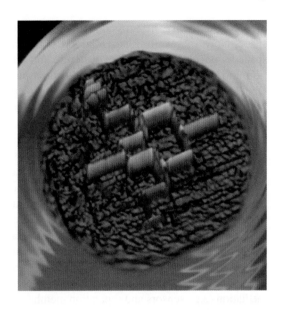

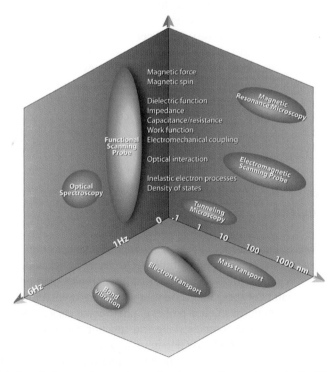

Fig. 11 Relation of advances in scanning probes to space, time and complexity. The future is in extending these probes into the regions in the center where spatial resolution, time resolution, and complexity can be simultaneously probed (Courtesy of D.A. Bonnell)

higher complexity in property functions and extended into time/frequency space. Now, the exciting potential exists to access regions in the center of this space by developing probes of structure and properties with high spatial resolution and time resolution. This would be a pathway to a vision articulated by Don Eigler [52]: 3D atomic resolution imaging of charge, spin, and vector properties at femtosecond time resolution. This vision is clearly in the center region of Fig. 11.

The next decade will realize exciting advances in electron beam instrumentation as well as innovative accessories and specimen stages for creative and meaningful experiments to decipher atomic- and molecular-scale structural, chemical, and electronic phenomena at the heart of nanoscience and nanotechnology. These are being coupled with dynamic and *in situ* capabilities extending the temporal scale to unprecedented limits down to nano- and femtosecond scattering [53, 54]. Such advances will be essential to achieving fundamental understanding of and engineering developments in materials, structures, and systems for next-generation technologies related to energy (e.g., improved photovoltaics via understanding of interfacial dynamics, and 3D architecture of fuel cell electrodes); environmental monitoring/mediation (e.g., sensors and separation membrane structures); biomedicine (e.g., 3D reconstruction of protein complexes and bio-nano interfaces in implants/devices); and fuel production in extreme environments, among others.

The 1999 Nobel Prize in Chemistry (http://nobelprize.org/nobel_prizes/chemistry/laureates/1999/) recognized the importance of using ultrafast lasers to reveal how atoms move during reactions. The trajectory of advances in laser technology will enable experimental configurations with spatial control, opening new opportunities in monitoring of femtosecond processes—the time scale of many nanoscale phenomena. Synchrotron facilities now offer femtosecond optical spectroscopy combined with x-ray radiation.

3.1 Probes of the Atomic Origin of Electromagnetic Phenomena

The physical interactions that yield function in solar cells, nanoelectronics, plasmonic sensors, etc., involve the interplay of multiple electromagnetic behaviors. Advancing energy systems, beyond-CMOS solutions in nanoelectronics, and optical devices require characterization of these interactions at increasing small, ultimately atomic scales. Simultaneous atomic resolution probes of photo conductance, resistance, work function, polarizability, and complex dielectric function are needed that operate over frequency ranges from static to GHz in order to access dynamic processes at the level of bond vibrations. Advances in integrating high frequency circuitry to scanning probe systems and combing multiple probes in one station is a path forward. This strategy would also result in new tools for *in situ* characterization of electronic and optical devices.

3.2 3D Atomic-Resolution Structure Determination with Chemical Specificity

Reports at the beginning of the NNI listed 3D atomic-resolution imaging as an important goal. It has required the tool advances of the last decade along several fronts to even begin to address this challenge. It is now possible to envision pathways toward this goal. Hybrid scattering near-field scanning optical microscopy (NSOM) has exhibited sub-100 nm resolution in a 3D holographic mode. Higher spatial resolution and easier experimental realization is necessary. Electron holography has made impressive advances and in the context of aberration-corrected instruments could be combined with *in situ* spectroscopies and energy filtering to increase spatial resolution and chemical identification. 3D characterization by x-ray tomography is also a promising approach to 3D structure determination.

3.3 Characterization of Dynamic Processes with High Spatial and Temporal Resolution

While there exist atomic resolution structural characterization tools and spectroscopies that access processes in the fête second time regime, new opportunities to combine these have been enabled by advances in the last decade. The synergistic combination of aberration-corrected electron microscopy coupled with innovative *in situ* specimen stages should facilitate a wide range of in-process observation/analysis of important phenomena at high spatial resolution (atomic/molecular scale) and range of temporal resolution (seconds→femtoseconds). The high brightness, high coherence beam lines offer opportunities to access processes in this regime with scattering tools. Using new scattering techniques to directly measure dynamic processes on time scales of attoseconds[2] will provide crucial information on the dynamic processes that drive replication, assembly, folding, and functioning in complex systems at the nanoscale. Combinations of scanning probes and laser spectroscopy or inelastic spectroscopies offer another approach.

3.4 Complexity in Biological and Soft Matter Systems

The potential of nanoscale probes for elucidating processes underlying protein function is considerable, but the experimental challenges are severe. Basic requirements are consistent video-rate (and above) imaging and analysis capability, simultaneous molecular resolution of many constituents, and control and manipulation of

---

[2] 10^{-18} of a second.

proteins in physiologically relevant environments. In addition to single-molecule/single-protein approaches, the behavior of wide networks of interacting biomolecules must be pursued. Combining near-field optical imaging, fluorescent microscopy, and scanning probes that image in liquid is one path forward. New tools such as ion conductance microscopy and higher spatial resolution patch clamp analysis would access cell processes. Increasing the spatial resolution and extending detection capability to higher complexity analysis by photo-activated localization microscopy (PALM) would access in-cell processes. Electron beam tools for soft matter should also be furthered. Continued developments are required on cryogenic preservation, encapsulated fluidic specimen cells, minimum-dose techniques, and detector technologies (e.g., phase plates).

3.5 In Situ and Multifunctional Tools

As outlined above, an enormous range of capability has emerged to characterize structure from different perspectives and to measure spatially resolved properties from bond vibrations to photoconductivity. Judicious combinations of *in situ* capabilities will be required to address the goals of the next decade. These considerations are compelling, for example, in "hybrid soft-hard" nanostructures, wherein hard-microscopy tools/techniques need to be merged with those in soft microscopy. A combination of SPM and SEM/FIB would allow researchers to create site-specific sections of important subsurface structure, patterns, or defects, which can be quantitatively measured by SPM techniques for local mechanical electronic properties. Similarly, given the remarkable range and innate quantitative nature of several x-ray scattering techniques, especially with synchrotron radiation, it should be possible to combine the best attributes of electron microscopy with (synchrotron) x-ray scattering, such that a given specimen or an experiment can be examined with diverse yet complementary techniques of x-ray scattering and electron microscopy/spectroscopy.

3.6 Tools for Nanomanufacturing

The next decade will see many of the early nanoscale science discoveries transition to manufacturing. The need for process metrology, quality control measurements, and associated standards is acute. These unmet requirements apply not only to manufacturing but also to the analysis of workplace safety, environmental impact, and life cycle calculations. In some areas, most notably the electronics industry, roadmapping exercises clearly articulate the needs. For example, additional R&D is required in next 5–10 years, particularly in the area of actinic mask inspection [55–57]. An effective actinic mask metrology tool that operates at a wavelength of 13.5 nm is required, as is actinic pattern inspection capability suitable for high-volume manufacturing to mitigate risks. In others, for example in the case of nanoparticles, nanotubes, etc., reliable fabrication and measurement technologies do not exist for all size regimes.

4 Scientific and Technological Infrastructure Needs

As noted in a recent report by the National Academy of Sciences, a strategic plan and infrastructure for mid-sized instrumentation are missing and/or problematic. The electron microscopes, scanning probe systems, and advanced optical systems that push the limits in performance are high-cost ($1–5 million) and require highly specialized staff to develop and maintain them. A new mechanism is required to support the human resource infrastructure. Universities are not able to subsidize these costs.

5 R&D Investment and Implementation Strategies

Success in pushing the frontiers in nanoscale measurement science requires effort sustained for appropriate time frames. Note that accomplishing the observation of single-electron charge and electron spin required over 10 years. It can be argued that the United States is falling behind in some nanocharacterization fields, perhaps as a consequence of short support time lines. To realize the ultimate implementation of nanotechnology, support mechanisms should be developed that account for the realities of relevant (i.e., longer-term) time frames.

The various types of centers of excellence supported by the National Science Foundation, the National Institutes of Health, and the Department of Energy have soundly demonstrated the positive impact of collaborative programs, particularly for topics that intersect several traditional fields. Investments in such centers and synergistic networks of centers should continue. New longer-term funding mechanisms should not deplete resources for individual investigator efforts, where significant innovation occurs.

Incentives should be instituted that encourage government labs, industry, and universities to collaborate in the development of standards for nanometrology. These mechanisms should not simply subsidize industry research and development but should facilitate coordinated advancement toward common research goals.

6 Conclusions and Priorities

Advanced capabilities in investigative, fabrication, and metrology tools will be a primary factor that enables new scientific discovery and the translation of nanoscience to nanotechnology. In the absence of these critical instruments and tools, the full potential of nanotechnology cannot be realized. The previous decade produced concepts and approaches that provide a platform for a new generation of localized measurement tools. The priorities for the next decade are to develop:

- Probes of the atomic origin of electromagnetic phenomena
- 3D atomic resolution structure determination with chemical specificity
- Characterization of dynamic processes with high spatial and temporal resolution
- *In situ* and multifunctional tools for device characterization
- Translation of fabrication metrology tools to facilitate manufacturing
- New strategies that allow instrument access to relevant research and communities
- Support of academic infrastructures that train the workforce

7 Broader Implications for Society

The implications of the research and development of experimental tools for the characterization and manipulation of nanoscale phenomena cannot be overstated. Discovery requires observation, so promoting the next generation of these measurement capabilities will produce the tools that produce new science. The manufacturing of nanodevices and systems cannot be accomplished without the development of new metrologies and standards. New horizons offered by biological systems will not be achieved without the extension of these characterization instruments into higher levels of complexity in broader environmental conditions. The concomitant development of a workforce trained in these metrologies will fuel national economic growth.

8 Examples of Achievements and Paradigm Shifts

Several examples here illustrate advances, in addition to those described above, that are transforming our ability to probe nanoscale phenomena.

8.1 Complex Dielectric Function of the Molecular Layer

Contact person: Dawn Bonnell, University of Pennsylvania

Various techniques have been developed for high-spatial-resolution probes of light interactions with surfaces, including photon-assisted STM, surface-enhanced Raman spectroscopy, and NSOM. A scattering-type near-field scanning optical microscopy (s-NSOM) offers the best opportunity for high spatial resolution. A sharp probe tip is positioned near a surface and illuminated with optical radiation. The field is enhanced at the tip, which acts as an optical antenna, [9, 10, 12, 13, 58], modulating the tip-sample distance in the enhanced field, detecting the reradiated light to distinguish variations in materials properties. Dielectric functions are an inherent component of the response of a tip-surface junction in this configuration.

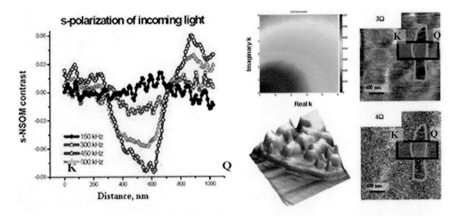

Fig. 12 The scattered NSOM signal profiles (*left*) and images (*right*) are compared to a model of the complex dielectric function of a molecular layer (*top, right*) and the topographic structure (*bottom, right*) [59]

Figure 12 illustrates that it is possible to determine the real and imaginary coefficient of the dielectric function for a single molecular layer [59]. Harmonics up to the fourth order and polarization-dependence of incident light enabling probes of dielectric properties monolayers of organic molecules on atomically smooth substrates. An analytical treatment of light/sample interaction using the NSOM tip was developed in order to quantify the dielectric properties. To date, the third harmonic provides the best lateral resolution and dielectric constant contrast, and the strength of s-NSOM contrast for s-polarized light is 10 times higher than that of p-polarized light for this configuration.

8.2 Single-Electron Spin Detection

Contact person: Dan Rugar, IBM

The continued development of magnetic resonance force microscopy (MRFM) by Rugar and colleagues (Fig. 13) has demonstrated unprecedented coherent control of individual electron spins, and has extended the sensitivity and spatial resolution of nuclear magnetic resonance (NMR) and electron paramagnetic resonance (EPR) spectrometry by many orders of magnitude.

8.3 Attosecond Processes with X-Ray Scattering

Contact persons: Peter Abbamonte, University of Illinois Urbana-Champaign; Dawn Bonnell, University of Pennsylvania

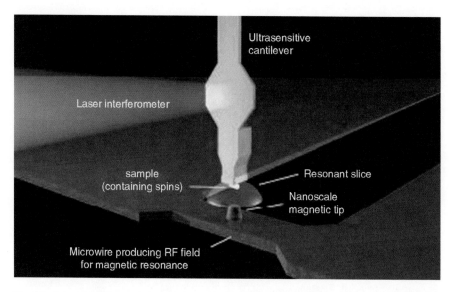

Fig. 13 Schematic diagram of magnetic resonance force microscopy (From Degen et al. [60])

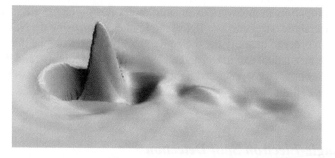

Fig. 14 Attosecond snapshots of electronic disturbances in water calculated to be produced by a diffusing gold ion based on inelastic x-ray scattering measurements

Current x-ray probes are capable of determining the electronic and structural dynamics of individual nanostructures (Fig. 14), whereas neutrons measure the dynamic response of assemblies of nanostructures—with the additional capability of separating a single particle from collective dynamics. Abbamonte et al. [61] show that the momentum flexibility of inelastic x-ray scattering may be exploited to invert its loss function, allowing real-time imaging of density disturbances in a medium. They show the disturbance arising from a point source in liquid water, with a resolution of 41.3 as (4.13×10^{-17} s) and 1.27 Å (1.27×10^{-8} cm). This result is used to determine the structure of the electron cloud around a photoexcited chromophore in solution, as well as the wake generated in water by a 9 MeV gold ion.

8.4 Electron Optics

Contact person: Vinayak Dravid, Northwestern University

Collectively, basic and aberration-corrected S/TEMs (and SEMs) coupled with accessories have taken the field of electron microscopy to new heights. These techniques have been applied to reveal atomic and molecular scale architecture, chemical partitioning, and varied aspects of electronic structures in materials and systems of great significance to modern society.[3] They range from understanding interface limitations in modern microelectronics devices to emerging opportunities beyond classical silicon technologies, as in multifunctional oxides. Though advanced and cutting-edge improvements in the graphical user interfaces (GUI), new ease of operation and diversity of applications have made electron microscopy tools, techniques, and associated accessories essential to understanding broad issues and limiting factors in energy/environment technologies, information technology devices, biomedical materials and interactions, and food-package interactions, to name a few broad applications.

8.4.1 Analysis of Soft Matter

Since 1999, the community of vendors/manufacturers and scientists has risen to the challenge and contributed to a portfolio of electron microscopy-based tools and instrumentation essential for imaging and analysis of soft matter.

Field-emission-gun SEM/S/TEM instruments capable of operating at ultralow voltages with minimally required beam current, coupled with innovative specimen stages (liquid-cell stage, cryo stage, and environmental stage), have greatly advanced the ability to image/analyze the surfaces and sub-surface phenomena of soft matter [76–79]. Vendors/manufacturers have identified critical applications to soft matter and configured modern S/TEMs to include variable energy (60–300 keV), cryo-preservation of biological and soft matter, thin sectioning methods, cryospecimen stages for S/TEM, and 3D tomographic reconstruction (Fig. 15) [80] via automated specimen tilting coupled with high-throughput image collection and analysis, among other related developments [77, 79, 81–86].

STEM imaging techniques, especially annular dark field (ADF) and high-angle annular dark field (HAADF) techniques, have been successfully developed and applied for quantitative imaging/analysis of soft matter [77, 81, 87, 88]. Specialized instruments have been conceived, designed, and developed for specific soft matter scientific issues, including cryo-bio nanoscale chemical analysis and mapping with

---

[3]For more information on these structures and their significance, resources include Muller et al. [30], Pennycook et al.[31], Smith [32], Zhu et al. [33], Prabhumirashi et al. [62], Dahmen et al. [63], Kisielowski et al. [37], Castell et al. [64], Girit et al. [34], Jia et al. [65], Meyer et al. [66], Midgley and Durkan [67], Nellist et al. [68], Oshima et al. [69], Hawkes and Spence [70], Rossell et al. [71], Suenaga et al. [72], Thomas [73], Varela et al. [74], and Voyles et al. [75].

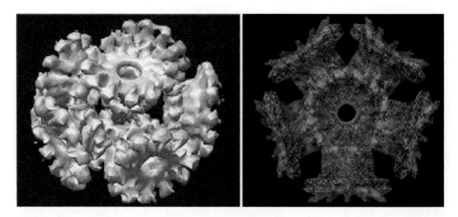

Fig. 15 Image reconstruction of biological ellipsoid nanoparticle of bacteriophage DNA packaging motor complex [80]

high sensitivity and overall stability. The key issues in microscopy of soft matter continue to revolve around consistent, reproducible, and viable specimen preparation means and methods, increasing automation, and computer-aided microscopy/ analysis.

Moving forward in the next decade, the analysis of soft matter will greatly benefit from aberration-corrected S/TEM, coupled with back-end innovative developments in specimen preparation techniques, novel multispecimen stages and specimen holders, as well as continued incorporation of automation, computer/ software interfaces, and pattern recognition and processing. Novel detectors and their configuration (e.g., phase plate) coupled with fast detection [53, 54] are being pursued; these promise significant improvements in ubiquitous contrast problems in soft matter with minimal dosage to reduce beam damage. Correlative microscopy [86], i.e., coupling and correlation of data and analysis from different microscopy (and spectroscopy) approaches, will provide unique opportunities for understanding the form-function relationship, which is the cornerstone of modern biology and nano-bio medicine. These advances in soft microscopy are critical for detailed understanding of complex biological structures and phenomena—especially at the interface with physical science and engineering. These developments will prove invaluable for improving health, environment, and sustainability in the context of nanomaterial life cycles.

8.4.2 3D Structure Determination

The decade 2000–2010 has seen innovative developments in electron microscopy tools and instrumentation and associated accessories (specimen holders) and techniques for 3D reconstruction and visualization of nanostructures [67, 89–92]. Three-dimensional tomographic reconstruction in biology has advanced significantly to allow for sub-nanometer-scale spatial resolution for 3D structure of

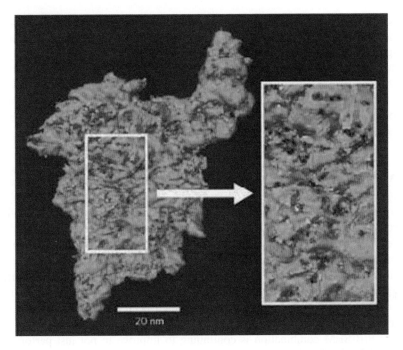

Fig. 16 3D tomographic reconstruction of a heterogeneous catalyst based on disordered mesoporous silica supporting bimetallic ruthenium–platinum nanoparticles [91]

biomolecular complexes ranging from protein clusters to biological tissues and hybrid organic-inorganic nanostructures [76–79, 87].

Specialized low-dosage exposure, high- and low-kV imaging with TEM and STEM imaging modalities, coupled with advances in specimen stages (high-tilt), specimen holders (axial full-tilt, cryo), and specimen support (hydrophilic surface treatment, SiNx window grids, graphene substrate, etc.) have all contributed to significant achievements in 3D rendering of nanostructures. Image processing algorithms, computational and processing power, and wide spectrum visualization tools and techniques have enabled improved appreciation of 3D rendering of nanostructures (Fig. 16), defects, and substructure.

The 3D tomography and analysis approaches are rapidly becoming mainstream, thanks to improved microscope stability, innovative specimen stage and holder designs, and much more integrated hardware (microscope, detectors, and holders) and software (auto-image capture, auto-focus, rapid archiving, etc.). Nanometer-scale resolution in most cases and sub-nanometer-scale resolution in emerging approaches is readily possible in modern commercial S/TEMs.

Here, too, the emergence of aberration-corrected S/TEM would prove indispensable in pushing the spatial (and depth) resolution in 3D reconstruction towards the molecular and atomic scales. The ability to open up pole-piece gaps in aberration-corrected S/TEM would allow researchers to develop innovative specimen stages, holders, and *in situ* experiments geared towards understanding biological processes under physiologically viable fluidic environment and external stimuli to probe complex dynamics of bio-nano interfaces. Such studies are vital not only for fundamental understanding of biological structures and processes but also for their tailoring and control for development of technologies for human health, environmental monitoring/remediation, and energy production, transfer, and storage using biological principles, among other developments that depend on 3D atomic and molecular structures and their dynamics.

8.4.3 Subsurface Metrology and Imaging Buried Structures/Phenomena

The remarkable advances in focused ion beam (FIB, including dual-beam SEM+FIB) instrumentation have enabled sectioning and fabrication for revealing buried and embedded structures and features [43, 93]. This has greatly advanced microelectronic metrology and defect analysis to the extent that significant improvement in yield and quality of final products has become possible in manufacturing. So-called dual-beam FIBs have proved to be a major boon to semiconductor/microelectronics manufacturing industries in their need for sub-surface metrology and defect analysis [40, 94].

The FIB-SEM combination is continuing to improve in not just performance figures (i.e., of beam size and current) but also in innovative *in situ* capabilities such as lift-off, micro/nano-manipulation, gas-injection/deposition, and related *in situ* material manipulation capabilities (Fig. 17).

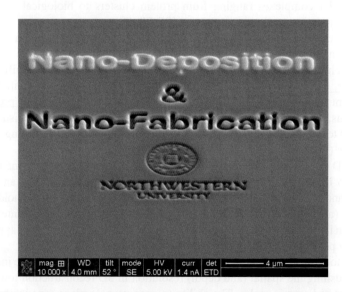

Fig. 17 The complementary nanoscale fabrication capabilities of modern FIB are depicted in this image, showing both deposition and fabrication of materials (Courtesy of B. Myers and V.P. Dravid)

These capabilities are rapidly advancing to enable multiple experiments and are being coupled with other instrumentations for synthesis, characterization, and measurements.

9 International Perspectives from Site Visits Abroad

The following are summaries from the International WTEC Nano2 workshops held in Germany, Japan, and Singapore, with a focus on international convergence in governance.

9.1 United States-European Union Workshop (Hamburg, Germany)

Panel members/discussants

Liam Blunt (co-chair), University of Huddersfield, UK
Dawn Bonnell (co-chair), University of Pennsylvania, United States
Richard Leach, National Physical Laboratory, UK
Clivia M. Sotomayor Torres, Centre d'Investigació en Nanociencia i Nanotecnologia, Spain
Malcolm Penn, Future Horizons, Ltd., UK

This group of scientists from the UK, Spain, and the United States met for several days at the Hamburg "Nano2" Workshop to assess the impact of tools that access nanoscale phenomena on scientific discovery and technology innovation. Vigorous discussion informed by the participants' expertise and knowledge of various European road-mapping and strategic planning exercises resulted in the following summary of opportunities for the next decade of nanotechnology. The discussions emphasized metrology needs for commercialization.

The last decade has seen a realignment of the perception of nanotechnology according to the Gartner Hype Cycle [95] moving from the early over-hyped "technology trigger phase" through the "trough of disillusionment phase," where the technology failed to meet early wild speculation. Finally we are now reaching the "slope of enlightenment phase" where realistic yet exciting possibilities in nanotechnology are becoming clear.

In the next decade, the design and production of nanostructures and materials will be of enormous importance in order to create materials with new and novel combinations of properties and functions. This will necessitate the convergence of the key enabling disciplines of chemistry, physics, and biology.

New disciplines have emerged from interdisciplinary mixes such as plasmonics. We are now at the point where relatively inexpensive and accessible methods of nano-patterning (micro-contact printing, nanoimprint lithography, thin layer deposition, etc.) allow research and product development in laboratories all over the world, facilitating fabrication of specialized complex devices that allow exploration of new phenomena.

The commercial availability of off-the-shelf nanoparticles and nanoporous materials in sizes ranging from 1 nm to hundreds on nanometers has enabled rapid advances in the fundamental science and application of particles ranging from medical therapeutics to hybrid nanoelectronics. Patterning processes by electron or ion beam as well as x-rays, offer the chance to control and modify the relevant materials parameters at the nanoscale in a research environment. These examples of patterning and available nanoparticles are key drivers towards up-scaling nanotechnology from the research field into the world of real industrial production. Key to developing nanoscale production are quality systems; at the heart of these systems is metrology.

Key needs for nanometrology in the future will be:

- Full 3-D metrology using reliable instrumentation
- Mastering the challenge of handling large volumes of metrology data in a production-like environment
- Multiscale measurement technology (resolution and range, nm to m)
- Hybrid measurements for materials (dimensional/properties)
- Hybrid measurements for nanostructures (dimensional/functional)
- Globally accepted standards for measurement
- Standard reference materials for nanoscale measurement
- Routine online measurements suitable for a production environment
- Metrology across the process chain
- Metrology for nano-bio processes
- Metrology for dynamic processes

Instruments that access structure and properties at the nanoscale and atomic length scales were, and continue to be, essential in the understanding of physics and chemistry of nanostructures and nanosystems. The ability to quantify nanoscale behavior is a limiting factor in understanding new physics and is a prerequisite to manufacturing. If it cannot be measured, it cannot be made or even understood in any quantifiable manner. If specifications cannot be characterized, reliable and reproducible manufacturing is not possible. Critical processes for up-scaling nanotechnologies into production are infrastructure development and cooperation with R&D organizations during the development phase, and of particular importance, transfer of knowledge to SMEs in the nanotechnology field.

The consequence of the advances in instrumentation science of the last decade is that we are now poised for a new generation of tools that will:

- Revolutionize the fundamental concepts of solids and biomolecules
- Allow new views of complex systems at small scales
- Probe dynamic processes and unexplored time scales

Novel concepts emanating from the understanding of the properties and interactions in the nanometer scale may provide new metrology methods. To enable nanoscale metrology requires instrumentation that is easy to use and fit for purpose in an industrial setting. However, the investment needed in research at this second stage is far greater than at the basic research stage. Unless this is recognized, nanometrology research will fail to expand to meet the future needs. Metrological standardization will require the development of routes for communication between standards bodies and industrial users at the global level.

9.2 United States-Japan-Korea-Taiwan Workshop (Tokyo/Tsukuba, Japan)

Panel members/discussants

Dae Won Moon (co-chair), Korea Research Institute of Standards and Science (KRISS)
Dawn Bonnell (co-chair), University of Pennsylvania, United States
Seizo Morita, Osaka University, Japan
Masakazu Aono, National Institute for Materials Science (NIMS), Japan
Mike B.C. Yao, Industrial Technology Research Institute (ITRI), Taiwan
Kunio Takayanagi, Tokyo Institute of Technology, Japan

A team of scientists from Korea, Japan, and the United States met for several days at the Tsukuba Nano2 Workshop to assess the impact of tools that access nanoscale phenomena on scientific discovery and technology innovation. Based on the expertise of the participants, discussion emphasized scientific achievements that offer opportunities for future technological breakthroughs.

Ten years ago, nanotechnology was an area of academic research exploring noble science and new applications in various disciplines. At present, nanotechnology demonstrates a strong potential to overcome critical challenges in many areas of electronics, energy, and medicine. This potential arises in a large part due to the advances in tools for measurement and atomic manipulation.

For example, in the last decade, scanning tunneling microscopy manipulation at low temperature for atom-by-atom assembly made several dramatic demonstrations, and its main application was in physical science. Recently, atomic force microscopy was used to interchange atoms with an AFM tip apex atom at room temperature. The ability for room-temperature atomic manipulation leads to practical strategies for the construction of new compounds by design. Multiple-tip scanning probe systems have been developed that quantify local transport properties in atomic and nanodevices, which was not possible previously.

Sub-Ångstrom resolution in TEM with aberration correction demonstrated in the last decade. An illustrative example of the potential of TEM is given by sub-70 pm spatial resolution imaging of Li atoms; a most stringent test of detectability and critical in energy applications. In addition, the capabilities of beamline-based characterization tools have evolved, with a ten orders of magnitude increase in brightness over the last 10 years. The fourth-generation beamlines promise advances in atomic dynamics and biomolecule analysis.

A shift from nanotechnology to "nanoarchitectonics" (from nanofunctionality to "nanosystem functionality") is a clear change moving forward. No full-fledged application of nanotechnology can be expected without this paradigm shift; conversely, if this paradigm shift is realized, it will bring invaluable advancement in a wide variety of technological fields. Control of self-assembly and -organization of atoms, molecules, and nanostructures (with the use of local external fields and advanced nanochemical control) with single-molecule-level ultra-high sensitivity and novel methods to measure electrical, optical, and magnetic properties at the nanoscale (including multi-probe SPMs) enable this vision.

Success in realizing any of the future advances in nanotechnology depends on the capability to manipulate matter at nanometer length scales and measure the relevant and often complex properties at the same scale. The ultimate goal of atomic/molecular imaging is the spatial, temporal, 3D, single-molecule identification on surfaces and in complex environment such as cells *in vivo* and *in vitro* for nanoscale biology and medical diagnosis. Successful approaches may involve combinations of scanning probe, electron, and optical microscopy/spectroscopies. Realizing this goal will simultaneously provide several instrumentation innovations required for similar advances in devices.

A prerequisite for transitioning from fundamental physics to applications is the ability to routinely manipulate matter at room temperature in order to fabricate new compounds/nanostructures by design. AFM- and STM-based approaches have been demonstrated, but this must become routine.

To realize the exciting potential in science, engineering, and applications, the requirements for scientific and technological infrastructure are:

- Mechanisms to maintain infrastructure for 5 years, which is a typical length of time for nanotechnology programs in many countries. The lack of continuity has particularly impact in the physical infrastructure required for characterization at the nanoscale.
- In order for nanomanfacturing to be realized at a large scale, new infrastructure will be required, and global partnerships will be critical.
- New educational systems are necessary to train an innovative workforce with the appropriate skill sets.

9.3 *United States-Australia-China-India-Saudi Arabia-Singapore Workshop (Singapore)*

Panel members/discussants

John Miles (co-chair) National Measurement Institute of Australia
Dawn Bonnell (co-chair), University of Pennsylvania, United States
Li-jun Wan, Institute of Chemistry, Chinese Academy of Sciences (CAS)
Yong Lim Foo, Institute of Materials Research & Engineering (IMRE), Singapore
Huey Hoon Hng, Nanyang Technological University (NTU), Singapore

Scientists and research agency directors from Australia, China, Singapore, and India met to determine the global impact of experimental tools that enable scientific advance and commercialization of nanotechnology. The outcome summarized here emphasizes developing global consensus in developing tools and metrology standards and places these needs in the context of opportunities for the next decade of nanotechnology.

The vision of nanotechnology has changed in the last 10 years to a quieter, more subdued one, with an understanding that the potential risks need to be addressed as much as the benefits. This has led to less funding from government and industry. Nanotechnology has penetrated into almost every research area and discipline in science and engineering, although the word "nanotechnology" has changed from a stand-alone term to one used now used in conjunction with specific applications, such as energy, water, medicine, etc. R&D has moved from individual nanostructures to nanosystems. A large number of commercial products have been developed.

The vision for the next 10 years is that most R&D will be motivated by the requirements of society (e.g., energy, water, health care). There will be an increasing need for the development of instruments capable of measuring nanomaterials *in situ* in matrices such as soil, water, food, and living organisms. The manipulation and manufacture of nanomaterials at the atomic level at room temperature and the simultaneous imaging of multiple complex processes in living systems are considered to be important, desirable, and bold objectives.

Generally, the main metrological advancements in the last 10 years are the significant increases in the use and capability (e.g., uncertainty, resolution, range, accuracy) of microscopy, analytical instruments, and spectroscopic and related techniques. These include the development of aberration correction for TEMs, the development of metrological AFMs, the measurement of single-charge, single-spin, inelastic vibrations, advances in energy-dispersive x-ray (EDX) detector technology and Si-drift detectors that can cope with higher count rates, allowing faster and more accurate mapping and more accurate chemical analyses, electron holography and magnetic induction maps, and multiprobe SPMs. The development of methods for creating reference standards for nanoparticles and for two-dimensional and three-dimensional nanoscale measurements is also an important advance.

The main goal for the next 5 years for general nanotechnology is to achieve large-scale applications of nanotechnology, with barriers being the scaling-up of processing/synthesis processes and safety issues of nanoparticles/nanomaterials. The main goals for the measurement and standards sector are

- Increased engagement and understanding between national measurement institutes, documentary standards developers, and the R&D community
- Further development of the linkages between national and international nanometrological systems
- Lower operating voltages (<35 kV) for TEM; 3D imaging with atomic resolution
- Publication of a large number of documentary standards for use by the nano community
- Bond-level, femtosecond-time frame measurement capability

- Development of new portable and cheap nanoscale instruments capable of operating in an industrial environment
- Development of instruments capable of measuring nanomaterials in matrices such as water, soil, food and living tissue, etc.

The scientific and technological infrastructure considered to be required includes the rather bold need for an international funding agency to pay for work in activities such as documentary standards development, toxicological testing, international regulatory reform, etc. These activities often rely on goodwill and are too important to be left to volunteers. More realistic is a strong need for the continued development of the international metrological infrastructure to support the development of nanotechnology. This includes the establishment of physical standards by national measurement institutes that provide measurements traceable to the SI (international system of units), reference materials, standardized methods, international comparisons, uncertainty analyses, and appropriate quality systems. The need for documentary standards for nanotechnology is still urgent and growing.

Instrumentation and sensing is seen as the core of the emerging topics and priorities for future nanoscale science and engineering research, particularly in the biomedical and energy sectors.

References

1. M.C. Roco, R.S. Williams, P. Alivisatos (eds.), *Nanotechnology Research Directions: IWGN [NSTC] Workshop Report: Vision for Nanotechnology R&D in the Next Decade* (International Technology Research Institute at Loyola College, Baltimore, 1999), Available online: http://www.nano.gov/html/res/pubs.html
2. D.A. Bonnell, Pushing resolution limits of functional imaging to probe atomic scale properties. ACS Nano **2**, 1753–1759 (2008). doi:10.1021/nn8005575
3. T. Kohsaka, C. Taylor, K. Fujita, A. Schmidt, C. Lupien, T. Hanaguri, M. Azuma, M. Takano, H. Eisaki, H. Takagi, S. Uchida, J.C. Davis, An intrinsic bond-centered electronic glass with unidirectional domains in underdoped cuprates. Science **315**, 1380–1385 (2007). doi:10.1126/science.1138584
4. C. Brun, J.C. Girard, Z.Z. Wang, J. Dumas, J. Marcus, C. Schlenker, Charge-density waves in rubidium blue bronze $Rb_{0.3}MoO_3$ observed by scanning tunneling microscopy. Phys. Rev. B **72**, 235119–235126 (2005). doi:10.1103/PhysRevB.72.235119
5. M.P. Nikiforov, A.F. Isakovic, D.A. Bonnell, Atomic structure and charge-density waves of blue bronze $K_{0.3}MoO_3(201)$ by variable-temperature scanning tunneling microscopy. Phys. Rev. B **76**, 033104 ((2007)
6. F.J. Giessibl, Advances in atomic force microscopy. Rev. Mod. Phys. **75**, 949–983 (2003)
7. F.J. Giessibl, H. Bielefeldt, Physical interpretation of frequency-modulation atomic force microscopy. Phys. Rev. B **61**, 9968–9971 (2000). doi:10.1103/PhysRevB.61.9968
8. Y. Sugimoto, P. Pou, M. Abe, P. Jelinek, R. Perez, S. Morito, O. Custance, Chemical identification of individual surface atoms by atomic force microscopy. Nature **446**, 64–67 (2007). doi:10.1038/nature05530
9. T. Eguchi, Y. Fujikawa, K. Akiyama, T. An, M. Ono, T. Hashimoto, Y. Morikawa, K. Terakura, T. Sakurai, M.G. Lagally, Y. Hasegawa, Imaging of all dangling bonds and their potential on the Ge/Si(105) surface by noncontact atomic force microscopy. Phys. Rev. Lett. **93**, 266102 (2004). doi:10.1103/PhysRevLett.93.266102

10. Y. Cho, R. Hirose, Atomic dipole moment distribution of Si Atoms on a Si(111)-(7×7) surface studied using noncontact scanning nonlinear dielectric microscopy. Phys. Rev. Lett. **99**, 186101–186105 (2007). doi:10.1103/PhysRevLett.99.186101
11. B.J. Rodriguez, S. Jesse, A.P. Baddorf, S.V. Kalinin, High-resolution electromechanical imaging of ferroelectric materials in a liquid environment by piezoresponse force microscopy. Phys. Rev. Lett. **96**, 237602 (2006)
12. B.C. Stipe, M.A. Rezaci, W. Ho, Single-molecule vibrational spectroscopy and microscopy. Science **280**(5370), 1372–1375 (1998)
13. J.I. Pascual, N. Lorente, Z. Song, H. Conrad, H.-P. Rust, Selectivity in vibrationally mediated single-molecule chemistry. Nature **423**, 525–528 (2003)
14. J.R. Hahn, W. Ho, Orbital specific chemistry: controlling the pathway in single-molecule dissociation. J. Chem. Phys. **122**, 244 (2005)
15. A.J. Heinrich, J.A. Gupta, C.P. Lutz, D.M. Eigler, Single-atom spin-flip spectroscopy. Science **306**, 466–469 (2004). doi:10.1126/science.1101077
16. A.F. Otte, M. Ternes, S. Loth, C.P. Lutz, C.F. Hirjibehedin, A.J. Heinrich, Spin excitations of a Kondo-screened atom coupled to a second magnetic atom. Phys. Rev. Lett. **103**, 107203–107207 (2009). doi:10.1103/PhysRevLett.103.107203
17. P. Han, A.R. Kurland, A.N. Giordano, S.U. Nanayakkara, M.M. Blake, C.M. Pochas, P.S. Weiss, Heads and tails: simultaneous exposed and buried interface imaging of monolayers. ACS Nano **3**, 3115–3121 (2009). doi:10.1021/nn901030x
18. H.M. Saavedra, T.J. Mullen, P.P. Zhang, D.C. Dewey, S.A. Claridge, P.S. Weiss, Hybrid strategies in nanolithography. Rep. Prog. Phys. **73**, 036501 (2010). doi:10.1088/0034-4885/73/3/036501
19. A.M. Moore, A.A. Dameron, B.A. Mantooth, R.K. Smith, D.J. Fuchs, J.W. Ciszek, F. Maya, Y. Yao, J.M. Tour, P.S. Weiss, Molecular engineering and measurements to test hypothesized mechanisms in single-molecule conductance switching. J. Am. Chem. Soc. **128**, 1959–1967 (2006). doi:10.1021/ja055761m
20. D.C. Coffey, D.S. Ginger, Time-resolved electrostatic force microscopy of polymer solar cells. Nat. Mat. **5**, 735–740 (2006). doi:10.1038/nmat1712
21. D.A. Bonnell, S. Kalinin, Local potential at atomically abrupt oxide grain boundaries by scanning probe microscopy, in *Solid State Phenomena*, ed. by O. Bonnaud, T. Mohammed-Brahim, H.P. Strulnk, J.H. Werner (SciTech Publishing, Uettikon am See, 2001), pp. 33–47
22. D.B. Williams, D.B. Carter, *Transmission Electron Microscopy*, 2nd edn. (Springer, New York, 2009)
23. M. Haider, S. Uhlemann, E. Schwan, H. Rose, B. Kabius, K. Urban, Electron microscopy image enhanced. Nature **392**(6678), 768–769 (1998). doi:10.1038/33823
24. H. Rose, Correction of aberrations. A promising means for improving the spatial and energy resolution of energy-filtering electron-microscopes. Ultramicroscopy **56**(1–3), 11–25 (1994)
25. A. Tonomura, T. Matsuda, J. Endo, H. Todokoro, T. Komoda, Development of a field emission electron microscope. J. Electron Microsc. Tokyo **28**(1), 1–11 (1979)
26. B. Kabius, H. Rose, *Novel Aberration Correction Concepts*. Advances in Imaging and Electron Physics, vol. 153 (Elsevier, San Diego, 2008), pp. 261–281
27. P.E. Batson, N. Dellby, O.L. Krivanek, Sub-angstrom resolution using aberration corrected electron optics. Nature **418**(6898), 617–620 (2002)
28. C. Kisielowski, B. Freitag, M. Bischoff, H. van Lin, S. Lazar, G. Knippels, P. Tiemeijer, M. van der Stam, S. von Harrach, M. Stekelenburg, M. Haider, S. Uhlemann, H. Müller, P. Hartel, B. Kabius, D. Miller, I. Petrov, E.A. Olson, T. Donchev, E.A. Kenik, A.R. Lupini, J. Bentley, S.J. Pennycook, I.M. Anderson, A.M. Minor, A.K. Schmid, T. Duden, V. Radmilovic, Q.M. Ramasse, M. Watanabe, R. Erni, E.A. Stach, P. Denes, U. Dahmen, Detection of single atoms and buried defects in three dimensions by aberration-corrected electron microscope with 0.5-Ångstrom information limit. Microsc. Microanal. **14**(5), 469–477 (2008)
29. O.L. Krivanek, G.J. Corbin, N. Dellby, B.F. Elston, R.J. Keyse, M.F. Murfitt, C.S. Own, Z.S. Szilagyi, J.W. Woodruff, An electron microscope for the aberration-corrected era. Ultramicroscopy **108**(3), 179–195 (2008)

30. D.A. Muller, L.F. Kourkoutis, M. Murfitt, J.H. Song, H.Y. Hwang, J. Silcox, N. Dellby, O.L. Krivanek, Atomic-scale chemical imaging of composition and bonding by aberration-corrected microscopy. Science **319**(5866), 1073–1076 (2008). doi:10.1126/science.1148820
31. S.J. Pennycook, M. Varela, C.J.D. Hetherington, A.I. Kirkland, Materials advances through aberration-corrected electron microscopy. MRS Bull. **31**(1), 36–43 (2006)
32. D.J. Smith, Development of aberration-corrected electron microscopy. Microsc. Microanal. **14**(1), 2–15 (2008)
33. Y. Zhu, H. Inada, K. Nakamura, J. Wall, Imaging single atoms using secondary electrons with an aberration-corrected electron microscope. Nat. Mater. **8**(10), 808–812 (2009). doi:10.1038/nmat2532
34. C.O. Girit, J.C. Meyer, R. Erni, M.D. Rossell, C. Kisielowski, L. Yang, C.-H. Park, M.F. Crommie, M.L. Cohen, S.G. Louie, A. Zettl, Graphene at the edge: stability and dynamics. Science **323**(5922), 1705–1708 (2009)
35. R.P. Feynman, There's plenty of room at the bottom. Talk given at the annual meeting of the American Physical Society at the California Institute of Technology, Dec 1959. Available online: http://www.its.caltech.edu/~feynman/plenty.html
36. R. Erni, M.-D. Rossell, C. Kisielowski, U. Dahmen, Atomic-resolution imaging with a sub-50-pm electron probe. Phys. Rev. Lett. **102**(9), 096101 (2009). doi:10.1103/PhysRevLett.102.096101
37. C. Kisielowski, C.J.D. Hetherington, Y.C. Wanga, R. Kilaas, M.A. O'Keefe, A. Thust, Imaging columns of the light elements carbon, nitrogen and oxygen with sub Ångstrom resolution. Ultramicroscopy **89**(4), 243–263 (2001)
38. B. Kabius, P. Hartel, M. Haider, H. Müller, S. Uhlemann, U. Loebau, J. Zach, H. Rose, First application of C_c-corrected imaging for high-resolution and energy-filtered TEM. J. Electron Microsc. **58**(3), 147–155 (2009)
39. O.L. Krivanek, M.F. Chisholm, V. Nicolosi, T.J. Pennycook, G.J. Corbin, N. Dellby, M.F. Murfitt, C.S. Own, Z.S. Szilagyi, M.P. Oxley, S.T. Pantelides, S.J. Pennycook, Atom-by-atom structural and chemical analysis by annular dark-field electron microscopy. Nature **464**(7288), 571–574 (2010)
40. R.M. Langford, A.K. Petford-Long, Broad ion beam milling of focused ion beam prepared transmission electron microscopy cross sections for high-resolution electron microscopy. J. Vac. Sci. Technol. A **19**(3), 982–985 (2001)
41. J. Li, D. Stein, C. McMullan, D. Branton, M.J. Aziz, J.A. Golovchenko, Ion-beam sculpting at nanometre length scales. Nature **412**(6843), 166–169 (2001). doi:10.1038/35084037
42. M. Marko, C. Hsieh, R. Schalek, J. Frank, C. Mannella, Focused-ion-beam thinning of frozen-hydrated biological specimens for cryo-electron microscopy. Nat. Meth. **4**(3), 215–217 (2007). doi:10.1038/nmeth1014
43. J. Mayer, L.A. Giannuzzi, T. Kamino, J. Michael, TEM sample preparation and FIB-induced damage. MRS Bull. **32**(5), 400–407 (2007)
44. O.G. Shpyrko, E.D. Isaacs, J.M. Logan, Y. Feng, G. Aeppli, R. Jaramillo, H.C. Kim, T.F. Rosenbaum, P. Zschack, M. Sprung, S. Narayanan, A.R. Sandy, Direct measurement of antiferromagnetic domain fluctuations. Nature **447**, 68–71 (2007)
45. M.F. Crommie, C.P. Lutz, D.M. Eigler, Confinement of electrons to quantum corrals on a metal surface. Science **262**, 218–220 (1993)
46. Y. Sugimoto, M. Abe, S. Hirayama, N. Oyabu, Ó. Custance, S. Morita, Atom inlays performed at room temperature using atomic force microscopy. Nat. Mater. **4**(2), 156–159 (2005). doi:10.1038/nmat1297
47. Lyding, J.W. Shen, T.C. Hubacek, J.S. Tucker, J.R. Abein, Nanoscale patterning and oxidation of H-passivated Si(100)-2×1 surfaces with an ultrahigh vacuum scanning tunneling microscope. Appl. Phys. Lett. **64**(15), 2010–2012 (1994). doi:10.1063/1.111722
48. K. Tanaka, Y. Kurihashi, T. Uda, Y. Daimon, N. Odagawa, R. Hirose, Y. Hiranaga, Y. Cho, Scanning nonlinear dielectric microscopy nano-science and technology for next generation high density ferroelectric data storage. Jpn. J. Appl. Phys. **47**(5), 3311–3325 (2008)

49. S. Kalinin, D. Bonnell, T. Alvarez, X. Lei, Z. Hu, J. Ferris, Q. Zhang, S. Dunn, Atomic polarization and local reactivity on ferroelectric surfaces: a new route toward complex nanostructures. Nano Lett. **2**, 589–594 (2002)
50. D.B. Li, M.H. Zhao, J. Garra, A.M. Kolpak, A.M. Rappe, D.A. Bonnell, J.M. Vohs, Direct in situ determination of the polarization dependence of physisorption on ferroelectric surfaces. Nat. Mater. **7**, 473–477 (2008)
51. P. Banerjee, P. Conklin, D. Nanayakkara, T.-H. Park, M.J. Therien, D.A. Bonnell, Plasmon-induced electrical conduction in molecular devices. ACS Nano **4**(2), 1019–1025 (2010)
52. D. Eigler, New Tools for Nanoscale Science and Engineering. Paper read at the workshop, international study of the long-term impacts and future opportunities for nanoscale science and engineering, Evanston, 9–10 Mar 2010
53. J.S. Kim, T. LaGrange, B.W. Reed, M.L. Taheri, M.R. Armstrong, W.E. King, N.D. Browning, G.H. Campbell, Imaging of transient structures using nanosecond *in situ* TEM. Science **321**(5895), 1472–1475 (2008). doi:10.1126/science.1161517
54. A.H. Zewail, Four-dimensional electron microscopy. Science **328**(5975), 187–193 (2010). doi:10.1126/science.1166135
55. F. Brizuela, S. Carbajo, A.E. Sakdinawat, Y. Wang, D. Alessi, B.M. Luther, W. Chao, Y. Liu, K.A. Goldberg, P.P. Naulleau, E.H. Anderson, D.T. Attwood Jr., M.C. Marconi, J.J. Rocca, C. S. Menoni, Improved performance of a table-top actinic full-field microscope with EUV laser illumination. *Proc SPIE* **7636** (2010)
56. S. Huh, L. Ren, D. Chan, S. Wurm, K. Goldberg, I. Mochi, T. Nakajima, M. Kishimoto, B. Ahn, I. Kang, J. Park, K. Cho, S.-I. Han, T. Laursen, A study of defects on EUV masks using blank inspection, patterned mask inspection, and wafer inspection. In *Extreme Ultraviolet (EUV) Lithography. Proceedings of SPIE*, vol 7636, ed. by B.M. La Fontaine (SPIE, San Jose, 2010).
57. K. Ushida, The future of optical lithography. Plenary talk, in *SPIE Advanced Lithography conference*, Santa Clara, 22 Feb 2010
58. Y. Kim, T. Komeda, M. Kawai, Single-molecule reaction and characterization by vibrational excitation. Phys. Rev. Lett. **89**, 126104–126108 (2002). doi:10.1103/PhysRevLett.89.126104
59. M.P. Nikiforov, S. Schneider, T.-H. Park, P. Milde, U. Zerweck, C. Loppacher, L. Eng, M.J. Therien, N. Engheta, D. Bonnell, Probing polarization and dielectric function of molecules with higher order harmonics in scattering–near-field scanning optical microscopy. J. Appl. Phys. **106**, 114307 (2009). doi:10.1063/1.3245392
60. C.L. Degen, M. Poggio, H.J. Mamin et al., Nanoscale magnetic resonance imaging. Proc. Natl. Acad. Sci. U.S.A. **106**(5), 1313–1317 (2009)
61. P. Abbamonte, K.D. Finkelstein, M.D. Collins, S.M. Gruner, Imaging density disturbances in water with a 41.3-attosecond time resolution. Phys. Rev. Lett. **92**, 237401 (2004). doi:10.1103/PhysRevLett.92.237401
62. P. Prabhumirashi, V.P. Dravid, A.R. Lupini, M.F. Chisholm, S.J. Pennycook, Atomic-scale manipulation of potential barriers at $SrTiO_3$ grain boundaries. Appl. Phys. Lett. **87**(12), 121917–121920 (2005)
63. U. Dahmen, R. Erni, V. Radmilovic, C. Ksielowski, M.-D. Rossell, P. Denes, Background, status and future of the transmission electron aberration-corrected microscope project. Philos. Trans. Math. Phys. Eng. Sci. **367**(1903), 3795–3808 (2009)
64. M.R. Castell, D.A. Muller, P.M. Voyles, Dopant mapping for the nanotechnology age. Nat. Mater. **2**(3), 129–131 (2003). doi:10.1038/nmat840
65. C.L. Jia, M. Lentzen, K. Urban, Atomic-resolution imaging of oxygen in perovskite ceramics. Science **299**(5608), 870–873 (2003)
66. J.C. Meyer, C.O. Girit, M.F. Crommie, A. Zettl, Imaging and dynamics of light atoms and molecules on graphene. Nature **454**(7202), 319–322 (2008)
67. P.A. Midgley, C. Durkan, The frontiers of microscopy. Mater. Today **11**, 8–11 (2009)
68. P.D. Nellist, M.F. Chisholm, N. Dellby, O.L. Krivanek, M.F. Murfitt, Z.S. Szilagyi, A.R. Lupini, A. Borisevich, W.H. Sides Jr., S.J. Pennycook, Direct sub-Ångstrom imaging of a crystal lattice. Science **305**(5691), 1741–1741 (2004)

69. Y. Oshima, Y. Hashimoto, Y. Tanishiro, K. Takayanagi, H. Sawada, T. Kaneyama, Y. Kondo, N. Hashikawa, K. Asayama, Detection of arsenic dopant atoms in a silicon crystal using a spherical aberration corrected scanning transmission electron microscope. Phys. Rev. B **81**(3), 035317–035322 (2010). doi:10.1103/PhysRevB.81.035317
70. P.W. Hawkes, J.C.H. Spence (eds.), *Science of Microscopy* (Springer, New York, 2007)
71. M.D. Rossell, R. Erni, M. Asta, V. Radmilovic, U. Dahmen, Atomic-resolution imaging of lithium in Al$_3$Li precipitates. Phys. Rev. B **80**(2), 024110 (2009). doi:10.1103/PhysRevB.80.024110
72. K. Suenaga, Y. Sato, Z. Liu, H. Kataura, T. Okazaki, K. Kimoto, H. Sawada, T. Sasaki, K. Omoto, T. Tomita, T. Kaneyama, Y. Kondo, Visualizing and identifying single atoms using electron energy-loss spectroscopy with low accelerating voltage. Nat. Chem. **1**(5), 415–418 (2009). doi:10.1038/nchem.282
73. S.J.M. Thomas, The renaissance and promise of electron energy-loss spectroscopy. Angew. Chem. Int. Ed Engl. **48**(47), 8824–8826 (2009)
74. M. Varela, S.D. Findlay, A.R. Lupini, H.M. Christen, A.Y. Borisevich, N. Dellby, O.L. Krivanek, P.D. Nellist, M.P. Oxley, L.J. Allen, S.J. Pennycook, Spectroscopic imaging of single atoms within a bulk solid. Phys. Rev. Lett. **92**(9) (2004). doi:10.1103/PhysRevLett.92.095502
75. P.M. Voyles, J.L. Grazul, D.A. Muller, Imaging individual atoms inside crystals with ADF-STEM. Ultramicroscopy **96**(3–4), 251–273 (2003)
76. A. Bartesaghi, P. Sprechmann, J. Liu, G. Randall, G. Sapiro, S. Subramaniam, Classification and 3D averaging with missing wedge correction in biological electron tomography. J. Struct. Biol. **162**(3), 436–450 (2008)
77. M.J. Costello, Cryo-electron microscopy of biological samples. Ultrastruct. Pathol. **30**(5), 361–371 (2006). doi:10.1080/01913120600932735
78. N. de Jonge, R. Sougrat, B.M. Northan, S.J. Pennycook, Three-dimensional scanning transmission electron microscopy of biological specimens. Microsc. Microanal. **16**(1), 54–63 (2010). doi:10.1017/S1431927609991280
79. R.I. Koning, A.J. Koster, Cryo-electron tomography in biology and medicine. Ann. Anat. **191**(5), 427–445 (2009). doi:10.1016/j.aanat.2009.04.003
80. F. Xiao, Y. Cai, J.C.-Y. Wang, D. Green, R.H. Cheng, B. Demeler, P. Guo, Adjustable ellipsoid nanoparticles assembled from reengineered connectors of the bacteriophage Phi29 DNA packaging motor. ACS Nano **3**(8), 2163–2170 (2009). doi:10.1021/nn900187k
81. A. Al-Amoudi, J.-J. Chang, A. Leforestier, A. McDowall, L.M. Salamin, L.P.O. Norlén, K. Richter, N. Sartori Blanc, D. Studer, J. Dubochet, Cryo-electron microscopy of vitreous sections. EMBO J. **23**(18), 3583–3588 (2004)
82. M. Beck, F. Förster, M. Ecke, J.M. Plitzko, F. Melchior, G. Gerisch, W. Baumeister, O. Medalia, Nuclear pore complex structure and dynamics revealed by cryoelectron tomography. Science **306**(5700), 1387–1390 (2004)
83. M. Cyrklaff, A. Linaroudis, M. Boicu, P. Chlanda, W. Baumeister, G. Griffiths, J. Krijnse-Locker, Whole cell cryo-electron tomography reveals distinct disassembly intermediates of *Vaccinia* virus. PLoS ONE **2**(5), e420 (2007). doi:10.1371/journal.pone.0000420
84. K. Grunewald, P. Desai, D.C. Winkler, J.B. Heymann, D.M. Belnap, W. Baumeister, A.C. Steven, Three-dimensional structure of herpes simplex virus from cryo-electron tomography. Science **302**(5649), 1396–1398 (2003)
85. S. Nickell, F. Beck, A. Korinek, O. Mihalache, W. Baumeister, J. Plitzko, Automated cryo-electron microscopy of "single particles" applied to the 26S proteasome. FEBS Lett. **581**(15), 2751–2756 (2007)
86. A. Sartori, R. Gatz, F. Beck, A. Rigort, W. Baumeister, J. Plitzko, Correlative microscopy: bridging the gap between fluorescence light microscopy and cryo-electron tomography. J. Struct. Biol. **160**(2), 135–145 (2007)
87. K. Aoyama, T. Takagia, A. Hirasec, A. Miyazawa, STEM tomography for thick biological specimens. Ultramicroscopy **109**(1), 70–80 (2008)
88. M.F. Hohmann-Marriott, A.A. Sousa, A.A. Azari, S. Glushakova, G. Zhang, J. Zimmerberg, R.D. Leapman, Nanoscale 3D cellular imaging by axial scanning transmission electron tomography. Nat. Meth. **6**(10), 729–732 (2009)

89. K.J. Batenburg, S. Bals, J. Sijbers, C. Kübel, P.A. Midgley, J.C. Hernandez, U. Kaiser, E.R. Encina, E.A. Coronado, G. Van Tendeloo, 3D imaging of nanomaterials by discrete tomography. Ultramicroscopy **109**(6), 730–740 (2009)
90. J.C. Gonzalez, J.C. Hernández, M. López-Haro, E. del Río, J.J. Delgado, A.B. Hungría, S. Trasobares, S. Bernal, P.A. Midgley, J.J. Calvino, 3D characterization of gold nanoparticles supported on heavy metal oxide catalysts by HAADF-STEM electron tomography. Angew. Chem. Int. Ed Engl. **48**(29), 5313–5315 (2009)
91. P.A. Midgley, R.E. Dunin-Borkowski, Electron tomography and holography in materials science. Nat. Mater. **8**(4), 271–280 (2009)
92. U. Ziese, C. Kübel, A. Verkleij, A.J. Koster, Three-dimensional localization of ultrasmall immuno-gold labels by HAADF-STEM tomography. J. Struct. Biol. **138**(1–2), 58–62 (2002)
93. D.J. Stokes, L. Roussel, O. Wilhelmi, L.A. Giannuzzi, D.H.W. Hubert, Recent advances in FIB technology for nano-prototyping and nano-characterisation, in *: Ion-Beam-Based Nanofabrication. MRS Proceedings,* Vol. 1020, Paper No. 1020-GG01-05, ed. by D. Ila, J. Baglin, N. Kishimoto, P.K. Chu. (Materials Research Society, Warrendale, 2007)
94. Y. Fu, N.K.A. Bryan, Fabrication of three-dimensional microstructures by two-dimensional slice by slice approaching via focused ion beam milling. J. Vac. Sci. Technol. B **22**(4), 1672–1678 (2004)
95. J. Fenn, *The Microsoft System Software Hype Cycle Strikes Again* (Gartner Group, Stamford, 1995)

Synthesis, Processing, and Manufacturing of Components, Devices, and Systems

Chad A. Mirkin and Mark Tuominen*

Keywords Fabrication • Patterning • Assembly • Scalability • Large-area • High-throughput • Nanomanufacturing • Carbon-based devices • Polymers • Roll-to-roll • Dendrimers • Forest products

1 Vision for the Next Decade

1.1 Changes of Vision over the Previous 10 Years

The last decade has been an exciting period of discovery in the synthesis and processing of nanostructures. Many new nanomaterials have emerged, along with new fabrication processes to generate them. The last decade has seen penetration of nanotechnology into almost every area and discipline in science and engineering. Nanotechnology has been used in commercial products, including nanostructured coatings, cosmetics, textiles and magnetic storage devices, among many others. While such products mark much more purpose-oriented use and application of

---

*With contributions from: Matthew R. Jones, Louise R. Giam, Richard Siegel, James Ruud, Fereshteh Ebrahimi, Sean Murdock, Robert Hwang, Xiang Zhang, John Milner, John Belk, Mark Davis, Tadashi Shibata.

C.A. Mirkin (✉)
Department of Chemistry and Director of the International Institute for Nanotechnology, Northwestern University, 2145 Sheridan Road, Evanston, IL 60208, USA
e-mail: chadnano@northwestern.edu

M. Tuominen
Department of Physics and Co-director of the Center for Hierarchical Manufacturing and MassNanoTech, University of Massachusetts, Amherst,
411 Hasbrouck Laboratory, Amherst, MA 01003, USA

nanostructures, there also has been important basic research concerning the toolkits for synthesis, fabrication, and patterning of nanostructures, in addition to bioinspired synthesis and directed self-assembly. Many of these advances show great promise for the development of new nanomanufacturing processes that will drive the creation of future nanosystems and devices. For example, the last ten years have seen the development of novel synthesis approaches for a range of nanoscale materials including aerosols, colloids, thin-films, nanocrystalline metals, ceramics, biomaterials, and nanoporous or nanocomposite structures. Importantly, several of these methodologies have improved upon industrially-relevant practices such as combustion, electrophoretic processes, electrodeposition, electrospinning, anodization, and sputtering. Over the same period of time, entirely new nanostructures, such as graphene, have been identified and their unique properties may lead to important technology advances.

Another important concept, in addition to synthesis, that has been heavily explored is the presence of long-range order in nanomaterial systems. The past decade has seen incredible advances to methods for controlling the placement of nanostructures in one-, two-, and three- dimensional arrays with extraordinary precision. For example, new concepts in bottom-up self-assembly have led to three-dimensional programmable superlattices using electrostatic, chemical, and biological interactions. Advances to the fundamental understanding of these forces and computational simulations have allowed for self-assembling nanostructures to created "by design". In addition, directed and hierarchical assembly has been achieved using various polymeric systems (e.g. block copolymers) which have already shown applications in data storage, nanoimprint lithography, and video displays. Using more conventional lithographic techniques that couple micro- and nanofabrication processes, new structures have been produced called metamaterials, which exhibit fascinating optical properties that manifest directly from the presence of periodicity in their physical architecture.

While typical inorganic devices and sensors have brought about much research into top-down lithography tools, which rely on high-energy destructive methods, there have been transformational developments for patterning surfaces with soft materials that would otherwise be damaged by such techniques (Fig. 1). The emergence of soft lithography and scanning probe-based methods that are high-throughput, low-cost, and amenable to arbitrary pattern formation and rely on the constructive delivery of materials mark a new set of tools that researchers can use for systematically investigating organic electronic device performance, biological interfaces, and chemical constructs. Moreover, as these top-down approaches meet fundamental resolution limits, it is necessary to look at bottom-up synthesis of materials at the single molecule level for higher order assemblies and material property testbeds.

Some of the main issues to consider for synthesis, assembly, and processing approaches include scalability, flexibility, producibility, predictability, low cost, safety, and the establishment of standards for human health and environmental protection. Breakthroughs with respect to the fundamental physics and chemistry of both inorganic and organic nanostructures have formed the new engine for

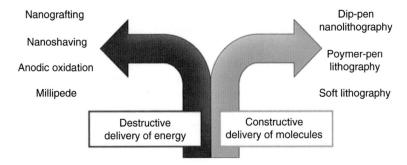

Fig. 1 Diagram showing the differences in two of the modern lithography strategies. A paradigm shift from energy to molecular delivery has taken place and enabled rapid advances in parallel scanning probe-based molecular printing techniques (From Braunschweig et al. [1])

exploration of materials properties, assembly mechanisms, and tool development. Many proof-of-concept devices have been demonstrated over the last decade; these have stimulated progress within fabrication technology.

1.2 Vision for the Next 10 Years

The exploration of next generation electronic devices is always a driving force in nanotechnology and will continue to be a focus for the next ten years. However, electronics are only part of the opportunities afforded by advances in nanotechnology over the last decade. Breakthroughs with respect to the fundamental physics and chemistry of both inorganic and organic nanostructures have formed the new engine for exploration of materials properties, assembly mechanisms, and tool development. These advances will largely act to drive the development of nanoscience over the next 10 years, in the context of crucial nanomanufacturing principles.

The challenge of printing features over large areas exceeding many square centimeters with sub-10 nm resolution and materials flexibility for both hard and soft matter requires the integration of many aforementioned strategies. It is expected that the combination of lithography techniques with supramolecular chemistry will enable the creation of nanostructures with a high degree of design flexibility. Moreover, directed self-assembly and self-alignment processes such as the use of block copolymers for high-density magnetic data storage and energy applications should be translated to commercially viable nanoscale patterning platforms in the coming decade. Other relevant bottom-up approaches include the fabrication of one-dimensional systems like gapped nanowires in on-wire lithography or to use programmable structures like DNA to assemble nanoparticles into ordered crystalline materials. Such systems not only allow one to probe molecular electronics, but also to understand the fundamental processes that govern crystallization. In addition to research on the fundamental aspects of nanotechnology, there likely will be a rise in emphasis on the manufacturing of useful products and devices. Of special importance is that these capabilities remain low-cost and can be scaled to high throughput roll-to-roll processes. While maintaining these considerations, it will be critical to develop nanomanufacturing processes amenable for integration with existing device fabrication technologies, which require contamination-free environments. The integration of large volume industry techniques with nanotechnology methods can expedite the production of devices that meet Moore's limit in the context of integrated electronics.

The ability to control the synthesis and assembly of devices will enable the emerging fields of plasmonic metamaterials, combinatorial catalysis, carbon-based electronics, and bio-inspired fabrication. Nanotechnology tools offer rapid approaches for making, screening, optimizing, and scaling potential catalysts for activity, selectivity, and environmental compatibility. In a single experiment, it would be possible to probe multiple questions as they pertain to physical, chemical, mechanical, electrical, optical, or biological properties. High-throughput methods for screening nanocompositions relevant in a given photonic, electronic, or biomedical application will become increasingly important as science enables smaller feature dimensions with increased sensitivity.

Furthermore, as integrated circuit device metrics demand more processing power, science and engineering could transition from the silicon to the carbon age where the predominant focus is on soft materials that are easy to manufacture and low cost in spite of the nanomanufacturing implementation. Driven by rapid improvements in understanding and manipulation of sp^2 nanocarbon materials, flexible and transparent electronics have become a point of focus for many device studies. Specifically, their use as transparent conducting electrodes may be a critical replacement for indium tin oxide, the current standard for flat panel displays and solar cells. Not only should researchers focus on better material performance for transistor applications, energy production, and biological sensors, but they should understand the importance of transferring revolutionary findings beyond the benchtop with scaling up and mass production of commercially applicable nanomaterials. Large-scale methods for making chemically well-defined graphene building blocks with desired dimensions and composition will be one thrust of such development, but it is critical to also advance fundamental science and develop tools capable of synthesizing and characterizing novel materials.

The nanomanufacturing tools and methods described herein will also be of particular importance to the growing interest and focus on bio-inspired and biomimetic assembly and synthesis. There have been developments and expansion in the areas of nanobiotechnology and nanomedicine for better disease diagnosis, drug delivery, and molecule detection. In the coming years, a number of efforts should be made to understand how to establish an effective biomolecule-electronic interface and advance the utilization of bio-inspired systems, fabricate implantable devices composed of biocompatible nanocomposites that are non-toxic and long-lasting, and an overarching goal of improving methods for inexpensive and high quality processing.

In conjunction with advances in fundamental materials research and fabrication tools for efficient manufacturing, there is a simultaneous need for high-resolution and high-sensitivity characterization techniques. Such advanced instruments may include improvements in cryo-transmission electron microscopy or cryo-scanning tunneling microscopy. It would be useful in the coming decade to achieve better in situ characterization of materials while the nanomanufacturing processes are occurring and to develop robust protocols for reproducibility. Within this area, significant progress should also be made to establish safety standards for human and environmental health.

2 Advances in the Last 10 Years and Current Status

Numerous new synthesis, processing and manufacturing methods have demonstrated proof-of-concept feasibility at the laboratory scale, of which an important subset has progressed to scale-up pilot plants and full-scale commercial production. Some examples represent the continued advancement of techniques developed over past decades, whereas other methods are new, providing the opportunity for manufacturing routes of unprecedented efficiency and entirely new applications. Manufacturing brings to bear a new range of issues that are typically well outside of the domain of lab-scale research: process development and modeling, scale-up, metrology, process control, tooling, standards, workforce, safety, and supply chain. To realize the beneficial economic and societal impact of nanotechnology, these issues must be addressed. Furthermore, because of its inherent relationship to commercial activity, nanomanufacturing requires productive cooperation between industry, academia, and government.

The following sections describe a specific set of laboratory-scale processes developed over the past decade. These methods, by their spatially localized, temporally sequential nature, and concomitant limitations, are presented to illustrate the needs for scalability, affordability, robustness, and environmental friendliness in new nanomanufacturing processes to be invented over the next 10 years for industry-scale production.

Figure 2 illustrates the timeline for several inorganic nanomaterials. Nanotechnology tools offer rapid approaches for making, screening, optimizing, and scaling potential catalysts for activity, selectivity, and environmental compatibility. In a single experiment, it would be possible to probe multiple questions that pertain to physical, chemical, mechanical, electrical, optical, or biological properties. High-throughput methods for screening nanocomposites relevant in a given photonic, electronic, or biomedical application will become increasingly important as science enables smaller feature dimensions with increased sensitivity.

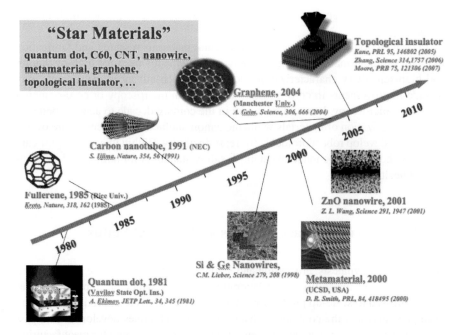

Fig. 2 Timeline for creating new nanostructured materials with current areas of research: carbon nanotubes, graphene, and metamaterials

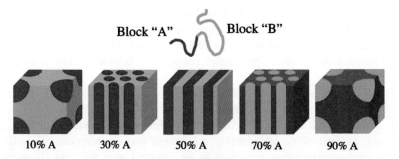

Fig. 3 Some key periodic nanostructure phases produced naturally by diblock copolymers as controlled by block volume fraction (Courtesy of M. Tuominen)

2.1 Block Copolymer Nanolithography

The use of block copolymers as a platform to create self-assembled templates for nanoscale patterning has developed rapidly over the last decade [3]. In general, block copolymers are molecules that have two or more immiscible polymer blocks that arrange thermodynamically [4] into periodic spherical, cylindrical, lamellar, or more complex array patterns (Fig. 3).

Microphase separated diblock copolymers led to early experimental demonstrations [5–8] of pattern transfer by subsequent deposition or etching that suggested the potential for large-scale production capabilities and hierarchical pattern control of nanoscale materials and devices, relevant to magnetic data storage, semiconductor devices, nanophotonics, and other technologies. One explicit device example is that of a flash electronic memory cell patterned via a diblock copolymer process [9]. Some projects demonstrate the conversion of block copolymer array structures into inorganic mesoporous structures or the formation of nanoparticles within the domains [10, 11]. Other research uses the co-assembly of block copolymers with nanoparticles to achieve phase-selective assembly/disassembly of nanoparticles in block copolymer thin films and produce nanotubes, hexapods, and other complex mesostructures by adjusting the relative size of nanoparticles [12, 13]. Diblock copolymers serve as a popular motif for directed self-assembly in which fields or pre-patterned surface features can control the orientation, alignment, and long-range order of the nanoscale polymer domain arrays [14–18, 20, 21]. Cylindrical microdomains as small as 3 nm have been produced, along with orientational control methods that suggest routes for low-cost continuous roll-to-roll manufacturing [22]. Recent work demonstrates the ability to use pre-patterned surfaces to coerce block copolymer assemblies into non-natural, somewhat arbitrary patterns more suitable for nanoelectronic devices – including square arrays, T-junctions, and bends ([23–27], [29]) (Fig. 4).

Quite clearly, the design rules for block copolymer nanolithography are rapidly becoming established and will continue to advance through this type of fundamental research. Multi-level alignment and 3D patterns are critical research targets for the future.

2.2 Scanning Probe-Based Lithography

2.2.1 Dip-Pen Nanolithography

Dip-pen nanolithography (DPN), polymer pen lithography (PPL), inkjet printing, transfer-printing techniques, and scanning probe block copolymer lithography over the past decade are illustrated in Fig. 5. Such tools have enabled the controlled synthesis and placement of nanomaterials and nanostructures on a surface with a broad range of materials and substrate compatibility.

The ability to pattern surfaces with sub-100 nm resolution has been a driving force in nanotechnology fueled by the semiconductor industry's desire to continually shrink the size of bulk materials, and by new capabilities for biological experiments made possible through high-density biomolecule arrays. In this respect, DPN [30–32] has become a commercial technique for direct-write molecular printing; it is capable of patterning surfaces with sub-50 nm feature size (Fig. 6). As a patterning tool, many applications have been explored, and DPN has been used for fundamental transport studies [33, 34], as a fabrication technique for photomasks

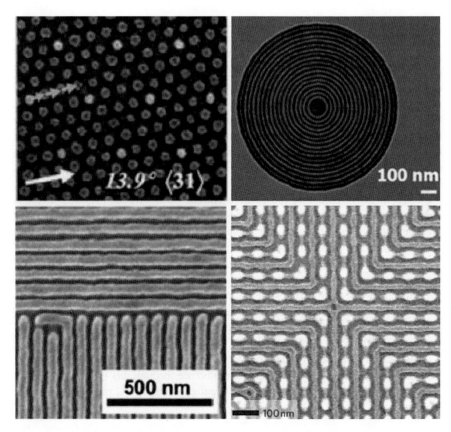

Fig. 4 Four examples of nanostructure patterns fabricated by directed self-assembly using diblock copolymers (Sources: *upper left*, Bita et al. [14]; *upper right*, Chai and Buriak [23]; *lower left*, Galatsis et al. [28]; *lower right*, Yang et al. [29])

Fig. 5 Timeline for scanning probe-based molecular printing tools such as dip-pen nanolithography [1]

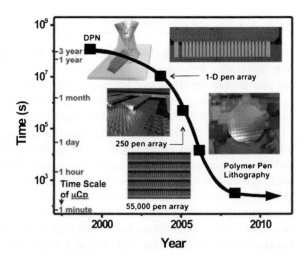

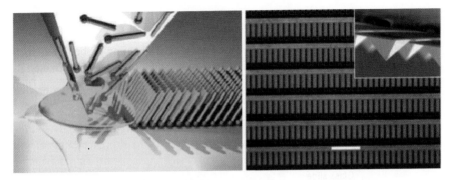

Fig. 6 (*Left*) Schematic of patterning alkanethiols on Au by DPN. (*Right*) Optical micrograph of 55,000 Si pen array; scale bar is 100 μm. (*Inset*) Scanning electron micrograph of the pens (From Salaita et al. [32])

[35], and as a method of creating biological screening devices, including an assay for HIV virus p24 antigen in serum samples [36].

One-dimensional (1D) arrays of cantilevers have been fabricated to increase throughput by a factor equal to the number of tips in the array [37, 38]. This idea of passive parallelization of DPN was further expanded to generate two-dimensional (2D) arrays consisting of 55,000 tips capable of increasing throughput by as much as four orders of magnitude with as many as 88 million dots fabricated in approximately 5 min (Fig. 6) [39]. To date, DPN arrays with as many as 1.3 million cantilevers have been fabricated. The massive parallelization of DPN opens up the possibility to fabricate combinatorial libraries of nanostructures with feature sizes two to three orders of magnitude smaller than current widely used microfabrication techniques such as photolithography, ink jet printing, and robotic spotting. The ability to shrink feature size to the nanoscale therefore allows researchers to print on the scale of biology. This feature size reduction not only increases the number of features per unit area but also allows single particle structures such as viruses and cells to be manipulated individually. Before combinatorial arrays generated by DPN become commonplace however, robust methods of inking and simultaneously transporting multiple different molecules to a substrate in multiplexed fashion must be developed.

A microfluidic inkwell platform capable of delivering multiple different inks to a 1D array of tips has been developed, allowing for the simultaneous deposition of up to eight different inks [40]. This technique, however, is not capable of addressing a 2D array of tips for massively parallel multiplexed DPN. Toward this goal, an inkjet printing technique has been developed for massively multiplexed parallelization, whereby tips within 1D and 2D arrays are inked with chemically distinct inks [102]. DPN may become in the next decade a general nanofabrication tool that combines high throughput, high resolution, and multiplexed deposition capabilities.

2.2.2 Polymer Pen Lithography

Polymer-pen lithography (PPL) merges the concepts of scanning probe lithography and contact printing to achieve high-throughput molecular printing of many materials (Fig. 7) [41]. PPL relies on an array of elastomeric tips to print features ranging from

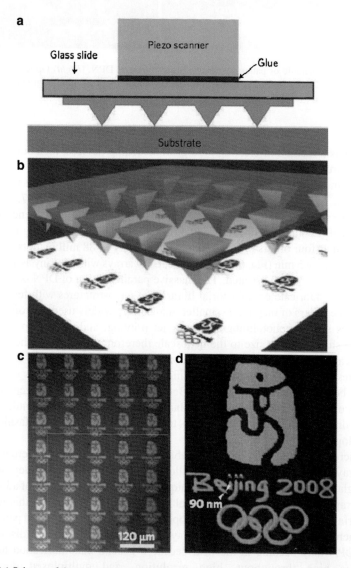

Fig. 7 (a) Scheme of the polymer pen array in a patterning instrument. (b) Scheme for fabricating the Beijing Olympic logo. (c) Optical micrograph of Au features showing high-throughput patterning. (d) Scanning electron micrograph where the smallest feature is 90 nm in diameter [41]

90 nm to more than 10 μm in a force- and time-dependent manner. As with conventional DPN, the deposited feature area increases linearly with tip-substrate contact time. In addition, feature size depends linearly on the amount of force applied to the pen arrays because they are elastomeric. The same polymer pen array may be used multiple times without reinking, and can be used to print patterns over large areas exceeding several square centimeters.

The same materials that can be patterned in a DPN experiment are applicable to PPL – alkanethiols, polymers, and proteins – to name a few. To generate multiplexed patterns of several materials, it is possible to use the same masters that acted as the polymer pen molds and deposit ink in the pyramidal pits, much like an inkwell [42].

2.2.3 Beam Pen Lithography

An extension of PPL termed beam pen lithography (BPL) uses the pen arrays to deliver light to a surface in a manner conventionally known as near-field scanning optical microscopy (Fig. 8) [43]. The polymer pen arrays are coated with a thin opaque layer such as Au and then brought in contact with an adhesive poly(methyl methacrylate) surface to fabricate micrometer-sized apertures or by focused ion beam lithography to generate nanometer-sized apertures. Light can then be exposed to the backside of the pen arrays and channeled through the apertures to a photosensitive surface for fabricating subdiffraction-limited features in a high-throughput manner. This tool could enable researchers to rapidly design and produce novel devices and marks an additional capability enabled by scanning probe-based lithographies.

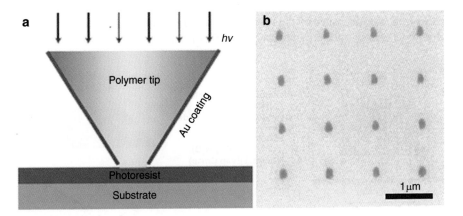

Fig. 8 (a) Scheme of beam pen lithography. (b) Scanning electron micrograph of 100 nm features made using 400 nm light [43]

2.3 1D Systems

2.3.1 On-Wire Lithography

1D systems (e.g., nanowires and nanorods) represent an area of intense research interest. Like their zero-dimensional (0D) counterparts (e.g., nanoparticles and quantum dots), there are now a variety of methods for synthesizing such structures. However, by transitioning from 0D to 1D systems, the design space available to the researcher and ultimate user significantly increases [44]. In addition to controlling the diameter, length, and composition of such structures, positive and negative architectural features can be introduced along the wire to realize structures with even greater functionality. In this regard, methods for nanowire fabrication and manipulation, analogous to the many powerful types of nanolithographies available to the materials researcher (e-beam lithography, nanoimprint lithography, and DPN), would be extremely useful.

On-wire lithography (OWL) is a powerful process that allows for the control of feature composition and size from the sub-5 nm to many micrometer length scales (Fig. 9). In the OWL process, anodic aluminum oxide films (either purchased from commercial vendors or fabricated in the lab) are used as templates to electrochemically deposit nanowires. Cylindrical, aligned pores permeate these templates and serve as discrete regions for nanowire growth. Anodic aluminum oxide films are available, with pores ranging in diameter from 400 to 13 nm. Deposition of materials into these pores is made possible by first evaporating a metal backing onto one side of the alumina. This evaporated film as a working electrode during wire synthesis (Fig. 9). "By electrochemically reducing metal ions from solution into the now half-closed pores of the template, nanowires can be grown with lengths and compositions corresponding to the applied current and metal ion precursor, respectively. Dissolution of the template and evaporated metal backing results in a suspension of billions of nanowires (Fig. 9). This suspension is sprayed onto a glass slide, and a backing layer is deposited on the wires by chemical or physical deposition methods. Both conducting (metals) and insulating (SiO_2) backing materials can be used.

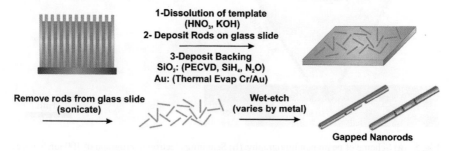

Fig. 9 The on-wire lithography (OWL) process [45]

With the backing layer in place, wire segments can be selectively dissolved on the basis of differences in chemical reactivity. This approach allows nanometer-scale gaps, disks, and disk arrays with precise control over not only the size of the gap, but also the thickness, composition, and periodicity of the disks to be prepared.

2.3.2 OWL-Based Encoding Materials

Encoded materials are used for many applications, including cryptography, computation, brand protection, covert tracking of material goods and personnel, and labeling in biological and chemical diagnostics [46]. The disk and gap structures made by OWL are particularly interesting for this class of materials because they are dispersible, allow for massive encoding on the basis of the length and location of individual chemical blocks within the structures, and can be easily functionalized using conventional surface chemistries. These properties of OWL-generated nanostructures have been used to create a library of optimized disk pair structures that vary in the number and position of the disk pairs along the silica backing [47, 48]. Each of these structures represents a unique nanodisk code label (Fig. 10).

By functionalizing the disk pairs with an oligonucleotide capable of binding a target DNA followed by hybridization of a chromophore labeled "reporter" strand, particular oligonucleotides can be captured and detected by means of hybridization to the structure. This sandwich assay design has been successfully implemented using the intensity of the Raman reporter as a measure of the concentration of the target to detect oligonucleotides with low pM sensitivity. These examples highlight the unique applications of OWL generated structures, which are made possible by the placement of SERS-active materials in a highly tailorable manner.

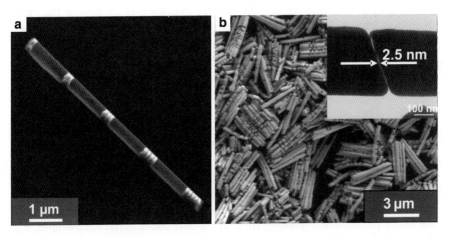

Fig. 10 Structures made by OWL. (**a**) Disk arrays (Adapted from [46]). (**b**) Zoomed out image demonstrating uniformity of rods [45]. (**b** *inset*) 2.5-nm gap produced by means of OWL [47, 48]

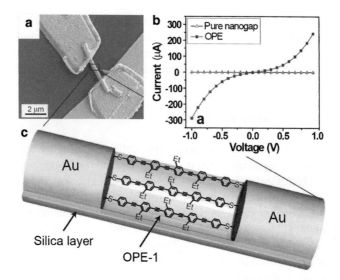

Fig. 11 (a) SEM image of a device prepared with an OWL-fabricated wire having a 3-nm gap. (b) Representative *I-V* response for 3-nm OWL-fabricated gaps before being modified with OPE-1 (Pure nanogap line) and after (OPE line). (c) A diagram of OPE-1 molecules spanning the 3-nm gap (Adapted from Chen et al. [50])

2.3.3 Test Bed for Molecular Electronics

In addition to plasmonic materials and SERS active substrates, the OWL process has been applied to synthesizing unique materials for understanding nanoscale transport phenomena [50]. Molecular electronics is a promising route to extremely compact, high-speed computing and data storage systems that are beyond the limits of conventional, solid-state circuitry [49]. Because OWL can be used to produce large quantities of high-quality nanowires with sub-5 nm gaps, it an ideal platform with which to study the charge transport properties of organic molecules that are designed to self-assemble across such gaps (Fig. 11).

2.4 DNA-Mediated Assembly of Gold Nanostructures

The ability to direct the placement of nanomaterials in three dimensions with a high degree of specificity and tailorability has been a goal of nanotechnology since its inception. By building complex nanostructures from the ground up with control over the lattice parameters, crystallographic symmetry, and material composition, it is thought that materials with new, emergent properties can be synthesized [51, 52]. Methods to arrive at an ordered nanocrystal superlattice typically rely on drying

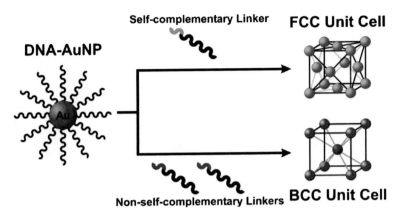

Fig. 12 Schematic illustration of the DNA-mediated assembly of gold nanoparticles. DNA linkers with short recognition sequences are hybridized to oligonucleotide functionalized gold nanoparticles. The sequence of the overhanging recognition unit dictates the assembly into FCC or BCC superlattices (Adapted from Park et al. [19])

effects or layer-by-layer assembly methods [53, 54]. Although these techniques have been used to demonstrate an impressive degree of crystallographic order over large areas, they do not afford a particularly high level of control over the lattice parameters of the resulting superlattices and cannot assemble particles independent of their size. An especially attractive alternative involves the use of DNA as a programmable linker that serves to direct the assembly of nanostructures in three dimensions with extraordinary control over crystallographic parameters [19]. DNA is an ideal material for the creation of nanoscale architectures, because its self-recognition capabilities allow for the assembly of materials with properties that can be varied via judicious DNA design. When DNA "linkers" that have short, self-complementary sticky ends are hybridized to oligonucleotide functionalized gold nanoparticles and allowed to induce assembly, a face-centered cubic (FCC) super-lattice is observed using small angle X-ray scattering (Fig. 12). Likewise, when non-self-complementary DNA linkers are hybridized to oligonucleotide function-alized particles, a body-centered cubic (BCC) superlattice is observed.

This difference in crystal symmetry arises from the driving force for DNA-gold nanoparticle conjugates to maximize the number of hybridization interactions in a given arrangement. For example, nanoparticles in a 1-component system (i.e., self-complementary linkers) have the greatest number of interparticle connections in an FCC configuration. Similarly, nanoparticles in a 2-component system (i.e., non-self-complementary linkers) have the greatest number of favorable hybridization interactions in a BCC configuration. Simply by changing the number of nucleobases in the linker oligonucleotides, scientists can systematically control the interparticle spacing of the resulting colloidal crystals anywhere from ~20 to ~55 nm [55]. Interestingly, these colloidal crystals initially assemble into a

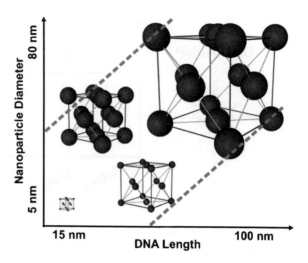

Fig. 13 Illustration of the "zone of crystallization" (denoted by *dashed lines*) in which certain combinations of nanoparticle size and DNA length are able to form ordered superlattices. Modeled unit cells are drawn to scale relative to one another [57]

disordered aggregate that subsequently undergoes a reorganization process ultimately resulting in an ordered superlattice [56].

One clear difference between solid-state atomic assembly and DNA-mediated nanoparticle assembly is the ability to control the properties of a nanostructure by changing its size, independent of its crystallographic arrangement. When oligonucleotide-functionalized gold nanoparticles of different sizes are used in the assembly process, a fascinating trend emerges in which only certain combinations of nanoparticle diameter and DNA length are able to reorganize into an ordered superlattice (Fig. 13) [57]. When the DNA length is too short relative to the nanoparticle diameter, the flexibility of the DNA cannot compensate for the size polydispersity of the nanoparticles (i.e., the elasticity of the oligonucleotides must be on the same order as the distribution in particle diameters). Likewise, if the DNA is too long relative to the nanoparticle diameter, the effective concentration of the linker sticky ends is decreased enough to suppress the "on rate" for linker-linker hybridization which prevents reorganization. Understanding these fundamental processes has helped to elucidate the design rules for nanoparticle crystallization and has carved out a "zone of crystallization" in which appropriate choice of particle size and DNA length can yield ordered superlattices.

Although the spherical particles used in DNA-mediated crystallization provide an interesting parallel to atoms in crystal lattices, they do not possess the directional bonding interactions present in solid-state systems, which account for the incredible diversity of crystal structures present in nature. One way to impart directional interactions in this colloidal assembly scheme is to replace the spherical cores with anisotropic nanostructures whose shape can template unique superlattice structures [58]. In the case of simple 1D nanorods, a 2D ordering of particles into extended, hexagonally packed sheets is observed (Fig. 14). This crystallographic arrangement is facilitated by the large number of hybridization

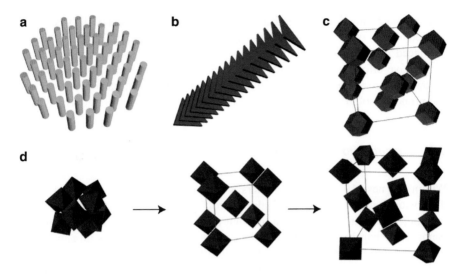

Fig. 14 Models of superlattices derived from DNA-mediated assembly of anisotropic nanostructures. (**a**) 2D hexagonal arrangement of nanorods. (**b**) 1D columnar ordering of triangular nanoprisms. (**c**) Face-centered cubic ordering of rhombic dodecahedra. (**d**) Illustration of the transition from disordered to BCC to FCC ordering with increasing linker length in the assembly of octahedra [58]

interactions perpendicular to the long axis of the rods, favoring a coplanar hexagonal packing. For the assembly of 2D triangular nanoprisms, a 1D columnar stacking of particles is observed, demonstrating that hybridization can be maximized by face-to-face association between prisms (Fig. 14). In the case of rhombic dodecahedra, both positional and orientation order is observed between neighboring particles in an FCC lattice. This configuration of particles allows every parallel face to be in registry with each other, maximizing the interaction area between them (Fig. 14). In the case of octahedra, a phase transition is observed as a function of DNA length between a disordered lattice, a BCC lattice, and an FCC lattice (Fig. 14).

3 Goals for the Next 5–10 Years: Barriers and Solutions

In the coming decade, synthesis work should endeavor to link itself with research on scale-up and advanced processing, in addition to the research that enables precise control of nanomaterial shape, structure, composition, and crystal facet. There is an enormous untapped potential in this area. The implementation of processes based on inherently scalable scientific phenomena, such as self-assembly, directed assembly, and bioinspired synthesis, will lower manufacturing costs and broaden the applicability of nanomaterials. Combined with a new generation of automation and manufacturing design optimization, advancements over the next 10 years have the potential to usher in a new age of sustainable manufacturing. Emphasis should be placed on the manufacturing science and engineering, which requires close cooperative efforts among industry, academia, and government. The section that follows discusses specific opportunities for nanofabrication, barriers to success, and proposed solutions to overcome the challenges and reach these goals.

3.1 Nanopatterning Tools

Within the next 10 years, researchers are expected to achieve two- and three-dimensional macroscopic materials control, with the ability to dictate where building blocks are placed down to 1-nm resolution, especially in the context of nanoprinting capabilities. Currently, there are no suitable methods with appropriate resolution and general materials compatibility. Moreover, it is challenging to maintain a narrow distribution in defect sizes. In this regard, the convergence of top-down and bottom-up chemistry or directed assembly approaches may enable high-resolution synthesis and control of nanoscale architectures. Such enabling work will require a shift in the emphasis from fundamental research to manufacturing in nanotechnology, which currently suffers from insufficient collaboration between academia and industry. Specifically, the scale-up of nanomaterials synthesis is disconnected from fabrication and manufacturing at the industrial scale level. Further financial incentives for industry leaders to engage and partner with academia will foster united efforts that address the challenges in nanomanufacturing science. Industry, academic, and government partnerships will enable all parties to run and assess pilot projects and manufacturing test beds. In such work, they must take laboratory benchtop proof-of-concept methods and develop approaches that address commercially relevant production volumes. This goal can be realized through developments in science, metrology, test data, tooling, and materials.

3.2 Plasmonic Metamaterials

Within the emerging field of plasmonic metamaterials, there are no accurate structure-property relationships or design rules due to the lack of complete scientific understanding for these systems. Experiments and data specifically geared toward obtaining accurate first principles scientific understanding of structure and resulting properties will be enabling for model predictions and property customization. Improved theoretical algorithms and better nanomanufacturing could allow researchers to tailor the optical properties of oriented metallic superstructures. These metamaterials could then be used in optoelectronic applications as light harvesting systems and, more specifically, affect the performance to cost ratio for energy conversion, storage, transmission, and efficiency.

3.3 Combinatorial Chemistry

Currently, high-sensitivity, high-throughput assays are only in development, even though many scientific fields could greatly benefit from rapid ways for making, screening, optimizing, and scaling nanoscale materials ranging from biological targets to catalysts for their activity, selectivity, and environmental compatibility. Clever combinatorial approaches will be needed to enable rapid screening of materials. In the context of nanoparticle catalysts, there needs to be greater nanoparticle design control, including shape, valency, composition, and other factors that provide recipes for designer functionality. It is important to establish design rules for multiscale synthesis and integration of nanoscale systems to assemble these particles into hierarchical structures.

3.4 Transition from Silicon- to Carbon-Based Devices

There are currently infrastructure barriers to growth for the transition from the silicon age to the carbon age for integrated electronic and photonic devices. Lower cost methods for making and integrating device components and taking advantage of component structural flexibility will help realize carbon-based devices. At the same time, it is important to realize the potential of silicon beyond electronics. It may be possible to use nanoscale silicon structures in the fields of photonics, catalysis, biotechnology, and energy applications. This work will involve the synthesis and purification of components, chemical stability (e.g., avoiding oxidation), and sometimes component integration.

3.5 Bioinspired and Biomimetic Devices

For the integration of biomolecules into functional bioinspired or biomimetic devices, it is important to synthesize, purify, and scale up these nano-based instruments and therapeutics to meet regulatory agency requirements. To do so, there should be reduced heterogeneity of nanostructures and improved understanding and engineering of biomolecules on nanostructured surfaces. Furthermore, it will be critical to establish an effective biomolecule-electronic interface, although currently there is an incomplete understanding of molecule-level interactions. As with nanopatterning tools, the convergence of top-down and bottom-up chemistry and directed assembly approaches are promising for this challenge. Advancing organic-inorganic nanomaterial hybrids will also require appropriate processing methods that preserve the properties of the organic component.

3.6 Nanomanufacturing Capabilities

The lack of the ability to introduce nanoscale discoveries in manufacturing, economically, with repeatability, and with full *in situ* instrumentation to monitor the processes, is a barrier that must be addressed in the future.

3.7 Nanomaterials in the Forest Products Industry

The forest products nanotechnology roadmap (www.nanotechforest.org) identifies the industry vision as "sustainably meeting the needs of present and future generations for wood-based materials and products by applying nanotechnology science and engineering to efficiently and effectively capture the entire range of values that wood-based lignocellulosic materials are capable of providing." The industry vision is well aligned with society's need for establishing a source of sustainable materials and products. Priority areas for nanotechnology in the forest product industry are: improving the strength-to-weight performance; liberating and using nanocellulose and nanofibrils naturally present in wood; and achieving a better understanding of water-lignocellulosic interactions with the aim of improving the dimensional stability and durability of wood-derived products.

4 Scientific and Technological Infrastructure Needs

Synthesis and manufacturing in nanotechnology has demanding infrastructure needs. Although some of the new nanofabrication and synthesis processes can be inherently inexpensive, their characterization is not. As processes make their way from the laboratory to the production plant, the demand on metrology is even greater, because characterization speed and throughput is often an issue. The cost of the production toolset increases as well, because high volume and high speed are often requirements. Successfully responding to the following infrastructure needs will help accelerate nanotechnology research and development:

- *In situ* or online characterization facilities for synthesizing and manufacturing high precision, high purity nanomaterials and nanostructures
- Greater availability of large scale expensive fabrication, characterization, and measurement facilities (e.g., electron-beam lithography, clean room foundries, synchrotron sources, neutron sources)
- Extensive integration of interdisciplinary research and activities for technological breakthrough and transition of nanoscience to practical products

- Development of nanomanufacturing education curricula as an integral part of such activities, with a strong emphasis on innovation education and manufacturing engineering principles
- National nanomanufacturing development roadmap, jointly developed by industry, academia, and government stakeholders
- Long-term strategic research focus on basic nanotechnology research
- Adequate national facilities for rapid and inexpensive screening of new nanomaterials for environmental health and safety impact
- A broader portfolio of standard reference nanomaterials for instrument and tool calibration
- Accessible databases and libraries with information on nanomaterial properties, nanomanufacturing process information, and safety

5 Research and Development Investment and Implementation Strategies

The United States must continue a strong base of fundamental research in nanotechnology but at the same time significantly increase activities in nanomanufacturing, product engineering, and innovation education. This combination serves best to generate societal and economic benefits via several main points:

- Foster collaboration among industry, universities, and research institutes. These relationships provide cost sharing, avoid redundant efforts, generate more effective facility use, and facilitate a more rapid commercialization. Efforts should be made to ensure that both large corporations and small start-up companies participate and benefit.
- Promote and fund a set of complementary regional clusters that work to accelerate nanotechnology development and commercialization. Each cluster should have a relatively narrow thematic focus as a genuine national center of excellence.
- Promote interdisciplinary work. Specifically, the challenges facing medicine need direct involvement of engineers, scientists, and clinicians.
- Build a robust value chain from raw materials to nanocomponents to final products. Fill weaknesses and gaps in the nanotechnology value chain to enable promising applications to grow and thrive. Strengthen the U.S. National Nanomanufacturing Network.
- Develop training programs and nanomanufacturing education curriculum. Increase investment in people who are trained in developing new analytical techniques. Such efforts can include postgraduate courses in nanoscience and nanotechnology and long-term positions for researchers. Promote multidisciplinary and integrative research.
- Increase fundamental research funding to generate new knowledge of phenomena and manipulation of matter at the nanoscale.
- Provide more investment in nanotoxicology to evaluate the safety of nanomaterials.

6 Conclusions and Priorities

In the coming decade, the research and development community must complement its ongoing fundamental research activities in synthesis, assembly, and processing by placing a stronger emphasis on the development of nanomanufacturing science and engineering. Nanomanufacturing (science based, reproducible, sustainable, and cost-effective) needs to be developed in conjunction with other areas such as nanobiotechnology and nanomedicine (e.g., diagnostics, drug delivery, and disease treatment), energy applications (e.g., conversion, storage, transmission, and efficiency), environmental fields (e.g., sensors, remediation, water purification), informatics (e.g., providing data, models, and information needed for efficient design, testing, development, and manufacturing), electronics (based on new architectures that specifically utilize the intrinsic properties and geometries of nanomaterials), and educational challenges (e.g., promoting the value of nanoscience and nanoengineering degrees; addressing lack of textbooks, emphasizing community college, undergraduate, and graduate education; and integrating partnerships with industry).

The following priorities have been identified for the next decade:

- Creating nanomaterials and systems "by design" is a main goal. Integration of fundamental research, modeling, simulation, processing, and manufacturing in a continuing R&D approach is essential.
- Development of a library of nanostructures (particles, wire, tubes, sheets, modular assemblies) of various compositions with industrial-scale quantities
- Investigation of new processes for large scale environmentally benign manufacturing of graphene and plasmonics materials
- Fundamental understanding of the pathways for self-assembly or controllable assembly of atoms or molecules into larger and stable nanostructures. Nanobiomanufacturing will expand to new approaches.
- Emulation of proven natural designs in nanosystem manufacturable architectures
- Scalable manufacturing processes using three-dimensional programmable assembly will be realized. Several "killer applications" are expected to emerge.
- Support manufacturing of large and flexible displays for mass use.
- Develop ability to print features over large areas exceeding many square centimeters with sub-10 nm resolution is needed in electronics and photonics. Materials flexibility for both hard and soft matter will require the integration of various strategies. The purpose is that nano-imprinting will be low-cost and in mass production. AFM lithography should be developed toward manufacturing use.
- Life-cycle environmentally-friendly nanomanufacturing technologies will increase as required by market.
- Develop nanoinformatics for nanomaterials and nanoscale devices.

7 Broader Implications of Nanotechnology Research and Development on Society

Synthesis and manufacturing are essential steps to be addressed for economical application of nanotechnology for societal needs as described in more detail in chapter "Innovative and Responsible Governance of Nanotechnology for Societal Development". Nanotechnology has the potential to benefit society in numerous areas from cosmetics to cars and from electricity to medicines. This impact can be readily seen with biomedical devices (e.g., diagnostics, drug delivery), electronic devices (e.g., mobile communication systems, portable data devices), efficient energy technologies (e.g., generation, conversion, transmission, and storage), and food industry products (e.g., production, packaging, safety) as they are translated from biology, chemistry, engineering, and materials science.

8 Examples of Achievements and Paradigm Shifts

8.1 Discovery of Graphene (See Also Chapter "Applications: Nanoelectronics and Nanomagnetics")

Contact person: Mark Tuominen, University of Massachusetts, Amherst

Hand-in-hand with explorations of material properties have existed efforts to synthesize them in a cost-effective, commercially relevant fashion. A good case study is the carbon nanomaterial, graphene – a single atomic layer of graphite. Originating with a process [59] that could not easily be scaled up (mechanical rubbing off a layer of graphite), graphene was produced so that its properties could be tested. Very quickly, the scientific community discovered that graphene has excellent electrical properties [60] that could be used in a variety of ways, including transistors and transparent conducting electrodes. Indeed, this excitement was ratified by the 2010 Nobel Prize in physics being awarded for the discovery of graphene. The exceptional properties of graphene drove the research community to develop new, commercially viable methods to produce graphene. Indeed, the International Technology Roadmap for Semiconductors added graphene and carbon nanotubes to the roadmap (Fig. 15).

Huge strides have been accomplished within the last 2 years in large-area [61], roll-to-roll production of graphene, and it has now been demonstrated for use as a transparent conducting electrode in prototype displays (Fig. 16) [62]. This is only one representative application, but one that solves a pressing issue: replacement of indium tin oxide as a transparent electrode when indium reserves are rapidly depleting. Many other uses for nanoscale carbon can be expected over the next decade.

Research and technology development schedule proposed for carbon-based nanoelectronics

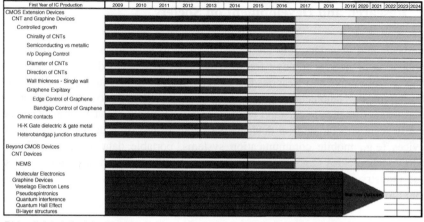

Fig. 15 Schedule for carbon nanoelectronics from the 2009 International Technology Roadmap for Semiconductors

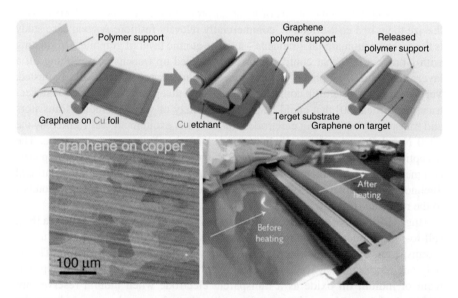

Fig. 16 Images of large scale production of graphene grown on copper foils and transferred to a polymer layer for use in large area display applications (From Li et al. [61] and Bae et al. [62])

8.2 The Opportunity for Atomically Precise Manufacturing

Contact person: John N. Randall, Zyvex Labs

Improved manufacturing precision is a proven path to increasing the efficiency, quality, and reliability of existing products. It is also a key enabler in developing new products and applications. As manufacturing precision approaches the atomic and molecular scale, researchers have the unique opportunity to exploit the quantized nature of matter and make precision absolute. For the first time, the possibility of making nanoscale objects that are not just similar but are in fact exact copies is within reach.

Zyvex Labs has a project to develop an atomically precise manufacturing technology [63]. The technical approach is an integration of two known experimental techniques: H depassivation lithography from Si (100) surfaces [64] with a scanning tunneling microscope (STM) and silicon atomic layer epitaxy [65] using disilane or other Si-H precursor gases to deposit Si where H has been removed (Fig. 17). This pattern and deposit cycle is repeated in ultra high vacuum to control the creation of 3D structures.

This digital fabrication process exploits the discrete nature of matter. Several key features differentiate this process from typical scanning probe manipulation of matter: The tip never contacts anything; the H atoms are removed into the gas phase; and the deposited material arrives from the gas phase. Although this fabrication process has not been demonstrated with atomic (absolute) precision, perfect patterning has. The principal challenge is demonstrating good quality epitaxy below 300°C that maintains atomic precision while creating 3D structures. The experimental results suggest that epitaxial growth is possible at 220°C where H mobility on Si surfaces is still very slow. The project also includes monolayer passivation of Si surfaces, improved STM tip technology, and MEMS (Micro electro mechanical systems) closed loop nanoscanners. This approach will enable applications for processes requiring only an extremely small volume of material. It has been

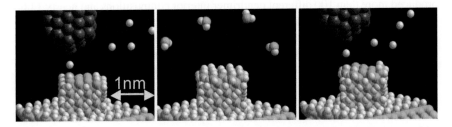

Fig. 17 (*Left*) H (*white*) atoms removed from Si (*blue*) surface by electron stimulated desorption by STM. (*Middle*) Si-H precursor gas selectively deposits on passivated Si where H has been removed (*middle*). (*Right*) Patterning process is repeated with another deposition cycle to create 3D structures

calculated, using a full cost of ownership model, that the cost of production would be approximately $2,100/µm^3. Many applications would be cost-effective even at this price point. Among these applications are nanopores for DNA sequencing, metrology standards, templates for nanoimprint lithography, nano electromechanical resonators for extremely low power radios, and Kane Qubit's for quantum computing. In the long term, with the sort of exponential increase in performance/cost seen in other technologies, and because this fabrication technology will be adaptable to many other material systems (semiconductors, metals, and insulators), it could become the foundry that enables widespread, cost-effective nanomanufacturing. Initial estimates predict that by 2020, Zyvex Lab consortium members will develop seven to nine key new technologies using this platform with a market valuation of $7–$600 million.

8.3 Dendrimers: 2010–2020

Contact person: Donald A. Tomalia, Central Michigan University

Dendrimer structures and processes (2000–2010). Dendrimers are synthetic coreshell, soft matter nanobuilding blocks that are recognized as members of the fourth major class of macromolecular architecture after linear, crosslinked, and branched polymers. They are derived from two or more dendrons attached to a common core. Architecturally, they possess onion-like topology that consists of a core, an interior (e.g., shells or generations), and terminal groups (e.g., surface chemistry).

They are synthesized by two major "bottom-up" strategies; namely: covalent divergent and covalent convergent syntheses [66] (Fig. 18). These pre-2000 processes produced precise nanostructure controlled dendrimers as a function of size, shape, surface chemistry, and flexibility/rigidity that rival structural regulation normally observed for biological nanoparticles such as proteins, DNA, and RNA [67]. In the last decade, substantial progress has occurred for covalent dendrimer synthesis on the basis of Sharpless-type "click chemistry." These covalent processes have produced more than 100 different dendrimer interior compositions and nearly 1,000 differentiated surface chemistries. More recently, a new class of "supramolecular dendrimers" has emerged based on self-assembly processes pioneered by Percec and colleagues (Fig. 19) involving amphiphilic dendrons [68].

Amphiphilic dendrons may be designed as a function of critical nanoscale design parameters such as size, shape, and surface chemistry to produce hollow/solid spherical dendrimers, as well as cylindrical-type dendrimers. These self-assembly principles/patterns have led to the first examples of Mendeleev-like nanoperiodic tables [69]. Such tables allowed a priori predictions of expected supramolecular dendrimer type with 85–90% accuracy based on a nanoperiodic concept reported by Tomalia [70, 71]. The current production cost limits the applications to high added value products.

Synthesis, Processing, and Manufacturing of Components, Devices, and Systems 135

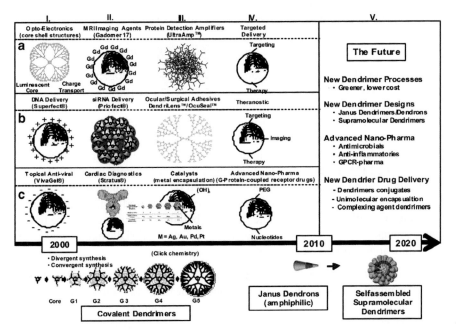

Fig. 18 Enhancement of nanostructure complexity for commercialized dendrimer-based products (2000–2010; Columns I–IV) and predicted future developments (Column V; 2010–2020)

Dendrimer applications/products (2000–2010). Dendrimer-based commercial products were first introduced in the late 1990s. Pre-2000 dendrimer commercial activity focused on the development of simple dendrimer compositions as a function of their nanoscale size and surface chemistry [71, 72]. The first products to emerge in 2000–2010 are described as follows (Fig. 18):

- Column I: (a) organic light emitting diodes (Cambridge Display/Sumitomo, Japan), (b) DNA gene vectors (e.g., Superfect®; Qiagen, Germany), and (c) antiviral topical nanopharma (e.g., VivaGel®; Starpharma, Australia). VivaGel® is presently in Phase IIa, FDA-based clinical trials.
- Column II: (a) magnetic resonance imaging agents (e.g., Gadomer-17®; Bayer/Schering Pharma AG, Germany), (b) siRNA delivery vector (e.g., Priofect, EMD and Merck), and (c) cardiac-diagnostics (e.g., Stratus; Siemens, Germany).
- Column III: (a) protein detection amplifiers (e.g., UltraAmp™, Affymetrix/Genisphere, Inc.), (b) ocular/surgical adhesives (e.g., DendriLens™/OcuSeal™; HyperBranch Medical Technology, Inc.), (c) and metal encapsulated/metal ligated dendrimer catalysts. More complex dendrimer-based nanodevices,
- Column IV a–c are focused on critical nanomedicine applications that include (a) targeted delivery of cancer therapies, (b) targeted delivery of cancer therapies with concurrent imaging capability and (c) advanced polyvalent

nanopharmaceutical prototypes referred to as G-protein-coupled receptor (GPCR) drugs, all of which are currently under development (e.g., National Institutes of Health and private sector) but not at a commercial stage [73].

Present market sizes/values for dendrimer-based biological/nanomedicine commercial applications are estimated to be more than $100 million/year. Large markets exceeding $1 billion/year are expected to emerge in the near term pending the final FDA approval of dendrimer-based topical, antiviral pharmaceuticals (e.g., VivaGel). These dendrimer-based microbicide agents are active against HIV, genital herpes, and human papillomaviruses and have been granted fast track status by the FDA. Essentially all critical behavior-influencing specific dendrimer applications and commercial products are based on intrinsic nanoscale properties associated with this broad architectural category of quantized building blocks [71]. These features distinguish dendrimer/dendron building blocks as one of the unique and dominant nanoscale platforms that allow systematic CNDP engineering/design for future applications and commercialization.

The Future (2010–2020). Activities expected to emerge in the next decade will focus on products of value to a variety of social, health, and economic areas. More efficient dendrimer processes (i.e., high atom efficiencies, lower recycle and byproducts; Column V) will allow larger volume value markets outside of the medical field: (a) manufacturing (additives for enhancing productivity, quality, or properties of commodity goods); (b) food production (agricultural products for enhanced crop production, controlled deliveries of herbicides, pesticides, and fertilizers); (c) environmental remediation (clean water and air, sequestering radioactive materials); (d) alternative energy (conversion and storage); (e) electronics [computing miniaturization, memory devices, illumination/displays (organic light-emitting diodes)]; and (f) miscellaneous (e.g., personal care products, antiaging, sensors, diagnostics, jet ink printing).

Completely new dendrimer/dendron structural designs (i.e., Janus dendrons/supramolecular dendrimers; Column V) will emerge based on "click chemistry" synthesis, poly(peptide) dendrimer constructions, region-specific dendrimer surface functionalizations, megamer (i.e., poly(dendrimer)) synthesis and dendrimer-based covalent/self-assembly hybridizations with other well-defined hard/soft nano building blocks to produce unprecedented new nanocompounds and assemblies. Several examples include, advanced dendrimer-based polyvalent nanopharma (Column V) in areas such as antimicrobials, anti-inflammatories, and GPCR drugs, and so forth, as well as dendrimer-based drug delivery vectors/excipients and advanced delivery targeting/strategies (Column V). These dendrimer-based vectors are designed to guide/target genetic or small molecule therapies to specific disease sites (e.g., cancer, diabetes) with minimal collateral damage to healthy tissue. They will be selected and synthesized based on the implementation of combinatorial libraries/techniques much as is used for traditional small molecule pharmaceutical screening [74].

8.4 Dendrimersomes

Contact person: Virgil Percec, University of Pennsylvania

Tiny bubbles and other nanostructures that form spontaneously when highly branched bifunctional compounds are put in water have been discovered and characterized (Fig. 19). The nanostructures may be more broadly useful for delivering drugs and other substances than similar nanostructures made from phospholipids or polymers. Researchers at the University of Pennsylvania reported in *Science* that a new family of vesicles, tubes, disks, and other shapes – which they call "dendrimersomes" – self-assemble from Janus dendrimers in water [76]. Dendrimersomes can act as hosts for a variety of guest molecules and could, therefore, have widespread applications as delivery vehicles for drugs, genes, imaging compounds, diagnostic agents, cosmetics, and other substances. Liposomes and polymersomes – synthetic vesicles made from phospholipids and polymers, respectively – have similar uses as delivery vehicles but several drawbacks: liposomes tend to be unstable and have short lifetimes, polymersome membranes are too thick to accommodate biological receptor and pore-forming proteins, and both liposomes and polymersomes adopt a wide range of sizes when they form and are difficult to derivatize. Dendrimersomes are stable for longer periods of time, are highly uniform in size, have the proper dimensions to accommodate membrane-spanning proteins, and are easily functionalized.

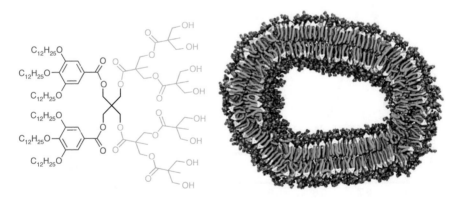

Fig. 19 (*Left*) Janus dendrimers are amphiphilic, with nonpolar (*left side*) and polar (*right*) ends. (*Right*) Nanostructure dendrimersome cross-section shows its cell-membrane-like bilayer [75]

8.5 New Paradigms in Nanoceramics

Contact person: Lynnette D. Madsen, National Science Foundation

Key breakthroughs in nanoscale ceramics in the last decade include the following: nanoporous materials and structures, strain-engineered complex oxides, and inorganic nanotubes and related materials.

Nanoporous materials and structures. Research has focused on nanoporous materials with tunable pore sizes and porous structures. Application areas are wide ranging, including capacitive energy storage, hydrogen storage, methane storage, gas separation, sorption of biomolecules, water desalination, and porous electrode materials for electrical energy applications.

Carbide-derived carbons (CDCs) produced by extraction of metals or metalloids from metal carbides have shown promise [76]. The method allows the synthesis of most known carbon structures. Further, CDC formation can be controlled to produce highly porous carbon materials with superior tribological properties. Precise control of porosity is required to maximize the material performance (Fig. 20). Research on the fundamental mechanisms governing the adsorption of cytokines by carbon materials could provide further insight and has the potential to save lives of people suffering from autoimmune diseases, severe sepsis, and multiple organ failure [78].

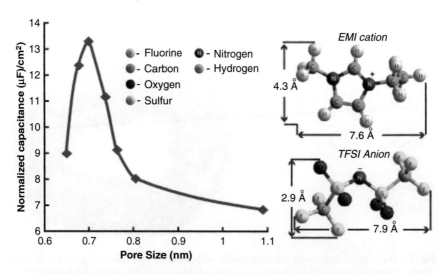

Fig. 20 Normalized capacitance change versus the pore size of the carbide-derived carbon samples tested in an ionic liquid electrolyte. Normalized capacitance is obtained by dividing the gravimetric capacitance by the specific surface area. HyperChem models of the structure of ethyl-methylimmidazolium (EMI) and bis(trifluoro-methane-sulfonyl)imide (TFSI) ions show a size correlation [77]

Researchers also focused on exploring predictive ways to synthesize closed-pore materials based on the use of surfactant micelles as templates to generate well-defined pore arrays. Surfactant templating has revolutionized the synthesis of nanoporous materials. Extending the scope of this powerful approach on new material compositions and on achieving more beneficial structural properties of the closed-pore materials is needed [79].

Similarly, electrically conducting zeolite-like or zeoate frameworks were targeted for improved electrical energy storage [80]. Some zeoate materials combine the properties of the well-known, but electrically insulating, microporous zeolites with an electronically active framework (Fig. 21). Based on fundamental science, novel energy storage and conversion mechanisms can be predicted and designed into new materials for many electrical energy applications from transportation to power-consuming electronics.

Atomic-scale engineering of ferroelectric and related materials. Strain has been used in epitaxial thin-film structures to induce or enhance ferroelectricity [81–83]. For example, ferroelectric capability was added to strontium titanate, $SrTiO_3$, to provide the first ferroelectric directly on silicon ([84]; Fig. 22). This ferroelectric material may eliminate the time-consuming booting and rebooting of computer operating systems and prevent losses from power outages.

Using the same principles, a new ferroelectric ferromagnet has been synthesized based on europium titanate ($EuTiO_3$) with a spontaneous magnetization × spontaneous polarization product at 4 K that is more than 100 times higher than any other known material [85]. First principles calculations were used to design a strong ferroelectric ferromagnet by exploiting strain in combination with spin-phonon coupling. These results open the door to higher temperature embodiments of strong ferromagnetic ferroelectrics, which would enable dramatic improvements in numerous devices, including magnetic sensors, energy harvesting, high-density multistate memory elements, wireless powering of miniature systems, and tunable microwave filters, delay lines, phase shifters, and resonators.

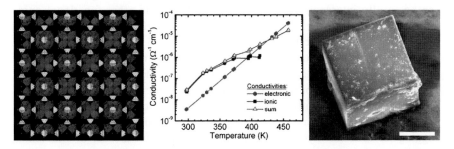

Fig. 21 (*Left*) Zeoate framework structure with [(As6VIV12VV3O51)–9Z] composition, where vanadium is *blue*, arsenic is *yellow*, and oxygen is *red*. The material crystallizes in a cubic space group with cell edge of 1,600 pm. (*Middle*) Plot of measured conductivities as a function of temperature (Data from [80]). (*Right*) Scanning electron micrograph of zeoate material from left panel; scale bar is 50 μm (Courtesy of V. Soghomonian, Virginia Polytechnic Institute and State University)

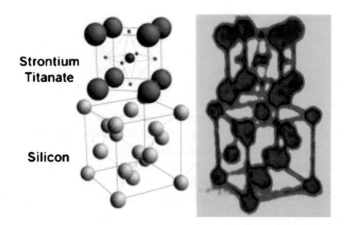

Fig. 22 (*Left*) The arrangement between atoms of a film of strontium titanate and single crystal of silicon on which it was made. When sufficiently thin, strontium titanate can be strained to match the atom spacing of the underlying silicon and becomes ferroelectric. (*Right*) The schematic at the left has been written into such a film using the ability of a ferroelectric to store data in the form of a reorientable electric polarization (Courtesy of J. Levy, University of Pittsburgh)

These and other advances have uncovered spectacular phenomena of fundamental scientific interest at oxide interfaces: self-generated ultrathin magnetic layers, orbitally reordered electronic systems, and conducting electron gases that can be patterned into nano-sized transistors [86]. In most cases, full implementation is impeded by several factors. Examples include room-temperature mobilities that are orders of magnitude below useful values. Limitations on mobilities could be caused by sample quality as a result of defects such as cation nonstoichiometries, oxygen defects, intermixing, and nonhomogeneities. An improved control and understanding of defects is necessary for significant advances in the fabrication of individual films as well as superlattice stacks containing several conducting interfaces that can be locally interconnected. Structures may include other functional oxide materials, such as ferroelectrics or strongly correlated materials. Samples that contain many tunable interfaces in parallel will demand and enable new transport studies. Other considerations in terms of moving toward devices are cost efficiency of fabrication and compatibility with existing semiconductor fabrication processes. In the long run, the possibility exists to create new artificial materials in bulk form. As Mannhart and Schlom [86] explained, "The theoretical modeling of interfaces in correlated systems is another area ripe with opportunity. This is a gold mine begging for materials-by-design solutions using the full arsenal of simulation, modeling, and theory."

Inorganic nanotubes. Inorganic nanotubes, such as nitride, sulfide, and oxide nanotubes, have the potential to be used for high-load, high-temperature, and high-pressure applications [87]. TiO_2 nanotubes have a role in photovoltaics, water purification, and CO_2 sequestration. Boron nitride nanotubes were predicted [88] and subsequently synthesized [89]. Their energy gap is independent of wall number, diameter, or chirality and thus they constitute desirable insulating structures for geometrically and electronically confining atomic, molecular, or nanocrystalline species [90]. In addition, boron–carbon–nitrogen systems have potential in terms of nanotubes and in other configurations [91]. Key factors in the development of various nanotubes seem to be (a) efficient and effective synthesis methods, (b) control of crystallinity and specific properties, (c) dispersion of nanotubes, and (d) the creation of composite materials with strong bridges between the different materials [92]. There are also opportunities in creating complex, hybrid (multicomponent) nanowires and nanotubes to build multifunctional nanomaterials. In the next decade, the creation of complex architectures with multiple nanoscale components assembled in a hierarchical fashion is likely.

8.6 Advanced Carbon Wiring

Contact person: Rick Ridgley, National Reconnaissance Office

The need for advanced wiring concepts can minimize ohmic losses and signal noise. The aerospace industry is constantly interested in reducing inactive spacecraft mass as a way to save payload costs and increase hardware efficiency. In general, the wire harness mass is approximately 10–15% of that of the total spacecraft. The harness mass includes the power distribution cables (~25%), data transfer cables (~55%), and mechanical fasteners and shielding (20%). If wire mass associated with spacecraft function, solar panel interconnects, and data transmission were decreased, there could be significant savings in spacecraft mass. In addition, if electronic failure mechanisms could be lessened through the use of advanced wires, with less arcing and shorting, there would be considerable benefit to enhance mission safety and lifetime (Fig. 23).

In high-power transmission lines, resistive losses consume about 7% of the energy produced. Reducing these losses to 6% would result in a national annual energy savings of 4×10^{10} kh (an annual energy savings roughly equivalent to 24 million barrels of oil annually or at \$80/barrel, \$1.92 billion annually). Advanced data transfer and low-voltage cables are another technology area that can reduce U.S. energy consumption – given that data centers consume annually 3% of the electricity, and the portion is growing at 12% a year.

All these technologies would benefit from advanced materials to improve electrical conductivity and mechanical stability. Historically, conductivity and mechanical stability have been obtained from a few common materials, such as steel, copper, and aluminum with sufficient, but not completely ideal, properties. Recently, the discovery of nanomaterials, such as carbon nanotubes (CNTs), opens up the possibility to push the frontier of materials development such that advanced wiring concepts can be realized to begin addressing many of these present-day challenges. Power cables, manufactured from CNTs, could be used to rewire electrical circuits in planes and even in the electrical transmission grid.

Fig. 23 Use of single walled carbon nanotube (SWCNT) ribbon photovoltaic interconnect (*top*). Image of a data center (*middle*). High-tension power lines (*bottom*)

Potential for carbon nanotubes. Single-wall carbon nanotubes (SWCNTs) can be envisioned as a graphene sheet rolled up into a seamless cylinder with fullerene caps. The van der Waals interaction between sidewalls leads to closely packed "bundles," which are an important physical property and the dimensions can be observed in a scanning electron micrograph (Fig. 24). The arrangement of

Fig. 24 Scanning electron micrograph of SWCNTs with insets. Cross-section of a wire (*top*); single-wall carbon nanotube ribbon cable (*middle*); carbon nanotubes thread (*bottom*). (Courtesy of Nanopower Research Labs at Rochester Institute of Technology: http://www.sustainability.rit.edu/nanopower/)

carbon-carbon bonding will determine the so-called "chirality" of the SWCNT and will determine whether the structure will be metallic or semiconducting. Depending on the chirality and physical bundling, carbon nanotubes can have outstanding electrical and thermal conductivities. Both properties are essential to wire and cable applications. An order of magnitude increase in conductivity exists for SWCNTs compared to copper when considering that SWCNT resistivity (r) is 1.3×10^{-6} Ω cm or conductivity is 7.7×10^5 S/cm. The bulk resistivity (r) for copper at room temperature is 1.7×10^{-6} Ω cm or a conductivity of 5.9×10^5 S/cm. This yields a specific conductivity for Cu (density is 8.92 g/cm^3) of 6.6×10^4 (S cm^2/g). Assuming a density of 0.8 g/cm^3 for SWCNTs yields a specific conductivity for the SWCNTs of 9.6×10^5 (S cm^2/g). Thus, there exists a nearly 15-fold improvement of the specific conductivity when comparing SWCNT wires to copper.

Another important attribute of SWCNTs is their current carrying capacity. SWCNTs have been shown to be ballistic conductors at room temperature, with mean free paths up to hundreds of microns [93]. Current densities have been measured as high as 10^7 A/cm^2 and have been predicted to be as high as 10^{13} A/cm^2 [94]. A lightweight material that can carry extremely high currents with superior strength and flexibility should be ideal for conventional wire applications. In addition, the extraordinary mechanical properties of these materials may ultimately improve spacecraft robustness and mission life as well as the lifetime and reliability of high-tension power transmission lines.

8.7 Fractal Nanomanufacturing: Multiscale Functional Material Architectures

Contact person: Haris Doumanidis, National Science Foundation

One of the manufacturing challenges arising from synthesis of nanoparticles and nanostructures is their multiscale integration into architectures with optimal transport functionality. Some efforts in this direction have, for example, succeeded in growing secondary carbon nanotubes on a substrate of multiwall carbon nanotubes (MWNT) in a branched pattern (Fig. 25) by ethylene decomposition on Ni nanoparticle catalysts on the original CNTs. Nanofiber/microfiber interconnected mesh networks made of cellulose acetate and other polymers have been developed by electrospinning at variable target distance and voltage settings through evaporation control of the solvent [96].

Flower-like architectures with random tree structures have been grown of various metal oxides (e.g., ZnO, SnO) and bimetallic oxides (e.g., silver molybdates; [103]), whereas branching alumina and titania nanotube ensembles have been developed by decomposition of anodized membranes (Fig. 26). Finally multiscale nano/microcrack networks in random fractal patterns have been achieved by controlled ultrasonic corrosion anodization of Ti and several other materials [104]. In these examples, the synthesis or fabrication of the nanoelements, such as CNTs, nanofibers, Al_2O_3 nanotubes, nanocrystals, and some with bifurcations and branching, was known by 2000, but their controlled integration into multiscale ensembles started in the past decade.

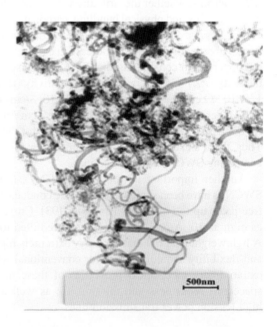

Fig. 25 Multiscale CNT network [95]

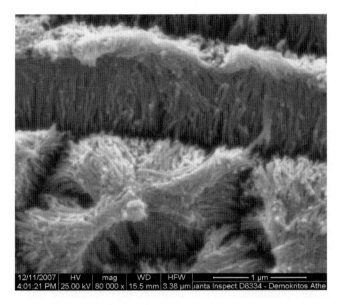

Fig. 26 Random branching alumina nanotubes [105]

Unlike their natural archetypes, such random fractal material networks with controlled stochastic features of the generator pattern pose several manufacturing challenges related to nanoscale and multi-scale integration. Top-down fabrication is primarily oriented toward deterministic, Euclidean designs rather than probabilistic patterns; production of whole branches (without assembly) is challenged by locally nondifferentiable nature of fractals; tool-part interference (visibility) constraints may arise in shaping recursive, often self-intersecting topologies; and layered manufacturing of internal vasculatures and pores may be limited by removal of overhang-supporting material.

Conversely, bottom-up synthesis of self-assembled structures yields morphologies inexorably determined by the physicochemical connectivity of their building blocks, therefore coupling geometric topology with material selection. In all cases, the multiscale complexity of self-similar forms challenges the workspace-resolution capabilities of any single process, additive or subtractive, or involves the tedium and cost of scaled lithographic masks and other patterned tools. Thus, to date, fabrication of branching structures is limited to nonscalable, ad hoc techniques, such as electrostatic discharge of a high-energy, beam-irradiated insulator (e.g., in a linear accelerator), producing fractal Lichtenberg patterns.

Innovation is needed in nanoscale manufacturing processes to address these limitations and bring forth the potential of random branching fractal materials to energy, health, and environmental applications. For the next decade, research needs in fractal nanomanufacturing will be oriented toward functional materials and vasculatures suitable for scalable industrial production. These include vascular scaffolds that mimic the internal structure of extracellular matrix; applications in scaffolds for tissue engineering on osseous materials and coatings; minimally invasive surgical patches; and soft tissues for artificial erythropoiesis for thalassemias and hemoglobinopathies. They are also tested as photoelectrodes in excitonic (dye-sensitized and hybrid polymer) solar cells; photocatalytic substrates for hydrogen production and oxidative removal of xenobiotics in wastewater treatment; and as water desalination membranes for reverse osmosis, electrodialysis, and capacitive deionization. Fractal nanomanufacturing is expected to follow a biomimetic approach, implementing processes inspired by natural patterning and integration (e.g., in lightning bolts, rivers, snowflakes, plant and animal tissues), optimizing transport of mass, energy, or information. The hypothesis of bioinspired fractal nanomanufacturing is that, as happens in nature, similar or related transformation and transport phenomena dominate and optimize both manufacture and operation processes of such structures. This promises to spur important contributions in experimental and computational science of the next decade bridging fractal mathematics with manufacturing process phenomena.

Fractal nanomanufacturing applications by 2020 are expected to cover areas such as internally vascularized scaffolds for tissue engineering; minimally invasive surgical patches; photoelectrodes in excitonic (dye-sensitized and hybrid polymer) solar cells; photocatalytic substrates for hydrogen production; reversible hydrogen storage and fuel cell membranes; nanostructured battery and supercapacitor materials; membranes for advanced oxidative removal of xenobiotics in wastewater treatment; and desalination membranes for reverse osmosis, electrodialysis, and capacitive deionization.

8.8 National Nanomanufacturing Network and Internano

Contact person: Mark Tuominen, University of Massachusetts, Amherst

The National Nanomanufacturing Network (NNN) is an alliance of academic, government, and industry partners that cooperate to advance nanomanufacturing strength in the United States. The goal of the NNN is to build a network of experts and organizations that facilitate and expedite the transition of nanotechnologies from core research and breakthroughs in the laboratory to production manufacturing. Partners and affiliates find value added through a range of services, including training and education, industrial vision and roadmap development, thematic conferences and workshops, contributions to standards development, and a comprehensive online information resource for nanomanufacturing, InterNano.

The NNN is funded and coordinated by the National Science Foundation Center for Hierarchical Manufacturing, a Nanoscale Science and Engineering Center (NSEC) at the University of Massachusetts Amherst. It works in close cooperation with the three other nanomanufacturing NSECs – the Center for High-Rate Manufacturing at Northeastern/University of Massachusetts Lowell/University of New Hampshire, the Center for Scalable and Integrated Nanomanufacturing at

UCLA/Berkeley, and the Center for Nanoscale Chemical–Electrical–Mechanical Manufacturing Systems at University of Illinois at Urbana-Champaign. The NNN also includes cooperative participation from the DOE Center for Integrated Nanotechnologies and the National Institute of Standards and Technology Center for Nanoscale Science and Technology and Precision Engineering Program and other institutions including Department of Defense, National Institutes of Health, National Institute for Occupational Safety and Health, and others.

The NNN strives to identify and help support critical components of the emerging nanomanufacturing value chain – each is an essential part of an overall nanomanufacturing system or enterprise (Fig. 27). The growth of new partnerships involving industry, government, academic, and nongovernmental organizations will advance the value chain by addressing current needs and challenges.

InterNano (http://www.internano.org), a service of the NNN, supports the information needs of the nanomanufacturing community by bringing together resources about the advances in applications, devices, metrology, and materials that will facilitate the commercial development or marketable application of nanotechnology. InterNano collects and provides access to information on nanomanufacturing centers, experts, and resources; proceedings of NNN thematic workshops and conferences; nanomanufacturing news, reviews, and events; and nanomanufacturing processes, test bed reviews, and research literature, with links to best practices. It is the central, online resource through which nanomanufacturing relevant information is archived and disseminated.

InterNano is actively building a suite of informatics tools for the nanomanufacturing community, beginning with three components: a smart company and institutional directory to find and establish connections with potential collaborators; a

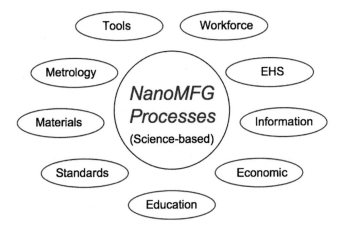

Fig. 27 Vital parts of a robust national nanomanufacturing enterprise

taxonomy to explore and discover nanomanufacturing content; and a nanomanufacturing process database to share standard nanomanufacturing processes and information on the properties of engineered nanomaterials across sectors. InterNano is also currently forging partnerships that will work toward the open exchange of data through federated databases, metadata standards, and the development of advanced analytical tools for evaluating trends within the nanomanufacturing community.

InterNano represents the crossroads where issues of lab-scale scientific research intersect with the industry-scale issues of environmental, health and safety (EHS), regulation, market research, and entrepreneurship. As such, the availability of consistent and reliable materials and process information, effective modeling and analysis tools, and trusted standards and protocols for information sharing with respect to intellectual property rights are all very relevant challenges that InterNano and the NNN are invested in addressing

9 International Perspectives from Site Visits Abroad

9.1 United States-European Union Workshop

Contact persons: Vasco Teixeira, University of Minho, Portugal; Mark Tuominen, University of Massachusetts, Amherst, United States

New techniques to generate, fabricate, and integrate novel materials architectures and devices have experienced substantial growth in the past 10 years. In addition, sophisticated analytical techniques (e.g., cryo-TEM) have emerged recently that enable new characterizations of nanostructures enabling time-resolved reactions that support their production.

In the past 10 years, researchers have started to understand the detailed science of production processes for nanomaterials and devices [97, 98]. Nanosynthesis tools highlighted include bioinspired and self-assembly processes. For example, nanoimprinting has emerged as method for nanofabrication with potentially broad impact. Scale-bridging multiscale modeling has experienced advances, but it still is not fully developed. Multiple nanolayering has made significant progress, including an improvement in the breadth of available materials for surface- and thin-films industries. Other thin-films technologies have advanced substantially, such as wet coating (as a lower cost thin-film production technique), vacuum coating, and roll-to-roll processing. Use for hierarchical and 3D nanofabrication is still under development. Functional thin-films are used today in many systems for tailored electronic, magnetic, optical, chemical, and thermal properties. Some nanotech-based products are already on the market (e.g., nanostructured coatings, cosmetics, textile industry, magnetic storage), but there is a strong potential for many more.

Concerning the use of nanomaterials and devices in nanomedicine, cost-effective lab-on-a-chip devices have been developed and commercially produced. In most cases, research on nanoparticles for medicine, diagnostics, and drug delivery has demonstrated principles of potential applications but has not reached widespread use.

Future nanomaterials, devices, and systems may lead to a new generation of high-added value and competitive products and services with superior performance across a range of applications. These new products will bring great impact and benefits throughout society and its activities, improving the quality of life of citizens. Breakthroughs in nanomaterials engineering depend on how precisely their nanoscale structure can be measured and controlled, achieving large-scale fabrication and "in-situ" measuring during processing. The short-term strategy for nanotechnology is to integrate nanotechnology with industry (including improving the materials of traditional industry) and to develop novel products with competitive quality and performance that would benefit consumers.

The expert group has discussed the synthesis, assembly, and processing of components, devices, and systems and concluded that there was a need to focus and strengthen the U.S. and European Union nanotechnology research effort in this area by building on past strengths and exploiting new opportunities. The main conclusions have been summarized as follows.

9.1.1 Vision for the Next 10 Years

- Cost-effective scale-up of nanofabrication methods
- Nanomaterials design and fabrication inspired by nature (e.g., biomimetics, tissue engineering)
- Adaptable, self-healing nanomaterial systems; increased and deeper knowledge of the self-assembling and self-adapting processes and mechanisms
- Encapsulation and active packaging for delivery of food, antimicrobial, fungicide, or nanoparticles for sensing (e.g., smart labels embedded in food packaging)
- Multifunctional nanoparticles (e.g., single photon sources for quantum information processing, devices that use spin and photons, photonics, nanodevices for drug delivery and biotracking)
- Carbon nanomaterials (e.g., nanotubes, graphene, nanodiamond, nonvisible emitters, carbon quantum dots)
- Higher precision of positional placement (e.g., quantum dots in cavities using a guiding field)
- *In situ* characterization for understanding and measuring improvements of processing (For industry to meet future requirements of properties, cost, and performance, it is necessary to understand the fundamental mechanisms of nanoscale processing by *in situ* analysis of reactions, structures, phase transformations, treatments, and 3D growth processes.)

9.1.2 Strategic Goals for the Next 5–10 Years

- Production of complex systems (3D nanosystems) (In the near future, complex (and multicomponent) 3D nanosystems will emerge that have a heterogeneous structure and different spatial properties.)
- Green nanotechnology as it applies to large-scale nanomaterials synthesis, low-cost, safer, and more environmentally friendly products or processes; replacement and recovery processes for limited raw materials (e.g., Indium; [62])
- Nanomedicine in the areas of diagnostics, drug delivery systems, and smart implants
- Organic-inorganic hybrid nanomaterials for flexible and transparent optoelectronics devices
- Low-cost and large-area printing and coating processes with nanoscale precision
- Novel material nanoarchitectures for high-efficiency energy conversion (e.g., substitution of electrodes in conventional solar cells for cost reduction and improved performance)
- Better understanding and control of nucleation processes that affect the quality of many materials
- Nanofabrication scale-up aiming at production of higher quality materials with lower cost (e.g., nanoimprint lithography), understanding the basics of production for different materials, and developing *in situ* large area characterization techniques for process improvements
- Safety of nanotechnology-based products in the human body and environment

9.1.3 Scientific and Technological Infrastructure in the Next 5–10 Years

- Enable a cross-cutting, world-class, competitive research and development infrastructure to support research in nanoscience and nanotechnology ensuring the best use of resources and take into account the needs of both research and industry organizations via combined centers for university and industry to share expensive equipment costs
- Provide fundamental research funding to studying phenomena and manipulation of matter at the nanoscale
- Train people in nanotechnology with the investment and creation of specific postgraduate courses on nanoscience and nanotechnology to promote researcher mobility and multidisciplinary research (Long-term positions for researchers are a concern for future strategic nanotechnology development. In addition to postgraduate programs in nanotechnology, the research and development policy should also move toward vocational training in nanotechnology for effective technician education in industry and research.)

9.1.4 Emerging Topics and Priorities for Future Nanoscale Science and Engineering Research and Education

- Energy and environment (e.g., development of novel material nanoarchitectures for high-efficiency energy conversion, water filtration and purification)
- Nanomedicine (e.g., nanodiagnostics, nanodelivery systems, smart bioimplants)
- Food industry (e.g., low-cost flexible sensors, bioactive packaging)
- Lab to pilot lines (more efficient transition into a nanomanufacturing pipeline)
- Integration of existing infrastructures, including providing a United States–European Union inventory of available facilities
- Postgraduate courses on nanoscience and nanotechnology
- Public acceptance of nanomanufacturing and generally nanotechnology is assumed to depend on confidence, which is created through information, education, openness, and debate.

9.2 United States-Japan-Korea-Taiwan Workshop

Contact persons: Tadashi Shibata, University of Tokyo, Japan; Chad Mirkin, Northwestern University, United States

The topics from this breakout session, with 25 people in attendance, were directly related to the bridging of basic discoveries and developments in nanoscience and technology to end-user products that allow society to benefit from nanotechnology. Five speakers presented their expert views from a variety of areas ranging from medicine, biology, chemistry, materials science, and nanofabrication technologies. The discussion focused on important nanotechnology breakthroughs during the last 10 years, important envisioned development in the next 10 years with specific goals for 2020.

9.2.1 Important Nanotechnology Breakthroughs During the Last 10 Years

- Development and study of two important nanotechnology materials: single-walled carbon nanotubes [99] and graphene [100] (These materials have high mechanical strength and excellent electronic properties, are lightweight and flexible, and can be scaled for large area production at low costs.)
- Establishment of synthetic methods for large volume production
- Enablement of large-area nanomanufacturing compatible with mass production by nanoimprinting and nanoinjection molding
- Combination of top-down and bottom-up approaches in block copolymer lithography (which is quite promising)
- Creation of supramolecular systems [101] for high customization and molecular structure organization

- Oral drug delivery of insulin through chitosan nanoparticles (which has had a great impact with *in vivo* experiments using a diabetic rat model)

9.2.2 Important Envisioned Nanotechnology Development in the Next 10 Years

On the basis of the achievements during the last 10 years, the direction for the nanotechnology development in the next 10 years is summarized in the following bullets:

- Memory applications, including resistive RAM and patterned media, are definitely important for nanopatterning because memory capacity could be dramatically enhanced.
- Carbon-based electronics based on single-walled nanotubes and graphene will allow for integrated circuits in main stream production beyond the Moore limits.
- Wafer-scale DPN technology would be a key tool for fabricating nanodevices and sensors. Enhancing the throughput by parallelizing writing probes is mandatory.
- Bottom-up self-assembly technologies merged with top-down processes is becoming increasingly important.
- Building single-crystal films at low temperatures will be important for building 3D very large scale integrated circuit (VSLI) chips.
- Bioinspired nanocatalytic systems mimicking biological enzymes are promising for "green" fuel production.

9.2.3 Goals for the Next 10 Years

- Ability to print over large areas with 1 nm resolution and the flexibility to work with many materials by merging scanning probe lithographies, nanoimprint, and block copolymer strategies
- Orally bioavailable, nanoparticle insulin delivery system with large-scale manufacturing of optimal nanostructure in good manufacturing practice manner
- Large-scale methods for making chemically well-defined graphene, nanotubes, nanoparticles, and nanowires of well-defined size and composition
- Nanomaterials that have well-defined composition and structure for optimum gas, ion, storage, and transport for improved electrodes
- High-throughput methods of screening compositions for a given photonic, electronic, or biomedical application
- Biocompatible implant devices that are nontoxic, noncorroding, and long lasting
- Development of contamination-free processes acceptable for integration with device fabrication lines
- Low-temperature and high-quality chemical vapor deposition to grow various single crystal thin films for integrated chips

9.3 United States-Australia-China-India-Singapore Workshop

Contact persons: Zhongfan Liu, Peking University, People's Republic of China; Chad Mirkin, Northwestern University, United States

The last decade has witnessed rapid development of nanotechnology. There has been a great stepping up of research emphasis from nanomaterials synthesis to purpose-oriented nanosystems. Many new "hot spots" have emerged, including 2D graphene, topological insulators, plasmonic metamaterials, and ZnO nanowire piezoelectronics. Synthesis and growth always play key roles for pushing emerging fields ahead, and numerous methods have been explored for achieving better control of shape, structure, and crystalline facets of nanomaterials. Great advances have also been achieved on scaling up and mass production of important nanomaterials, such as carbon nanotubes, luminescent semiconducting quantum dots, which have stimulated applications on reinforced composite materials, supercapacitors for energy storage, and medical labeling and nanodiagnosis. Bioinspired and biomimetic synthesis have gained wide attention for obtaining new nanostructures.

Several Challenges and Opportunities Exist for the Coming Decade:

- Transition from silicon to carbon-based devices
- 2D and 3D control of materials arrangement
- Synthesis, screening, and optimization of catalysts for their activity, selectivity, and environmental compatibility
- Synthesis and distribution of safe nanotechnology-based therapeutics
- Shift from fundamental nanoscience to large volume manufacturing of nanotechnology
- Integration of biomolecules with electronic interfaces
- Developing and understanding the structure–property relationships of plasmonic metamaterials and their applications as light harvesting systems

Several kinds of infrastructures are essential in which *in situ* or online characterization facilities would be of great help for synthesizing and manufacturing high-precision, high-purity nanomaterials and nanostructures. Broader availability of large-scale expensive fabrication and characterization facilities would determine how far researchers can go with nanostructured electronic devices. Extensive integration of cross-disciplinary research and activities are essential for technological breakthroughs and the transition of nanoscience to practical nanotechnology. Furthermore, the development of nanomanufacturing education curriculum should be an integral part of all such activities, with a strong emphasis on innovation and engineering principles.

References

1. A. Braunschweig, F. Huo, C. Mirkin, Molecular printing. Nat. Chem. **1**(5), 353–358 (2009)
2. President's Council of Advisors on Science and Technology (PCAST), *Report to the President and Congress on the Third Assessment of the National Nanotechnology Initiative,*

Assessment and Recommendations of the National Nanotechnology Advisory Panel (Office of Science and Technology Policy, Washington, DC, 2010)
3. J. Bang, U. Jeong, D.Y. Ryu, T.P. Russell, C.J. Hawker, Block copolymer nanolithography: translation of molecular level control to nanoscale patterns. Adv. Mater. **21**, 1–24 (2009)
4. F.S. Bates, G.H. Fredrickson, Block copolymer thermodynamics: theory and experiment. Annu. Rev. Phys. Chem. **41**, 525–557 (1990)
5. C.T. Black, K.W. Guarini, K.R. Milkove, S.M. Baker, M.T. Tuominen, T.P. Russell, Integration of self-assembled diblock copolymers for semiconductor capacitor fabrication. Appl. Phys. Lett. **79**, 409 (2001)
6. K.W. Guarini, C.T. Black, Y. Zhang, H. Kim, E.M. Sikorski, I.V. Babich, Process Integration of self-assembled polymer templates into silicon nanofabrication. J Vac. Sci. Technol. B **20**, 2788 (2002)
7. M. Park, C. Harrrison, P.M. Chaikin, R.A. Register, D.H. Adamson, Block copolymer lithography: periodic arrays of 1011 holes in 1 square centimeter. Science **276**, 1401 (1997)
8. T. Thurn-Albrecht, J. Schotter, G.A. Kaestle, N. Emley, T. Shibauchi, L. Krusin-Elbaum, K. Guarini, C.T. Black, M.T. Tuominen, T.P. Russell, Ultrahigh-density nanowire arrays grown in self-assembled diblock copolymer templates. Science **290**, 2126 (2000)
9. C.T. Black, R. Ruiz, G. Breyta, J.Y. Cheng, M.E. Colburn, K.W. Guarini, H.C. Kim, Y. Zhang, Polymer self assembly in semiconductor microelectronics. IBM J. Res. Dev. **51**, 605 (2007)
10. R.A. Pai, R. Humayun, M.T. Schulberg, A. Sengupta, J.N. Sun, J.J. Watkins, Mesoporous silicates prepared using preorganized templates in supercritical fluids. Science **303**, 507 (2004)
11. N. Sivakumar, M. Li, R.A. Pai, J.K. Bosworth, P. Busch, D.M. Smilgies, C.K. Ober, T.P. Russell, J.J. Watkins, An efficient route to mesoporous silica films with perpendicular nanochannels. Adv. Mater. **20**, 246 (2008)
12. S.C. Warren, F.J. Disalvo, U. Wiesner, Nanoparticle-tuned assembly and disassembly of mesostructured silica. Nat. Mater. **6**, 156 (2007)
13. S.C. Warren, F.J. Disalvo, U. Wiesner, Erratum: nanoparticle-tuned assembly and disassembly of mesostructured silica hybrid. Nat. Mater. **6**, 248 (2007)
14. I. Bita, J.K.W. Yang, Y.S. Jung, C.A. Ross, E.L. Thomas, K.K. Berggren, Graphoepitaxy of self-assembled block copolymers on 2D periodic patterned templates. Science **321**, 939 (2008)
15. C.T. Black, O. Bezencenet, Nanometer-scale pattern registration and alignment by directed diblock copolymer self-assembly. IEEE Trans. Nanotechnol. **3**, 412–415 (2004)
16. J.Y. Cheng, A.M. Mayes, C.A. Ross, Nanostructure engineering by templated self-assembly of block copolymers. Nat. Mater. **3**, 823–828 (2004)
17. J.Y. Cheng, C.T. Rettner, D.P. Sanders, H.C. Kim, W.D. Hinsberg, Dense self-assembly on sparse chemical patterns: rectifying and multiplying lithographic patterns using block copolymers. Adv. Mater. **20**, 3155–3158 (2008)
18. S. Park, B. Kim, O. Yavuzcetin, M.T. Tuominen, T.P. Russell, Ordering of PS-b-P4VP on patterned silicon surfaces. ACS Nano **2**, 1363 (2008)
19. S.Y. Park, A.K.R. Lytton-Jean, B. Lee, S. Weigand, G.C. Schatz, C.A. Mirkin, DNA-programmable nanoparticle crystallization. Nature **451**(7178), 553–556 (2008)
20. R. Ruiz, H.M. Kang, F.A. Detcheverry, E. Dobisz, D.S. Kercher, T.R. Albrecht, J.J. de Pablo, P.F. Nealey, Density multiplication and improved lithography by directed block copolymer assembly. Science **321**, 936 (2008)
21. R.A. Segalman, H. Yokoyama, E.J. Kramer, Graphoepitaxy of spherical domain block copolymer films. Adv. Mater. **13**, 1152–1155 (2001)
22. S. Park, D.H. Lee, J. Xu, B. Kim, S.W. Hong, U. Jeong, T. Xu, T.P. Russell, Macroscopic 10-terabit–per–square-inch arrays from block copolymers with lateral order. Science **323**, 1030 (2009)
23. J. Chai, J.M. Buriak, Using cylindrical domains of block copolymers to self-assemble and align metallic nanowires. ACS Nano **2**, 489 (2008)

24. Y.S. Jung, J.B. Chang, E. Verploegen, K.K. Berggren, C.A. Ross, A path to ultranarrow patterns using self-assembled lithography. Nano Lett. **10**, 1000 (2010)
25. S.M. Park, G.S.W. Craig, Y.H. La, H.H. Solak, P.F. Nealey, Square arrays of vertical cylinders of PS-b-PMMA on chemically nanopatterned surfaces. Macromolecules **40**, 5084–5094 (2007)
26. C.B. Tang, E.M. Lennon, G.H. Fredrickson, E.J. Kramer, C.J. Hawker, Evolution of block copolymer lithography to highly ordered square arrays. Science **322**, 429–432 (2008)
27. G.M. Wilmes, D.A. Durkee, N.P. Balsara, J.A. Liddle, Bending soft block copolymer nanostructures by lithographically directed assembly. Macromolecules **39**, 2435–2437 (2006)
28. K. Galatsis, K.L. Wang, M. Ozkan, C.S. Ozkan, Y. Huang, J.P. Chang, H.G. Monbouquette, Y. Chen, P. Nealey, Y. Botros, Patterning and templating for nanoelectronics. Adv. Mater. **22**, 769–778 (2010)
29. J.K.W. Yang, Y.S. Jung, J.-B. Chang, R.A. Mickiewicz, A. Alexander-Katz, C.A. Ross, K.K. Berggren, Complex self-assembled patterns using sparse commensurate templates with locally varying motifs. Nat. Nanotechnol. **5**, 256 (2010)
30. A. Braunschweig, A. Senesi, C. Mirkin, Redox-activating dip-pen nanolithography (RA-DPN). J. Am. Chem. Soc. **131**(3), 922–923 (2009)
31. R. Piner, J. Zhu, F. Xu, S. Hong, C.A. Mirkin, "Dip-pen" nanolithography. Science **283**(5402), 661–663 (1999)
32. K. Salaita, Y. Wang, C.A. Mirkin, Applications of dip-pen nanolithography. Nat. Nanotechnol. **2**(3), 145–155 (2007)
33. L. Giam, Y. Wang, C. Mirkin, Nanoscale molecular transport: the case of dip-pen nanolithography. J. Phys. Chem. A **113**, 3779–3782 (2009)
34. S. Rozhok, R. Piner, C.A. Mirkin, Dip-pen nanolithography: what controls ink transport? J. Phys. Chem. B **107**(3), 751–757 (2003)
35. R. Jae-Won Jang, R.G. Sanedrin, A.J. Senesi, Z. Zheng, X. Chen, S. Hwang, L. Huang, C.A. Mirkin, Generation of metal photomasks by dip-pen nanolithography. Small **5**(16), 1850–1853 (2009)
36. K.-B. Lee, E.-Y. Kim, C.A. Mirkin, S.M. Wolinsky, The use of nanoarrays for highly sensitive and selective detection of human immunodeficiency virus type 1 in plasma. Nano Lett. **4**(10), 1869–1872 (2004)
37. S. Minne, S.R. Manalis, A. Atalar, C.F. Quate, Independent parallel lithography using the atomic force microscope. J. Vac. Sci. Technol. B **14**(4), 2456–2461 (1996)
38. K. Salaita, S.W. Lee, X. Wang, L. Huang, T.M. Dellinger, C. Liu, C.A. Mirkin, Sub-100 nm, centimeter-scale, parallel dip-pen nanolithography. Small **1**(10), 940–945 (2005)
39. K. Salaita, P. Sun, Y. Wang, H. Fuchs, C.A. Mirkin, Massively parallel dip-pen nanolithography with 55000-pen two-dimensional arrays. Angew. Chem. Int. Ed Engl. **45**(43), 7220–7223 (2006). doi:10.1002/anie.200603142
40. D. Banerjee, A. Nabil, S. Disawal, J. Fragala, Optimizing microfluidic ink delivery for dip pen nanolithography. J. Microlith. Microfab. Microsys. **4**(2), 023014 (2005). doi:10.1117/1.1898245
41. F. Huo, Z. Zheng, G. Zheng, L.R. Giam, H. Zhang, C.A. Mirkin, Polymer pen lithography. Science **321**(5896), 1658–1660 (2008). doi:10.1126/science.1162193
42. Z. Zheng, W.L. Daniel, L.R. Giam, F. Huo, A.J. Senesi, G. Zheng, C.A. Mirkin, Multiplexed protein arrays enabled by polymer pen lithography: addressing the inking challenge. Angew. Chem. Int. Ed Engl. **48**(41), 7626–7629 (2009). doi:10.1002/anie.200902649
43. F. Huo, G. Zheng, X. Liao, L.R. Giam, J. Chai, X. Chen, W. Shim, C.A. Mirkin, Beam pen lithography. Nat. Nanotechnol. **5**, 637–640 (2010). doi:10.1038/nnano.2010.161
44. G.A. Ozin, A.C. Arsenault, *Nanochemistry: A Chemical Approach to Nanomaterials* (RSC Publishing, Cambridge, 2005)
45. L.D. Qin, S. Park, L. Huang, C.A. Mirkin, On-wire lithography. Science **309**, 113–115 (2005)
46. L.D. Qin, S. Zou, C. Xue, A. Atkinson, G.C. Schatz, C.A. Mirkin, Designing, fabricating, and imaging Raman hot spots. Proc. Natl. Acad. Sci. U.S.A. **103**, 13300–13303 (2006). doi:10.1073/pnas.0605889103

47. L.D. Qin, J.W. Jang, L. Huang, C.A. Mirkin, Sub-5-nm gaps prepared by on-wire lithography: correlating gap size with electrical transport. Small **3**, 86–90 (2007)
48. L.D. Qin, M.J. Banholzer, J.E. Millstone, C.A. Mirkin, Nanodisk codes. Nano Lett. **7**, 3849–3853 (2007)
49. A. Nitzan, M.A. Ratner, Electron transport in molecular wire junctions. Science **300**, 1384–1389 (2003)
50. X. Chen, Y.-M. Jeon, J.-W. Jang, L. Qin, F. Huo, W. Wei, C.A. Mirkin, On-wire lithography-generated molecule-based transport junctions: a new testbed for molecular electronics. J. Am. Chem. Soc. **130**(26), 8166–8168 (2008). doi:10.1021/ja800338w
51. Z. Nie, A. Petukhova, E. Kumacheva, Properties and emerging applications of self-assembled structures made from inorganic nanoparticles. Nat. Nanotechnol. **5**(1), 15–25 (2010)
52. D.V. Talapin, J.-S. Lee, M.V. Kovalenko, E.V. Shevchenko, Prospects of colloidal nanocrystals for electronic and optoelectronic applications. Chem. Rev. **110**(1), 389–458 (2009). doi:10.1021/cr900137k
53. M.-H. Lin, H.-Y. Chen, S. Gwo, Layer-by-layer assembly of three-dimensional colloidal supercrystals with tunable plasmonic properties. J. Am. Chem. Soc. **132**(32), 11259–11263 (2010)
54. E.V. Shevchenko, D.V. Talapin, N.A. Kotov, S. O'Brien, C.B. Murray, Structural diversity in binary nanoparticle superlattices. Nature **439**(7072), 55–59 (2006). http://www.nature.com/nature/journal/v439/n7072/abs/nature04414.html - a1
55. H.D. Hill, R.J. Macfarlane, A.J. Senesi, B. Lee, S.Y. Park, C.A. Mirkin, Controlling the lattice parameters of gold nanoparticle FCC crystals with duplex DNA linkers. Nano Lett. **8**(8), 2341–2344 (2008). doi:10.1021/nl8011787
56. R.J. Macfarlane, B. Lee, H.D. Hill, A.J. Senesi, S. Seifert, C.A. Mirkin, Assembly and organization processes in DNA-directed colloidal crystallization. Proc. Natl. Acad. Sci. U.S.A. **106**(26), 10493–10498 (2009)
57. R. Macfarlane, M.R. Jones, A.J. Senesi, K.L. Young, B. Lee, J. Wu, C.A. Mirkin, Establishing the design rules for DNA-mediated programmable colloidal crystallization. Angew. Chem. Int. Ed Engl. **49**(27), 4589–4592 (2010)
58. M. Jones, R.J. Macfarlane, B. Lee, J. Zhang, K.L. Young, A.J. Senesi, C.A. Mirkin, DNA-nanoparticle superlattices formed from anisotropic building blocks. Nat. Mater. **9**, 913–917 (2010). doi:10.1038/nmat2870
59. K.S. Novoselov, A.K. Geim, S.V. Morozov, D. Jiang, Y. Zhang, S.V. Dubonos, I.V. Grigorieva, A.A. Firsov, Electric field effect in atomically thin carbon films. Science **306**, 666–669 (2004)
60. A.K. Geim, K.S. Novoselov, The rise of graphene. Nat. Mater. **6**, 183–191 (2007)
61. X. Li, W. Cai, J. An, S. Kim, J. Nah, D. Yang, R. Piner, A. Velamakanni, I. Jung, E. Tutuc, S.K. Banerjee, L. Colombo, R.S. Ruoff, Large-area synthesis of high-quality and uniform graphene films on copper foils. Science **324**, 1312–1314 (2009)
62. S. Bae, H. Kim, Y. Lee, X. Xu, J.-S. Park, Y. Zheng, J. Balakrishnan, T. Lei, H.R. Kim, Y.I. Song, Y.-J. Kim, K.S. Kim, B. Özyilmaz, J.-H. Ahn, B.H. Hong, S. Iijima, Roll-to-roll production of 30-inch graphene films for transparent electrodes. Nat. Nanotechnol. **5**(8), 574–578 (2010)
63. J.N. Randall, J.B. Ballard, J.W. Lyding, S. Schmucker, J.R. Von Her, R. Saini, H. Xu, Y. Ding, Atomic precision patterning on Si: an opportunity for a digitized process. Microelectron. Eng. **87**(5–8), 955–958 (2010)
64. J.N. Randall, J.W. Lyding, S. Schmucker, J.R. Von Ehr, J. Ballard, R. Saini, H. Xu, Y. Ding, Atomic precision lithography on Si. J. Vac. Sci. Technol. B **27**(6), 2764 (2009). doi:10.1116/1.3237096
65. Y. Suda, N. Hosoya, K. Miki, Si submonolayer and monolayer digital growth operation techniques using Si_2H_6 as atomically controlled growth nanotechnology. Appl. Surf. Sci. **216**(1–4), 424–430 (2003)
66. D.A. Tomalia, J.M.J. Fréchet (eds.), *Dendrimers and Other Dendritic Polymers* (Wiley, Chichester, 2001)
67. D.A. Tomalia, Birth of a new macromolecular architecture: dendrimers as quantized building blocks for nanoscale synthetic polymer chemistry. Prog. Polym. Sci. **30**(3–4), 294–324 (2004)

68. M. Peterca, V. Percec, M.R. Imam, P. Leowanawat, K. Morimitsu, P.A. Heiney, Molecular structure of helical supramolecular dendrimers. J. Am. Chem. Soc. **130**(44), 14840–14852 (2008)
69. B.M. Rosen, D.A. Wilson, C.J. Wilson, M. Peterca, B.C. Won, C. Huang, L.R. Lipski, X. Zeng, G. Ungar, P.A. Heiney, V. Percec, Predicting the structure of supramolecular dendrimers via the analysis of libraries of AB3 and constitutional isomeric AB2 biphenylpropyl ether self-assembling dendrons. J. Am. Chem. Soc. **131**(47), 17500–17521 (2009). doi:10.1021/ja806524m
70. D.A. Tomalia, In quest of a systematic framework for unifying and defining nanoscience. J. Nanopart. Res. **11**(6), 1251–1310 (2009)
71. D.A. Tomalia, Dendrons/dendrimers: quantized, nano-element like building blocks for soft-soft and soft-hard nano-compound synthesis. Soft Matter **6**(3), 456–474 (2010)
72. V. Marx, Poised to branch out. Nat. Biotechnol. **26**(7), 729–732 (2008)
73. A.R. Menjoge, R.M. Kannan, D.A. Tomalia, Dendrimer-based drug and imaging conjugates: design considerations for nanomedical applications. Drug Discov. Today **15**(5–6), 171–185 (2010)
74. C.C. Lee, J.A. MacKay, J.M.J. Fréchet, F.C. Szoka, Designing dendrimers for biological applications. Nat. Biotechnol. **23**(12), 1517–1526 (2005). doi:10.1038/nbt1171
75. V. Percec, D.A. Wilson, P. Leowanawat, C.J. Wilson, A. Hughes, M.S. Kaucher, D.A. Hammer, D.H. Levine, A.J. Kim, F.S. Bates, K.P. Davis, T.P. Lodge, M.L. Klein, R.H. DeVane, E. Aqad, B.M. Rosen, A.O. Argintaru, M.J. Sienkowska, K. Rissanen, S. Nummelin, J. Ropponen, Self-assembly of Janus dendrimers into uniform dendrimersomes and other complex architectures. Science **328**(5981), 1009–1014 (2010). doi:10.1126/science.1185547
76. F. Beguin, E. Frackowiak (eds.), Carbide-derived carbon and templated carbons, in *Carbons for electrochemical energy storage and conversion systems,* ed. by T. Kyotani, J. Chmiola, Y. Gogotsi (CRC Press/Taylor and Francis, Boca Raton, 2009), pp. 77–114
77. C. Largeot, C. Portet, J. Chmiola, P.-L. Taberna, Y. Gogotsi, P. Simon, Relation between the ion size and pore size for an electric double-layer capacitor. J. Am. Chem. Soc. **130**(9), 2730–2731 (2008). doi:10.1021/ja7106178
78. S. Yachamaneni, G. Yushin, S.H. Yeon, Y. Gogotsi, C. Howell, S. Sandeman, G. Phillips, S. Mikhalovsky, Mesoporous carbide-derived carbon for cytokine removal from blood plasma. Biomaterials **31**(18), 4789–4794 (2010)
79. M. Kruk, C.M. Hui, Thermally induced transition between open and closed spherical pores in ordered mesoporous silicas. J. Am. Chem. Soc. **130**(5), 1528–1529 (2008)
80. V. Soghomonian, J.J. Heremans, Characterization of electrical conductivity in a zeolite like material. Appl. Phys. Lett. **95**(15), 152112 (2009)
81. K.J. Choi, M. Biegalski, Y.L. Li, A. Sharan, J. Schubert, R. Uecker, P. Reiche, Y.B. Chen, X.Q. Pan, V. Gopalan, L.-Q. Chen, D.G. Schlom, C.B. Eom, Enhancement of ferroelectricity in strained $BaTiO_3$ thin films. Science **306**(5698), 1005–1009 (2004). doi:10.1126/science.1103218
82. J.H. Haeni, P. Irvin, W. Chang, R. Uecker, P. Reiche, Y.L. Li, S. Choudhury, W. Tian, M.E. Hawley, B. Craigo, A.K. Tagantsev, X.Q. Pan, S.K. Streiffer, L.Q. Chen, S.W. Kirchoefer, J. Levy, D.G. Schlom, Room-temperature ferroelectricity in strained $SrTiO_3$. Nature **430**(7001), 758–761 (2004). doi:10.1038/nature02773
83. D.G. Schlom, L.-Q. Chen, C.-B. Eom, K.M. Rabe, S.K. Streiffer, J.-M. Triscone, Strain tuning of ferroelectric thin films. Annu. Rev. Mater. Res. **37**(1), 589–626 (2007)
84. M.P. Warusawithana, C. Cen, C.R. Sleasman, J.C. Woicik, Y. Li, L.F. Kourkoutis, J.A. Klug, H. Li, P. Ryan, L.-P. Wang, M. Bedzyk, D.A. Muller, L.-Q. Chen, J. Levy, D.G. Schlom, A ferroelectric oxide made directly on silicon. Science **324**(5925), 367–370 (2009)
85. J.H. Lee, L. Fang, E. Vlahos, X. Ke, Y.W. Jung, L.F. Kourkoutis, J.-W. Kim, P.J. Ryan, T. Heeg, M. Roeckerath, V. Goian, M. Bernhagen, R. Uecker, P.C. Hammel, K.M. Rabe, S. Kamba, J. Schubert, J.W. Freeland, D.A. Muller, C.J. Fennie, P. Schiffer, V. Gopalan, E. Johnston-Halperin, D.G. Schlom, A strong ferroelectric ferromagnet created by means of spin-lattice coupling. Nature **466**(7309), 954–958 (2010)
86. J. Mannhart, D.G. Schlom, Oxide interfaces – an opportunity for electronics. Science **327**(5973), 1607–1611 (2010)

87. H. Chen, Y.A. Elabd, G.R. Palmese, Plasma-aided template synthesis of inorganic nanotubes and nanorods. J. Mater. Chem. **17**(16), 1593–1596 (2007)
88. A. Rubio, J.L. Corkill, M.L. Cohen, Theory of graphitic boron nitride nanotubes. Phys. Rev. **B 49**(7), 5081 (1994)
89. A. Zettl, Non-carbon nanotubes. Adv. Mater. **8**(5), 443–445 (1996)
90. X. Blase, A. Rubio, S.G. Louie, M.L. Cohen, Stability and band gap constancy of boron nitride nanotubes. Europhys. Lett. **28**(5), 335 (1994). doi:10.1209/0295-5075/28/5/007
91. L. Ci, L. Song, C. Jin, D. Jariwala, D. Wu, Y. Li, A. Srivastava, Z.F. Wang, K. Storr, L. Balicas, F. Liu, P.M. Ajayan, Atomic layers of hybridized boron nitride and graphene domains. Nat. Mater. **9**(5), 430–435 (2010). doi:10.1038/nmat2711
92. D. Golberg, Y. Bando, C.C. Tang, C.Y. Zhi, Boron nitride nanotubes. Adv. Mater. **19**(18), 2413–2432 (2007)
93. E. Brown, L. Hao, J.C. Gallop, J.C. Macfarlane, Ballistic thermal and electrical conductance measurements on individual multiwall carbon nanotubes. Appl. Phys. Lett. **87**(2), 023107 (2005)
94. P.G. Collins, A. Phaedon, Nanotubes for electronics. Sci. Am. **283**(6), 62–69 (2000)
95. P.G. Savva, K. Polychronopoulou, R.S. Ryzkov, A.M. Efstathiou, Low temperature catalytic decomposition of ethylene into H2 and secondary carbon nanotubes over Ni/CNTs. Appl. Catal. **B 93**(3–4), 314 (2010)
96. T. Christoforou, C. Doumanidis, Biodegradable cellulose acetate nanofiber fabrication via electrospinning. J. Nanosci. Nanotechnol. **10**(9), 1–8 (2010)
97. European Commission, Toward a European strategy for nanotechnology (Office for Official Publications of the European Communities, Luxembourg, 2004). Available online: http://ec.europa.eu/nanotechnology/pdf/nano_com_en_new.pdf
98. National Science and Technology Council, Committee on Technology, Subcommittee on Nanoscale Science, Engineering, and Technology, *National Nanotechnology Initiative: Research and Development Supporting the Next Industrial Revolution* (National Nanotechnology Initiative, Washington, DC, 2003). Available online: www.nano.gov/html/res/fy04-pdf/fy04%20.../NNI-FY04_front_matter.pdf
99. S. Iijima, Helical microtubules of graphitic carbon. Nature **354**(6348), 56–58 (1991)
100. F. Schwierz, Graphene transistors. Nat. Nanotechnol. **5**(7), 487–496 (2010)
101. W.-S. Li, T. Aida, Dendrimer porphyrins and phthalocyanines. Chem. Rev. **109**(11), 6047–6076 (2009)
102. Y. Wang, L.R. Giam, M. Park, S. Lenhert, H. Fuchs, C.A. Mirkin, A self-correcting inking strategy for cantilever arrays addressed by an inkjet printer and used for dip-pen nanolithography. Small **4**(10), 1666–1670 (2008). doi:10.1002/smll.200800770
103. D. Stone, J. Liu, D.P. Singh, C. Muratore, A.A. Voevodin, S. Mishra, C. Rebholz, Q. Ge, S.M. Aouadi, "Layered atomic structures of double oxides for low shear strength at high temperatures." Scripta Materialia **62**(10), 735–738 (2010)
104. C.C. Doumanidis, "Nanomanufacturing of random branching material architectures." Microelectronic Engineering **86**(4-6): 467–478
105. M. Kokonou, C. Rebholz, K.P. Giannakopoulos, C.C. Doumanidis, Low aspect ratio porous alumina templates, Microelectron. Eng. **85**(2008) 1186

Nanotechnology Environmental, Health, and Safety Issues

André Nel, David Grainger, Pedro J. Alvarez, Santokh Badesha,
Vincent Castranova, Mauro Ferrari, Hilary Godwin, Piotr Grodzinski,
Jeff Morris, Nora Savage, Norman Scott, and Mark Wiesner

Keywords Nano EHS • Nanomaterial properties • Hazard, risk reduction • *In-vitro, in-vivo* • Predictive methods safe-by-design approach • Role of industry • International perspective

1 Vision for the Next Decade

The environmental, health, and safety (EHS) of nanomaterials has been defined as "the collection of fields associated with the terms 'environmental health, human health, animal health, and safety' when used in the context of risk assessment and

---

A. Nel (✉)
Department of Medicine and California NanoSystems Institute, University of California, 10833
Le Conte Avenue, 52–175 CHS, Los Angeles, CA 90095, USA
e-mail: anel@mednet.ucla.edu

D. Grainger
Department of Pharmaceutics and Pharmaceutical Chemistry, University of Utah,
30 South 2000 East, Room 301, Salt Lake City, UT 84112–5820, USA

P.J. Alvarez
Department of Civil and Environmental Engineering, Rice University, Houston, TX
77251–1892, USA

S. Badesha
Xerox Corporation, PO Box 1000, Mail Stop 7060–583, Wilsonville, OR 97070, USA

V. Castranova
Centers for Disease Control, National Institute for Occupational Safety and Health, Health
Effects Laboratory Division, 1095 Willowdale Road, Morgantown, WV 26505–2888, USA

M. Ferrari
The University of Texas Health Science Center, 1825 Pressler Street, Suite 537D Houston, TX
77031, USA

H. Godwin
Pubic Health–Environmental Health Science, University of California, 951772, Los Angeles,
CA, 90095, USA

risk management" ([1], p. 2). In this chapter, the term "nano-EHS" is used for convenience to refer specifically to environmental, health, and safety research and related activities as they apply to nanoscale science, technology, and engineering. This chapter outlines the major advances in nano EHS over the last 10 years and the major challenges, developments, and achievements that we can expect over the next 10 years without providing comprehensive coverage or a review of all the important issues in this field.

1.1 Changes in the Vision over the Last 10 Years

Although exposure to engineered nanomaterials (ENMs) in the workplace, laboratory, home, and the environment is likely more widespread than previously perceived, no specific human disease or verifiable environmental mishap has been ascribed to these materials to date. Perceptions of ENM hazard have evolved from "small is dangerous" to a more realistic understanding that ENM safety should best be considered in terms of the specific-use contexts, applications, exposures, and the specific properties of each nanomaterial.

Because organic, inorganic, and hybrid materials can be produced in various sizes, shapes, surface areas, surface functionalities, and compositions, and because of their widely tunable compositions and structures that can be dynamically modified under different biological and environmental use conditions, most ENMs cannot be described as a uniform molecular, chemical, or materials species. One major conceptual advance in nano-EHS assessment has been the recognition that these dynamic material properties play a determination role in ENM conditioning, dissemination, exposure, and hazard generation at the nano-bio

---

P. Grodzinski
Office of Technology and Industrial Relations, National Cancer Institute, Building 31, Room 10, A49 31 Center Drive, Mail Stop 2580, Bethesda, MD 20892–2580, USA

J. Morris
Ronald Reagan Building and International Trade Center, U.S. Environmental Protection Agency, Room 71184 1300 Pennsylvania Avenue NW, Washington, DC 20004, USA

N. Savage
U.S. Environmental Protection Agency Office of Research and Development, National Center for Environmental Research, 1200 Pennsylvania Avenue NW, Mail Stop 8722F, Washington, DC 20460, USA

N. Scott
Biological and Chemical Engineering, Cornell University, 216 Riley Hall, Ithaca, NY 14853–5701, USA

M. Wiesner
Department of Civil and Environmental Engineering, Duke University, 90287, 120 Hudson Hall, Durham, NC 27708–0287, USA

interface¹ [2–6]. Thus, it has become clear that since a large number of novel materials and material properties are continuously being introduced, it is imperative to develop a robust scientific platform to understand the relationship of these properties to EHS outcomes [3, 7–9]. Because this knowledge generation will require time and consensus building, rational decision-making in nano-EHS is likely to be incremental. However, this process could be accelerated by implementation of high-throughput and rapid ENM screening platforms [10–13], as well as exploiting computational methods to assist in risk modeling and hazard assessment.

We have come to recognize that, because of the diverse and unique properties of engineered nanomaterials, safe implementation of nanotechnology is a multidisciplinary exercise that goes beyond traditional hazard, exposure, and risk assessment models. In addition to properties research, the nano-EHS community requires information about the commercial uses of ENMs, their fate and transport, bioaccumulation, and lifecycle analysis, all of which demand careful coordination and incremental and adaptive decision making to guide safe implementation of nanotechnology. The need for data and information collection is now understood to be essential for researchers, producers, consumers, and regulators of ENM products to allow the formulation of adequate regulatory policy for engineered nanomaterials.

The National Science Foundation has established a research program solicitation with a focus on nanoscale processes in the environment beginning with August 2000. The Environmental Protection Agency has a research program solicitation on nanotechnology EHS since 2003, and the National Institute for Environmental Health Sciences in 2004.

1.2 Vision for the Next 10 Years

Due to the rapid pace at which nanotechnology is expanding into society via its many applications, as well as to the likelihood that significant human, animal, and ecosystem exposures are already occurring [9, 14–16], it is necessary to develop an integrated, validated scientific platform for assessment of hazards, exposures, and risks at a scale commensurate with the growth of this technology. Instead of performing the nano-EHS exercise one material at a time, rapid-throughput and high-content screening platforms will emerge to survey large batches of nano-phased materials in parallel [10–13].

Thus, the vision for the next 10 years includes the discovery and development of ENM property–activity relationships, high-volume data sets, and computational methods used to establish knowledge domains, risk modeling, and nano-informatics

---

¹ The nano-bio interface is defined here as the dynamic physicochemical interactions, kinetics, and thermodynamic exchanges between nanomaterial surfaces and the surfaces of biological components such as proteins, membranes, phospholipids, endocytic vesicles, organelles, DNA, and biological fluids.

capabilities to reliably assist decision-making. This information needs to be integrated into predictive science [8, 9, 17, 18] and risk management platforms that relate specific materials and ENM properties to hazard, fate and transport, exposure, and disease outcomes. Ensuring safe implementation of nanotechnology over the next decade also requires the development of new, sensitive analytical methodologies, tools, and accepted protocols for screening, detection, characterization, and monitoring of ENM exposure in the workplace, laboratory, home, and the environment [19, 20]. We also need to develop effective monitoring, containment, and nanomaterial removal methods for waste disposal systems. New data and knowledge gathering will lead to the design of safer materials and green manufacturing that could transform nanotechnology into a cornerstone of sustainability [19]. Safe implementation of nanotechnology requires close cooperation between academia, industry, government, and the public, all of whom have a stake in seeing this technology succeed for the benefit of society, the economy, and the environment.

2 Advances in the Last 10 Years and Current Status

Ten years ago nanotechnology was recognized for its enormous potential to produce revolutionary advances in electronics, low-cost solar cells, next-generation energy storage, and smart anti-cancer therapeutics, among other fields of application. The first collective efforts in nano-EHS awareness commenced early after the National Nanotechnology Initiative (NNI) was established in 2000, including organization of several workshops that addressed the environment, nanobiotechnology, and societal implications [21]; nevertheless, it required considerable time to comprehend and integrate all the scientific disciplines that are necessary to understand the possible impact of this disruptive new technology on humans and the environment. Some of the early steps in the awareness/integration process were the following:

- In 2003, the National Toxicology Program considered first tests on nanoparticles, nanotubes, and quantum dots, and Environmental Protection Agency (EPA) has the first program announcement on nano-EHS.
- In December 2004, the Nanoscale Science, Engineering, and Technology (NSET) Subcommittee of the Committee on Technology of the National Science and Technology Council published the NNI Strategic Plan for the Federal fiscal years (FY) 2006–2010 in which environmental science and technology were well represented.
- Several coordinated academic centers emerged early in the decade that began to focus on nano-EHS, such as the Center for Biological and Environmental Nanotechnology (CBEN) at Rice University and the University of California Nanotoxicology Research Training Program.

The pace of research and implementation of nano-EHS regulatory policy began to speed up by 2008, at which point the number of peer-reviewed publications addressing nano-EHS risk assessment increased rapidly, amounting to >250 papers in 2009 as compared to ~50 in 2004. Concerns about ENM safety also led to a steady increase in the number of regulatory interventions by Federal agencies, as well as an increase in the U.S. Federal budget for nano-EHS research from $67.9 million in FY 2008 to a requested amount of $116.9 million in FY 2011. (Budget considerations will be covered in Sect. 3.)

From an EHS standpoint, researchers have made some progress in developing toxicological screening for the most abundant ENMs in their primary form, and new data have emerged on the importance of several material properties that may pose a hazard at the nanoscale level [5, 9, 20, 22]. This has elicited new concerns about possible hazard, fate and transport, exposure, and bioaccumulation. The significant challenge now is the standardization, harmonization, and implementation of nano-EHS monitoring and screening, data collection, streamlined risk reduction procedures, and a coordinated governance strategy to ensure safe implementation of this technology. The imminent introduction of active nanosystems and nano-engineered devices, including integrated assemblies of multiple different nanomaterials that perform more complex functions than those of individual materials, will necessitate the development of additional nano-EHS procedures for composite materials.

2.1 Data Gathering, Monitoring, and Governance of Carbon Nanotubes

Carbon nanotubes (CNTs) are one example of an important industrial class of ENMs for which considerable nano-EHS data collection is now available [23–29]. CNT inhalation exposure in the workplace is a potential concern, as a result of the widespread use of CNTs in manufacturing, their high volume of production, and ready aerosolization by activities such as packaging, dispensing, vortexing, acting, grinding, and vessel transfer. Extensive current CNT production and distribution capabilities, together with expanding product and user bases, have led to a significant increase in the number of studies and guidance procedures. Several acute toxicity studies with rodents that have been completed since 2003 support some likelihood that certain types of single- and multiwalled CNTs pose hazards to the lung or mesothelial surfaces under experimental exposure conditions [23–27, 29]. One scenario is the potential for CNTs to induce granulomatous airway inflammation or interstitial fibrosis in the alveolar region of the lung, depending on the dispersal state of the carbon nanotubes. Another possible hazard emerging from these studies is granulomatous inflammation in the mesothelial lining after peritoneal instillation in mice. This could be a precursor to mesothelioma, as demonstrated in disease outcome in p53 knockout mice exposed to CNTs [30].

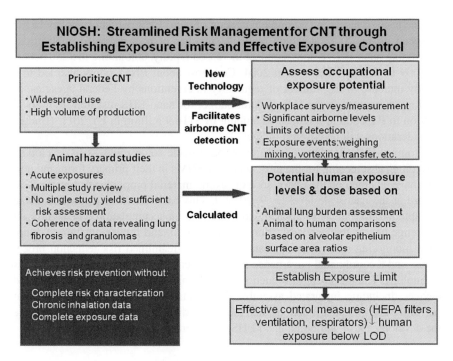

Fig. 1 NIOSH scheme for streamlined risk management for CNTs (Courtesy of A. Nel)

While there is no clinical evidence to date that CNT exposure is responsible for pulmonary fibrosis or mesothelioma in humans, the U.S. National Institute of Occupational Safety and Health (NIOSH) has concluded after its review of multiple rodent studies that the collective evidence points to the possibility that human CNT exposures in the workplace may indeed pose a hazard to the human lung (Fig. 1). NIOSH set up a new generation of airborne particle counters to monitor and quantify airborne CNTs in the workplace (further reviewed in Sect. 8.5). Not only did the occupational surveys demonstrate significant airborne levels in response to specific workplace procedures, but that monitoring could also establish limits of detection (LOD). Utilizing animal lung burden assessments and extrapolation of those data to humans using alveolar epithelial surface area ratios, NIOSH has established an exposure limit and demonstrated that control measures such as ventilation, respirators, and HEPA filters can effectively decrease workplace exposure to below the LOD. NIOSH has also published guidelines for worker safety and recommends that companies working with ENMs implement a safe risk management program as outlined in Sect. 8.5. The NIOSH risk management scheme for CNTs is outlined in Fig. 1.

It is important to emphasize that the generic NIOSH guidelines for CNTs do not imply that all CNT formulations are harmful. There is a burgeoning literature demonstrating in animal studies that CNTs can be functionalized and used safely as imaging and drug delivery agents [31–33]. Thus, it is important to distinguish the properties of CNTs in their as-prepared states (e.g., carbon allotrope with substantial surface-adsorbed contaminants and associated synthetic by-products) from their purified and functionalized forms, which appear to be more benign.

In step with scientific developments and occupational guidelines, CNTs have also come under increased scrutiny from the EPA. In October 2008, the EPA issued a formal notice of the agency's interpretation of the inventory status of CNT under the Toxic Substances Control Act (TSCA), and announced a plan to enforce that interpretation, beginning in March 2009. EPA's position is that CNTs are not equivalent to graphite or carbon allotropes for TSCA purposes, and therefore it is illegal for companies to import or manufacture CNTs in any amount for non-exempt commercial purposes until after a TSCA pre-manufacture notice (PMN) for the CNT has been submitted to EPA and the 90-day review period has expired [34].

2.2 Data Gathering, Monitoring, and Governance of TiO_2, ZnO, and Silica Nanoparticles

The CNT example is just one of a number of ENM decision-making approaches emerging from data gathering. Titanium dioxide (TiO_2), zinc oxide (ZnO), and silica nanoparticles also represent mature, relatively well-characterized materials in terms of available information and readiness of regulators to address risk and hazard concerns [35–38]. For instance, TiO_2 has been used as a pigment for decades and has been studied in its nano-particulate form since the 1980s. Not only is there an extensive literature, but NIOSH has established effective risk management strategies for TiO_2 practices in the workplace. These guidelines have been made available through portals like NIOSH's report and website, *Approaches to Safe Nanotechnology* [39], and DuPont and Environmental Defense Fund's *NANORisk Framework* report and website [40] (see also Sect. 8.4). Moreover, extensive research into the use of TiO_2 and ZnO in sunscreens and cosmetics has demonstrated that the actual consumer risks are low, even prompting the nongovernmental organization the Environmental Working Group, previously critical of nanoparticle use in sunscreens, to make a statement that, "many months and nearly 400 peer-reviewed studies later, we find ourselves drawing a different conclusion and recommending some sunscreens that may contain nano-sized ingredients" [41]. While there remain a number of unanswered questions about the end-of-life risk of TiO_2, there is no evidence that the spread of these particles to water treatment systems or the environment pose any greater risks than the more widespread micron-scale pigment-grade materials. Currently, nano-structured TiO_2 is still officially regarded as "potentially harmful" to the environment [42]. It should be clarified that the end-of-life risk for nano-ZnO may be different from that of TiO_2, as it is regarded in the literature as being "extremely toxic" in the environment [42]. EPA's current inventory approach is that new nanoscale forms of TiO_2 and ZnO are not considered new chemicals requiring reporting under Section 5 of TSCA [34]. However, EPA is developing a Significant New Use Rule (SNUR) to require reporting and filing a 90-day PMN for new nanomaterials based on existing chemical substances.

2.3 Data Gathering, Monitoring, and Governance of Nanostructured Silver

Researchers and regulators are looking more closely at nano-silver, because it is one of the most commonly cited ENMs in "nano"-branded products. Because products containing nano-structured silver often make pesticidal claims for antimicrobial activity, EPA has been evaluating nano-silver under its Federal Insecticide, Fungicide, and Rodenticide Act (FIFRA) statute (7 U.S.C. §136 *et seq.*; [43]). From a toxicological perspective, most of the concern is not directed as much to the apparent modest risk to workers and consumers as to the hazard potential in the environment, especially for aquatic life forms [44, 45].

Policymakers from around the world have indicated that insufficient data have emerged to implement rational changes to existing frameworks for risk management of chemicals and nanomaterials. After a relatively long period of inactivity, national and international governments have begun to collaborate and are now more proactive on the regulatory front. Major regulatory activities include more deliberate data-gathering efforts, global standardization, and coordination of risk assessment to enable regulatory agencies to formulate policy. Examples include the data collection programs and risk management best practices initiatives from organizations such as the Organization for Economic Co-operation and Development (OECD)[2] and the International Organization for Standardization (ISO).[3]

A number of key additional nano-EHS advances over the past 10 years are worth mentioning here and will be discussed in more detail elsewhere, namely advances related to environmental remediation (Sect. 6.1), green chemistry (Sect. 6.1), and improved water and food safety and supplies (Sects. 6.1 and 6.3).

---

[2] For examples, see the OECD department website on Safety of Manufactured Nanomaterials http://www.oecd.org/department/0,3355,en_2649_37015404_1_1_1_1_1,00.html

[3] For examples, see the ISO catalog website for standards devised by its Technical Committee 229 on Nanotechnologies: http://www.iso.org/iso/iso_catalogue/catalogue_tc/catalogue_tc_browse.htm?commid=381983

3 Goals, Barriers, and Solutions for the Next 5–10 Years

3.1 Develop Validated Nano-EHS Screening Methods and Harmonized Protocols that Promote Standardized ENM Risk Assessment at Levels Commensurate with the Growth of Nanotechnology

While some progress has been made in developing toxicological screening for abundantly produced ENMs, there is still a lack of standardized methods and protocols to assess and manage nano-EHS issues. This has resulted in contradictory and even irreproducible ENM hazard assessment that has sparked considerable debate on how best to conduct toxicity screening for risk assessment and regulatory purposes [17, 46]. One significant barrier to the development of validated and harmonized screening protocols is insufficient knowledge about which physicochemical properties of ENMs are relevant to transport, exposure, dose calculation, and hazard assessment. Other obstacles include lack of standardized nomenclature for nanomaterials classification, lack of standard reference materials to use as controls, the high rate at which materials with new properties are being introduced, and ongoing debate about whether *in vitro* and *in vivo* testing best constitute a valid approach to reliable, predictive hazard screening [17, 46, 47]. To address these barriers, a number of solutions are likely to emerge in the next 10 years. These include the following:

- Development of validated hazard assessment strategies and protocols that consider the correct balance of *in vitro* and *in vivo* testing, of biologically relevant screening platforms, and of high-throughput methods. Both *in vitro* and *in vivo* testing are important for knowledge generation about hazardous material properties [17, 48–50]. *In vitro* studies at the molecular and cellular level allow for rapid knowledge generation, but the relevance of this screening must be carefully connected to a desired, validated toxicological outcome *in vivo* to make the screening predictive [17]. This connectivity establishes the relevance of using cellular and biomolecular endpoints to collect primary screening data that can then be used to prioritize animal testing, where fewer observations are possible and mechanistic studies are difficult (see Fig. 2). This approach could limit the extent, volume, and cost of animal testing. (Examples of the use of *in vitro* screening efforts that could be regarded as predictive of *in vivo* pathology or disease outcome are reviewed in Sect. 8.1.) Important considerations for the design of *in vitro* cellular assays include the choice of representative cell lines, their phenotypic fidelity, stability in culture, appropriate use of single- versus multi-parametric response tracking, reporting for acute versus chronic effects, use of an extensive dose range that assesses lethal and graded sub-lethal response outcomes in the linear part of the dose–response curve, and the ability to adapt high-content and rapid-throughput screening approaches to speed up and multiplex hazard data collection [46, 51]. To assist these screening efforts, an important

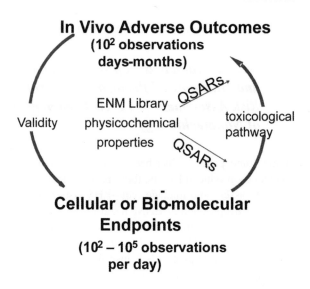

Fig. 2 Differences in the rate of knowledge generation *in vitro* and *in vivo* show the utility of using both approaches but the necessity of validating biomolecular events *in vivo* to establish a predictive toxicological outcome (Courtesy of A. Nel)

goal is to develop and validate harmonized protocols that lead to standardized testing; one example is the efforts by the International Alliance for NanoEHS Harmonization (http://www.nanoehsalliance.org), in which a number of leading international scientists seek to establish and validate protocols anticipated to become useful for toxicological testing of representative nanoparticles in round-robin experiments. Interlaboratory tests are designed to validate the reliability and reproducibility of the protocols as practiced in representative laboratories. At present we do not have databases adequately reporting and tracking data reliability and reproducibility. Yet without quantitative measures of error, uncertainty and sensitivity it is not possible to rationally design nanomaterial, or to evaluate a nanomaterial's health, safety or environmental risk.

- Development of appropriate ENM dosimetry tools that go beyond the traditional mass-dose, particle number, and surface area-dose (SAD) considerations. While traditionally chemical dose levels are determined based on what the organism ingests, dosimetry for nanoparticles is often calculated based on quantities added to the exposure medium, which is conceptually incorrect. The appropriate considerations should be for the bioavailable dose at the site of injury: To relate the toxicology of ENMs to physicochemical properties that are responsible for injury, it is critical to take into consideration cationic charge, surface reactivity, redox activity, surface shedding of metal ions, dissolution chemistry and morphological changes, and the effect of chemisorbed chemical substances, stabilizers, and capping agents [12, 36, 52–54]. To make valid comparisons between *in vitro* short-term mechanistic observations and *in vivo* toxicity and pathology as a result of toxicologically relevant ENM exposures, it is essential to perform dosimetry experiments in the linear region of the ENM dose–response curve. Examples of progress being made in dosimetry assessment in the field of pulmonary toxicology include the tiered

assessment of cellular oxidative stress in response to abiotic and biotic oxygen radical production as well as relating SAD to pulmonary inflammation as reflected in bronchoalveolar lavage (BAL) polymorphonuclear (PMN) cell counts [52, 53, 55, 56].
- *Improved technology to track the presence, fate, and transport of nanomaterials and improve exposure assessment.* Tracking, sensing, detecting, and imaging of nanoscale materials in environmental, biomedical, and biological systems require new analytical technologies that require the same level of technical sophistication as the design of ENMs. Rapid progress is foreseen in technologies that detect and characterize ENMs in aerosols, comparable to the progress discussed above for detection of CNTs in the workplace. Similar advances are being made for detection of other types of nanoparticles in the workplace. Improvements in new, sensitive instrumentation that can detect ENMs in complex biological environments are reviewed in Sect. 4. Technological requirements to assess the presence, spread, and bioavailability of ENM in complex environmental media such as agricultural products and wastewater systems are discussed in Sects. 4 and 6.
- *Life cycle analysis.* An analysis of the energy consumption and materials usage throughout the value chain of ENM production, use, and disposal is essential to understand the overall environmental impact of emerging nanotechnology industries [57]. Similarly, an assessment of the wastes generated by nanotechnology production processes is needed and should include attention to waste streams coming from nanomaterial production facilities as well as conventional waste streams that may impose new pressures on environmental systems (see Sect. 8.3). This life cycle assessment of ENMs should be accompanied by a value-chain analysis that begins with estimates and projections of nanomaterials production. Such estimates are needed to obtain quantitative estimates of expected nanomaterial exposures. Important factors to be identified in evaluating potential nanomaterial exposure are the format in which nanomaterials will be present in commercial products, the potential for these materials to be released to the environment, and the transformations that those materials may undergo that affect their transport and potential for exposure.

3.2 Develop Risk Reduction Strategies that Can Be Implemented Incrementally Through Commercial Nanoproduct Data Collection, Regulatory Activity, and EHS Research Directly Linked to Decision Making

A major barrier to performing comprehensive risk characterization (Fig. 3) is the lack of sufficient knowledge about ENM hazard, fate and transport, dosimetry, and how to perform ENM exposure assessment [5]. This precludes rational implementation of a mature and comprehensive risk management strategy for most ENMs.

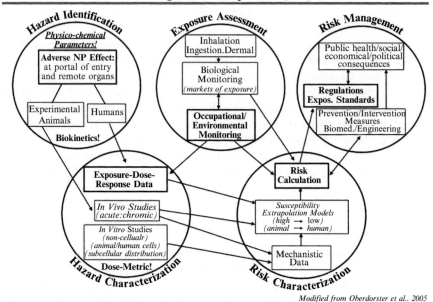

Fig. 3 Risk assessment and risk management paradigm for engineered nanoparticles (Adapted with permission from Oberdörster et al. [5])

However, to mitigate perceived risk and promote widespread public acceptance of nanotechnology, it is necessary to develop safe implementation strategies using current capabilities and infrastructure that are presently at our disposal [5]. We can then proceed with risk reduction strategies that inform the community and the public and also help prevent unanticipated negative EHS consequences of nanotechnology implementation.

To manage risks associated with ENMs, commercial use data must be collected and made public to enable independent EHS researchers to conduct life cycle and exposure analyses [57]. This includes information about the chain of commerce, quantities, and types of ENMs being used in commercial applications. Although both Federal and state agencies (e.g., the California Environmental Protection Agency) have existing authorities dictating how and what data will be collected, improved NNI coordination can play a critical role in fostering the political will to collect commercial use data. Regulatory agencies worldwide are gearing up to fill the major knowledge data gaps about commercial use of nanotechnology by making changes to existing regulations or enactment of new policy to assist the data collection. Current and forecasted policies of regulatory agencies in the United States, Canada, and the European Union (EU) appear in Table 1. Of particular note are the enactment of the significant new use rule (SNUR) by the U.S. EPA and the EU decision to classify specific nanomaterials

Table 1 Current and forecasted regulatory policies of United States, Canada, and EU regulatory agencies

Agency/Law	Jurisdiction	Current stance	Future prospects
Environmental Protection Agency (EPA)	U.S.	TSCA does not require registration and testing for ENMs already in its inventory, but it considers ENMs with novel molecular structures as new materials (e.g., carbon nanotubes)	Rather than labeling ENMs as new substances, the EPA is currently using tools like SNUR to restrict uses of particular nanomaterials if they are expected to present risks. TSCA reform is being considered
Food and Drug Administration (FDA)	U.S.	FDA considers its current practices sufficient to cover NMs, but the agency will issue guidance on data to be included in submissions, including size	Emerging scientific information suggests that certain NMs do present EHS risks. The FDA will modify its policy on a case-by-case basis
Consumer Product Safety Commission (CPSC)	U.S.	CPSC considers its current policies sufficient for NMs until more information is known	CPSC will consider modifications on a case-by-case basis depending upon evidence
Occupational Safety and Health Administration (OSHA)	U.S.	OSHA considers its current policies sufficient for NMs until more information is known	OSHA will consider modifications on a case-by-case basis depending upon evidence
REACH (Registration, Evaluation and Authorization of Chemicals)	EU	REACH identifies chemicals by CAS registry numbers, which identify molecular structure but not particle size	Pending new data, the European Commission through REACH may classify specific NMs as SVHC, similar to EPA's SNUR, to limit or restrict nanomaterial usage in lieu of more concrete regulations
Canadian Environmental Protection Act (CEPA)	Canada	Through CEPA, the Canadian government in 2009 mandated that companies working with ENM must submit usage and toxicity data	Further legislation is under consideration in 2010 requiring notification of significant new activity, risk assessment procedures, and establishment of a public inventory for nanotechnology and ENMs

Source: Adapted with permission from *The Nanotech Report* [35]

as "Substances of High Concern" (SVHC) under the Registration, Evaluation and Authorisation CHemicals (REACH) regulation of the European Chemicals Agency, both of which decisions put use of specific nanomaterials under close scrutiny and regulatory procedures.

EHS research should be driven by the need to make informed decisions on hazard and risk management as well as regulatory decision making. To date, U.S. interagency cooperation has not facilitated effective linkage of risk research to decision making; this disconnect has resulted in actions and strategies that do not fully address policy needs. At the moment, individual agencies are independently establishing connections between research and decision making. Similar efforts are needed at an interagency level to ensure that risk assessment and evidence-based decision making are addressed collectively. Finally, it is also important to mention the possible contribution of *in silico* methods for risk ranking and risk modeling.

3.3 Develop a Clearly Defined Strategy for Nano-EHS Governance that is Compatible with Incremental Knowledge Generation and Stepwise Decision Making

There are a number of divergent positions among different international stakeholders regarding regulatory policy for engineered nanomaterials, as indicated in Table 2, divided roughly into the positions of policymakers, business, academia, and civil society organizations (CSOs). While an integrated strategy for nano-EHS governance currently does not exist in the United States, the trend appears to be shifting from that shown in the second row of Table 2 to the position shown in the third row—that is, toward an across-the-board more precautionary and proactive approach to the regulation of ENMs. While putatively the best position will be evidence-based decision making, there are a number of barriers that preclude this goal, including insufficient knowledge about ENM hazard, dosimetry, exposure, and how to best perform risk assessment.

Attributes of a desirable nano-regulatory process that most stakeholders could possibly agree upon include the following [58]:

- Responsible development of nanotechnology should be accomplished without hampering innovation and commercial growth
- Governance and regulation of nanotechnologies is a dynamic exercise that needs to be continuously adapted
- Appropriate regulation of nanomaterials requires constant implementation of state-of-the-art knowledge, methods, and monitoring
- Timely and appropriate response is needed to address the data gaps and challenges that are continuously being generated in a dynamically changing field
- Global agreement is necessary to promote commercialization

Table 2 Regulatory policy for ENMs among stakeholders around the world

Position/opinion	Policymakers	Business	Researchers	CSOs
The existing regulatory situation is adequate. In the case that scientific evidence indicates a need for modification; the regulatory framework will be adapted	+	+		
Specific guidance and standards must be developed to support existing regulations when dealing with N&N, but the existing regulatory situation is generally adequate	++	++	++	
Regulation should be amended (on a case-by-case basis) for specific N&N, above all when a high potential risk is identified. A precautionary approach is envisaged	++	+	++	+
The existing regulatory position is not adequate at all. Nanomaterials should be subject to mandatory, nano-specific regulations				++

Source: Adapted with permission from FramingNano Governance Platform ([58], p. 69)
Note: *CSO* civil society organizations, *N&N* nanoscience and nanotechnologies.

- All relevant stakeholders and the interested public must be engaged in the development of new policies and regulation of ENMs
- Cooperation between government, industry, academia, and the public is essential in developing the knowledge base required for evidence-based governance

On the basis of these principles as well as the perception that knowledge generation about essential nano-EHS domains is likely to be incremental, the goal for the next 10 years could be to follow an adaptive, iterative approach to nano-regulatory policy (left side of Fig. 4). According to this approach we should identify current knowledge and capacity, and use the statutes and governance infrastructure currently in place, but make it more effective through coordination of data gathering and informatics efforts as well as by involving all stakeholders. This could be done by adjusting and improving the oversight procedures as the knowledge base and capacity increases. Thus, short-term approaches could include information collection, implementation of safe practices in the workplace and laboratory, use of best practices, streamlined risk management for specific ENMs (e.g., the NIOSH guidelines for CNTs), as well as augmenting current statutes to obtain more complete product information (e.g., the SNUR by the EPA or SVHC in the EU).

In the long term, this approach to nano-regulatory policy should shift to a risk prevention paradigm (right side of Fig. 4) in which the emphasis becomes the use of hazard, exposure, and lifecycle data to provide proof of risk reduction through the implementation of safe management and best practices. The long-term goal

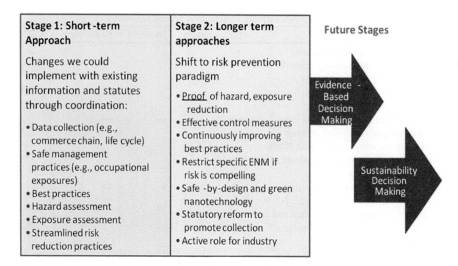

Fig. 4 An example of an adaptive iterative approach to nano-regulatory policy that considers what can be accomplished immediately within our current framework and regulations and where we should aim to move to next as more data and information become available. This could lead us to evidence-based and ultimately sustainability-based decision making (Courtesy of A. Nel)

should be to utilize the information gathered through high-throughput property–activity relationships and computational analysis to develop safe design guidelines for ENM along with implementation of green manufacturing. This could ultimately evolve into evidence-based decision making and policy that promotes sustainability (see Sect. 6).

3.4 Develop Computational Analysis Methods Capable of Providing In Silico Modeling of Nano-EHS Risk Assessment and Modeling

Challenging barriers to evidence-based decision making in nano-EHS include the complexity of environmental and mammalian systems, the large number of variables engineers employ to design nanomaterials, absence of critical knowledge of how to perform risk analysis, the rapid rate of expansion of nanotechnology, inability to deal with large databases, and the lack of a standardized nomenclature to codify engineered nanomaterials [17]. As a result, it is apparent that the cornerstone of research in nano-EHS must be systematic, quantitative studies designed to inform and promote the use of accurate, predictive models and reliable, relevant simulations [59]. Such models must effectively and rigorously address diverse nanomaterial types, including their dissemination and interactions with a multitude of complex and diverse environmental and biological systems. Judicious application of models that ultimately should incorporate predictive power can accelerate safe commercialization of nanotechnology throughout its innovation pipeline. Models that describe the nano–environmental interface will enable engineers to devise "safe-by-design" nanosystems and will also equip companies to design and create containment and waste treatment strategies to minimize nanomaterial exposures. Quantitative adaptive graphical and accessible simulations of nanomaterial transport, interactions at the nano–bio interface, lifecycle analysis, and risk modeling can provide information not currently obtainable from experiments. Some concepts of what these models might look like are available in cutting-edge cell and developmental biology, where virtual environments are being designed to mimic complex biological responses to various stimuli [60].

At present there is no equivalent of the Protein Data Bank (PDB) for ENMs. The PDB serves as a repository for protein structures to archive published molecular structures and conformations for proteins cited in the published literature, and serves as a focus for annotation, curation and validation of those structures. The lack of a repository for ENM structures is critical. Correlations of nanomaterial structure with their physico-chemical, biological, toxicological and biomedical data are being performed without knowledge of the differences in structure of the ENMs in the experimental samples or of the sensitivity of the experimental results to those differences. Nanomaterial is, in general, both polydisperse and polymorphic and modeling efforts may require structural models for several different subpopulations. The Collaboratory for Structural Nanobiology (CSN http://nanobiology.utalca.cl or http://nanobiology.ncifcrf.gov) has been developed and is being used to prototype tools to construct and validate molecular models, to obtain realistic user requirements for a repository from practitioners across the disciplines relevant to nanotechnology, and to explore nanobioparticle data storage, retrieval and analysis in the context of nanobiological studies.

Although it is likely that computational models will need to be trained or fed with valid experimental input data to be valuable for predicting actual behaviors, predictive models can be used to great effect in determining the sensitivity of ENM properties to changes in their environment and structure. With the advent of new online environments, key databases are being developed by consortia such as nanoHUB.org or the caBIG(R) Integrative Cancer Research Nanotechnology Working group at the National Cancer Institute (http://www.nanoehsalliance.org/index.php). Another example of how machine-learning analysis is being used to provide predictive modeling is the framework developed by the University of California Center for Environmental Implications of Nanotechnology (UC CEIN) [61]. (This aspect is further discussed in Sect. 8.2.)

The OECD has published a set of guidelines for the validation of quantitative structure-activity relationship (QSAR) models for regulatory purposes [62]. These guidelines focus on five main concepts:

- Defined endpoint
- Use of an unambiguous algorithm
- Defined applicability domain

- Measures of goodness-of-fit, robustness, and predictability
- Mechanistic interpretation of the model whenever possible

QSAR modeling requires the computation of structural and chemical descriptors as well as large experimental databases of physicochemical properties. In contrast with QSARs for chemicals, the nano-QSAR concept is still in its early development [63]. Due to high variability in the molecular structures and different mechanisms of action, one goal could be to group ENMs into classes and model individual classes of ENM separately. In each case, the applicability domain of the models should be carefully validated. Successful development of new nano-QSARs needs reliable experimental data and requires experimentalists to work together with the nano-informatics community. (Sect. 4 discusses the new capabilities, instruments, and tools that are required for nano-informatics.)

3.5 Develop High-Throughput and High-Content Screening as a Universal Tool for Studying Nanomaterial Toxicity, Ranking Hazards, Prioritizing Animal Studies and Nano-QSAR Models, and Guiding the Safe Design of Nanomaterials

Major barriers in the assessment of ENM hazard potential include the lack of capacity to perform safety screening on large batches of nanomaterials, lack of data on core structure-activity relationships that predict toxicity, inability to cover all of the potentially hazardous materials or material properties in a single experiment, inability to prioritize the execution of costly animal experiments, and the limitations of using single-response parameters (e.g., lethality) without considering a full range of sublethal and lethal dose–response parameters.

One possible solution to these problems is the use of high-content screening methodologies that have facilitated understanding of biological phenomena in cells as well as improved drug screening [10–13]. Rapid-throughput multiparametric cellular screening recently has been shown to be an important tool for toxicology. The goal over the next 10 years is to develop and implement new screening tools to enhance the efficiency and rate at which ENM hazard profiling can be performed. A considerable amount of exploration is required to produce appropriate, dose-dependent responses at the nano-bio interface that can be used for high-throughput screening [51, 55].

Examples of possible applications of high-content screening include assessments of toxicological injury pathways, signaling pathways, membrane damage, organelle damage, apoptosis and necrosis pathways, DNA damage, and mutagenicity [17]. Rapid pathway-based cellular screening studies that utilize one or more of these endpoints allow the establishment of property-activity relationships in which material properties such as size, shape, dissolution, band-gap, surface charge, and so forth are varied to test the biological consequences [12]. This requires the development of ENM libraries that include specific properties for testing (as outlined in

Sect. 8.2), as well as microplate optical reader–based assays that rely on fluorescence, fluorescence polarization, time-resolved fluorescence, luminescence, or absorbance [61]. High-throughput or high-content screening can also help to identify hazardous material properties that could be used for safe-by-design approaches to nanomaterials synthesis [12].

While much of the knowledge about current ENM cellular toxicity has been generated by single-readout cellular screening assays, the major drawback is that each assay represents only a single specific reaction to a toxic stimulus and thus is of limited predictive value. The use of multiparametric screening assays allows for the elucidation of connected molecular pathways or biochemical events [12]. Thus, understanding the initial mechanism of injury and time-sequence information is gained. An estimation of the severity of the insult (e.g., lethal vs. sublethal) is possible through the use of dose–response relationships captured during high-throughput screening.

Cytotoxicity screening as a stand-alone exercise has several limitations, and the true toxicological significance of a cellular injury response can only be determined if it is correlated with adverse biological effects in intact organisms and animals [46]. For the *in vitro* screening to be a truly predictive toxicological tool, the *in vitro* injury response should be directly and unequivocally connected to an *in vivo* injury response or adverse health effect. The duration and intensity of exposure (i.e., acute vs. chronic) must also be considered. Thus, *in vitro* screening assays constitute just one of multiple steps required for ENM safety assessment and validation.

3.6 Improve Safety Screening and Safe Design of Nanomaterials Used for Therapeutics and Diagnostics

Nanotechnology has made major inroads in medicine and looks poised to transform many of nanomedicine's traditional components (see chapter "Applications: Nanobiosystems, Medicine, and Health"). The expected advances encompass improvements in the delivery of therapeutic molecules through systemic injection and locally implanted devices, contrast agents for all modalities of radiological imaging, and innovation in laboratory diagnostics and screening methodologies [64, 65]. In addition to improvements over existing approaches in health care, nanotechnology offers truly transformative opportunities as a necessary enabler of key aspects of personalized medicine [66], regenerative medicine, and reformulation of basic tenets of biological and medical sciences [67].

Nanomedicine is pervasive throughout contemporary medicine, with nanostructured drugs and contrast agents widely available in the clinic [67]. Since the approval of the first nanotechnology-based drugs in 1994, this sector has grown into a $6 billion market as of 2006 in the United States alone. Because of direct and deliberate human exposure in their use, the safety of nanoscale devices is of prime importance and can benefit from some of the same discovery platforms as discussed above for exploration of the nano–bio interactions leading to toxicity. There is relatively little known currently about the safety of the nanoscale devices used increasingly for drug delivery, imaging, or theranostics [68]. There is a lack of detailed information about hazardous nanoscale properties that could require novel safety testing procedures currently not included in the traditional drug screening approaches. However, all current clinically available nanotechnology-enabled agents of therapy and imaging contrasts have obtained regulatory approval from the U.S. Food and Drug Administration (FDA) and other similar agencies worldwide. No adverse event has been reported to date for nanoparticles being used in the clinic. Similarly, there is no literature evidence of health hazards or adverse impact on personnel working in manufacturing, transportation, disposition, storage, medical administration, or dispersion of clinically used nanoparticles. Thus, current concerns about the safe design of nanomaterials largely relate to hypothetical problems that may arise for future generations of nanoscale drugs and devices.

Rationalizing the regulatory process guiding the use of ENM in nanomedicine is a major challenge. The FDA's current position is that nanostructured drugs and nanoparticulate imaging agents and theranostics can be regulated without special consideration of the nanoscale [69]. Demonstrated safety and efficacy of the therapeutic platform is the most important requirement, and experience to date indicates that the drugs being delivered by the nanoparticles are generally much more toxic than the ENM carriers being used. (Parenthetically, it is noted that a nontoxic chemotherapeutic drug would be like a blunt surgical scalpel that would not have any efficacy against cancer.) Thus it may be argued that it is not the lack of toxicity that is the objective, but the balance between risk and benefit for the patients and the community at large [70]. At this time, the regulatory approval pathway is the same as for any other drug or contrast agent; the position of the FDA [71] is that intact nanoparticles as drugs or agents are to be tested as a unit rather than as a combination of individual components.

Some consumer groups and nongovernmental organizations (NGOs) question whether the regulatory processes for nanoscale therapeutics are sufficient, given the lack of comprehensive knowledge about reactions to ENM properties in the body. Moreover, the field of nanotechnology-enabled biomedicine is advancing rapidly and may yield more complicated biomedical nanostructures in the future. These uncertainties are further compounded by the observation that the FDA has not taken any specific actions with respect to monitoring ENM use in foods and cosmetics, despite perceived risks. The FDA does address the safety of drugs and devices before allowing their entering the market and their use in healthcare. On the other hand, the FDA authority for what relates to foods and cosmetics does not include a requirement for premarket authorization, but only a monitoring of post-market safety with the authority to mandate removal of unsafe products.

The FDA has been considering safe design principles for nanoparticles in medicine [69]. Design practices have been sufficient to date to avoid undue safety risks from medical nanoparticles; however, novel design paradigms are emerging under rubrics such as "rational design of nanoparticles" [72, 73]. This has yielded "design maps" to attempt to assess the biological properties of nanomaterials according to their design parameters. Among the biological properties changed by

ENM redesign are the shape characteristics that allow disk-shaped nanoparticles to selectively move to the flow margins in blood vessels, firmly adhere to the vessel walls, and then slip through some vascular fenestrations where they enter cells. Although these methods were primarily intended to optimize biodistributions and therapeutic indices, it is expected that they may also be the foundational cornerstone for a rigorous, quantitative modeling exercise useful to promote nanoparticle safety. Additional research that is ongoing in many laboratories and industries worldwide is directed at the development of "safe" nanoscale vectors that can optimize delivery of therapeutic and contrast agents to intended sites in the body and then disintegrate in full without leaving any trace behind them, *in situ* or systemically. This research has to take into account accessing the body's metabolic and excretory pathways.

3.7 Consideration of Safety Assessment of Increasing More Complex ENMs that Are Being Introduced in a Functionalized, Embedded or Composite Material Format

While to date most of the efforts in hazard and risk assessment have concentrated on primary ENMs such as nanoparticles, nanotubes, and nanofibers, increasingly more complex, composite, embedded, hybrid, and functionalized materials are being introduced. Such new ENMs will necessitate adaptation of study approaches and deciding which materials and commercial products to prioritize for testing. Examples include a number of active nanostructures that are being introduced as "second-generation" nanomaterials as well as "third and fourth-generation materials" that will be obtained through guided assembly, assembly of hierarchical architectures, development of nano-composites, and organic–inorganic hybrids. In addition to the current OECD efforts that focus on high volume or high tonnage primary materials, we should expect in the next decade to see materials such as platinum/palladium nanoparticles in auto catalysts, organic-modified inorganic systems, nanostructured protective coatings, nanostructured reinforced materials, designed microstructures, nanostructured composites, nano-reinforced metallics, and nano/bio-soft-condensed matter. Thus, hazard and risk assessment tools will also have to incorporate methodology and approaches to deal with these novel material characteristics. This will necessitate studies that can assess commercial products and embedded nanoscale materials in their as-produced form as well following their disintegration, shedding or emission of ENMs in the environment or human living space. This introduces another level of complexity that initially should involve data collection about material use, lifecycle analysis, and ultimate disposal in the environment. Industry will play an important role in the data generation and research into the safety of these products. These data are important for definition of the potential exposure scenarios, from which the hazard and risk assessment approaches could evolve, such as use of environmental mesocosms for ENM deposition and aging studies, collection of wear and tear particulates from car tires or vehicular emissions on or close to freeways, looking at the release of embedded nanoscale materials during combustion, erosion, grinding, sanding of composite materials, and assessment of nanomaterial release from building materials.

4 Scientific and Technological Infrastructure Needs

4.1 Development of Advanced Instrumentation and Analytical Methods for More Competent and Reliable ENM Characterization, Assessment, and Detection in Complex Biological and Environmental Media

Relatively few techniques are able to interrogate nanoparticles properties directly with sufficient chemical or physical sensitivity in complex environmental, agricultural, or biological milieus, in real time, or under batch-processed analytical formats. Rapid, sensitive and definitive tools to identify amounts and types of ENMs in complex samples remain a challenge and a need. New characterization tools that directly detect small ENM amounts in "real" biological and exposure environments (i.e., tissue slices, food, environmental samples, blood) are necessary to better evaluate the dynamics of nanomaterials interactions at the biological interface for nano-EHS research [20]. Examples of tools and capabilities that have recently emerged or are in development for characterizing biomaterials include the following:

- Improved tracking of cellular and tissue uptake of ENM using scanning electron microscopy (SEM) and transmission electron microscopy (TEM). Standard SEM and TEM approaches are useful for imaging electron-dense (e.g., metallic nanoparticles), but not soft materials (e.g., dendrimers and liposomes). However, several recent advances improve the utility of TEM in studying the nano–bio interface. TEM cryomicroscopy is now used routinely to image intercellular structures and unstained biomolecules at the sub-nanometer level. When combined with data processing, this technique allows the molecular topographies of single biomolecules to be visualized in conformational states that are not accomplishable with X-ray diffraction [74]. Thus, it is now standard practice to obtain three-dimensional reconstructions of nanoscale biovolumes using eucentric-tilting goniometers [75, 76]. Moreover, sub-Ångstrom, aberration-corrected TE[A]M instruments have been developed to directly image the volumes and surface edge atomic structures of nanoparticles using both transmission (TEM) and scanning transmission modes (STEM) [76]. STEM holds great promise in enhancing the contrast of biostructures when combined with energy-filtered TEM imaging [33, 76]. In addition, a new generation of low voltage electron microscopes are now becoming available to take advantage of the high contrast of biological materials at low energies and which permit multimode (ED, SEM, TEM, STEM) operation in a desk-top instrument (http://www.lv-em.com). This development

makes it possible to place an electron microscope on a manufacturing floor as well as in the field.
- *Improved techniques to resolve nanoscale particles in very large biomaterial volumes.* One approach is correlative microscopy: using optical techniques to identify targets, transferring the sample and grid coordinates to a TEM, and automatically navigating those targets to obtain high-resolution images, while maintaining the sample in a frozen, hydrated state [77–79]. An alternative technique uses a dual-beam instrument—an ion beam to cut a cross-section in the bulk biomaterial, and SEM to record it. By automating the cutting and recording of the image, data can be processed to provide tomographic representations of the volume [80]. This approach has applications in the fully automated analyses of bulk materials with site-specific targeting, where the structure is recognized through the different rates of sublimation of its cellular components. This technique has been used to site-specifically remove artifact-free, thin lamellae of frozen tissue for TEM cryomicroscopy [81].
- *New approaches to fluorescence imaging.* Fluorescently labeled nanoparticles and related imaging techniques (e.g., confocal microscopy) suffer potential problems such as label instability, altered physicochemical properties, and photobleaching from laser exposure. Ideally, novel imaging techniques are required to visualize local populations of nanoparticles at nanometer resolution in real time within cells without structural damage. A promising development is live cell confocal microscopy, which is ideal for high-resolution imaging of movement through intracellular environments, including endo-exocytosis, vesicle tracking, particle transport, and nuclear-cytosol membrane mechanisms [82].
- Advances in Coherent Anti-Stokes Raman (CARS) scattering now permit Raman spectroscopy to be used as a "chemical microscope", mapping the detailed structure of cells and organelles according to the chemical composition at each point in three-dimensional space. without the use of dyes. (e.g., Broadband-CARS (B-CARS) (http://www.sciencedaily.com/releases/2010/10/101014121156.htm) and Femtosecond Adaptive Spectroscopic Techniques for CARS (FAST-CARS) (http://www.ncbi.nlm.nih.gov/pmc/articles/PMC123198/)).
- *Surface-enhanced Raman scattering (SERS).* Another technique being used increasingly for bioimaging of cells and intact animals is SERS [83], which measures the enhanced Raman scattering of molecules adsorbed onto (e.g., nanotextured) metal surfaces. With enhancement factors as high as 10^{15}, this technique is sufficiently sensitive to detect single molecules (e.g., PEGylated Au and Ag nanoparticles). Recent tumor imaging with radio-labeled single-walled carbon nanotubes (SWCNTs) suggests that SERS may be a promising molecular imaging technique in living subjects [24, 84].

In similar fashion, new characterization tools/techniques such as the following are emerging to evaluate the structure and dynamics of the environmental interface:
- *Liquid chromatography-atmospheric pressure photoionization-mass spectrometry (LC-APPI-MS).* This can be used to determine aqueous concentrations of

ENMs with positive electron affinity at relatively low levels (e.g., 0.15 pg detection limit for C_{60}).
- *Spectroscopic techniques.* Techniques such as x-ray absorption fine structure (XAFS), including x-ray absorption near-edge spectroscopy (XANES) and extended x-ray absorption fine-structure spectroscopy (EXAFS), can be used in conjunction with electron microscopy to determine the chemical state and local atomic structure of inorganic ENMs and assess their chemical transformations. However, these methods often require a synchrotron beamline, which is expensive, non-routine, and often inaccessible for most needs.
- *The environmental scanning electron microscope (ESEM).* This allows a gaseous environment in the specimen chamber, whereas other electron microscopes operate under vacuum. ESEM allows imaging of wet specimens and can be useful for detecting nanomaterials in the environment. Hydrated specimens can be examined, because any pressure greater than 609 Pa allows water to be maintained in its liquid phase for temperatures above 0°C, in contrast to the SEM, where specimens are desiccated by the vacuum condition. Moreover, electrically nonconductive specimens do not require the preparation techniques used in SEM, such as the deposition of a thin gold or carbon coating.

The informatics infrastructure for nanotechnology should incorporate web-enabled websites and forums to advance collaboration in gathering user requirements for use cases employing advanced instrumentation deployed in realistic settings, instrument prototyping, and partnering in production of these new instruments which promise to rapidly advance our understanding of the behavior of ENMs in biological environments.

4.2 Development of Computational Models, Algorithms, and Multidisciplinary Resources for Increasingly Sophisticated Predictive Modeling

The importance of computational and predictive modeling in advancing the goals of nano-EHS has been noted previously. The technological infrastructure required for these developments includes new computational tools that transcend traditional analytical methods, which often assess a single material under specific use conditions. The new computational methods and tools allow forecasting (of variable materials, diverse uses, and new hazards), construction of quantitative structure–activity relationships (nano-QSARs), fuzzy logic, self-learning, neural networks, and artificial intelligence. Important nano-informatics requirements are summarized in Table 3.

Also needed is a systematic nomenclature to codify engineered nanostructures for computational analysis. The current lack of a coherent nomenclature confounds the interpretation of data sets and hampers the pace of progress and risk assessment.

Table 3 Important nano-informatics requirements

Data collection and curation	Tools/methods for discovery, innovation, communication, and management	Social dimensions to information sharing
Lab automation for high-throughput collection	Data mining	Defining and addressing sociological issues
Tools for literature data collection	Machine learning	Overcoming education and perception barriers
Databases and data sharing	Visual analytics	Determining and establishing rational governance parameters
Tracking error, uncertainty, and sensitivity in ENM data		
Interoperability	Semantic search and analysis	Instituting terms of use
Metadata standards	Literature analysis	
Nanomaterial property data	Quality control	
Ontologies	Standards development	
Taxonomies	Open source	
Open access		
ENM molecular structure	CSN	
Advance instrumentation		
Collaboratories	Predictive model development for risk and ENM design	

The International Union of Physical and Applied Chemistry (IUPAC) has developed a nomenclature for organic, inorganic, biochemical, and macromolecular chemistries [85], and the Chemical Abstracts Service (CAS) has developed a cataloging system for reagents and new substances [86]. However, neither of these nomenclature systems is appropriate for nanostructures. For nanostructures, a systematic nomenclature based on material composition and nanoscale properties such as size, shape, core/surface chemistry, and solubility may be particularly relevant to nano-EHS activities.

4.3 Development of Workforce Capacity Through Interdisciplinary Education and Training, Particularly in the Nano-EHS Field, Where a Large Number of Research Areas Are Converging

The market for nanotechnology-enabled products was estimated at $254 billion in 2009 and is projected to increase to $2.5 trillion by 2015 (Lux [87]). A corresponding increase in the number of individuals trained to work in the various sectors involved in the development and production of nanotechnology-enabled products is essential to maintaining a competitive edge in this area and for harvesting the benefits of this effort. As is the case for many cutting-edge areas of science and technology, the future of nanotechnology is inextricably linked to interdisciplinary education and training. To create new "smart" ENMs for medical applications, researchers must be well versed not only in materials science, chemistry, and physics, but also in biological sciences, physiology, pharmacology, and engineering.

NSF has played a critical role in building a pipeline of multidisciplinary researchers and engineers through its programs Nanoscale Science and Engineering Education (NSEE), Nanotechnology Undergraduate Education (NUE), Research Experience for Teachers (RET), and Research Experience for Undergraduates (REU). These programs allow for the development of education modules in nanotechnology that can be used in a broad range of settings—from K–12 through graduate education—and hence have the ability to impact the education of a broad spectrum of our society. The National Institutes of Health (NIH) has its T32 and R25 Institutional Research Training/Education Grants programs for emerging technologies, which have had a similar impact on interdisciplinary graduate education related to nanotechnology, including the use of a multiple principal investigator (PI) mechanism. Funding targeted at interdisciplinary educational programs (e.g., NSF's Integrative Graduate Education and Research Traineeship program) and interdisciplinary research centers (e.g., The National Cancer Institute's Centers of Cancer Nanotechnology Excellence and NSF's Nanoscale Science and Engineering Centers) plays a critical role in enabling interdisciplinary nanotechnology education and training programs in ways that individual-investigator funding cannot.

These types of interdisciplinary education and training are absolutely essential to meet the significant challenges presented by nano-EHS as an emerging field. In the short term, the development of guidelines for safe handling of ENMs requires engagement of researchers and practitioners of both industrial hygiene and public health. Dissemination of these practices across the communities of scientists and engineers who develop and work with ENMs will require collaboration not only between industrial hygienists and nanoscience-focused researchers and engineers, but also with members of the education community. Likewise, the development of risk management practices and appropriate policy and regulatory strategies for nano-EHS will require basic researchers in the nanoscience community to engage and partner with individuals working in the fields of EHS, risk management, public policy, and law. Because toxics policy is currently under broad review in the United States, an investment in activities that foster this cross-fertilization is likely to help drive stability within the field of nanotechnology as well as to inform decision making on toxics policy.

In the long term, movement beyond risk management and toward a risk prevention strategy that embraces the concepts of inherently safer design and predictability should develop the best models for correlating physicochemical ENM properties with their biological and environmental impacts and robust decision-making tools based on these models. Such tools will require coordination of research and data collection from a broad variety of disciplines. Programs such as the Centers for Environmental Implications of Nanotechnology being coordinated from the University of California, Los Angeles (UCLA), and Duke

University with funding by the NSF and EPA [61, 88] play a critical role in driving this agenda. (The multidisciplinary research integration in the UC CEIN is summarized in Sect. 8.2.)

5 R&D Investment and Implementation Strategies

5.1 Increase the Role of Industry in Nano-EHS R&D Funding

Sufficient Federal funding is required during the early discovery and incubation phases of nano-EHS research, but funding for nano-EHS research and development should also be shared by the private sector, where implementation of nano-EHS knowledge should lead to improved products with enhanced commercial value. Industry has a particular responsibility as a partner in establishing standardized testing and development of safe nanomaterials. Moreover, nano-EHS research will also contribute to new nanotechnology-based green technologies and innovative environmental cleanup strategies, with their intrinsic commercial value added. In order to contribute to sustainability, it is important that nano-EHS funding be implemented as an integral part of new product design and manufacturing rather than as a *post facto* add-on, safety mandate, or as an imposed cleanup cost.

5.2 Increase Federal Focus on Building an Accessible Infrastructure for Understanding ENM Toxicity

A key question at present is whether the U.S. Federal support for nano-EHS efforts is sufficient to build the capacity required for safe implementation of nanotechnology. EHS spending in the United States amounted to 2.8–5.4% of the total Federal spending on nanotechnology between 2006 and 2010. The FY2011 Federal budget proposes $1.8 billion spending on nanotechnology, with $117 million, or 6.6% of the total, earmarked for research on EHS considerations. It remains to be seen whether this budget allocation is sufficient to allow for the implementation of all the research and knowledge generation needed in terms of new methods development, coordinated ENM candidate screening, data collection, model development, risk assessment, and effective end-stage commercialization. It is important to consider that some of the Federal money for nano-EHS research has been allocated in the past to general ENM characterization and methods development and validation, rather than for specific research directed at understanding nanomaterial toxicology. There is currently insufficient funding for extensive research and analysis of the possible health consequences of ENM in food, agricultural products, and industrial processes such as printing. Another key investment should be in technological infrastructure and new instrumentation to address the diverse analytical needs for nano-EHS.

While instrumentation and tools have been addressed in previous sections, it is important to highlight the need for shared user facilities where industry, academia, and government can coordinate nano-EHS research. While several national laboratories and academic institutions have outstanding facilities and infrastructure to conduct general or applied nanotechnology research, there are no shared use facilities for nano-EHS research. As a result, there is little or no transfer of knowledge and protocols. This has contributed to a lack of cooperation and disclosure, and guarded secrecy about nano-EHS efforts in the private sector, including in the food and cosmetics industries. Moreover, in food and agricultural research, the materials to be investigated are often "dirty" and demand dedicated equipment for analysis of composition and synthesis of what are more complex test systems.

5.3 Promote Cross-Sector Partnerships in Nano-EHS R&D Efforts

The promotion of collaborative partnerships between academia, government, and industry is essential for successful creation, design, development, and value capture of nanotechnology advancements, including widespread public acceptance. These partnerships are critical not only for harvesting knowledge but for enabling investment options by creating needs-driven knowledge. Dialogue is needed to overcome the reticence of industry to actively participate in nano-EHS efforts, particularly in the formative stages of strategic program development. In this regard, it is helpful to examine the efforts by some industry sectors and corporations that have promoted safety in nanotechnology development (examples are discussed in Sect. 8.4), as well as the reasons for selective non-participation in industrial surveys by other industry sectors and businesses. Important issues that have surfaced in surveys to date include the current lack of standardized ENM screening protocols, uncertainty about the regulatory environment, possibilities of inviting unnecessary scrutiny, cost-benefit factors, and public perceptions. Industry needs to see that government and universities are listening to and addressing these kinds of concerns.

Even as we are moving to more regulation of nanotechnology-based products as a result of knowledge gaps, it is highly desirable to establish private-public partnerships to change the dynamics of the current dialogue. The best R&D partnerships involve government, industry, and academia, each playing to its own strengths. Ideally, the data collection on the safe use of nanotechnology in commercial products and industrial processes should be a position of consensus rather than of unilateral enforcement. While it is currently still possible for industry to withhold data because of fear of disclosing confidential business information, a continuing reluctance to share information could prompt a change in the environmental statutes and laws to essentially demand disclosure—a situation not conducive to fostering collaboration and trust. This possibility is evident in the recent U.K. House of Lords [89] recommendation that noncompliance with respect to the use of ENMs in foods could serve as the basis for exclusion of food products from the marketplace.

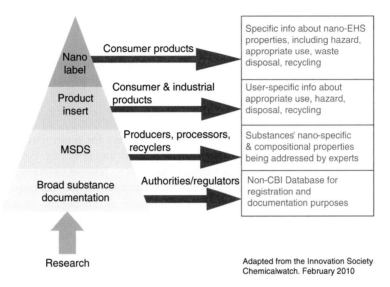

Fig. 5 This nano-information pyramid illustrates development of an incremental information-sharing collaboration between government, academia, and industry (Adapted with permission from Widmer [90])

An example of what may be achieved through product information disclosure is encapsulated in the proposed nano-information pyramid shown in Fig. 5. This illustrates that the first tier of information disclosure could involve broad substance information collection for registration and documentation purposes. This could transition at the next level to material substance data sheets (MSDS) handled by producers, processors, and recyclers, which depending on the level of risk, may require at the next level product inserts or labeling that provide specific information about hazard, safe handling, disposal, and recycling [3]. Another solution is an incentive-based system in which voluntary business disclosure of nano-EHS data in collaboration with academia and government agencies would facilitate safety profiling that would make it easier to move to the marketplace as compared to when there is no safety information available. Combined public-and-private research efforts can also help to develop, optimize, and validate *in vitro* and *in vivo* safety assessment protocols. Specifically, nano-EHS consensus in food, medicine, and cosmetic safety requires continuous industry participation and assistance in policy formation.

Private-public partnerships could also help develop the high-throughput methods, property-activity relationships, and computational methods necessary to understand any risks and hazards, and produce safer nanotechnology-based products. Not only will this promote sustainability but it will also deliver a chain of superior products capable of returning the up-front investment in nano-safety. If this kind of cross-sector interaction is established early, including dedicated funding to make it an integral part of the nanotechnology development enterprise, it will provide a precedent and strong incentive for ongoing industry participation in nano-EHS R&D. Examples of several successful private–public partnerships are highlighted in Sect. 8.4.

6 Emerging Topics and Priorities

6.1 Role of Nanotechnology in Promoting Environmental Remediation and Sustainability, Including Through Green Manufacturing

Nanomaterials have potentially beneficial applications for future environmental remediation or as active transforming agents, sensors, and detectors. For example, iron nanoparticles can serve as powerful reductants to remove oxidized contaminants from soil and ground water as sorbents. Nanomaterials and nanodevices can be exploited for pollution prevention by functioning as components in advanced biosensors, monitors, adsorption surfaces for toxic chemicals, and new filtration membranes [91, 92]. Noteworthy examples include:

- Use of natural and manufactured nanostructured clays and zeolites for filtration of undesirable compounds from air or water
- Removal from groundwater of trichloroethylene (TCE) by reductive dechlorination and hexavalent chromium (Cr(VI) or Cr^{6+}) by reductive immobilization using zero-valent Fe and adsorption of nanostructured TiO_2, zeolites, nanomagnetite, or dendrimers.
- Use of polymeric membranes impregnated with silver/zeolite or photosensitive ENM to improve resistance to bio-fouling in structures in aquatic environments without the use of biocides

In addition to remediation, ENMs can help meet a growing need for point-of-use water treatment and reuse. Advancements in decentralized water treatment and reuse alleviate dependence on major infrastructure, avoid degradation of water quality within distribution networks, exploit alternative water sources for a growing population (e.g., recycled grey water), and reduce energy consumption. Future urban systems will increasingly rely on high-performance nanotechnology-enabled water monitoring, treatment, and reuse systems that target a wide variety of water pollutants, are affordable, easy to operate, and contribute toward a zero discharge paradigm, which is the ultimate goal of sustainable urban water management. Examples of ENMs that can enable this vision are summarized in Table 4.

Green nanoscience aims to create and apply design rules proactively for greener nanomaterials as well as for developing efficient and reproducible synthetic strategies to produce materials with defined composition, structure, and purity [19]. As such, green nanoscience incorporates the 12 well-known principles of green chemistry in the design, production, and use of ENMs (Table 5).

Green nanomaterials/processes can substitute for dangerous materials and processes shown to pose more risk. Nanotechnology-inspired production is likely to also lead to more efficient use of materials and lower energy needs, thereby decreasing the environmental footprint. Nonetheless, the entropic penalties associated with creating order at the atomic scale set boundaries on the possible gains

Table 4 Example of opportunities for ENM in water treatment and reuse

Desirable ENM properties	Examples of ENM-Enabled technologies
Large surface area to volume ratio	Superior sorbents with high, irreversible adsorption capacity (e.g., nanomagnetite to remove arsenic and other heavy metals)
Enhanced catalytic properties	Hypercatalysts for advanced oxidation (TiO_2 & fullerene-based photocatalysts) and reduction processes (Pd/Au to dechlorinate TCE)
Antimicrobial properties	Disinfection without harmful disinfection by-products (e.g., enhanced UV and solar disinfection by TiO_2 and derivatized fullerenes)
Multi-functionality (antibiotic, catalytic, etc.)	Fouling-resistant (self-cleaning), functionalized filtration membranes that inactivate virus, fungal, and bacterial threats, and destroy organic contaminants
Self-assembly on surfaces	Surface structures that decrease bacterial adhesion, biofilm formation, and corrosion of water distribution and storage systems
High conductivity	Novel electrodes for capacitive deionization (electro-sorption) and low-cost, energy-efficient desalination of high-salinity water
Fluorescence	Sensitive sensors to detect pathogens and other priority pollutants

Table 5 Applying green chemistry principles to the practice of green nanoscience

Green chemistry principles	Designing greener NMs and NM production methods	Practicing green nanoscience
P1: Prevent waste P2: Atom economy	Design of safer NMs (P4, P12)	Determining the biological impacts of nanoparticle size, surface area, surface functionality; utilize this knowledge to design effective safer materials that possess desired physical properties; avoid incorporation of toxic elements in nanoparticle compositions
P3: Less hazardous chemical synthesis P4: Designing safer chemicals	Design for reduced environmental impact (P7, P10)	Study NM degradation and fate in the environment; design material to degrade to harmless subunits of products. An important approach involves avoiding the use of hazardous elements in nanoparticle formulation; the use of hazard-less, bio-based nanoparticle feedstocks may be a key
P5: Safer solvents/reaction media P6: Design for energy efficiency	Design for waste reduction (P1, P5, P8)	Eliminate solvent-intensive purification by utilizing selective nanosyntheses—resulting in great purity and nanodisparity; develop new purification methods (e.g., nanofiltration) that minimize solvent use, utilize bottom-up approaches to enhance material efficiency and eliminate steps

(continued)

Table 5 (continued)

Green chemistry principles	Designing greener NMs and NM production methods	Practicing green nanoscience
P7: Renewable feedstocks P8: Reduce derivatives	Design for process safety (P3, P5, P7, P12)	Design and develop advanced syntheses that utilize more benign reagents and solvents than used in the "discovery" preparations; utilize more benign feedstocks, derived from renewable sources, if possible; identify replacements for highly toxic and pyrophoric reagents
P9: Catalysis P10: Design for degradation; design for end of life	Design for material efficiency (P2, P5, P9, P11)	Develop new, compact synthetic strategies; optimize incorporation raw material in products through bottom-up approaches; use alternative reaction media and catalysis to enhance reaction selectivity; develop real-time monitoring to guide process control in complex nanoparticle syntheses
P11: Real-time monitoring and process control P12: Inherently safer chemistry	Design for energy efficiency (P6, P9, P11)	Pursue efficient synthetic pathways that can be carried out at ambient temperature rather than elevated temperatures; utilize noncovalent and bottom-up assembly method near ambient temperature; utilize real-time monitoring to optimize reaction chemistry and minimize energy costs

Source: Adapted from Anastas and Warner [93], p. 30. With permission from Oxford University Press)
Note: The principles are listed, in abbreviated form, along with the general approaches to designing greener nanomaterials and nanomaterial production methods and specific examples of how these approaches are being implemented in green nanoscience. Within the figure, PX, where $X = 1–12$, indicates the applicable green chemistry principle [19]

that are achievable by applying ENMs to solve environmental problems. For example, theoretical gains in adsorptive efficiency using nanostructured iron oxides for arsenate oxo-anion removal are more than outweighed by their necessary energy investments and associated costs when compared with conventional ferric chloride salts.

6.2 Safe-By-Design Approaches to Promote Sustainable Implementation of Nanotechnology

The awareness of safe-by-design ENM approaches is moving the nano-EHS field toward thinking about the possible proactive implications of specific applications of nanotechnology at the design and development stages, rather than waiting to reactively consider impacts until after the technology has been matured and deployed [19, 20]. An understanding of hazardous ENM properties is essential for safe design from a biological and lifecycle perspective. While there is no single design feature that currently fits this description, possible approaches that might contribute to this area are being identified. It is important to note that redesign of some of these properties may affect ENM performance characteristics (e.g., electrical conductivity, thermal conductivity, or magnetic properties) that are essential for technology or product development. Thus, while the potential impact on product performance must be properly explored, it is possible that certain compromises may result.

Focusing on ENM exposure control rather than on suppressing ENM intrinsic reactivity that contributes to toxicity might be a useful compromise strategy [20]. Thus, risk abatement options worthy of consideration include tailored coatings that reduce bioavailability or mobility. The modern chemical industry has demonstrated that some substances can be reengineered to create safer, greener, and yet efficient products [19]. Encouraging examples include the substitution of branched alkylbenzene sulfonate surfactants that cause excessive foaming in the environment, with biodegradable linear homologues [94]. It is therefore important to discern the specific critical functionalities and physicochemical properties that make ENMs harmful, then reengineer these properties to achieve safer products.

Another route to mitigate ENM toxicity is to exploit the tendency of nanoparticles to aggregate in natural and biological media, which naturally decreases their bioavailability and possible bio-reactivity [20]. Colloidal stabilizers with kinetic degradation in certain conditions allow initial ENM dispersion as desired but with a programmed loss of their dispersibility and resulting aggregation over time, controlling their nano-specific properties. Surface coating is a design feature being exploited to improve nanoparticle safety by preventing undesired bio-reactivity. For instance, TiO_2, ZnO, and Fe_2O_3 nanoparticles within cosmetic formulations (e.g., sunscreen lotions) are often coated with a hydrophobic polymer (e.g., poly[methyl vinyl ether]/maleic acid) to reduce direct contact with the human skin [95]. Coating of nanoscale zero-valent iron (NZVI) with polyaspartate not only prevents particle aggregation to enhance nanoparticle mobility in contaminated groundwater so as to reach and reductively dechlorinate trichloroethylene, but it also mitigates NZVI toxicity to indigenous bacteria, enhancing their possible co-participation in the cleanup process [96, 97]. This also suggests that artificial as well as natural coatings (e.g., dissolved natural organic matter) can be used to mitigate ENM toxicity and alter impacts on microbial ecosystem services.

An extension of this principle is the use of polymer and detergent coatings that reduce eukaryotic cell particle contact and uptake by steric hindrance. Many such coatings are environmentally labile or degradable. Thus, an initially nontoxic material may become hazardous after shedding its coating if resulting aggregation does not reliably remove it from the system. An important design feature would be to enhance the stability of coating materials or design them originally to prevent adverse biological responses. Coating nanoparticles with protective shells (i.e., core-shell systems) can also reduce the dissolution and release of toxic ions [98], while also providing a physical barrier against cellular uptake if undesired. Suitable shell materials include biocompatible organic or inorganic substances such as PEG-SiO$_2$, gold, and biocompatible polymers [99].

Altered dissolution rates and limited metal ion leaching could also be deliberately achieved by material doping (e.g., doping of ZnO with Fe$_3$O$_4$, leading to decreased cellular and zebrafish toxicities) [12, 20]. Modification of surface charge is another approach towards reducing nanoparticle toxicity [100, 101]. For example, layer-by-layer coatings of polyelectrolytes on gold nanorods decrease their cellular uptake via modified surface charge and functionality. For the safe design of materials that form bio-persistent fibers (e.g., CNTs), it is important to consider aspect ratios, hydrophobicity, and stiffness [26]. Chemical functionalization of short (<5 μm) multiwalled carbon nanotubes (MWCNTs) can provide stable dispersions of individual tubes in physiological media, thereby allowing their safe use as imaging and drug-delivery devices [32]. Functionalization with small hydrophilic groups is a safety feature allowing the formation of stable dispersions with high excretion rates [32]. Thus, ENM coatings and surface properties can produce diverse properties that enhance or diminish certain types of exposure, depending on the application, chemistry, design, and ENM properties. Identifying desired and undesired specific ENM functions and possible risks, and applying safe-by-design principles to realize these properties while mitigating risks, represent attractive objectives for this strategy.

Finally, consideration should also be given to material disposal, life cycle fate, and containment. Several priority research areas can inform the ecologically responsible design and disposal of ENMs. As a first step to understanding potential impacts resulting from incidental or accidental releases of nanomaterials and evaluating the need for ENM interception, containment, or treatment technologies, we should understand sources and the scale of potential discharges into various environmental compartments (including ENMs leaching from commercial products during their entire life cycles) [94]. This requires having an inventory of the magnitude of ENM use within defined spatial domains and the possible flow of ENMs across domains. Quantification of potential fluxes to the environment from both point and non-point sources is also a priority that can only be accomplished after developing appropriate analytical tools or identifying sentinel species that can be monitored to detect environmental presence and pollution by ENMs.

Furthermore, ENM waste will enter protective environmental infrastructures such as sewage treatment, air filters, bag houses, and landfill liners. It is unknown how accidental or deliberate ENM releases may affect the performance of such processes (e.g., toxicity to probiotic bacteria essential in activated sludge) and how effective barrier technologies (e.g., landfill liners) would be at intercepting and containing ENMs. Knowledge about the flows of ENMs from different stages in their life cycles to waste-handling institutions will provide a basis for prioritizing research on this topic. An example of impactful research in this area appears in Sect. 8.3. Distinct properties that make ENMs so useful in a vast spectrum of products are also those that may challenge their recyclability. Specific guidelines and possibly product labeling are needed to safely and responsibly dispose of and recycle the waste products that contain ENMs.

6.3 Role of Nanotechnology in Agriculture and Food Systems, Including Enhancement of Food Safety as well as Ability to Demonstrate that ENMs in Foods Are Safe

Nanotechnology has an important role in creating a safer food system [102–105]. The food supply chain can and will be affected by the utilization of nanotechnology at each point in the system along the supply chain—from production through domestic consumption [106]. While the advances and technological impacts of nanotechnology on agriculture and food systems in the past 5–6 years are limited due to its relative "newness" in this sphere, some encouraging results have been obtained in the various agri-food sectors discussed in chapter "Nanotechnology for Sustainability: Environment, Water, Food, Minerals, and Climate". From the perspective of food safety, nanotechnology has much to offer, including:

- Carbon nanotube and surface-enhanced Raman spectroscopy (SERS) nanosensor arrays can help ensure the safety of the food supply by identifying the presence of pathogens, toxins, and bacteria, and actively eliminating their impact
- Edible nanoparticle sensors can detect food quality and safety
- DNA barcoding methods are a simple and low-cost way to detect the presence of bacteria and other pathogens in foods
- New biosensors can detect the presence of avian influenza virus
- Nano-sensing formats can be used for food packaging security, freshness, and sustainability

In order to promote and expand the role of nanotechnology in food and agriculture, it will be necessary to address the inherent safety of nanomaterials that enter the food chain as well as articulate the benefits of nanotechnology in this sector to the public. In addition to concern about the health and safety issues of possible new "nano-produced" or "nano-monitored" foods, there is a concern among some NGOs about broad social and ethical issues. One such concern is that nanotechnology will become concentrated within multinational corporations and that this could impact the livelihood of the poor. These areas of health and safety and the impact on agriculture infrastructure are currently an area of intense interest and much debate, mirroring similar concerns and issues in some previous emerging technology situations.

Some public skepticism can be influenced by factors such as a fear of novel risks, trust or lack of trust in the regulatory process, and wider social and ethical concerns. A recent study by Britain's House of Lords [89] offers several recommendations to build public confidence and trust: (1) there should be increased research on toxicological impacts of nanomaterials, particularly in areas relating to risks posed by ingesting nanomaterials; (2) a definition of nanomaterials should be added to food legislation to ensure that all nanomaterials that interact differently with the body as the result of their small size be assessed for risk before they are allowed on the market; (3) food regulators and the food industry should collaborate to develop a database of information about nanomaterials in development to anticipate future risk needs; and (4) food regulatory agencies should create and maintain a list of products containing nanomaterials as they enter the market, to promote transparency.

Issues of perceptual risk and social and ethical concerns might be addressed with a number of steps: (1) develop a broad coalition of scientists, engineers, farmers, food processors, and manufacturers, interested NGOs, government agencies, and consumers to engage in discussions that will promote common understanding and agendas; (2) develop comprehensive interactions with the FDA and EPA to discuss whether regulations are required; (3) develop public–private partnerships in which agricultural and food companies interact with universities, the USDA, EPA, and FDA; and (4) offer increased opportunities for the public to participate in open forums to help create an intelligent understanding of concerns and benefits.

6.4 The Following Key Priorities Have Been Identified for the Next Decade

- Develop validated nano-EHS screening methods and harmonized protocols that promote standardized ENM risk assessment at levels commensurate with the growth of nanotechnology
- Obtain active industry participation and NGOs in nano-EHS, including hazard and risk assessment, lifecycle analysis, non-confidential product information disclosure to assess exposure scenarios, and use of nanomaterial property-activity relationships to implement safe-by-design for product life cycle strategies
- Introduce environmentally benign nanomanufacturing methods and using nanotechnology to replace commonly used processes, compounds and products with adverse effects to human health and the environment
- Develop risk reduction strategies that can be implemented incrementally through nano-EHS research, commercial nanoproduct data collection and the use of streamlined decision-making tools
- Develop high-throughput approaches, nanoinformatics and *in silico* decision-making tools that can help model and predict nanomaterial hazard, risk assessment, and safe design of nanomaterials as an integral part of new program development.
- Develop clearly defined strategies for nano-EHS governance that takes into consideration knowledge gathering and stepwise decision-making that ultimately leads to evidence-based and sustainability-enhancing decision-making.

7 Broad Implications for Society

Although academia, industry, and government deal with real risk issues, the public is more prone to react to unproven perceived risks, and their views are often shaped by often-unsubstantiated reports coming from popular news media and NGOs [107]. As long as nano-EHS data gaps remain, threats of perceived risks, despite lack of evidence, will persist, potentially hindering market and technology development. NGOs are continuously pushing for concrete regulations (Table 2), and some like the Natural Resources Defense Council and Friends of the Earth continue to argue that voluntary data collection programs should be mandatory. A key issue therefore for academia, industry, and government is to effectively communicate, inform and involve the participation of the public in the dialogue on the beneficial implications of nanotechnology, the potential for risk, and what is being done to ensure safe implementation of the technology. Due to the complex and multidisciplinary nature of nanoscience and nanotechnology, knowledge transfer and public education has not been effective and needs urgent attention. Strategies for communication and education of the public are discussed in chapter "Developing the Human and Physical Infrastructure for Nanoscale Science and Engineering".

Closely associated with the issue of perceived risks is the safety of common consumer products that contain nanotechnology-based ingredients; these include such products as sunscreens, soaps, toothpastes, clothing, food, and cosmetics.[4] Greater transparency is required in disclosing the presence of nanomaterials in these products, including why their addition and use provides a better product, as well as specific technical data on their compounding and formulation. Due to perceived risks, some information about nanotechnology-based products is deliberately withheld instead of being disclosed, which in the long run could be counterproductive to credibility, transparency, perception, and image. The nano-information pyramid and proactive recommendations about how package inserts or labeling may be introduced with care and forethought could help remove such uncertainty (Fig. 5). It is also important to explain to the public that nanotechnology can play an important role in promoting food safety, environmental remediation, better medical therapies, and product enhancements.

8 Examples of Achievements and Paradigm Shifts

8.1 Examples of Predictive Toxicological Paradigms that Connect In vitro Hazard Assessment to In vivo Injury in Intact Animals

Contact person: André Nel, University of California, Los Angeles (UCLA)

Both the National Toxicology Program and the National Research Council (NRC) in the U.S. National Academy of Sciences (NAS) have recommended that toxicological testing in the 21st century evolve from a predominantly observation science

---

[4] For examples of specific nanotechnology-based products, see the Project for Emerging Nanotechnologies (PEN) consumer products inventory at http://www.nanotechproject.org/inventories/consumer.

at the level of disease-specific models to predictive science models focused on broad inclusion of target-specific, mechanism-based biological observations [8, 18]. Predictive toxicology is an essential tool for successful drug development because it is crucial to identify and exclude new drug candidates with unfavorable safety profiles as early as possible [108]. Predictive toxicology has recently been introduced to industrial chemical toxicity and is also relevant to the assessment of ENM hazard [109]. A predictive toxicological approach for ENM hazard screening could, for instance, include the assessment of injury at cellular and molecular levels as a way to predict adverse biological effects and health outcomes *in vivo* [12, 17, 49, 52, 53]. Evidence that such a mechanistic approach is possible emerged from the study of the adverse health effects of ambient particulate matter [36, 110, 111]. The physicochemical properties of ambient ultrafine particles (UFP), including their small size, large surface area, and high content of redox-cycling organic chemicals and transition metals, are instrumental in these particles' pro-inflammatory effects in cellular targets such as macrophage, epithelial, endothelial, and dendritic cells [112]. Similar responses in the lung and cardiovascular system likely play a role in the pathogenesis of inflammatory disease states such as allergic airway inflammation and atherosclerosis.

While no definitive disease processes have emerged as a result of ENM exposure in humans [113], a number of research studies have shown correlation between toxicological effects at the cellular level and organ injury at the intact animal level. Becher et al. [114] showed good correlation between the pro-inflammatory effects (IL-6, TNF-α and MIP-2) of stone particles (e.g., quartz, feldspar, and mylonite) in macrophages and epithelial cells and their ability to generate polymorphonuclear (PMN) inflammation in the lungs of rats. Sayes et al. [47] failed to demonstrate a correlation between the cellular and *in vivo* results when comparing carbonyl iron, crystalline silica, amorphous silica, nano-ZnO, and fine-sized ZnO in a well-designed dose–response study. This included measurement of lactate dehydrogenase (LDH) release, metabolic activity (MTT assay), and cytokine production (IL-6, TNF-α and MIP-2) in rat lung epithelial cells and alveolar macrophages versus measurement of PMN cell count or LDH values in the BAL fluid of rats. However, upon reanalysis of the previous data set, Rushton et al. [49] demonstrated that there was indeed a positive correlation if the particle mass was converted to SAD and the analysis performed at the steepest slope of the dose–response curve. Thus, the picture that emerged in the reanalysis was a good correlation between MIP-2 levels in cells versus the PMN response in the lung or LDH release from cells versus the PMN response in the BAL fluid.

The conclusion was that it is possible to show *in vitro/in vivo* predictions when using a surface area-normalized response metric. The Oberdörster laboratory [49] independently demonstrated through cell-free and cell-based measurement of reactive oxygen species (ROS) production, LDH release, and the use of an IL-8 promoter-luciferase reporter assay that there is a correlation between *in vitro* and acute pulmonary inflammation (PMN levels) in rats being challenged by intratracheal instillation of seven distinct particle types (Au, nano-TiO_2, fine TiO_2, NH2-PS, Ag, elemental carbon, and Cu). In addition, Ken Donaldson's laboratory (Edinburgh,

Scotland) has demonstrated that IL-8 production in A549 cells, exposed to a panel of low toxicity (e.g., TiO_2, carbon black) versus highly reactive quartz and metal (e.g., Ni, Co) nanoparticles, correlates with BAL polymorphonuclear cell counts in Wistar rats [52, 53]. This group also demonstrated that the expression of the particles' SAD versus PMN counts in the BAL fluid yields a shallow dose–response curve for low-toxicity particles, whereas highly reactive materials produced a steeper dose–response curve due to a high "surface reactivity." Thus, although such predictive modeling and correlations remain at an early stage, it appears that if appropriate response metrics are chosen and corrected to appropriate dose metrics, it is possible to develop reliable scientific paradigms that allow cellular screening to predict *in vivo* hazard potential [49].

Even if a link is established between *in vitro* and *in vivo* toxicological outcomes, human disease pathogenesis is dependent on real-life exposures at toxicologically relevant doses and distinguishes between dose-dependent acute versus chronic exposures. Fate and transport as well as exposure assessments are key ingredients that are not included in the predictive toxicological paradigm but are important ingredients for proper risk assessment. There are also chronic toxicological scenarios that involve a series of initiation and promoter events that cannot be simulated by a one-step toxicological paradigm. An example is the oncogenesis that is required to transform chronic granulomatous peritoneal inflammation into a mesothelioma in response to asbestos fibers [26, 30]. Although a screening assay for "frustrated phagocytosis" in response to long and biopersistent fibers may predict chronic mesothelial inflammation, this response profiling will not shed light on the mutagenic events that are required for development of a mesothelioma. This may require another event such as p53 gene knockout to elucidate the secondary event [30].

8.2 Example of the Use of Multidisciplinary Research in the University of California Center for the Environmental Impact of Nanotechnology Leading to the Establishment of Knowledge for the Safe Implementation of Nanotechnology in the Environment

Contact person: André Nel, University of California, Los Angeles (UCLA)

The mission of the University of California Center for the Environmental Impact of Nanotechnology (UC CEIN) is to develop a broad-based predictive scientific model [17, 61] premised on ENM properties and behavior that determine ENM spread to the environment, bioaccumulation, trophic transfer, and catalysis of potentially hazardous interactions at cellular, tissue, organism, and ecosystem levels (Fig. 6). The key components of this multidisciplinary model include the following: (1) the construction of well-characterized compositional and combinatorial ENM libraries to reflect the most abundant materials in the

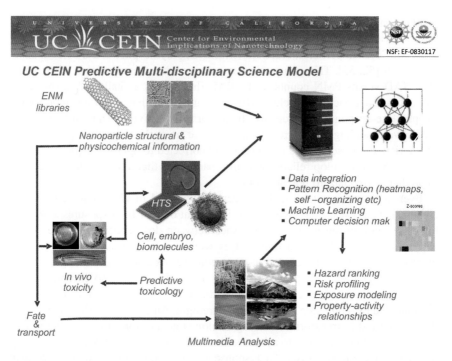

Fig. 6 UC CEIN uses a predictive multidisciplinary model for hazard ranking and risk profiling (Courtesy of V. Castranova)

marketplace; (2) the fate and transport of ENM, including methods of release and physicochemical and transport properties that could lead to interactions with biological substrates; (3) biomolecular and cellular injury mechanisms that relate to bio-physicochemical interactions at the nano-bio interface [20]; (4) use of injury mechanisms and bio-physicochemical interactions at the nano–bio interface to perform high-throughput screening in tissue culture cells, bacteria, yeast, and embryos; (5) use of the *in vitro* relationships to understand the possible harm to different strata or trophic life forms in freshwater, seawater, and terrestrial environments, including the identification of sentinel species to screen for ENM hazard in the environment; and (6) computational decision-making tools that utilize data capture and processing in the center for machine learning and provide a series of modeling predictions (Fig. 6).

Predictive science as practiced at UC CEIN refers to each scientific discipline performing research that predicts or informs every other discipline what those investigators may expect to find if they utilize a common set of compositional ENM libraries as well as materials that are made to systematically vary property or property sets to study biological effects at cellular, organism and population level. An attempt is made to elucidate cellular, bacterial, yeast or embryo stress responses, including through high throughput screening, that are also relevant to whole organisms that are being studied at increasing trophic level in freshwater, seawater and terrestrial mesocosms.

Fate and transport assessment as well as multi-media modeling are performed to determine how the alteration of the primary material properties in response to real-life environmental media may contribute to ENM spread, exposure, bio-accumulation and bio-processing. Computational biological and computerized decision tool are involved in data integration for purpose of hazard ranking, exposure modeling, risk profiling, and construction of property–activity relationships. These research activities are being combined with educational programs to inform the public, future generations of scientists, Federal and state agencies, and industrial stakeholders of the importance of safe implementation of nanotechnology in the environment.

Since its founding in September 2008, the UC CEIN (http://cein.ucla.edu) has successfully integrated the expertise of engineers, chemists, colloid and material scientists, ecologists, marine biologists, cell biologists, bacteriologists, toxicologists, computer scientists, and social scientists into a synergistic research program that has demonstrated the feasibility of using well-designed and well-characterized metal oxide libraries (TiO_2, CeO2, and ZnO) as well as property variations (e.g., size, shape, dissolution, and band gap tuning) to study ENM behavior in different environmental media and under different biological conditions [115–117]. The implementation of this research is being facilitated by the development of protocols to harmonize particle suspension, dispersal, and initiation of experiments under freshwater, seawater, and tissue culture conditions [118]. This illustrates the importance of multidisciplinary collaboration and harmonization efforts at national and international levels.

Collaborative research at the UC CEIN has identified the key material properties that lead to aggregation and sedimentation of the metal oxides in seawater, freshwater, and groundwater environments, and has also illustrated the ease with which these nanoparticles can be stabilized by capping agents under freshwater conditions, including the likelihood of inhibiting or averting spread to wastewater treatment plants and storm-water runoffs [115, 119, 120]. The availability of the nanoparticle libraries has facilitated the implementation of rapid-throughput screening studies that utilize a robotized and automated high-throughput screening laboratory, epifluorescence microscopy, and reporter cell lines to perform hazard ranking and analysis of the property–activity relationships at the cellular level that may predict *in vivo* toxicity [12]. The differential toxicity at cellular level has been further reflected by similarities and differences in the toxicity of these materials in bacteria, algae, phytoplankton, germinating seeds, sea urchins, and zebrafish embryos [117, 121].

There are other illustrations of the importance of the UC CEIN multidisciplinary approach to generating knowledge about nano-EHS. Mesocosm studies being carried out in collaboration with dynamic energy budget modeling have demonstrated that specific ENM properties contribute to the environmental impact at the population level and bioaccumulation at higher trophic levels in terrestrial and freshwater environments. The UC CEIN has obtained strong confirmation of the high toxicity of ZnO in primary producers in aquatic environments and could ascribe that to particle dissolution and the release of toxic Zn^{++}. This relationship was confirmed by high-throughput screening and property-activity analyses that have allowed the synthesis of less toxic ZnO nanoparticles through Fe doping [12]. The accompanying change in the particle matrix decreased Zn^{++} shedding, thereby lowering toxicity in cellular assays, bacteria, zebrafish embryos, and rodent lungs. Another potentially useful procedure for exposure reduction involved ENM removal from the experimental aqueous systems through optimal pH destabilization, coagulant dosing, sedimentation, and ultrafiltration. This research also allowed computerized modeling to study nanoparticle aggregation under various environmental and experimental conditions. Data capturing and analysis in the computerized expert system allow the development of novel feature selection algorithms to screen and rank nanoparticle properties to establish quantitative property–structure relationships. In summary, the integration at UC CEIN of multidisciplinary scientific platforms has been a particularly fruitful pathway to better understanding of environmental, health, and safety aspects of nanotechnology.

8.3 Quantitative Assessment of Environmental Exposure to Engineered Nanomaterials from Wastewater Systems

Contact person: Paul Westerhoff, Arizona State University, Tempe

Wastewater treatment plants (WWTPs) are major sources of ENM introduction into aquatic and terrestrial ecosystems. With more than 16,000 WWTPs in the United States alone that serve more than 75% of the population, WWTPs serve as interceptors of materials from residential, commercial, and industrial sources. The commercial introduction of engineered nanomaterials is already leading to a detectable footprint in sewage at WWTPs, such that it is possible to differentiate ENMs from natural colloids containing similar elements. Studies of these systems are beginning to demonstrate how properties of ENMs affect their removal from biological wastewater treatment and lead to their distribution in liquid effluent discharged to lakes and rivers or biosolid sludges that are often applied to land-based disposal sites such as agriculture crops.

Commercial products containing ENMs have been widely used for more than a decade. Titanium dioxide is an example of an ENM used for many years. Several toothpaste products that are being disposed into sewage systems were analyzed and observed to contain aggregates (200 nm–500 nm in size) of near-spherical primary TiO_2 nanoparticles (30–50 nm in size) suspended in an organic matrix [122]. Electron microscopy imaging and elemental composition were greatly enhanced by removing the organic background matrix, by applying hydrogen peroxide, and by heating to 60°C. Other larger-sized titanium materials were identified in wastewater, including angular micron-sized titanium dioxide being mined and used in paints as well as nanostructured silver. The Westerhoff group has demonstrated that products containing nanostructured silver (e.g., some fabrics, shampoos, detergents, towels, and toys) release ionic and nano-size silver during use, some of which finds its way

Table 6 Titanium concentrations across a wastewater treatment plant

Sampling location	Titanium concentration (µg/L)	Biosolids concentration (µg/g-solids)
Raw sewage	180±51	
After primary settling	113±63	
After activated sludge and secondary settling	50	
After tertiary filtration	39	
Biosolids from primary settling		257
Biosolids from secondary settling		8,139

into sewage systems [14, 15, 123]. Likewise, it was demonstrated that fullerenes released from cosmetic products can be washed into sewage systems [15].

This experimental work helps confirm estimates that predict ENM release as part of lifecycle assessments. These models predict that TiO_2 will occur at the highest levels among several types of ENMs. Results from sampling at one WWTP are shown in Table 6. Overall, the facility removed nearly 80% of the influent titanium. Titanium was accumulated in the biosolids (settled bacterial materials). Titanium dioxide nanoparticles were imaged in the samples: (1) liquid effluent contained primarily nanoscale nearly spherical TiO_2 and (2) biosolids contained spherical nanoscale TiO_2, angularly shaped micron-sized TiO_2, and micron-sized sediment containing titanium, silicates, and other elements. Sampling at a dozen other WWTPs is showing similar trends and indicate that the type of wastewater treatment (e.g., fixed vs. attached bacteria, or sedimentation vs. membrane bioreactors) affects the potential to remove ENMs such as TiO_2.

Because ENMs other than TiO_2 are not yet used in high enough quantities, Westerhoff's group has developed laboratory batch experiments to compare ENM removal capabilities. Batch sorption tests between ENMs and wastewater bacteria show that different types of ENMs exhibit different affinities for bacterial surfaces (Fig. 7). They have shown that standard protocols of EPA's Office of Pollution Prevention and Toxics (OPPT) used to evaluate organic chemical pollutant removal during wastewater treatment are not suitable for ENMs, and new protocols are required (unpublished data). Separate long-term operational experiments that simulate WWTPs indicate that ENMs in mg/L quantities in sewage have negligible effects on the biological function (nutrient removal) of WWTPs [125].

While Westerhoff's group has made significant progress in understanding the fate of ENMs during wastewater treatment and their likelihood to enter aquatic systems (river and streams), they are just beginning to understand the fate of ENMs in biosolids that may be land-applied, incinerated, or otherwise disposed. Improved analytical techniques are required to differentiate ENMs from natural or non-engineered forms of colloids of similar composition (e.g., titanium as discussed earlier, or silver from silver chloride). National reconnaissance monitoring projects should be conducted to assess current levels of ENMs in wastewaters (raw sewage and effluents), biosolids, and rivers receiving wastewater. The beneficial effects of ENM removal at WWTPs could be greatly enhanced by understanding the fundamental interaction of

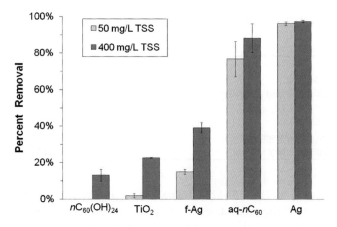

Fig. 7 Propensity of ENMs to biosorb to wastewater bacteria [124]

ENMs of different size, charge, and composition with the surfaces of different types of bacteria (gram negative or positive, filamentous, etc.).

8.4 Public-Private Partnerships for Nano-EHS Awareness and Risk Reduction Strategies

Contact persons: David Grainger, The University of Utah, Salt Lake City; Santokh Badesha, Xerox Corporation

Industries interested in commercializing emerging nanotechnologies face the usual market risks of any new product development, but these risks are compounded by the uncertainties of worker and consumer safety, unknown regulatory restrictions, and possible public backlash in the current era of disinformation and unknowns [3, 37, 58, 89, 107, 113]. Additionally, for companies operating multinationally, there is high probability that any regulations imposed on nanotechnology's use and dissemination will be highly variable across borders (Lux [35]). Private-public partnerships (PPPs) can help provide both the structures and conduits for information flow to and from nanotechnology stakeholders to stymie reaction and stigma that might otherwise unfairly plague this developing industry at this early stage [126]. A variety of PPP research models are available, including those of the U.S. domestic microelectronics organization Sematech (http://www.sematech.org/corporate), the European Union's Sixth and Seventh Framework Programmes (FP6 and FP7; [127]), and the U.S. government's National Institute of Standards and Technology (NIST) Advanced Technology Partnership program [128]. Nonetheless, there are currently few readily recognized or known PPPs that link governments or NGOs with private companies to jointly produce risk governance, best practices, and safety guidelines in nanotechnology and commercialization. Yet, it is likely to be in the best interests of businesses to foster an open dialogue with the various public and private constituents involved in the current discussions of the risks and benefits of nanotechnologies. The PPP mechanism is well suited to promote stakeholder interests and transparency in developing nano-EHS risk governance.

One example of a working nano-safety partnership is the DuPont and Environmental Defense NANORisk Framework [40], an open information-gathering system to generate data to help support decisions and practices concerning the safe production and use of nanomaterials. Under development since late 2005, the pioneering program also offers guidance on how to communicate information and decision processes to key stakeholders. The intent of the Framework is to "promote responsible development of nanotechnology products, facilitate public acceptance, and support the development of a practical model for reasonable government policy on nanotechnology safety." The Framework strategy seeks to "define a systematic and disciplined process for identifying, managing, and reducing the risk of unintended consequences from engineered nanomaterials across all stages of a product's 'lifecycle'" [129].

Significantly, as a model, DuPont's private-public partnering efforts to address questions about nanomaterials extend to other working relationships it has with NGOs, including its involvement with the OECD. Through OECD's Business Industry Advisory Committee and related activities in OECD's Working Party on Manufactured Nanomaterials, DuPont helps to provide information on potential nanotechnology-related health and environmental issues. DuPont is also involved with the American Chemistry Council Nanotechnology Panel, providing information and recommendations to the U.S. EPA and the chemical industry on safety, health, and environmental issues and regulatory guidelines for nanomaterials. DuPont was the first company to provide product information under the EPA's voluntary Nanomaterials Stewardship Basic Program (http://www.epa.gov/oppt/nano/stewardship.html). As a member of the European Chemical Industry Association and subordinate nanotechnology working groups, DuPont is helping to develop similar industry recommendations in Europe, representing the European Chemical Industry Council on the European Competent Authority working group reviewing nanotechnology in the context of the REACH chemicals regulation. DuPont has made a commitment [130] to participate actively in the ISO framework [131] for comprehensively evaluating and addressing potential environmental, health, and safety risks of nanomaterials and their applications.

DuPont has supported research at Rice University's Center for Biological and Environmental Nanotechnology and is a founding member of ICON, the International Council on Nanotechnology at Rice (http://icon.rice.edu). ICON represents industry, academia, regulatory agencies, and NGOs seeking to "assess, communicate, and reduce nanotechnology environmental and health risks while maximizing its societal benefit" [130].

While DuPont figures prominently as a current and past nanotechnology PPP participant and catalyst, other operational examples (e.g., ICON, NOSH, and OECD) currently foster PPPs in the nanotechnology risk-benefit dialog, and best practices are emerging. The EU FP7 Framework has recently announced a renewal of PPP targets for new research programming (see http://cordis.europa.eu/fp7/dc/index.cfm).

Generally speaking, all nanotechnology commercialization efforts should follow principles of good product stewardship and good risk management strategies in the design and manufacturing of products made with engineered nanomaterials.[5] To accommodate commercialization strategies and motives, industry response to emerging public attitudes and NGO positions on nanotechnology need to be based on facts and realism, with a rational and rapid recognition that nanotechnology as a young, dynamic field requires active, ongoing learning, rather than *post facto* reactions. Mutual stakeholder education essential to establishing public-private credibility and trust would be accelerated through open sharing of emerging experiences and data on a global basis. This is best facilitated through an open-exchange PPP mechanism that promotes active exchange of information with other industries, academia, public, and government agencies by enabling public disclosure of testing and possible risks of nanomaterials as the field and new products develop. Industry, governments, NGOs, and other stakeholders must openly collaborate to lay the proper foundation for imminent regulatory actions and to assess the potential for international voluntary agreements. To avoid backlash from relative positions of ignorance, stakeholders must be reassured that their respective concerns are considered and that private and public risk management institutions assigned to risk governance are held to accountability and articulated good practices.

It is likely that industry will advocate a system of voluntary risk governance and compliance rather than a unilaterally imposed legal regulatory enforcement (e.g., see http://www.cefic.be/en/Legislation-and-Partnership.html). Therefore, voluntary risk governance systems might best be proactively developed via a PPP mechanism to consider (1) development of standards and good practice guidelines encompassing basic research all the way to product testing and tracking, with methods for assessing hazards and exposure as a priority; (2) development of occupational safety guidelines, best practice scenarios, and information disclosure programs for consumers; and (3) establishment of transparent reporting processes and expectations, particularly for new data and events disclosures relevant to risk management. Nonetheless, voluntary reporting quality that assures adequate participation and transparency is difficult to achieve, and thus the desired watchdog function can be weak. Regardless of voluntary or mandatory governance, industries maintain concerns about protecting their intellectual property rights and intrinsic competitive advantage. Additionally, voluntary self-policing systems can often result in a "lowest common denominator" outcome, and as such, may not impose a sufficient incentive to those who prefer to operate outside of the voluntary system or choose not to comply.

Through PPP operations, emerging industry should try to expediently adopt preemptive, credible, and comprehensive self-regulations, which are often implemented more rapidly and efficiently than most governmental regulations. A continued

---

[5] See http://www.nanoandme.org/downloads/The Responsible Nano Code.pdf for examples of responsible risk management strategies.

and consistent focus on "best practices for risk governance" should be a priority. As a stakeholder, industry requires the continued capability to ensure technology leadership, harmonized global standards for risk assessment that ensure workplace and consumer safety and health, and a validated scientific base for efficient, appropriate adoption of regulations by engaging with academic scientific teams, policymakers, and NGOs as credible dialogue partners.

8.5 NIOSH Guidelines for Occupational Safety, Including the Use of Monitoring Equipment to Survey the Workplace

Contact person: Vincent Castranova, NIOSH

There has been a dramatic increase in production of ENMs—including CNTs—in recent years. Although aerosolization during handling of nanoparticles is feasible, data are lacking on the exposure levels in workplaces where nanoparticles are synthesized, packaged, used, or disposed. In addition, data concerning the effects of exposure to various types of nanoparticles are incomplete. The National Institute for Occupational Safety and Health (NIOSH) is conducting a multidisciplinary research program to (1) develop methods to monitor airborne levels of nanoparticles in the field; (2) determine airborne levels of nanoparticles in various workplaces and link peak exposures to certain work processes; (3) identify respiratory and systemic effects of pulmonary exposure of laboratory animals to various nanoparticles; (4) determine dose response, time course, mechanisms of action, and structure–function relationships; (5) develop models to relate responses in animal models to those in humans; (6) conduct risk assessment; and (7) evaluate the effectiveness of control technology and personal protective equipment. The NIOSH research plan is published in the Strategic Plan for NIOSH Nanotechnology Research and Guidance: Filling the Knowledge Gaps [132]. Progress reports are published by NIOSH on a regular basis (Progress towards Safe Nanotechnology in the Workplace; [133]). As a result of available data, NIOSH [39] has published Approaches to Safe Nanotechnology: Managing the Health and Safety Concerns Associated with Engineered Nanomaterial. This document suggests that in the absence of complete information, companies either manufacturing or working with nanomaterials should follow the precautionary principle and implement a risk management program in the workplace in order to minimize the risk of worker exposure to these materials. Critical elements of such a program include the following:

- Capability to anticipate new and emerging risks (hazard determination) and whether they are linked to changes in the manufacturing process, equipment, or introduction of new materials
- Installation and evaluation of engineering controls (e.g., exhaust ventilation and dust collection systems)
- Evaluation of the effectiveness of controls through monitoring of airborne nanoparticles in the workplace

- Education and training of workers in the proper handling of nanomaterials (e.g., in safe work practices)
- Selection and use of personal protective equipment (e.g., clothing, gloves, and respirators)

NIOSH has evaluated the air environment of several nanotechnology worksites using a sophisticated array of particle analysis instrumentation to determine particle size distribution, mass concentration, number concentration, mass median aerodynamic diameter, count median aerodynamic diameter, and particle surface area. Examples of the instrumentation are shown in Fig. 8. Since this instrumentation is bulky and not commonly available to industrial hygienists, NIOSH has also developed the Nanoparticle Emission Assessment Technique that uses common, handheld, real-time monitors to evaluate workplace levels of airborne nanoparticles [134].

In addition, NIOSH is reviewing existing toxicology data and conducted risk assessment to recommend exposure limits to selected nanoparticles. A Current Intelligence Bulletin [135] evaluated tumor induction data in rats after long-term inhalation of fine or nano-sized TiO_2 and will recommend an exposure limit for the nanosized form that is an order of magnitude lower than for the fine form. This document is in the final stages of review before release. NIOSH is also drafting a Current Intelligence Bulletin [136] that notes the congruence of data from the number of animal studies for granulomatous inflammation or fibrosis in response to SWCNTs and MWCNTs and will conduct risk assessments from these data to recommend an exposure limit. This development is illustrated in Fig. 1.

Fig. 8 Example of field application of instruments needed for real-time measurement of number, mass, size distribution, and surface area of engineered nanomaterials (Courtesy of A. Nel)

9 International Perspectives from the Overseas Workshops

9.1 United States-European Union Workshop (Hamburg, Germany)

Panel members/discussants
Bengt Fadeel (co-chair), Karolinska Institutet, Stockholm, Sweden
André Nel (co-chair), University of California, Los Angeles (UCLA), United States
Peter Dobson, Oxford University, United Kingdom
Rob Aitken, Institute of Occupational Medicine, Edinburgh, United Kingdom
Kenneth Dawson, University College Dublin, Ireland
Wolfgang Kreyling, Helmholtz Centre, Munich, Germany
Lutz Mädler, University of Bremen, Germany
George Katalagarianakis, European Commission
Ilmari Pykkö, University of Tampere, Finland
Jean-Christophe Schrotter, Anjou Recherche, Water Research Center of Veolia Water, France

Overall, there has been a huge increase in activity in the nano-EHS field in the past decade, but emphasis has been on hazard assessment, and less progress has been made on exposure issues. Therefore, the available information is insufficient for predictions of *in vivo* effects or effects on human health. Moreover, the toxicological results generated to date do not allow for comprehensive conclusions on nanomaterial safety, due to conflicting data related to issues of physico-chemical characterization of materials but also due to the fact that the sheer numbers of different nanomaterials that are currently being produced and explored, with tunable compositions and structures, make it challenging to address EHS outcomes. More systematic research is thus needed. There is an awareness that nanomaterials need to be studied on a case-by-case basis in order to discern associations between specific material properties and hazardous effects. Hence, research on EHS issues pertaining to nanomaterials is an interdisciplinary exercise involving researchers in material sciences, biology, (eco)-toxicology, medicine, and so on. Moreover, paradigms have emerged to support our understanding of the interaction and/or interference of nanomaterials with biological systems.

The panel members agreed that there should be more focus on "nanosafety" instead of addressing only "nanotoxicology." In other words, safety assessment of engineered nanomaterials should not be a barrier to development but rather should enable the safe and sustainable development of nanotechnology. The concepts of "safety-by-design" (i.e., intelligent material design to mitigate adverse effects on human health and the environment) and proactive risk management of ENMs were also promoted by workshop participants. To this end, more systematic research is needed in the field of EHS, making use of high-throughput screening (HTS) and systems biology approaches. The implementation of such technologies could also aid in the reduction of the number of animal experiments by serving as a triage system for ENMs. The panel highlighted the need to focus on the following emerging topics:

- Development of new methods for detection and characterization of nanomaterials *in situ*, i.e., in living systems and relevant environmental matrices
- Standardization and validation of test methods for the assessment of hazards of nanoparticles as well as more complex nano-systems
- Understanding bio-nano interactions, including the behavior and fate of engineered nanomaterials *in vivo*, e.g., navigation of nanoparticles into the brain and other organs
- Long-term *in vivo* toxicity studies of selected nanomaterials, applying realistic doses, with assessment of genotoxicity end-points
- Monitoring of human as well as environmental/ecological exposure to nanoparticles to allow for risk assessment of these materials
- Development of HTS platforms and QSARs
- Systems biology approaches for profiling/fingerprinting of categories of ENMs
- Implementation of a "safety culture," i.e., a system of certified testing, labeling, etc., to manage the risk of nanomaterials

Overall, the emerging concept in the field of environmental, health, and safety (EHS) of nanomaterials assessment is "safety by design" as a result of the development of reliable and predictive test methods. Fostering international cooperation (as in the recent joint EU–US call on modeling) will be important, as will be sharing of research facilities and infrastructures (as in the European NanoSafety Cluster of FP6 and FP7 projects) [137]. Moreover, interdisciplinary education of the next generation of nanosafety researchers is also needed.

9.2 United States-Japan-Korea-Taiwan Workshop (Tsukuba, Japan)

Panel members/discussants

Tatsujiro Suzuki (co-chair), University of Tokyo; Japan Atomic Energy Commission, Japan,
André Nel (co-chair), University of California (UCLA), United States
Masafumi Ata, National Institute of Advanced Industrial Science and Technology (AIST), Japan
Masashi Gamo, Research Center for Chemical Risk Management, AIST, Japan
Satoshi Ishihara, Japan Science and Technology Agency (JST), Japan
Chin-Chung Tsai, Chiao Tung University, Taiwan
Chung-Shi Yang, Center for Nanomedicine Research, National Health Research Institutes, Taiwan

The following is a summary of the key points discussed during the session.

9.2.1 Changes in the Vision over the Last 10 Years

- The need to address the potential risks by this new science is now widely acknowledged and has changed nanotechnology from a business and science dream to an inclusive societal feature.
- NGOs have advanced the codes of conduct, ethics, etc., of nanotechnology.
- Governments have begun to address regulatory issues, are viewing existing regulations, and looking at where there may be nanotechnology-specific issues.

9.2.2 Advances in Last 10 Years

- Many EHS studies have been conducted in all Asian countries, and some regulatory bodies have taken action, but no long-term policy or strategy has been established. Large companies are more active in addressing EHS issues than are small companies/startups.

9.2.3 Vision for the Next 10 Years

- In Asia, future funding trends for EHS research is uncertain. While steady funding in Taiwan is expected for next 5 years, it is not clear what will happen after the national program ends. Taiwan is spending about 10% of total R&D expenditure on EHS, while Japan is spending less than 2% on average; however, this expenditure fluctuates every year.
- While the United States is expecting that predictive toxicology for ENMs using computer and simulated modeling can play an important role, in Asia it is currently viewed as being of questionable value.
- All participants agreed that public involvement/outreach is critically needed in addressing potential EHS concerns.
- While there are new efforts on technology assessment in Japan, a new focus has emerged on the use of "distributed governance," which is premised on collective knowledge dissemination and not necessarily associated with government agencies. In Taiwan, the "nanoMark program" promotes best practices; this program has been successful in actively involving consumer groups.
- International collaboration in EHS research is considered a key factor. Japan and Korea are both involved in the Working Party on Manufactured Nanomaterials (WPMN) of the OECD, and Taiwan is interested in joining the WPMN sponsorship program. Japanese scientists also participate in the voluntary International Association of Nano Harmonization (IANH). The ISO Nanotechnologies Technical Committee (TC229) standards activities are also important.

9.2.4 Goals for 2020

- International harmonization is an important goal, for instance, in terms of development of standardized methods, risk evaluation, and risk assessment and management protocols. Korea has proposed in ISO/TC229 the use of nano-MSDSs, while the Taiwanese equivalent of the U.S. EPA uses a policy similar to TSCA in the United States.
- Classification of some ENMs as toxic substances should be considered an important goal.
- Institutionalization of technology assessment should be realized. It means that such activity needs to be an embedded function of societal efforts. The funding source should be stable and should be routinely carried out by an independent agency.
- Vision to develop tools and processes for public engagement is needed to assure responsible development of nanotechnologies. This vital effort should involve all stakeholders, such as the scientific community, public, government, industry, and media.
- International collaboration is important in sharing best practices for public engagement. The U.S. Centers for the Environmental Implications of Nanotechnology (CEIN) is an interesting model that could be followed in Asia.
- There is a strong need for information sharing, common databases, and for research that uses standard protocols to generate comparable data. Three Taiwan agencies are developing common databases. There is also a need to encourage industry to share data and information about the use of nanotechnology in its products. For example, there is a need to know what products have nanotechnology in them in order to assess exposure, hazard, and risks.

9.3 United States-Australia-China-India-Singapore Workshop (Singapore)

Panel members/discussants

Yuliang Zhao (co-chair), Chinese Academy of Sciences (CAS); CAS Key Laboratory for Biomedical Effects of Nanomaterials and Nanosafety, China
André Nel (co-chair), University of California, Los Angeles (UCLA), United States
Graeme Batley, Commonwealth Scientific and Industrial Research Organization, Australia
Graeme Hodge, Monash University, Australia
Joachim Loo, Nanyang Technological University, Singapore
Yiyan Yang, Institute of Bioengineering and Nanotechnology, Singapore
Yong Zhang, National University of Singapore

The following is a summary of the key points discussed during the session.

9.3.1 Changes in the Vision over the Last 10 Years

- The need to address the potential risks by this new science is now widely acknowledged and has changed nanotechnology from a business and science dream to an inclusive societal feature.
- NGOs have advanced the codes of conduct, ethics, etc. for nanotechnology.
- Governments have begun to address regulatory issues, are viewing existing regulations, and are looking at nanotechnology-specific issues.

9.3.2 Advances in Last 10 Years

- Over the last 10 years, more than 20 ENMs have been tested for potential toxicity. Although the initial toxicological data were inconsistent due to insufficient material characterization, current data reported in the literature are more consistent and reproducible due to more stringent characterization and harmonized test efforts
- The number of nanotechnology characterizations and definitions has narrowed, among which the recently published ISO definition of "nanotechnology" represents a significant advance.
- Rapid- as well as high-throughput screening techniques for assessment of potential toxicity of nanomaterials have been proposed, and implementation has begun.

9.3.3 Vision for the Next 10 Years

- Exciting advances in nanotechnology applications will occur, enabled by continuous incremental progress in nano-EHS issues.
- Since more is understood about the mechanisms and properties leading to nanomaterial hazard, safe implementation and design of nanomaterials have become possible.
- There is a need to establish nanotechnology-specific regulatory procedures for risk assessment of nanomaterials, including for governance.
- Guidelines must be developed for safe use of nanomaterials/nanotechnology in applications, and there should be continued development of self-regulation.

9.3.4 Goals for 2020

- **Goal 1:** Knowledge-generation about nanomaterial properties that could pose hazard at the biological level at a rate commensurate with the expansion of nanotechnology and new products.
  - Barrier: One-material-at-a-time analysis is impractical, given the large number of properties and the many new materials being produced.

- Solution: Large-scale implementation of high-throughput screening techniques.
- **Goal 2:** To consider the safety of ENMs at the initial stages of their development and the development of the products incorporating them.
  - Barrier: Inability to predict whether ENMs with potentially hazardous properties may pose biological hazards.
  - Solution: Develop safe ENMs "by design" using principles similar to those of "green chemistry," e.g., coated materials to reduce/eliminate toxicity.
- **Goal 3:** The development of public confidence in/acceptance of nanotechnology as a result of all of the above.

9.3.5 Infrastructure Needs

- Funding for EHS research, integrated with applications development
- Nanomaterial reference libraries
- Databases of properties
- Instrumentation
- Environment detection poses a "grand challenge"

9.3.6 R&D Strategies

- Standardized assays and methodologies, validated and internationally accepted
- International cooperation, leveraging, e.g., OECD, WPMN, ISO, and others
- Industry participation (including funding support) and role in nano-EHS efforts

9.3.7 Emerging Topics and Priorities

- Occupational safety studies and defining of LOD and minimal exposure thresholds
- Mechanisms of nanotoxicity (important for predictable and designable nanotechnology), reliable ADME/Tox data (important for development of safety assessments)
- Modeling of risk assessment, fate and transport, and QSARs
- Nano-ESH methodology development
- Nano-informatics
- Addressing the current lack of knowledge concerning impacts on the environment, fate and transport, bioaccumulation, trophic transfer, etc.
- Knowledge translation: making "nano" accessible to the general public, increase the public's trust in science

References

1. National Science, Engineering, and Technology (NSET) Subcommittee of the Committee on Technology of the National Science and Technology Council, Environmental, health, and safety research needs for engineered nanoscale materials (NSET, Washington, DC, 2006), Available online: http://www.nano.gov/html/res/pubs.html
2. V.L. Colvin, The potential environmental impact of engineered nanomaterials. Nat. Biotechnol. **21**(10), 1166–1170 (2003)
3. A.D. Maynard, R.J. Aitken, T. Butz, V. Colvin, K. Donaldson, G. Oberdörster, M.A. Philbert, J. Ryan, A. Seaton, V. Stone, S.S. Tinkle, L. Tran, N.J. Walker, D.B. Warheit, Safe handling of nanotechnology. Nature **444**(7117), 267–269 (2006)
4. A.E. Nel, T. Xia, L. Madler, N. Li, Toxic potential of materials at the nanolevel. Science **311**(5761), 622–627 (2006)
5. G. Oberdörster, A. Maynard, K. Donaldson, V. Castranova, J. Fitzpatrick, K. Ausman, J. Carter, B. Karn, W. Kreyling, D. Lai, S. Olin, N. Monteiro-Riviere, D. Warheit, H. Yang, ILSI Research Foundation/Risk Science Institute Nanomaterial Toxicity Screening Working Group., Principles for characterizing the potential human health effects from exposure to nanomaterials: elements of a screening strategy. Fibre Toxicol. **2**, 8 (2005). doi:10.1186/1743-8977-2-8
6. A. Seaton, L. Tran, R. Aitken, K. Donaldson, Nanoparticles, human health hazard and regulation. J. R. Soc. Interface **7**, S119–S129 (2010)
7. National Institute of Environmental Health Sciences (NIEHS), Toxicology in the 21st century: the role of the National Toxicology Program (NIEHS, Research Triangle Park, 2004), Available online: http://ntp.niehs.nih.gov/ntp/main_pages/NTPVision.pdf
8. National Research Council, *ToxicityTesting in the 21st Century: A Vision and a Strategy* (National Academies Press, Washington, DC, 2007), Available online: http://www.nap.edu/catalog.php?record_id=11970#toc or http://dels.nas.edu/resources/static-assets/materials-based-on-reports/reports-in-brief/Toxicity_Testing_final.pdf
9. N. Walker, J.R. Bucher, A 21st century paradigm for evaluating the health hazards of nanoscale materials? Toxicol. Sci. **110**, 251–254 (2009)
10. V.C. Abraham, D.L. Taylor, J.R. Haskins, High-content screening applied to large-scale cell biology. Trends Biotechnol. **22**, 15–22 (2004)
11. V.C. Abraham, D.L. Towne, J.F. Waring, U. Warrior, D.J. Burns, Application of a high-content multi-parameter cytotoxicity assay to prioritize compounds based on toxicity potential in humans. J. Biomol. Screen. **13**, 527–537 (2008)
12. S. George, S. Pokhrel, T. Xia, B. Gilbert, Z. Ji, M. Schowalter, A. Rosenauer, R. Damoiseaux, K.A. Bradley, L. Mädler, A.E. Nel, Use of a rapid cytotoxicity screening approach to engineer a safer zinc oxide nanoparticle through iron doping. ACS Nano **4**, 15–29 (2010)
13. R.F. Service, Nanotechnology: can high-speed tests sort out which nanomaterials are safe? Science **321**(5892), 1036–1037 (2008)
14. T.M Benn, B. Cavanagh, B.K. Hristovski, J. Posner, P. Westerhoff, The release of (nano) silver from consumer products used in the home. J. Environ. Qual., published online 12 July 2010. doi:10.2134/jeq2009.0363
15. T.M Benn, P. Westerhoff, P. Herckes, Detection of fullerenes (C60 and C70) in commercial cosmetics. Environ. Pollu. **159**(5), 1334–1342 (2011) http://www.sciencedirect.com/science/article/
16. D.B. Warheit, C.M. Sayes, K.L. Reed, K.A. Swain, Health effects related to nanoparticle exposures: environmental, health, and safety considerations for assessing hazards and risks. Pharmacol. Ther. **120**, 35–42 (2008)
17. H. Meng, T. Xia, S. George, A.E. Nel, A predictive toxicological paradigm for the safety assessment of nanomaterials. ACS Nano **3**, 1620–1627 (2009)
18. National Toxicology Program (NTP), Toxicology in the 21st century: the role of the National Toxicology Program (Department of Health and Human Services, NIEHS/NTP, Research Triangle Park, 2004), Available online: http://ntp.niehs.nih.gov/ntp/main_pages/NTPVision.pdf

19. J.E. Hutchinson, Greener nanoscience: a proactive approach to advancing applications and reducing implications of nanotechnology. ACS Nano **2**, 395–402 (2008)
20. A.E. Nel, L. Madler, D. Velegol, T. Xia, E.M.V. Hoek, P. Somasundaran, F. Klaessig, V. Castranova, M. Thompson, Understanding biophysicochemical interactions at the nano-bio interface. Nat. Mater. **8**, 543–557 (2009)
21. M.C. Roco, Environmentally responsible development of nanotechnology. Environ. Sci. Technol. **39**(5), 106A–112A (2005). doi:10.1021/es053199u
22. M. Lundqvist, J. Stigler, G. Elia, I. Lynch, T. Cedervall, K.A. Dawson, Nanoparticle size and surface properties determine the protein corona with possible implications for biological impacts. Proc. Natl. Acad. Sci. U. S. A. **105**, 14265–14270 (2008)
23. C.W. Lam, J.T. James, R. McCluskey, R.L. Hunter, Pulmonary toxicity of single-wall carbon nanotubes in mice 7 and 90 days after intratracheal instillation. Toxicol. Sci. **77**, 126–134 (2004)
24. Z. Liu, *In vivo* biodistribution and highly efficient tumour targeting of carbon nanotubes in mice. Nat. Nanotechnol. **2**, 47–52 (2007)
25. R. Mercer, R.J. Scabilloni, L. Wang, E. Kisin, A.R. Murray, D. Schwegler-Berry, A.A. Shvedova, V. Castranova, Alteration of deposition pattern and pulmonary response as a result of improved dispersion of aspirated single-walled carbon nanotubes in a mouse model. Am. J. Physiol. Lung Cell. Mol. Physiol. **294**, L87–L97 (2008)
26. C.A. Poland, R. Duffin, I. Kinloch, A. Maynard, W.A.H. Wallace, A. Seaton, V. Stone, S. Brown, W. MacNee, K. Donaldson, Carbon nanotubes introduced into the abdominal cavity of mice show asbestos-like pathogenicity in a pilot study. Nat. Nanotechnol. **3**, 423–428 (2008)
27. D.W. Porter, A.F. Hubbs, R.R. Mercer, N. Wu, M.G. Wolfarth, K. Sriram, S. Leon, L. Battelli, D. Schwegler-Berry, S. Friend, M. Andrew, B.T. Chen, S. Tsuruoka, M. Endo, V. Castranova, Mouse pulmonary dose- and time course-responses induced by exposure to multi-walled carbon nanotubes. Toxicology **269**(2–3), 136–147 (2010). 10
28. A.A. Shvedova, V. Castranova, E.R. Kisin, D. Schwegler-Berry, A.R. Murray, V.Z. Gandelsman, A. Maynard, P. Baron, Exposure to carbon nanotube material: assessment of nanotube cytotoxicity using human keratinocyte cells. J. Toxicol. Environ. Health A **66**, 1909–1926 (2003)
29. A.A. Shvedova, E.R. Kisin, R. Mercer, A.R. Murray, V.J. Johnson, A.I. Potapovich, Y.Y. Tyurina, O. Gorelik, S. Arepalli, D. Schwegler-Berry, A.F. Hubbs, J.S. Antonini, D.E. Evans, B.K. Ku, D. Ramsey, A. Maynard, V.E. Kagan, V. Castranova, P. Baron, Unusual inflammatory and fibrogenic pulmonary responses to single-walled carbon nanotubes in mice. Am. J. Physiol. Lung Cell. Mol. Physiol. **289**, L698–L708 (2005)
30. A. Takagi, A. Hirose, T. Nishimura, N. Fukumori, A. Ogata, N. Ohashi, S. Kitajima, J. Kanno, Induction of mesothelioma in p53+/− mouse by intraperitoneal application of multiwall carbon nanotube. J. Toxicol. Sci. **33**, 105–116 (2008)
31. N.W.S. Kam, M. O'Connell, J.A. Wisdom, H. Dai, Carbon nanotubes as multifunctional biological transporters and near-infrared agents for selective cancer cell destruction. Proc. Natl. Acad. Sci. U. S. A. **102**, 11600–11605 (2005)
32. K. Kostarelos, The long and short of carbon nanotube toxicity. Nat. Biotechnol. **26**, 774–776 (2008)
33. A.E. Porter, Direct imaging of single-walled carbon nanotubes in cells. Nat. Nanotechnol. **2**, 713–717 (2007)
34. U.S. Environmental Protection Agency (U.S. EPA), TSCA inventory status of nanoscale substances: general approach (2008), Available online: http://www.epa.gov/oppt/nano/nmsp-inventorypaper2008.pdf
35. L. Research, *The Nanotech Report*, 5th edn. (Lux Research, New York, 2007)
36. T. Xia, M. Kovochich, M. Liong, L. Mäedler, B. Gilbert, H. Shi, J.I. Yeh, J.I. Zink, A.E. Nel, Comparison of the mechanism of toxicity of zinc oxide and cerium oxide nanoparticles based on dissolution and oxidative stress properties. ACS Nano **2**, 2121–2134 (2008)
37. D.B. Warheit, T.R. Webb, C.M. Sayes, V.L. Colvin, K.L. Reed, Pulmonary instillation studies with nanoscale TiO_2 rods and dots in rats: toxicity is not dependent upon particle size and surface area. Toxicol. Sci. **91**, 227–236 (2006)

38. D.D. Zhang, M.A. Hartsky, D.B. Warheit, Time course of quartz and TiO_2 particle: induced pulmonary inflammation and neutrophil apoptotic responses in rats. Exp. Lung Res. **28**, 641–670 (2002)
39. National Institute for Occupational Safety and Health (NIOSH), Approaches to safe nanotechnology: managing the health and safety concerns associated with engineered nanomaterials (DHHS (NIOSH) publication 2009–125, Washington, DC, 2009), Available online: http://www.cdc.gov/niosh/topics/nanotech/safenano
40. Environmental Defense Fund (EDF), NANO risk framework (2007), Available online: http://nanoriskframework.com/page.cfm?tagID=1083
41. Environmental Working Group (EWG), Nanotechnology and sunscreens: EWG's 2009 sunscreen investigation Sect. 4 (2009), Available online: http://www.ewg.org/cosmetics/report/sunscreen09/investigation/Nanotechnology-Sunscreens
42. A. Kahru, H.-C. Dubourguier, From ecotoxicology to nanoecotoxicology. Toxicology **269**, 105–119 (2010)
43. U.S. Environmental Protection Agency (U.S. EPA), Federal insecticide, fungicide, and rodenticide act *(FIFRA)* (1996), Available online: http://www.epa.gov/oecaagct/lfra.html
44. P.V. Asharani, Y.L. Wu, Z. Gong, S. Valiyaveettl, Toxicity of silver nanoparticles in zebrafish models. Nanotechnology **19**, 255102–255110 (2008)
45. N.C. Mueller, B. Nowack, Exposure modeling of engineered nanoparticles in the environment. Environ. Sci. Technol. **42**, 4447–4453 (2008)
46. C.F. Jones, D.W. Grainger, *In vitro* assessments of nanomaterial toxicity. Adv. Drug Deliv. Rev. **61**, 438–456 (2009)
47. C.M. Sayes, K.L. Reed, D.B. Warheit, Assessing toxicity of fine and nanoparticles: comparing *in vitro* measurements to *in vivo* pulmonary toxicity profiles. Toxicol. Sci. **97**, 163–180 (2007)
48. K. Donaldson, P.J. Borm, G. Oberdörster, K.E. Pinkerton, V. Stone, C.L. Tran, Concordance between *in vitro* and *in vivo* dosimetry in the proinflammatory effects of low-toxicity, low-solubility particles: the key role of the proximal alveolar region. Inhal. Toxicol. **20**, 53–62 (2008)
49. E. Rushton, J. Jiang, S. Leonard, S. Eberly, V. Castranova, P. Biswas, A. Elder, X. Han, R. Gelein, J. Finkelstein, G. Oberdörster, Concept of assessing nanoparticle hazards considering nanoparticle dosemetric and chemical/biological response-metrics. J. Toxicol. Environ. Health A **73**, 445–461 (2010)
50. T.M. Sager, D.W. Porter, V.A. Robinson, W.G. Lindsley, D.E. Schwegler-Berry, V. Castranova, Improved method to disperse nanoparticles for *in vitro* and *in vivo* investigation of toxicity. Nanotoxicology **1**, 118–129 (2007)
51. J.G. Teeguarden, P.M. Hinderliter, G. Orr, B.D. Thrall, J.G. Pounds, Particokinetics *in vitro*: dosimetry considerations for *in vitro* nanoparticle toxicity assessments. Toxicol. Sci. **95**, 300–312 (2007)
52. R. Duffin, L. Tran, D. Brown, V. Stone, K. Donaldson, Proinflammogenic effects of low-toxicity and metal nanoparticles *in vivo* and *in vitro*: highlighting the role of particle surface area and surface reactivity. Inhal. Toxicol. **19**, 849–856 (2007)
53. C. Monteiller, L. Tran, W. MacNee, S. Faux, A. Jones, B. Miller, K. Donaldson, The proinflammatory effects of low-toxicity low-solubility particles, nanoparticles and fine particles, on epithelial cells *in vitro*: the role of surface area. Occup. Environ. Med. **64**, 609–615 (2007)
54. T. Xia, M. Kovochich, M. Liong, J.I. Zink, A.E. Nel, Cationic polystyrene nanosphere toxicity depends on cell-specific endocytic and mitochondrial injury pathways. ACS Nano **2**, 85–96 (2008)
55. G. Oberdörster, E. Oberdörster, J. Oberdörster, Concepts of nanoparticle dose metric and response metric. Environ. Health Perspect. **115**, A290 (2007)
56. T. Xia, M. Kovochich, J. Brant, M. Hotze, J. Sempf, T. Oberley, C. Sioutas, J.I. Yeh, M.R. Wiesner, A.E. Nel, Comparison of the abilities of ambient and manufactured nanoparticles to induce cellular toxicity according to an oxidative stress paradigm. Nano Lett. **6**, 1794–1807 (2006)

57. S. Chellam, C.A. Serra, M.R. Wiesner, Life cycle cost assessment of operating conditions and pretreatment on integrated membrane systems. J. Am. Water Works Assn. **90**(11) 96–104 (1998)
58. M. Widmer, C. Meili, E. Mantovani, A. Porcari, The framing nano governance platform: a new integrated approach to the responsible development of nanotechnologies (FP7: FramingNanoProject Consortium, 2010), Available online: http://www.framingnano.eu/index.php?option=com_content&task=view&id=161&Itemid=84
59. A. Barnard, How can *ab initio* simulations address risks in nanotech. Nat. Nanotechnol. **4**, 332–335 (2009)
60. E.C. Butcher, E.L. Berg, E.J. Kunkel, Systems biology in drug discovery. Nat. Biotechnol. **22**, 1253–1259 (2004)
61. H.A. Godwin, K. Chopra, K.A. Bradley, Y. Cohen, B. Herr Harthorn, E.M.V. Hoek, P. Holden, A.A. Keller, H.S. Lenihan, R. Nisbet, A.E. Nel, The University of California Center for the Environmental Implications of Nanotechnology. Environ. Sci. Technol. **43**, 6453–6457 (2009)
62. Organisation for Economic Co-operation and Development (OECD), The UN principles for responsible investment and the OECD guidelines for multinational enterprises: complementarities and distinctive contributions. Annex II-A4, in *Annual Report on the OECD Guidelines for Multinational Enterprises* (OECD, Paris, 2007)
63. T. Puzyn, D. Leszczynska, J. Leszczynski, Toward the development of "nano-QSARs": advances and challenges. Small **5**, 2494–2509 (2009)
64. M. Ferrari, Cancer nanotechnology: opportunities and challenges. Nat. Rev. Cancer **5**(3), 161–171 (2005)
65. K. Riehemann, S.W. Schneider, T.A. Luger, B. Godwin, M. Ferrari, H. Fuchs, Nanomedicine – Challenge and perspective. Angew. Chem. Int. Ed Engl. **48**(5), 872–897 (2010)
66. J.H. Sakamoto, A.L. van de Ven, B. Godin, E. Bianco, R.E. Serda, A. Grattoni, A. Ziemys, A. Bouamrani, T. Hu, S.I. Ranganathan, E. De Rosa, J.O. Martinez, C.A. Smid, R.M. Buchanan, S.-Y. Lee, S. Srinivasan, M. Landry, A. Meyn, E. Tasciotti, X. Liu, P. Decuzzi, M. Ferrari, Enabling individualized therapy through nanotechnology. Pharm. Res. **62**(2), 57–89 (2010)
67. M. Ferrari, Frontiers in cancer nanomedicine: directing mass transport through biological barriers. Trends Biotechnol. **28**(4), 181–188 (2010)
68. S.E. McNeill, Nanotechnology for the biologist. J. Leukoc. Biol. **78**, 585–594 (2005)
69. W.R. Sanhai, J. Spiegel, M. Ferrari, A critical path approach to advance nanoengineered medical products. Drug Discov. Today Technol. **4**(2), 35–41 (2007)
70. M. Ferrari, M. Philibert, W. Sanhai, Nanomedicine and society. Clin. Pharmacol. Ther. **85**(5), 466–467 (2009)
71. Food and Drug Administration (FDA), Fact sheet: FDA nanotechnology task force report outlines scientific, regulatory challenges (2007), Available online: http://www.fda.gov/ScienceResearch/SpecialTopics/Nanotechnology/NanotechnologyTaskForce/ucm110934.htm. Also, the Nanotechnology task force report to which it refers, http://www.fda.gov/ScienceResearch/SpecialTopics/Nanotechnology/NanotechnologyTaskForceReport2007/default.htm
72. P. Decuzzi, M. Ferrari, Design maps for nanoparticles targeting the diseased microvasculature. Biomaterials **29**(3), 377–384 (2008)
73. P. Decuzzi, R. Pasqualani, W. Arap, M. Ferrari, Intravascular delivery of particulate systems. Pharm. Res. **2**(1), 235–243 (2008)
74. X. Yu, L. Jin, Z.H. Zhou, A structure of cytoplasmic polyhedrosis virus by cryo-electron microscopy. Nature **453**, 415–419 (2008)
75. W. Baumeister, A voyage to the inner space of cells. Protein Sci. **14**, 257–269 (2005)
76. B. Carragher, D. Fellmann, F. Guerra, R.A. Milligan, F. Mouche, J. Pulokas, B. Sheehan, J. Quispe, C. Suloway, Y. Zhu, C.S. Potter, Rapid routine structure determination of macromolecular assemblies using electron microscopy: current progress and further challenges. J. Synchrotron Radiat. **11**, 83–85 (2004)

77. V. Lucic, A.H. Kossel, T. Yang, T. Bonhoeffer, W. Baumeister, A. Sartori, Multiscale imaging of neurons grown in culture: from light microscopy to cryo-electron tomography. J. Struct. Biol. **160**, 146–156 (2007)
78. A. Sartori, R. Gatz, F. Beck, A. Kossel, A. Leis, W. Baumeister, J.M. Plitzko, Correlative microscopy: bridging the gap between fluorescence light microscopy and cryo-electron tomography. J. Struct. Biol. **160**, 135–145 (2007)
79. A.C. Steven, W. Baumeister, The future is hybrid. J. Struct. Biol. **163**, 186–195 (2008)
80. J.A. Heymann, M. Hayles, I. Gestmann, L.A. Giannuzzi, B. Lich, S. Subramaniam, Site-specific 3D imaging of cells and tissues with a dual beam microscope. J. Struct. Biol. **155**, 63–73 (2006)
81. M. Marko, Focused-ion-beam thinning of frozen-hydrated biological specimens for cryo-electron microscopy. Nat. Methods. **4**, 215–217 (2007)
82. D.J. Stephens, V.J. Allan, Light microscopy techniques for live cell imaging. Science **300**, 82–86 (2003)
83. X. Qian, X.-H. Peng, D.O. Ansari, Q. Yin-Goen, G.Z. Chen, D.N. Shin, L. Yang, A.N. Young, M.D. Wang, S. Nie, *In vivo* tumor targeting and spectroscopic detection with surface-enhanced Raman nanoparticle tags. Nat. Biotechnol. **26**, 83–90 (2008)
84. S. Keren, C. Zavaleta, Z. Cheng, A. de la Zerda, O. Gheysens, S.S. Gambhir, Noninvasive molecular imaging of small living subjects using Raman spectroscopy. Proc. Natl. Acad. Sci. U. S. A. **105**, 5844–5849 (2008)
85. D.J. Gentleman, W.C.W. Chan, A systematic nomenclature for codifying engineered nanostructures. Small **5**, 426–431 (2009)
86. R.J. Rowlett, An interpretation of Chemical Abstracts Service indexing policies. J. Chem. Inf. Comput. Sci. **24**, 152–154 (1984)
87. L. Research, *The Recession's Ripple Effect on Nanotech: State of the Market Report* (Lux Research, New York, 2009)
88. M.R. Wiesner, G.V. Lowry, K.L. Jones, M.F. Hochella, R.T. Di Guilio, E. Casman, E.S. Bernhardt, Decreasing uncertainties in assessing environmental exposure, risk, and ecological implications of nanomaterials. Environ. Sci. Technol. **43**, 6458–6462 (2009)
89. House of Lords of the UK Parliament, Science and Technology Committee, Nanotechnologies and food. 1st Report of Session 2009–10, vol I, HL Paper 22-I (The Stationery Office Limited, London, 2010), Available online: http://www.publications.parliament.uk/pa/ld/ldsctech.htm
90. M. Widmer, The "Nano Information Pyramid" as an approach to the "no data, no market" problem of Nanotechnologies (The Innovation Society, St. Gallen, 2010), Available online: http://www.innovationsgesellschaft.ch/index.php?newsid=265&section=news&cmd=details
91. Q. Li, S. Mahendra, D.Y. Lyon, L. Brunet, M.V. Liga, D. Li, P.J.J. Alvarez, Antimicrobial nanomaterials for water disinfection and microbial control: potential applications and implications. Water Res. **42**, 4591–4602 (2008)
92. M.A. Shannon, P.W. Bohn, M. Elimelech, J.G. Georgiadis, B.J. Mariñas, A.M. Mayes, Science and technology for water purification in the coming decades. Nature **452**, 301–310 (2008)
93. P.T. Anastas, J.C. Warner, *Green Chemistry: Theory and Practice* (Oxford University Press, New York, 1998)
94. P.J.J. Alvarez, V. Colvin, J. Lead, V. Stone, Research priorities to advance eco-responsible nanotechnology. ACS Nano **3**, 1616–1619 (2009)
95. W.A. Lee, N. Pernodet, B. Lin, C.H. Lin, E. Hatchwell, M.H. Rafailovich, Multicomponent polymer coating to block photocatalytic activity of TiO_2 nanoparticles. Chem. Commun. Camb **45**, 4815–4817 (2007)
96. T.L. Kirschling, K.B. Gregory, E.G. Minkley, G.V. Lowry, R.D. Milton, Impact of nanoscale zero valent iron on geochemistry and microbial populations in trichloroethylene contaminated aquifer materials. Environ. Sci. Technol. **44**, 3474–3480 (2010)
97. Z. Xiu, Z. Jin, T. Li, S. Mahendra, G.V. Lowry, P.J.J. Alvarez, Effects of nano-scale zero-valent iron particles on a mixed culture dechlorinating trichloroethylene. Bioresour. Technol. **101**, 1141–1146 (2010)

98. C. Kirchner, T. Liedl, S. Kurdera, T. Pellegrino, A. Muño Javier, H.E. Gaub, S. Stölzie, N. Fertig, W.J. Parak, Cytotoxicity of colloidal CdSe and CdSe/ZnS nanoparticles. Nano Lett. **5**, 331–338 (2005)
99. T.K. Jain, M.A. Morales, S.K. Sahoo, D.L. Leslie-Pelecky, V. Labhasetwar, Iron oxide nanoparticles for sustained delivery of anticancer agents. Mol. Pharm. **2**, 194–205 (2005)
100. T.S. Hauck, A.A. Ghazani, W.C. Chan, Assessing the effect of surface chemistry on gold nanorod uptake, toxicity, and gene expression in mammalian cells. Small **4**, 153–159 (2008)
101. J.A. Khan, B. Pillai, T.K. Das, Y. Singh, S. Maiti, Molecular effects of uptake of gold nanoparticles in HeLa cells. Chembiochem **8**, 1237–1240 (2007)
102. N.R. Scott, H. Chen, *Nanoscale Science and Engineering for Agriculture and Food Systems*. Roadmap report of the national planning workshop, 18–19 November 2002 (USDA/CSREES, Washington, DC, 2003), Available online: http://www.nseafs.cornell.edu/web.roadmap.pdf
103. P.R. Srinivas, M. Philbert, T.Q. Vu, Q. Huang, J.K. Kokini, E. Saos, H. Chen, C.M. Petersen, K.E. Friedl, C. McDade-Nguttet, V. Hubbard, P. Starke-Reed, N. Miller, J.M. Betz, J. Dwyer, J. Milner, S.A. Ross, Nanotechnology research: applications to nutritional sciences. J. Nutr. **140**, 119–124 (2009)
104. T. Tarver, Food nanotechnology: a scientific status summary synopsis. Food Technol. **60**(11), 22–26 (2006)
105. J. Weiss, P. Takhistov, J. McClement, Functional materials in food nanotechnology. J. Food Sci. **71**(9), R107–R116 (2006)
106. N.R. Scott, Impact of nanoscale technologies in animal management, in *Animal Production and Animal Science Worldwide*, ed. by A. Rosati, A. Tewolde, C. Mosconi (Wageningen Academic Publishers, Wageningen, 2007), pp. 283–291
107. N. Pidgeon, B. Herr Harthorn, K. Bryant, T. Rogers-Hayden, Deliberating the risks of nanotechnologies for energy and health applications in the United States and United Kingdom. Nat. Nanotechnol. **4**, 95–98 (2009)
108. H.S. Rosenkrantz, A.R. Cunningham, Y.P. Zhang, H.G. Claycamp, O.T. Macina, N.B. Sussman, S.G. Grant, G. Klopman, Development, characterization and application of predictive-toxicology models SAR. QSAR Environ. Res. **10**, 277–298 (1999)
109. R. Benigni, T.I. Netzeva, E. Benfenati, C. Bossa, R. Franke, C. Helma, E. Hulzebos, C. Marchant, A. Richard, Y.-T. Woo, C. Yang, The expanding role of predictive toxicology: an update on the (Q)SAR models of mutagens and carcinogens. J. Environ. Sci. Health C **25**, 53–97 (2007)
110. B. Fubini, Surface reactivity in the pathogenic response to particulates. Environ. Health Perspect. **105**, 1013–1020 (1997)
111. V. Vallyathan, S. Leonard, P. Kuppusamy, D. Pack, M. Chzhan, S.P. Sanders, J.L. Zweir, Oxidative stress in silicosis: evidence for the enhanced clearance of free radicals from whole lungs. Mol. Cell. Biochem. **168**, 125–132 (1997)
112. A. Nel, Atmosphere. Air pollution-related illness: biomolecular effects of particles. Science **308**, 804 (2005)
113. T. Xia, N. Li, A.E. Nel, Potential health impact of nanoparticles. Annu. Rev. Public Health **30**, 21.1–21.14 (2009)
114. R. Becher, R.B. Hetland, M. Refsnes, J.E. Dahl, H.J. Dahlman, P.E. Schwarze, Rat lung inflammatory responses after *in vivo* and *in vitro* exposure to various stone particles. Inhal. Toxicol. **13**, 789–805 (2001)
115. A. Keller, X. Wang, D. Zhou, H. Lenihan, G. Cherr, B. Cardinale, R.J. Miller, Stability and aggregation of metal oxide nanoparticles in natural aqueous matrices. Environ. Sci. Technol. **44**(6), 1962–1967 (2010)
116. M.L. López-Moreno, G. de la Rosa, J.A. Hernández-Viezcas, J.R. Peralta-Videa, J.L. Gardea-Torresdey, XAS corroboration of the uptake and storage of CeO_2 nanoparticles and assessment of their differential toxicity in four edible plant species. J. Agric. Food Chem. **58**, 3689–3693 (2010)

117. R.J. Miller, H.S. Lenihan, E.B. Muller, N. Tseng, S.K. Hanna, A.A. Keller, Impacts of metal oxide nanoparticles on marine phytoplankton. Environ. Sci. Technol (online publication 14 May 2010). doi:10.1021/es100247x
118. Z. Ji, X. Jin, S. George, T. Xia, H. Meng, X. Wang, E. Suarez, H. Zhang, E.M.V. Hoek, H. Godwin, A.E. Nel, J.I. Zink, Dispersion and stability optimization of TiO_2 nanoparticles in cell culture media. Environ. Sci. Technol (online publication 10 June2010). doi: 10.1021/es100417s
119. P. Wang, A. Keller, Natural and engineered nano and colloidal transport: role of zeta potential in prediction of particle distribution. Langmuir **25**(12), 6856–6862 (2009)
120. P. Wang, Q. Shi, H. Liang, D. Steuerman, G. Stucy, A.A. Keller, Enhanced environmental mobility of carbon nanotubes in the presence of humic acid and their removal from aqueous solution. Small **4**(12), 2166–2170 (2008)
121. J. Priester, P. Stoimenov, R. Mielke, S. Webb, C. Ehrhardt, J. Zhang, G. Stucky, P. Holden, Effects of soluble cadmium salts versus CdSe quantum dots on the growth of planktonic *Pseudomonas aeruginosa*. Environ. Sci. Technol. **43**(7), 2589–2594 (2009)
122. M.A. Kiser, P. Westerhoff, T. Benn, Y. Wang, J. Pérez-Rivera, K. Hristovski, Titanium nanomaterial removal and release from wastewater treatment plants. Environ. Sci. Technol. **43**, 6757–6763 (2009)
123. T.M. Benn, P. Westerhoff, Nanoparticle silver released into water from commercially available sock fabrics. Environ. Sci. Technol. **42**, 4133–4139 (2008)
124. A. Kiser, H. Ryu, G. Jang, K. Hristovski, P. Westerhoff, Biosorption of nanoparticles on heterotrophic wastewater biomass. Water Res. **44**(14), 4105–4114 (2010). doi:10.1016/j.watres.2010.05.036
125. P. Westerhoff, G. Song, K. Hristovski, M.A. Kiser, Occurrence and removal of titanium at full scale wastewater treatment plants: Implications for TiO2 Nanomaterials, J. Environ.Moni. DOI: 10.1039/C1EM10017C (2011)
126. ChemicalWatch, A range of tools are needed to communicate the risks of nanomaterials through the value chain. *Monthly Briefing* (CW Research, Shrewsbury, 2010), Available at: http://chemicalwatch.com/3311
127. P. Aguar, J.J. Murcia Nicolás, EU Nanotechnology R&D in the Field of Health and Environmental Impact of Nanoparticles (European Commission Research Directorate-General (FP6/7), Brussels, 2008), Available online: ftp://ftp.cordis.europa.eu/pub/nanotechnology/docs/final-version.pdf
128. National Institute of Standards and Technology (NIST), Advance Technology Program (ATP) economic studies, survey results, reports and working papers (2009) (online index), Available online: http://www.atp.nist.gov/eao/eao_pubs.htm
129. Environmental Defense Fund (EDF), DuPont nano risk framework (2008), Available online: http://innovation.edf.org/page.cfm?tagID=30725
130. DuPont, Position statement: DuPont NanoScale Science & Engineering (NS&E) (2010), Available online: http://www2.dupont.com/Media_Center/en_US/position_statements/nanotechnology.html
131. International Organization for Standardization (ISO), Web site of ISO Technical Committee 229 (Nanotechnologies) (2010), http://www.iso.org/iso/iso_technical_committee?commid=381983
132. National Institute for Occupational Safety and Health (NIOSH), Strategic plan for NIOSH nanotechnology research and guidance: filling the knowledge gaps (DHHS/CDC, Atlanta, 2008), Available online: http://www.cdc.gov/niosh/topics/nanotech/strat_plan.html
133. National Institute for Occupational Safety and Health (NIOSH), Progress toward safe nanotechnology in the workplace (DHHS/CDC, Atlanta, 2007), Available online: http://www.cdc.gov/niosh/docs/2007-123
134. M. Methner, L. Hodson, C. Geraci, Nanoparticle emission assessment technique (NEAT) for the identification and measurement of potential inhalation exposure to engineered nanomaterials – Part A. J. Occup. Environ. Hyg. **7**, 127–132 (2010)
135. National Institute for Occupational Safety and Health (NIOSH), NIOSH current intelligence bulletin: evaluation of health hazards and recommendations for occupational exposure to

titanium dioxide, in Final Policy Clearance for Full Publication (NIOSH, NIOSH Docket #100, Washington, DC, 2005), Available online: http://www.cdc.gov/niosh/review/public/tio2

136. National Institute for Occupational Safety and Health (NIOSH), NIOSH current intelligence bulletin: occupational exposure to carbon nanotubes and nanofibers. Draft being evaluated for policy clearance for placement online on the NIOSH Web site for public comment. Approval anticipated by the end of 2010

137. M. Riedicker, G. Katalagarianakis (eds.), Compendium of projects in the European NanoSafety Cluster (2010), Available online: ftp://ftp.cordis.europa.eu/pub/nanotechnology/docs/compendium-nanosafety-cluster2010_en.pdf

Nanotechnology for Sustainability: Environment, Water, Food, Minerals, and Climate*

Mamadou Diallo and C. Jeffrey Brinker

Keywords Nanomaterials • Water filtration • Clean environment • Food and agricultural systems • Minerals • Climate change • Transportation • Biodiversity • Green manufacturing • Geoengineering • International perspective

The global sustainability challenges facing the world are complex and involve multiple interdependent areas. Chapter "Nanotechnology for Sustainability: Environment, Water, Food, Minerals, and Climate" focuses on sustainable nanotechnology solutions for a clean environment, water resources, food supply, mineral resources, green manufacturing, habitat, transportation, climate change, and biodiversity. It also discusses nanotechnology-based energy solutions in terms of their interdependence with other sustainability target areas such as water, habitat, transportation, and climate change. Chapter "Nanotechnology for Sustainability: Energy Conversion, Storage, and Conservation" is dedicated to energy resources.

---

*With contributions from: André Nel, Mark Shannon, Nora Savage, Norman Scott, James Murday.

M. Diallo (✉)
Environmental Science and Engineering, Division of Engineering and Applied Science,
California Institute of Technology, 1200 East California Boulevard, Mail Stop 139-74,
Pasadena, CA 91125, USA
and
Graduate School of Energy, Environment, Water and Sustainability (EEWS),
Korea Advanced Institute of Science and Technology (KAIST), 291 Daehak-ro, Yuseong-gu,
Daejeon 305-701, Republic of Korea
e-mail: Diallo@wag.caltech.edu, mdiallo@kaist.ac.kr

C.J. Brinker
Department of Chemical and Nuclear Engineering, University of New Mexico,
1001 University Boulevard SE, Albuquerque, NM 87131, USA
and
Department 1002, Sandia National Laboratories, Self-Assembled Materials,
Albuquerque, NM 87131, USA

1 Vision for the Next Decade

1.1 Changes in the Vision Over the Last 10 Years

Brundtland's Commission placed sustainability at the intersection of social, economic, and environmental factors (Fig. 1) where, "sustainable development is that which meets the needs of the present without compromising the ability of future generations to meet their own needs" [1]. Sustainability entails considerations of people, the environment, and the economy. To achieve sustainability, it is vital to take into account the complex linkages between the "social system" (i.e., the institutions that support human existence on Earth), the "global system" (i.e., the Earth's ecosystems that support human life) and the "human system" (i.e., all the other factors that impact the health and well being of humans) [2]. Every human being needs food, water, energy, shelter, clothing, healthcare, employment, etc., to live and prosper on Earth. One of the greatest challenges facing the world in the twenty-first century is providing better living conditions to people while minimizing the impact of human activities on Earth's ecosystems and global environment.

In 2000, the nanotechnology research agenda was primarily focused on the discovery, characterization, and modeling of nanoscale materials and phenomena. As nanotechnology continues to advance, the agenda is increasingly focused on addressing two key questions related to sustainability over the next 10 years:

- How can nanotechnology help address the challenges of improving global sustainability?
- Can nanotechnology be developed in a sustainable manner?

Soon after the inception of the National Nanotechnology Initiative (NNI), it was envisioned that nanotechnology could provide more sustainable solutions to the global challenges related to providing and protecting water, energy, food and shelter habitat, mineral resources, clean environment, climate, and biodiversity. Indeed, sustainability has been a goal of the NNI from the outset: "Maintenance of industrial sustainability by significant reductions in materials and energy use, reduced sources of pollution, increased opportunities for recycling" was listed as an

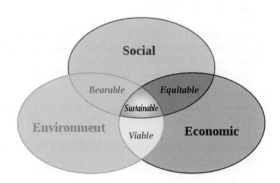

Fig. 1 The three pillars of sustainability [1]

important goal of the NNI in the 1999 Nanotechnology Research Directions Report [85]. Also, in a subsequent address at the Cornell Nanofabrication Center on September 15, 2000, Mike Roco [3] of the National Science Foundation (then co-chair of the Nanoscale Science and Technology Subcommittee of the National Science and Technology Council's Committee on Technology), discussed how nanotechnology could improve agricultural yields for an increased population, provide more cost-efficient and cost-effective water treatment and desalination technologies, and enable the development of renewable energy sources, including highly efficient solar energy conversion systems. Continuing in that vein, he has also noted that nanotechnology promises to "extend the limits of sustainable development… For example, nanoscale manufacturing will provide the means for sustainable development: less materials, less water, less energy, and less manufacturing waste for manufacturing, and new methods to convert energy and filter water …" ([4], pp. 181, 185).

1.2 Vision for the Next 10 Years: A World in Balance

Although Earth has experienced many cycles of significant environmental change, during which civilizations have arisen, developed, and thrived, the planet's environment has been stable during the past 10,000 years [5]. This stability is now threatened as the world's population will reach about seven billion in 2012 [6], and industrial output per capita continues to increase around the world.

Since the Industrial Revolution, human actions have become the main drivers of global environmental change, and could put the "Earth System" (Fig. 2) outside a stable state, with significant or catastrophic consequences. This thesis was proposed by group of investigators from the Resilience Alliance [5]. They defined the Earth System as the set of coupled and interacting physical, chemical, biological, and socioeconomic processes that control the environmental state of Planet Earth. Rockström and colleagues [5] proposed a new conceptual framework, "planetary boundaries," for "estimating a safe operating space for humanity with respect to the functioning of the Earth System." They suggested planetary boundaries in nine areas underlying global sustainability: climate change, rate of biodiversity loss (terrestrial and marine), interference with the nitrogen and phosphorus cycles, stratospheric ozone depletion, ocean acidification, global freshwater use, change in land use, chemical pollution, and atmospheric aerosol loading (see Fig. 2 and Table 1).

Rockström et al. [5] argue that humanity must stay within defined planetary boundaries for a range of key ecosystem processes to avoid catastrophic environmental changes; they maintain we have already transgressed three of these of nine boundaries: (1) atmospheric CO_2 concentration, (2) rate of biodiversity loss, and (3) input of nitrogen into the biosphere. In the case of global freshwater, they believe that "the remaining safe operating space for water may be largely committed already to cover necessary human water demands in the future"(http://www.ecologyandsociety.org/vol14/iss2/art32/).

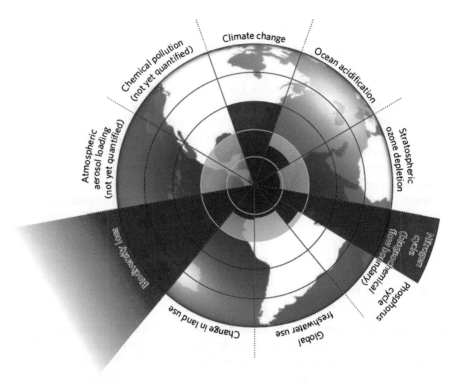

Fig. 2 Planetary boundaries: The inner (*green*) shaded nonagon represents the safe operating space with proposed boundary levels at its outer contour. The extent of the wedges for each boundary shows the estimate of current position of the control variables as highlighted in Table 1 [5]

Arguably, nanotechnology has the potential to address each of these areas of global sustainability, where metrics could be developed to quantify the impact of, for example, the ability of nanotechnology to ameliorate a critical value of one or more control variables for the planetary boundaries, such as carbon dioxide concentration. The following discussion highlights several topical areas in sustainability where nanotechnology is likely to have the greatest impact over the next decade. Efficiency—the amount of energy, water, and natural resources consumed per unit of goods produced or performance achieved—is the most important metric for gauging progress in achieving sustainability.

2 Advances in Last 10 Years and Current Status

To address the two interrelated questions about sustainability raised above, the following subsections provide a background discussion (the status) for each of the nine sustainability goals, an assessment of how nanotechnology may advance the goal, and quantitative metrics against which progress can be measured.

Table 1 Planetary boundaries with proposed boundary and current values of the control variables

Planetary boundaries				
Earth-system process	Parameters	Proposed boundary	Current status	Pre-industrial value
Climate change	(i) Atmospheric carbon dioxide concentration (parts per million by volume)	350	387	280
	(ii) Change in radiative forcing (watts per metre squared)	1	1.5	0
Rate of biodiversity loss	Extinction rate (number of species per million species per year)	10	>100	0.1–1
Nitrogen cycle (part of a boundary with the phosphorus cycle)	Amount of N_2 removed from the atmosphere for human use (millions of tonnes per year)	35	121	0
Phosphorus cycle (part of a boundary with the nitrogen cycle)	Quantity of P flowing into the oceans (millions of tonnes per year)	11	8.5–9.5	~1
Stratospheric ozone depletion	Concentration of ozone (Dobson unit)	276	283	290
Ocean acidification	Global mean saturation state of aragonite in surface sea water	2.75	2.90	3.44
Global freshwater use	Consumption of freshwater by humans (km^3 per year)	4,000	2,600	415
Change in land use	Percentage of global land cover converted to cropland	15	11.7	Low
Atmospheric aerosol loading	Overall particulate concentration in the atmosphere, on a regional basis	To be determined		
Chemical pollution	For example, amount emitted to, or concentration of persistent organic pollutants, plastics, endocrine disrupters, heavy metals and nuclear waste in, the global environment, or the effects on ecosystem and functioning of Earth system thereof	To be determined		

Source: Rockström et al. [5]

2.1 Sustainable Water Supply: Provide Clean Water for the Planet

The United States and many regions of the world face multiple challenges in sustainably supplying potable water for human use and clean water for agriculture, food processing, energy generation, mineral extraction, chemical processing, and industrial manufacturing. Demand for water is increasing due to population growth at the same time as water supplies are being stressed by the increasing contamination and salinization of fresh waters, the depletion of groundwater aquifers, and loss of snowpacks and water stored in glaciers, due to climate change. Figure 3 is a map of the sources of freshwater in the United States. The red regions of Fig. 3 correspond to stressed aquifers that experienced declines in water level of more than 60 ft between 1980 and 1999. It should be noted that as aquifers are drawn down to great depths, their salinities increase significantly. Aquifer salinization is a growing problem along the Gulf Coast and the southern Atlantic and Pacific coasts of the United States. The salinization of rivers and lakes also is increasing due to increase discharges of pollutants and nutrients (nitrates and phosphates) from surface runoff.

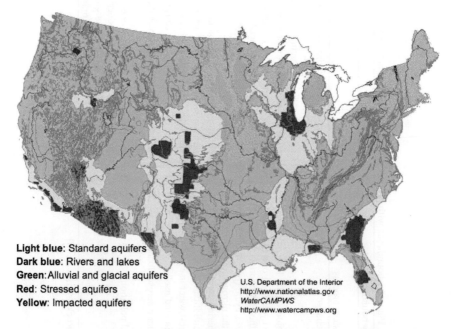

Fig. 3 Sources of freshwater in the United States, including all rivers, lakes, and standard and "fossil" groundwater aquifers. Over-pumping can stress aquifers and can impact water supplies. Estimates are shown of stressed (*red*) and impacted (*yellow*) aquifers throughout the United States [7]

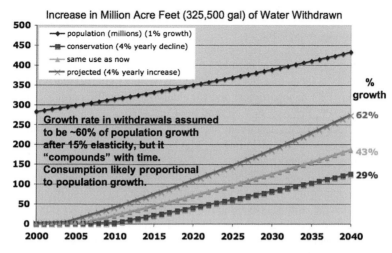

Fig. 4 Predictions of water withdrawals in the United States [7]. (*Top line*) The average overall increase in population of the United States, assuming a 1% increase (between the low and high estimates). Estimates for the growth in water supplies needed to sustain the population growth, assuming projected changes in per capita consumption, as follows: higher use of energy and economic expansion, with current technologies, yielding a 4% per annum increase in per capita use (*62% by 2040, second line from top*); maintenance of current levels (*43% by 2040, third line from top*); and a drop in per capita use of 4% per annum due to increased conservation and efficiency (*20% by 2040, bottom line*) of 20%. The conservation projection for 2040 requires 60% less water in domestic use, 30% less for energy production, and 20% less for agriculture and livestock. Achieving this will require new technologies

As water supplies are decreasing, growth in population, energy use, and economic expansion are driving demand for more water. Figure 4 shows predicted water withdrawals in the United States during the next two decades.

Many other areas of the world also are experiencing water stresses. The United Nations Environment Program (UNEP) predicts that freshwater will become scarcer in many regions of the world by 2020 [8].

Nanotechnology has the potential to provide efficient, cost-effective, and environmentally sustainable solutions for supplying potable water for human use and clean water for agricultural and industrial uses.

2.1.1 Status

During the last 10 years, there has been significant progress in the development and applications of nanotechnology-based solutions in the areas of water treatment, desalination, and reuse ([9–13]). Relevant examples are:

- Nanosorbents with high capacity and selectivity that can remove cations, anions, and organic solutes from contaminated water, including (1) nanoclays; (2) metal

oxide nanoparticles; (3) zeolites; (4) nanoporous carbon fibers; and (5) nanoporous polymeric adsorbents
- Nanocatalysts and redox active nanoparticles that can convert toxic organic solutes and oxyanions into harmless byproducts, including (1) titanium dioxide (TiO_2) photocatalysts that can be activated by visible light; (2) redox active zero valent iron (Fe^0) nanoparticles; and (3) catalytic bimetallic particles, including Fe^0/Pd^0, Fe^0/Pt^0, Fe^0/Ag^0, Fe^0/Ni^0 and Fe^0/Co^0
- Nanobiocides that can deactivate bacteria in contaminated water without generating toxic byproducts, including (1) MgO nanoparticles; (2) Ag^0 nanoparticles; and (3) bioactive dendrimers
- Nanostructured filters and reactive membranes for water treatment, desalination, and reuse, including (1) carbon nanotube filters that can remove bacteria/viruses [14]; (2) reverse osmosis (RO) membranes with enhanced water flux, including zeolite nanocomposite membranes [15] and carbon nanotube membranes [16]; and (3) polymeric nanofibrous membranes with enhanced separation efficiency and water flux [86]
- Nanoparticle-based filtration systems and devices, including (1) a dendrimer-enhanced ultrafiltration system that can remove ions from aqueous solutions using low-pressure membrane filtration [17, 18], and (2) a nanofluidic seawater desalination system [19] (see the example in Sect. 8.1)

2.2 Food Security and Sustainability: Feed the Planet

In 2008, the total amount spent for all food consumed in the United States was $1,165 billion [20]. Another food-related cost is that of food-borne illness, estimated at $152 billion a year in the United States [21]. Godfray and colleagues [22] suggested that the world will face major challenges in meeting the global demand for adequate food over the next 40 years as the world population reaches approximately nine billion by 2050. Viable solutions to this challenge will require a radical transformation of agriculture by growing more food while (1) minimizing the environmental impact of the agriculture and food industries; (2) managing the impact of global climate change; and (3) ensuring the safety and security of the food supply. Advances in nanotechnology could result in major improvements in the technologies used to grow, process, store, and distribute food [23–25].

2.2.1 Status

The application of nanotechnology to agriculture and food systems became part of the U.S. research agenda after a workshop report held in Washington D.C. was

Nanotechnology for Sustainability

published in November 2002 [26]. Several potential applications of nanotechnology to agriculture and food systems were discussed in the report, including:

- Disease diagnosis and treatment delivery systems
- New tools for molecular and cellular breeding
- Development and modification of new food products (smaller, more uniform particles, heat-resistant chocolate, powder suspension, etc.)
- New food packaging materials and systems

To date, a number of significant advances have been made in areas of materials (nanoparticles, nanoemulsions, and nanocomposites), food safety and biosecurity (nanosensors and nanotracers), products (delivery systems and packaging), and processing (nanobiotechnology). Relevant examples include:

- Molecular imprinted polymers to recognize plant and insect viruses
- DNA nanobarcodes to track bacteria in agriculturally important microbial environments
- Nanocomposite materials to detect pore-forming toxins
- Surface-enhanced Raman spectroscopy (SERS) nanosensor array systems to detect food-borne bacteria and toxins
- Edible nanoparticles as sensors of food quality and safety
- "Nutraceutical" nanocomposites utilizing engineered edible films with controlled release morphology
- Nanofluidic arrays for detection of pathogens and bacteria

2.3 Sustainable Habitats: Provide Human Shelter

In the United States, commercial and residential buildings consume 40% of the total energy and account for 39% of the total emissions of CO_2 [27]. Heating, cooling, and lighting are responsible for about 50% of the total energy consumed by commercial and residential buildings [27]. Thus, increases in the energy efficiency of buildings could result in a significant reduction of the environmental footprint of edifices. The U.S. Department of Energy (DOE) has initiated a broad range of research activities to develop and demonstrate zero energy buildings (ZEB) [28]. The underlying premise of the ZEB concept is that buildings can be designed and constructed to achieve zero net energy consumption and zero carbon emissions annually.

2.3.1 Status

Nanotechnology is emerging as a versatile platform technology for producing key ZEB components, including (1) super-insulating aerogels (see chapter "Applications: High-Performance Materials and Emerging Areas"), and (2) more efficient solid-state lighting and heating systems (see chapter "Nanotechnology for Sustainability: Energy Conversion, Storage, and Conservation").

2.4 Sustainable Transportation: Build "Greener" Automobiles and Trucks

In the United States, transportation is responsible for approximately 33% of CO_2 emissions, 66% of oil consumption, and 50% of urban air pollution [29–31]. The United States and many other countries have initiated a broad range of research and development (R&D) programs to build greener (e.g., electric and hybrid) automobiles and heavy-duty trucks and passenger vehicles with 50% improvements in fuel efficiency (e.g., see the DOE $187 Million Super Truck R&D Program website. http://www.energy.gov/8506.htm). Nanotechnology is emerging as an enabling platform technology for producing key components of the next generation of sustainable transportation systems, including automobiles, aircrafts, and ships.

2.4.1 Status

A study by the National Research Council [32] found a linear relationship between vehicle fuel consumption and weight (Fig. 5). The NNI program envisioned very early that nanotechnology could lead to "materials that are ten times stronger than steel, but a fraction of the weight" ([33], p. 14). Subsequent measurements have shown that the density-normalized modulus and strength of single-walled nanotubes (SWNTs) are, respectively, 19 and 56 times greater than that of steel wires [34] and thus SWNTs provide clear opportunities for significantly improved new vehicle materials. Critical technical issues for producing lighter materials with enhanced mechanical properties include achieving uniform dispersions of SWNTs within a

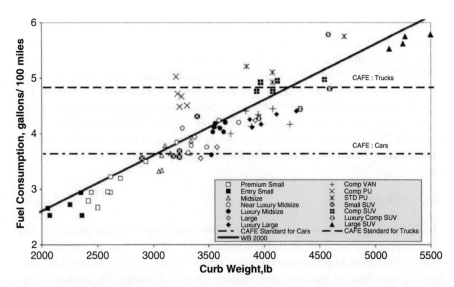

Fig. 5 Relationship between vehicle weight and fuel consumption. For example, by reducing the weight of a vehicle by 20%, fuel consumption also is reduced by 15% [32]

host matrix (e.g., polymer) as well as tailoring interfaces to control adhesion and stress transfer. During the last 10 years, significant advances have been made in the development of carbon nanotube polymer nanocomposites (PNCs) and clay PNCs with enhanced stiffness, strength, and toughness [35–36].

2.5 Mineral Resources: Establish Sustainable Mineral Extraction and Use

Natural resources derived from the Earth's lithosphere, hydrosphere, biosphere, and atmosphere are the key building blocks of a sustainable human society. Like energy and water, the availability of minerals is critical to the United States and the world economy. In 2006, the United States National Research Council (NRC) estimated the added value of processed nonfuel minerals to the U.S. economy to exceed $2.1 trillion [9].

In situ mining (ISM) is emerging as a more efficient and environmentally sound alternative mining technology for valuable metal ions [36]. *In situ* leaching (ISL) is an ISM process that involves extracting a valuable metal/element (e.g., U(VI), Cu(II) and Au(I)) by injection of a leaching solution (commonly referred to as lixiviant) into the ore zone of a subsurface formation (Fig. 6). Because ISL enables

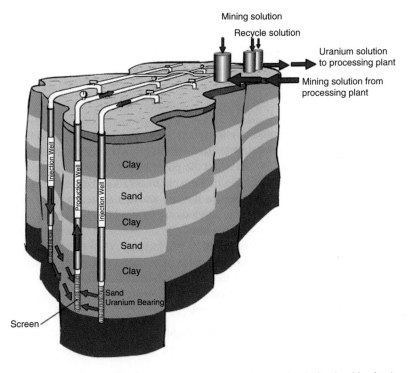

Fig. 6 Uranium mining by ISL (http://www.uraniumsa.org/processing/insitu_leaching.htm)

Table 2 End uses of rare earth elements

End use	Percentage
Automotive catalytic converters	32
Metallurgical additives and alloys	21
Glass polishing and ceramics	14
Phosphors (television, monitors, radar, lighting)	10
Petroleum refining catalysts	89
Permanent magnets	2
Other	13

Source: Committee on Critical Mineral Impacts of the U.S. Economy, Committee on Earth Resources, National Research Council [9]

the mining and recovery of the minerals or elements of interest without excavating the underlying subsurface formation, it has emerged at the technology of choice for mining uranium from permeable subsurface deposits such as sandstones [32]. About 20% of the world uranium production is generated by ISL.

Although the United States is one of the world's largest producers of minerals, it imports more than 70% of its needs for important minerals, many of which have been listed as *critical minerals* by the U.S. National Research Council ([9]). In addition to uranium used in nuclear power generation, many key segments of U.S. industry (Table 2) use significant amounts of minerals such as copper, manganese, lithium, titanium, tungsten, cobalt, nickel, chromium, platinum group metals (e.g., platinum, palladium, and ruthenium), and rare earth elements (e.g., europium, cerium, neodymium, gadolinium, and terbium) ([9]).

Many of these minerals will be required to build nanotechnology-based materials, devices, and systems for sustainable energy generation and storage (Table 3; see also chapter "Nanotechnology for Sustainability: Energy Conversion, Storage, and Conservation") [37].

2.5.1 Status

The application of nanotechnology to mineral discovery, mining, extraction, and processing has thus far received little attention. In a white paper published by the Foresight Institute, Gillett [38] surveyed the potential applications of nanotechnology to the extraction and processing of valuable minerals and elements from ores. Although the U.S. Geological Survey has an active nanotechnology research program, most of its current research activities are focused on (1) bacteria-mediated synthesis of nanoparticles; and (2) characterization of the environmental impact of engineered nanomaterials (http://microbiology.usgs.gov/nanotechnology.html). During the last 10 years, significant advances have been made in the development of nanoscale supramolecular hosts that can serve as high-capacity, selective, and recyclable ligands and sorbents for extracting valuable metal ions from solutions and mixtures; advances include the following:

- Dendrimer-based chelating agents for valuable metal ions, such as Cu(II), Ni(II), Zn(II), Fe(III), Co(II), Pd(II), Pt(II), Ag(I), Au(I), Gd(III), or U(VI) ([87]; [18])

Table 3 Selected applications of nanomaterials in power and energy systems

Start-up	Technology	Power segment	Applications	Nanomaterial used
A_{123} Systems	Lithium iron phosphate (LFP)	Utility applications	Frequency regulation	LFP nanoparticles
Altair Nanotechnologies	Lithium titanate (LTO)	Utility applications	Frequency regulation	LTO nanoparticles
GeoBattery	LFP	Utility applications	Grid storage	LFP nanoparticles i
A_{123} Systems	LFP	Portable power applications	Power tools	LFP nanoparticles
American lithium energy	Lithium nickel cobalt oxide (LNCO), LFP	Portable power applications	E-bikes	LFP nanoparticles
Anzode	Nickel zinc	Portable power applications	E-bikes	Nanoporous zinc
CFX battery	Lithium primary	Portable power applications	Military	Carbon nanotubes, nanoporous carbon, graphene
China BAK battery	Lithium ion, including LCO, nickel manganese cobalt (NMC), lithium manganese spinel (LMS), and LFP	Portable power applications	E-bikes, scooters, power tools	LFP nanoparticles
International battery	Lithium ion, including NMC and LFP	Portable power applications	Military portable power	LFP nanoparticles
K_2 energy solutions	LFP	Portable power applications	Power tools, garden tools, e-bikes, scooters	LFP nanoparticles
NanoeXa	LFP and NMC	Portable power applications	Power tools	Nanostructured LFP, Nanostructured NMC
Pihsiang energy technology	LFP, Lithium polymer	Portable power applications	Medical, e-bikes, scooters	Carbon coated LFP nanoparticles
Planar energy devices	Lithium ion	Portable power applications	Military and specialty handsets	Nano-enabled composite

Source: Lux [37]

- Dendrimer-based separation systems for recovering metal ions from aqueous solutions [17, 18] (see Sect. 8.3)
- Nanosorbents based on self-assembled monolayers on mesoporous supports for recovering metal ions, such as Cu(II), Ni(II), Zn(II), Fe(III), Co(II), Pd(II), Pt(II), Ag(I), Au(I), Gd(III), or U(VI) [39]

2.6 Sustainable Manufacturing: Reduce the Environmental Footprint of Industry

Industrial manufacturing has a heavy environmental footprint. First, it requires a significant amount of materials, energy, and water. Second, it generates a lot of wastes (gaseous, liquid, and solid) and toxic by-products that need to be disposed of or converted into harmless products. Thus, many industries spend a significant amount of financial and human resources in waste treatment and environmental remediation. Green manufacturing encompasses a broad range of approaches that are being used to:

- Design and synthesize environmentally benign chemical compounds and processes (green chemistry)
- Develop and commercialize environmentally benign industrial processes and products (green engineering)

Nanotechnology is emerging as an enabling platform for green manufacturing in the semiconductor, chemical, petrochemical, materials processing, pharmaceutical, and many other industries [40].

2.6.1 Status

The Semiconductor Research Corporation, through the Engineering Research Center for Environmentally Benign Semiconductor Manufacturing at the University of Arizona, is exploring the use of nanotechnology to reduce the environmental footprint of the semiconductor industry [41]. This includes the development of new methods for layering microchips with nanofilms (e.g., selective deposition). Carbon nanotubes and nanoclays also are being evaluated as flame-retardant additives for polymeric materials. The hope is that these nanoparticles can someday replace toxic brominated fire-retardant additives [42]. Fe-based nanocatalysts are providing new opportunities to synthesize valuable chemicals with high yield (~90%) and reduced waste generation. Zeng and colleagues [43] have developed recyclable Fe_3O_4 magnetic nanoparticles that can catalyze the coupling of aldehydes, alkynes, and amines to produce bioactive intermediates such as propargylamines. They were able to recover the Fe_3O_4 nanocatalysts by magnetic separation and reuse them 12 times without activation.

2.7 Sustain a Clean Environment: Reduce the Impact of Pollution

Green manufacturing is arguably the most efficient way to reduce and (eventually) eliminate the release of toxic pollutants into the soil, water, and air. However, large-scale implementation by industry will take decades. Thus, more efficient and cost-effective technologies are critically needed in the short term to (1) detect and monitor pollutants (environmental monitoring), (2) reduce the release of industrial pollutants (waste treatment), and (3) clean polluted sites (i.e., undertake environmental remediation). The 2010 Deepwater Horizon oil spill in the Gulf of Mexico (http://www.energy.gov/open/oil_spill_updates.htm) suggests that more efficient technologies are needed to monitor and clean up oil spills in marine ecosystems.

2.7.1 Status

Application of engineered nanomaterials to sensing and detection devices has enabled the development of a new generation of advanced monitoring and detection concepts, devices, and systems for various environmental contaminants [44–48]. Nanoscale sensors are less energy-intensive than conventional ones, use fewer material resources, and are often reusable. The devices can combine a variety of sensing and detection modalities, such as chemical (e.g., molecular recognition), optical (e.g., fluorescence), and mechanical (e.g., resonance). Potential applications of engineered nanomaterials for environmental monitoring include detection of various compounds in gaseous, aquatic, or soil media; sampling and detecting in biological media (cells, organs, tissues, etc.), and monitoring of physical parameters (pressure, temperature, distance, etc.). Nanotechnology-enabled sensors can enable rapid and accurate detection of harmful compounds using only minute concentrations of analytes.

Advances in nanotechnology are enabling development of more efficient and cost-effective waste treatment and environmental remediation technologies [12, 49]. Nanoscale zero valent iron (NZVI) particles have proven to be very efficient redox-active media for the degradation of organic contaminants, especially chlorinated hydrocarbons [50–52]. Dendritic nanomaterials, which consist of highly branched nanoscale polymers, have been successfully employed to "encapsulate" environmental pollutants. Dendritic nanomaterials are often recyclable and water-soluble and have demonstrated great potential for removing inorganic pollutants, heavy metals, biological, and radiological compounds [53–56]. Another key advance is the development of nanowire membranes with tunable wettability ranging from superhydrophobic to superhydrophilic [73] these have good potential as oil-spill cleanup media (see Sect. 8.4).

2.8 Sustain Earth's Climate: Reduce and Mitigate the Impacts of Greenhouse Gases

Global climate change has emerged as the most daunting challenge facing the world in the twenty-first century [57]. During the last two decades, a consensus has gradually emerged that increasing emissions of carbon dioxide (CO_2) from the combustion of fossil fuels (e.g., coal and petroleum) are the key drivers of global climate change [57]. Currently, fossil fuels provide approximately 80% of the energy used worldwide [58]. Although many non-CO_2-emitting energy sources are being developed (chapter "Nanotechnology for Sustainability: Energy Conversion, Storage, and Conservation"), the world will continue to burn significant amounts of fossil fuels in the foreseeable future. Thus, carbon capture and storage (CCS) is emerging as a viable short- to medium-term alternative for reducing the amounts of anthropogenic CO_2 released into the atmosphere [58].

2.8.1 Status

The efficient and selective separation of CO_2 from gas mixtures is the critical step of any CCS technology. Nanotechnology has the potential to provide efficient, cost-effective, and environmentally acceptable sorbents for CO_2 separation (capture and release) from the flue gases of fossil-fuel-fired power plants and relevant industrial plants (Fig. 7). During the last 5 years, there has been significant progress in

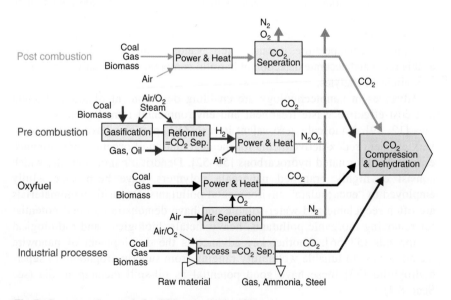

Fig. 7 Gas separation needs for CCS technologies [58]

the development of high-capacity and selective nanosorbents for the CO_2 separation from gaseous mixtures. Relevant examples include:

- Nanoporous silica particles with covalently attached amine groups [59]
- Nanoscale metal organic frameworks (MOFs) [60]
- Nanoscale zeolitic imidazolate frameworks (ZIF) [61, 62] (see Sect. 8.5)

2.9 Sustain Earth's Natural Capital: Preserve the Biodiversity of Earth's Ecosystems

Human beings depend heavily on Earth's natural capital—its biological resources and ecosystems—to live and prosper. Earth's biological resources are extraordinarily diverse. They include a myriad of species of plants, animals, and microorganisms, and they provide a significant fraction of mankind's food, agricultural seeds, pharmaceutical intermediates, and wood products [63]. Earth also possesses a variety of ecosystems (e.g., wetlands, rainforests, oceans, coral reefs, and glaciers) that provide critical services such as (1) water storage and release; (2) CO_2 absorption and storage; (3) nutrient storage and recycling; and (4) pollutant uptake and breakdown. Preservation of the biodiversity of Earth's ecosystems is critical to human life and prosperity.

2.9.1 Status

- The application of nanotechnology to biodiversity has so far received little attention [64]. An integrated research and development program has yet to be formulated.

3 Goals, Barriers, and Solutions for the Next 5–10 Years

3.1 Sustainable Supplies of Clean Water

The availability of clean water has emerged as one of the most critical problems facing society and the global economy in the twenty first century [12, 13]. As previously stated, the United States and many regions of the world are already experiencing higher demands for clean water while freshwater supplies are being stressed. Energy and water issues are strongly coupled (Fig. 8). Energy generation requires reliable and abundant sources of clean water. Conversely, the production and delivery of clean water require a lot of energy. Thus, we expect that increasing demands of clean water for energy generation [65] will cause additional stresses to our freshwater resources globally.

A report published by the Intergovernmental Panel on Climate Change [66] suggests that global climate change will adversely impact the world's freshwater

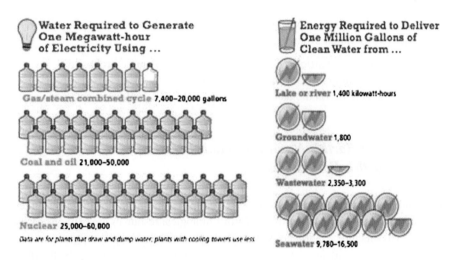

Fig. 8 Energy and water are strongly linked: both energy and water are in short supply; thus, both crises needed to be solved together [65]

resources in several ways: (1) increase the frequency of droughts and floods; (2) decrease the amount of water stored in snowpack and glaciers; and (3) decrease the overall water quality due to salinity increase and enhanced sediment, nutrient, and pollutant transport in many watersheds throughout the world. Thus, a significant increase is needed in the amount of clean water produced from impaired water (e.g., wastewater, brackish water, and seawater) to meet the growing demand throughout the world in the next decade—and beyond.

- Approximately 70–90% of the water used in agriculture and industry and for human consumption is returned to the environment as wastewater [7]. Wastewaters contain 25 MJ/kg of dry weight in organics, including nutrients and compounds with fuel values [7]. Current treatment technologies spend energy to destroy (i.e., mineralize) organic compounds in wastewater; instead, efficient ways must be found to extract clean water, energy, nutrients, and valuable organic compounds from wastewater.
- Seawater and brackish water from saline aquifers constitute ~97% of the water on Earth [13]. Approximately 2.58–4.36 KWh of energy is need to produce 1 m^3 of clean water from saline water [65]. Development of low-energy desalination technologies must be a priority for extracting clean water and valuable minerals (e.g., lithium) from brackish water and seawater.

A convergence between nanotechnology, chemical separations, biotechnology, and membrane technology will lead to revolutionary advances in water desalination and reuse technologies, including the following:

- Solar-powered electrochemical or photo-catalytic systems, which can extract clean water and produce energy (e.g., hydrogen) from wastewater (Sect. 8.2).

- Separation systems (e.g., membranes and sorbents) that can produce clean water and valuable products through selective extraction and release of valuable compounds (e.g., organics and nutrients) from industrial and municipal wastewater.
- Low-energy membranes and filtration systems that selectively reject and reversibly bind and release ions from brackish water and seawater with very high water recovery (>90%) and minimum environmental impact (e.g., reduced brine generation).
- Solar-powered and high-performance deionization systems that can desalinate brackish water and seawater at lower cost and reduced environmental impact (e.g., brine generation).

3.2 Sustainable Agriculture and Food Production

It is envisioned that the convergence between nanotechnology, biotechnology, plant science, animal science, crop, and food science/technology will lead to revolutionary advances in the next 5–10 years, including

- "Reengineering" of crops, animals and microbes at the genetic and cellular level
- Nanobiosensors for identification of pathogens, toxins, and bacteria in foods
- Identification systems for tracking animal and plant materials from origination to consumption
- Development of nanotechnology-based foods with lower calories and with less fat, salt, and sugar while retaining flavor and texture
- Integrated systems for sensing, monitoring, and active response intervention for plant and animal production
- Smart field systems to detect, locate, report and direct application of water
- Precision- and controlled release of fertilizers and pesticides
- Development of plants that exhibit drought resistance and tolerance to salt and excess moisture
- Nanoscale films for food packaging and contact materials that extend shelf life, retain quality and reduce cooling requirements

3.3 Sustainable Human Habitats

During the next 5–10 years, it is expected that nanotechnology will continue to be an enabling technology for zero energy commercial and residential buildings. Anticipated major advances include:

- More efficient organic light-emitting diodes (chapter "Nanotechnology for Sustainability: Energy Conversion, Storage, and Conservation")
- Super-insulating and self-cleaning windows

- More efficient roof-top photovoltaic systems (chapter "Nanotechnology for Sustainability: Energy Conversion, Storage, and Conservation")
- More efficient sensors for monitoring and optimizing energy usage in buildings

3.4 Sustainable Transportation

It is envisioned that in the next 5–10 years nanotechnology will become a key enabling platform technology for the next generation of transportation systems. Significant advances are expected in:

- More efficient and lighter materials for automotive and aircraft systems
- High-performance tires for automobiles
- Efficient and non-platinum-based catalytic converters
- Novel and more efficient fuel and power sources (see chapter "Nanotechnology for Sustainability: Energy Conversion, Storage, and Conservation")

3.5 Sustainable Raw Mineral Extraction and Use

China had 97% of the market of rare-earth elements (REE) in 2010 and its export limitations [67, 68] have disrupted the sustainable supply of *critical minerals* (Table 2). Nanotechnology is emerging as key platform technology for solving the global challenges in sustainable supply of *critical minerals* such as REE. It is envisioned that in the next 10 years the convergence between nanotechnology, geosciences, synthetic biology, biotechnology, and separations science and technology will lead to major advances and significant improvements in mineral extraction, processing and purification technologies, including:

- Development of non-acidic microbial strains that can selectively leach valuable metal ions including platinum group metals, rare-earth elements and uranium from mineral ores without extensive dissolution of the surrounding rock matrices
- Development of more efficient and environmentally benign leaching solutions for in-situ leach (ISL) mining (Fig. 6) and hydrometallurgical processing of ores containing *critical minerals*
- Development of more efficient, cost-effective, and environmentally acceptable separation systems (e.g., chelating ligands for solvent extraction, ion-exchange media, and affinity membranes) for recovering valuable minerals and elements such as rare earth elements from mine tailings, leaching/hydrometallurgical solutions and wastewater from mineral/metallurgical extraction and processing plants

Nanotechnology could also lead to significant reductions in the consumption of *critical minerals* through the efficient use of mineral resources and the development of nontoxic and cost-effective substitutes to rare-earth elements for the applications listed in applications in Table 2. For example, a group of nanotechnology researchers from the Universities of Delaware and Nebraska have received a $4.5 million grant from the Advanced Research Projects Agency for Energy (ARPA-E) to develop magnetic nanomaterials (20–30 nm) with higher magnetic strength than those of Nd2Fe14B, which is the "world's strongest magnet" [69]. A key goal of this research is the development of "REE-free high anisotropy and high magnetization compounds in doped Fe-, Co- or Mn- rich materials" [69].

3.6 Sustainable Manufacturing

In the next 10 years, a convergence between nanotechnology, green chemistry, and green engineering will enable society to build the sustainable products, processes, and industries of the twenty-first century, including:

- Environmentally benign building blocks and manufacturing processes for the semiconductor, chemical, petroleum, metal/mineral, and pharmaceutical industries
- High-performance nanocatalysts for the chemical, petroleum, and pharmaceutical industries
- More efficient nanotechnology-based consumer products such as nanotechnology-enabled, high performance, and environmentally sustainable "green cars" (Fig. 9)

3.7 Sustaining a Clean Environment

It is envisioned that in the next 5–10 years, small-scale and ubiquitous sensors will be developed and deployed that can be customized to perform "real time" monitoring of environmental systems, including air, water, and soils. For example, "smart" and

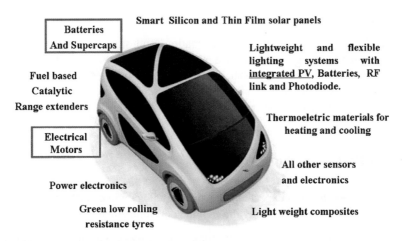

Fig. 9 Nanomaterials as components of the next generation of electric cars (Courtesy of Pietro Perlo, Fiat, http://www.gennesys2010.eu/)

"ubiquitous" nanosensor devices—with the ability to perform a specified action upon detection of a compound—could be placed in surface water systems or subsurface environments to track contaminant migration and implement preventive measures to keep critical compounds from contaminating local water sources. Reduced size coupled with an increase in computational power and speed will make these sensors particularly effective. We also envision the continued development of cost-effective environmental cleanup and remediation technologies for emerging contaminants, including pharmaceuticals, household products, and nanomaterials.

3.8 Sustaining the Climate: Reducing the Impact of Greenhouse Gases

Nanoscale metal organic frameworks (MOFs) and zeolite imidazolate frameworks (ZIFs) are promising CO_2 sorbents with high adsorption capacity, selectivity, and reversibility. However, the first generation of nanoscale MOFs and ZIFs can only perform a single function, i.e., CO_2 separation. Thus, their use alone might not lead to the revolutionary advances needed to significantly decrease the atmospheric release of greenhouse gases by capturing CO_2 and converting it to useable products (e.g., fuels and chemicals). The vision for the 5–10 year timeframe is that convergence between nanotechnology, chemical separations, catalysis, and systems engineering will lead to revolutionary advances in CO_2 capture and conversion technologies, including:

- Nanoscale sorbents containing functionalized size- and shape-selective molecular cages that can capture CO_2 and convert it to useable products
- Nanoporous fibers and/or membranes containing functionalized size- and shape-selective molecular cages that can capture CO_2 and convert it to useable products

In addition to CO_2 capture, transformation, and storage, geoengineering is being considered as a potential climate mitigation technology. The ultimate goal of geoengineering is to reduce global warming by developing and deploying large scale "cooling" systems in the stratosphere. Sect. 8.6 discusses the use of nanotechnology as enabling technology platform for geoengineering.

3.9 Sustaining Biodiversity

In the next 10+ years, it is expected that nanotechnology will contribute significantly to the preservation of biodiversity through the development and implementation of:

- Advanced sensors and devices for monitoring ecosystem health (e.g., soil/water composition, nutrient/pollutant loads, microbial metabolism, and plant health)

- Advanced sensors and devices for monitoring and tracking animal migration in terrestrial and marine ecosystems
- Cost-effective and environmentally acceptable solutions to the global sustainability challenges, including energy, water, environment, and climate change as described in this report

4 Scientific and Technological Infrastructure Needs

During the past 10 years, there has a gradual shift from the discovery, characterization and modeling of nanoscale materials toward the development of nano-enabled systems, devices, and products. In the next 10 years, it will be necessary to harness the power of nanotechnology to develop and commercialize the next generation of sustainable products, processes, and technologies. Key science and technological infrastructure needs and R&D investment needs include:

- Holistic investigations of all interdependent aspects of sustainable development, including cost-benefit environmental risk assessments
- Nanomaterial scale-up and manufacturing facilities/hubs for sustainability applications
- Dedicated nanomaterial characterization facilities for sustainability applications
- Computer-aided nanomaterials modeling and process design tools for sustainability applications
- Test-beds for nanotechnology-enabled sustainability technologies

5 R&D Investment and Implementation Strategies

Because sustainability entails considering social, economic, and environmental factors, it is critical in all cases to integrate fundamental science (e.g., materials synthesis, characterization, and modeling) with engineering research (e.g., system design, fabrication, and testing), commercialization (e.g., new products), and societal benefits (e.g., new jobs and cleaner environment) as scientists and engineers pursue the research priorities outlined in this report. Thus, nanotechnology solutions for sustainable development cannot simply be addressed at the level of small- and single-investigator funded research grants. Sustainability R&D has to be integrated with broader research goals and included from the beginning in large interdisciplinary programs to be carried out by interdisciplinary teams of investigators and/or dedicated Federally funded research and development centers. To achieve these objectives, it will be necessary to:

- Establish centers to develop and implement nanotechnology solutions tied to key aspects of sustainability, as well as to develop suitable open-source databases and partnerships

- Develop new funding mechanisms to advance promising early-stage research projects, e.g., automatic supplemental funding for projects with commercial potential
- Involve industry at the outset of programs
- Accelerate knowledge and technology transfer from academic/government laboratories (e.g., academic spin-off companies)

6 Conclusions and Priorities

The global sustainability challenges facing the world are complex and involve multiple interdependent areas. Energy generation/usage, water usage/delivery, CO_2 emissions, and industrial manufacturing are strongly coupled. Nanotechnology has the potential to provide breakthrough solutions for sustainable development, particularly in the areas of energy generation, storage, and usage (See chapter "Nanotechnology for Sustainability: Energy Conversion, Storage, and Conservation"), clean water resources, food/agriculture resources, green manufacturing, and climate change. The following key priorities have been identified for the next decade:

- A comprehensive study of all interdependent factors affecting sustainability and how nanotechnology solutions can extend the limits of sustainable development must be undertaken and updated each year.
- Solar-powered photocatalytic systems and separation systems (e.g., nanoporous membranes with ion-channel mimics) that extract clean water, energy and valuable elements (e.g., nutrients and minerals) from impaired water including wastewater, brackish water and seawater.
- Multifunctional sorbents/membranes that can capture CO_2 from flue gases and transform it into useful products (e.g., chemical feed stocks) must be optimized and then scaled up for industrial scale use.
- Development of more efficient, cost-effective, and environmentally acceptable separation systems (e.g., chelating ligands for solvent extraction, ion-exchange media, and affinity membranes) for recovering *critical minerals* such as rare earth elements (REE) from mine tailings, leaching/hydrometallurgical solutions and wastewater from mineral/metallurgical extraction and processing plants.
- Green manufacturing technologies to (1) develop nontoxic and cost-effective substitutes for REE and (2) reduce and (eventually) eliminate the release of toxic pollutants into the environment (soil, water, and air).

7 Broader Implications for Society

Every human being needs adequate food, water, energy, shelter, clothing, healthcare, and employment to live and prosper on Earth. One of the greatest challenges facing the world in the twenty-first century is continuing to provide better living conditions to people while minimizing the impact of human activities on Earth's ecosystems and the global environment. Nanotechnology offers the potential to expend the limits of sustainability in all critical areas of human development on the Earth. However, it is critical to make sure that any potential adverse effects on humans and the environment are effectively assessed and addressed before the large-scale deployment of nanoscience-enabled sustainability technologies.

8 Examples of Achievements and Paradigm Shifts

8.1 Nanofluidic Water Desalination System

Contact person: Jongyoon Han, Massachusetts Institute of Technology

Figure 10 shows a new promising nanotechnology for desalinating water at the point of use. This nanofluidic device desalinates water with energy consumption that approaches that of a large-scale reverse osmosis desalination system [19]. However, no high pressure is needed to operate the system. Instead, low-pressure and electricity are used to drive seawater through a microchannel containing a nanojunction, consisting of an ion-selective nanoporous Nafion membrane, to connect two microchannels. This causes an ion concentration polarization that separates the seawater stream into an ion "depleted" stream (freshwater) and ion "rich" stream (concentrate).

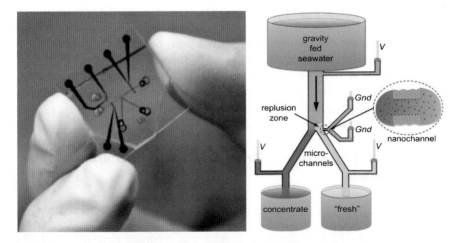

Fig. 10 A chip-sized nanofluidic device (*left*) desalinates water at an energy use approaching that of a large-scale reverse osmosis desalination system Kim et al. [19]

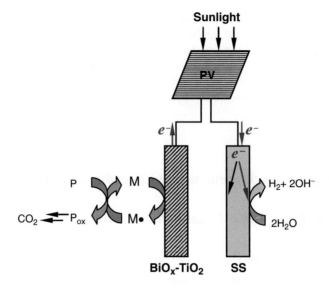

Fig. 11 Solar-powered photocatalytic systems for sustainable water reuse

8.2 Solar-Driven Photocatalytic and Electrochemical Systems for Sustainable Water Reuse

Contact person: Michael Hoffmann, California Institute of Technology

The need for alternative energy sources with reduced carbon footprints is growing. Solar-powered electrochemical or photo-catalytic systems (Fig. 11), which produce hydrogen via water splitting using organic pollutants as sacrificial electron donors, provide a possible solution to achieve two objectives: generation of energy and production of clean water [70–72]. Hybridization of a BiO_x-TiO_2/Ti anode with stainless steel or functionalized metal cathodes, powered by photo-voltaic (PV) arrays, has been shown to achieve simultaneous water purification coupled with H_2 generation. Note that hydrogen generation can be suppressed by purging the reactor system with air. A variety of other hetero-junction, mixed-metal nanooxide semiconductors have also been shown to be highly efficient electro-chemical catalysts for both anodic and cathodic electron transfer.

8.3 Recovery of Metal Ions from Solutions by Dendrimer Filtration

Contact person: Mamadou Diallo, California Institute of Technology

Diallo et al. ([17–18]) have developed a dendrimer-enhanced ultrafiltration (DEF) systems that can remove dissolved cations from aqueous solutions ions using low-pressure membrane filtration (Fig. 12). DEF works by combining dendrimers

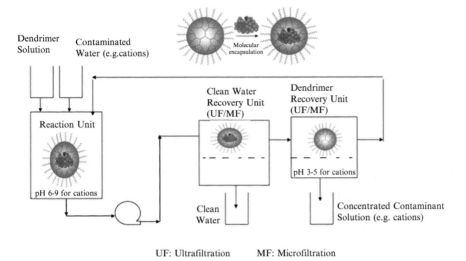

Fig. 12 Recovery of metal ions from aqueous solutions by dendrimer filtration (Adapted from Diallo [18])

with ultrafiltration or microfiltration membranes. Functionalized and water-soluble dendrimers with large molar mass are added to an incoming aqueous solution and bind with the target ions. For most metal ions, a change in solution acidity and/or salinity causes the dendrimers to bind or release the target metal ions. Thus, a two-stage filtration process can be used to recover and concentrate a variety of dissolved ions in water including Cu(II), Ag(I), and U(VI). A key feature of the DEF process is the combination of dendritic polymers with multiple chemical functionalities with the well-established separation technologies of ultrafiltration (UF) and microfiltration (MF). This allows a new generation of metal ion separation processes to be developed that are flexible, reconfigurable, and scalable. The flexibility of DEF is illustrated by its modular design approach. DEF systems will be designed to be "hardware invariant" and thus reconfigurable in most cases by simply changing the "dendrimer formulation" and "dendrimer recovery system" for the targeted metal ions of interest. The DEF process has many applications including the recovery of valuable metal ions including platinum group metals, rare-earth metals and actinides from mineral/hydrometallurgical processing solutions, *in situ* leach mining solutions and industrial wastewater solutions.

8.4 Superhydrophobic Nanowire Membranes for Oil-Water Separation

Contact person: Francesco Stellaci, Massachusetts Institute of Technology

Yuan et al. [73] have developed a new generation of nanoporous membranes with tunable wettability (Fig. 13). These membranes consist of manganese oxide nanowires that self-assemble into an open porous network. By coating the membranes

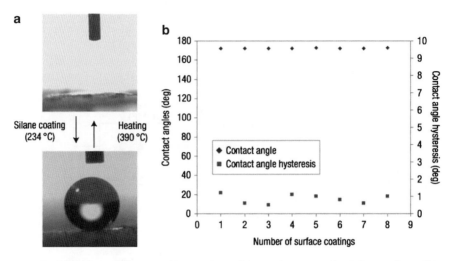

Fig. 13 Nanowire membranes with tunable surface wetting properties [73]. (**a**) Reversible transition between superhydrophilic (*top*) and superhydrophobic (*bottom*); (**b**) contact angles and hysteresis measurements between cycles

Fig. 14 "Seaswarm" is a nanotechnology-enabled oil-spill-cleaning robot developed by a group of MIT engineers (*TechNews Daily*, August 25, 2010). The conveyor belt of the robot is coated with a mesh of superhydrophobic nanowires (Developed by Yuan et al. [73])

with silicone, Yuan et al. [73] were able to fabricate oil-water separation membranes to "selectively absorb oils up to 20 times the material's weight in preference to water, through a combination of superhydrophobicity and capillary action" systems. Recently, a team of MIT engineers used the superhydrophobic nanowires as components to build a robot (Fig. 14) that can clean oil spills in water (*TechNews Daily*, August 25, 2010).

8.5 MOFs and ZIFs for CO_2 Capture and Transformation

Contact person: Omar Yaghi, University of California, Los Angeles

Nanoscale zeolitic imidazolate frameworks (ZIFs; Fig. 15; [61, 62]) and metal organic frameworks (MOFs) with multivariate (MTV) functionalities [74] exhibit

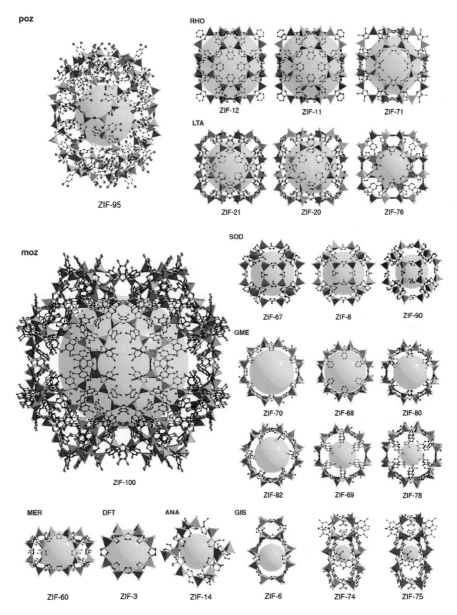

Fig. 15 Crystal structures of zeolitic imidazolate frameworks grouped according to their topology (three-letter symbol). The ball indicates the space within the cage [62]

very large CO_2 sorption capacity, selectivity, and reversibility [62]. ZIFs possess larger pores (2–3 nm) than zeolites and can be processed into adsorbents and gas separation membranes [75]. MTV-MOFs incorporate multiple functionalities on their linking groups. This can lead to a significant enhancement of their CO_2 selectivity. ZIFs and MTV-MOFs have high chemical stability and can be readily reacted with various organic moieties to produce industrial-scale quantities of materials. Thus, nanoscale ZIFs and MTV-MOFs have great potential as building blocks for the next generation of CO_2 capture and transformation media.

8.6 Nanotechnology and Geoengineering

Contact persons: Jason Blackstock and David Guston, Arizona State University

As concerns about climate change continue to mount, "geoengineering" concepts for intentionally altering part of the climate system in order to counteract greenhouse gas–induced changes are receiving rapidly increasing scientific and public attention (*Science Daily* 2010). Among the most prominent concepts is the notion of artificially reducing the amount of sunlight absorbed by the Earth's atmosphere and surface by injecting reflective nanoparticles into the stratosphere. Until recently, the material and design specifications for such reflective nanoparticles for "solar radiation management" (SRM) had only been the subject of speculative discussions and very limited theoretic calculations. But in recent years, calls for developing serious SRM research programs (Fig. 16)—which would include specific focus on the "engineering" of suitable nanoparticles for potential megaton-quantity atmospheric dispersion—have emerged in the scientific [78] and public literature [79]. Recently, the first field-experiment dispersing very small quantities of sulfur-dioxide aerosol particles into the environment were conducted in Russia [80].

The most common reflective nanoparticulates considered for SRM are those of sulfur-dioxide [81, 82], because such particles mimic those naturally produced by volcanoes and could have their dispersed size distribution optimized/engineered for "optimal" radiative properties. However, dielectric and metallic materials predicted to have unique radiation properties or even magnetic or photophoretic levitation characteristics (potentially useful for moderating altitudinal and geographic dispersion of nanoparticulate "mists") have also been seriously proposed in recent literature [83, 84]. Plans for expanded small- to medium-scale field tests of nanoparticle-based climate-modifying technologies are now being discussed seriously and could be underway as soon as 2015. The development of and potential for large-scale atmospheric dispersion of such nanoparticles should be managed under a framework of responsible nanotechnology development.

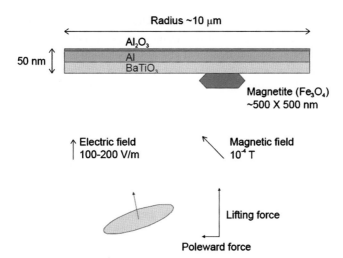

Fig. 16 Conceptual design of a self-levitated nano-disk for SRM geoengineering [76, 77]. This nano-disk configuration is theoretically designed to have three properties: (1) the aluminum layer ~25 nm thick would be reflective in the visible and transparent in the infrared portion of the electromagnetic spectrum; (2) the nano-disks would self-levitate in the stratosphere due to the interaction of the barium titanate ($BaTiO_3$) with the natural electric-field (100–200 V/m) in the stratosphere; and (3) the nano-disks would "tilt" due to the off-center presence of the magnetic iron oxide (Fe_3O_4) interacting with the earth's magnetic field in the stratosphere, leading to the nano-disks being pushed by Brownian forces in the direction of the earth's magnetic poles and concentrating the particles over the Arctic. All of the material layers and size scales indicated in this figure could be produced using existing nanofabrication techniques

9 International Perspectives from Site Visits Abroad

9.1 U.S.–European Union Workshop (Hamburg, Germany)

Panel members/discussants

Antonio Marcomini (co-chair), University of Venice, Italy
Mamadou Diallo (co-chair), California Institute of Technology, USA
C. Jeffrey Brinker, University of New Mexico and Sandia National Laboratory, USA
Inge Genné, VITO, Belgium
Karl-Heinz Haas, Fraunhofer-Institut fuer Silicatforschung, Germany
John Schmitz, NXP,
Udo Weimar, Institut für Physikalische und Theoretische Chemie, University of Tübingen, Germany

During the last 10 years, nanotechnology has emerged as a suitable platform technology for addressing global sustainability challenges, especially in terms of energy, water, food, habitat, transportation, mineral resources, green manufacturing, clean environment, climate change, and biodiversity. Key advances include the development of (1) more efficient renewable energy generation and storage technologies, (2) improved water treatment and desalination membranes, (3) improved food safety (detection and tracking) systems, (4) more efficient separation systems for recovering valuable metals, (5) miniaturized sensors that can detect and monitor pollutants more efficiently, and more effective environmental remediation technologies, (6) more efficient CO_2 capture media, and (7) microchip fabrication processes that use less materials, energy, and water.

In the next 10 years, the power of nanotechnology should be used to develop and commercialize the next generation of sustainable products, processes, and technologies. Priority areas will include (1) water desalination and reuse, (2) CO_2 capture and transformation to useful products and (3) green manufacturing. Current reverse osmosis (RO) desalination membranes have limited water recovery (e.g., 40–70%) and require high pressure (e.g., 10–70 bar) to operate and generate significant amounts of liquid wastes that need to be disposed. Thus, a key priority in water desalination and reuse is to develop the next generation of low-pressure membranes (e.g., 0.5-5 bar) with high water recovery (>95%) and capability to:

- Reject dissolved anions and cations from saline water (e.g., brackish water and seawater)
- Reversibly and selectively bind and release dissolved anions and cations in saline water
- Extract clean water, nutrients, and other valuable compounds from domestic and industrial wastewater

Current CO_2 capture media only perform a single function, i.e., separate CO_2 from flue gases. Thus, a key priority in climate change mitigation and reduction of greenhouse gas emissions is to develop:

- Size- and shape-selective media with nanocages that can capture CO_2 and convert it to useable products
- Membrane reactors with embedded nanocages that can capture CO_2 and convert it to useable products

Green manufacturing is arguably the most efficient way to fabricate high-performance products at lower cost while reducing and (eventually) eliminating the release of toxic pollutants in the environment. Nanotechnology is a key enabling platform technology for green manufacturing. Priority research areas include the development of:

- Nano-enabled, environmentally benign manufacturing processes and products for the semiconductor industry
- High-performance nanocatalysts for the chemical, petroleum and pharmaceutical industries
- More efficient nanotechnology-based consumer products (e.g., nano-enhanced "green" cars)

Potential negative impacts on human health and the environment should not be overlooked. In order to mitigate the potential environmental impact from nanomaterials and products, it is important to design them in a sustainable, life-cycle-oriented way, balancing their environmental performance with technical, cost, cultural, and legal requirements. Suitable design strategies for pollution prevention and resource conservation include product/material life extension, reduced material intensiveness, and process management. In order to achieve this, appropriate environmental analysis tools should be applied.

The sustainable development of nanotechnology also requires the education and training of a new generation of scientists, engineers, entrepreneurs, policymakers, and regulators. This education should not be limited to natural sciences or engineering, but, being more interdisciplinary in nature, it should also develop societal awareness and cover practical communication, entrepreneurial and management skills. Today, there is growing recognition that the field is in urgent need for trained workforce and any delay in launching suitable educational initiatives would ultimately impede innovation.

Finally, although nanotechnology holds great promise for solving global sustainability challenges, there is growing awareness that some nanomaterials could pose environmental and health hazards. Thus, we have to make sure that any potential adverse effects on humans and the environment are effectively assessed and addressed before large-scale deployment of nano-enabled sustainability technologies.

9.2   U.S.–Japan–Korea–Taiwan Workshop (Tokyo/Tsukuba, Japan)

Panel members/discussants

Chul-Jin Choi (co-chair), Korea Institute of Materials and Science, Korea
James Murday (co-chair), University of Southern California, USA
Tomoji Kawai , Osaka University, Japan
Chuen-Jinn Tsai, National Chiao Tung University, Taiwan
Jong Won Kim, Higher Education Research Data Collection Office, Korea
Kohei Uosaki, National Institute for Materials Science , Japan

During the last 10 years, nanotechnology has emerged as an ideal platform technology for solving global sustainability challenges in energy, water (see Fig. 17), food, habitat, transportation, mineral resources, green manufacturing, clean environment, climate change and biodiversity. Key advances in 10 years has been achieved in many areas of sustainability including: (1) new sensing concepts and devices to detect nanomaterials and monitor their toxicity in a workplace environment; (2) smart (self-cleaning) windows and walls; (3) nanoabsorbents, nanocatalysts, nanobiocides, and nanostructured filters/membranes; (4) nanofluidic arrays for detection of pathogens; and (5) more efficient nanostructured building insulation materials.

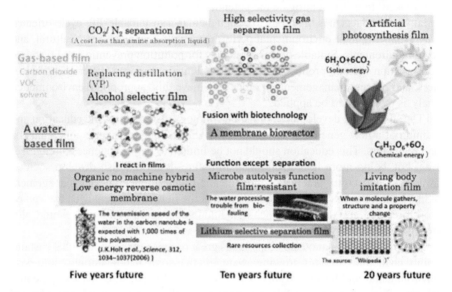

Fig. 17 Suggested research agenda for nano-enabled sustainability (Courtesy of Tomoji Kawai, Osaka University)

In the next 10 years, priority areas will include (1) CO_2 capture and conversion, (2) water desalination and reuse, e.g., more effective desalination membranes; (3) artificial photosynthesis and (4) green manufacturing, e.g., change in feedstock from petroleum/naphtha to biomass/methane enabled by nano catalysts, and (5) recycling and recovery of limited materials from semiconductor plants and electronic devices (Fig. 17).

9.3 U.S.–Australia–China–India–Saudi Arabia–Singapore Workshop (Singapore)

Panel members/discussants

Murali Sastry (co-chair), Tata Chemicals Innovation Center, India
James Murday (co-chair), University of Southern California, USA
Rose Amal, University of New South Wales, Australia
Calum Drummond, Commonwealth Scientific and Industrial Research Organization, Australia
Craig Johnson, Department of Innovation, Industry, Science, and Research, Australia
Subodh Mhaisalkar, Nanyang Technological University, Singapore

During the last 10 years, there has been growing shift from nanoscale science/engineering discovery to nano-enabled processes, technologies, and products. However, the development of nanotechnology solutions and products for sustainability applications has received limited attention. In the next 5–10 years, the emphasis should on the development/commercialization of:

- Green manufacturing and nanotechnologies
- Improved water filtration technologies
- Functional coatings that enable self-healing and wear/ice/bio-fouling resistance
- Solution-based processing of nanostructures (as opposed to vacuum-based processing)
- Nano-enabled recycling with minimal waste, including scavenging

The R&D investment and implementation strategies will continue to evolve toward greater focus on topics of regional and national interest, with collaboration between industry, academia, and research institutes. With the growing emphasis on near-term commercial goals, attention must be paid to a balance with long-term strategic goals and knowledge generation.

References

1. H. Brundtland, Towards sustainable development. (Chapter 2 in A/42/427). *Our common Future: Report of the World Commission on Environment and Development* (1987), Available online: http://www.un-documents.net/ocf-02.htm
2. D.J. Rapport, Sustainability science: an ecohealth perspective. Sustain. Sci. **2**, 77–84 (2007)
3. M.C. Roco, From vision to the implementation of the U.S. National Nanotechnology Initiative. J. Nanopart. Res. **3**(1), 5–11 (2001)
4. M.C. Roco, Broader societal issues of nanotechnology. J. Nanopart. Res. **5**, 181–189 (2003)
5. J. Rockström, W. Steffen, K. Noone, Å. Persson, F.S. Chapin III, E.F. Lambin, T.M. Lenton, M. Scheffer, C. Folke, H.J. Schellnhuber, B. Nykvist, C.A. de Wit, T. Hughes, S. van der Leeuw, H. Rodhe, S. Sörlin, P.K. Snyder, R. Costanza, U. Svedin, M. Falkenmark, L. Karlberg, R.W. Corell, V.J. Fabry, J. Hansen, B. Walker, D. Liverman, K. Richardson, P. Crutzen, J.A. Foley, A safe operating space for humanity. Nature **461**, 472–475 (2009)
6. U.S. Census Bureau, World POPClock Projection (2010), Available online: http://www.census.gov/ipc/www/popclockworld.html
7. M.A. Shannon, Net energy and clean water from wastewater. ARPA-E Workshop (2010), Available online: http://arpae.energy.gov/ConferencesEvents/PastWorkshops/Wastewater.aspx
8. United Nations Environment Programme (UNEP), *Challenges to International Waters—Regional Assessments in a Global Perspective* (UNEP, Nairobi, 2006)
9. Committee on Critical Mineral Impacts of the U.S. Economy, Committee on Earth Resources, National Research Council, *Minerals, Critical Minerals, and the U.S. Economy* (National Academies Press, Washington, 2008). ISBN 0-309-11283-4
10. T. Hillie, M. Munshinghe, M. Hlope, Y. Deraniyagala, n.d. Nanotechnology, water, and development, Available online: http://www.merid.org/nano/waterpaper
11. Organisation for Economic Co-operation and Development (OECD), Global challenges: nanotechnology and water (2008). Report of an OECD workshop on exposure assessment. Series on the safety of nanomanufactured materials #13 (2008), DSTI/STP/NANO(2008) 14
12. N. Savage, M. Diallo, Nanomaterials and water purification: opportunities and challenges. J. Nanopart. Res. **7**, 331–342 (2005)

13. M.A. Shannon, P.W. Bohn, M. Elimelech, J.G. Georgiadis, M.J. Mariñas, A.M. Mayes, Science and technology for water purification in the coming decades. Nature **452**, 301–310 (2008)
14. A. Srivastava, O.N. Srivastava, S. Talapatra, R. Vajtai, P.M. Ajayan, Carbon nanotube filters. Nat. Mater. **3**(9), 610–614 (2004)
15. B.H. Jeong, E.M.V. Hoek, Y. Yan, X. Huang, A. Subramani, G. Hurwitz, A.K. Ghosh, A. Jawor, Interfacial polymerization of thin film nanocomposites: a new concept for reverse osmosis membranes. J. Memb. Sci. **294**, 1–7 (2007)
16. J.K. Holt, H.G. Park, Y. Wang, M. Stadermann, A.B. Artyukhin, C.P. Grigoropoulos, A. Noy, O. Bakajin, Fast mass transport through sub-2-nanometer carbon nanotubes. Science **312**, 1034–1037 (2006)
17. M.S. Diallo, S. Christie, P. Swaminathan, J.H. Johnson Jr., W.A. Goddard III, Dendrimer-enhanced ultrafiltration. 1. Recovery of Cu(II) from aqueous solutions using Gx-NH2 PAMAM dendrimers with ethylene diamine core. Environ. Sci. Technol. **39**(5), 1366–1377 (2005)
18. M.S. Diallo, Water treatment by dendrimer enhanced filtration. U.S. Patent 7,470,369, 30 Dec 2008
19. S.J. Kim, S.H. Ko, K.H. Kang, J.Y. Han, Direct seawater desalination by ion concentration polarization. Nat. Nanotechnol. **5**, 297–301 (2010)
20. U.S. Department of Agriculture, Economic Research Service (USDA-ERS), Table 1 – Food and alcoholic beverages: total expenditures (2009), Available online: http://www.ers.usda.gov/briefing/CPIFoodAndExpenditures/Data/Expenditures_tables/table1.htm
21. R.L. Scharff, Health-related costs from foodborne illness in the United States (Produce Safety Project, Washington, DC, 2010), Available online: http://www.producesafetyproject.org/media?id=0009
22. H.C.J. Godfray, J.R. Beddington, I.R. Crute, L. Haddad, D. Lawrence, J.F. Muir, J. Pretty, S. Robinson, S.M. Thomas, C. Toulmin, Food security: the challenge of feeding 9 billion people. Science **327**, 812–818 (2010)
23. P.R. Srinivas, M. Philbert, T.Q. Vu, Q. Huang, J.K. Kokini, E. Saos, H. Chen, C.M. Petersen, K.E. Friedl, C. McDade-Nguttet, V. Hubbard, P. Starke-Reed, N. Miller, J.M. Betz, J. Dwyer, J. Milner, S.A. Ross, Nanotechnology research: applications to nutritional sciences. J. Nutr. **140**, 119–124 (2009), Available online: http://www.foodpolitics.com/wp-content/uploads/NanotechReview.pdf
24. T. Tarver, Food nanotechnology, a scientific status summary synopsis. Food Technol. **60**(11), 22–26 (2006), Available online: http://members.ift.org/IFT/Pubs/FoodTechnology/Archives/ft_1106.htm
25. House of Lords of the United Kingdom, Nanotechnologies and food (2010). Science and Technology Committee.1st Report of Session 2009–10, Vol. I. HL Paper. 22-I, Available online: http://www.publications.parliament.uk/pa/ld/ldsctech.htm
26. N.R. Scott, H. Chen, Nanoscale science and engineering for agriculture and food systems. Roadmap report of the national planning workshop (Washington DC, 2003), Available online: http://www.nseafs.cornell.edu/web.roadmap.pdf. Accessed 18–19 Nov 2002
27. U.S. Department of Energy (DOE), Buildings energy data book (2009), Available online: http://buildingsdatabook.eere.energy.gov
28. P.A. Torcellini, D.B. Crawley, Understanding zero energy buildings. Am. Soc. Heat. Refriger. Air Cond. Eng. J. **48**(9), 62–69 (2006)
29. Committee on State Practices in Setting Mobile Source Emissions Standards, National Research Council, *State and Federal Standards for Mobile-Sources Emissions* (National Academies Press, Washington, 2006)
30. S.C. Davis, S.W. Diegel, R.G. Boundy, *Transportation Energy Data Book*, 27 ORNL-6981st edn. (Oak Ridge National Laboratory, Oak Ridge, 2008)
31. U.S. Environmental Protection Agency (EPA), *Inventory of U.S. Greenhouse Gas Emissions and Sinks: 1990–2006* (EPA, Washington, 2008)
32. Committee on the Effectiveness and Impact of Corporate Average Fuel Economy (CAFE) Standards, National Research Council., *Effectiveness and Impact of Corporate Average Fuel Economy (CAFE) Standards* (National Academies Press, Washington, 2002)

33. National Nanotechnology Initiative (NNI), The initiative and its implementation plan (2000), Available online: http://www.nano.gov/html/res/nni2.pdf
34. R.H. Baughman, A.A. Zakhidov, W.A. de Heer, Carbon nanotubes-the route toward applications. Science **297**, 787–792 (2002)
35. J.N. Coleman, U. Khan, Y. Gun'ko, Mechanical reinforcement of polymers using carbon nanotubes. Adv. Mater. **18**, 689–706 (2006)
36. Committee on Technologies for the Mining Industry, Committee on Earth Resources, National Research Council., *Evolutionary and Revolutionary Technologies for Mining* (National Academies Press, Washington, 2002)
37. Lux Research, The governing green giants: makers of cleantech nanointermediates on the Lux Innovation Grid. Paper LRNI-R-09-07 (Lux Research Nanomaterials Intelligence service, New York, 2010)
38. S.L. Gillett, Nanotechnology: clean energy and resources for the future. White paper for the Foresight Institute (2002), Available online: http://www.foresight.org/impact/whitepaper_illos_rev3.pdf
39. Pacific Northwest National Laboratory (PNNL), SAMMS technical summary (2009), Available online: http://samms.pnl.gov/samms.pdf
40. K.F. Schmidt, Green nanotechnology: it is easier than you think. Project on Emerging Nanotechnologies (Pen 8) (Woodrow Wilson International Center for Scholars, Washington, DC, 2007)
41. F. Shadman, Environmental challenges and opportunities in nano-manufacturing. Project on Emerging Nanotechnologies (Pen 8) (Woodrow Wilson International Center for Scholars, Washington, DC, 2006), Available online: http://www.nanotechproject.org/file_download/58
42. T. Kashiwagi, E. Grulke, J. Hilding, R. Harris, W. Awad, J. Douglas, Thermal degradation and flammability properties of poly(propylene)/carbon nanotube composites. Macromol. Rapid Commun. **23**, 761–765 (2002)
43. T. Zeng, W.-W. Chen, C.M. Cirtiu, A. Moores, G. Song, C.-J. Li, C-J. Fe_3O_4 nanoparticles: a robust and magnetically recoverable catalyst for three-component coupling of aldehyde, alkyne and amine. Green Chem. **12**, 570–573 (2010)
44. C.L. Aravinda, S. Cosnier, W. Chen, N.V. Myung, A. Mulchandani, Label-free detection of cupric ions and histidine-tagged proteins using single poly(pyrrole)-NTA chelator conducting polymer nanotube chemiresistive sensor. Biosens. Bioelectron. **24**, 1451–1455 (2009)
45. Z.Y. Fan, D.W. Wang, P.C. Chang, W.Y. Tseng, J.G. Lu, ZnO nanowire field-effect transistor and oxygen sensing property. Appl. Phys. Lett. **85**, 5923–5925 (2004)
46. D.G. Rickerby, M. Morisson, Nanotechnology and the environment: a European perspective. Sci. Technol. Adv. Mat. **8**, 19–24 (2007)
47. A. Vaseashta, D. Dimova-Malinovska, Nanostructured and nanoscale devices, sensors, and detectors. Sci. Technol. Adv. Mat. **6**, 312–318 (2005)
48. B. Wang, A.P. Cote, H. Furukawa, M. O'Keeffe, O.M. Yaghi, Colossal cages in zeolitic imidazolate frameworks as selective carbon dioxide reservoirs. Nature **453**, 207–211 (2008)
49. P.G. Tratnyek, R.L. Johnson, Nanotechnologies for environmental cleanup. Nano Today **1**(2), 44–48 (2006)
50. Y. Liu, S.A. Majetich, R.D. Tilton, D.S. Sholl, G.V. Lowry, TCE dechlorination rates, pathways, and efficiency of nanoscale iron particles with different properties. Environ. Sci. Technol. **39**, 1338–1345 (2005)
51. G.V. Lowry, K.M. Johnson, Congener-specific dechlorination of dissolved PCBs by microscale and nanoscale zero-valent iron in a water/methanol solution. Environ. Sci. Technol. **38**(19), 5208–5216 (2004)
52. H. Song, E.R. Carraway, Reduction of chlorinated ethanes by nanosized zero-valent iron kinetics, pathways, and effects of reaction conditions. Environ. Sci. Technol. **39**, 6237–6254 (2005)
53. R.M. Crooks, M.Q. Zhao, L. Sun, V. Chechik, L.K. Yeung, Dendrimer-encapsulated metal nanoparticles: synthesis, characterization, and application to catalysis. Acc. Chem. Res. **34**, 181–190 (2001)

54. L. Balogh, D.R. Swanson, D.A. Tomalia, G.L. Hagnauer, E.T. McManus, Dendrimer-silver complexes and nanocomposites as antimicrobial agents. Nano Lett. **1**(1), 18–21 (2001)
55. E.R. Birnbaum, K.C. Rau, N.N. Sauer, Selective anion binding from water using soluble polymers. Sep. Sci. Technol. **38**(2), 389–404 (2003)
56. M.S. Diallo, S. Christie, P. Swaminathan, L. Balogh, X. Shi, W. Um, C. Papelis, W.A. Goddard, J.H. Johnson, Dendritic chelating agents. 1. Cu(II) binding to ethylene diamine core poly(amidoamine) dendrimers in aqueous solutions. Langmuir **20**, 2640–2651 (2004)
57. Intergovernmental Panel on Climate Change (IPCC), in *Climate Change 2007: The Physical Science Basis*, ed. by S. Solomon, D. Qin, M. Manning, Z. Chen, M. Marquis, K.B. Averyt, M. Tignor, H.L. Miller (Cambridge University Press, Cambridge, 2007)
58. Intergovernmental Panel on Climate Change (IPCC), in *Carbon Dioxide Capture and Storage*, ed. by Metz Bert, Davidson Ogunlade, Heleen de Coninck, Loos Manuela, Meyer Leo (Cambridge University Press, Cambridge, 2005)
59. F.R. Zheng, R.S. Addleman, C. Aardahl, G.E. Fryxell, D.R. Brown, T.S. Zemanian, Amine functionalized nanoporous materials for carbon dioxide (CO_2) capture, in *Environmental Applications of Nanomaterials*, ed. by G.E. Fryxell, G. Cao (Imperial College Press, London, 2007), pp. 285–312
60. D. Britt, H. Furukawa, B. Wang, T.G. Glover, O.M. Yaghi, Highly efficient separation of carbon dioxide by a metal-organic framework with open metal sites. Proc. Natl. Acad. Sci. U.S.A. **106**, 20637–20640 (2009)
61. R. Banerjee, A. Phan, B. Wang, C.B. Knobler, H. Furukawa, M. O'Keeffe, O.M. Yaghi, High-throughput synthesis of zeolitic imidazolate frameworks and applications to CO_2 capture. Science **319**, 939–943 (2008)
62. A. Phan, C.J. Doonan, F.J. Uribe-Romo, C.B. Knobler, M. O'Keeffe, O.M. Yaghi, Synthesis, structure, and carbon dioxide capture properties of zeolitic imidazolate frameworks. Acc. Chem. Res. **43**, 58–67 (2010)
63. Convention on Biological Diversity (CBD), *Sustaining Life on Earth* (Secretariat of the Convention on Biological Diversity, Montreal, 2010). ISBN 92-807-1904-1
64. Global Biodiversity Sub-Committee (GBSC), Nanotechnology and biodiversity: an initial consideration of whether research on the implications of nanotechnology is adequate for meeting aspirations for global biodiversity conservation (2009). Paper GECC GBSC (09)14, Available online: http://www.jncc.gov.uk/page-4628
65. M.E. Webber, Energy versus water: solving both crises together. *Scientific American Earth* (October Special Edition) (2008), pp. 34–41
66. Intergovernmental Panel on Climate Change (IPCC), Climate change and water, in *Technical Paper of the Intergovernmental Panel on Climate Change*, ed. by B.C. Bates, Z.W. Kundzewicz, S. Wu, J.P. Palutikof (IPCC Secretariat, Geneva, 2008)
67. A. Aston, China's rare-earth monopoly. MIT Technology Review (2010), Available online: http://www.technologyreview.com/energy/26538/?p1=A2 15 Oct 2010
68. C. Hurst, China's rare earth elements industry: what Can the West Learn (Institute for the Analysis of Global Security, Potomac, 2010), Available online: http://www.iags.org/reports.htm
69. Advanced Research Projects Agency-Energy (ARPA-E), High energy permanent magnets for hybrid vehicles and alternative energy (2010), Available online: http://arpa-e.energy.gov/ProgramsProjects/BroadFundingAnnouncement/VehicleTechnologies.aspx
70. J. Choi, H. Park, M.R. Hoffmann, Effects of single metal-ion doping on the visible-light photoreactivity of TiO_2. J. Phys. Chem. C **114**(2), 783 (2010)
71. H. Park, C.D. Vecitis, M.R. Hoffmann, Electrochemical water splitting coupled with organic compound oxidation: the role of active chlorine species. J. Phys. Chem. C **113**(18), 7935–7945 (2009)
72. L.A. Silva, S.Y. Ryu, J. Choi, W. Choi, M.R. Hoffmann, Photocatalytic hydrogen production with visible light over Pt-interlinked hybrid composites of cubic-phase and hexagonal-phase CdS. J. Phys. Chem. C **112**(32), 12069–12073 (2008)
73. J. Yuan, X. Liu, O. Akbulut, J. Hu, S.L. Suib, J. Kong, F. Stellacci, Superwetting nanowire membranes for selective absorption. Nat. Nanotechnol. **3**, 332–336 (2008)

74. H. Deng, C.J. Doonan, H. Furukawa, R.B. Ferreira, J. Towne, C.B. Knobler, B. Wang, O.M. Yaghi, Multiple functional groups of varying ratios in metal-organic frameworks. Science **327**, 846–850 (2010)
75. Y. Liu, E. Hu, E.A. Khan, Z. Lai, Synthesis and characterization of ZIF-69 membranes and separation for CO_2/CO mixture. J. Memb. Sci. **353**, 36–40 (2010)
76. D.W. Keith, Photophoretic levitation of aerosols for geoengineering. Geophy Res Abstr 10, EGU2008-A-11400 (European Geophysical Union, Vienna, 2008a)
77. D.W. Keith, Photophoretic levitation of stratospheric aerosols for efficient geoengineering. Paper read at *Kavli Institute for Theoretical Physics Conference: Frontiers of Climate Science*, Santa Barbara, 2008b, Available online: http://online.itp.ucsb.edu/online/climate_c08/keith
78. D.W. Keith, E. Parson, M.G. Morgan, Research on global sun block needed now. Nature **463**, 426–427 (2010)
79. T. Homer-Dixon, D. Keith, Blocking the sky to save the earth, Op-Ed. *New York Times,* 19 Sept 2008
80. YuA Izrael, V.M. Zakharov, N.N. Petrov, A.G. Ryaboshapko, V.N. Ivanov, A.V. Savchenko, AVYuV Andreev, YuA Puzov, B.G. Danelyan, V.P. Kulyapin, Field experiment on studying solar radiation passing through aerosol layers. Russ. Meteorol. Hydrol. **34**, 265–274 (2009)
81. J.J. Blackstock, D.S. Battisti, K. Caldeira, D.M. Eardley, J.I. Katz, D.W. Keith, A.A.N. Patrinos, D.P. Schrag, R.H. Socolow, S.E. Koonin, Climate engineering responses to climate emergencies (2009), Available online: http://arxiv.org/pdf/0907.5140
82. P.J. Rasch, P.J. Crutzen, D.B. Coleman, Exploring the geoengineering of climate using stratospheric sulfate aerosols: the role of particle size. Geophys. Res. Lett. **35**, L02809 (2008)
83. D.W. Keith, Geoengineering the climate: history and prospect. Annu. Rev. Energ. Environ. **25**, 245–284 (2000)
84. E. Teller, L. Wood, R. Hyde, Global warming and ice ages: I. Prospects for physics-based modulation of global change. University of California Research Laboratory Report UCRL-JC-128715 (Lawrence Livermore National Laboratories, Berkeley, Aug 1997)
85. M.C. Roco, R.S. Williams and P. Alivisatos, Nanotechnology Research Directions: Vision for Nanotechnology in the Next Decade, IWGN Workshop Report, U.S. National Science and Technology Council, (1999), Washington, D.C. Also published by Springer (previously Kluwer), (2000), Dordrecht. Available on line on http://www.wtec.org/loyola/nano/IWGN.Research.Directions/
86. S. Kaur, R. Gopal, W.J. Ng, S. Ramakrishna, T. Masuura, Next-Generation Fibrous Media for Water Treatment, MRS Bulletin. **33**(1), 21–26 (2008),
87. Tomalia, D.A., Henderson, S.A. Diallo, M.S. Dendrimers - An Enabling Synthetic Science To Controlled Organic Nanostructures. Chapter 24. Handbook of Nanoscience, Engineering and Technology. 2nd Edition. 2007. Second Edition; Goddard, W.A. III.; Brenner, D.W.; Lyshevski, S.E. and Iafrate, G.J.; Eds.; CRC Press: Boca Raton

Nanotechnology for Sustainability: Energy Conversion, Storage, and Conservation

C. Jeffrey Brinker and David Ginger

Keywords Photovoltaics • Solar cells • Batteries • Capacitors • Solid state lighting • Thermoelectric • Hydrogen storage • Thermal insulation • Lightning, Green building • International perspective

1 Vision for the Next Decade

Increasing standards of living and rising population numbers are leading to inevitable increases in global energy consumption. Worldwide energy usage is on track to increase by roughly 40% in the next 20 years (Fig. 1) and to nearly double by 2050. This demand could be met, in principle, from fossil energy resources, particularly coal. However, the cumulative nature of CO_2 emissions in the atmosphere demands that holding atmospheric CO_2 levels to even twice their pre-anthropogenic values by midcentury will require invention, development, and deployment of schemes for carbon-neutral energy production on a scale commensurate with, or larger than, the entire present-day energy supply from all sources combined [1, 2]. In addition to the negative climate impacts associated with burning fossil fuel, significant worldwide competition for these limited resources, and increases in the prices of energy-intensive commodities like fertilizer, are likely to have significant geo-political and social consequences, making energy an issue of national security.

C.J. Brinker (✉)
Department of Chemical and Nuclear Engineering, University of New Mexico,
1001 University Boulevard SE, Albuquerque, NM 87131, USA
and
Department 1002, Sandia National Laboratories, Self-Assembled Materials,
Albuquerque, NM 87131, USA
e-mail: jbrinker@unm.edu

D. Ginger
Department of Chemistry, University of Washington, Box 351700,
Seattle, WA 98195-1700, USA

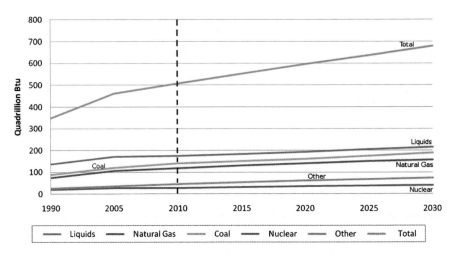

Fig. 1 World total energy consumption in quadrillion BTUs (1990–2030) (From U.S. Energy Information Administration, http://www.eia.doe.gov/oiaf/forecasting.html)

For these reasons, the president's goal has been that 10% of electricity generated should be derived from renewable sources by 2012 and 25% by 2025.[1]

Meeting the energy needs of the world's growing population in an environmentally and geopolitically sustainable fashion is arguably the most important technological challenge facing society today [1, 2]: addressing issues related to climate change, air and water pollution, economic development, national security, and even poverty and global health all hinge upon our ability to provide clean, low-cost, sustainable energy in a manner free of geopolitical conflict.

Even in mature economies like the United States where growth in energy consumption will be less rapid than in rapidly growing economies (Fig. 2), the challenge is enormous in scope. Successfully meeting this challenge will ultimately demand restructuring nearly 85% of the United States' primary energy supply. Thus, we will need to deploy *terawatts* of sustainable domestic energy generation capacity, find new ways to store this energy efficiently on truly massive yet sustainable scales, and develop new ways to separate unwanted by-products efficiently from waste streams. Furthermore we must accomplish these goals using abundant, low-cost materials in sustainable processes.

As a whole, present technologies are either too limited in terms of resources, too inefficient, or too expensive to deploy on the massive scale that will be necessary in the coming decades. It is in this context that nanoscience and nanotechnology are poised to play a transformative role in providing clean and sustainable energy from secure domestic resources.

---

[1]This goal was part of the president's campaign agenda, e.g., see http://change.gov/agenda/energy_and_environment_agenda/.

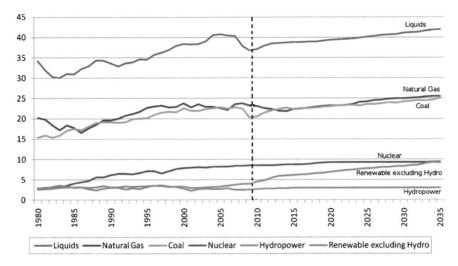

Fig. 2 U.S. total energy consumption in quadrillion BTUs according to fuel type (1980–2035) (From U.S. Energy Information Administration, http://www.eia.doe.gov/ oiaf/forecasting.html)

The past decade has shown that the technological challenges of making energy conversion and storage more efficient and more affordable are intimately tied to our understanding and control of nanoscale phenomena. In the next decade, we envision that research in nanoscience and nanotechnology will enable realization of new technologies such as low-cost photovoltaics for solar power generation, new classes of batteries for both transportation and grid-connected energy storage, efficient low-cost methods of converting both solar and electrical energy into chemical fuels, new catalysts and catalyst systems enabling artificial photosynthesis and utilization of CO_2 as a feedstock, ultrahigh surface area materials for energy storage, new membranes for applications such as water and gas purification, and improved thermoelectric and electrocaloric devices for more efficient conversion between heat and electricity.

2 Advances in the Last 10 Years and Current Status

Research in the period from roughly 2000 to 2010 has shown that nanotechnology is a powerful tool for a host of processes in support of efficient, sustainable energy conversion, storage, and conservation, in terms of:

- Tailoring the interaction of light with materials and enabling the processing of low-cost semiconductors into devices such as photovoltaics
- Making more efficient photocatalysts for converting sunlight into chemical fuels
- Developing new materials and membranes for the separations needed in many energy applications (see also Chapter "Nanotechnology for Sustainability: Environment, Water, Food, Minerals, and Climate")

- Converting chemical fuels into electrical energy (and vice versa)
- Improving energy and power density in batteries
- Improving efficiency in areas from displays and solid state lighting to thermoelectrics and friction

There are many promising research areas from the last decade could, with proper support, become transformative technologies in the coming decade. Advances in nanomaterials synthesis, integration into devices, and characterization, along with modeling and understanding of nanoscale physical phenomena, have all contributed to significant accomplishments in these areas. At the applications level, selected examples of the progress made in the last decade include those discussed in the subsections below.

2.1 Nanostructured Organic (Plastic) Photovoltaics

Solar energy is perhaps the most abundant and attractive long-term renewable energy source. However, new technology breakthroughs are needed to make solar energy conversion more cost-effective and more readily deployable on large scales. In the past decade, low-cost, nanostructured organic solar cells made from polymers like plastics have emerged as one possibility. Organic photovoltaics do not rely on conventional single p-n junctions for their function. Instead, a nanostructured donor/acceptor interface is used to dissociate excitons, while providing co-continuous transport paths for positive holes and negative electrons. The generation of photocurrent comprises four successive steps: generation of excitons by photon absorption, diffusion of excitons to the heterojunction, dissociation of the excitons into free charge carriers, and transport of these carriers to the contacts. Advances in the optimization of nanoscale phase separation in polymer blends and the use of new nanomaterials (e.g., Fig. 3) have enhanced the efficiency of each of these steps.

In addition, during the past decade, polymer-based solar cells have advanced from being subjects of basic research interest with less than ~1% power conversion efficiency to viable devices with demonstrated power conversion efficiency exceeding 8% in 2010 (e.g., [3, 4]). This has stimulated the formation of a number of startup companies, such as Konarka, Plextronics, and Solarmer, and interest from major corporations. Organic photovoltaics products based on these R&D advances are now reaching the market in consumer electronics applications. If investments in both basic and applied research can lead to similar improvements in performance and lifetime in the next decade, then organic photovoltaics will be poised to impact energy generation on the terawatt scale. Nanoscience and nanotechnology could offer improved performance through better control over the active layer morphology via self-assembly, an understanding and reduction of the various loss mechanisms that occur in processes from optical excitation to charge collection [5], better lifetimes with nanostructured barrier coatings, better light harvesting using nanostructured

Fig. 3 The past decade has shown that controlling nanoscale film morphology is key to the performance of plastic solar cells (*Physics Today* cover, May 2005)

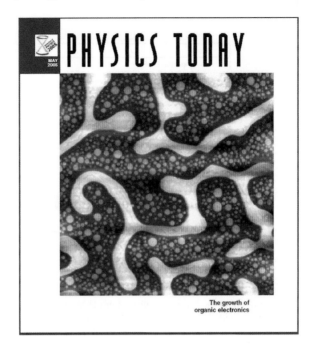

light-trapping strategies, and use of new materials to harvest a wider range of the solar spectrum, including nanostructured hybrid organic/inorganic composites. In combination, these advances should lead to efficiencies on the order of 10% in the near future.

2.2 Nanostructured Inorganic Photovoltaics

In the last decade, advances in colloidal synthesis have facilitated the use of inorganic nanoparticles as precursors for low-cost, solution-phase deposition of thin-film solar cells. Nanoscale size control offers the ability to tailor optical absorption and energy band alignments, as well as the potential to utilize more exotic phenomena such as carrier multiplication to increase photovoltaic performance (Fig. 4).

The theoretical efficiency limit for even an optimal single–band gap solar conversion device is 31%, because photons having energies lower than the absorption threshold of the active photovoltaic (PV) material are not absorbed, whereas photons having energies much higher than the band gap rapidly release heat to the lattice of the solid and have useful internal energy equal to that of the bandgap. Improvements in efficiency above the 31% theoretical limit are possible if the constraints that are incorporated into the so-called Shockley-Queisser theoretical efficiency limit are relaxed [1, 7]. Regardless, since no solar cell can be made more than 100% efficient, these performance increases must come at the same time costs are brought further down.

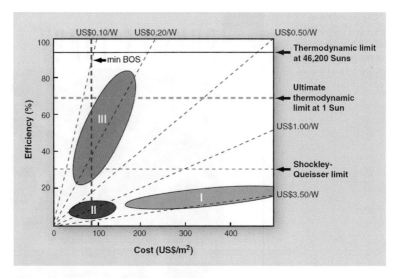

Fig. 4 Photovoltaic (PV) power costs ($/W) as function of module efficiency and areal cost. The cost figure of merit for PV cell modules ($/W) is determined by the ratio of the module cost per unit area divided by the maximum electric power per unit area. *Dashed lines* are constant $/W. *Highlighted regions* refer to Generation I, Generation II (*thin-film PV*), and Generation III (advanced/future PV) solar cells (From Green [6])

Nanostructuring has also been demonstrated as a means to increase the efficiency of current extraction by decreasing the distance charges have to travel, which would allow the use of lower-cost, lower-purity inorganic materials for photovoltaic (PV) applications without sacrificing performance. In the next decade, we see these advances poised to enable alternative thin-film technologies with lower costs and, critically, faster energy payback times.

2.3 *Artificial Photosynthesis*

Photosynthesis provides a blueprint for solar energy storage in fuels. Indeed, all of the fossil fuel-based energy consumed today is a product of sunlight harvested by photosynthetic organisms. Fuel production via natural or artificial photosynthesis requires three main nanoscale components: (1) a reaction center complex that absorbs sunlight and converts the excitation energy to electrochemical energy (redox equivalents); (2) a water oxidation complex that uses this redox potential to catalyze conversion of water to hydrogen ions, electrons stored as reducing equivalents, and oxygen; and (3) a second catalytic system that uses the reducing equivalents to make fuels such as carbohydrates, lipids, or hydrogen gas.

During the past decade a number of research groups have prepared synthetic analogues of the principal nanoscale photosynthesis components and have developed artificial systems that use sunlight to produce fuel in the laboratory. For example artificial reaction centers—where electrons are injected from a dye molecule into the conduction band of nanoparticulate titanium dioxide on a transparent electrode, coupled to catalysts such as platinum or hydrogenase enzymes—can produce hydrogen gas. Oxidizing equivalents from such reaction centers can be coupled to iridium oxide nanoparticles that can oxidize water. This coupled nanoscale system demonstrates the possibility of using sunlight to split water into oxygen and hydrogen fuel; however, efficiencies are low, and an external electrical potential is required. Dramatic improvements in efficiency, durability, and nanosystems integration are needed in the next decade to advance artificial photosynthesis as a practical technology for energy harvesting and conversion [8].

2.4 Nanostructures for Electrical Energy Storage

Along with energy production, renewable energy systems such as solar or wind require the ability to store energy for reuse on many different scales. Electrical energy, which offers the greatest potential for meeting future energy demands as a clean and efficient energy source, can be stored by electrically pumping water into reservoirs, transforming it to potential energy and back. However, this is only possible for very large-scale localized storage. As recently outlined in a workshop report from the U.S. Department of Energy [7], the use of electricity generated from renewable sources, such as water, wind, or sunlight, requires efficient distributed electrical energy storage on scales ranging from public utilities to miniaturized portable electronic devices. This can be accomplished with chemical storage (i.e., batteries) or capacitive storage (i.e. electrical capacitors). Nanostructuring can increase the efficiency of both storage, release of electrical energy, and the stability of electrode materials against swelling-induced damage from ion uptake.

Battery technologies face issues of internal surface area, electronic and ionic conduction, and phase-stability/reversibility that can benefit from the use of nanostructures. In the year 2000, there were still questions about the viability of batteries for transportation applications, hybrid vehicles were still met with widespread skepticism by most of the public and the automobile industry, and Li-ion battery energy and power densities were in the range of 100 Wh/kg–200 W/kg, respectively. In 2010, several vehicles rely heavily on battery technology (e.g., Toyota Prius, one of the best-selling cars, and the Chevy Volt and Nissan Leaf). Automotive lithium-ion cells have performance in the range of 150 Wh/kg–3,000 W/kg with 10 year predicted lifetimes, and Li-ion batteries are making inroads into large-scale grid storage. During the next decade we might expect development of new classes of nanostructured battery systems. For example, the Li–S battery has been under intense scrutiny for over two decades, as it offers the possibility of high gravimetric capacities and theoretical energy densities ranging up to a factor of five beyond conventional Li-ion systems. Recently a highly ordered, nanostructured carbon–sulphur cathode was reported enabling reversible capacities up to 1,320 mA h g^{-1}

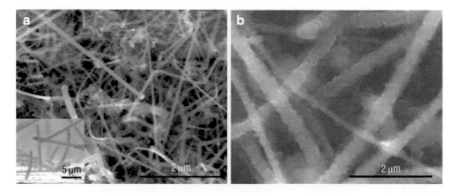

Fig. 5 Nanostructured materials, such as these Si nanowires used as an anode in a Li ion battery before (**a**) and after (**b**) electrochemical cycling (From Chan et al. [10]) have the potential to increase the performance of batteries by ten times or more

to be attained [9]. Figure 5 shows an example of Si nanowires that are being explored as high surface area anodes in Li batteries. Low dimensional nanowires can reversibly and rapidly intercalate Li without degradation.

Supercapacitors can provide a high power density (several kW/kg) and moderate energy density [11, 12]. Energy storage scales as $½ CV^2$, where C is the capacitance and V is the operating voltage across the electrodes. Because C is proportional to the surface area of the electrode accessible to ions, nanostructured electrodes can store hundreds or thousands of times more charges than conventional electrolytic capacitors. Nanostructuring also enhances so-called pseudocapacitance in which the charge storage is not electrostatic but occurs via charge transfer coming from a surface or bulk redox reaction. A decade ago RuO_2 was identified to have an enormous specific capacitance of nearly 1,000 F/g; however, Ru is expensive and rare, and during the last decade alternative nanostructured materials incorporating cheap and available MnO_2, NiO, and V_2O_5 have been developed. MnO_2 has a theoretical specific capacitance of 1,380 F/g, but so far only 30% of this value can be achieved [11]. Further improvements over the next decade will require the design and fabrication of nanoarchitectures or composites that integrate cheap transition metal oxide nanoparticles within porous carbon or polymeric matrices.

2.5 Nanotechnology for Hydrogen Storage

In 1997 it was reported that single walled carbon nanotubes (SWNTs) with diameters of about 1.2 nm synthesized by co-evaporation of cobalt and graphite in an electric arc exhibit a hydrogen storage capacity of ~10 wt% at room temperature under modest pressure [13]. These promising results were followed quickly by others, suggesting carbon nanotubes could fulfill the requirements of on-board energy storage, ~6 wt%, set by the U.S. Department of Energy (DOE). During the last decade, reported hydrogen uptake of carbon nanotubes (CNTs) has been the subject of much controversy, and recent results show that under a pressure of 12 MPa and at room temperature, the hydrogen storage capacity of CNTs is less than 1.7 wt% [14], which is far below the DOE benchmark. Overall during the last decade, significant improvements in nanoparticle purity and characterization techniques have allowed the field to arrive at a consensus that it is no longer worth investigating hydrogen uptake in pure CNTs for on-board storage applications. It is anticipated that the next decade will see new types of ultrahigh-surface-area nanoscale materials, like metal organic frameworks (MOFs) designed and developed for more efficient hydrogen storage. Continued improvements in battery technology are likely to place increasing pressure on hydrogen as an energy storage medium.

2.6 Nanotechnology for Improved Lighting

Lighting consumes roughly 22% of U.S. electricity, at a cost of $50 billion/year to U.S. consumers. Solid-state lighting is an emerging technology with the potential to achieve luminous efficacies that span an estimated range from $\sim 3 \times$ (163 lm/W) to $\sim 6 \times$ (286 lm/W) higher than that of traditional lighting technology and to reduce energy consumption proportionately. Much progress has been made over the past decade. High-color-rendering quality in commercial solid-state lighting presently achieves a luminous efficacy of about 59 lm/W when driven under moderate current densities. Today's state of the art in solid-state-lighting technology is based on highly nano-engineered InGaN/GaN (indium gallium nitride/gallium nitride) layered structures [15]. The structures are initially grown on sapphire substrates and undergo a complex series of processing steps: film deposition and removal, layer bonding and removal, and lateral patterning.

It is important to note that current state-of-the-art solid-state-lighting technology with characteristics necessary for widespread adoption in general illumination has a luminous efficacy much lower than the aggregate luminous efficacy of current artificial lighting (a mix of incandescent, fluorescent, and high-intensity discharge). Luminous efficacies for high-color-rendering indexes and for operating conditions (higher drive currents), which are in the long term necessary to minimize lifetime ownership costs, are only 23 lm/W. Thus, the targets and estimates for potential luminous efficacies are 5X and 10X away from current artificial lighting efficiencies. The magnitude of these gaps is such that a concerted effort is necessary to explore and understand the fundamental mechanisms and processes by which energy conversion occurs in photonic structures: the desired routes, conversion of injected charge carriers (electrons and holes in solids) to useful light (photons in free space), as well as the parasitic routes. Achieving such a fundamental understanding will accelerate our ability [2] to rationally design solid-state lighting materials and structures [16] and to tailor experimental conditions to control losses in the light emission process—the two grand challenges of the DOE Office of Science *Basic Research Needs for Solid-State Lighting* workshop report [17].

2.7 Nanotechnology for Thermoelectrics

The vast majority (~90%) of the world's power is generated by heat energy, typically operating at 30–40% efficiency. Thus, roughly 15 terawatts of heat is lost annually to the environment. Devices based on the thermoelectric effect—the direct conversion of temperature differences to electric voltage and vice versa—have long been recognized as a potentially transformative energy conversion technology due to their ability to convert heat directly into electricity. Despite this potential, thermoelectric devices are not in common use because of their low efficiency, and today they are only used in niche markets where reliability and simplicity are more important than performance [18, 19].

Thermoelectric materials are ranked by a figure of merit, ZT, defined as $ZT = S^2\sigma T/k$, where S is the thermopower or Seebeck coefficient, σ is the electrical conductivity, k is the thermal conductivity, and T is the absolute temperature. ZT determines the fraction of the Carnot efficiency that can be theoretically attained by a thermoelectric material. The quantity $S^2\sigma$ is called the *power factor* (PF) and is the key to achieving high performance. A large PF means that a large voltage and a high current are generated during power generation. There is no thermodynamic upper limit to ZT, however, currently available thermoelectrics have a ZT < 1. To be competitive with conventional refrigerators and generators, one must develop materials with $ZT > 3$ [18]. The challenge lies in the fact that S, σ, and k are interdependent; changing one alters the others, making optimization extremely difficult. Recently, however, the ability to create nanostructured thermoelectric materials has led to significant progress in enhancing thermoelectric properties, making it plausible that thermoelectrics could start being used in new settings in the near future.

A decade ago, the main strategy for reducing k without affecting S and σ was to use semiconductors of high atomic weight such as Bi_2Te_3 and its alloys with Sb, Sn, and Pb. High atomic weight reduces the speed of sound in the material and thereby decreases the thermal conductivity. But by 1999 there were promising reports that layered nanoscale structures might have a more profound effect on phonon transport, resulting in significant reductions in thermal conductivity either through phonon confinement [20] or phonon scattering mechanisms [21]. Soon, thin-film superlattice structures of PbTe-PbSe (grown with molecular beam epitaxy, MBE) and Bi_2Te_3-Sb_2Te_3 (grown with chemical vapor deposition) with nanoscale features were reported to have extremely low lattice thermal conductivities [22, 23]. These reports motivated the investigation of additional systems, particularly bulk materials, because superlattice structures were perceived to be too costly and fragile for real-world devices.

Although recent evidence suggests that the lattice thermal conductivities of superlattices are not as low as originally claimed [24], investigations of bulk materials that followed the thin-film reports did produce systems with extremely low lattice thermal conductivities and correspondingly high ZT (>1) [25–27]. Since then, rather dramatic enhancements in figure of merit in bulk thermoelectric materials have come from further reductions in lattice thermal conductivity rather than from improvement in power factors, giving rise to a next generation of bulk thermoelectric materials with

ZT ranging from 1.3 to 1.7. These enhancements have resulted from optimization of existing materials using nanoscale inclusions and compositional inhomogeneities, which can dramatically suppress the lattice thermal conductivity. Thus, through several avenues of nanotechnology research, the potential has expanded for achieving significantly higher efficiencies in thermoelectric power-generation schemes.

2.8 Nanotechnology for Thermal Insulation

Based on recent DOE Annual Energy Outlook reports, residential and commercial buildings account for 36% of the total primary energy use in the United States and 30% of the total U.S. greenhouse gas emissions. About 65% of the energy consumed in the residential and commercial sectors is for heating (46%), cooling (9%), and refrigeration (10%). In addition to developing new renewable sources of energy for heating and cooling, nanotechnology can play an important role in energy conservation. Nanoscale titania low-emissivity coatings made by sputtering or chemical vapor deposition are now commonplace on commercial and residential insulating glass units (IGU). Porous and particulate nanoscale materials are also crucial to advanced thermal insulation. Silica aerogels are exceptional thermal insulators because they minimize the three methods of heat transfer (conduction, convection, and radiation). Their ultra-low density (as low as 1.9 mg/cm^3 (http://eetd.lbl.gov/ECS/Aerogels/sa-thermal.html) and weakly connected, fractal silica framework reduce conductive heat transfer. Additionally because their pore size is less than the mean free path of gas at ambient pressure, they suppress convective diffusion. Finally, silica aerogel strongly absorbs infrared radiation. In combination, these properties result in extremely low thermal conductivity: from 0.03 W/m·K down to 0.004 W/m·K [28] upon moderate evacuation. This corresponds to thermal R-values of 14–105 for 8.9 cm thickness. For comparison, typical wall insulation is 13 for 8.9 cm thickness. During the last decade commercial aerogel manufacturing has become possible by avoidance of supercritical processing through adaptation of processing strategies developed by Brinker and Smith in the mid-1990s [29, 30]. In the next decade, it is expected that nanoporous aerogel and nanoparticulate fumed silica will be implemented more widely in vacuum insulation panels, because their nanostructure allows enables high R values to be obtained at very modest levels of vacuum achievable with advanced nanoscale diffusion barrier coating systems [31].

3 Goals, Barriers, and Solutions for the Next 5–10 Years

In the next decade, nanoscience research and nanotechnology will be key enablers of a variety of energy technologies at many scales. Examples of key goals where nanotechnology is likely to play a critical role in next decade include:

- Terawatt-scale solar energy generation at cost approaching that of fossil energy
- Economical catalysts to convert electricity and/or sunlight into chemical fuels

- Electrical storage for electrified transportation and grid-connected renewables
- Nanostructured materials for improved lighting efficiency
- Nanostructured thermoelectric materials for economical energy conversion

3.1 Terawatt-Scale Solar Energy at Fossil Energy Costs

From a long-term perspective, solar photovoltaics are perhaps the most attractive renewable energy source. Photovoltaics can convert sunlight directly to electricity with no moving parts or noise. Photovoltaics produce more power per unit land area than any other renewable energy technology. Indeed, the DOE has estimated that existing rooftop capacity is already sufficient to generate roughly 20% of all current U.S. energy consumption (not limited to just electricity) [27]; this is an important consideration given the often considerable public opposition to the land use that many large renewable projects are currently facing. Perhaps even more importantly, photovoltaics are inherently scalable: they can power small electronics, residential installations, and utility-scale projects. The photovoltaics industry has been growing at an annualized average rate of ~40+%/year over the past decade, making it one of the fastest growing businesses worldwide [32]. Over that same time period, China, Europe, and Japan have become the dominant worldwide photovoltaics manufacturers. However, existing photovoltaic technologies are not presently capable of rapidly delivering terawatts of new photovoltaic (PV) capacity in a cost-effective manner, creating a major environmental incentive to pursue alternative technologies. Since the worldwide photovoltaics market in 2009 exceeded $50 billion [32], and since photovoltaics are on track to become the largest semiconductor business in the world, there are also major economic incentives to invest in nanoscale photovoltaics research.

If photovoltaics are to become major contributors to our energy supply, photovoltaic technologies are needed that can deliver electricity at or below grid parity, with short energy pay-back times, using abundant materials and scalable manufacturing processes to achieve annual production capacities of many tens of gigawatts. Nanoscale phenomena may provide the key to unlocking the potential of economical, large-scale photovoltaics over the next decade. Almost all photovoltaic technologies would benefit from new strategies for light-trapping and light-harvesting using nanostructures. Likewise, the full range of organic, hybrid, and inorganic photovoltaic technologies would benefit from improved control of charge and energy transfer at heterogeneous nanoscale interfaces; improved methods to synthesize nanostructured materials; and better tools for probing nanoscale charge generation, transport, and recombination processes.

Specific selected examples of how nanotechnology could benefit next-generation photovoltaics include:

- New nanoscale chemistry and materials science for the synthesis and self-assembly of bulk heterojunction (plastic) solar cells with higher performance and uniformity possibly through ordered phase separation

- New nanoscale chemistry and materials science for the low-cost production of inorganic thin films with controlled stoichiometry and low defect density
- Self-assembly for the low-cost fabrication of nanorod structures for high-aspect-ratio electrodes
- Use of low-cost chemically synthesized quantum confined structures as tunable bandgap absorbers
- Use of nanoparticles and quantum dots in carrier multiplication and hot carrier collection strategies to circumvent the Schockley-Queisser limit in thin-film devices
- Development of new nanostructures for light harvesting—including plasmonic nanostructures, nanostructures for up-conversion, and fluorescent nanoparticles for use in fluorescent concentrators
- New transparent conductors based on inorganic nanowires, carbon nanotubes, graphene sheets, and related materials for use as solar cell substrates and superstrates to replace expensive transparent conductive oxides
- New imaging and metrology tools to characterize performance and manufacturing defects in next-generation nanostructured solar cells
- New nanostructured barrier coatings to improve environmental resilience of low-cost thin film semiconductor materials

3.2 Economical Catalysts to Convert Electricity and Sunlight into Chemical Fuels

The economical conversion of electrical energy into high-energy-density fuels for land, sea, and air transportation would increase the viability of alternative energy sources, including solar photovoltaic, concentrating solar thermal, wind, hydroelectric, and nuclear power. Converting sunlight directly to chemical fuel in an artificial manner analogous to plant photosynthesis has also captured the imaginations of many scientists. Both approaches share common challenges that must be addressed in the coming decade, including accomplishing complex, multi-electron redox chemistry with electrodes that exhibit excellent long-term stability with low-cost materials. Photocatalysts must also achieve efficient broadband light absorption.

New approaches combine multiple materials in nanoscale composite systems to achieve separate steps such as light absorption from redox chemistry with the aim of improving the performance of low-cost materials. Close synergies with investigations into nanostructured heterogeneous catalysts may emerge to allow faster turnover rates, increased active site densities, improved lifetimes, and better resilience to contaminants. Like photovoltaics, photocatalysts would also benefit from the use of nanostructures to manipulate light absorption in nanoscale structures. Biomimetics and synthetic biology may borrow concepts from natural biological systems, while greatly exceeding their performance.

3.3 Electrical Storage for Electrified Transportation and Grid-Connected Renewables

Improved storage is crucial if renewable energy sources such as solar and wind are to contribute more than ~20% of our electricity supply [33]. In the next 10 years, improved electrical energy storage will remain critical for mobile electronics and become increasingly important in the electrification of transportation applications. Advances in battery technology will make a hybrid option available for all vehicles and allow millions of plug-in hybrid electric vehicles (PHEVs) to operate on roadways. By 2020, next-generation "beyond-lithium" chemistry will need to be ready for a transition to the marketplace. An electric vehicle with a 200-mile range will require cell performance of 400 Wh/kg to 800 Wh/L, at a price point of < $100/kWh. Reaching such levels of performance will require the use of nanotechnology: nanostructures enable efficient diffusion and efficient displacement reactions in batteries and can be used to produce multifunctional and hybrid materials that combine mixed electronic/ionic conduction. Nanotechnology could also lead to improved electrolytes that use nanostructured block copolymers or nanostructured hybrid organic/inorganic composites. Self-assembling junctions, subassemblies, and bipolar electrochemical junctions could improve performance while reducing overall cost. These advances could allow the widespread adoption of batteries not just for transportation, but also for high-power, short-term storage in the electrical grid, and for integrated local storage in renewable solar and wind energy installations. With sufficient market penetration, the battery capacity of the millions of PHEVs on the road could provide large-scale scalable storage and load balancing options for a smart electrical grid.

3.4 Nanotechnology for Improved Lighting Efficiency

State-of-the-art light-emitting diodes (LEDs) are sophisticated and represent the culmination of a decade of evolving R&D. Further progress, however, is likely to be slowed by two technology challenges underlying the current five to tenfold performance gap between solid-state and current artificial lighting discussed above. The first technology challenge is the green-yellow-orange gap. Although InGaN/GaN layered structures can emit relatively efficiently at wavelengths in the blue and purple wavelengths, they emit relatively inefficiently at longer green, yellow, and orange wavelengths. Phosphors, which down-convert light to longer wavelengths, can ameliorate this to some extent, but at the expense of significant Stokes-shift (energy) losses. An ideal high-color-rendering-index, high-efficiency white light source would be composed of four wavelengths (463, 530, 573, and 614 nm) emitted directly from semiconductors [34], but efficient sources have been achieved to date at only the first of these wavelengths.

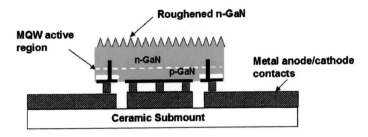

Fig. 6 State-of-the-art thin-film flip-chip (TFFC) InGaN-GaN light-emitting diode

The second technology challenge is the high-power-density rollover. Although InGaN/GaN layered structures emit relatively efficiently at modest input-power densities, they emit much less efficiently at higher input-power densities. However, higher input-power densities are necessary to achieve the highest lumen output per device, and thus a low per-lumen lifetime ownership cost.

At a superficial level, it is tempting to explain the origins of these technology challenges as the result of defect-mediated or many-body (e.g., Auger) processes of various kinds. However, at a deeper level there is little that is positively known about their origins. Consider the consequences of injecting electrons and holes into the LED structure illustrated in Fig. 6. The electrons and holes would interact to form excitons, with the subsequent fate of the excitons dependent on the density of both the excitons themselves and on other entities (defects, phonons, photons) that may interact with them. Interactions within LED nanostructures are not fully predictable and under some circumstances can lead to exciton decay—thus energy loss—whereas under other circumstances they do not, generating much controversy in the research community regarding materials and experimental conditions. This points to the ongoing requirement for fundamental as well as applied research to achieve higher-efficiency LEDs.

At low exciton and photon densities, exciton-exciton and exciton-photon interactions are relatively weak. Exciton decay through electron–hole recombination and spontaneous emission of photons can occur and is, of course, one preferred energy conversion route. But because this route can be slow, excitons can have time to interact with other entities such as defects and phonons. Some of these routes lead to emission of photons and thus light, but other parasitic routes (with respect to solid-state lighting) do not. Because there are a wide range of possible defects (vacancies, interstitials, impurities), most of which are difficult to identify with existing characterization techniques, little is positively known about these routes to exciton decay—knowledge that could provide a motivation to control particular defects through tailored growth of novel nanostructured materials and structures.

At high exciton or photon densities, exciton-exciton and exciton-photon interactions are relatively strong. Additional exciton decay routes open up, including stimulated emission, polariton formation, many-body Auger processes, and defect- or phonon-assisted Auger processes. In the presence of metals, excitons may even interact with plasmons. Again, some of these routes lead to emission of photons, but others are parasitic in that they do not. And again, because there are a number of routes with ambiguous signatures under common experimental conditions, the routes that dominate for particular materials structures and experimental conditions are currently quite controversial [35].

3.5 Economically Processable Thermoelectrics

The most promising approach for fabrication of practical thermoelectric materials is bulk nanocomposites. Figure 7 shows a typical $Si^{80}Ge^{20}$ nanocomposite, along with several critical length scales. Because the nanostructure has a grain size smaller than the phonon mean free path but greater than the electron or hole mean free path, phonons are more strongly scattered by grain boundaries than by electrons or holes, yielding a net increase in ZT [19]. At present, two major types of bulk nanostructured materials are emerging [36]: (a) materials with self-formed inhomogeneities on the nanoscale driven by phase segregation phenomena such as spinodal decomposition and nucleation and growth, and (b) materials that have been processed (e.g., ground) so they are broken up into nanocrystalline pieces, which then are sintered or pressed into bulk objects. Compared to superlattices, either of these bulk approaches is considerably cheaper and quicker to produce, and this concept has been extended to many bulk compositions, including $BixSb_2-xTe_3$ [37] and $AgPbmSbTe_2+m$, which has shown ZT~1.7 at 600 K [38]. Improvements in ZT over the last decade have come almost exclusively from strategies that reduce the lattice thermal conductivity, but it is presently difficult to imagine how k_{LAT} can be reduced much lower. Future improvements will have to be based on big jumps in the power factor from current levels [36].

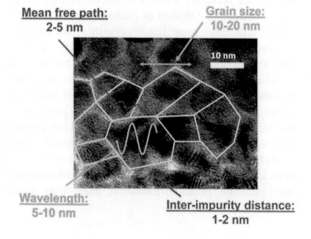

Fig. 7 A TEM image of a heavily doped $Si^{80}Ge^{20}$ nanocomposite along with some important numerically calculated characteristic lengths [19]

4 Scientific and Technological Infrastructure Needs

The diversity of energy-related research, from batteries to solar power, will necessarily require a diverse portfolio of scientific and technological infrastructure; nevertheless, some common requirements emerge:

- Support for both development and acquisition of university instrumentation and national shared user facilities to probe fundamental phenomena of charge transport, generation, and recombination, as well as light absorption and energy transfer in chemically heterogeneous, nanostructured materials with high resolution in time and space.
- More Federal support for instrumentation and computing staff at universities to maintain infrastructure investments over the long term at a time when both private and state-supported universities are cutting staff.
- An increase in certified performance measurement facilities (ranging from solar cell performance and lifetime, to grid-scale storage).
- Expanded computational resources (both hardware and software) for tackling multiscale problems that span atomistic processes to device-level performance.
- Scale-up facilities for testing multiscale manufacturing, ranging from thin-film deposition and coatings to colloidal syntheses.
- More energy-related codes in terms of computation and theory, maintained by topical experts.
- Infrastructure to accelerate development and adoption of nanotechnologies for targeted energy applications. Examples include hubs and consortia for evaluating efficient building technologies, fuel conversion, solar and battery manufacturing, and real-world performance/lifetime testing.

5 R&D Investment and Implementation Strategies

In terms of investment and implementation strategies, broadly speaking, the funding of energy-related research at both the basic sciences and applied/translational levels needs to be commensurate with the scale of this challenge over a stable, long-term period, reflecting the long-term nature of the problem. Specific recommendations for improving the effectiveness of research dollars spent on energy include:

- *Continue expansion of both student and postdoctoral fellowship programs.* Nanoscience and nanotechnology fellowships in energy will train the next generation of scientific leaders, while enabling the best students to choose the most innovative projects and to explore creative research with more freedom than would be possible if the student/postdoc was funded through an individual principle-investigator (PI) or center grant.
- *Create a precompetitive "Energy Research Corporation."* Borrowing ideas and "best practices" from the Semiconductor Research Corporation model used by the semiconductor industry could create a framework for exploring ways to

strengthen precompetitive research in energy science and technology while also building closer ties between academia and industry to accelerate the movement of ideas from the lab to real-world implementation.
- *Realize improved synergy, cooperation, and integration between Federal agencies.* Due to the crosscutting nature of these problems, energy research will require less-exclusive "ownership" of ideas and programs than the historical norm from Federal sponsors. Leveraging support from multiple sources to achieve major programmatic goals should be encouraged.
- *Increase both average award sizes and overall success rates.* Many recent "special programs" for energy research have had success rates approaching 1%. On the positive side, this reflects the large number of ideas and untapped potential for addressing the energy challenge at all levels of science; however, the 1–10% success rates typical of many new programs result in significant wasted effort by proposers and reviewers and make it difficult for program administrators to pick the best proposals in an environment with a low-signal/noise ratio. Within Federal agencies, subunits that fund disproportionate amounts of energy research should be targeted for appropriate shares of any increases in Federal research dollars.
- *Sustain support for national energy research centers and hubs.* Many large energy initiatives have recently been funded by DOE and other Federal agencies. These centers have the potential to achieve transformative breakthroughs, but they will need stable/predictable support over periods longer than a single funding cycle to achieve their ambitious long-term goals.
- *Fund more small team awards over longer periods.* The energy challenge is interdisciplinary, and building collaborations takes time. Small team awards can promote collaboration to tackle new ideas in a nimble fashion. Awards that encourage real connections and convergence between computation and experiment have good potential for high impact.

6 Conclusions and Priorities

Meeting our energy needs in an environmentally and geopolitically sustainable fashion is likely the most important technological challenge facing society today. Aside from the impacts of climate change, the economic damages from domestic fossil fuel use already approach $120 billion annually [34]. Over the next 10 years, energy will likely emerge as a preeminent application for nanotechnology, and energy applications will be a prime driver for nanoscience research. Nanoscience and nanotechnology will be primary enablers for a broad range of technologies that achieve efficient energy use, abundant low-cost alternative energy generation, and efficient energy storage. Key priorities are:

- *Solar energy generation*: The energy challenge will only be solved by sustained research and development investment in many new technologies. The problem is both too large, and our technologies too primitive, to pick clear winners for the next decade at this stage. However, solar energy is the most abundant, and arguably the most attractive *long-term* renewable energy source. The current high relative cost of solar energy compared with fossil fuels and other alternatives means solar also stands to benefit disproportionately from advances in nanotechnology (deploying terawatts of solar energy with existing technologies would require trillions of dollars of capital, and is not something we can subsidize our way out of). We thus identify solar photovoltaics, (and to a lesser extent solar photoelectrochemical cells), as a primary target for significantly increased strategic research funding: the development of terawatt-scale solar energy at a cost lower than fossil fuels would be perhaps the most significant achievement one could make in energy research. In any circumstance, solar photovoltaics are already on track to become the largest semiconductor business in the world. The U.S. can either invest in technologies to compete in this space, or cede economic strength in this market to others.

- *Storage*: Since storage will be essential for multiple renewable technologies, electrical storage for electrified transportation, and grid-connected renewables we also identify electrical storage as a second critical research target in nanoscience and nanotechnology. While batteries appear to have a technological lead at present, efficient catalysts for interconverting electricity and chemical fuel could also play an important role if fundamental scientific challenges were overcome.
- *Efficiency*: While it is critical to develop new renewable energy generation technologies such a solar, it will remain important to make the best use of the limited resources currently available. New technologies for efficient energy use, including solid state lighting, waste heat harvesting via thermoelectrics, and better insulation for buildings should represent a third core research goal for nanotechnology.
- *Human Resource Development*: Today's graduate students and postdocs will be the ones to found the next "Google of clean energy." The best and brightest students are currently drawn to energy research in chemistry, physics, biology, mathematics, and all branches of engineering. Increasing direct research opportunities for PhD and postdoctoral students would allow these scientists to meaningfully contribute to the field in the short- to-medium term with fewer bureaucratic hurdles than are often associated with major new research initiatives. That said, new integrated research initiatives are needed that are commensurate with the size, scope, and time-scale of the energy challenge.

7 Broader Implications for Society

If nanotechnology research can be transitioned to applications to provide low-cost affordable energy conversion, storage, and efficient use, it will become a central pillar in supporting economic growth, sustainable development, and national security. Renewable energy is also intimately connected with water supply, both through energy-related water usage and possible impacts of climate change on precipitation. Low-cost renewable energy is also tied intimately to global health in terms of the impacts of air and water pollution, the availability of refrigeration and sanitation in the developing world, and the possibility of new patterns of disease transmission due to habitat and climate change. The ability to positively affect the energy landscape with nanotechnology will provide direct tangible examples that can be understood by students and the general public; this may ultimately exceed nanomedicine and nanoelectronics in increasing the general public's awareness of the benefits of nanotechnologies.

8 Examples of Achievements and Paradigm Shifts

8.1 Plastic Photovoltaics and Nanostructured Photovoltaics

Plastic solar cells, which have currently reached roughly 8% efficiency in the lab and ~4–5% efficiency in prototype module production, could transform the economics of solar energy not only with lower materials costs, but lower capital costs for manufacturing and lower balance-of-system costs through the use of flexible, lightweight form factors, as in the example on the right side of Fig. 8 [4, 39, 40].

In the last 15 years, the power conversion efficiency of polymer bulk heterojunction solar cells has increased significantly, and achieved efficiencies have evolved from less than 1% in the poly(phenylene vinylene) (PPV) system in 1995 [41] to 4–5% in the poly(3-hexylthiphene) (P3HT) system in 2005 [42], to around 6%, as reported recently [43], and over 8% as reported in 2010. However, the efficiency of polymer solar cells is still lower than their inorganic counterparts, such as silicon, CdTe and CIGS, which is a limitation to practical applications at large scales.

As recently reported by Park et al. [43] achieving efficiencies in bulk heterojunction (BHJ) devices in excess of 6% will depend on a systems approach, employing several classes of nanoscale materials engineered to optimize both the total number of absorbed photons within the solar spectrum and the internal quantum efficiency

Fig. 8 *Left*: Organic solar cells are inherently nanostructured devices, as this nanoscale image of the photocurrent from a prototype cell illustrates. *Right*: Nanostructured flexible plastic solar cell produced by Konarka incorporated into a soft-sided bag for charging portable electronics sold commercially by a German company. Nanotechnology-enabled advances in solar-cell efficiencies and lifetimes could enable organics to be utilized for large-scale commercial power generation (Image from http://www.konarka.com)

(IQE) of the device [40]. Using a new class of alternating low-band-gap copolymers incorporating a fullerene derivative along with a nanoscale titania optical spacer and hole-blocking layer, they demonstrated more efficient harvesting of the solar energy spectrum and a higher open circuit voltage. The IQE is determined by a three-step process consisting of: (1) migration/diffusion of the photogenerated excitons to the bulk heterojunction interface; (2) exciton dissociation and charge separation at the interface; and (3) charge collection at the electrodes. Because of step (1), the nanoscale phase separation of the BHJ was designed to be less than 20 nm, because the exciton diffusion length is generally less than 10 nm [44, 45], and smaller-scale phase separation creates larger-area donor–acceptor interfaces where charge separation can take place. The combined nanoscale features of their device resulted in over a 6% power conversion efficiency and an internal quantum efficiency of nearly 100%, implying that essentially every absorbed photon results in a separated pair of charge carriers and that all photo-generated carriers are collected at the electrodes.

Although the previously described work is promising, integrated module efficiencies in the 8–10% range with lifetimes of roughly 7 years have been cited as a target needed for organic solar cells to compete with commercial power generation [44]. Future efforts will require improvements of the polymer, the morphology of the BHJ interface, and the overall device architecture. Important in this regard is the recent report of a new class of thieno[3,4-b]thiophene and benzodithiophene polymers (PTBs) used in BHJ polymer/fulleride solar cells. These polymers have a low-lying highest occupied molecular orbital (HOMO) energy level that provides a large open-circuit voltage (Voc) and a suitable lowest unoccupied molecular orbital (LUMO) energy level that provides enough offset for charge separation. Furthermore by judicious use of a cosolvent, the film morphology was controlled at the scale of the exciton diffusion length, resulting altogether in a power conversion efficiency of 7.4% [4]. It is expected the effective translation of these lab-scale efficiencies to large-scale module production will hinge on even greater understanding and control of the nanoscale film morphology [45]. Potentially, block copolymers that self-assemble to form highly ordered bulk heterojunctions with controlled, thermodynamically defined three-dimensional interfaces could allow engineering of optimized bulk heterojunction interfaces (Fig. 9).

Organic solar cells are not the only type of photovoltaics where nanotechnology could have a significant impact. Wadia et al. [16] recently surveyed the supply of raw materials for a variety of inorganic photovoltaics (Fig. 10). Their conclusion was that a number of materials have the potential to exceed the annual electricity production of crystalline silicon with reduced materials costs. However, utilizing many of these materials in efficient, low-cost photovoltaics will likely require nanotechnology solutions—through tailoring the semiconductor band gaps through quantum confinement effects; overcoming recombination losses with nanoscale charge collection strategies; improving light harvesting in thin films; and/or processing films with appropriate stoichiometry, surface chemistry, and defect passivation using nanoscale colloids.

Nanostructuring is also viewed to be important in increasing the efficiency of devices above the so-called Shockley-Queisser theoretical efficiency limit of 31%.

Fig. 9 Self-assembled bicontinuous BHJ morphology prepared from block copolymers [46]

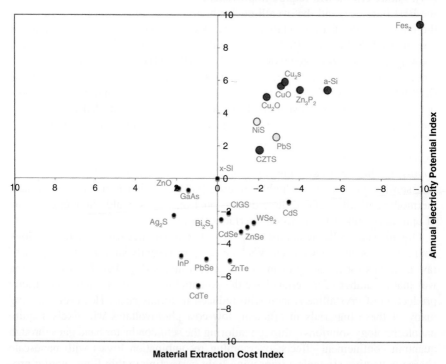

Fig. 10 Various inorganic materials ranked by extraction costs and electricity production potential for photovoltaic applications from the paper by Wadia et al. [16] Materials in *red* are those the authors identified with the greatest long-term potential. Nanotechnology to enhance light absorption, facilitate thin-film processing, improve the ratio of carrier collection to recombination loss, and tune bandgaps could be beneficial for many materials combinations

This limit arises because photons having energies lower than the absorption threshold of the active PV material are not absorbed, whereas photons having energies in excess of the band gap rapidly release heat to the lattice of the solid. It has been suggested that improvements in efficiency above the 31% theoretical limit could be achieved by generation of multiple electron–hole pairs (excitons) by high-energy single photons. This effect, known as carrier multiplication (CM), was first reported for quantum confined semiconductor nanocrystals in 2004 [47] based on a distinct dynamical signature of multiexcitons associated with their fast Auger recombination. However recently CM has been debated and in some cases claimed not to exist or that the CM yield in nanocrystals is less than in the parent solid [48]. In a recent review of this topic, discrepancies in CM measurements have been attributed to photoinduced charging of nanocrystals [49]. It is concluded that CM is enhanced in nanocrystals compared to the bulk and due to the fact that appreciable CM efficiencies are obtained for a significantly greater band gap, CM in nanocrystals is of greater potential utility in photovoltaics and photocatalysis compared to CM in parental bulk solids. No matter what the actual yield of CM effects, their use in increasing PV device efficiency will no doubt require advanced nanoscale architectures to harvest multiexcitons on a time scale faster than competing Auger recombination.

8.2 Nanostructured Batteries

The success of hybrid vehicles over the past decade perhaps best embodies the nanotechnology-based paradigm shift that has taken place in the energy storage field over the last 10 years. As late as the year 2000, many questioned the appropriateness of batteries for transportation applications. There were outstanding safety issues, battery fires were in the news, millions of lithium-ion laptop batteries were recalled, and hybrid vehicles were met with widespread skepticism by most of the public and the automobile industry. Ten years later, the Toyota Prius is one of the best-selling cars in the world, and almost all major automobile manufacturers make and sell hybrid vehicles. Nanotechnology has the potential to improve battery power and energy density as well as battery lifetime by increasing the efficiency of diffusion and displacement reactions in new battery materials, leading to the development of new hybrid electronic/ionic conduction and storage mechanisms, novel electrolytes, and new electrochemical assemblies.

8.3 New Solid-State Lighting Architectures

8.3.1 Beyond 2D—Wires, Dots, and Hybrid Structures

Layered, planar 2D heterostructures (illustrated above in Fig. 6) dominate the current architecture for solid-state lighting. This architecture can be highly nano-engineered, and its performance will surely continue to improve over time.

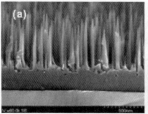 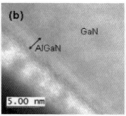 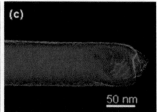

Fig. 11 Images of nanowires with approximately 2.5-nm thick AlN or AlGaN shell layers surrounding GaN cores. (**a**) SEM image, (**b**) high-resolution cross-section TEM image, and (**c**) reconstructed 3D scanning TEM tomograph showing AlN at edges, GaN core, and embedded Ni catalyst at the tip [50, 51]

This architecture, however, carries with it two severe constraints. From a materials synthesis point of view, strain cannot relax elastically in the two lateral dimensions, leading to well-known limitations on the alloy compositions and thicknesses of the various layers. From a physics point of view, confinement of charge carriers (whether quantum or classical) occurs only in one dimension, leading to less localization and weaker carrier–carrier interactions than in architectures with confinement in more than one dimension. Thus, wires and dots are expected to be important solid-state lighting structures in the coming decade. Figure 11 shows GaN (core) / InGaN (shell) (followed by additional possible shells) radial heterostructures. These structures basically constitute a nanowire, grown vertically first and then horizontally outwards. InGaN forms a quantum well structure where electron–hole pairs, i.e., excitons, are localized, thus increasing their radiative recombination. Radial nanowire heterostructures maximize the surface area of the active regions that emit light relative to the nanowire volume, and hence they could serve as a general architecture for nanowire light emitters that emit efficiently in the yellow and red wavelengths.

Although lower dimensionality can enhance strain accommodation, widen accessible alloy composition ranges, and confine electrons, thus leading to size-tuning of emission energies over an incredible range as well as with high radiative efficiencies, it can make carrier injection and transport much more difficult. These competing advantages and disadvantages suggest that the highest-performance solid-state lighting architectures may someday be coupled subsystems of different dimensionalities, where advantage is taken of each dimension's best attributes. A particularly promising approach might be one in which an exciton in the quantum well is converted into an exciton in a quantum dot overlayer via nonradiative (e.g., dipole-dipole, or Förster) coupling [52], followed by subsequent photon re-emission. Klimov recently demonstrated [53] the potential of this alternative as a means to create white light (see Fig. 12).

8.3.2 Strong Light-Matter Interactions

The regime in which light-matter interactions are weak may not be the best long-term paradigm for solid-state lighting. LEDs are based on pairs interacting with the

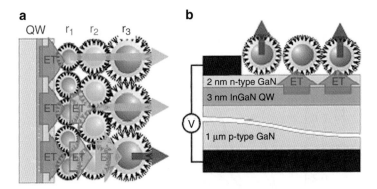

Fig. 12 (a) Schematic of a potential white light emitter: nanocrystals with emission at 480 nm (r_1), 540 nm (r_2), and 630 nm (r_3). (b) Schematic of an electrically driven, single-color energy transfer device (Adapted from Achermann et al. [53])

weakest of all electromagnetic fields, the vacuum field. When fields are this weak, the conversion of energy from electron–hole pairs to photons is slow, and for that conversion to be efficient, all other parasitic processes must be even slower. Lasers are based on stimulated emission in which electron–hole pairs now interact with an amplified and resonant electromagnetic field, so the transfer of energy from electron–hole pairs to photons is faster. Indeed, the highest-efficiency electroluminescent devices at any wavelength are high-power InGaAs/GaAs semiconductor diode lasers operating in the infrared at around 900 nm: these lasers are now 80%-or-so efficient!

Beyond this, one can imagine photonic structures, like resonant microcavities, in which electromagnetic fields are so strong that energy flow between electron–hole pairs and the electromagnetic field become continuous and cyclic. Rather than the typical irreversible decay of an exciton into a spontaneously emitted photon, the exciton and photon exchange energy resonantly at the so-called Rabi frequency. Under these conditions, the strongly coupled exciton and photon states are split into so-called upper and lower polariton states, each with mixed exciton and photon character, with a separation equal to the Rabi frequency. In semiconductors, this would represent a new regime in light-matter interactions, one that would reorder the relative importance of all the various energy conversion routes, and perhaps cause those that ultimately lead to free-space photons to become dominant.

8.4 Increasing Figure of Merit of Thermoelectrics: $ZT > 1$

As discussed above, dramatic improvements in the thermoelectric figure of merit, $ZT = S^2\sigma T/k$ (where S is the thermopower or Seebeck coefficient, σ is the electrical conductivity, k is the thermal conductivity, and T is the absolute temperature) over the last decade have come almost exclusively from strategies that reduce the lattice

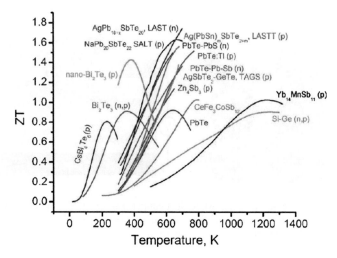

Fig. 13 Current state of the art in bulk thermoelectric materials. All the *top* performing (high ZT) materials are nanostructured. ZT 3 is needed for many practical applications of thermoelectrics [36]

thermal conductivity (see Fig. 13). Future improvements will have to be based on big jumps in the power factor from today's levels [36].

Based on recent reviews [19, 36], there are several promising nanostructuring strategies to improve the power factor $S2\sigma$, as follows:

- *Increasing the mobility.* One way in which electrical properties could be improved is simply by reducing the impact of grain boundaries on electron transport. For heavily doped nanostructured n-type $Si^{80}Ge^{20}$, the experimental mobility is about 40% lower than the theoretical expectation at room temperature, meaning that if the mobility could be restored to closer to the theoretical value, ZT could be increased by up to 40% for this case.
- *Energy filtering.* Rather than grain boundaries reducing the mobility, it is also possible for grain boundaries to play a positive role through energy filtering. The concept of energy filtering is that low energy electrons reduce the Seebeck coefficient because this portion of the Seebeck distribution is negative. This suggests an energy filtering strategy whereby an additional scattering mechanism is introduced that preferentially scatters low-energy electrons, minimizing their contribution to transport properties and thus increasing the Seebeck coefficient. In nanocomposites, the obvious scattering mechanism is electron grain boundary scattering, which does reduce the mobility but also preferentially scatters electrons with energies less than the barrier height.
- *Resonant levels to enhance the density of states.* A recent result, which has resulted in a remarkable increase in ZT, is the use of impurity band energy levels to increase the Seebeck coefficient [54]. Here the energy level created by the impurity can actually lie in the conduction or valence bands, creating a resonant

level and a local maximum in the electronic density of states. If the Fermi level is close to the distortion, it is theoretically expected that the Seebeck coefficient will increase. According to Minnich et al., [19]. It should be possible to create a material that uses both the resonant level and the nanocomposite concepts: one simply needs to add Tl to PbTe and generate the nanocomposite in the usual manner. The resulting material should have a lower thermal conductivity than the bulk value but significantly improved electrical properties due to the Tl doping.

- *Reducing the bipolar effect.* The major theme of nanocomposites for improving thermoelectric figure of merit has been a reduction of the lattice thermal conductivity. However, nanocomposites now have such a low lattice thermal conductivity that the other contributions to the thermal conductivity, viz. the electronic thermal conductivity and the bipolar thermal conductivity, can be comparable in magnitude. The bipolar effect is due to the fact that there are two types of carriers, electrons and holes. The nanocomposite strategy to reduce the bipolar effect is to create a nanostructure that preferentially scatters minority carriers but not the majority carrier. This concept has been demonstrated, but the reason why electrons are preferentially scattered in a $Bi_xSb_{2-x}Te_3$ nanocomposite alloy [37] is currently not understood. Further enhancing this preferential scattering requires detailed knowledge of the electron and hole scattering processes at grain boundaries.
- *Reducing the electronic thermal conductivity.* Another strategy to enhance ZT would be to reduce the electronic thermal conductivity, which would seem impossible, since all charge carriers must travel through the thermoelectric material, and when they do so they inevitably carry heat. While some theoretical work has shown potential situations where the Wiedemann–Franz law relating electrical and thermal conductivity is violated, no nanocomposite strategy has been developed.
- *Optimization of the physical characteristics of the nanocrystal.* Theoretical calculations on thermoelectric nanocomposite materials consisting of granular regions [55] suggest that by changing the physical characteristics of the nanocrystal, such as potential barrier height, width, and the distance between them, it is possible to increase the mean energy per carrier to obtain an enhanced power factor for improved thermoelectric performance. The model is promising and can be generic because it can be applied to other nanocomposites by incorporating the appropriate electronic structure parameters [36].

8.5 *Nanotechnology and Other Thermal Properties*

Solid-state conversion of heat to energy has been a long-sought after technology for a broad array of applications. The thermoelectric effect, magnetocaloric effect, thermionic emission [56], and recently, laser cooling [57] have all been employed as the basis for solid-state energy conversion. In spite of some niche applications, solid-state conversion is a minor player in the nation's energy technologies. Part of the reason for this limited role, discussed in connection with the thermoelectric effect above, was emphasized by the recent commentary, "An inconvenient truth about thermoelectrics" [58], where the importance of the thermodynamic efficiency of any solid-state system was emphasized. For other technologies such as the magnetocaloric effect, recent work suggests that materials may already be close to optimal without new physical mechanisms being discovered [59]. In order to make a significant contribution to the nation's energy use and greenhouse gas generation, a path to efficiencies competitive with and exceeding those of existing mechanical heat engines must be envisioned.

Recent improvements in the electrocaloric effect (ECE) have underscored the potential for using the ECE for thermal applications [60–63]. The electrocaloric effect is the inverse of the pyroelectric effect. Adiabatically applying an electric field to an electrocaloric material raises its temperature as the entropy of the polarization field is reduced, while removing the field lowers its temperature. Similarly, changing the temperature of electrocaloric materials generates electrical power. While the pyroelectric effect has been the basis of thermal sensor technologies, the ECE has not been utilized for power generation or in heat engines.

The recent improvement in the electrocaloric coefficient has occurred for nanostructured thin films that can be made with higher material quality and thus higher performance than bulk electrocalorics. Publications since 2005 have documented this improvement for thin films of $PbZr_{0.95}Ti_{0.05}O_3$ [62] and poly(vinylidene fluoride) (PVDF) and related copolymers [63]. Even more recently, researchers reported that (PVDF)–based copolymer and terpolymer films support a reversible adiabatic temperature change of 21 K with an applied voltage of 27 V [15].

As described by Epstein and Malloy. [60], electrocaloric thin films form the basis for a new class of high-performing heat engines. Four factors make electrocaloric thin films a potential breakthrough energy technology: The first is that previous use of the bulk ECE (and magnetocaloric effect) was limited by the long time required to move heat out of the bulk [60]. Using nanoscale films of electrocaloric material increases the surface-to-volume ratio and accelerates heat transport. Second, the improvements in the electrocaloric coefficient come with improvements in materials quality at the nanoscale as enabled by thin-film growth and processing techniques. In particular, understanding and control of nanoscale polymer crystallization has proven very important [61]. Third, thin films reduce the voltages required for the ECE. Finally, PVDF and related materials are environmentally friendly because they contain no lead or other heavy metals and minimal amounts of greenhouse gas chemicals.

As described previously, the efficiency of thermoelectrics and of Peltier/Seebeck devices are constrained by the competition between thermal and electrical conductivity. Vining [58] points out that a new thin-film heat engine technology could be developed based on electrocaloric thin films if a controllable thin-film thermal conductivity existed.

One example of a nanoscale mechanism for controllable thermal conductivity is based on the properties of some liquid crystal systems. It has long been known that the thermal conductance of some liquid crystals is strongly anisotropic [64–68]. In some measured rod-like or calamitic liquid crystals, the thermal conductivity is more than three times greater along the molecular director than perpendicular to it. This thermal conductivity anisotropy of liquid crystals can be harnessed to make thin-film heat switches. A thin-film heat switch could then consist of a thin layer of liquid crystal and a mechanism for changing the orientation of liquid crystal's director from parallel to perpendicular to the plane of the film. When the director is mainly perpendicular to the film, the thermal conductivity across the film is enhanced compared with the director lying in the plane of the film. The orientation of the liquid crystal directors can be controlled by electric fields applied across the liquid crystal, thus effectively acting as a "heat switch."

Carbon nanotubes also possess very high thermal conductivity anisotropies [69–72] and could be the basis for improving the switching contrasts in liquid crystals. Other mechanisms such as electrowetting would also lead to thermal switches. Performance is determined by nanoscale phenomena such as interfacial phonon density of states and fast switching speeds; high thermal contrasts are required.

8.6 Radiation-Tolerant Metals for Next-Generation Nuclear Reactors

Designing nuclear materials that can sustain extreme amounts of radiation damage is an important challenge for next-generation nuclear reactors. During years of service, radiation-induced point defects (interstitials and vacancies) are created [73], that can aggregate to form interstitial clusters, stacking fault tetrahedra, and voids [74]. Eventually, this can lead to swelling, hardening, amorphization, and embrittlement, which are primary causes of material failure [75]. Over the past decade, nanocrystalline materials containing a large fraction of grain boundaries (GB), such as nickel and copper [76], gold [77], palladium and ZrO_2 [78], and $MgGa_2O_4$ [79] have shown improved radiation resistance compared to their polycrystalline counterparts.

These experimental results are suggestive of a mechanism whereby grain boundaries absorb defects, as evidenced in large-grained polycrystalline materials by regions within 10–100 nm from grain boundaries that are denuded of extended defects following irradiation. Therefore, for materials with a grain size smaller than the material's characteristic defect-denuded zone width, extended defect formation should be minimal in the grain interiors. Very recently the mechanistic details of this nanoscale radiation tolerance mechanism have been revealed through atomic simulation [80]. Bai et al. found that grain boundaries have a surprising "loading-unloading" effect. Upon irradiation, interstitials are loaded into the boundary, which then acts as a source, emitting interstitials to annihilate vacancies in the surrounding bulk. This unexpected recombination mechanism has a much lower energy barrier than conventional vacancy diffusion and is efficient for annihilating immobile vacancies in the nearby bulk, resulting in self-healing of the radiation-induced damage. The "loading-unloading" mechanism of interstitial emission may help explain the experimental observations that nanocrystalline (NC) materials have better or worse radiation tolerance than polycrystalline (PC) materials, depending on the conditions [76–79].

8.7 Nanostructuring in Fuel Cells

The ever-increasing demand for powering portable devices has generated a worldwide effort for development of high-energy-density power sources. Although advancements in lithium-ion battery technology in recent years have provided higher-power devices, this progress has not kept pace with the portable technologies, leaving a so-called power gap that is widely expected to grow in coming years. Micro fuel cell (MFC) technology that has been under development for some time has the potential to bridge this power gap. The energy density of the fuels used in MFCs exceeds that of the batteries by an order of magnitude. However, efforts to harvest this high energy density have been hampered by issues concerning MFCs' fabrication, performance, reliability, size, and cost.

At the heart of the issues is the use of polymer membranes, which exhibit both low proton conductivity at low humidity and a large volumetric size change with humidity that is a major source of failure and integration difficulties. Improved membrane materials and configurations have been widely sought for decades and if discovered would represent a key advancement in low-temperature fuel cell technology. In addition, development of a membrane compatible with the manufacturing infrastructure within the semiconductor and microelectromechanical systems (MEMS)–based silicon-processing industries would be a major technological breakthrough. To achieve both of these objectives, Moghaddam et al. [81] recently introduced the concept of a surface nano-engineered fixed-geometry PEM, enabling nearly constant proton conductivity over a wide humidity range with no changes in volume. Key to achieving these advantages was the fabrication of a silicon membrane with ~5–7 nm pores, deposition of a self-assembled molecular monolayer on the pore surface, and then capping the pores with a layer of porous silica. The silica layer formed by plasma-assisted atomic layer deposition (a self-limiting layer-by-layer deposition process) [82] reduces the diameter of the pores and ensures their hydration, resulting in a proton conductivity of 2–3 orders of magnitude higher than that of Nafion at low humidity. An MEA constructed with this proton exchange membrane delivered an order of magnitude higher power density than that achieved previously with a dry hydrogen feed and an air-breathing cathode.

9 International Perspectives from Site Visits Abroad

9.1 United States-European Union Workshop (Hamburg, Germany)

Panel members/discussants

Bertrand Fillon (co-chair), Alternative Energies and Atomic Energy Commission (CEA), France
C. Jeffrey Brinker (co-chair), University of New Mexico, United States

Nanotechnology for Sustainability: Energy Conversion, Storage, and Conservation

Udo Weimar, University of Tübingen, Germany
Vasco Teixeira, University of Minho, Portugal
Liam Blunt, University of Huddersfield, United Kingdom
Lutz Maedler, University of Bremen, Germany

The past decade has shown that the technological challenges of making energy conversion and storage more efficient and more affordable are intimately tied to our understanding and control of nanoscale phenomena. But the improvements in realizing energy efficiency have been incremental, with research focusing primarily on discovery of nanoscale phenomena and development and experimental validation of appropriate theoretical frameworks; this has been the case for third-generation PV cells, fuel cell membranes, and thermoelectric devices. A major change in strategy in the next decade will be taking a systems approach to nanotechnology that couples multiple nanoscale phenomena and architectures with an end goal of optimized device performance—in contrast to focusing on individual nanoscale components (e.g., the fuel cell instead of only the membrane of the fuel cell, the global batteries management system instead of the electrode of batteries, etc.). Also there will be a focus on the global process chain from the material sources to the devices through the processes. This will allow the development of lower-cost processes like roll-to-roll processing with the use of flexible substrates to produce energy devices (e.g., printed electrodes for batteries, use of printed solar cells instead of vacuum deposition processes, etc.), new device architectures based on multi-nanolayering, heterostructuring, etc. Some research topics have came back with new ideas like Li-ion batteries and energy harvesting based on thermoelectricity, and other research topics have turned out to be impractical, like hydrogen storage in carbon nanotubes.

Over the next 10 years, energy will likely emerge as a preeminent application for nanotechnology, and energy applications will be a prime driver for nanoscience research. Nanostructured materials offer high potential in the area of energy technology, provided that they are well understood and tailored to exactly the right size and structure on the nanometer scale, and that powerful tools are available for nanoscale manufacturing and characterization. The vision is to develop adaptable nanostructured/nanotextured devices for better energy efficiency. Self-regenerating systems will be needed (e.g., new membranes for artificial photosynthesis—reverse engineering of nature). We expect nanoscience and nanotechnology to be primary enablers for a broad range of low-cost technologies that achieve highly efficient energy use, are composed of abundant low-cost carbon-neutral materials, enable alternative energy generation, and exhibit highly efficient energy storage. Generally, the main target will be low-cost energy devices obtained with high efficiency and low-cost processing, meaning that the emphasis will shift to large-scale manufacturing with high efficiency and low cost. Tailoring energy needs to different scales and lifetimes will also be important. Life cycle analysis will be a growing and obligatory demand.

Major nanoscience and technology research priorities for the next decade should focus on the following topics:

- Develop scalable, low-cost manufacturing methods for energy device fabrication, e.g., roll–to-roll, through self-assembly (for photovoltaic, fuel cell, batteries,

etc.), and other bottom-up approaches utilizing non precious, abundant, and "impure" material resources (e.g., FeS_2).
- Develop adaptable nanotextured, nanostructured devices for better energy efficiency.
- Improve the efficiency of photovoltaic devices through enhanced light coupling, engineered interfaces, and bandgap engineering.
- Improve safe battery power and energy density as well as lifetimes for electrified transportation and grid-connected renewables.
- Develop economical catalysts and scalable nanoscale architectures to convert electricity and/or sunlight to chemical fuels with efficiencies exceeding natural photosynthesis
- Increase significantly the figure of merit for thermoelectric and electrocaloric materials and devices that convert heat directly to electricity by decoupling electrical and thermal conductivity.
- Continue development of nanostructured materials for improved energy efficiency in everything from buildings to industrial separations.

At present, scientific inquiry in the energy arena is scattered and diversified, with many research groups working separately toward different pursuits without a clear roadmap to a better energy future. Today there is a need to facilitate and simplify access to large-scale laboratories, and for an international inventory of what facilities are available in different countries, including pilot and scale-up equipment/facilities.

A breakthrough in energy technology demands for nanomaterials science is on the way through the complete process chain: synthesis, characterization, phenomena and properties modeling. This new science era will ensure economical and at the same time safe, durable, reliable, and environmentally friendly energy conversion systems for the future. This science is still in its infancy; in order to make revolutionary progress, which is of utmost importance for the world, the following advice could be followed:

- Conduct an inventory of facilities (infrastructures, pilot lines, etc.)
- Develop an international roadmap "nano for energy applications"
- Identify energy needs in relation to applications
- Analyze the existing institutionally supported project portfolios
- Propose a strategic research agenda on nano/energy

9.2 United States-Japan-Korea-Taiwan Workshop (Tokyo/Tsukuba, Japan)

Panel members/discussants

Wei-Fang Su (co-chair), National Taiwan University, Taiwan
James Murday (co-chair), University of Southern California, United States
Chul-Jin Choi, Korea Institute of Materials Science, Korea

Participants agreed that meeting the energy needs of the world's growing population in an environmentally and geopolitically sustainable fashion is arguably the most important technological challenge facing society today. Island/peninsular/mountainous countries such as Japan, Taiwan, and Korea face the additional challenge of minimal landmass to direct toward solar energy conversion. It is in this context that nanoscience and nanotechnology are poised to play a transformative role in providing clean and sustainable energy using each country's resources.

The workshop delegates from Japan, Taiwan, Korea, and the United States proposed that major nanoscience and technology research priorities for the next decade should focus on the following topics:

- Develop high-power conversion efficiency (>40%) solar cells through enhanced light coupling, engineered interfaces, and bandgap engineering.
- Develop scalable, low-cost manufacturing methods for long-life photovoltaic fabrication (<0.5 \$/Wp >15 year) through self-assembly and other bottom-up approaches utilizing non-precious and abundant materials resources such as hydrocarbons. Instead of burning them for fuel and generating harmful CO_2, we should use them to fabricate solar cells. Low-cost, flexible non-Indium-based transparent electrodes, perhaps carbon-based, should be pursued for this kind of solar cell.
- Improve battery power and energy density as well as lifetime for electrified transportation and grid-connected renewables through the science of nanoscale interfacial interactions and nanostructures.
- Design and develop super capacitors for high-efficiency energy storage.
- Develop economical catalysts and scalable nanoscale architectures to convert electricity and/or sunlight to chemical fuels with efficiencies exceeding natural photosynthesis.
- Improve the efficiency, stability, and life of fuel cells by working on the system level rather than on single components such as electrodes. A Li-ion version is highlighted in Fig. 14. A mid-range power capacity fuel cell is the most needed

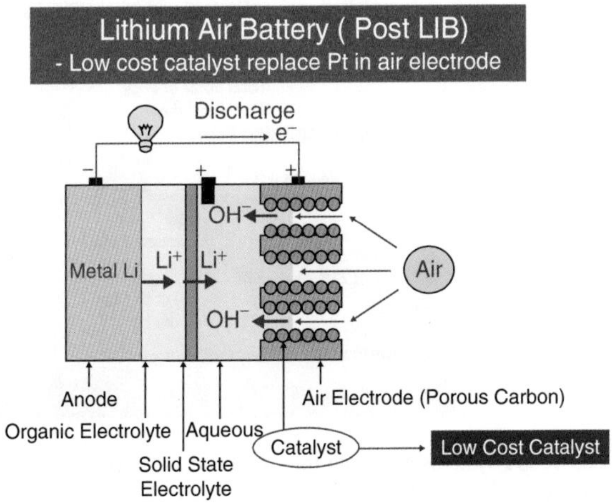

Fig. 14 Organic and aqueous electrolytes, separated by a solid state electrolyte, are used to create a new lithium-air battery, resulting in a large capacity (50,000 mAh/g on air electrode basis) [83]

for Taiwan, China, and other Asian countries to operate motorcycles and/or portable appliances.
- Increase significantly the figure of merit for thermoelectric and electrocaloric materials and devices that convert heat directly to electricity. New materials and niche markets will be needed.
- Design predictive models for the development of nanomaterials and nanostructures for applications in energy generation and conservation.
- Work out the details of total system cost of the various approaches for generation and storage of energy. One such system is illustrated in Fig. 15. The life-cycle issues need to be addressed.
- Continue development of 3D nanostructured materials for improved energy efficiency in everything from buildings to industrial separations.

Three kinds of infrastructure needs for the advancement of nanoscience and nanotechnology in the next 10 years are proposed: (1) the establishment of user facilities of measurements down to the atomic scale, (2) the establishment of data centers and top-notch scientific teams to develop predictive models for research in novel new nanomaterials and nanostructures, and (3) the establishment nanomanufacturing user facilities by interfacing with industry in the various applications.

Emerging topics for future nanoscience and nanoengineering research and education are proposed: (1) beginning in grade school, the discussion of energy utilization/conservation in our daily life, such as drying clothes with sunlight rather than an electric dryer, (2) understanding the true cost of energy consumption incorporated into products, and (3) discussion of nanotechnology's role in reducing energy costs.

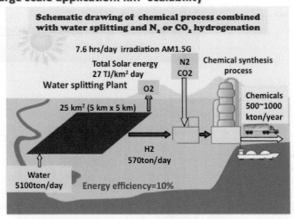

Fig. 15 A systems approach for combined solar harvesting, hydrogen production, and chemical production from a CO_2 feedstock (Courtesy of K. Domen, University of Tokyo)

In terms of R&D strategies, integrated research initiatives commensurate with the size, scope, and time-scale of the energy challenge are needed. The R&D institutions and standards groups should involve industry for energy-application-focused R&D. However, basic research funding in the nanoscience and nanoengineering should not be reduced. The R&D investments will focus on the establishment of shared-instrumentation user facilities that should include the costs of operating expenses/maintenance/local expertise, and staying at the state-of-art.

Energy impacts everything in our daily life; the continued efforts in the R&D of nanoscience and nanotechnology will resolve the issues in energy generation and energy consumption for the coming decade.

9.3 United States-Australia-China-India-Saudi Arabia-Singapore Workshop (Singapore)

Panel members/discussants

Subodh Mhaisalkar (co-chair), Nanyang Technological University, Singapore
James Murday (co-chair), University of Southern California, United States
Huey-Hoon Hng, Nanyang Technological University, Singapore
Scott Watkins, Commonwealth Scientific and Industrial Research Organisation, Australia

> China will soon publish a plan for what it called "newly developing energy industries" that will involve 5 trillion yuan ($739 billion) in investment through 2020.
>
> China Securities Journal, 8 Aug 2010

> ...no single issue is as fundamental to our future as energy...
>
> President Barack Obama, 26 January 2009

Rising global population and living standards, concerns over climate change, secure and safe low-carbon energy supplies, and the fact that the energy-water nexus represents an unparalleled threat (and opportunity) to business-as-usual, all have made the issue of energy a global priority. Over the next 40 years, in order to sustain life and standards of living to which we have grown accustomed, we must develop solutions for massively scaling *terawatts* of affordable sustainable energy *and* develop means to reduce our CO_2 emissions (Fig. 16).

The three largest greenhouse gas-emitting areas are China, the United States, and the European Union; together they produce 44% of the world's emissions (Wikipedia, "Kyoto Protocol") and are representative of the problems faced by both developing and developed countries. For the United States, the Electric Power Research Institute estimates the need to reduce greenhouse gas emissions by more than 80% over 2005 levels. Only drastic measures fuelled by breakthroughs in every aspect of science, technology, business, and industry will enable us to limit the ocean temperature rise to less than 3°C and provide renewable energy sources

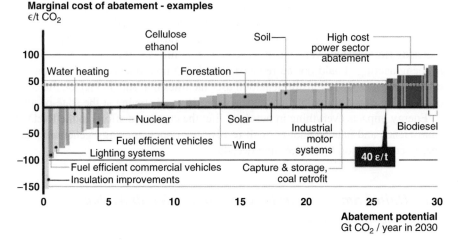

Fig. 16 Cost of greenhouse emissions abatement versus abatement potential of various technologies and systems. Items below the zero €/t CO_2 line on the *left-hand side* of the graph represent abatement obtained by increased efficiency

that will guarantee a sustainable future for the generations to come. It is in this context that nanoscience and nanotechnology are poised to play a transformative role in areas encompassing energy generation, conversion, transmission, harvesting, storage, recovery, and transportation, as well as carbon capture, storage, and conversion.

Over the past decade the largest contribution of nanoscience and nanotechnology has been towards catalysis and microelectronics processing technologies that enable nanofilms and nanofabrication that use less materials, energy, and water. In the next decade, nanotechnology will undoubtedly make a significant impact in a number of areas (see Fig. 17), as discussed below.

9.3.1 Generation and Conversion

- *Renewables and other alternatives*: It has been well recognized that nanotechnology can play a key enabling role in areas of sustainable energy generation that include solar photovoltaics (Gen I: crystalline silicon, Gen II: thin-film solar cells, Gen III: polymer, dye sensitized). Solar thermal energy enabled by active photothermal nanomaterials would be key in not only solar-thermal-based energy systems but also in energy efficiency applications. Nanotechnologies, such as nanocomposites and nanocoating, will play a critical role in wind and marine energy opportunities. The entire field of nuclear energy, which has been ignored for the past 20 years, has clearly come through a renaissance; needs

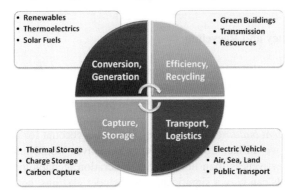

Fig. 17 Nanotechnology and nanomaterials can impact all areas of the energy/sustainability cycle

from a nanotechnology perspective exist in nuclear fuels, reactor designs, and construction, as well as waste separation/immobilization.

- *Thermoelectrics, piezoelectrics*: The opportunity to harvest waste heat and waste kinetic energy is tremendous; nanotechnology can play a significant role in increasing the energy efficiency/figure of merit for these systems
- *Solar and synthetic fuels*: Nanotechnology can be used to generate effective catalysts and nanostructures that will enable photocatalysis, carbon capture, and conversion to liquid hydrocarbons. The high energy density of hydrocarbons and their existing distribution system make this approach particularly important for those regions of the world without other options for renewable energy sources.

9.3.2 Capture and Storage

- *Charge storage*: The need is to improve battery energy and power densities and lifetime and also establish schemes for sustainable battery materials and also for battery recyclability. Nanomaterials and nanocarbons (graphene, CNT, amorphous carbons) are expected to spearhead the next breakthrough that will support mobility (transportation) as well as stationary energy solutions (e.g., peak shaving).
- *Thermal storage*: In temperate climates this will be critical for applications from domestic hot water applications to industrial applications. In tropical conditions, solar thermal can be used in combination with nanotechnology-enabled phase change materials to reduce thermal loading.

- *Carbon capture*: Nanomaterials and nanofiber-based separation systems and methodologies for carbon capture and storage are needed.

9.3.3 Efficiency and Recycling

- *Transmission*: Electrical grid transmission accounts for 25% of energy losses. New materials to reduce transmission losses, including use of high-temperature superconductors, would provide a critical breakthrough in this area.
- *Energy-efficient buildings* (see Fig. 18): Buildings account for up to 60% of the energy that we utilize. Energy efficiency initiatives would have much higher payback effectiveness than new energy generation technologies. Areas where nanomaterials can play a significant role include: glazings, heat rejection coatings, phase-change materials, high reflectivity coatings, and alternate building construction materials (e.g., high-performance concrete reinforced by nanomaterials). New concepts in air conditioning (e.g., magnetocaloric cooling, absorption/adsorption chillers), dehumidification, and solid-state lighting will also be enabled by research in nanomaterials.
- *Resources*: Recycling and syngas production would benefit from nanocatalyst breakthroughs. Efforts should also be continued in improving energy effi-

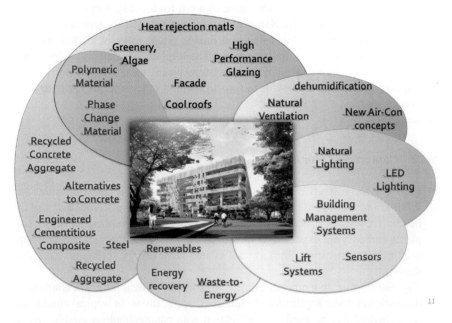

Fig. 18 Green building components

ciency of petroleum-based products—purification as well as combustion. Other opportunities may readily be identified in fields of biofuels, hydrogenated fuels (also from waste animal products), and nanotechnology for cellulosic biofuels.

9.3.4 Transportation and Logistics

- *Electric vehicles*: Transportation in general accounts for up to 20% of the CO_2 generated in megacities, and any move towards electric vehicles will play a very big role in improving energy efficiency and reducing atmospheric CO_2 (presuming low-CO_2 emissive electrical power plants). Keys to the success of electric vehicles are battery technologies and novel nanomaterials for the cathode, anode, and electrolyte materials.
- *Air, sea, land transportation*: Similar to electric vehicles for land transportation, electrification of sea transportation with fuel-cell-powered ships would also improve energy efficiency and CO_2 emission reduction. Nanocomposites in both air and sea transport will help with fuel efficiencies. Scrubber technologies and new catalyst technologies will improve with nanotechnological breakthroughs.

9.3.5 Education and Training

- The training of new generation of scientists and engineers skilled in multidisciplinary sciences is a priority. A closer public-private-academia partnership will allow the students to be world-wise and provide nanotechnology-driven solutions to address the multitude of challenges that we face from energy and sustainability perspectives.

References

1. N.S. Lewis, Toward cost-effective solar energy use. Science **315**(5813), 798–801 (2007)
2. N.S. Lewis, D.G. Nocera, Powering the planet: chemical challenges in solar energy utilization. Proc. Natl. Acad. Sci. U.S.A. **103**(43), 15729–15735 (2006)
3. Business Wire, Solarmer Energy, Inc. breaks psychological barrier with 8.13% OPV efficiency (2010), Available online: http://www.businesswire.com/news/home/20100727005484/en/Solarmer-Energy-Breaks-Psychological-Barrier-8.13-OPV. 27 July 2010
4. Y. Liang, Z. Xu, J. Xial, S.T. Tsai, Y. Wu, G. Li, C. Ray, L. Yu, For the bright future – bulk heterojunction polymer solar cells with power conversion efficiency of 7.4%. Adv. Mater. **22**, 1–4 (2010). doi:10.1002/adma.200903528
5. P. Heremans, D. Cheyns, B.P. Rand, Strategies for increasing the efficiency of heterojunction organic solar cells: material selection and device architecture. Acc. Chem. Res. **42**(11), 1740–1747 (2009)

6. M.A. Green, *Third Generation Photovoltaics: Advanced Solar Energy Conversion* (Springer, Berlin, 2004)
7. U.S. Department of Energy Office of Basic Energy Sciences (DOE/BES), Basic research needs for solar energy utilization. Report of the Basic Energy Sciences Workshop on Solar Energy Utilization, 18–21 April 2004 (U.S. Department of Energy Office of Basic Energy Sciences, Washington, DC, 2005), Available online: http://www.sc.doe.gov/bes/reports/files/SEU_rpt.pdf
8. D. Gust, T.A. Moore, A.L. Moore, Solar fuels via artificial photosynthesis. Acc. Chem. Res. **42**(12), 1890–1898 (2009)
9. X. Ji, T.L. Kyu, L.F. Nazar, A highly ordered nanostructured carbon–sulphur cathode for lithium–sulphur batteries. Nat. Mater. **8**, 500–506 (2009)
10. C.K. Chan, H. Peng, G. Liu, K. McIlwrath, X.F. Zhang, R.A. Huggins, Y. Cui, High-performance lithium battery anodes using silicon nanowires. Nat. Nanotechnol. **3**, 31–35 (2008). doi:10.1038/nnano.2007.411
11. C. Xu, F. Kang, B. Li, H. Du, Recent progress on manganese dioxide supercapacitors. J. Mater. Res. **25**(8), 1421–1432 (2010). doi:10.1557/JMR.2010.0211
12. L. Zhang, X.S. Zhao, Carbon-based materials as supercapacitor electrodes. Chem. Soc. Rev. **38**(9), 2520–2531 (2009)
13. A.C. Dillon, K.M. Jones, T.A. Bekkedahl, C.H. Kiang, D.S. Bethune, M.J. Heben, Storage of hydrogen in single-walled carbon nanotubes. Nature **386**, 377–379 (1997)
14. C. Liu, Y. Chen, C.-Z. Wu, S.-T. Xu, H.-M. Cheng, Hydrogen storage in carbon nanotubes revisited. Carbon **48**, 452–455 (2010)
15. O.B. Shchekin, J.E. Epler, T.A. Trottier, D.A. Margalith, High performance thin-film flip-chip InGaN–GaN light-emitting diodes. Appl. Phys. Lett. **89**, 071109 (2006)
16. C. Wadia, A.P. Alivisatos, D.M. Kammen, Materials availability expands the opportunity for large-scale photovoltaics deployment. Environ. Sci. Technol. **43**(6), 2072–2077 (2009)
17. U.S. Department of Energy Office of Basic Energy Sciences (DOE/BES), Basic Research Needs for Solid-State Lighting. Report of the Basic Energy Sciences Workshop on Solid-State Lighting 22–24 May 2006 (U.S. Department of Energy Office of Basic Energy Sciences, Washington, DC, 2006), Available online: http://www.sc.doe.gov/bes/reports/files/SSL_rpt.pdf
18. A. Majumdar, Materials science: enhanced thermoelectricity in semiconductor nanostructures. Science **303**(5659), 777–778 (2004)
19. A.J. Minnich, M.S. Dresselhaus, Z.F. Ren, G. Chen, Bulk nanostructured thermoelectric materials: current research and future prospects. Energy Environ. Sci. **2**(5), 466–479 (2009)
20. A. Balandin, K.L. Wang, Effect of phonon confinement on the thermoelectric figure of merit of quantum wells. J. Appl. Phys. **84**(11), 6149–6153 (1998)
21. G. Chen, Thermal conductivity and ballistic-phonon transport in the cross-plane direction of superlattices. Phys. Rev. B **57**(23), 14958 (1998)
22. T.C. Harman, P.J. Taylor, M.P. Walsh, B.E. LaForge, Quantum dot superlattice thermoelectric materials and devices. Science **297**(5590), 2229–2232 (2002)
23. R. Venkatasubramanian, E. Siivola, T. Colpitts, B. O'Quinn, Thin-film thermoelectric devices with high room-temperature figures of merit. Nature **413**(6856), 597–602 (2001)
24. Y.K. Koh, C.J. Vineis, S.D. Calawa, M.P. Walsh, D.G. Cahill, Lattice thermal conductivity of nanostructured thermoelectric materials based on PbTe. Appl. Phys. Lett. **94**(15), 153101–153103 (2009)
25. G. Chen, M.S. Dresselhaus, G. Dresselhaus, J.P. Fleurial, T. Caillat, Recent developments in thermoelectric materials. Int. Mater. Rev. **48**, 45–66 (2003)
26. H.-K. Lyeo, A.A. Khajetoorians, L. Shi, K.P. Pipe, R.J. Ram, A. Shakouri, C.K. Shih, Profiling the thermoelectric power of semiconductor junctions with nanometer resolution. Science **303**(5659), 816–818 (2004). doi:10.1126/science.1091600
27. U.S. Department of Energy Office of Energy Efficiency and Renewable Energy (DOE/EEaR), Solar energy technologies program: Multi-Year Program *Plan* 2007–2011, 2006 (DOE) (2006), Available online: http://www1.eere.energy.gov/solar/pdfs/set_myp_2007-2011_proof_2.pdf

28. J. Fricke, A. Emmerling, Aerogels. J. Am. Ceram. Soc. **75**(8), 2027–2036 (1992). Available online: http://eetd.lbl.gov/ECS/Aerogels/sa-thermal.html
29. R. Deshpande, D.W. Hua, D.M. Smith, C.J. Brinker, Pore structure evolution in silica-gel during aging drying. 3. Effects of surface-tension. J. Non. Cryst. Solids **144**(1), 32–44 (1992)
30. S.S. Prakash, C.J. Brinker, A.J. Hurd, S.M. Rao, Silica aerogel films prepared at ambient-pressure by using surface derivatization to induce reversible drying shrinkage. Nature **374**(6521), 439–443 (1995)
31. R. Baetens, B.P. Jelle, J.V. Thue, M.J. Tenpierik, S. Grynning, S. Uvsløkk, A. Gustavsen, Vacuum insulation panels for building applications: a review and beyond. Energy Build **42**, 147–172 (2010)
32. A. Jaeger-Waldau, PV Status Report 2009: research, solar cell production, and market implementation of photovoltaics (European Commission Joint Research Centre Institute for Energy, Ispra, 2009), Available online: http://re.jrc.ec.europa.eu/refsys/pdf/PV-Report2009.pdf
33. N.R. Council, *Electricity from Renewable Resources: Status, Prospects, and Impediments* (National Academy of Sciences, Washington, DC, 2010)
34. J. Johnson, Fossil fuel costs. Chem. Eng. News **87**(43), 6 (2009)
35. J. Hader, J.V. Moloney, B. Pasenow, S.W. Koch, M. Sabathil, N. Linder, S. Lutgen, On the importance of radiative and Auger losses in GaN-based quantum wells. Appl. Phys. Lett. **92**, 261103 (2008). doi:10.1063/1.2953543
36. M.G. Kanatzidis, Nanostructured thermoelectrics: the new paradigm? Chem. Mater. **22**(3), 648–659 (2009)
37. B. Poudel, Q. Hao, Y. Ma, Y. Lan, A. Minnich, B. Yu, X. Yan, D. Wang, A. Muto, D. Vashaee, X. Chen, J. Liu, M.S. Dresselhaus, G. Chen, Z. Ren, High-thermoelectric performance of nanostructured bismuth antimony telluride bulk alloys. Science **320**(5876), 634–638 (2008). doi:10.1126/science.1156446
38. K.F. Hsu, S. Loo, F. Guo, W. Chen, J.S. Dyck, C. Uher, T. Hogan, E.K. Polychroniadis, M.G. Kanatzidis, Cubic $AgPb_mSbTe_{2+m}$: bulk thermoelectric materials with high figure of merit. Science **303**(5659), 818–821 (2004). doi:10.1126/science.1092963
39. Business Wire, To cap off a magnificent year, Solarmer achieves 7.9% NREL Certified Plastic Solar Cell Efficiency (2009), Available online: http://www.businesswire.com/news/home/20091201005430/en/Cap-Magnificient-Year-Solarmer-Achieves-7.9-NREL
40. R. Gaudiana, Third-generation photovoltaic technology – the potential for low-cost solar energy conversion. J. Phys. Chem. Lett. **1**(7), 1288–1289 (2010). doi:10.1021/jz100290q
41. G. Yu, J. Gao, J.C. Hummelen, F. Wudl, A.J. Heeger, Polymer photovoltaic cells: enhanced efficiencies via a network of internal donor-acceptor hetrojunctions. Science **270**, 1789–1791 (1995)
42. G. Li, V. Shrotriya, J.S. Huang, Y. Yao, T. Moriarty, K. Emery, Y. Yang, High-efficiency solution processable polymer photovoltaic cells by self-organization of polymer blends. Nat. Mater. **4**, 864–868 (2005). doi:10.1038/nmat1500
43. S.H. Park, A. Roy, S. Beaupré, S. Cho, N. Coates, J.S. Moon, D. Moses, M. Leclerc, K. Lee, A.J. Heeger, Bulk heterojunction solar cells with internal quantum efficiency approaching 100%. Nat. Photonics **3**(5), 297–302 (2009)
44. G. Dennler, M.C. Scharber, C.J. Brabec, Polymer-fullerene bulk-heterojunction solar cells. Adv. Mater. **21**(13), 1323–1338 (2009). doi:10.1002/adma.200801283
45. R. Giridharagopal, D.S. Ginger, Characterizing morphology in bulk heterojunction organic photovoltaic systems. J. Chem. Phys. Lett. **1**(7), 1160–1169 (2010)
46. E.J.W. Crossland, M. Kamperman, M. Nedelcu, C. Ducati, U. Wiesner, D.-M. Smilgies, G.E.S. Toombes, M.A. Hillmyer, S. Ludwigs, U.O. Steiner, H.J. Snaith, A bicontinuous double gyroid hybrid solar cell. Nano Lett. **9**(8), 2807–2812 (2009). doi:10.1021/nl803174p
47. R.D. Schaller, V.I. Klimov, High efficiency carrier multiplication in PbSe nanocrystals: implications for solar energy conversion. Phys. Rev. Lett. **92**, 186601 (2004). doi:10.1103/PhysRevLett.92.186601
48. G. Nair, M.G. Bawendi, Carrier multiplication yields of CdSe and CdTe nanocrystals by transient photoluminescence spectroscopy. Phys. Rev. B **76**, 081304(R) (2007)

49. J.A. McGuire, M. Sykora, J. Joo, J.M. Pietryga, V.I. Klimov, Apparent versus true carrier multiplication yields in semiconductor nanocrystals. Nano Lett. **10**, 2049–2057 (2010)
50. I. Arslan, A.A. Talin, G.T. Wang, Three-dimensional visualization of surface defects in core-shell nanowires. J. Phys. Chem. C **112**, 11093 (2008)
51. G.T. Wang, A.A. Talin, D.J. Werder, J.R. Creighton, E. Lai, R.J. Anderson, I. Arslan, Highly aligned, template-free growth and characterization of vertical GaN nanowires on sapphire by metal-organic chemical vapor deposition. Nanotechnology **17**, 5773 (2006)
52. V.M. Agranovich, D.M. Basko, G.C. La Rocca, F. Bassani, New concept for organic LEDs: non-radiative electronic energy transfer from semiconductor quantum well to organic overlayer. Synth. Met. **116**(1–3), 349–351 (2001)
53. M. Achermann, M.A. Petruska, S. Kos, D.L. Smith, D.D. Koleske, V.I. Klimov, Energy-transfer pumping of semiconductor nanocrystals using an epitaxial quantum well. Nature **429**(6992), 642–646 (2004)
54. J.P. Heremans, V. Jovovic, E.S. Toberer, A. Saramat, K. Kurosaki, A. Charoenphakdee, S. Yamanaka, G.J. Snyder, Enhancement of thermoelectric efficiency in PbTe by distortion of the electronic density of states. Science **321**(5888), 554–557 (2008). doi:10.1126/science.1159725
55. A. Popescu, L.M. Woods, J. Martin, G.S. Nolas, Model of transport properties of thermoelectric nanocomposite materials. Phys. Rev. B **79**(20), 205302 (2009)
56. Y. Hishinuma, T.H. Geballe, B.Y. Moyzhes, T.W. Kenny, Refrigeration by combined tunneling and thermionic emission in vacuum: use of nanometer scale design. Appl. Phys. Lett. **78**(17), 2572–2574 (2001)
57. D.V. Seletskiy, S.D. Melgaard, S. Bigotta, A. Di Lieto, M. Tonelli, S.-B. Mansoor, Laser cooling of solids to cryogenic temperatures. Nat. Photonics **4**(3), 161–164 (2010). doi:10.1038/nphoton.2009.269
58. C.B. Vining, An inconvenient truth about thermoelectrics. Nat. Mater. **8**(2), 83–85 (2009)
59. V.I. Zverev, A.M. Tishin, M.D. Kuz'min, The maximum possible magnetocaloric Delta T effect. J. Appl. Phys. **107**(4), 043907–043903 (2010)
60. R.I. Epstein, K.J. Malloy, Electrocaloric devices based on thin-film heat switches. J. Appl. Phys. **106**(6), 064509–064507 (2009)
61. P.F. Liu, J.L. Wang, X.J. Meng, J. Yang, B. Dkhil, J.H. Chu, Huge electrocaloric effect in Langmuir-Blodgett ferroelectric polymer thin films. New J. Phys. **12**, 023035 (2010). doi:10.1088/1367-2630/12/2/023035
62. A.S. Mischenko, Q. Zhang, J.F. Scott, R.W. Whatmore, N.D. Mathur, Giant electrocaloric effect in thin-film $PbZr_{0.95}Ti_{0.05}O_3$. Science **311**(5765), 1270–1271 (2006). doi:10.1126/science.1123811
63. B. Neese, B. Chu, S.-G. Lu, Y. Wang, E. Furman, Q.M. Zhang, Large electrocaloric effect in ferroelectric polymers near room temperature. Science **321**(5890), 821–823 (2008). doi:10.1126/science.1159655
64. T. Kato, T. Nagahara, Y. Agari, M. Ochi, High thermal conductivity of polymerizable liquid-crystal acrylic film having a twisted molecular orientation. J. Polym. Sci. B Polym. Phys. **44**(10), 1419–1425 (2006)
65. M. Marinelli, F. Mercuri, U. Zammit, F. Scudieri, Thermal conductivity and thermal diffusivity of the cyanobiphenyl (nCB) homologous series. Phys. Rev. E **58**(5), 5860 (1998)
66. J.R.D. Pereira, A.J. Palangana, A.C. Bento, M.L. Baesso, A.M. Mansanares, E.C. da Silva, Thermal diffusivity anisotropy in calamitic-nematic lyotropic liquid crystal. Rev. Sci. Instrum. **74**(1), 822–824 (2003). doi:10.1063/1.1519677 DOI:dx.doi.org
67. F. Rondelez, W. Urbach, H. Hervet, Origin of thermal conductivity anisotropy in liquid crystalline phases. Phys. Rev. Lett. **41**(15), 1058 (1978)
68. W. Urbach, H. Hervet, F. Rondelez, Thermal diffusivity in mesophases: a systematic study in 4-4[prime]-di-(n-alkoxy) azoxy benzenes. J. Chem. Phys. **78**(8), 5113–5124 (1983)
69. I. Dierking, G. Scalia, P. Morales, Liquid crystal-carbon nanotube dispersions. J. Appl. Phys. **97**(4), 044309–044305 (2005)

70. J. Lagerwall, G. Scalia, M. Haluska, U. Dettlaff-Weglikowska, S. Roth, F. Giesselmann, Nanotube alignment using lyotropic liquid crystals. Adv. Mater. **19**(3), 359–364 (2007). doi:10.1002/adma.200600889
71. M.D. Lynch, D.L. Patrick, Organizing carbon nanotubes with liquid crystals. Nano Lett. **2**(11), 1197–1201 (2002)
72. W. Song, I.A. Kinloch, A.H. Windle, Nematic liquid crystallinity of multiwall carbon nanotubes. Science **302**(5649), 1363 (2003). doi:10.1126/science.1089764
73. G.D. Watkins, EPR Observation of close Frenkel pairs in irradiated ZnSe. Phys. Rev. Lett. **33**(4), 223 (1974)
74. B.D. Wirth, Materials science: how does radiation damage materials? Science **318**(5852), 923–924 (2007). doi:10.1126/science.1150394
75. T. Diaz de la Rubia, H.M. Zbib, T.A. Khraishi, B.D. Wirth, M. Victoria, M.J. Caturla, Multiscale modelling of plastic flow localization in irradiated materials. Nature **406**(6798), 871–874 (2000)
76. N. Nita, R. Schaeublin, M. Victoria, Impact of irradiation on the microstructure of nanocrystalline materials. J. Nucl. Mater. **329–333**(Part 2), 953–957 (2004)
77. Y. Chimi, A. Iwasea, N. Ishikawaa, M. Kobiyamab, T. Inamib, S. Okuda, Accumulation and recovery of defects in ion-irradiated nanocrystalline gold. J. Nucl. Mater. **297**(3), 355–357 (2001). doi:10.1016/S0022-3115(01)00629-8
78. M. Rose, A.G. Balogh, H. Hahn, Instability of irradiation induced defects in nanostructured materials. Nucl Instrum. Meth. B **127–128**, 119–122 (1997)
79. T.D. Shen, S. Feng, M. Tang, J.A. Valdez, Y. Wang, K.E. Sickafus, Enhanced radiation tolerance in nanocrystalline $MgGa_2O_4$. Appl. Phys. Lett. **90**(26), 263115–263113 (2007). doi:10.1063/1.2753098
80. X.-M. Bai, A.F. Voter, R.G. Hoagland, M. Nastasi, B.P. Uberuaga, Efficient annealing of radiation damage near grain boundaries via interstitial emission. Science **327**, 1631–1634 (2010)
81. S. Moghaddam, E. Pengwang, Y.-B. Jiang, A.R. Garcia, D.J. Burnett, C.J. Brinker, R.I. Masel, M.A. Shannon, An inorganic–organic proton exchange membrane for fuel cells with a controlled nanoscale pore structure. Nat. Nano **5**(3), 230–236 (2010). doi:10.1038/nnano.2010.13
82. Y.B. Jiang, N.G. Liu, H. Gerung, J.L. Cecchi, C.J. Brinker, Nanometer-thick conformal pore sealing of self-assembled mesoporous silica by plasma-assisted atomic layer deposition. J. Am. Chem. Soc. **128**(34), 11018–11019 (2006)
83. H. Zhou, Y. Wang, Development of a new-type lithium-air battery with large capacity. Advanced Industrial Science and Technology (AIST) Press Release (2009), Available online: http://www.aist.go.jp/aist_e/latest_research/2009/20090727/20090727.html

Applications: Nanobiosystems, Medicine, and Health*

Chad A. Mirkin, André Nel, and C. Shad Thaxton

Keywords Nanotechnology • Nanodiagnostic • Nanotherapeutics • Theranostics
• Translational Nanotechnology, Imaging • Drug delivery • Cancer treatment
• Tissue regeneration • Synthetic biology • Sensors to monitor human health
• International perspective

1 Vision for the Next Decade

1.1 Changes in the Vision over the Last 10 Years

Over the past decade, nanomedicine and nanobiology have undergone radical transformations from fantasy to real science. The days of discussing advances in this area in the context of "nanobots" are over, and systems and nanomaterials have emerged that provide major analytical or therapeutic advantages over conventional molecule-based structures and approaches. We have come to recognize that much of biology is executed at the nanoscale level, therefore providing a rational approach to using

---

*With contributions from: Barbara A. Baird, Carl Batt, David Grainger, Sanjiv Sam Gambhir, Demir Akin, Otto Zhou, J. Fraser Stoddart, Thomas J. Meade, Piotr Grodzinski, Dorothy Farrell, Harry F. Tibballs, Joseph De Simone.

C.A. Mirkin (✉)
Department of Chemistry, Northwestern University, 2145 Sheridan Road,
Evanston, IL 60208, USA
e-mail: chadnano@northwestern.edu

A. Nel
Department of Medicine and California NanoSystems Institute, University of California,
10833 Le Conte Avenue, 52-175 CHS, Los Angeles, CA 90095, USA

C.S. Thaxton
Institute for Bionanotechnology in Medicine, Northwestern University,
Robert H. Lurie Building, 303 E Superior Street, Room 10-250,
Chicago, IL 60611, USA

the structure and function of engineered nanomaterials at the nano-bio interface for interrogation of disease, diagnosis, treatment, and imaging at levels of sophistication not possible before [1]. Fabrication of a host of nanostructures has been coupled with advanced chemical manipulation in order to impart biological recognition and interaction capabilities. Often, chemical manipulation results in nanomaterials that provide performance enhancement of therapeutics, imaging agents, diagnostics, and materials for tissue engineering and for basic science applications.

Thus, the foundation has been laid for creating systems that can revolutionize the fields of medicine and biology. Early work has provided significant evidence that the properties afforded by nanostructures offer not only different but also better ways of detecting, managing, treating, and in certain cases, preventing disease. Analytical tools have been invented that allow imaging and manipulation of biological structures in ways that would have been viewed as science fiction just a few years ago. These tools are dramatically accelerating the fundamental understanding of complex biological systems and providing a basis for understanding the rapid translational advances being made on the nanomedicine front. Through the application of nanomaterials in *in vitro* and *in vivo* biological systems, and with sophisticated tools to monitor such nano-bio interactions, it has become increasingly appreciated that such interactions are complex and warrant directed evaluation as we move forward in the next 10 years. This chapter describes some representative examples of nanotechnology-based tools, materials, and systems that are having major impacts in both biology and medicine. The final several sections provide an international perspective on the impact of nanotechnology and nanobiosystems in health and medicine.

1.2 Vision for the Next 10 Years

Bio-organic and synthetic chemistry are contributing to a new interfacial science that is of great importance to better understand biology at the nano-bio interface and to specifically design engineered nanomaterials or nanoscale systems that can fundamentally impact human health [1]. (For a unique perspective on synthetic biology and nanotechnology see Sect. 8.2) The nano-bio interface is defined as the dynamic physicochemical interactions between nanomaterial surfaces and the surfaces of biological components such as proteins, membranes, phospholipids, endocytotic vesicles, organelles, DNA, and biological fluids. The outcomes of these interactions determine nanoparticle uptake, bioavailability, and the possibility to carry out bio-physicochemical reactions that could be of therapeutic and diagnostic use. This includes the dynamic interactions that determine the formation of a protein corona that is specific to individual biological compartments, particle wrapping at the cell surface membrane, endocytotic uptake, and lysosomal as well as mitochondrial perturbation that can affect drug delivery and the safety of nanomaterials [1]. Interactions between biological and nanoparticle components could, in reverse, lead to phase transformation, surface reconstruction, dissolution, and release of

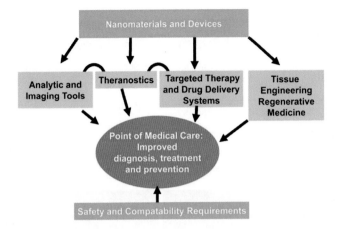

Fig. 1 The cornerstones of nanomedicine (Courtesy of A. Nel)

the surface energy from the nanoparticles. Probing these nano-bio interfaces at the biological level permits prediction of structure-activity outcomes based on nanomaterial properties such as size, shape, surface chemistry, surface charge, hydrophobicity, roughness, and surface coatings.

Through a more comprehensive understanding of the bio-nano interface, nanomedicine will mature into a higher-throughput and more predictable endeavor. This new branch of medicine will revolutionize the way medicine is practiced, create a new pipeline of diagnostic and therapeutic capabilities for the pharmaceutical industry, and catalyze extraordinary advances in molecular and cell biology. The cornerstones of nanomedicine are depicted in Fig. 1, which delineates that substantially improved diagnosis, imaging, and treatment can be obtained through the use of nanomaterials and devices that are capable of performing analytical and imaging functions, targeted therapy and drug delivery, imaging plus delivery functions (theranostics), as well using nano-enabled material surfaces for tissue engineering.

The most promising future nanoscience-based applications in medicine (Fig. 2) are ultrasensitive and selective multiplexed diagnostics, drug delivery, targeted treatment of cancer and other diseases, body imaging, tissue/organ regeneration, and gene therapy. All of these applications combine engineering advances with improved strategies for manipulating biological systems. New approaches for drug delivery, imaging, and diagnostics will be refined and developed, and more sophisticated nano-therapeutics and diagnostics will supplement those already in clinical use or in clinical trials. To facilitate this development, it will be necessary to implement new manufacturing approaches. All new products must address stringent safety and compatibility standards that are being challenged by the novel properties of engineered nanomaterials and the potential that these may introduce new biohazards [3].

Nanotechnology is enabling the development of highly accurate *in vitro* and *in vivo* sensors, novel imaging contrast agents, and platforms for localized

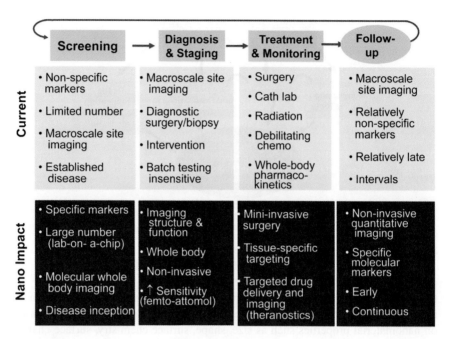

Fig. 2 Proposed impact of nanotechnology on future clinical care (Modified from the European Science Foundation Forward Look report [2])

therapy [4] (see also Sect. 8.3). Current sensors are largely *in vitro*; a reduction in sample size and multi-analyte detection capabilities continue to improve guidance for traditional therapies. In the future, diagnostics will provide earlier and more accurate bioanalyte detection capabilities, and localized nanotherapy will provide more effective treatments. Furthermore, advances in nanodiagnostic capabilities will allow for a more robust realization of personalized medicine, because disease phenotyping biomarkers will be used to direct appropriate patient- and disease-targeted therapies. Quantitative *in vivo* nanosensors will allow a leap in systemic diagnostics at multiple sites, and this capability together with localized and targeted nanotherapy will greatly increase treatment efficacy and serve to minimize side effects. Although *in vivo* diagnostics hold the greatest promise for the long run, *in vitro* diagnostic approaches are more quickly translated to clinical applications, for several reasons: these include a less complex regulatory environment, greater familiarity and success with current sensor platforms, and more direct access to relevant clinical samples. The next step will be *in vivo* sensors that do not require systemic injection. Ultimately, as knowledge of specific disease biomarkers increases and therapeutics become more highly refined, the demand and necessity for systemic *in vivo* sensors will increase. Finally, continued advances in bioengineering and at the bio-nano interface will provide robust means to guide stem cell fate and to regenerate tissues.

Nanotechnology is also enabling novel, specialized treatments for cancer. This will be a high-impact area of nanomedicine in the next 10 years (see Sect. 8.1).

In spite of historic advances in cancer biology and the development of ever more sophisticated chemotherapeutics, the outlook for patients with advanced cancers is still grim; for some cancers (e.g., brain, ovarian, lung, and pancreatic cancers), there have been no appreciable changes in life expectancy for the past 50 years. Several nanomedicine pathways can improve this outlook through: (1) fuller realization of the opportunities in personalized medicine for the identification and targeted treatment of sub-sets of patients with given cancer types, and (2) development of targeted therapeutics with higher efficacy and reduced side-effect profiles. Accompanying the increasing capabilities for nanoscale therapeutics to target treatments and limit adverse effects for cancer patients over the next 10 years should be regulatory changes that make the most promising candidates available to patients more rapidly. In fact, opportunities for nanotechnology in cancer treatment are many fold and are expected to have a significant impact. Over the next 10 years it should be possible to:

- Develop point-of-care nanodevices for early diagnosis and therapeutic response monitoring capable of using unprocessed bodily fluids with multiplexing and rapid analysis capabilities
- Develop diagnostic and post-therapy monitoring nanodevices for the detection and interrogation of circulating tumor cells and circulating tumor initiating cells
- Conduct successful clinical trials for nanoparticle delivery of siRNA molecules and other nucleic acid therapeutics
- Demonstrate novel nanoparticle-based drug formulations with significant improvement in targeting therapeutic windows as compared to free drug delivery
- Design particles to enable penetration of the blood-brain-barrier and enable more effective treatment of brain tumors
- Leverage nanotechnology-based studies of cell migration and cell motility for the development of anti-metastatic drugs
- Leverage nanotechnology tools to enable patient stratification for more personalized medicine
- Develop nanoparticle-based techniques to overcome the multidrug-resistance (MDR) mechanism
- Develop nanoparticle constructs capable of probing tumor microenvironments for tumor recognition and/or triggered drug release
- Implement nanotechnology-based techniques for intraoperative monitoring during surgery
- Develop theranostic multifunctional nanoscale platforms capable of interrogating the tumor microenvironment, subsequently administering therapy, and providing a readout of therapeutic efficacy

The impact of nanotechnology on cancer and other diseases depends on the design and construction of devices to diagnose, treat, and monitor disease at all stages. In addition, new tools and devices are needed to understand the processes behind the development and spread of a disease and to reverse or alter the progress of the disease.

In summary, the overall impact of nanotechnology on future point-of-clinical-care delivery will be multifaceted, with significant advances in patient screening, diagnosis, staging, treatment, and monitoring.

2 Advances in the Last 10 Years and Current Status

Nanobiosystems and nanomedicine are the most exciting and fastest-growing areas in modern nanotechnology research. Indeed, nanostructures are now being used in the development of powerful tools for manipulating and studying cellular systems, in ultrasensitive *in vitro* diagnostics that can track indications of disease at very early stages, in *in vivo* imaging agents that provide better contrast and more effective targeting compared to molecular systems, and in novel therapeutics for debilitating diseases such as cancer, heart disease, and regenerative medicine. Over the past decade, research and advances in these areas have occurred at an extraordinary pace, and looking to the future, these subfields of nanotechnology will experience a revolution in new capabilities. Major milestones over the past decade have included:

- FDA-approvals for the first nanotherapeutics [5]
- FDA-clearances for the first nano-enabled *in vitro* diagnostic tools [6]
- The first siRNA human trials involving nanomaterial delivery systems [7]
- Development of intracellular probes for measuring the genetic content of a cell as well as its metabolic activity [8–10]
- Imaging of the sites of therapeutic targeting (theranostics) [11–15]
- Development of mechanized nanoparticles that are capable of stimulus-responsive release of guest molecules [16–22]

Although there are many nanotechnologies that have had significant impact in biomedicine, a few important examples are included below for illustrative purposes.

2.1 In Vitro Diagnostics

2.1.1 The Bio-Barcode Assay

The medical field is increasingly anxious to adopt novel technologies that enable the accurate and early detection of disease-specific biomarkers so that timely and personalized treatments can be delivered. High priorities, moving in parallel with the development of new technologies to detect existing biomarkers, are the identification and validation of "low-abundance" biomarkers. Decxxtection of disease—or patient-specific biomarkers at an early stage requires technology capable of measuring markers when in low abundance. The bio-barcode assay, pioneered by the Mirkin Group at Northwestern University, is a detection technology that

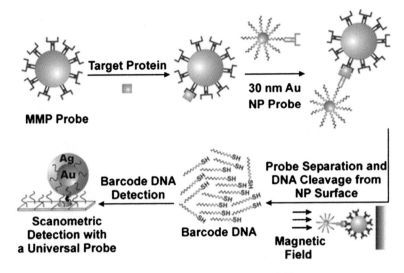

Fig. 3 The bio-barcode assay for protein detection (Nam et al. [26])

provides significant sensitivity advantages over conventional methodologies such as the enzyme-linked immunosorbent assay (ELISA), without requiring costly, time-consuming, and technically demanding enzymatic amplification steps (e.g., those required by polymerase chain reaction, PCR, techniques) [23–29]. The bio-barcode assay and variants of it are being developed for a number of disease processes where early detection is of paramount importance; it is suitable for subsequent adaptation to any number of others.

The bio-barcode assay is a sandwich assay involving two probe species that collectively function to detect and isolate a targeted biomarker and then amplify its presence at the time of assay readout without using enzymatic signal amplification (Fig. 3) [26].

Illustrated in Fig. 3 is the concept of protein detection [26]; however, the bio-barcode assay can be adapted to detect any molecular target, for example nucleic acids [25], amenable to "sandwich" capture between the two assay probes. The first particle element is a micron-sized magnetic particle (MMP) surface decorated with target-specific recognition elements. Monoclonal antibodies and short complementary oligonucleotides are used for protein and nucleic acid biomarker targets, respectively. Mixing the MMPs with a solution (e.g., serum) containing the biomarker of interest results in specific target binding to the surface of the MMP. Magnetic separation isolates the bound target. Next, a solution containing gold nanoparticles (AuNP) is added to the MMP-target hybrids.

Importantly, the AuNP probes have two important functional elements. First, a recognition element is bound to the AuNP probe, which acts to sandwich the targeted biomolecule. Antibodies are used for proteins and oligonucleotide probes for targeted nucleic acids. Second, the AuNP has surface-immobilized bio-barcode sequences that are arbitrarily chosen as surrogate markers for the presence of the targeted biomolecule. Typically, the bio-barcode is 20, or so, nucleotides in length, providing approximately 4^{20} unique barcode identities. The barcode DNA does not participate in target recognition, but it is ultimately responsible for target amplification, identification, and quantification.

Following AuNP probe addition, the hybrid structures sandwiching the targeted biomolecule are separated using a magnet, the solution is washed free of unbound AuNP probes, and then the bio-barcodes are chemically released from the surface of the AuNPs. Released bio-barcode DNA is then added to a DNA microarray, where the barcodes can be sorted to specific array addresses and then detected using AuNPs surface-functionalized with oligonucleotides complementary to a universal bio-barcode tag. In a final step known as the Scanometric assay, the optical signature provided by AuNPs bound to bio-barcode DNA is greatly amplified through the electroless reduction of silver and/or gold catalyzed by the surface-bound AuNP probes [30, 31]. Ultimately, the biomarker presence and quantity in the original solution is reflected by the presence and intensity of staining at the corresponding bio-barcode array address.

Initially demonstrated for the prostate cancer biomarker, prostate specific antigen (PSA),[1] the bio-barcode assay can be up to 10^6 times more sensitive than conventional ELISA assays for the same target [26]. The bio-barcode assay sensitivity is mainly derived from the following:

- Target capture is highly efficient due to the homogeneous nature of the assay and because a large concentration of target-binding elements are present in the assay
- In addition to the recognition elements, AuNP probes carry with them hundreds of bio-barcode DNA strands, which lead to direct amplification of signal
- The scanometric assay, the first FDA-cleared nanoassay and the final step of the bio-barcode assay [31], has high sensitivity based upon the catalytic amplification of the optical signature of individual bio-barcode-bound AuNPs

A significant amount of work has been done to move from the initial proof-of-concept PSA nanoassay to clinical applications. As a marker for the early detection of prostate cancer, PSA suffers from a lack of specificity for prostate cancer and is controversial in this clinical setting [33, 34]. However, following surgical removal of the prostate gland (i.e., prostatectomy) for cancer, the presumed sole source of serum PSA has been removed. In the post-prostatectomy setting, assuming organ-confined disease, PSA serum levels fall to values that are undetectable using commercial ELISA-based technologies. Unfortunately, up to 40% of patients who undergo radical prostatectomy have prostate cancer recurrence, often first detected by a PSA that increases from undetectable to detectable and rising [32, 35–37]. Here, clinical data demonstrates that earlier salvage radiation delivered at the lowest possible serum PSAs leads to improved patient survival [38].

Accordingly, and building upon previous work demonstrating that more sensitive testing for PSA post-prostatectomy could provide substantial lead times

---

[1] PSA is a serum biomarker used to screen individuals for prostate cancer and also as a marker for prostate cancer recurrence following primary and secondary prostate cancer intervention [32].

in the diagnosis of prostate cancer recurrence [39], the Mirkin group utilized the bio-barcode assay to assess whether increased sensitivity would provide clinically useful information in the setting of prostate cancer recurrence [40]. In a pilot study, serial serum samples were collected prospectively from patients following prostatectomy. The serum samples had been tested using conventional assays prior to testing with the bio-barcode assay. The study included a mix of patients with and without prostate cancer recurrence, first demonstrated by a detectable and then rising PSA. Serum samples for interrogation with the bio-barcode assay had undetectable (<0.1 ng/mL PSA) PSA values as measured with conventional tests. However, the bio-barcode assay is over 300 times more sensitive than conventional ELISA assays, with a lower limit of detection of 0.3 pg/mL. Using the bio-barcode assay, PSA was detected in the serum of all patients, whereupon they could be stratified into three groups:

1. Low and non-rising bio-barcode PSA values (Fig. 4a)
2. Initially low bio-barcode PSA values, but then rising values (Fig. 4b), in some cases, that became detectable with conventional PSA assays (Fig. 4c)

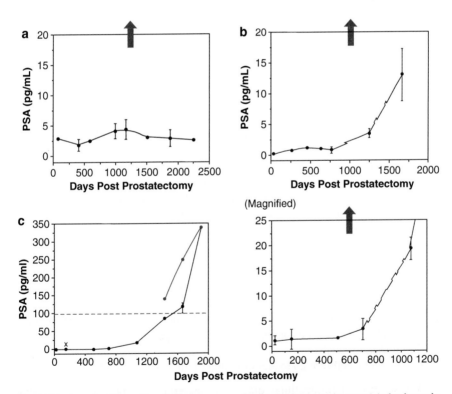

Fig. 4 Bio-barcode PSA profiles. Following surgery ($T=0$), the bio-barcode assay (■) clearly tracks PSA response to treatment at values well below the threshold of detection of current commercial PSA immunoassays (100 pg/mL, ●). *Arrows* (magnified plot) or *dotted line* indicate that values shown are below the limit of clinical detection of 1 ng/mL (From Thaxton et al. [40])

3. Low bio-barcode values following secondary prostate cancer interventions that drove PSA values to undetectable levels as measured with conventional tests (not shown)

Overall, this pilot study concluded that by using the more sensitive nanotechnology-enabled analytical method, prostate cancer patients could be more accurately and more quickly risk-stratified following prostatectomy and closely monitored after undergoing secondary treatment.

Most recently, a commercial-research-use-only nanoparticle assay for PSA (Verisense PSA, Nanopshere, Inc., Northbrook, IL) with a comparable sensitivity to the bio-barcode assay was used in a similar analysis. That study was designed to validate the results of the pilot study and was conducted retrospectively using samples from over 400 patients who had undergone prostatectomy for clinically localized prostate cancer. Although these data are not yet published, the results of this much larger retrospective study corroborate those of the pilot study, and a significant case is being made for the initiation of a large, multi-institution prospective clinical trial to validate the technology and to assess its clinical value in the setting of prostate cancer recurrence.

2.1.2 Gold Nanoparticle-Based Nano-flares: Intracellular Probes

Discovery of the polymerase chain reaction (PCR) revolutionized nearly all aspects of basic science research and has profoundly impacted clinical medicine. Similarly, more contemporary quantitative PCR (qPCR) techniques allow for the detection and quantification of mRNA targets of interest such that relative changes in mRNA transcript levels can be assessed under myriad conditions and from a wide variety of different cells and tissues. Due to the sensitivity and dynamic range of the technique, quantitative analysis can be performed on either single cells or large numbers of cells. However, one requirement for mRNA analysis using PCR or qPCR approaches is that the cells or tissue under study must be destroyed. That is, PCR requires cell lysis for solubilizing intracellular mRNA species. This severely and fundamentally limits qPCR as follows:

1. Removing mRNA targets from the friendly confines of the cell lipid bilayer membrane makes them susceptible to degradation by abundant and highly effective nucleases (e.g., RNase). Meticulous care is required for experimental success.
2. Cell lysis precludes the simultaneous assessment of protein and/or other small molecule changes in the context of ongoing changes at the mRNA level in real time and in live cells.
3. Experimental cells are no longer available for downstream applications.

A new class of intracellular diagnostic materials has been developed that addresses the above limitations [8–10]. The synthesis of gold nanoparticles surface-functionalized with DNA oligonucleotides is shown in Fig. 5. DNA AuNPs have unique properties that make them superior probes for complementary oligonucleotides of interest. In an extracellular diagnostic context, they demonstrate increased

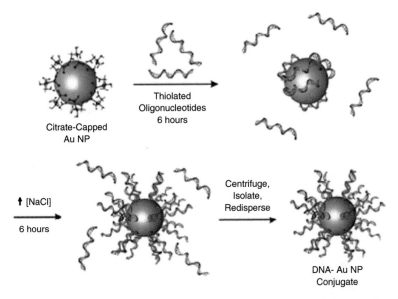

Fig. 5 Synthesis of DNA AuNPs. Citrate-capped gold nanoparticles are mixed with alkyl-thiol modified oligonucleotides for 6 h. The thiol adsorbs to the AuNP surface resulting in a loose association of DNA with the particle surface. Through a process of salt (NaCl) additions, the particles are 'aged' over 6 h to achieve a dense loading of alkyl-thiol oligonucleotides to the surface of the AuNP. Centrifugation isolates the DNA AuNP conjugates from unreacted components (Courtesy of Chad Mirkin)

binding constants for targeted oligonucleotides, i.e., increased *sensitivity* [41] and exquisitely sharp melting transitions between perfectly matched and mismatched targets, i.e., increased *selectivity* [42–44]. These properties enable point-of-care diagnostics for both proteins and nucleic acids.

Commercialized by Nanosphere, Inc., a number of FDA-cleared pharmacogenomic tests that detect single base changes in drug-metabolizing enzymes are providing real opportunities for personalized medicine at points of care. However, extracellular *in vitro* diagnostics are just a first glimpse at the impact that these unique probes will have in biomedicine. DNA AuNPs have also shown extraordinary promise with regard to intracellular applications, including in the detection of target molecules and as a potentially new and potent class of therapeutics [45, 46]. Here, DNA AuNPs have demonstrated:

- Universal and >99% uptake into cultured cells (over 40 cell lines and primary cells have been tested to date)
- Cellular uptake without the use of cytotoxic transfection agents
- DNA AuNP uptake is not inherently cytotoxic
- The unique properties of the particles stabilize surface-modified oligonucleotides against nuclease degradation [47]
- Nanoparticle bound oligonucleotides can be used to both *detect* and *regulate* target genes at the mRNA level [9, 46, 48]

Nano-flare technology (Fig. 6), pioneered by the Mirkin Group, is an elegant example of an intracellular application of these unique nanomaterials [8, 9]. When bound to an AuNP, the fluorescent emission from a molecular fluorophore "flare" is effectively quenched by the close proximity to the AuNP [49]. Effective fluorescence quenching and the increased stability of the oligonucleotides on the surface of the DNA AuNPs [47] leads to significantly lower background signal upon transfection (*vida infra*). Upon transfection into cells, the fluorophore-labeled oligonucleotide flares dissociate from AuNP surface-immobilized complements upon specific mRNA target binding, removing the quenching effects of the AuNP and resulting in a detectable increase in intracellular fluorescence. Because the increase in fluorescence correlates with the presence of specific mRNA targets (Fig. 7),

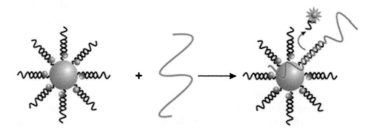

Fig. 6 Nano-flares. AuNPs (*large spheres*) are densely surface-functionalized with thiolated oligos. Bound to the AuNP oligos are fluorophore (*small sphere*)-labeled short oligos (*flare*). Hybridization orients the molecular fluorophore sufficiently close to the AuNP such that effective quenching of the fluor is observed. Upon encountering an intracellular mRNA target (*wavy line*), the flare is displaced and the molecular fluorophore exhibits bright fluorescence (Courtesy of Chad Mirkin)

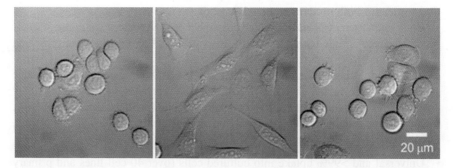

Fig. 7 Intracellular nano-flares. A human breast cancer cell line (SKBR3, *left and right panels*) is shown after transfection with nano-flare particles specifically targeting the survivin mRNA sequence known to be expressed by SKBR3 cells. On the *left*, bright fluorescence is observed as the nano-flare sequence is complementary to the targeted survivin mRNA sequence. On the *right*, a control scrambled nano-flare sequence was added and minimal fluorescence is observed. The *middle panel* demonstrates the lack of survivin nano-flare fluorescence following transfection in a cell line that does not express survivin (mouse endothelial, C166) (Courtesy of Chad Mirkin)

DNA AuNPs conjugated to fluorophore-labeled oligonucleotides act as intracellular target mRNA probes for assessing mRNA target changes which result from directed cellular perturbations [48].

Intracellular binding of target mRNA is further augmented by the unique properties of the DNA AuNPs [42–44]. Due to the ability for the nano-flare probes to readily enter all cell types tested to date, the Mirkin group proposes that the nano-flare agents can be used as a universal live cell target mRNA quantification platform with the potential for simultaneous protein or small molecule detection. Because cells are not destroyed, flow cytometric and light microscopy coupled with nano-flares may replace qPCR, while preserving cells for downstream applications. Importantly, using either flow cytometry or confocal microscopy, there is no loss of sensitivity with reference to qPCR (i.e., single-cell analysis is possible with confocal microscopy), while also maintaining the analytical range to perform ensemble measurements averaged from the measurements of hundreds to thousands of measurements made on individual cells (flow cytometry).

2.2 In Vivo Imaging

2.2.1 Instrumentation: Carbon Nanotube-Based X-Ray Systems for Diagnostic Imaging and Radiotherapy

X-ray radiation is widely used today for *in vivo* cancer detection and for radiotherapy. For example, mammography is the most common modality for breast cancer screening, and over 50% of all cancer patients in the United States undergo radiation therapy. For x-ray based imaging and radiotherapy devices, there are constant demands to increase the resolution so that tumors can be detected at an earlier stage, to minimize the imaging dose, improve the accuracy of dose delivery, and minimize normal tissue damage during radiotherapy. The new carbon nanotube–based x-ray technology enables new imaging and radiotherapy devices with improved performance in these respects.

Utilizing the recent advances in nanomaterials, a new x-ray source technology has been developed (Fig. 8). Carbon nanotubes (CNTs) instead of the conventional thermionic filaments are used as the "cold" electron sources for x-ray generation. The technology is capable of generating both temporally and spatially modulated x-ray radiation that can be readily gated and synchronized with physiological signals. The spatially distributed x-ray source array technology opens up new possibilities for designing *in vivo* imaging systems, especially tomography systems, with increased resolution and imaging speed, and expanded functionalities. By distributing the x-ray power over a large area, the technology can generate a significantly higher dose rate for certain radiotherapy applications. Since its invention, this nanotechnology-enabled x-ray source technology has moved from being a simple academic curiosity to commercial production. Its applications for cancer detection and treatment are being actively investigated in academic institutions and in industry.

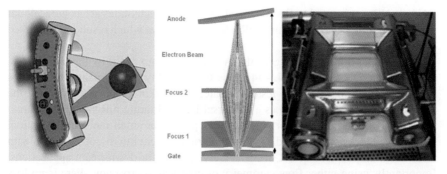

Fig. 8 (*Left*) Schematics showing a CNT x-ray source array; (*right*) a square-geometry nanotube x-ray source array with 52 individually controllable x-ray beams, manufactured by XinRay Systems (Courtesy of Otto Zhou, University of North Carolina)

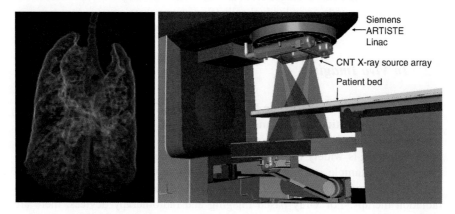

Fig. 9 *Left*: Prospective-gated micro-CT image of a mouse lung tumor model (UNC data. Mouse model from Dr. B. Kim). *Right*: An illustration of a CNT x-ray source array mounted on a radiotherapy machine (Image courtesy of J. Maltz of Siemens and P. Lagani of XinRay Systems)

Utilizing the electronic programmable capability of the CNT x-ray source, a physiologically gated, microcomputed tomography scanner has been developed. By synchronizing x-ray exposure and data collection with the non-periodic respiratory and cardiac motions, high-resolution CT (computed tomography) images with minimum motion blurs can be obtained from free-breathing mice. The scanner is now used routinely by a large number of cancer researchers at the University of North Carolina (UNC) (Fig. 9).

Siemens and XinRay Systems developed a high-speed tomosynthesis scanner to provide real-time image guidance for radiation therapy using the distributed CNT x-ray source array technology. The device will enable oncologists to "see" tumors in real time during treatment and will allow more accurate radiation delivery. The scanner has been integrated with the Siemens Artiste treatment system. It is currently under testing at the UNC Cancer Hospital. Clinical trials are scheduled for 2010.

2.2.2 Molecular Imaging

Macroscale anatomical and physiological information acquired from diagnostic imaging techniques has successfully guided clinician prognosis. However, imaging at this scale overlooks the molecular processes that underlie most diseases. Molecular imaging has emerged in an attempt to exploit molecular markers for early diagnosis and tracking of disease progression with the goal to significantly improve patient care. Molecular imaging can be defined as "the characterization and measurement of biologic processes in living animals, model systems, and humans at the cellular and molecular level by using remote imaging detectors" [50].

Over the past decade versatile surface functionalization of nanomaterials has enabled the realization of additional functions, including the development of multimodal probes to utilize more than one imaging modality, and theranostic probes to achieve simultaneous drug delivery and diagnostic imaging. Some of the strategies in the design of new nanomaterials include development of synthetic procedures and surface functionalization with molecules of interest.

Controlling reaction conditions can generate nanomaterials of similar composition with distinct physical properties for a variety of applications. For example, a one-pot thermal decomposition reaction in the presence of polyethylene glycol and oleylamine yields water-soluble Fe_3O_4 magnetic "nanoflowers" of 39, 47, and 74 nm in size with high crystallinity [110]. The nanoflower consists of a cluster of multiple Fe_3O_4 nanoparticles. The synergistic magnetism of the Fe_3O_4 nanoparticles increases the magnetic resonance image (MRI) contrast effect to be significantly higher than that of commercially available agents. Through modification of the reaction conditions in the nanoflower synthesis (e.g., stabilizing surfactants, reagent concentrations, reaction time, and temperature), ultrasmall Fe_3O_4 magnetic nanoparticles with tunable core size are produced [110]. The synthesized 3–5 nm ultrasmall nanoparticles have comparable MRI contrast effects to commercial agents. In contrast to commercial agents, the ultrasmall nanoparticles are expected to have the desirable pharmacokinetics for escaping the reticuloendothelial system (RES) and increasing their circulation lifetime for molecular targeting [51].

The properties of nanomaterials can be manipulated through surface conjugation to molecules with desired functions. Frequently, the cooperative effects between the nanomaterials and the attached molecules enhance the nanomaterial performance, such as in the conjugation of molecular gadolinium(III) complexes to nanodiamonds [52]. The magnetic gadolinium(III) complex and non-magnetic nanodiamond are conjugated to yield a highly sensitive MRI contrast agent [Gd(III)-ND]. The immobilized Gd(III) complexes on the nanodiamond surface exhibit a tenfold increase in MR signal enhancement over the free molecular Gd(III) complex, representing the second most potent Gd(III) contrast agent reported to date (58.8 $mM^{-1}s^{-1}$). The exceptional enhancement of the MRI contrast is likely attributed to the slower tumbling rate of the conjugated Gd(III) complexes, clustering of Gd(III)-ND in solution, and the unique hydration environment around the nanodiamond surface.

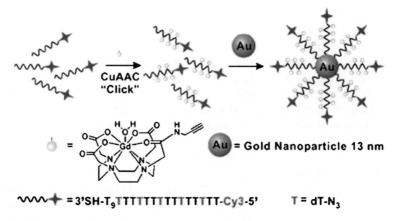

Fig. 10 Schematic of Cy3-DNA–GdIII@AuNP conjugates. Gd(III) complexes are covalently attached to fluorescent DNA using click chemistry and conjugated to Au nanoparticles. The resulting nanoconjugate is stable, multimodal (MR, fluorescence, and CT), and cell permeable (Song et al. [14])

Utilizing the extensive surface modification potential of nanomaterials, complex systems with multimodal imaging capabilities have been achieved. The reported Cy3DNA–GdIII@AuNP multimodal probe consists of a gold nanoparticle with immobilized DNA on the surface decorated with Cy3 fluorescent dyes and Gd(III) complexes (Fig. 10). The co-existence of magnetic Gd(III) complexes and Cy3 fluorescent dye enables the detection of Cy3DNA–GdIII@AuNP conjugates by MRI and confocal microscopy (Fig. 11). The high sub-cellular resolution of confocal microscopy verifies the intracellular delivery of Cy3DNA–GdIII@AuNP also visualized by MRI. Since AuNPs serve as CT contrast agents, these nanoconjugates are promising multimodal probes for MR, fluorescence, and CT. This system is one of the latest examples of nanomaterials that possess multiple characteristics for molecular imaging, including increased MR contrast sensitivity, intracellular delivery, and multimodal capability.

The nanomaterial field has made great strides in the development of MR contrast agents for molecular imaging. Advances in synthetic procedures and complex surface modifications are among the many strategies to generate nanomaterials with desired properties. Simultaneous achievement of high sensitivity probes with optimal pharmacokinetics, intracellular delivery, multimodality, and theranostic capabilities is becoming a reality.

2.2.3 Nanoimaging for Cancer

Nanoscience applied to cancer research is critical to the future of eliminating cancer and is making a significant impact on cancer diagnosis and management in revolutionary ways already. With its capacity to provide enormous sensitivity, throughput, and flexibility, nanotechnology has the potential to profoundly impact cancer patient

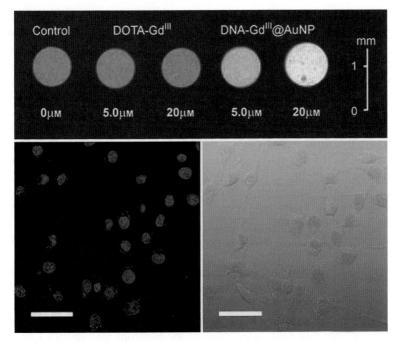

Fig. 11 *In vitro* MR and confocal fluorescence micrographs of NIH/3 T3 cells incubated with Cy3-DNA-Gd(III)@AuNP conjugates. MR images (*top*) were acquired at 14.1 T and 25°C on cells incubated with 5 μM and 20 μM nanoconjugate with comparison to a molecular control agent, Gd(III)-DOTA. For confocal imaging (*bottom*), the cells were incubated with 0.2 nm nanoconjugate. (*Bottom left*) merged DAPI and Cy3-DNA-Gd(III)@AuNP channels. (*Bottom right*) Overlay with the transmitted light image. Scale bar is 50 μm (From Song et al. [14])

management. *Ex vivo* diagnostics used in conjunction with *in vivo* diagnostics can markedly impact future cancer patient management by providing a synergy that neither strategy alone can offer. Nanotechnology can significantly advance both *ex vivo* diagnostics through genomic/proteomic nanosensors and *in vivo* diagnostics through nanoparticles for molecular imaging. A technology that is amenable to detection of small tumors is based on the Raman Effect, inelastic scattering of light by molecules. The Raman Effect is very weak, producing low signals. However, the signal to noise ratio is typically very high due to the lack of autofluorescence. This relatively inexpensive and easy to use imaging technology can achieve better depth and spatial information than existing diagnostic imaging methods. To improve the utility of this technology further, gold nanoparticles coated with small molecules that enhance the Raman effect by increasing inelastic scattering of light are being developed (surface-enhanced Raman scattering, SERS). These same nanoparticles are also functionalized with molecules that allow them to home in on the tumor [12, 15, 53]. Small animal imaging instruments have already been developed to detect these nanoparticles in living subjects, and this technology currently is going through FDA evaluation for its use as an endoscopic Raman imaging and screening tool for human colorectal cancer.

The sensitivity of the technique is a 100 times higher than that of fluorescence using quantum dots. Depending on the molecule on the surface of a gold nanoparticle, each nanoparticle produces a unique Raman signal, allowing for simultaneous detection of 10–40 different signals from a living subject; hence, the Raman technique's pM sensitivity and multiplexing capabilities are unsurpassed. However, its limited depth of light penetration hinders direct clinical translation. Therefore, a more suitable way to harness its attributes in a clinical setting would be to couple Raman spectroscopy with endoscopy. It was recently reported that flat lesions in the colon were five times more likely to contain cancerous tissue than polyps detected by conventional colonoscopy. The use of an accessory Raman endoscope in conjunction with locally administered (e.g., via spraying), tumor-targeting Raman nanoparticles during a routine colonoscopy could offer a new way to sensitively detect dysplastic flat lesions and very small tumors that otherwise would be missed. Topical application of SERS nanoparticles in the colon appears to minimize their systemic distribution, thus avoiding potential toxicity and supporting the clinical translation of Raman spectroscopy as an endoscopic imaging tool.

It is not currently expected that targeting tumors with nanoparticles will entirely replace current procedures used to detect cancer, but the hope is that the combination of blood biomarkers will indicate the presence of early-stage disease, and an imaging study will follow in the hope that it will lead to detection of relevant disease. Towards this goal, researchers at the Stanford University's Center for Cancer Nanotechnology Excellence Focused on Therapy Response (CCNE-TR) have worked together for 5 years to examine panels of biomarkers that are relevant and predictive of disease in ovarian, prostate, lung, and pancreatic cancer and now functionalizing their nanoparticle-based imaging agents to detect these biomarkers. While all of these technologies are still in their infancy, the next generation of instruments—for example, those that would replace mammograms using photoacoustic imaging [54]—will be seen imminently.

In vivo molecular imaging has a great potential to impact cancer by detecting disease in early stages, identifying extent of disease, personalizing treatments by theranostics or targeted therapy, and measuring molecular effects of treatment.

2.3 Therapeutics

2.3.1 Targeted Nanoparticle siRNA Delivery in Humans

Work leading to the first targeted nanoparticle delivery of siRNA in humans, the drug CALAA-01, was begun by the Davis Group at the California Institute of Technology in 1996 [55]. Their goal was to develop a multifunctional targeted cancer therapeutic that would enable the systemic administration of nucleic acids. Their self-stated design methodology was a systems approach to a multifunctional colloidal particle—a rationally derived nanoparticle therapeutic employed before the term "nanoparticle" was widely used. The initial drug schematic (Fig. 12) presented

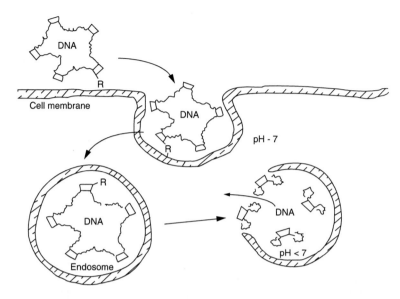

Fig. 12 Envisioned nanoparticle delivery strategy that eventually became CALAA-01 [55]

the desired components of an envisioned therapeutic: (1) a cyclodextrin-containing polymer (CDP) core that spontaneously self-assembles with nucleic acids yielding small colloidal particles less than 100 nm in diameter, (2) a targeting ligand providing for tumor cell specificity and uptake, and (3), an appreciation of endosomal acidification as a mechanism for particle disassembly and endosomal escape of the therapeutic nucleic acid [55]. This CDP system was initially envisioned for use with plasmid DNA; however, because the association of the nucleic acid with CDP is based upon electrostatic interactions, it was understood that the approach could be a general one for candidate nucleic acid therapeutics. In addition to platform generality, the platform components were also chosen due to their amenability to scale-up and manufacture.

CALAA-01 is an embodiment of this initial vision. CALAA-01 ultimately evolved to include a number of key components that spontaneously assemble into therapeutic nanoparticles ~70 nm in diameter [55]. In addition to the CDP particle core and the siRNA payload, the surface of the formed nanoparticles is decorated with (1) adamantane-polyethylene glycol (AD-PEG) for drug stabilization in biological matrices, (2) adamantane-PEG-transferrin (AD-PEG-Tf) for tumor-specific targeting and cellular uptake, and (3) imidazole residues to titrate the decrease in endosomal pH upon cellular uptake and promote the endosomal escape of the otherwise sequestered nucleic acid drug. Adamantane is a small molecule that binds tightly and forms an inclusion complex with cyclodextrin on the surface of the formed nanoparticles, thus, displaying PEG and Tf.

Shortly after the conjugate therapeutic was developed, investigators found that the nanoparticles could be formed by self-assembly through simultaneous component

Fig. 13 Formulation strategy for CALAA-01 [55]

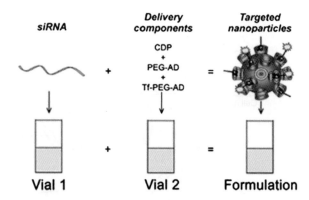

reconstitution and mixing [56, 57]. This finding gave rise to a unique two-vial formulation strategy (Fig. 13) which allows for the rapid and straight forward self-assembly of the nanoparticle delivery system components (CDP, AD-PEG, AD-PEG-Tf) with siRNA, at the point of care [56].

This formulation provides for siRNA solvation immediately prior to reconstitution with the nanoparticle delivery system components. This is an advantage because siRNA is highly unstable when solvated, but, following the self-assembly process, is protected from nuclease degradation. In addition, this two-vial formulation allows for the molecular delivery of system components of CALAA-01 to be separated from siRNA and tested separately for safety in animal models prior to introduction in humans [58, 59]. The composition of the formed nanoparticle therapeutic following mixing of the separated components has been well-characterized from the standpoint of size and molecular composition [56].

The initial *in vitro* and *in vivo* demonstration of the CDP-based siRNA delivery system was published in 2005 using a disseminated murine model of Ewing's sarcoma [60]. In these studies, the siRNA targeted the breakpoint of the EWS-FLI1 fusion gene, which is an oncogenic transcriptional activator in TC71 cells positive for EWS-FLI1 and for the transferrin receptor [60]. In addition to *in vitro* inhibition of their targeted gene product, TC71 cells transfected with firefly luciferase and injected into NOD/SCID mice served as a model system of metastatic Ewing's sarcoma where tumor dissemination and treatment efficacy could be assessed using bioluminescent imaging. In this murine model, investigators administered their targeted, CDP-based siRNA delivery particle and demonstrated anti-tumor effects and target-specific mRNA down-regulation [60]. Further studies provided evidence that the CDP-based delivery system does not illicit an innate immune response [60], and that active targeting to the transferrin receptor enhances tumor cell uptake [61]. With these promising animal results, a solid foundation was provided for moving the CDP-based siRNA therapeutic platform to the clinic.

In the case of CALAA-01, siRNA-targeting ribonucleotide reductase subunit 2 (RRM2) was identified [58, 59]. In addition to potency, siRNA-targeting RRM2 demonstrated complete sequence homology in mouse, rat, monkey, and humans, which allowed for a single siRNA to be used for conducting preliminary studies in all animal models. Targeting RRM2, Davis and colleagues confirmed effective protein knockdown with concomitant reduction in tumor cell growth potential in a subcutaneous mouse model of neuroblastoma [62].

Building upon the above experiments and drawing closer to the initiation of clinical drug testing, Davis and colleagues performed the first study showing that multidosing of siRNA, in the context of the CDP-based nanoparticles, could be done safely in a nonhuman primate [58, 59]. This study demonstrated that dosing parameters that were well tolerated were similar to those that had demonstrated anti-tumor efficacy in mouse models. Furthermore, reversible toxicity was observed in the form of mild renal impairment at high dose; however, extrapolations from mouse model efficacy studies suggested that the therapeutic dosing window would be large.

In May of 2008, CALAA-01 became the first targeted delivery of siRNA in humans [55]. Details of this study, and others focused on nanoparticle therapeutics, can be found at http://www.clinicaltrials.gov. In Phase I studies, patients were administered CALAA-01 to assess drug safety. Patients were administered CALAA-01 by way of intravenous infusion on days 1, 3, 8, and 10 of a 21 day cycle. Importantly, in a recent study by the Davis group [7], the authors demonstrate in tumor tissue taken from patients with melanoma after systemic administration of CALAA-01that CALAA-01 effectively targets RRM2 through an RNAi mechanism of action. Certainly, the updated results of clinical trials with CALAA-01 are eagerly anticipated.

Many more nanoparticle therapeutics are on the horizon and will follow CALAA-01 into the clinical setting. As a pioneering nanoparticle therapeutic, much can be learned from CALAA-01. For instance, scrupulous detail to the design, fabrication, characterization, and quality control of CALAA-01 provided a solid platform for moving the drug forward in preclinical and clinical trials. In addition, the choice of a species-generic but target-specific siRNA sequence targeting RRM2 provided direct access to multiple animal models, ultimately for translation to humans. Finally, testing in a wide variety of preclinical animal models provided the data necessary to anticipate safe dosing parameters in humans and dictated drug-specific safety monitoring protocols in humans. Building upon CALAA-01, it is anticipated that what is now an "infant" field will ultimately provide new therapeutics, based upon advances in nanotechnology that will provide for significant improvements in human health management.

2.3.2 Nucleic Acid Functionalized Gold Nanoparticles for Therapeutic Use

The DNA AuNP platform described above with regard to *in vitro* extra – and intra-cellular diagnostic applications is also being developed as the next generation of therapeutic nanoparticles for regulation of gene expression. As previously noted, the AuNP serves as a platform for oligonucleotide attachment (functionalization) and allows for a high density of oligonucleotides to be loaded onto the surface of each gold nanoparticle (approximately 80 strands on a 13 nm particle) [63, 64] (see Fig. 5, above). The conjugate can include a number of sequence modifiers, including releasable fluorophores, for easy detection, as was demonstrated in the nano-flare example. This high local density of DNA and the ability of the nanoparticles to engage in multivalent interactions are hypothesized to be the origin of a number of unique properties associated with this class of materials, including their target-binding properties and propensity for cellular transfection [45, 64]. These particles are readily internalized by cells [64], demonstrate enhanced binding to complementary targets [45], are stable to nuclease degradation [45, 46], and are not inherently cytotoxic. The conjugates remain intact and function to inhibit protein expression for at least several days in an intracellular environment [45, 46]. Accordingly, these conjugates act as *single-entity agents* capable of simultaneous transfection and gene regulation. Furthermore, DNA AuNPs bound to complementary targets exhibit unusually sharp melting transitions, in some cases over 1°C, when compared with unlabeled or fluorophore-labeled DNA probes [42–44, 65]. This greatly increases the selectivity and specificity for perfectly matched target sequences versus those with nucleotide mismatches.

Introduction of genetic material (e.g., DNA and RNA) into cells and tissues holds significant promise for therapeutic and diagnostic applications [66]. However, development of nucleic acids into viable intracellular diagnostic or therapeutic agents has faced challenges with regard to (1) stable cellular transfection, (2) entry into diverse cell types, (3) toxicity, (4) agent stability, and (5) efficacy [66]. Initial evaluation of DNA AuNPs as therapeutic agents was driven by the unique properties of the nanoconjugates realized in the context of extracellular diagnostic applications. For intracellular therapeutic use, DNA-AuNPs were discovered to be an innovative way to efficiently transfect antisense oligonucleotides for regulating gene expression; importantly, the particle itself acts as the antisense agent and is not just a carrier of nucleic acids ([46]; Fig. 14). Furthermore, AuNPs surface-functionalized with siRNA were also synthesized and demonstrated to regulate intracellular

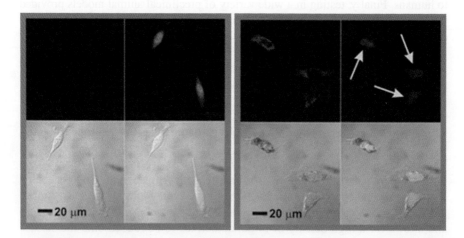

Fig. 14 Antisense DNA AuNPs. (*Left-hand group of images*) Mouse endothelial C166 cells expressing green fluorescent protein (GFP; light areas, untreated). (*Right-hand group of images*) Cy5-labeled antisense GFP-AuNPs transfected into C166 cells downregulate the expression of GFP [46]

gene expression [45]. Significantly, through the course of these experiments, the collective properties of the nucleic acid–functionalized AuNP was investigated in the context of addressing the problems associated with conventional gene regulation techniques. Ultimately, the dense loading of DNA on the surface of the AuNPs, so vital in the context of extracellular diagnostics, provides significant advantages of the DNA AuNP system over conventional intracellular gene regulation strategies.

Currently, the siRNA- or DNA-AuNP therapeutic platform is being applied to a wide variety of candidate diseases from brain cancer to heart disease, where the unique properties to the platform may overcome the significant hurdles standing in the way of successful disease treatment. In order to more rapidly commercialize this technology and drive it from the bench to the bedside, Professors Mirkin and Thaxton of Northwestern University co-founded AuraSense, LLC.

2.3.3 Biomimetic High-Density Lipoproteins

With consideration given to mortality and worldwide prevalence, the significance of atherosclerosis is profound [67, 68]. Atherosclerosis is a chronic infiltrative and inflammatory disease of the systemic arterial tree caused by excess circulating cholesterol [69, 70]. Cholesterol is not soluble in the aqueous milieu of the human body, thus it travels by way of dynamic nanoparticle carriers known as lipoproteins (LPs) [69]. The main LP carriers of cholesterol are low-density lipoprotein (LDL) and high-density lipoprotein (HDL). High LDL levels promote atherosclerosis and are associated with an increased risk of cardiovascular disease [71, 72]. Therapeutic LDL lowering has been shown to reduce cardiovascular disease mortality [73–75]. Conversely, HDL is well-known to promote reverse cholesterol transport (RCT) from sites of peripheral deposition (macrophage foam cells) to the liver for excretion [76–78]. Accordingly, high HDL levels inversely correlate with cardiovascular disease risk [76].

There is intense interest in therapeutic strategies to harness the beneficial effects of HDL [78–82] in order to address the substantial cardiovascular disease burden that exists despite LDL-lowering therapies. HDL is a dynamic serum nanostructure that matures from nascent form to a mature spherical form; in both forms, the surface components are (1) APOA1, (2) phospholipids, and (3) free cholesterol. In the maturation process from nascent to spherical forms of HDL, free cholesterol is esterified through the action of lecithin:cholesterol acyl transferase (LCAT) where the cholesteryl esters are added to the core of HDL due to increased hydrophobicity. This process increases the size of the particle, which then takes on a spherical shape. In order to overcome the biological steps required to fabricate a biomimetic HDL nanostructure from the bottom-up, biology needed to be replaced, to some extent, by nanotechnology. Specifically, by using a 5 nm diameter AuNP as the core of the biomimetic HDL, the biological steps required to fabricate biomimetic HDLs with the surface chemistry of natural mature spherical HDL could be bypassed [83]. As proof of concept toward the development of nanoparticles as novel tools and therapeutic agents for studying and treating atherosclerosis, a synthetic HDL biomimetic nanostructure (HDL AuNP) was fabricated collaboratively

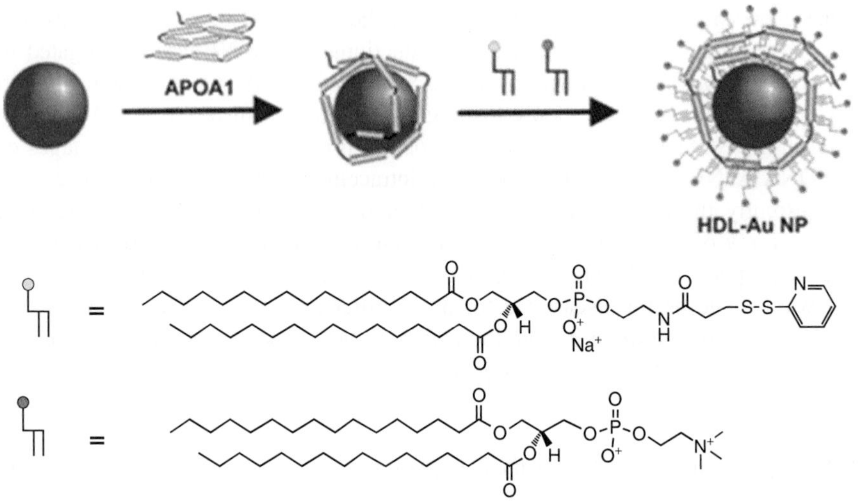

Fig. 15 Synthesis for templated, spherical HDL AuNPs [83]

Table 1 Hydrodynamic diameter of HDL conjugates

Particles	Hydrodynamic diameter (nm)
Au NP (5 nm diameter)	9.2 ± 2.1
Au NP + APOA1	11.0 ± 2.5
Au NP + APOA1 + phospholipids	17.9 ± 3.1

by the Thaxton and Mirkin research groups, and its cholesterol-binding properties were measured [83] (Fig. 15). The size, shape, and surface chemistry of the synthetic version closely mimicked that of natural mature spherical HDL (Table 1).

As discussed, transport of cholesterol to the liver by HDL is a key mechanism by which HDL protects against the development of atherosclerosis [78]. Thus, determining if HDL-AuNPs bind cholesterol is important for determining the potential of these structures as therapeutic agents. By using a cholesterol analog, the measured binding constant for cholesterol to HDL AuNPs was measured and found to be 3.8 nm (Fig. 16). Interestingly, there is little information regarding the K_d (dissociation constant) of natural HDL for comparison purposes. Accordingly, HDL AuNPs are key biomimetic nanostructures that can be used as reference points for further comparison. Current research activity is centered around the biology of HDL AuNPs, both *in vitro* and *in vivo*.

2.3.4 Mechanized Nanoparticles for the Delivery of Therapeutic Agents

Arguably, chemical systems are at their best when they are robust and "smart." Imagine a device for the delivery of an anticancer drug specifically targeted to cancer cells that involves a rugged nanoscale container endowed with nanoscale

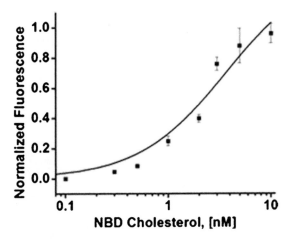

Fig. 16 Binding isotherm of a fluorescent cholesterol analogue to HDL-AuNPs. NBD cholesterol was titrated into a 5 nm solution of HDL-AuNPs. The fluorescence vs. NBD-cholesterol concentration was used to calculate the K_d of HDL-AuNPs (Courtesy of C.S. Thaxton)

antennae and associated machinery. One type of delivery system developed by the Stoddart Group at Northwestern University in collaboration with Professors Zink, Nel, and Tamanoi at the University of California Los Angeles (UCLA) consists of mesoporous silica nanoparticles 200–500 nm in diameter. Interspersed on the surface of the nanoparticles are different models of nanomachinery that can be actuated chemically (via pH change), biochemically (via enzyme action) or photophysically (via light) [16–22]. Silica is an attractive material to incorporate into such devices on account of it being rigid, chemically inert, and optically transparent.

Historically, molecular machinery very often has been based on bistable rotaxanes [84, 85], which are mechanically interlocked molecules that consist of a ring component that can be induced to move along a stalk component attached to the outside of the nanoparticle at one end and terminated by a large stopper that prevents the ring from leaving the stalk at the other end. These stalks are bestowed with a recognition element that is close to the surface of the mesoporous nanoparticle. Thus, when a macrocycle such as cyclodextrin or cucurbit[n]uril is complexed to this recognition element near the surface, it acts as a gatekeeper and retains the cargo (i.e., a drug or imaging agent) in the mesoporous channels of the nanoparticle. When it is moved away from the surface, the gates are opened and the contents of the pores (cargo) can be released into the surrounding medium. The macrocycle gatekeeper can be moved from the surface of the nanoparticles by either of two basic mechanisms: (1) nanovalve systems that incorporate a second recognition element in the stalk of the rotaxane that allows the gatekeeper macrocycle to slide away from nanoparticle surface, allowing the cargo to escape from the pores, or (2) snap-top systems that incorporate a cleavage point in the stalk of the rotaxane such that the stopper can be removed in response to a specific stimulus, allowing the gatekeeper macrocycle to slip off the stalk entirely, thus freeing the cargo from

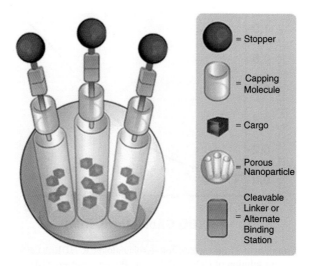

Fig. 17 Graphical representation of a mesoporous silica nanoparticle (SiO_2, average diameter of 200 nm) adorned with a layer of [2] rotaxanes on the nanoparticle surface. Cargo can be released from the pores of the nanoparticles by either forcing the capping molecule to move to another binding site of the rotaxane stalk (nanovalves), or by cleaving the rotaxane stoppers off the stalks completely (snap-tops) (Courtesy of F. Stoddart and A. Nel)

the pores of the mesoporous silica nanoparticles (MSNP). Fig. 17 shows the basic design of the nanovalve/snap-top mechanized nanoparticle system.

Recently, nanovalve systems have been developed that can release cargo by either raising the pH [86] or lowering the pH [87] of the system, which shows potential for use in a biological setting. Various snap-top systems have been developed also, including one that can release its cargo upon exposure to an esterase enzyme [88]. Currently, the Stoddart group is working on developing additional snap-top systems that can release cargo in response to specific enzymes that are over-expressed by tumor tissue, which will potentially provide new diagnostic and treatment options for various cancer types.

Drs. Zink, Nel, and Tamanoi at the Nano Machine Center at the California NanoSystems Institute at UCLA have developed a pH-sensitive MSNP that is capable of delivering doxirubicin to an acidifying intracellular compartment in cancer cells from where drug release to the nucleus takes place when the pH drops below 6 [20, 21]. The drug is retained in the nanoparticle at physiological pH (7.4) of the blood, which theoretically means that drug release can be controlled to prevent systemic side effects while being used to target its delivery to cancer cells, e.g., by the attachment of surface ligands such as folic acid [17].

In addition to the mechanized features of the MSNP that allow on-demand drug delivery, this particle platform has been endowed with additional design features to emerge as a multifunctional therapeutics and imaging platform. First, attachment of ligands such as folic acid has allowed the particles to be targeted to breast or pancreatic cancer cells; second, super-paramagnetic iron oxide nanocrystals were encapsulated inside the MSNP, allowing magnetic resonance imaging, with the possibility of using the particles for theranostics [17]. Third, the UCLA group has recently demonstrated that it is possible to deliver doxirubicin from the pores simultaneous with the release of a siRNA species that targets the Pgp (permeability glycoprotein) transporter that is involved in doxirubicin and other chemotherapeutic agent resistance [20, 21]. This was accomplished by attaching polyethylenimine to the negatively charged particle surfaces, allowing secondary but stable attachment of siRNA while still keeping the pores available for electrostatic attachment of doxirubicin. This has allowed this dual delivery system to be used for partial restoration of doxirubicin sensitivity in a squamous carcinoma cell line that has been selected for doxirubicin resistance due to overexpression of Pgp (Fig. 18). Fourth, additional surface functionalization with polyethylene glycol (PEG) and other steric hindrants has allowed modification of MSNP biodistribution, with reduced reticuloendothelial system (RES) uptake but with enhanced electron paramagnetic resonance (EPR) spectra in subcutaneous xenografts in mice. All of the above features, combined with a demonstration of safety during intravenous injection, in

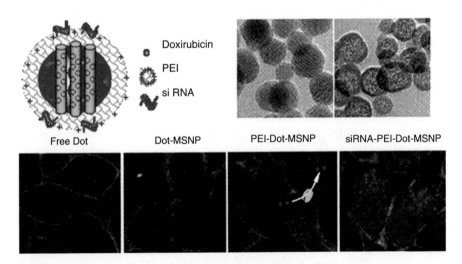

Fig. 18 Contemporaneous drug-plus-siRNA delivery to a squamous carcinoma cell line (HELA), which exhibits doxirubicin resistance due to overexpression of the Pgp drug transporter, which also expels doxorubicin [20, 21]. By utilizing mesoporous silica nanoparticles (*MSNP*; TEM image in *top middle*) coated with a cationic polyethyleneimine (*PEI*) polymer (*top right*), it is possible to attach Pgp siRNA to the particle surface (*top left*). This arrangement leaves the MSNP pores open for electrostatic binding of doxirubicin. The HELA confocal series (*bottom images*) shows the fluorescent doxirubicin located in the particles (*dots*) after uptake into acidifying endosomes. This series further demonstrates that whereas doxirubicin trapped in the particles is taken up in larger amounts than the free drug, essentially all of the doxirubicin being released from within the endosome by protons (interfering with the drug's electrostatic attachment) is expelled by the Pgp transporter in the image of the PEI-Dot-MSNP superimposed on the confocal picture. If, however, the drug is delivered concurrent with the siRNA, which knocks out Pgp expression, the released drug is retained in sufficient quantities to enter the nucleus where it induces apoptosis as shown in the image labeled siRNA-PEI-Dot-MSNP. Thus, dual delivery can overcome cancer drug resistance (Courtesy of A. Nel and F. Stoddart)

addition to recovering most of the injected silica in the urine and feces within days, establishes this system as a versatile drug delivery platform [19].

Functionalized MSNPs show great promise for being the preferred method of administering therapeutics in the not-too-distant future. Perhaps the most attractive property of these delivery systems is their inherent modularity that allows easy interchange between the various components of the functionalized MSNP. This arises from the piece-wise synthesis that is employed to assemble the delivery systems, so that various epitopes can be attached, such as targeting moieties or fluorescent tags, to further tailor the properties of the nanoparticles. Furthermore, MSNPs are known [18] to be capable of being endocytosed by live cells to deliver a cargo of imaging agents or therapeutics to the lysosomal compartment of the cell [20, 21]. The endocytotic uptake is dramatically enhanced by surface ligation or attachment of a cationic group [17, 22]. Moreover, long-aspect-ratio MSNPs show dramatically enhanced uptake by macropinocytosis through an apparent outside-in as well inside-out communication system that involves filipodia and the cellular cytoskeleton. *In vivo* studies have also shown that these delivery systems are nontoxic due to mostly unreactive components such as the silica nanoparticles and the cyclodextrin capping molecule; *in vitro* studies have also not shown any adverse effects upon exposure to these vectors [19–22]. This evidence supports the notion that functionalized MSNPs can and will be used as weapons in the fight against deadly diseases.

2.3.5 PRINT Nanoparticles as Therapeutics

The DeSimone Group at the Carolina Center of Cancer Nanotechnology Excellence at UNC invented PRINT (*Pattern Replication In Non-wetting Templates*) technology, which enables the design and manufacture of precisely engineered nanoparticles with respect to particle size, shape, modulus, chemical composition, and surface functionality (Fig. 19) [89–91].

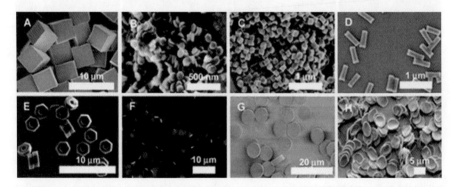

Fig. 19 PRINT is an innovative approach to fabricate nanoparticles with precise control of particle size, shape, composition, cargo, and surface properties, all of which affect how nanoparticles loaded with an anticancer agent will behave in the body. (**A**)–(**H**) show nanoparticles of different sizes, shapes, and compositions, made using PRINT (From Gratton et al. [89])

PRINT particles are presently being designed for use as new therapies in cancer prevention, diagnosis, and treatment. The ability to control the size, composition, and shape of the particles on the nanoscale permits construction of specific materials that are ideally suited for many specialized diagnostic and drug-delivery applications.

2.4 Tissue Regeneration

There is a major need in the medical field for tissue-regeneration technologies that can contribute to potential applications such as bone, cartilage, vascular, bladder, and neural regeneration. Each year almost 500,000 patients worldwide receive hip implants while about the same number need bone reconstruction due to injuries or congenital defects. Moreover, 16 million Americans lose teeth annually and require dental implants. The projected annual market for medical implant devices in the U.S. alone is estimated to be > $20 billion and is expected to grow by 10% a year for the next decade. Unfortunately, medical implant devices have been associated with a variety of side effects, including wear, immunogenicity, inflammation, and fibrosis. Contributing to these side effects is poor tissue integration that leads to loosening of implants and mechanical damage to the surrounding host tissues.

There is a growing consensus that nanostructured implants have potential advantages over conventional materials and that the use of nano-architectures and synthetic pro-morphogens (bioactive analogs of growth factors) can dramatically enhance tissue regeneration by, for instance, mimicking natural extracellular matrix (ECM). A key design feature of these types of engineered nanoparticles appears to be their surface topography, especially with regard to surface features that affect cell attachment, growth signals, and ability to induce genes involved in cellular differentiation in cell types such as endothelial cells, vascular smooth muscle cells, chondrocytes, osteoblasts, neurons, and embryonic stem cells.

Although there is a consensus that regenerative medicine could have a major impact on the healthcare system, cost is an issue. Research must focus not only on addressing major chronic disease processes with poor clinical solutions but on high healthcare costs that can be lowered by nano-enabled materials. Thus, the major impact of nanotechnology in the field of regenerative medicine will likely be in the areas of wound healing, urinary incontinence, osteoarthritis of major joints, diabetes, coronary artery disease and cardiac failure, stroke, Parkinson's disease, spinal cord injury, and renal failure. The competitive scenario is that while orthopedics, dental implants, and wound healing are the easiest markets to penetrate, significant advances are taking place in addressing cardiovascular disease, stem cell therapy, and spinal cord injury. These advances are made possible through research advances that are moving the field of regenerative medicine from inert polymers that mimic biomechanical properties of native tissue to the construction of bioactive materials that promote tissue self-healing.

For instance, because most ECM features naturally operate at the nanoscale level, advanced bio-inspired materials are incorporating nano-architecture features into multifunctional ECM analogs. This work necessitates understanding how cells detect biomaterials as well as understanding the downstream pathways that could lead to inflammation, fibrosis, and immune rejection. There is also is a high clinical demand for therapeutic tools that control inflammation and jumpstart ECM production by endogenous or transplanted cells. High-throughput screening (HTS) approaches can help to answer some of these questions by using cellular endpoints and nanomaterial libraries with property variations, allowing study of cellular biology responses that can assist tissue generation. HTS can also be used to detect potential adverse outcomes that result from potentially hazardous nanomaterial properties that can be redesigned to improve material safety [1, 3, 16]. This aspect is described in more detail in chapter "Synthesis, Processing, and Manufacturing of Components, Devices, and Systems".

Two major classes of multifunctional ECM analogues are emerging, namely nano-architectured materials and synthetic pro-morphogens. Nano-architectured bio-mimics require the development of 3D engineered or self-assembling scaffolds that can be achieved by nanofabrication techniques such as electrospinning and soft lithography. Functionalized nanowires, nanoparticles, and CNTs are good platforms for constructing multitasking engineered nanomaterials (ENMs). Other work is being done to engineer self-adaptive materials that modify their response properties in relation to environmental changes. Biocompatible ENM components could also be designed to be sensitive to on-demand stimuli such as chemical, electric, mechanistic, photonic, and thermal triggers. One example is a polymer substance [poly(N-isopropylacrylamide)] that undergoes a temperature-dependent transformation from hydrophobic to hydrophilic upon incubation of the coated tissue culture dishes at 20°C rather than 37°C (body temperature). This hydrophilic transformation allows primary human tissue culture cells such as myocardiocytes or corneal epithelial cells to detach from the surface as intact sheets of cells that maintain an intact ECM [92, 93]. When, for instance, this approach is used to harvest sheets of myocardial cells (Fig. 20), the intact ECM allows the sheets to adhere to themselves, thereby forming a spontaneously contracting multilayer of cells that can be used as myocardial patches for the heart, e.g., dyskinetic post-myocardial infarct segments. Cell sheet engineering offers the promise of means to patch up damaged cornea and other tissue as well as heart tissue. Moreover, the technique can possibly also be used to grow an entire organ, such as a heart or a bladder, on a series of 3D templates.

Synthetic pro-morphogens drive tissue regeneration through the provision of biochemical cues. Examples include nanostructured biomaterials that incorporate morphogens into their structure to provide extracellular or intracellular cues in an appropriate 3D context, laying down gradients and/or timing the delivery. Examples include dendrimer systems presenting bio-ligands, nanoparticles carrying growth factor analogues, and molecularly driven nanogel gradients and nanopatches.

Increasingly, stem cells are being proposed as agents for cell-based therapies due to available isolation techniques, plasticity, and capacity for *ex vivo* expansion.

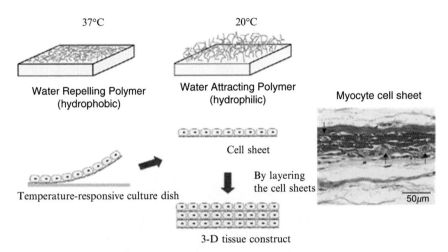

Fig. 20 Myocardial cells sheet engineering. Cell sheet technology is based on the use of thermo-responsive polymers, poly(N-isopropylacrylamide), which are hydrophobic at 37°C, allowing primary tissue culture cells to lay down an ECM, which remains intact when the temperature of the culture dish is brought down to 20°C and the cell sheet released due to the fact that the polymer then becomes hydrophilic. Harvesting of numerous layers of myocyte cell sheets allows layering and the formation of a 3D spontaneously beating myocardial-like tissue construct that can be used for patching real-life myocardial defects as well as offering the possibility to reconstruct an entire myocardial tube (Adapted from Masuda et al. [94])

However, to date the available technologies are hindered by a lack of reproducible control of the desired differentiation pathway, lack of control over the production volume, and the low percentage of cells reaching the target site. Nanotechnology can contribute to two types of products that enhance cell therapy, namely (1) a delivery system for cell transplantation and (2) appropriately differentiated cells. Delivery vehicles include polymeric and biocompatible devices that act as cellular reservoirs that also provide immune protection, enhance cellular survival, and provide controlled release of recruitment and growth factors at the site of carrier degradation. An example would be a material able to surround pancreatic islets during transplantation and grafting via the portal vein. Alternatively, nanomaterials can be used to construct 3D multifunctional scaffolds that provide constant mechanical stability and structural integrity comparable to the native tissue. During the culture, the cell-biomaterial compound can be provided with physicochemical cues that mimic *in vivo* tissue growth and maturation conditions. Work by the Stupp Laboratory at Northwestern University illustrates this in the context of spinal cord injury and repair [112]. As shown in Fig. 21, peptide amphiphiles (Pas) are unique materials that self-assemble into cylindrical nanofibers that display surface peptides known to promote axon growth [111]. In a mouse model of spinal cord injury, significant differences in axon regeneration were observed between untreated and those animals treated with peptide amphiphile-generated nanofiber scaffolds for axonal regeneration [112].

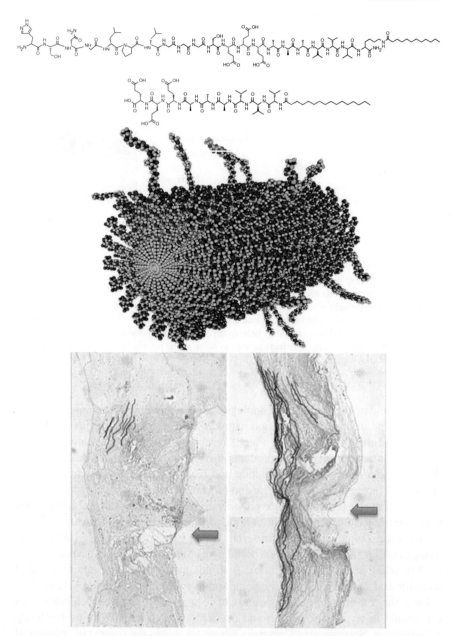

Fig. 21 Peptide amphiphiles (PAs) and spinal cord injury. (*Top*) Two peptide PA molecules are shown. The top PA has a peptide growth factor extending from the core PA that is depicted underneath. (*Middle*) Mixing PAs in aqueous saline solution results in the self-assembly of PA nanofibers which display peptide growth factors. (*Bottom*) In a mouse model of spinal cord injury, injection of PAs at the site of injury (*arrow*) promotes axonal outgrowth (*colored lines*) and bridging of the area of injury (*right*) versus untreated mice (*left*) (Courtesy of S. Stupp)

Cellular therapeutics also requires engineering and manufacturing of patient-specific rather than off-the-shelf cell products. Because culturing of cells or tissue-engineered products is still characterized by high variability and poor efficiency, economically viable tissue culture manufacturing systems need to be developed. Nanomaterials are likely to play a role in this development through their ability to provide physicochemical environmental cues for 3D culture systems and bioreactor designs.

2.5 Nanobiotechnology and Cell Biology

The nanostructures that are fundamental to life, including proteins, DNA, and lipids, operate collectively in cells that may be considered the units of complex biosystems that build from cells to tissues to organs to organisms. Efforts to analyze and engage with complex systems, such as in human medicine, are often targeted at the cellular level. For example, cancer diagnostics methods seek biomarkers on cell surfaces or secreted by cells; drugs must be delivered to certain types of cells and must negotiate cellular barriers and pathways; successful tissue engineering and implants require compatible cellular interactions and growth. Progress in nanomedicine will always be aided—or limited—by researchers' understanding of how cells operate, in exquisite detail. Similarly their ability to make smart hybrid devices for medical or for other purposes will benefit enormously from the intricate tools that have evolved in nature to sustain complicated cellular systems. In many ways, cells lie at the heart of nanobiotechnology's promise.

From a biological perspective, cells are hierarchical organizations of diverse biomolecular constituents that enable cells to maintain homeostasis in a particular biological environment and to respond appropriately to external signals. Nanobiotechnology offers micro- and nanofabricated materials, structures, and devices to examine and engage with cellular systems on the subcellular and molecular level. Other sections in this chapter describe nanotechnology-based analytical methods that can now determine the composition of individual cells. This information can be considered in the context of increasing biological understanding of the cellular processes that the components represent. For example, many laboratories have investigated signal transduction pathways that are stimulated in a large variety of physiologically relevant cell types. By incorporating genetic, biochemical, and physical approaches, key enzymes as well as regulatory structures and mechanisms have been defined. However, the necessary spatial arrangements and regulation have been very difficult to approach because of the heterogeneity of cells and the nanoscale resolution required. Advances in microfabrication and nanofabrication have opened new opportunities to investigate complicated questions of cell biology in ways not before possible. In particular, the spatial regulation of cellular processes can be examined by engineering the chemical and physical environment at the level at which the cell responds. Lithographic methods and selective chemical modification schemes provide biocompatible surfaces that control cellular interactions on the micron and submicron scales on which cells are organized. Combined with fluorescence microscopy and other approaches of cell biology, a widely expanded toolbox has become available.

Responsive interactions of cells with surfaces are fundamental to physiology, but if dysregulated they can lead to pathological conditions such as tumor growth. Surface patterning with microcontact printing has made it possible to examine systematically the importance of the spatial dimensions over which these chemical interactions occur [95–99]. In the last several years a large number of studies have taken advantage of patterned surfaces to study spatial control of cell surface interactions and address specific questions of morphology and adhesion, dendritic branching, migration, and mechanotransduction (how cells convert mechanical signals into chemical responses). Quantitative measurements have been made of mechanical interactions between cells and microfabricated surfaces containing needle-like posts made of elastomeric PDMS (polydimethylsiloxane) [100, 101].

Cells respond to chemical messages (e.g., antigens, growth factors, cytokines) in the environment by means of specific cell surface receptors. Binding of these molecular stimuli to receptors stimulates transmembrane and intracellular signaling events that result in a global cellular response. Spatial and temporal targeting of signaling components in the membrane on micron and submicron scales is critical for overall efficiency and regulation of the cellular response. If the environment is engineered, then the spatially regulated response of the cell to defined stimuli can be investigated by monitoring redistributions of selectively labeled components or alterations in signaling pathways. Patterned surface approaches lend themselves readily to signaling that involves specialized regions or interactions such as membrane domains formed by clustered receptors or synapses formed at a cell–cell interface. These approaches have proven valuable for examining immune system cells, which are stimulated by clustering of antigen receptors. One example is the receptor (FcεRI) for immunoglobulin E (IgE), which plays a central role in the allergic immune response.

Spatial regulation arises early in the signaling process at two stages: (1) ligand-dependent cross-linking of the IgE-FcεRI complexes required for cell activation and (2) selective targeting of signaling proteins to the region of the activated receptors. Structurally defined ligands can be used to examine structural constraints in signaling pathways. The first stage of spatial regulation is examined with a bottom-up approach of nanobiotechnology: synthesizing multivalent ligands with defined architectural features that control the manner in which the IgE-FcεRI come together when cross-linked to activate the cells. Bivalent ligands with antigenic groups on both ends and flexible spacers of poly(ethylene) glycol in the range of 10 nm cross-link the two binding sites of IgE intramolecularly. These serve as potent inhibitors of intermolecular IgE cross-linking and cell activation and provide a potential lead for allergy therapeutics. Signaling mechanisms stimulated by cross-linked IgE-FcεRI were investigated with bivalent ligands containing rigid spacers of double-stranded DNA (dsDNA), which has a persistence length of ~50 nm. These ligands stimulated low-level degranulation and also revealed a length dependence. Branched dsDNA constructions were found to be more effective because trimeric and larger cross-linking of IgE receptors causes significantly higher levels of cellular responses.

Trivalent Y-DNP$_3$ ligands trigger robust signaling responses in two distinguishable pathways, one that depends on ligand length and another that is independent of length. This spatial differentiation provides direct support for physical coupling of proteins as a key mechanistic step in one pathway, and it also reveals branching of pathways in signaling events.

Patterned antigens have proven to be valuable for examining membrane-compartmentalized signaling. Spatial targeting of signaling components at the plasma membrane has been a subject of considerable interest, and accumulating evidence points to the participation of membrane domains. However, direct methods to investigate spatial rearrangements in the membrane and targeting of components were initially limited by typical heterogeneity of the cells and stimuli, as well as by the diffraction limit of light. Surfaces were patterned with lipid bilayers containing receptor-specific ligands using a parylene lift-off method. These microfabricated substrates allowed observation by fluorescence microscopy of labeled cellular components co-redistributing with the receptors clusters in the same pattern under physiological conditions. Substrates with patterned lipid bilayers lend themselves to other optical measurements. For example, fluorescence photobleaching recovery can been used to evaluate the dynamic nature of the proteins concentrating in the regions of patterned ligands and clustered IgE-FcεRI (Fig. 22).

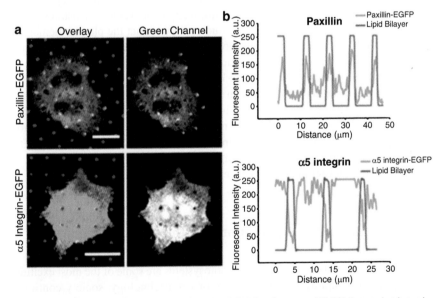

Fig. 22 Paxillin co-redistributes with clustered IgE-FcRI and patterned lipid bilayers, but integrins avoid these regions. Confocal micrographs show cells expressing paxillin-EGFP or integrin α5-EGFP (*green*) after incubation with patterned lipid bilayers (*red dots*) containing receptor-specific ligands. (**a**) Paxillin is recruited with IgE-FcεRI in the regions of patterned ligands, and α5-integrin is excluded (*scale bars*: 20 μm). (**b**) Intensity line profiles from images in A confirm that paxillin accumulation and integrin exclusion correspond exactly to where the patterned bilayers are localized (From Torres et al. [102])

Patterned substrates also reveal selective targeting of higher-order cellular structures to clustered, activated receptors or elsewhere. One study showed that cell surface integrins preferentially bind to the silicon surface, whereas particular cytoskeleton binding proteins (e.g., paxillin) cluster with the receptors, independent of the integrins. Complementary biochemical studies with siRNA showed that paxillin participates in signaling initiated by the clustered receptors. Other studies monitored stimulated membrane trafficking and revealed that recycling endosomes are targeted to clustered receptors, whereas secretory lysosomes are targeted elsewhere.

Nanoapertures provide the means to examine plasma membrane dynamics with single-molecule resolution. Although surfaces patterned with micron-size features have proven valuable in providing new information about spatial regulation of cell signaling that occurs proximal to the plasma membrane, ultimate elucidation of cellular homeostasis and stimulated responses requires investigations of molecular dynamics and interactions that occur on the length scale of 100 nm. Although it has been possible to pattern these features as small as 50 nm, visualization by fluorescence microscopy with live cells has not been practical below 600 nm. New super-resolution fluorescence methods, including PALM, STORM, and STED, are beginning to overcome these limitations. Fabrication techniques, including zero-mode waveguides (ZMWs), can also overcome the diffraction limit of light. ZMWs were initially used to examine diffusion of fluorescent lipid probes on model membranes and single-enzyme turnover of fluorescent substrates with micromolar concentrations. Subsequent studies showed that the plasma membrane of live cells can be probed with high sensitivity with these optical nanostructures. Characterization of mast cell interactions with ZMW using a combination of fluorescence and electron microscopy showed that plasma membranes from live cells penetrate these nanostructures in a cytoskeleton-dependent manner. This positioning of the plasma membrane into the apertures allows high-resolution examination of fluorescently labeled molecules with fluorescence correlation spectroscopy. These optical nanostructures offer the unique advantage of studying single-molecule dynamics in the plasma membrane, or within close proximity, at physiologically relevant molecular concentrations.

3 Goals, Barriers, and Solutions for the Next 5–10 Years

While the science and engineering of nanobiosystems are some of the most exciting, challenging, and rapidly growing sectors of nanotechnology, society continues to face daunting challenges in biomedicine and healthcare delivery. Over the next 5–10 years, nanotechnology will play a significant role in overcoming these challenges. Several specific goals are described below, the current barriers to success are discussed, and then solutions are proposed where nanotechnology can overcome the barriers and reach these goals.

3.1 Biomarkers

Successful treatment of human disease requires accurate diagnosis and effective treatments. Successful realization of personalized care requires that markers exist both for the disease and for identifying subsets of patients who will respond to specific treatments. Thus, improved diagnostics are needed to detect disease as early as possible, even to the point of detecting single defective cells or biomarkers that predict the future onset of a disease. Currently, there is insufficient knowledge of disease-specific biomarkers; further, the sensitivity of current diagnostic tests may not allow for the detection and quantification of the most informative low-abundance biomarkers. Finally, it will likely be the case that a panel of biomarkers is most accurate with regard to disease detection, and diagnostic platforms are needed that can detect multiple targets rapidly, cheaply, and at the point of care. Advances in nanotechnology will provide both *in vitro* and *in vivo* diagnostics using nanoscale materials that are capable of specifically binding with target biomarkers and amplifying the detection response for identification and quantification. Importantly, adapting these technologies so that they can be made to identify low-abundant unknown biomarkers, and then identify which ones are specific to a given disease state, is possible with exquisitely sensitive nanotechnologies coupled with imaging modalities that relay morphological, structural, and functional features of disease; devices that use multiplexing for *in vitro* diagnoses; and implantable devices that monitor analytes and therapies.

3.2 In Vitro Diagnostics

Improved diagnostics that achieve PCR-like sensitivity through the use of nanoscale operations that assess specific disease biomarkers at points of care are a goal for future development. However, the development of such platforms is hindered by insufficient knowledge about specific and non-specific biological interactions, inadequate affinity agents and amplification methods, and the lack of appropriate and validated biomarkers. To overcome these barriers, it will necessary to achieve high sensitivity and specificity in nanoscale processes, coupled with a more robust knowledge of the bio–nano interface affecting nanomaterial probes and target analytes. The surface functionality and modification processes of nanomaterials will be tailored to better resist non-specific biomolecular interactions and foster specific ones. Finally, sample processing and autocatalytic nanotechnologies will provide unprecedented purification and signal generation upon binding to target analytes. Taken together, advances in nanotechnology and a more thorough understanding of the bio-nano interface will provide robust, multiplexed diagnostic assays where specific signal(s) generated from only a few target molecules can be readily distinguished from background noise.

3.3 Nanotherapeutics

The pharmaceuticals industry is a sector seeking radical innovation. A significant goal for nanotechnology over the next 5–10 years is developing nanoscale pharmaceuticals that offer new therapeutic approaches. Significant barriers exist with regard to both the scientific and the business aspects of these endeavors. Development and translation of nanotherapeutics to patient use are expensive, time-consuming, dependent upon the approval of regulatory agencies, and require appropriate reimbursement. Over the next decade, significant financial input from pharmaceutical companies and the Federal government is expected to more fully realize the promise of developing nanotherapeutics, especially at a time when venture capital financing of small businesses has diminished. In addition, regulatory agencies are realizing that novel approaches are needed with respect to nanotherapeutic testing in humans and human clinical trial design. With regard to product development, nanotherapeutics are facing significant challenges with regard to demonstrating favorable pharmacokinetics, biodistribution profiles, targeted drug delivery, tissue penetration, drug release, and the ability to provide an image signature of location and efficacy. Each of these challenges is being met head-on with development of novel methods of nanoparticle fabrication, increasingly sophisticated nanoparticle architectures, a better understanding of the bio-nano interface, and better ways of developing hierarchical nanotherapeutics.

3.4 Nanotechnology and Stem Cells

Stem cells interact with their physical and chemical environments, and these interactions dictate stem cell maintenance and differentiation. At present, insufficient knowledge of stem cell regulation, including the biology of the multi-potent state and underlying biological regulation mechanisms, are limiting the directed manipulation of stem cells. Nanotechnology can be harnessed to more thoroughly understand the complex stem cell-environment interactions in order to more formally monitor stem cell responses and then to systematically understand those features which dictate stem cell differentiation and phenotype. Nanoarray technology, nanoparticles used to deliver differentiating signals to stem cells, and a robust understanding of the bio-nano interface will provide ways of building and manipulating appropriate scaffolds for directing stem cell fate.

3.5 Tissue Engineering

Engineering functional tissues that repair or replace damaged tissues and organs holds significant promise for treating some of the most devastating and debilitating human neurological and musculoskeletal diseases. Currently, there exist no off-the-shelf technologies or means to rapidly and reproducibly build appropriate tissue architectures. Nano-architectures and synthetic pro-morphogens (bio-active analogues of growth factors) that recapitulate natural extracellular matrix and provide growth signals can be fabricated using a number of techniques, including (1) electrospinning and molecular printing strategies (e.g., dip pen nanolithography, contact printing) to develop 3D engineered or self-assembled scaffolds; (2) functionalized nanoparticles, nanowires, and carbon nanotubes to build composite multi-tasking engineered nanomaterials; and (3) dendrimers presenting bio-ligands or nanobeads carrying growth factor analogues. Such materials and technologies will permit rapid growth in the field of tissue regeneration and repair.

3.6 Economics of Nanomedicine

The goal of nanomedicine is to reduce costs, provide more accurate and earlier diagnoses, deliver effective and personalized care, and simplify and standardize healthcare delivery. This is especially true with regard to chronic diseases (e.g., cardiovascular disease and diabetes) that have high healthcare costs. However, the development of nanotechnologies is expensive, there is competition from generic companies, the environment for reimbursement is difficult, regulatory agencies have not previously focused on complex and often multidisciplinary drugs and/or devices, and the funding environment is quite restricted. Ultimately, nanotechnology can deliver low-cost, accurate, quantitative, reliable, and accessible materials for diagnosis and treatment, and these cost savings will emerge through innovative plans to fund, regulate, and commercialize nanotechnologies. It is difficult to predict the overall economic impact of the nanobiomaterials discussed in this report and of the others that will be translated into use in the medical arena over the next 5–10 years. Many estimates exist for the overall market size of nanotechnologies. One recent review [103] cited three different sources and estimated that the overall market size would be between $2.6 and 2.95 trillion by 2014 or 2015. NSF has estimated a market size of $1 trillion by 2015, $180 billion of which will be attributed to nanopharmaceuticals [103]. Regardless of the estimate, and despite the challenges to nanobiosystems development noted throughout this chapter, nanotechnologies will most certainly have a substantial impact in health and medicine over the next 5–10 years.

4 Scientific and Technological Infrastructure Needs

In order to fully realize the promise of nanotechnology in biomedicine, significant improvements in scientific and technological infrastructure are needed. Investment in innovative nanotechnological research is imperative. Centers of excellence have proven to be very effective tools for bringing together many disciplines within medicine, science, and engineering to tackle important problems in the field. A significant emphasis should be on establishing such centers on a disease-by-disease basis. This will require increased and targeted funding approaches, but importantly, such centers will centralize and galvanize new opportunities for commercialization and business development in the nanomedicine arena. An infrastructure for doing clinical trials based upon the pipeline of new diagnostic and therapeutic tools emerging from the field must be established. The U.S. Food and Drug Administration (FDA) must build its own internal infrastructure to handle the evaluation of these important innovations in a timely, efficient, and safe manner. In addition, thought must be given to how the adoption of these technologies by physicians can be facilitated through targeted education, participation in the clinical trials, and ultimately, reimbursement policies.

Investment and innovation in education is required from the K–12 levels up through physicians and professors to enhance matriculation to careers in science and math fields and to improve the adoption of emerging nanotechnologies in the clinic. Enhanced communication is required between those on the front lines of academic nanomedicine research, clinicians, and individuals in the pharmaceutical and biotechnology industries. Focused understanding of the most significant problems facing patients, and then focused energies on developing collaborative solutions to key problems will certainly facilitate and expedite translation of the most promising technologies.

In order to fully realize the potential of nanotechnology in biomedicine, reliable production of high-quality nanomaterials and their characterization is imperative. This will enable reproducible fabrication of high-quality materials platforms and answering of directed questions with regard to their use in *in vitro* diagnostics, as therapeutics, as tissue scaffolds, and as materials for directing stem cell fate, among others. Furthermore, broad access to new tools for characterizing such materials and their interactions with biological systems is needed to transform what has been heretofore an empiric exercise into a predictive one. Establishing scientific consensus as to the array of *in vitro* and *in vivo* tests needed for nanomaterials would provide scientists with a framework for evaluating novel agents.

5 R&D Investment and Implementation Strategies

The challenges described above demonstrate that substantial investment in nanobiomedicine is required; however, the return on investment will be substantial with regard to jobs created, the U.S. competitive position in the global economy, and improvements in human health that directly result from nanotechnology. Often small start-up companies are the initial entities making strides toward product development. One way to substantially enhance translation of their efforts to commercial success would be a system of financial incentives available to both the start-up company and to large biotechnology and pharmaceutical companies who choose to invest in them. Another way, with a more direct impact on small start-up success, would be to increase government spending on small start-up nanotechnology companies in order to facilitate their ability to attract more substantial investment from the private sector.

Next, it is critical to identify and address the most substantial problems in biomedicine. Identifying key issues is not a straightforward process. Nanoscience research is often done by chemists, materials scientists, and the like, who undertake technology development based upon the known novel properties of nanostructures; however, they may not focus on specific applications in biomedicine. Increasing the communication between practicing health professionals and pharmaceutical industry experts, for example, with those in academia may be a direct way to augment nanotechnology development efforts in all areas of biomedicine in order to facilitate early commercial and clinical interest and, therefore, enhance the likelihood of translation. As evidence continues to be generated and nanotechnology moves from an empiric science to one based upon well-characterized and predictable nanosystems, it will be imperative to broadly disseminate this information so as to avoid redundancy in R&D and more clearly focus on developing nanomaterials and systems for the next generation of *in vitro* diagnostics, nanotherapeutics, imaging agents and technologies, combination theranostics, and materials to manipulate stem cells and regenerate human tissue.

6   Conclusions and Priorities

Over the last 10 years there have been substantial developments in nanobiosystems. Currently, nanotechnology-enabled diagnostic, therapeutic, and imaging agents are being used in patients; these agents are being closely scrutinized with regard to the benefits that they provide in relation to conventional technologies and patient health outcomes. Furthermore, substantial advances have been made for developing nanomaterials that can manipulate stem cells and engineer cells and tissues. Finally, research is honing in on the bio-nano interface and how changes in nanomaterials and nanoengineered surfaces result in specific and nonspecific bio-nano interactions.

Building upon these advances, much work needs to be done over the next 5–10 years to more fully realize the promise of nanobiosystems. First, *in vitro* diagnostic capabilities need to be adapted to not only detect vanishingly small quantities of known target biomarkers, but also to identify new biomarkers specific for given diseases. Biomarker discovery efforts may well reveal multiple biomarkers, and nanotechnologies are needed that can robustly perform multiplexed detection. For therapeutics, novel methods are required to efficiently target nanoparticles to specific cells and tissues harboring disease. These nanostructures should be capable of carrying multifunctional cargo to enable drug delivery (e.g., small molecules or nucleic acid therapeutics), target-specificity, tissue penetration, and the capability for imaging both location and molecular mechanism of action. In the special case of RNAi, nanoparticle delivery strategies may well be absolutely required to realize the massive potential of this entire class of therapeutic molecules. Next, the bio-nano interface needs to be more fully explored to develop nanomaterials that both drive stem cell fate and regenerate tissues. In addition, a more complete understanding of the bio-nano interface is necessary to begin to predict the *in vivo* behavior of nanomaterials so as to more rationally develop nanomaterials as diagnostic, therapeutic, imaging, and theranostic agents, and also to predict toxicity.

Collectively, these advances are possible with continued and sustained investment in nanotechnology and nanobiosystems over the next decade. Innovative mechanisms are needed to provide funding for research laboratories and small start-up companies. Because advances in nanobiomedicine have largely been the result of robust collaborations that incorporate the expertise of individuals skilled in nanomaterial synthesis and characterization, biologists, engineers, and medical doctors, such collaborations should be encouraged. Finally, the value of the advances of the past 10 years, and the promise that nanotechnology has for the future of biosystems, medicine, and health care, should not be underestimated.

7 Broader Implications for Society

Society stands to reap substantial benefits from advances in nanobiotechnology. New nanotechnologies already are in the clinic and benefitting patients. The future will provide ever smarter and more advanced nanotechnologies that identify disease-specific biomarker profiles to enable early detection and monitoring of disease. Nanoparticle therapeutics will be tailored to individual patient profiles, realizing the promise of personalized medicine, targeting diseased tissue, and eradicating disease without harming normal nearby cells and tissue. Furthermore, nanomaterials hold the keys to treating some of the most debilitating and devastating human diseases through stem cell and tissue engineering approaches. Overall, continued investment in nanobiotechnology R&D is expected to dramatically improve health care, create jobs, help build a new high-tech industrial base, and drive a new era of U.S. prosperity.

8 Examples of Achievements and Paradigm Shifts

8.1 National Cancer Institute: Alliance for Nanotechnology in Cancer

Contact person: Piotr Grodzinski, National Institutes of Health/National Cancer Institute

Although the past 50 years have seen enormous advances in our understanding of the mechanisms and biology of cancer occurrence and spread, the standard regime for cancer treatment and management—surgical resection followed by chemo- and/or radiotherapy and periodic screenings to search for recurrence or metastasis—is largely the same now as it was in 1960. Although highly heterogeneous in its causes and characteristics, cancer is a disease of cells, typically localized to specific tissues or organs in its early stages; however, its treatment and management have been to a large extent highly invasive, extended across the whole body, and uniform across patient populations. These blunt-instrument approaches to the disease result in treatments with a burden of side effects that can limit or even terminate treatment, screening procedures that are high in false positives, and unnecessary costs and suffering, yet they are unable to detect the most dangerous and aggressive cancers early enough for effective intervention. The ability of nanotechnology to improve this situation is already registering, through nanoformulations that use tumor-specific drug delivery to localize treatments and improve the therapeutic index of highly toxic chemotherapeutic agents, and the use of nanomaterials with novel physical (optical, magnetic, and mechanical) properties that enable more sensitive measurement of cancer biomarkers and more reliable and specific *in vivo* imaging for cancer detection. The combination of advanced imaging with traditional surgical techniques for intraoperative guidance enables more successful resection of cancerous growths, which is still the most effective cure available for many cancers. These new technologies are beginning to shift cancer care to more targeted, less destructive treatment strategies that recognize and act only on cancerous cells, coupled with multiplexed, multimodal imaging and screening technologies that can detect, stratify, and monitor disease accurately and in real time.

Solid tumors are typically supported by a rapidly grown vasculature that is porous and lacking in sufficient lymphatic drainage. The fenestrations in this vasculature allow nanoscale materials (50–250 nm) to extravasate into the tumor, and the inadequate drainage allows these materials to accumulate in the tumor. This enhanced permeability and retention effect leads to higher tumor loading by large-molecular-weight therapeutics and nanoparticles than by small molecules. Increased loading of imaging agents and therapeutics in tumors reduces systemic distribution, reducing off-target effects of drugs and background signal in imaging. The large surface area and encapsulated volume of nanoparticles also enables delivery of multiple cargoes for combined imaging and drug delivery. Surface attachment of cancer cell-specific ligands, peptides, and antibodies leads to increased cellular internalization of cargo loaded into tumors. Intracellular delivery of drugs results in higher therapeutic efficacy and provides access to a greater variety of drug targets, including genes for silencing by RNA interference therapy. Early evidence also indicates that the endocytosis-mediated internalization of nanomaterials can evade efflux pump mechanisms of drug resistance in cancer cells.

As the use of nanocarriers to deliver already approved imaging agents and therapeutics is being clinically established, two further advances in nanoparticle-mediated treatment of cancer present themselves. The first is the exploitation of novel nanomaterial properties for therapeutic and/or diagnostic effect. Examples include the use of externally applied alternating magnetic fields to heat magnetic nanoparticles at a cancer site, resulting in hyperthermia and cell death, and the exploitation of the photothermal properties of gold nanoparticles/rods/spheres to detect and destroy cancerous tissue by exciting fluorescence or heating using infrared radiation. These physically mediated, localized cancer therapies promise low-side-effect profiles and imperviousness to cellular mechanisms of therapy resistance. The second advance is the introduction of functional nanosystems. These systems include particles and devices that release therapeutics in response to biochemical signals detected in the tumor or blood, while also reporting on the chemical and physical tumor microenvironment, providing therapy and monitoring in one package. An example would be a collective of magnetic nanoparticles with environmentally mediated aggregation, in which both drug release and MRI contrast are controlled by aggregation state. The ability to decorate the surface of nanoparticles with multiple reporter molecules and to load multiple nanoparticles into a single device will also enable multimodal and multiplexed molecular imaging and more informative monitoring. Such functional molecular imaging can act as optical biopsies, with tumors being continuously typed and staged.

The development and validation of targeting moieties directed at both cancer cell surfaces and tumor vasculature and microenvironment is necessary to increase the efficacy of the cancer nanotechnology strategies outlined above. Further enhancement of intratumoral and intracellular delivery will also result from the careful choice of nanocarrier shape in addition to size and rational design of nanocarrier properties. Well-designed nanocarriers can control drug release by time, response to internal environment (e.g., pH, presence of cancer specific enzymes or receptors) or response to external stimuli (e.g., radiation, ultrasound).

Improvements in cancer detection, therapy, and monitoring will be amplified by advances in *in vitro* assay technology. More rapid and sensitive assays will enable identification of and validation of new cancer biomarkers, including prognostic markers of tumor metabolism, growth, and dormancy as well as type and stage. Many of these technologies will rely on both the innate characteristics of nanomaterials and their large surface area–to-volume ratio to increase assay sensitivity. Examples include the magnetic nanoparticle-based magnetoresistive sensors, nanoscale cantilever resonance-based molecular detectors, and signal amplification through surface functionalization. Non-protein-based biomarkers will be essential for the development of cheap, reliable *in vitro* assays. Low-cost genomic and proteomic profiling of tumor types will also be crucial to choosing optimal care strategies for individual patients. Microfluidics will be a backbone technology for many of these advances, especially cancer cell detection strategies.

The use of nanotechnology for cancer prevention is currently an underdeveloped area. Nanoformulations of chemopreventives are currently under investigation, as are methods to identify biomarker profiles for at-risk patients. More ambitious research will incorporate the cancer detection technologies discussed above into the standard of care for high-risk populations, and even integrate treatment delivery for triggered release upon disease discovery.

8.2 Perspective on Nanobiotechnology and Synthetic Biology

Contact person: Carl Batt, Cornell University

The history of biotechnology, dating back to ancient times, has been highlighted by development of processes, identification of causative microorganisms, and in recent years, deliberate improvement of these microorganisms. Although the Egyptians more than 5,000 years ago did not comprehend the microbiological basis for the conversion of bread into beer, modern biotechnology has given rise to processes for producing the rarest of biomolecules for treating complex diseases. Genetic engineering has been used to deliberately improve single and multiple traits important for the individual products; however, genetic engineering is limited by the host background and the competing metabolic activities within the host cell. In the next 10 years significant advances will be made in synthetic biology, and these will be enhanced by nanotechnology-derived tools.

Over the past 50 years, advances in the broad area of biotechnology have been supported by the products of nanotechnology. Progress in nucleic acid sequencing and synthesis has enabled new applications, which in turn has driven demand. Improved speed, accuracy, and reduced cost of reading and creating nucleic acid sequences (Fig. 23) have been realized through nanotechnology-based components.

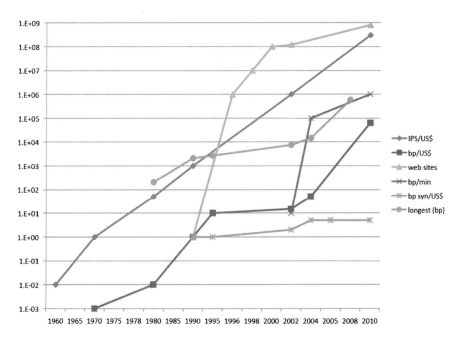

Fig. 23 The change in lengths of time for advances in DNA synthesis, sequencing costs, speed and gene synthesis as compared to advances in computing power and numbers of web sites. Instructions per second (IPS/US$), base pairs sequenced (BP/US$), base pair synthesized (BP syn/US$), sequencing speed (BP/min) and reported longest single DNA sequence (longest BP) are from reported values and or estimates derived from reported values [104])

Deconstruction of complex biological systems and then reconstruction of them will at some point largely be replaced by *de novo* approaches that rely upon our knowledge base and the ability to predict outcomes, where components are created without reference to existing systems. Synthetic biology is in essence an effort to totally create complex biological systems from scratch. To accomplish this, a concerted effort is being mounted that will drive the field of nanobiotechnology more than previously witnessed, when the technology was being used more to understand than to create.

The reduction of costs of DNA synthesis and sequencing parallel those seen with Moore's Law demonstrating the power of the demand for better tools for biotechnology (Fig. 24). The rate of improvement in both DNA synthesis and sequencing has perhaps exceeded that predicted by Moore. The rationale in part due to the more direct linkage between speed, cost and ultimate benefits that can arise from an enhance ability to manipulate nucleic acid.

Nanotechnology has enabled significant improvements in both nucleic acid sequencing and synthesis. In the field of nucleic acid sequencing, nanometer-scale electrophoresis platforms allow for separation of nucleic acid fragments with single-base-pair resolution and enhanced sensitivity by reducing the focal volume of the eluate. Technologies that have advanced beyond the industry-wide Sanger (chain-termination) sequencing coupled to capillary electrophoresis separation format have been enabled by nanotechnology. For example, advances in nucleic acid sequencing have been realized by moving from a relatively linear singular format to a massively parallel platform. The parallel sequencing format has been achieved using nanoscale beads and microfluidic handling that facilitate the reading of massive numbers of DNA molecules at the same time.

In a similar fashion, nucleic acid synthesis based upon well-refined phosphoramidite chemistry was similarly advanced using nanotechnology, and the formulation of highly controlled solid-state supports has pushed the limits of the length of

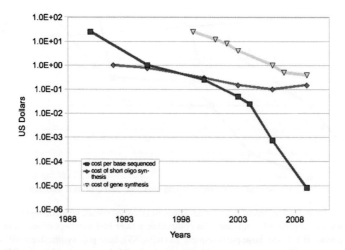

Fig. 24 Reduction in cost of DNA synthesis, 1990–2006 (From R. Carlson, http://www.synthesis.cc)

sequences that can be made, along with cost reductions. The relatively lower cost of DNA synthesis is primarily a matter of the market and the feasible commercial costs of set up, shipping, and handling. However, the most dramatic increase is observed for the length of DNA sequences that can be synthesized and assembled. Now advanced methods of chemical synthesis of DNA, coupled to facilitated assembly, provides a means to build long stretches of nucleic acid.

Combined advances in nucleic acid sequencing and synthesis have created a robust platform for biotechnology and the engineering of biological systems to create an array of products. Entire microorganisms have been recreated: a first test of a relatively simple one, *Mycoplasma*, was reported by [105]. Over the next 10 years the ability to synthesize longer and longer DNA sequences will be enhanced, and routine genome construction will begin with a design and end with a complete genome. While advances in genome synthesis will continue, use of genomes to prime the creation of new organisms will have significant challenges, as outlined below.

Synthetic biology is a view of biological systems from a *de novo* design perspective, incorporating knowledge gathered over the past 100 years to build potentially better systems. In the next 10 years, the available tools and advances driven by the demands of the entire biotechnology field will coalesce with nanotechnology, pushing the emerging field of synthetic biology forward.

Initial demonstrations of bottom-up construction of viruses have evolved into the *de novo* creation of simple single-cell organisms. Nanotechnology offers the tools and processes to further create hybrid biological systems that do not rely solely on natural biomolecules (nucleic acids, proteins, lipids, carbohydrates) but incorporate non-biological molecules, including semiconductor materials. Self-replicating hybrid biological systems that include what are not traditionally thought of as biological materials are already on exhibition in nature. Diatoms and other similar organisms incorporate silicon onto protein templates to create rigid exostructures. This biomineralization process suggests that synthetic biology extended to include semiconductor-based structures could be replicated through harnessing those biomineralization mechanisms used by diatoms and encoded in their genomes. The expansion of the materials palette compatible with biological systems is already exemplified by the development of components for the incorporation of unnatural amino acids [106]. This expanded amino acid repertoire enables a number of post-translational chemistries to be used, further increasing the spectrum of functionality in biological systems. For example the incorporation of amino acid analogs with an azide moiety could be used to carry out a subsequent "click" reaction [107]; the most simplistic application would be to immobilize a protein to a surface, but more elaborate schemes of mimicking various post-translational modifications (glycosylation) could be imagined. For nanotechnology to advance and be a vital part of synthetic biology, it will be necessary to overcome challenges that exist in interfacing the various elements and ensuring that processes are compatible, especially with regard to the relative fragility of biomolecules. Nevertheless, the resiliency of biomolecules is surprising, and life in extreme environments (pH, temperature, ionic strength) suggests routes to obtain a set of biological components compatible with the nanotechnology tools and processes.

The biggest challenges in the future relate to the *de novo* design of systems and an ability to accurately predict the performance of these synthetic systems. Mimicking biological systems is not pushing the limits of potential and imagination. From the standpoint of *de novo* constructed systems, the most immediate challenge is "booting up the system" where the only inputs are nucleic acids. These systems have developed over eons of evolution, and the generally recognized path starts with catalytic nucleic acid molecules. There is a potential analog in booting up the system; nanotechnology may provide a basis through the design and fabrication of devices that have the precision to orient parts, much as RNA polymerase helps to transcribe the DNA sequences and ribosomes help to translate this information into proteins. The debate about mechanosynthesis and its ultimate feasibility and/or utility will continue [108].

In essence, the future of synthetic biology will take a chapter from evolution but replace the laws of natural selection with ones that are derived from theoretical models. Since these models will be constructed and validated based upon observations made over the past 100 or more years, there will still be elements of evolution, but the process will be replaced by a few rounds of computation.

8.3 Paradigm Changes in Use of Nanoscale Sensors to Monitor Human Health/Behavior

Contact person: Harry Tibbals, University of Texas Southwestern Medical Center

Nanotechnology-based sensors for measurements in the human body have advanced significantly since 2000. In addition to incremental improvements in the technologies and types of applications envisioned 10 years ago, a number of developments have created paradigm shifts in methods and approaches. Significant advances for *in vivo* biomedical nanosensors have been achieved in the following areas: physical and physiological sensors, imaging, biomarkers and diagnostics, and integrated nanosensor systems [109].Measurement capability for biological physiology has improved in the past decade by application of nanotechnology in blood pressure, tissue pressures, pH, and electrophysiological monitoring of brain activity. Tables 2 and 3 and Fig. 25 give selected examples of key nanotechnology-based advances in medicine since the year 2000.

A number of research directions can be successfully addressed by 2020:

- Many of the medical technologies for monitoring the human body are just now at the "proof-of-concept" stage. The next 10 years will see them move into prototyping of devices and systems for medical application, followed by clinical trials. Those that pass tests of safety and effectiveness will rapidly appear in medical practice.
- We will continue to see convergent integration of sensors, communications, and therapeutics into systems that detect, monitor, and apply therapies in response to changes in health status of individuals and populations.

Table 2 Paradigm changes in use of nanoscale sensors for monitoring human health and behavior

Pre-2000	2000–2010	2010–2020	2020–2030	Medical applications
Pulse oximetry				
Attached, wired, 1–2 cm^3	Attached wired <1 cm^3	Adhered, wearable, wireless reporting 0.1 cm^3	Implanted, embedded, wireless, remote 0.001 cm^3	Cardiovascular, ICU, intraoperative surgical monitoring
Accelerometry				
External, attached, wired, 1–2 cm^3	Attached wired <1 cm^3	Adhered, implanted, wireless reporting 0.001 cm^3	Implanted, embedded, wireless, remote <0.001 cm^3	Home monitoring, post-operative, geriatrics, orthopedics, neurology (gait analysis), cardiovascular
Pressure				
External, attached, wired, 2–4 cm^3	Embedded wireless <1 cm^3	Implanted wireless 0.001 cm^3	Implanted, ingestible wireless, remote <0.001 cm^3	Ob-Gyn (uterine), intra- & post-operative monitoring, gastro-, intestinal, orthopedics, prosthetics, cardiovascular
Humidity				
External, attached, wired, 3–4 cm^3	Attached wired <2 cm^3	Wireless/remote optical sensing 0.001 cm^3	Implanted, ingestible wireless, remote <0.001 cm^3	Ob-Gyn (uterine), intra- & post-operative monitoring, gastro-, intestinal, urology, orthopedics, cardiovascular
Galvanic potential (skin)				
External, attached, wired, 3–4 cm^3	Attached wired <2 cm^3	Wireless/remote optical sensing 0.001 cm^3	Implanted, ingestible wireless, remote <0.001 cm^3	Ob-Gyn (uterine), intra- & post-operative monitoring, gastro-, intestinal, urology, orthopedics, cardiovascular

(continued)

Table 2 (continued)

	Pre-2000	2000–2010	2010–2020	2020–2030	Medical applications
Impedance (internal)	External, attached, wired, 3–4 cm^3	Attached wired <2 cm^3	Wireless/remote optical sensing 0.001 cm^3	Implanted, ingestible wireless, remote <0.001 cm^3	Ob-Gyn (uterine), intra- & post-operative monitoring, gastro-, intestinal, urology, orthopedics, cardiovascular
pH	External, attached wired 3–4 cm^3	Attachable wireless <1 cm^3	Wireless, implantable ingestible 0.001 cm^3	Implantable, ingestible wireless, remote <0.001 cm^3	Gastrointestinal, intra- & post-operative monitoring, urology, wound healing
Glucose sensors	External sampling	Sampling through skin with nano-needle arrays, implantable MEMS	Long-term implantable MEMS/NEMS with wireless monitoring and feedback for insulin delivery via wearable micropump	Incorporation of nanosensors in integrated artificial pancreas and/or encapsulated live-cell bioreactor	Diabetes, endocrinology, immunology
EMG: electromagnetoencephalography	External, room-sized (SQUID)	External, cabinet (SQUID)	Attachable, wireless 1,000 cm^3 (MEMS)	Wearable, wireless <1,000 cm^3 (NEMS)	Neurology, neurosurgery (with magnetic brain induction) prosthetics (brain interface)
Fluorescence: nanodots for biomarkers	External (in cell cultures and lab samples)	Internal: IV, injectable; (detection with fiberoptic probes, endoscopy) (cancer)	Internal: IV, Injectable, (detection with wireless spectroscopic probes) (other diseases)	Internal, IV, ingestible Injectable, ingestible, (detection with wireless spectroscopic probes, external IR through-skin monitors) (wide variety of diagnostics)	Cancer, infectious diseases, inflammatory, degenerative, genetic diseases

Ultrasound and opto-acoustic-enhancement nanoparticles			
External (in cell cultures and lab samples)	Internal: IV, injectable; (detection with fiberoptic probes, endoscopy) (cancer)	Internal: IV, injectable, (detection with wireless spectroscopic probes) (other diseases)	Internal, IV, ingestible injectable, (detection w/ wireless spectroscopic probes, external IR through-skin monitors) (wide diagnostic variety)	Cancer, infectious diseases, inflammatory, degenerative, genetic diseases
FTIR and Raman			
External (in cell cultures and lab samples)	External (single-cell detection)	Attachable, implantable MEMS for internal single-cell wireless monitoring	Attachable, implantable injectable, ingestible, NEMS for internal single-cell resolution screening, monitoring	Cancer, infectious diseases, inflammatory, degenerative, genetic diseases
MS: mass spectrometry			
External (cell cultures, extracts, samples)	1st single-molecule MS, external, lab (experimental proof-of-concept)	Insertable, attachable single-molecule MS for monitoring	Implantable «« Etc.	Cancer, infectious diseases, inflammatory, degenerative, genetic diseases
Nanosensors in tissue scaffolding			
Concept	Experimental guidance of stem cell and autologous cell growth	Proof-of-concept; 1st prototypes; clinical trials; 1st approved uses	Use in medical practice; on-demand generation of live-cell tissue scaffolds	Surgery, nerve, brain, and bone regeneration, joint and cartilage, heart, burns, wounds

(continued)

Table 2 (continued)

Pre-2000	2000–2010	2010–2020	2020–2030	Medical applications
Nanosensors in neurosensory prosthetics				
Concept, wireless operation	Experimental prototypes for improved retinal prostheses and cochlear implants with nanosensors	Reduction in size, power requirements; use of power harvesting via nanogenerators	Nano-surface engineering of sensor interfaces for improved bio-compatibility, longer lifetime; live-cell integration	Vision, hearing
Integrated nanodevices for targeted therapeutics				
Concept, theranostic targeted nanoparticles for drug and RF therapy delivery	Development of a number of specific targeted NPs for imaging, thermo- and radio-therapy, drug delivery	Platforms for customizable targeted nano-therapeutics based on individual genotypes and phenotypes	Implantable, integrated nanosystems for monitoring and responding to pathological conditions	Infectious diseases, cancer, degenerative diseases, chronic metabolic conditions, traumatic injury, stress-induced conditions

Table 3 Advances in medical nanotechnology for measurements in the human body

Technology	Application area	Qualitative/quantitative advances	Stage of development				
			Proof of concept	Prototype	Pre-clinical	Clinical trials	Application
MEMS pressure monitoring:	Blood pressure, intra-arterial (cardioMEMs)	Wireless, min-invasive, continuous, ×3 accuracy	2001	2002	2003	2005	2006
MEMS gastric and esophageal Monitoring	Gastric acid reflux, pre-cancer: pH, Impedance	Wireless, min-invasive, ×5 sensitivity	Bravo: 2002; UTA: 2004	Bravo: 2003 UTA: 2005	Bravo: 2003; UTA: 2007	Bravo: 2005	Bravo: 2007
MEG: MEMS Magneto-encephalography	Brain monitor; prosthetics	Min-invasive; sensitivity = ×100 to ×1,000	NIST, Bell Labs: 2003–2004	2005	2006–2007		
Gold nanodots and nanorods	Diagnostic imaging: x-ray and ultrasound enhancement	×1,000 in intensity, contrast, selectivity	2000–2005	2003–2006	2004–2008	2004–2009	2005–
Si, ceramic dye-doped Nanoparticles	Detection of biomarkers, in vivo diagnostics	Selectivity, stability, sensitivity = ×100–1,000	2001–2005	2003–2006	2005–2008		
MRI nano-engineered particles	Contrast enhancement for MRI imaging	Contrast enhancements = ×50	2005–2006	2006–2007	2007–2009		
Nanostructures for surface-enhanced Raman spectra	Detection of biomarkers on skin, in tissue, in breath, body fluids	Detection sensitivity increase of ×10^{14}–10^{15}	2000–2005	2003–2006	2005–2009		

(continued)

Table 3 (continued)

Technology	Application area	Qualitative/ quantitative advances	Stage of development				
			Proof of concept	Prototype	Pre-clinical	Clinical trials	Application
Label-free nano- engineered lab-on-chip arrays based on CNTs, Si nanowires, BioFETs, etc.	Detection of DNA, proteins, biomarkers, for early diagnosis of cancer, degenerative diseases, personalized medicine	Reductions of 100× in detection time, 20× in detection costs, 100× in device costs	2001–2005	2004–2007	2006–2008	2009–2010	2010
Nanotech for DNA sequencing	Personalized medicine; targeted therapeutics	400 million base pairs per 40 min.; entire human genome in 15 min (est. by 2013); Following Moore's law: doubling capacity every 18 months	2000	2001–2005	2005–2009	2005–2009	2009–2010
Wearable/ implantable medical sensors with wireless communication	Monitoring for post treatment, preventative care, health maintenance, remote checkup and disease management (e.g., diabetes, cardiovascular)	Following Moore's law	2002–2005	2003–2006	2005–2009	2007–2010	2010

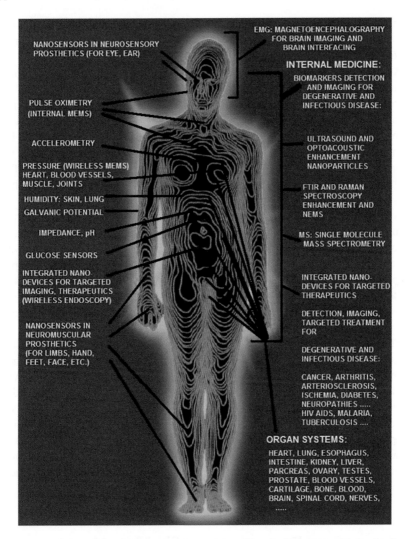

Fig. 25 Sensors for monitoring human health and behavior; see Tables 2 and 3 for details and time frame for application (Adapted from http://www.csm.ornl.gov/SC98/viz/viz4.html)

- Several new areas of smart integrated nanosensors and nanomaterials are especially exciting for their future potential to regenerate tissue affected by brain and nerve injury, cardiovascular ischemia, and degenerative diseases such as Alzheimer's disease and ALS. One is the application of protein nanoscience to the epigenetic reprogramming of cells. Cellular-level nanotechnology is being applied to create stem cells from a person's own mature cells, eliminating problems associated with immune rejection, embryonic stem cells, and genetic engineering of cells for medical treatments. Smart nano-implant nanotechnology will be applied to the guidance of cell growth, differentiation, and

migration; this will be important in making stem cell therapy more successful than it is currently, by providing nano-engineered tissue scaffolding to support stem cell implants.
- Another area that will develop in the next decade in increased use of sophisticated nanosensors to guide surgery and medical interventions such as tissue transplants, including the generation of live tissue scaffold implants on demand in the operating room. These custom implants will be generated in real time in response to sensor evaluation of tissue injuries, immunologies, and DNA matching, by three-dimensional tissue printing with nano-engineering rapid prototyping systems that assemble implants with biopolymers and cells to fit the shape, size, and structural requirements of a specific surgical operation while it is taking place.
- Another area is the increase in use of low-cost wearable and implantable intelligent nanosensors to give early warning of incipient health conditions, to monitor the progress of treatment, and to apply therapies by nano-activation of energy or chemical medications in the body, either automatically in response to embedded control systems, or on command from healthcare providers who can monitor the patient remotely. These will be used increasingly for remote examination and evaluation of patients by healthcare providers, and for organization of collaborative and consultative care [109].
- The exponentially growing development of nanotechnology-based biosensor capabilities is contributing to more effective and lower-cost approaches to health care. These approaches are needed to deal with economic, demographic, environmental, and epidemiological challenges. Nanosensor capabilities for biomedicine are a potentially disruptive technological force that will realign the tools and delivery modes of healthcare—a realignment that will represent a strategic approach to changing healthcare needs. Whether this challenge is met will depend on continued commitment to basic research and development as well as further work to move demonstrated new technical capabilities into practical medical application.

9 International Perspectives from Site Visits Abroad

9.1 United States-European Union Workshop (Hamburg, Germany)

Panel members/discussants

Chrit Moonen (co-chair), CNRS/University Bordeaux 2, France
Andre Nel (co-chair), University of California, Los Angeles, United States
Costas Kiparissides, Aristotle University of Thessaloniki, Greece
Simone Sprio, Institute of Science and Technology for Ceramics (ISTEC-CNR), Italy

Günter Gauglitz, University of Tübingen, Germany
Milos Nesladek, Hasselt University, Belgium
Peter Dobson, Oxford University, UK
Wolfgang Kreyling, Helmholtz Centre Munich, Germany

Contributors: Chad Mirkin, Northwestern University; C. Shad Thaxton, Northwestern University

This summary builds largely on the abstract by Chad Mirkin, Andre Nel, and C. Shad Thaxton of the March 9–10, 2010, workshop held in Evanston, Illinois, USA. The group working on the Nanobiosystems, Medicine, and Health topic at the Hamburg meeting fully agreed with the outcome of the Evanston meeting (Part 1 below). Most of the discussion in Hamburg then concentrated on the very important role of regulation. This is summarized in Part 2.

9.1.1 Part 1: Abstract of the Evanston Meeting

Nanostructures are now being used in the development of powerful tools for manipulating and studying cellular systems; ultrasensitive *in vitro* diagnostics that can track indications of disease at very early stages; *in vivo* imaging agents that provide better contrast and more effective targeting compared to molecular systems; and novel therapeutics for debilitating diseases such as cancer, heart disease, and regenerative medicine. These advances are empowering scientists and clinicians, and in the process they are changing the lives of citizens in very profound and meaningful ways. While the science and engineering of nanobiosystems is one of the most exciting, challenging, and rapidly growing sectors of nanotechnology, society continues to be faced with daunting challenges in biomedicine and healthcare delivery. Significant challenges that need to be addressed over the next decade include:

- Identification of the molecular and cellular origin of disease processes such as cancer, cardiovascular disease, diabetes mellitus, aging, and tissue/organ regeneration
- Health monitoring and early detection of disease through the identification of novel biomarkers, diagnostic tool kits, and advanced imaging technologies
- Establishment of scientific bases for regenerative medicine, biocompatible prosthetics, and stem cell engineering
- Development of a quantitative understanding of how biological nanosystems and subcellular constituents work in synchrony to establish systems biology that can be interrogated at the nanoscale level
- Synthesis of new devices or systems that provide targeted and on-demand drug delivery that also provides an imaging component
- Fostering an understanding of the tenets of nanoscience in the biomedical community, with the potential to accelerate bench-to-bedside implementation of new technologies
- Realization of personalized medicine and point-of-care (POC) healthcare delivery that is rapid, efficient, and cost-effective

In order to address these challenges through translational advances in biomedical nanotechnology, significant investment is imperative in biomedical nanotechnology research, small businesses aiming to commercialize nanotechnologies, and science and math education at all grade levels. Furthermore, regulatory changes are needed with regard to more rapid evaluation of new technologies for early approval and safe delivery to patients by educated healthcare providers. Nanomedicine has raced forward at a staggering pace with significant investment and focused collaborative energy. As a result, some paradigm shifts are now becoming reality; these include approaching cancer from a holistic perspective based upon early diagnosis and treatment, as well as personalized medicine and cancer prevention. The overall impact of nanotechnology on biomedicine will be multifaceted and consist of advanced screening, diagnosis and staging, personalized treatment and monitoring, and follow-up. Rapid success depends upon focused collaborative energies, future investments, and wide spread education about the realities and promise of nanomedicine.

9.1.2 Part 2: Regulatory Aspects of Nanomedicine and Nanobiosystems Development

It is evident that regulatory issues have a large impact on the translation to the clinic, on the impact of nanotechnologies on health, and on the creation of companies. There is a need for improved EU-U.S. collaboration on issues important for regulatory bodies, such as reproducibility, good manufacturing processes, bio-distribution, toxicity, immunogenicity, etc. Scientists working in medical nanotechnologies should be involved with regulatory agencies at an early stage. A major objective is the creation of a streamlined, effective, regulatory pathway. A secondary objective is the integration of research funding approaches to underpin facilitated regulatory approval. A third objective is to address the funding gap "from science to market" in view of the regulatory issues.

9.2 United States-Japan-Korea-Taiwan Workshop (Tokyo/Tsukuba, Japan)

Panel members/discussants

Kazunori Kataoka (co-chair), University of Tokyo, Japan
Chad A. Mirkin (co-chair), Northwestern University, United States
Andre Nel, University of California, Los Angeles, United States
Yoshinobu Baba, Nagoya University, Japan
Chi-Hung Lin, National Yang-Ming University, Taiwan
Teruo Okano, Tokyo Women's University, Japan
Joon Won Park, Pohang University of Science and Technology, Korea
Yuji Miyahara, National Institute for Materials Science (NIMS), Japan

Applications: Nanobiosystems, Medicine, and Health

9.2.1 Changes in the Vision of Nanotechnology in the Last 10 Years

Nanotechnology has definitely accelerated integration of established scientific fields, thereby accelerating the breaking down of boundaries between classic disciplines. Eventually, this culminated in establishing the new interdisciplinary science of nanotechnology and its associated biology branch known as "nanobiotechnology."

The visions emerging from nanobiotechnology over the last 10 years have progressively focused on the establishment of nanomedicine, including drug delivery systems, regenerative medicine, molecular imaging, and diagnostic devices. Nanomedicine enables minimally invasive or noninvasive diagnosis and targeted therapy by breakthroughs in sensing, processing, and operating modalities, as shown in Fig. 26.

In addition, there is significant research training for blending the above capabilities into single-platform diagnostic and therapeutic nanobiodevices (nanotheranostics).

9.2.2 Vision of Nanobiotechnology for the Next 10 Years

Five specific topics were featured in the panel discussion to delineate visions of what may happen in nanobiotechnology over the next 10 years:

- Ultra-sensitive, highly specific, low-invasive, and reliable (robust) detection technologies for diagnosis, prognosis, and prevention
- Multifunctional nanobiodevices for the integration of diagnosis and therapy
- Accelerating the translation of nanobiotechnology-based methods and devices to points of clinical care
- Innovation in nanophysiology through intravital and nanoscaled observation of operations in living organisms through the use of multifunctional nanodevices

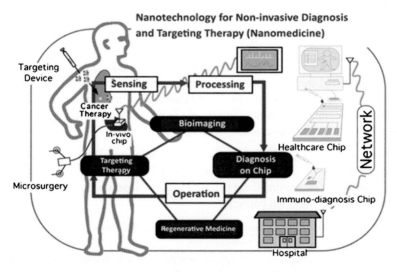

Fig. 26 Vision for nanotechnology in medicine (Courtesy of The University of Tokyo Center for NanoBio Integration, http://park.itc.u-tokyo.ac.jp/CNBI/e/index.html)

- Nanobiotechnology applications to ensure the implementation of environmentally friendly nanofabrication processes, and detection and monitoring of harmful chemicals and organisms

9.2.3 Main Scientific/Engineering Advancements and Technological Impacts of Nanobiotechnology in the Last 10 Years

The panel selected the following major advancements in the field of nanobiotechnology that have had high impact over the last 10 years:

- Development of nanocarrier systems for cancer chemotherapy premised on the EPR effect has yielded new cost efficient therapeutics with higher efficacy and lower side effects.
- Cell sheet engineering for tissue repair and regeneration has been implemented to treat cardiac failure and diseases of the cornea, with additional potential to be used for tissue repair after the surgery of esophageal epithelial cancer. In these three areas, cell sheet engineering therapy has already started to treat human patients.
- Ultrasensitive and highly specific biodetection systems by integrated bottom-up and top-down nanotechnologies led to significant impact in terms of protein biomarker detection and clinical diagnosis.

9.2.4 Goals for the Next 5–10 years: Opportunities, Barriers, and Solutions

The panel discussed the following seven goals in terms of opportunities, barriers, and solutions.

Goal 1: *Development of multifunctional nanobiodevices for single-platform bioimaging and targeting therapy (theranostic nanodevices).*

Barriers: A major challenge is in the development of nanocarriers with high extravasation potential through transcytosis. This may enable transport of probes and drugs through the blood–brain barrier, and thus opens the door to nanotechnology-enabled diagnosis and therapy of intractable brain diseases, including Alzheimer's disease and Parkinson's disease.

Solutions: Recent progress in phage display technology has highlighted the potential to use a variety of peptide sequences to facilitate transcytosis in endothelial cells. Coupling or combining these peptides to nanocarriers is a promising approach to crossing the blood–brain barrier.

Goal 2: *Development of ultrasensitive, highly specific, low-invasive, and reliable (robust) multiplexed detection technologies.*

This will enable diagnosis through the ability to detect disease processes at their inception, e.g., a single diseased cell or pathophysiology at the molecular level. This methodology will allow early cancer detection.

Barriers: Development of systems with the sensitivity of PCR techniques but without their complexity. The system should detect a variety of analytes, including nucleic acids, proteins, small molecules, metal ions, and should be capable of multiplexing and amenable to point-of-care use.

Solutions: Bar-code technology, lab-on-a-chip diagnostics. This requires close collaboration between the medical and nanotechnology communities. It will be important to demonstrate the utility of this approach as a key breakthrough in the major disease target to demonstrate the clinical utility of smart nanosensing devices.

Goal 3: *Remote disease monitoring through the ability of nanotechnology to provide online sensing and information relay.*

Barriers: Relative shortage of healthcare personnel and POC delivery in underserved and rural communities. Safety and regulation issues.

Solutions: Nanosensor devices with portability. Establish regulatory policy and educate people in government regulatory agencies.

Goal 4: *Construction of tissue growth facilitating structured cell sheets for organ repair and replacement.*

Barriers: Lack of organ donors and the availability of off-the-shelf tissue and organ resources.

Solutions: Use of nano-architectures, synthetic morphogens, and extracellular matrix mimics to induce cellular growth and reconstitution of tissues and organs. Preservation technology.

Goal 5: *Application of biological nanostructures (i.e., nucleic acids, proteins) in bioelectronics and environmentally friendly nanofabrication processes.*

Barriers: Creating viable strategies for fabricating and integrating.

Solutions: Research on interfacing biological structures with inorganic materials. Creating design rules for system-wide construction.

Goal 6: *Single-cell interventions and diagnostics, including stem cell growth and differentiation and cellular-level genomics and proteomics.*

Barriers: Sensitivity, transfection, and limited examples requiring this capability.

Solutions: Develop nano-bio tools with high sensitivity and more examples and applications.

Goal 7: *Stem cell differentiation and site-specific delivery by devices that allow protective delivery and can serve as growth templates that include nano-architectured surfaces.*

Barriers: Understanding chemical and physical stimulation to cell differentiation. Understanding signal pathways.

Solutions: Nanostructured interfaces, 3D nano-architectures, and synthetic morphogens.

9.2.5 R&D Investment and Implementation Strategies

Implementation of a strategy that facilitates multidisciplinary research through the creation of open innovation centers for "nano-bio" equipped with cutting-edge instruments and animal facilities. Particularly, R&D investment to establish nanomedicine centers in medical schools and hospitals with the ability to bring nanotechnology-based therapeutics, diagnostics, devices, and systems to the clinic.

9.2.6 Emerging Topics and Priorities for Future Nanoscale Science Research and Education

The panel reached the consensus that priority should be given to research for the understanding of "life" at the single molecular and cellular level through nanotechnology. This technology should also be applied to *in vivo* nanoscale biodevices that can perform smart functions through the combination of molecular self-assembly and nanofabrication processes. These nanodevices will also be helpful to establish the new scientific fields of "nano-physiology" and nano-systems biology that can be used for theranostics approaches in nanomedicine.

9.2.7 The Impact of Nanotechnology R&D on Society

Significant progress in nanomedicine could allow us to cure cancers at an early stage and therefore enhance human longevity. The combination of nanomedicine and information technology enables diagnosis and therapy with high quality of life regardless of age, wealth, country, and occupation, contributing to continuous development in human welfare. Furthermore, understanding the behavior of individual cells deepens our understanding of life on earth, with the potential to induce a paradigm shift in the entire area associated with the biosphere, including health science and medical practice.

9.3 United States-Australia-China-India-Saudi Arabia-Singapore Workshop (Singapore)

Panel members/discussants

Yi Yan Yang (co-chair), Institute of Bioengineering and Nanotechnology, Singapore
Chad Mirkin (co-chair), Northwestern University, United States

The panel made the following main points in answering the questions related to health and nanobiotechnology in a broad context. They focused primarily on the future.

Applications: Nanobiosystems, Medicine, and Health

9.3.1 How has the Vision of Nanotechnology Changed in the Last 10 Years?

- Transition from hype to realization (diagnostics, *in vivo* imaging, and therapeutics) – FDA-cleared systems and first human nanodrug trials
- Realization that nanotherapeutics will be a new drug-delivery platform
- Transition from individual components to more sophisticated multifunctional systems

9.3.2 Where is the Field Likely to Go?

- Imaging live systems with molecular resolution on the time scale of chemical reactions and biological processes
- Noninvasive pathology
- Biological drug (peptides, proteins, nucleic acids) and cell delivery systems
- Overcoming and preventing drug resistance
- Preventing, reversing, or eliminating metastatic cancer
- Atomic resolution of integral membrane proteins with a view to identifying drug targets.
- Imbedded and pervasive, true point-of-care molecular diagnostic systems.

9.3.3 What are the Main Scientific/Engineering Advances and Technological Impacts in the Last Decade?

- FDA cleared nano diagnostic systems
- First human trials with siRNA nanodelivery systems
- Order of magnitude improved resolution in *in vivo* imaging enabled by better tools assisted by nanoparticle contrast agents.

9.3.4 What are the Goals for the Next 5–10 Years: Opportunities, Barriers, Solutions?

- Imaging live systems with molecular resolution on the time scale of chemical reactions and biological processes
    - *Barrier*: Resolution, background, and timescale
    - *Solution*: Financial resource; understanding of contributions to background; probe sensitivity, integration of imaging modalities; overcoming computation and modeling deficiencies.
- Noninvasive Pathology
    - *Barrier*: Lack of tools and insufficient biomarker data base
    - *Solution*: Suite of *in vitro* molecular diagnostic and *in vivo* imaging tools, combined with an aggressive approach to identifying and validating molecular and structural markers for specific pathology and pathophysiology. New nanoscale

contrast agents with enhanced sensitivity and specificity for disease states. Resolving sampling statistical issues.

- Biological drug (peptides, proteins, nucleic acids) and cell delivery systems
  - *Barrier*: Drug and cell instability, environment compatibility, effective targeting schemes that overcome drug resistance.
  - *Solution*: Nanomaterials that encapsulate, stabilize, deliver, and release therapeutic agents at the point of disease.
- Overcoming and preventing drug resistance
  - *Barrier*: Drug and cell instability, environment incompatibility.
  - Solution: Nanoscale dual drug/gene regulation systems; stealth delivery and entry that bypasses drug resistant mechanisms. Contemporaneous multidrug and adjuvant delivery.
- Preventing, reversing, or eliminating metastatic cancer
  - *Barrier*: Lack of understanding and the inability to effectively intervene in metastatic pathways. Surreptitious nature of metastatic disease.
  - *Solution*: Delivery of biological signal that stops the budding of metastatic cancer cells; nanoparticle delivery systems; encapsulation metastatic cells
- Atomic resolution of integral membrane proteins with a view to identifying drug targets.
  - *Barrier*: Lack of understanding of nucleation and crystallization phenomena of integral membrane proteins. Tracking protein conformational changes in a membrane environment.
  - *Solution*: Better EM and related high resolution imaging methods; amphiphilic nanostructures that mimic membrane structure and dynamics.
- Embedded and pervasive, true point-of-care molecular diagnostic systems.
  - *Barrier*: Increasing sensitivity and specificity; Molecular structural and functional imaging; Molecular monitoring of response to therapy; Non-invasive imaging and minimally invasive imaging; Automation (sample prep, imaging and identification); Molecular single cell imaging and tomography; Single cell molecule extraction; Multi-analyte extraction (DNA, RNA, Protein, metabolites) from same few cells
  - *Solution*: Multiplexed nano-enabled multifunctional platforms

9.3.5 What Is the Impact of Nanotechnology R&D on Society?

- Transformative
- Economy building
- Improves quality of life
- Addresses global challenges such as personalized medicine, food security, climate change, water security, and sustainability.

References

1. A.E. Nel, L. Madler, D. Velegol, T. Xia, E.M.V. Hoek, P. Somasundaran, F. Klaessig, V. Castranova, M. Thompson, Understanding biophysicochemical interactions at the nano-bio interface. Nat. Mater. **8**(7), 543–557 (2009). doi:10.1038/nmat2442
2. European Science Foundation (ESF), *Nanomedicine: An ESF – European Medical Research Council Forward Look Report* (ESF, Strasbourg, France, 2005), Available online: http://www.esf.org/publications/forward-looks.html
3. A. Nel, T. Xia, L. Madler, N. Li, Toxic potential of materials at the nanolevel. Science **311**(5761), 622–627 (2006). doi:10.1126/science.1114397
4. P. Grodzinski, M. Silver, L.K. Molnar, Nanotechnology for cancer diagnostics: promises and challenges. Expert Rev. Mol. Diagn. **6**(3), 307–318 (2006)
5. F. Alexis, E.M. Pridgen, R. Langer, O.C. Farokhzad, Nanoparticle technologies for cancer therapy. Handb. Exp. Pharmacol. **197**, 55–86 (2010)
6. S.X. Tang, J. Zhao, J.J. Storhoff, P.J. Norris, R.F. Little, R. Yarchoan, S.L. Stramer, T. Patno, M. Domanus, A. Dhar, C. Mirkin, I.K. Hewlett, Nanoparticle-based biobarcode amplification assay (BCA) for sensitive and early detection of human immunodeficiency type 1 capsid (p24) antigen. J. Acquir. Immune Defic. Syndr. **46**(2), 231–237 (2007). doi:10.1097/QAI.0b013e31814a554b
7. M.E. Davis, J.E. Zuckerman, C.H. Choi, D. Seligson, A. Tolcher, C.A. Alabi, Y. Yen, J.D. Heidel, A. Ribas, Evidence of RNAi in humans from systemically administered siRNA via targeted nanoparticles. Nature **464**(7291), 1067–1070 (2010). doi:10.1038/nature08956
8. A.E. Prigodich, D.S. Seferos, M.D. Massich, D.A. Giljohann, B.C. Lane, C.A. Mirkin, Nano-flares for mRNA regulation and detection. ACS Nano **3**(8), 2147–2152 (2009). doi:10.1021/nn9003814
9. D.S. Seferos, D.A. Giljohann, H.D. Hill, A.E. Prigodich, C.A. Mirkin, Nano-flares: probes for transfection and mRNA detection in living cells. J. Am. Chem. Soc. **129**(50), 15477–15479 (2007). doi:10.1021/ja0776529
10. D. Zheng, D.S. Seferos, D.A. Giljohann, P.C. Patel, C.A. Mirkin, Aptamer nano-flares for molecular detection in living cells. Nano Lett. **9**(9), 3258–3261 (2009). doi:10.1021/nl901517b
11. A. De la Zerda, C. Zavaleta, S. Keren, S. Vaithilingham, S. Bodapati, Z. liu, J. Levi, B.R. Smith, T. Ma, O. Oralkan, Z. Cheng, X. Chen, H. Dai, B.T. Kuri-Yakub, S.S. Gambhir, Carbon nanotubes as photoacoustic molecular imaging agents in living mice. Nat. Nanotechnol. **3**(9), 557–562 (2008). doi:10.1038/nnano.2008.231
12. S. Keren, C. Zavaleta, Z. Cheng, A. de la Zerda, O. Gheysens, S.S. Gambhir, Noninvasive molecular imaging of small living subjects using Raman spectroscopy. Proc. Natl. Acad. Sci. U. S. A. **105**(15), 5844–5849 (2008). doi:10.1073/pnas.0710575105
13. J.L. Major, T.J. Meade, Bioresponsive, cell-penetrating, and multimeric MR contrast agents. Acc. Chem. Res. **42**(7), 893–903 (2009). doi:10.1021/ar800245h
14. Y. Song, X. Xu, K.W. MacRenaris, X.Q. Zhang, C.A. Mirkin, T.J. Meade, Multimodal gadolinium-enriched DNA-gold nanoparticle conjugates for cellular imaging. Angew. Chem. Int. Ed Engl. **48**(48), 9143–9147 (2009)
15. C. Zavaleta, A. de la Zerda, Z. Liu, S. Keren, Z. Cheng, M. Schipper, X. Chen, H. Dai, S.S. Gambhir, Noninvasive Raman spectroscopy in living mice for evaluation of tumor targeting with carbon nanotubes. Nano Lett. **8**(9), 2800–805 (2008). Available online: http://www.adelazerda.com/NanoLetters_08.pdf
16. S. George, S. Pokhrel, T. Xia, B. Gilbert, Z. Ji, M. Schowalter, A. Rosenauer, R. Damoiseaux, K.A. Bradley, L. Madler, A.E. Nel, Use of a rapid cytotoxicity screening approach to engineer a safer zinc oxide nanoparticle through iron doping. ACS Nano **4**(1), 15–29 (2010)
17. M. Liong, J. Lu, M. Kovochich, T. Xia, S.G. Ruehm, A.E. Nel, F. Tamanoi, J.I. Zink, Multifunctional inorganic nanoparticles for imaging, targeting, and drug delivery. ACS Nano **2**(5), 889–896 (2008). doi:10.1021/nn800072t

18. J. Lu, M. Liong, S. Sherman, T. Xia, M. Kovochich, A. Nel, J. Zink, F. Tamanoi, Mesoporous silica nanoparticles for cancer therapy: energy-dependent cellular uptake and delivery of paclitaxel to cancer cells. Nanobiotechnology **3**(3), 89–95 (2007). doi:10.1007/s12030-008-9003-3
19. J. Lu, M. Liong, Z. Li, J.I. Zink, F. Tamanoi, Biocompatability, biodistribution, and drug-delivery efficiency of mesoporous silica nanoparticles for cancer therapy in animals. Small **6**(16), 1794–1805 (2010)
20. H. Meng, M. Liong, T. Xia, Z. Li, Z. Ji, J.I. Link, A.E. Nel, Engineered design of mesoporous silica nanoparticles to deliver doxorubicin and Pgp siRNA to overcome drug resistance in a cancer cell line. ACS Nano **4**(8), 4539–4550 (2010). doi:10.1021/nn100690m
21. H. Meng, M. Xie, T. Xia, Y. Zhao, F. Tamanoi, J.F. Stoddart, J.I. Zink, A. Nel, Autonomous *in vitro* anticancer drug release from mesoporous silica nanoparticles by pH-sensitive nanovalves. J. Am. Chem. Soc. **132**(36), 12690–12697 (2010). doi:10.1021/ja104501a
22. T. Xia, M. Kovochich, M. Liong, H. Meng, S. Kabahie, S. george, J.I. Zink, A. Nel, Polyethyleneimine coating enhances the cellular uptake of mesoporous silica nanoparticles and allows safe delivery of siRNA and DNA constructs. ACS Nano **3**(10), 3273–3286 (2009)
23. D.G. Georganopoulou, L. Chang, J.W. Nam, C.S. Thaxton, E.J. Mufson, W.L. Klein, C.A. Mirkin, Nanoparticle-based detection in cerebral spinal fluid of a soluble pathogenic biomarker for Alzheimer's disease. Proc. Natl. Acad. Sci. U. S. A. **102**(7), 2273–2276 (2004). doi:10.1073/pnas.0409336102
24. C.A. Mirkin, C.S. Thaxton, N.L. Rosi, Nanostructures in biodefense and molecular diagnostics. Expert Rev. Mol. Diagn. **4**(6), 749–751 (2004)
25. J.M. Nam, S.I. Stoeva, C.A. Mirkin, Bio-bar-code-based DNA detection with PCR-like sensitivity. J. Am. Chem. Soc. **126**(19), 5932–5933 (2004). doi:10.1021/ja049384+
26. J.M. Nam, C.S. Thaxton, C.A. Mirkin, Nanoparticle-based bio-bar codes for the ultrasensitive detection of proteins. Science **301**(5641), 1884–1886 (2003). doi:10.1126/science.1088755
27. S.I. Stoeva, J.S. Lee, C.S. Thaxton, C.A. Mirkin, Multiplexed DNA detection with bio-barcoded nanoparticle probes. Angew. Chem. Int. Ed Engl. **45**(20), 3303–3306 (2006). doi:10.1002/anie.200600124
28. C.S. Thaxton, H.D. Hill, D.G. Georganopoulou, S.I. Stoeva, C.A. Mirkin, A bio-bar-code assay based upon dithiothreitol-induced oligonucleotide release. Anal. Chem. **77**(24), 8174–8178 (2005)
29. C.S. Thaxton, N.L. Rosi, C.A. Mirkin, Optically and chemically encoded nanoparticle materials for DNA and protein detection. MRS Bull. **30**(5), 376–380 (2005)
30. D. Kim, W.L. Daniel, C.A. Mirkin, Microarray-based multiplexed scanometric immunoassay for protein cancer markers using gold nanoparticle probes. Anal. Chem. **81**(21), 9183–9187 (2009). doi:10.1021/ac9018389
31. T.A. Taton, C.A. Mirkin, R.L. Letsinger, Scanometric DNA array detection with nanoparticle probes. Science **289**(5485), 1757–1760 (2000). doi:10.1126/science.289.5485.1757
32. C.R. Pound, A.W. Partin, M.A. Eisenberger, D.W. Chan, J.D. Pearson, P.C. Walsh, Natural history of progression after PSA elevation following radical prostatectomy. JAMA **281**(17), 1591–1597 (1999)
33. G.L. Andriole, R.L. Grubb III, S.S. Buys, D. Chia, T.R. Church, M.N. Fouad, E.P. Gelmann, P.A. Kvale, D.J. Reding, J.L. Weissfeld, L.A. Yokochi, E.D. Crawford, B. O'Brien, J.D. Clapp, J.M. Rathmell, T.L. Riley, R.B. Hayes, B.S. Kramer, G. Izmirlian, A.B. Miller, P.F. Pinsky, P.C. Prorok, J.K. Gohagan, C.D. Berg, Mortality results from a randomized prostate-cancer screening trial. N. Engl. J. Med. **360**(13), 1310–1319 (2009)
34. F.H. Schroder, J. Hugosson, M.J. Roobol, T.L.J. Tammela, S. Ciatto, V. Nelen, M. Kwiatkowski, M. Lujan, H. Lilja, M. Zappa, L.J. Denis, F. Recker, A. Berenguer, L. Määttänen, C.H. Bangma, G. Aus, A. Villers, X. Rebillard, T. van der Kwast, B.G. Blijenberg, S.M. Moss, H.J. de Koning, A. Auvinen, ERSPC Investigators, Screening and prostate-cancer mortality in a randomized European study. N. Engl. J. Med. **360**(13), 1320–1328 (2009)
35. W.J. Catalona, D.S. Smith, 5-year tumor recurrence rates after anatomical radical retropubic prostatectomy for prostate cancer. J. Urol. **152**(5), 1837–1842 (1994)

36. T.L. Jang, M. Han, K.A. Roehl, S.A. Hawkins, W.J. Catalona, More favorable tumor features and progression-free survival rates in a longitudinal prostate cancer screening study: PSA era and threshold-specific effects. Urology **67**(2), 343–348 (2006). doi:10.1016/j.urology.2005.08.048
37. J.G. Trapasso, J.B. deKernion, R.B. Smith, F. Dorey, The incidence and significance of detectable levels of serum prostate specific antigen after radical prostatectomy. J. Urol. **152**(5), 1821–1825 (1994)
38. B.J. Trock, M. Han, S.J. Freedland, E.B. Humphreys, T.L. DeWeese, A.W. Partin, P.C. Walsh, Prostate cancer-specific survival following salvage radiotherapy vs observation in men with biochemical recurrence after radical prostatectomy. JAMA **299**(23), 2760–2769 (2008)
39. H. Yu, E.P. Diamandis, A.F. Prestigiacomo, T.A. Stamey, Ultrasensitive assay of prostate-specific antigen used for early detection of prostate cancer relapse and estimation of tumor-doubling time after radical prostatectomy. Clin. Chem. **41**(3), 430–434 (1995)
40. C.S. Thaxton, R. Elghanian, A.D. Thomas, S.I. Stoeva, J.S. Lee, N.D. Smith, A.J. Schaeffer, H. Klocker, W. Horninger, G. Bartsch, C.A. Mirkin, Nanoparticle-based bio-barcode assay redefines "undetectable" PSA and biochemical recurrence after radical prostatectomy. Proc. Natl. Acad. Sci. U. S. A. **106**(44), 18437–18442 (2009). doi:10.1073/pnas.0904719106
41. A.K. Lytton-Jean, C.A. Mirkin, A thermodynamic investigation into the binding properties of DNA functionalized gold nanoparticle probes and molecular fluorophore probes. J. Am. Chem. Soc. **127**(37), 12754–12755 (2005)
42. R. Elghanian, J.J. Storhoff, R.C. Mucic, R.L. Letsinger, C.A. Mirkin, Selective colorimetric detection of polynucleotides based on the distance-dependent optical properties of gold nanoparticles. Science **277**(5329), 1078–1081 (1997). doi:10.1126/science.277.5329.1078
43. J.J. Storhoff, A.D. Lucas, V. Garimella, Y.P. Bao, U.R. Muller, Homogeneous detection of unamplified genomic DNA sequences based on colorimetric scatter of gold nanoparticle probes. Nat. Biotechnol. **22**(7), 883–887 (2004)
44. J.J. Storhoff, S.S. Marla, P. Bao, S. Hagenow, H. Mehta, A. Lucas, V. Garimella, T. Patno, W. Buckingham, W. Cork, U.R. Muller, Gold nanoparticle-based detection of genomic DNA targets on microarrays using a novel optical detection system. Biosens. Bioelectron. **19**(8), 875–883 (2004). doi:10.1016/j.bios.2003.08.014
45. D.A. Giljohann, D.S. Seferos, A.E. Prigodich, P.C. Patel, C.A. Mirkin, Gene regulation with polyvalent siRNA-nanoparticle conjugates. J. Am. Chem. Soc. **131**(6), 2072–2073 (2009). doi:10.1021/ja808719p
46. N.L. Rosi, D.A. Giljohann, C.S. Thaxton, A.K.R. Lytton-Jean, M.S. Han, C.A. Mirkin, Oligonucleotide-modified gold nanoparticles for intracellular gene regulation. Science **312**(5776), 1027–1030 (2006). doi:10.1126/science.1125559
47. D.S. Seferos, A.E. Prigodich, D.A. Giljohann, P.C. Patel, C.A. Mirkin, Polyvalent DNA nanoparticle conjugates stabilize nucleic acids. Nano Lett. **9**(1), 308–311 (2009). doi:10.1021/nl802958f
48. D.S. Seferos, D.A. Giljohann, N.L. Rosi, C.A. Mirkin, Locked nucleic acid-nanoparticle conjugates. Chem. Biochem **8**(11), 1230–1232 (2007). doi:10.1002/cbic.200700262
49. N. Nerambourg, R. Praho, M.H.V. Werts, D. Thomas, M. Blanchard-Desce, Hydrophilic monolayer-protected gold nanoparticles and their functionalisation with fluorescent chromophores. Int. J. Nanotechnol. **5**(6–8), 722–740 (2008). doi:10.1504/IJNT.2008.018693
50. T. Meade, Seeing is believing. Acad. Radiol. **8**(1), 1–3 (2001)
51. D. Neuberger, J. Wong, Suspension for intravenous injection: image analysis of scanning electron micrographs of particles to determine size and volume. PDA J. Pharm. Sci. Technol. **59**(3), 187–199 (2005)
52. L.M. Manus, D.J. Mastarone, E.A. Waters, X.-Q. Zhang, E.A. Schultz-Sikma, K.W. MacRenaris, D. Ho, T.J. Meade, Gd(III)-nanodiamond conjugates for MRI contrast enhancement. Nano Lett. **10**(2), 484–489 (2010). doi:10.1021/nl903264h
53. C.L. Zavaleta, B.R. Smith, I. Walton, W. Doering, G. Davis, B. Shojaei, M.J. Natan, S.S. Gambhir, Multiplexed imaging of surface enhanced Raman scattering nanotags in living mice using noninvasive Raman spectroscopy. Proc. Natl. Acad. Sci. U. S. A. **106**(32), 13511–13516 (2009). doi:10.1073/pnas.0813327106

54. S.M. van de Ven, N. Mincu, J. Brunette, G. Ma, M. Khayat, D.M. Ikeda, S.S. Gambhir, Molecular imaging using light-absorbing imaging agents and a clinical optical breast imaging system–a phantom study. Department of Radiology, Stanford University Medical Center, Stanford, CA, USA. Mol Imaging Biol. Apr;**13**(2):232–238 (2011)
55. M.E. Davis, The first targeted delivery of siRNA in humans via a self-assembling, cyclodextrin polymer-based nanoparticle: from concept to clinic. Mol. Pharm. **6**(3), 659–668 (2009). doi:10.1021/mp900015y
56. D.W. Bartlett, M.E. Davis, Physicochemical and biological characterization of targeted, nucleic acid-containing nanoparticles. Bioconjug. Chem. **18**(2), 456–468 (2007). doi:10.1021/bc0603539
57. S.H. Pun, N.C. Bellocq, A. Liu, G. Jensen, T. Machemer, E. Quijano, T. Schluep, S. Wen, H. Engler, J. Heidel, M.E. Davis, Cyclodextrin-modified polyethylenimine polymers for gene delivery. Bioconjug. Chem. **15**(4), 831–840 (2004). doi:10.1021/bc049891g
58. D.J. Heidel, J.D. Heidel, J. Yi-Ching Liu, Y. Yen, B. Zhou, B.S.E. Heale, J.J. Rossi, D.W. Bartlett, M.E. Davis, Potent siRNA inhibitors of ribonucleotide reductase subunit RRM2 reduce cell proliferation *in vitro* and *in vivo*. Clin. Cancer Res. **13**(7), 2207–2215 (2007). doi:10.1158/1078-0432.CCR-06-2218
59. D.J. Heidel, Z. Yu, J. Yi-Ching Liu, S.M. Rele, Y. Liang, R.K. Zeidan, D.J. Kornbrust, M.E. Davis, Administration in non-human primates of escalating intravenous doses of targeted nanoparticles containing ribonucleotide reductase subunit M2 siRNA. Proc. Natl. Acad. Sci. U. S. A. **104**(14), 5715–5721 (2007). doi:10.1073/pnas.0701458104
60. S. Hu-Lieskovan, J.D. Heidel, D.W. Bartlett, M.W. Davis, T.J. Triche, Sequence-specific knockdown of EWS-FLI1 by targeted, nonviral delivery of small interfering RNA inhibits tumor growth in a murine model of metastatic Ewing's sarcoma. Cancer Res. **65**(19), 8984–8992 (2005). doi:10.1158/0008-5472.CAN-05-0565
61. D.W. Bartlett, H. Su, I.J. Hildebrandt, W.A. Weber, M.E. Davis, Impact of tumor-specific targeting on the biodistribution and efficacy of siRNA nanoparticles measured by multimodality *in vivo* imaging. Proc. Natl. Acad. Sci. U. S. A. **104**(39), 15549–15554 (2007). doi:10.1073/pnas.0707461104
62. D.W. Bartlett, M.E. Davis, Impact of tumor-specific targeting and dosing schedule on tumor growth inhibition after intravenous administration of siRNA-containing nanoparticles. Biotechnol. Bioeng. **99**(4), 975–85 (2008). doi:10.1002/bit.21668
63. L.M. Demers, C.A. Mirkin, R.C. Mucic, R.A. Reynolds III, R.L. Letsinger, R. Elghanian, G. Viswanadham, A fluorescence-based method for determining the surface coverage and hybridization efficiency of thiol-capped oligonucleotides bound to gold thin films and nanoparticles. Anal. Chem. **72**(22), 5535–5541 (2000)
64. D.A. Giljohann, D.S. Seferos, P.C. Patel, J.E. Millstone, N.L. Rosi, C.A. Mirkin, Oligonucleotide loading determines cellular uptake of DNA-modified gold nanoparticles. Nano Lett. **7**(12), 3818–3821 (2007)
65. C.A. Mirkin, R.L. Letsinger, R.C. Mucic, J.J. Storhoff, A DNA-based method for rationally assembling nanoparticles into macroscopic materials. Nature **382**(6592), 607–609 (1996). doi:10.1038/382607a0
66. I. Lebedeva, C.A. Stein, Antisense oligonucleotides: promise and reality. Annu. Rev. Pharmacol. Toxicol. **41**, 403–419 (2001)
67. G.S. Getz, C.A. Reardon, Nutrition and cardiovascular disease. Arterioscler. Thromb. Vasc. Biol. **27**(12), 2499–2506 (2007)
68. R. Josi, S. Jan, Y. Wu, S. MacMahon, Global inequalities in access to cardiovascular health care. J. Am. Coll. Cardiol. **52**(23), 1817–1825 (2008)
69. A.J. Lusis, Atherosclerosis. Nature **407**(6801), 233–241 (2000)
70. L.G. Spagnoli, E. Bonanno, G. Sangiorgi, A. Mauriello, Role of inflammation in atherosclerosis. J. Nucl. Med. **48**(11), 1800–1815 (2007). doi:10.2967/jnumed.107.038661
71. W.B. Kannel, W.P. Castelli, T. Gordon, P.M. McNamara, Serum cholesterol, lipoproteins, and the risk of coronary heart disease: the Framingham study. Ann. Intern. Med. **74**(1), 1–12 (1971)
72. W.B. Kannel, P.W.F. Wilson, An update on coronary risk factors. Med. Clin. North Am. **79**(5), 951–971 (1995)

73. T.C. Andrews, K. Raby, J. Barry, C.L. Naimi, E. Allred, P. Ganz, A.P. Selwyn, Effect of cholesterol reduction on myocardial ischemia in patients with coronary disease. Circulation **95**(2), 324–328 (1997)
74. C.M. Ballantyne, J.A. Herd, J.K. Dunn, P.H. Jones, J.A. Farmer, A.M. Gotto Jr., Effects of lipid lowering therapy on progression of coronary and carotid artery disease. Curr. Opin. Lipidol. **8**(6), 354–361 (1997)
75. A. Zambon, J.E. Hokanson, Lipoprotein classes and coronary disease regression. Curr. Opin. Lipidol. **9**(4), 329–336 (1998)
76. H.B. Brewer, Increasing HDL cholesterol levels. N. Engl. J. Med. **350**(15), 1491–1494 (2004)
77. E.M. Degoma, R.L. Degoma, D.J. Rader, Beyond high-density lipoprotein cholesterol levels: evaluating high-density lipoprotein function as influenced by novel therapeutic approaches. J. Am. Coll. Cardiol. **51**(23), 2199–2211 (2008)
78. T. Joy, R.A. Hegele, Is raising HDL a futile strategy for atheroprotection? Nat. Rev. Drug Discovery **7**(2), 143–155 (2008)
79. P. Conca, G. Franceschini, Synthetic HDL as a new treatment for atherosclerosis regression: has the time come? Nutr. Metab. Cardiovasc. Dis. **18**(4), 329–335 (2008)
80. A. Kontush, M.J. Chapman, Antiatherogenic small, dense HDL – guardian angel of the arterial wall? Nat. Clin. Pract. Cardiovasc. Med. **3**(3), 144–153 (2006). doi:10.1038/ncpcardio0500
81. I.M. Singh, M.H. Shishehbor, B.J. Ansell, High-density lipoprotein as a therapeutic target – A systematic review. JAMA **298**(7), 786–798 (2007)
82. G.F. Watts, P.H.R. Barrett, D.C. Chan, HDL metabolism in context: looking on the bright side. Curr. Opin. Lipidol. **19**(4), 395–404 (2008)
83. C.S. Thaxton, W.L. Daniel, D.A. Giljohann, A.D. Thomas, C.A. Mirkin, Templated spherical high density lipoprotein nanoparticles. J. Am. Chem. Soc. **131**(4), 1384–1385 (2009). doi:10.1021/ja808856z
84. Y. Klichko, M. Liong, E. Choi, S. Angelos, A.E. Nel, J.F. Stoddart, F. Tamanoi, J.I. Zink, Mesostructured silica for optical functionality, nanomachines, and drug delivery. J. Am. Ceram. Soc. **92**(s1), s2–s10 (2009). doi:10.1111/j.1551-2916.2008.02722.x
85. S. Saha, E. Johansson, A.H. Flood, H.-R. Tseng, J.I. Zink, J.F. Stoddart, A photoactive molecular triad as a nanoscale power supply for a supramolecular machine. Chemistry **11**(23), 6846–6858 (2005). doi:10.1002/ chem.200500371
86. S. Angelos, Y.W. Yang, K. Patel, J.F. Stoddart, J.I. Zink, pH-responsive supramolecular nanovalves based on cucurbit[6]uril pseudorotaxanes. Angew. Chem. Weinheim. Bergstr. Ger. **47**(12), 2222–2226 (2008). doi:10.1002/anie.200705211
87. N.M. Khashab, M.E. Belowich, A. Trabolsi, D.C. Friedman, C. Valente, Y. Lau, H.A. Khatib, J.I. Zink, J.F. Stoddart, pH-responsive mechanised nanoparticles gated by semirotaxanes. Chem. Commun. Camb. **36**, 5371–5373 (2009). doi:10.1039/B910431C
88. K. Patel, S. Angelos, W.R. Dichtel, A. Coskun, Y.-W. Yang, J.I. Zink, J.F. Stoddart, Enzyme-responsive snap-top covered silica nanocontainers. J. Am. Chem. Soc. **130**(8), 2382–2383 (2008). doi:10.1021/ja0772086
89. S.E. Gratton, S.S. Williams, M.E. Napier, P.D. Pohlhaus, Z. Zhou, K.B. Wiles, B.W. Maynor, C. Shen, T. Olafsen, E.T. Samulski, J.M. DeSimone, he pursuit of a scalable nanofabrication platform for use in material and life science applications. Acc. Chem. Res. **41**(12), 1685–1695 (2008). doi:10.1021/ar8000348
90. L.E. Euluss, J.A. DuPont, S. Gratton, J. DeSimone, Imparting size, shape, and composition control of materials for nanomedicine. Chem. Soc. Rev. **35**(11), 1095–1104 (2006). doi:10.1039/B600913C
91. R.A. Petros, P.A. Ropp, J.M. DeSimone, Reductively labile PRINT particles for the delivery of doxorubicin to HeLa cells. J. Am. Chem. Soc. **130**(15), 5008–5009 (2008)
92. I. Elloumi-Hannachi, M. Yamato, T. Okano, Cell sheet engineering: a unique nanotechnology for scaffold-free tissue reconstruction with clinical applications in regenerative medicine. J. Intern. Med. **267**(1), 54–70 (2010)
93. T. Shimizu, H. Sekine, M. Yamato, T. Okano, Cell sheet-based myocardial tissue engineering: new hope for damaged heart rescue. Curr. Pharm. Des. **15**(24), 2807–2814 (2009)

94. S. Masuda, T. Shimizu, M. Yamato, T. Okano, Cell sheet engineering for heart tissue repair. Adv. Drug Deliv. Rev. **60**, 277–285 (2008)
95. C.S. Chen, M. Mrksich, S. Huang, G.M. Whitesides, D.E. Ingber, Micropatterned surfaces for control of cell shape, position, and function. Biotechnol. Prog. **14**(3), 356–363 (1998). doi:10.1021/bp980031m
96. J. James, E.D. Goluch, H. Hu, C. Liu, M. Mrksich, ubcellular curvature at the perimeter of micropatterned cells influences lamellipodial distribution and cell polarity. Cell Motil. Cytoskeleton **65**(11), 841–852 (2008). Available online: http://www.mech.northwestern.edu/medx/web/publications/papers/196.pdf
97. M. Mrksich, C.S. Chen, Y. Xia, L.E. Dike, D.E. Ingber, G.M. Whitesides, Controlling cell attachment on contoured surfaces with self-assembled monolayers of alkanethiolates on gold. Proc. Natl. Acad. Sci. U. S. A. **93**(20), 10775–10778 (1996)
98. M. Mrksich, L.E. Dike, J. Tien, D.E. Ingber, G.M. Whitesides, Using microcontact printing to pattern the attachment of mammalian cells to self-assembled monolayers of alkanethiolates on transparent films of gold and silver. Exp. Cell Res. **235**(2), 305–313 (1997)
99. M. Mrksich, G.M. Whitesides, Using self-assembled monolayers to understand the interactions of man-made surfaces with proteins and cells. Annu. Rev. Biophys. Biomol. Struct. **25**, 55–78 (1996)
100. S. Heydarkhan-Hagvall, C.H. Choi, J. Dunn, S. Heydarkhan, K. Schenke-Layland, W.R. MacLellan, R.E. Beygui, Influence of systematically varied nano-scale topography on cell morphology and adhesion. Cell Commun. Adhes. **14**(5), 181–194 (2007)
101. J.H. Silver, J.C. Lin, F. Lim, V.A. Tegoulia, M.K. Chaudhury, S.L. Cooper, Surface properties and hemocompatibility of alkyl-siloxane monolayers supported on silicone rubber: effect of alkyl chain length and ionic functionality. Biomaterials **20**(17), 1533–1543 (1999). doi:10.1016/S0142-9612(98)00173-2
102. A.J. Torres, L. Vasudevan, D. Holowka, B.A. Baird, Focal adhesion proteins connect IgE receptors to the cytoskeleton as revealed by micropatterned ligand arrays. Proc. Natl. Acad. Sci. U. S. A. **105**(45), 17238–17244 (2008). doi:10.1073/pnas.0802138105
103. D.W. Hobson, Commercialization of nanotechnology Wiley Interdiscip. Rev. Nanomed. Nanobiotechnol. **1**(2), 189–202 (2009). doi:10.1002/wnan.28
104. J. Shendure, R.D. Mitra, C. Varma, G.M. Church, Advanced sequencing technologies: methods and goals. Nat. Rev. Genet. **5**(5), 335–344 (2004). doi:10.1038/nrg1325
105. D. Gibson, G.A. Benders, C. Andrews-Pfannkoch, E.A. Denisova, H. Baden-Tillson, J. Zaveri, T.B. Stockwell, A. Brownley, D.W. Thomas, M.A. Algire, C. Merryman, L. Young, V.N. Noskov, J.I. Glass, J.C. Venter, C.A. Hutchison III, H.O. Smith, Complete chemical synthesis, assembly, and cloning of a Mycoplasma genitalium genome. Sci. Signal. **319**(5867), 1215–1220 (2008)
106. C. Noren, S. Anthony-Cahill, M. Griffith, P. Schultz, A general method for site-specific incorporation of unnatural amino acids into proteins. Science **244**(4901), 182–188 (1989). doi:10.1126/science.2649980
107. C. Gauchet, G. Labadie, C. Poulter, Regio- and chemoselective covalent immobilization of proteins through unnatural amino acids. J. Am. Chem. Soc. **128**(29), 9274–9275 (2006)
108. R. Baum, Drexler and Smalley make the case for and against 'molecular assemblers. Chem. Eng. News **81**(48), 37–42 (2003). Available online: http://pubs.acs.org/cen/coverstory/8148/8148counterpoint.html
109. H.F. Tibbals, *Medical Nanotechnology and Nanomedicine* (CRC Press, Boca Raton, 2010)
110. F. Hu, K.W. MacRenaris, E.A. Waters, E.A. Schultz-Sikma, A.L. Eckermann, T.J. Meade, Highly dispersible, superparamagnetic magnetite nanoflowers for magnetic resonance imaging. Chem. Commun. Camb. **46**(1), 73–75 (2010). doi:10.1039/b916562b
111. R.M. Shah, N.A. Shah, M.M. Del Rosario Lim, C. Hsieh, G. Nuber, S.I. Stupp, Supramolecular design of self-assembling nanofibers for cartilage regeneration. Proc. Natl. Acad. Sci. **107**(8), 3293–3298 (2010)
112. V.M. Tysseling-Mattiace, V. Sahni, K.L. Niece, D. Birch, C. Czeisler, M.G. Fehlings, S.I. Stupp, J.A. Kessler, Self-assembling nanofibers inhibit glial scar formation and promote axon elongation after spinal cord injury. J. Neurosci. **28**(14), 3814–3823 (2008)

Applications: Nanoelectronics and Nanomagnetics

Jeffrey Welser, Stuart A. Wolf, Phaedon Avouris, and Tom Theis

Keywords Nanoelectronics • Nanomagnetics • CMOS • FET • MRAM • Graphene • Flash • Self-assembly • Spintronics • Carbon nanotubes • Devices • International perspective

1 Vision for the Next Decade

1.1 Changes in the Vision Over the Last 10 Years

In the last 10 years, the state of the art in nanoelectronics, including nanomagnetics, has rapidly gone from devices at or above 100 nm in size to the realm of 30 nm and below, with a well-defined pathway to devices (including transistors for logic and memory) of about 15 nm. In the process of reaching this size, the thickness of the critical layers in many structures is approaching 1 nm; the threshold voltage of a metal-oxide semiconductor field effect transistor (MOSFET) device is now controlled by fewer than 100 atoms, and the line edge roughness requirements are a few nanometers. All of these advances have resulted in an increased demand for near-atomic-level control for deposition, patterning, and characterization.

J. Welser (✉)
Semiconductor Research Corporation, 1101 Slater Road, Suite 120, Durham, NC 27703, USA
and
IBM Almaden Research Center, 650 Harry Road, San Jose, CA 95120, USA
e-mail: jeff.welser@src.org

S.A. Wolf
NanoStar, University of Virginia, 395 McCormick Road, Charlottesville, VA 22904, USA

P. Avouris
IBM T.J. Watson Center, P.O. Box 218, Yorktown Heights, NY 10598, USA

T. Theis
IBM, 1101 Kitchawan Road, Yorktown Heights, NY 10598, USA

1.2 Vision for the Next 10 Years

In the coming decade the research community must focus on what happens at 10 nm and below for all devices. Moreover, there needs to be an increased focus on how to utilize the new physics offered at the nanoscale to increase device functionality. It is ironic that in the past 10 years, nanoelectronics has arguably been the primary driver (both economically and technologically) for increasing our ability to control material at the nanoscale, but the nanoelectronics community has largely used these increasing capabilities to shrink existing devices while attempting to maintain the microscale physics that has enabled exponential progress in miniaturization for the past half century. In the sub-10 nm world, there will be no choice but to embrace the new nanoscale phenomena and focus on how to utilize them for new functionality beyond the complementary metal oxide semiconductor (CMOS) device. This will not only mean taking advantage of new nanoelectronic phenomena, but increasingly it will also mean manipulating the magnetic, spintronic, and other properties of matter at the nanoscale as state variables for computation and new forms of data storage. Moreover, this new functionality will not only improve our existing products, but it will open up new application areas, including new sensors, new ultralow power devices, and new flexible electronics. The result will be ever more ubiquitous applications of semiconductor nanoelectronics—beyond those in today's laptops and cell phones and those that are an increasingly integral part of appliances, cars, homes, and healthcare—in products and industries yet to be conceived by a new generation of entrepreneurs.

The expansion of nanoelectronic application areas will require an increased focus on research that considers from the beginning not just the device, but the circuit, architecture, and end application, as well as the societal implications of devices that may be embedded in our everyday world. While focused research on new materials and new scientific phenomena for novel devices will clearly need to continue, there will also be a shift toward increased research on nanosystems that not only takes advantage of the properties of the devices but also considers how to exploit the novel interactions between multiple devices in new architectures. This will drive an increased need for design tools and methodologies that can bridge from the materials level to the system level, as well as drive increased interaction between researchers in the basic sciences and those focused on applications. Hence, the next 10 years will see even more need for multidisciplinary teams to do the nanoelectronic and nanomagnetic research for the future.

2 Advances in the Last 10 Years, and Current Status

There have been many significant advances in nanoelectronic materials and devices during the first decade of the twenty first Century, but nowhere has the impact of shrinking devices to the nanoscale been more dramatic than in the semiconductor industry. For over 30 years, the industry has been able to double the number of

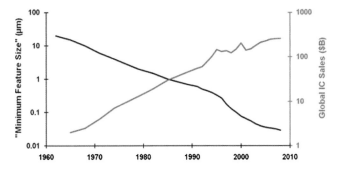

Fig. 1 Impact of transistor scaling (*purple curve*) on semiconductor revenue (*green curve*; Courtesy of R. Doering, Texas Instruments; data from Semiconductor Industry Association, http://www.sia-online.org)

field-effect transistors (FETs) on a chip every 18–24 months, a trend that has come to be known as "Moore's Law" [1]. The resulting exponential increase in the information-processing capability per unit area on the chip—or more importantly, per dollar—has meant not only that existing chip-based products have become faster and/or cheaper each year, but also that the number of products that use semiconductor chips has greatly expanded, from supercomputers to cell phones to toasters. The result can clearly be seen in the exponential increase in semiconductor revenues as the transistor size has shrunk (Fig. 1). Over the past decade, the industry has nearly doubled in size, and is expected to hit a record high of $300B in 2010, with the U.S. market share being approximately 50% of that. It is a driver of most economies of the developed world.

These advances have been won with increasing difficulty in recent years, and continuing them will require ever increasing utilization of nanoscale structures and properties. According to the long-established rules for transistor scaling [2], the operating voltage should be reduced in proportion to the reduction in the critical dimensions of the FET. Following this prescription, the speed of the FET went up while the area occupied by the device and power used by the device went down, so that the areal power density remained constant. In the past 10 years, as devices shrank below 100 nm in gate length, the rule governing reduction of voltage had to be abandoned, primarily because a minimum gate voltage swing is necessary to switch an FET from an "off" (low-current) state to an "on" (high-current) state. If that swing is too small, the device will have excessive leakage current (high passive power dissipation) in the nominally "off" state or low current (slow circuits) in the nominally "on" state. While it seems that devices can continue to be shrunk for the next 5–10 years, the designs utilizing them must increasingly trade density for speed in order to mitigate the increasing on-chip power density [3]. This will in limit the ability of further size reduction to achieve the full historical benefits of increasing computation/second/dollar/watt that has driven the industry up to now. Moreover, while the power-density challenge is affecting all applications (including large servers in datacenter environments), the overall energy consumption is even more problematic for the increasing number of mobile devices. These require not only finding a device with low operation energy, but also ultra-low (or zero) passive and standby energy needs. The trade-off between performance and leakage current is especially difficult in these applications, and finding novel non-volatile devices—potentially even for the logic devices—would offer major advantages. In addition, finding ultra-low power transistors not only increases traditional battery life, but opens the door to more exotic energy scavenging techniques to power these devices in the future.

In addition, as FETs have been shrunk to the tens of nanometers size range, rather than benefiting from new phenomena at the nanoscale, their performance has been challenged by increases in tunneling currents and the need for almost atomic-scale precision in fabrication. At the 90 nm node in the first part of the decade, for example, the gate insulator was approaching 1.5 nm in thickness—just a few atomic layers of silicon oxy-nitride—which not only resulted in unacceptably high tunneling leakage currents but also required monolayer control across a 200 mm wafer. The solution was to introduce an insulator with higher dielectric constant, which allowed for an increased thickness but in most cases also required changing the gate electrode process to a combination of polysilicon and metal [4]. The result was decreased electrical thickness, which allowed further shrinking of the gate length. However, even with high-K material, the insulator thickness is still only a few nanometers, and with a lowered barrier height. Further improvements in gate dielectric materials will probably be limited, and therefore further reduction of gate length may depend on introduction of new device structures as discussed below.

The gate length for modern transistors is now about 30 nm. This is exceedingly difficult to achieve with current optical lithography, which utilizes 193 nm wavelength light to pattern the features. It is only possible through many tricks, including chemically active resists, restricted design rules, extensive use of optical proximity correction on the masks, and most recently, the introduction of immersion techniques to reduce the effective refractive index for the exposure [5]. Moreover, since many of the electrical properties of the FET are exponentially dependent on gate length (including leakage currents), these dimensions must be controlled to just a few nanometers. Variations in line width across a wafer, and increasingly across a single die, are one of the major sources of variability that limit ultimate chip yield and performance, so it is crucial that new patterning solutions are found. The current approach for the industry is to push optical lithography to the extreme ultraviolet (EUV) range, but this continues to be a huge challenge, largely in finding suitable mask and resist materials [6], and significant research in this area is still required. Beyond optical lithography, emerging patterning processes such as nanoimprint lithography [7], scanning probe lithography [8], and various forms of self-assembly ([9, 10]) are being explored. These are key areas of research for the coming decade.

It is important to note that both the gate oxide and lithography improvements were only possible due to the research in nanoscale fabrication and characterization that has taken place over the past 10 years. Continued scaling will require further fundamental advances in lithographic and other processes for precise fabrication of nanometer-scale structures. Advances in metrology will be necessary to attain the required precision. Furthermore, future devices must take direct advantage of nanoscale effects as well. One interesting example that has already reached the market is the use of nanocrystal floating gates [11] for nonvolatile FLASH memory [12, 13]. Utilizing nanocrystals rather than continuous floating gates for storing charge should allow the use of thinner insulators and improve the scaling potential for future FLASH devices. Recent advances in nanoscale materials and device physics are also behind rapid progress in two other non-volatile memory devices that are entering the commercial arena. Phase Change Memory (PCM) has been in development for decades and is now considered to be a potential successor to silicon FLASH memory because of its superior potential to be scaled to small size. Extensive materials research shows that the phase change behavior is well defined in films as thin as 1 nm and exploratory device research suggests that devices with active material volumes of just a few cubic nanometers are feasible [14]. Looking forward, advances in the understanding of oxygen vacancy transport in metal oxides and other effects which can induce resistance changes in material stacks may find potential application in other resistive memory devices, including novel "memristor" devices [15–17], that may be useful in memory [18], storage, and even circuits which mimic the synaptic function of the brain [19, 20]. The other emerging non-volatile memory technology is Magnetic Random Access Memory (MRAM). It has been in commercial development since the mid-nineties, and promises much greater speed, lower power, and far better durability than either FLASH memory or PCM. However, a path to very small devices became evident only after a scientific breakthrough—the experimental demonstration of the spin-torque transfer effect [21]. See the discussion below of Nanomagnetics and Spintronics.

For FETs used in digital logic circuits, the most promising directions forward involve reduced-dimension structures, starting with ultrathin silicon-on-insulator (SOI) and double-gate or FinFET devices [22], ultimately leading to completely 1D structures such as nanowires [23] and nanotubes [24]. All of these structures enable improved gate control of the FET channel current, allowing shorter channels and better performance. Carbon nanotubes (CNTs) in particular have been studied intensely for this over the past 10 years, and many interesting demonstrations have been done that take advantage of their high carrier mobility, high thermal conductivity, and unique physics [24]. The key remaining obstacle to the use of nanotubes and nanowires in high performance electronics is the lack of fabrication methods that allow dense packing of uniform diameter tubes/wires across a large area with low defects and identical electrical behavior. For flexible electronics or applications that currently utilize thin film transistors (TFTs), the requirements are greatly relaxed. Hence the first large-scale application of CNTs may be as a mesh structure for moderate-performance flexible electronics—an area of growing importance in the future.

Regardless of the material or structure used, all of the above-mentioned FETs are approaching some hard limits due to the basic physics of the device operation. Thus there is a strong need to explore new device concepts that circumvent these limitations. Similar motivations drove the development of both the first solid state transistors, based on bipolar technology, when vacuum tubes and mechanical switches were reaching similar power constraints in the late 1940s, and the current

FET, which replaced bipolar transistors in the majority of semiconductor applications in the late 1980s. The potential for yet another major device transition was recognized at the beginning of the decade, and the Semiconductor Research Corporation (SRC) and the National Science Foundation (NSF) jointly organized a set of industry-academia-government workshops [25–27] to study the problem. In parallel, the Technology Strategy Committee of the Semiconductor Industry Association (SIA) conducted several workshops whose objective was to identify research initiatives to advance integrated circuit technology beyond currently identified scaling limits. These activities ultimately defined key research vectors considered to be important components of the search for the next switch and resulted in the formation of the Nanoelectronics Research Initiative (NRI, http://nri.src.org) by the SIA in 2005. Managed by the SRC, the NRI's mission is to demonstrate novel computing devices capable of replacing the CMOS FET as a logic switch in the 2020 timeframe. These devices must show significant advantage over ultimate FETs in power, performance, density, and/or cost, and most importantly must enable the semiconductor industry to extend the historical cost and performance trends for information technology [28].

The NRI's primary goal is to circumvent the power density issues that are currently limiting CMOS technology. To go beyond this limit, the physics of the device or the mode of its operation must be fundamentally changed so that much less energy is dissipated in each switching operation. To accomplish this, the primary focus is on switches that utilize alternative state variables to represent information (compared to the FET, which relies on the dissipative movement of charge), as well as on the corresponding interconnects and circuits needed to perform logic. Research is also being done on nanoscale phonon engineering, both to manage heat and potentially to isolate systems from thermal noise, and on directed self-assembly of key device structures. The challenge—and urgency—of finding a new switch to enable continued progress in nanoelectronics beyond 2020, similar to what has driven the economy for the past half century, should not be underestimated and should be one of the primary foci of nanoelectronics work in the coming decade.

While many of the nanoelectronic advances in the past decade have already had impact on mainstream electronics, many other breakthroughs show great promise for new innovations in the coming decade. A few examples include:

- The discovery of graphene and development of methods for controlled growth or synthesis [29].
- Elucidation of the electronic, optical and thermal properties of carbon nanotubes and graphene and establishment of a new class of electronic materials – carbon electronics.
- Discovery of magnetoelectric and multiferroic materials which promise voltage-control of magnetism [30].
- The experimental realization of spin torque switching [21].
- Discovery of the Spin Hall Effect [31–33].
- Demonstrations of spin injection into and spin readout from semiconductors.
- The discovery of topological insulators, a new and topologically distinct electronic state in matter with unique collective transport properties [34–36].

- Advances in the understanding and development of dilute magnetic semiconductors [37, 38].
- Fundamental understanding of the chemistry and physics of semiconductor nanowire growth.
- The exploration of nanoscale solid-state electrochemistry for a variety of potential applications in device technology [39].
- Some of these breakthroughs have already developed into major new fields of technology exploration, with two of the largest areas currently being graphene electronics and spintronics.

2.1 Graphene Electronics

One of the most interesting developments for future nanoelectronics has been the rediscovery of graphene—a single monolayer of graphitic carbon—as a potential substrate with unique physics. This material has almost all of the same advantages of CNTs (as well as other attributes), but it is a planar material, making device fabrication more straightforward. However, producing the initial substrate of graphene is still very challenging. While the electronic structure of graphene was calculated back in 1945 [40], and ways to produce graphene on metals [41], SiC [42], and graphene oxide [43] were demonstrated early on, research on the electronic properties of graphene did not start earnestly until 2004, when single-layer graphene was exfoliated from graphite and deposited on SiO_2 [29]. Early experimental and theoretical work was focused on the 2D electron gas properties, particularly on the study of graphene's quantum Hall effect [44, 45] and its minimum conductivity [46]. Transport measurements firmly established its excellent electrical properties: mean-free paths of hundreds of nanometers to a micrometer and mobilities of the order of 10,000 cm^2/Vs in the supported state and over 200,000 cm^2/Vs in the suspended state [46, 47]. Additionally, graphene possesses extremely high current-carrying capability and excellent thermal conductivity and mechanical strength.

Graphene was incorporated early on as the channel of field-effect transistors, and the device current could be modestly modulated by the gate field due to the linear variation of the graphene density-of-states with energy [48]. However, because of the lack of a bandgap (graphene is a zero bandgap semiconductor), the achieved current on/off ratios have been, in general, below 10. Thus, although, graphene was hailed in the popular press as a successor to Si-MOSFETs, the lack of a bandgap currently precludes its use in digital electronics. On the other hand, in bilayer graphene with the layers stacked in the so-called AB or Bernal stacking configuration, the interaction between the two layers was shown theoretically in 2006 to lead to the opening of a bandgap on application of a strong vertical electrical field [49]. Early experiments failed to reveal a significant bandgap opening [50], most likely because of the quality of the bilayer sample used. In more recent experiments using a better dielectric gate stack, the opening of an electrical bandgap >140 meV was demonstrated [51], and further improvements are expected.

The initial graphene experiments were performed using exfoliated graphite. Soon, however, synthetic techniques for growing monolayer and multilayer graphenes appeared. Currently, the key synthetic techniques involve the thermal decomposition of SiC [52] and chemical vapor deposition on metals such as Ni [53] or Cu [54].

While digital electronic applications are not currently possible for graphene, its extremely high carrier mobility at room temperature, ultimate body thinness, high transconductance, and modest field tuning of the current recommend graphene for high-frequency analog electronic applications. In particular, radio-frequency field-effect transistors (RF-FETs) could find use in wireless communications, radar, security and medical imaging, vehicle navigation systems, and a host of other applications. Currently, the record cutoff frequency (fT) achieved by RF graphene transistors based on exfoliated graphite stands at 50 GHz [55]. Of course, for commercial applications, wafer-scale graphene is required. This can be achieved using the above-mentioned graphene synthetic techniques. In particular, the SiC-based approach has yielded wafer-sized samples of graphene. Graphene wafers from the Si-face of SiC show good layer thickness control and morphology and have mobilities that currently are in the range of 1,000–3,000 cm^2/Vs. C-face SiC yields graphene with higher mobilities but with thicker layers and less controlled morphology. RF transistors have been fabricated on Si-face SiC-derived 2 in. graphene wafers [55, 56]. The current record involves 240 nm gate length transistors from IBM with cutoff frequencies of 100 GHz [57]. This is already remarkable since Si-CMOS transistors with the same gate length achieve cutoff frequencies of 40 GHz. The currently achieved f_T x (gate length) product of 22 GHz · μm is higher than that of most semiconductors, and both scaling of the devices and improvements in the graphene mobility are expected to yield much higher-frequency performance (Fig. 2).

Another possible area for applications of graphene that can utilize its exceptional transport properties for both electrons and holes, as well as its strong optical absorption over a very wide wavelength range (about 2% per atomic layer), is optoelectronics. Ultrafast (>40 GHz) metal-graphene-metal photodetectors have been demonstrated [58], and these photodetectors have been used recently to detect reliably optical data streams at 10 Gbits/s [59]. THz radiation emitters from monolayer and bilayer graphene are also very likely to emerge soon.

Finally, beyond the traditional electron transport and optical devices, the physics of graphene could offer new possibilities for devices with unique functionality. The two-dimensional honeycomb lattice of graphene gives rise to a conical band structure, which leads to electrons behaving as massless Dirac fermions. Some proposed devices take advantage of the resulting photon-like behavior of electrons in graphene and utilize p-n junctions to create programmable interconnects or Veselago lens devices [60].

A completely different approach is taken by another new transistor concept, the bilayer pseudospin FET (BiSFET) [61]. In this device, two metal oxide gates sandwich two separately contacted graphene monolayers separated by a tunnel oxide. This device takes advantage of the pseudospin property [62], which predicts that under certain gate conditions an exciton condensate forms between the graphene layers, leading to the possibility of a collective many-body current between the two layers [63]. This could potentially enable a very low-energy switch, even at room temperature [61].

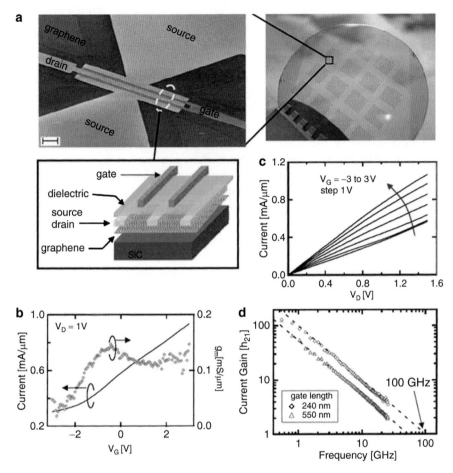

Fig. 2 (a) Scanning electron microscope image and the schematic cross-sectional view of a top-gated graphene field-effect transistor. The optical image of the 2 in. graphene/SiC wafer with arrays of graphene devices is on the right. The transistors possess dual-gate channels to increase the drive current and lower the gate resistance. The scale bar is 2 μm. (b) Drain current I_D of a 240 nm gate length graphene transistor as a function of gate voltage V_G at drain bias of 1 V. The current shown was normalized with respect to the total channel width. The device conductance $g_m = dI_D/dV_G$ is shown on the right axis. (c) Measured drain current I_D as a function of drain bias of a graphene FET with a gate length L_g of 240 nm for various top-gate voltages. (d) Measured small signal current gain $|h_{21}|$ as a function of frequency f for a 240 nm gate and a 550 nm gate graphene FET, represented by (◊) and (Δ), respectively. The current gain for both devices exhibits 1/f dependence, where a well-defined cutoff frequency f_T can be determined to be 53 GHz and 100 GHz for the 550 nm and 240 nm devices, respectively [57]

2.2 Nanomagnetics and Spintronics

While electronics and semiconductors have traditionally formed the basis for the chip industry, magnetics has formed the basis for the storage industry. During the past 10 years, though, advances in nanometer-sized magnetic devices as well as new methods for controlling spin properties in general have enabled increased usage of nanomagnetics, including on chip. There have been significant advances in spintronics, particularly in areas based on a nanoscale multilayer structure, that is, a magnetic tunnel junction (MTJ) [64, 65]. Magnetic tunnel junctions are already ubiquitous in that they are the key component in the sensor that reads the information stored on a magnetic disk. In 2006, a new type of nonvolatile, infinite-endurance computer memory was introduced to the marketplace called magnetic random access memory or MRAM [66]. This memory is now manufactured and sold by a spinoff of Freescale Semiconductor called Everspin, which has been introducing MRAM into more and more markets over the last few years. MRAM is also going to be a key component in defense systems that require radiation-hard, nonvolatile memory; Honeywell has teamed with Everspin to provide just such a product that is now going into production. The MTJ can also be used as a sensor, and this has found several applications in industrial sensing as well as in cell phones to provide directional information (a 3D magnetic compass).

Several key discoveries led to the successful development of MRAM, including a novel method of switching the bit, called toggle-switching [67], and the discovery of a new tunnel barrier (MgO) that enabled an order-of-magnitude increase in the tunneling magnetoresistive ratio (TMR) [68]. In the beginning of the last decade, another major hurdle for MRAM was overcome with the discovery of magnetization reversal due to scattering of a spin-polarized current [21]. This discovery, called spin torque transfer (ST or STT) switching, will allow MRAM to scale to well under 10 nm, dimensions that traditional switching methods using current-generated magnetic fields will never allow. Spin torque transfer also enables the generation of spin wave radiation due to the rapidly precessing magnetization of the "free layer" of a tunnel junction in a magnetic field. These nano-oscillators may find many uses in signal generation and processing as well as in phased arrays for RF detection and radiation. Finally, spin torque transfer provides a means to move magnetic domain walls in a magnetic nanowire [69]. This provides an alternate path for information storage and perhaps even processing. Prior to the discovery of spin torque transfer, magnetic domain walls were moved using changing magnetic fields, which are cumbersome and power-hungry.

Another major direction for spintronics was enabled by the realization that traditional semiconductors could be made magnetic by adding dilute amounts of magnetic ions, particularly manganese [70]. These materials, by virtue of the fact that in most cases the magnetism is mediated by carriers—and carrier concentrations can be controlled electrically—offer the unprecedented ability for the magnetism to be controlled with an electric field. The field of semiconductor spintronics has grown exponentially in the last decade but has been hindered by the absence of a magnetic semiconductor with a Curie temperature well above room temperature. However, proof-of-principle demonstrations of various spin-injection and FET-like devices show that if the right material is found, then adding the spin degree of freedom to semiconducting devices will enable a broad spectrum of applications spanning many diverse fields of electronics and photonics [70].

3 Goals, Barriers, and Solutions for the Next 5–10 Years

While there are many fruitful areas to be explored in nanoelectronics in the next 10 years, six primary goals and directions are highlighted as particularly important in the areas of fabrication, devices, and architectures, as described below.

3.1 Fabrication: Achieve 3D Near-Atomic-Level Control of Reduced Dimensional Materials

This includes not only the growth of nanotubes and nanowires with control of diameter, chirality, and placement across a wafer, but also the growth of large-area monolayer (or multilayer) graphene. It also includes the production of layered structures and interfaces, key to some emerging areas like topological insulators, multiferroic materials, and complex metal oxides, as well as more complex structures such as lattices of ferroelectric material with embedded ferromagnetic particles to create artificial magneto-electric structures. While there are specific barriers involved with different materials, all of these require increased control of both thickness and lateral dimensions during growth and/or etching, as well as metrology tools with sufficient resolution in multiple dimensions to measure the structures. It also requires continued advances in predictive modeling of materials to guide the design of new materials and interfaces with unique functionality.

3.2 Fabrication: Combine Lithography and Self-Assembly to Pattern Semi-Arbitrary Structures Down to 1 nm Precision

While self-assembly can potentially offer solutions for creating exceedingly small patterns, the number of defects is currently too high to be useful for most applications, and simple self-assembly processes can only produce repetitive structures. Defect rates in self-assembly must be improved, and efficient mechanisms for integrating self-assembly with lithographic patterning (templated self-assembly) must be developed to allow arbitrary pattern formation [71]. Given the current rate and pace of scaling, 1 nm precision will soon be necessary for fabrication of devices in complex circuits and systems. It is very important, therefore, to continue research investments in the science and engineering of precision patterning. At the same time, architectures and circuits must also evolve to work with imperfect structures and more regular layouts of devices. More communication between the various research areas will be necessary to achieve balance.

3.3 Devices: Discover Devices for Logic and Memory that Operate with Greatly Reduced Energy Dissipation

Power is the key barrier to continuing scaling of almost all nanoelectronic devices, so finding devices that can operate at room temperature with as little energy dissipation as possible is imperative. Logic devices are particularly challenging, since they must maintain adequate speed to allow continued scaling of overall computation/second/area/watt. It is unlikely this can be achieved solely through geometric scaling of the FET, as has been done in the past, so the focus must be on devices and device concepts that maintain distinguishability of logic state with less stored energy than the FET, or on devices that can do more complex functions and/or multibit calculations at a given energy dissipation and area. Beyond the speed consideration, the biggest barrier to low-energy devices of any kind is robustness to thermal noise. To address these challenges, collective effects and alternative state variables must be considered for future device candidates. This means finding new methods of representing information with material properties or parameters that can be manipulated into distinguishable states. Condensed matter theory offers many materials with broken symmetries that result in distinguishable regions characterized by different order parameters. These include ferroelectric, antiferroelectric, ferromagnetic, antiferromagnetic, ferrotorodic, ferroelastic, and ferrimagnetic materials, among others [72]. In many cases, these order parameters are coupled together by the atomic structure of the crystalline lattice structure. Examples of order coupling include magnetoelectric, piezoelectric, piezomagnetic, electrostrictive, and magnetostrictive effects. The key is to identify which of these can be manipulated at reasonable speed at low energy—while remaining thermally robust—to do computations at room temperature. In addition, consideration must also be given to the using these devices in circuits that can operate out of equilibrium with the ambient for a period of time or that can recover the switching energy as part of the computation cycle. This leads to the need for research also in the area of phonon engineering, to better control thermal interactions with nanodevices. Finally, for mobile applications, it's important to pay particular attention to reducing (or eliminating) standby power and leakage, potentially with non-volatile device solutions.

3.4 Devices: Exploit Spin for Memory, Logic, and New Functionality

While many potential state variables should be explored, an area that is particularly ripe for exploitation over the coming decade is spintronics. Spin and magnetics have already proven valuable for storage devices and increasingly are showing promise for memory, but they offer many other potential opportunities in sensors, oscillators, and logic devices as well. Finding energy-efficient ways to manipulate spin—potentially with electric fields using multiferroic and dilute magnetic

Applications: Nanoelectronics and Nanomagnetics

semiconductor materials—is imperative to this, as is studying the behavior of both single-spin and collective-spin states. This requires advances both in the materials and in the characterization tools for measuring individual spins and nanoscale magnets—and in particular, the dynamics of their behavior in novel materials.

3.5 Architecture: Integrate Architecture and Nanoscale Device Research for Unique Computation Functionality

While the past 10 years have largely seen demonstrations of individual nanoelectronic devices and new science phenomena, the next ten must focus on how these are integrated on a large scale and how they can be used to perform a useful function. For integration, that includes research on large-scale, reproducible fabrication of nanostructures; interconnects—potentially non-charge-based—between devices (and the impact they have on power and speed); consideration of nanostructure-environment and nanostructure-nanostructure interactions; and dealing with stochastic processes, noise, and thermal management. On the function side, rather than forcing new devices to fit into the current charge-based, Boolean systems, architecture researchers should look for novel architectures to take advantage of the unique functionality of a given device to implement important algorithms or applications, such as pattern matching, fast Fourier transform (FFT) computation, encryption, etc. In particular, they should consider how to work with the realities of most nanoscale devices (e.g., dense, slow, sea-of-gate structures with high levels of defects) while taking advantage of some of their assets (e.g., the prospect of merging memory and logic or making programmable architectures by taking advantage of the nonvolatile switches). Conversely, architects should also be thinking abstractly about entirely different approaches to doing even general-purpose computation, with a particular eye towards stochastic or "almost-right" computation, and giving feedback to device researchers on what kind of functionality would be needed in a new device to enable implementation of the architecture.

3.6 Architecture: Increase Focus on Emerging, Non-IT Applications

The major future market for electronics is likely to be outside of the traditional information technology (IT) space and instead geared for embedded applications in vehicles, appliances, homes, and even human bodies. Yet the research is still largely focused on traditional computer chip applications. Breakthroughs in the areas of low-power devices, sensors, transmitters, receivers, and flexible substrates are key components for these embedded applications, and to enable this, more research is needed on nanostructures specifically. At the same time, an increased emphasis is needed on understanding the environmental impacts of nanoelectronic components as they become more ubiquitous. Since each of these non-IT applications has unique requirements (e.g., high temperature, ultra-low power, biologically friendly packaging, etc.) it is important that research be more focused on the final application from the beginning, rather than just on the general problem of improving computation or memory in traditional environments.

4 Scientific and Technological Infrastructure Needs

The primary need to enable the nanoelectronic and nanomagnetic research necessary to achieve the goals listed in Sect. 3 is readily accessible tools—at low cost— to fabricate, characterize, and model active structures at the atomic scale. For fabrication, the ability to pattern reliably down to the 1 nm scale is crucial and will likely require a combination of tools for integrating both top-down (lithographic) and bottom-up (self-assembly) techniques. For characterization, the primary needs are for tools that can measure static structures (vertically and laterally) down to an atomic layer, as well as tools that can characterize the growth and patterning of these structures *in situ* to enable better understanding and control of the dynamics of the material growth. In addition, the ability to characterize the dynamics of single-carrier transport, spin precession, magnetic and ferroelectric domain reorientation, and a host of other material state transitions will become increasingly important as alternative state variables are explored. For modeling, an increased focus on tools that can scale-up atomic, first-principles simulations to macroscopic, device- (and circuit-) relevant structures is a major goal. Finally, the infrastructure and tools development for these areas should focus not only on tools that can accomplish this resolution for "single devices" in a research lab, but also on tools that enable the scale-up of the work to the "billion devices" needed for cost-effective nanomanufacturing. And this is not limited to just scaling up the fabrication tools, but also finding efficient, noninvasive characterization techniques for key parameters to use in monitoring and controlling inline manufacturing processes, and developing multiscale modeling tools that can facilitate rapid device and circuit design for product development while not sacrificing accuracy.

Although the number of materials being incorporated expands every year, the majority are still inorganic semiconductors, metals, and insulators, including an increasing number of ferroelectric, ferromagnetic, and even multiferroic materials. Therefore, the required tools are similar to what is already available in the national nanofabrication and characterization facilities, both at universities and at NIST and the DOE national laboratory user facilities. The National Nanotechnology Infrastructure Network (NNIN; http://www.nnin.org/) is a good model for this, but it needs to be greatly expanded both in terms of the number of facilities it supports across the country as well as in terms of updating the tools themselves. The NNIN facilities were largely built in the 1980s and 1990s for the microelectronic generations, so they aren't necessarily suitable for the nanoelectronics era. In addition, the current NNIN model, which largely invests only in the initial tool purchase and relies on user fees to cover the operations costs, is not practical for many of the modern tools. Many of these tools can cost upwards of $5–10 million, and typical warranty costs are 10% of the tool cost per year; in addition, there are the costs of a staff person dedicated to operating and maintaining the tool. Since user fees, particularly for university researchers or start-up companies, need to be kept to a level that can be contained in typical grants and angel funding, a significant portion of this ongoing operating cost should be built into the NNIN program.

The focus on tool investments in the universities should be on tools that are needed almost daily for effectively fabricating nanoscale devices. While many of these tools exist at the national labs and at a handful of universities in the NNIN, it is unrealistic to assume graduate students will be able to either travel to these as often as needed or to relocate there for the extended time needed to build a complicated structure. It is much more valuable for them to be working with their professors and fellow graduate students daily, so expanding the number of NNIN locations is vital. The tools needed include not just the fabrication tools, such as advanced e-beam, nano-imprint lithography, and integration tools, but also the characterization tools such as aberration-corrected TEM and nanoscale focused ion beam (FIB) instruments. It does no good to have a lithography and etching tool that is capable of sub-1 nm precision if it is necessary to fly back and forth to another location to measure the structure after each step in the process. With the increased focus on studying magnetic properties, advanced tools are increasingly necessary for magnetometry, magnetic force microscopy, and for studying susceptibility and transport in a magnetic field. Lastly, there should be increased support for experimental tool development, in some cases in conjunction with the tool vendors themselves, as is common in the semiconductor industry guided by the International Technology Roadmap for Semiconductors (ITRS). The focus at universities to support new tool development should be on areas where breakthroughs could enable a whole new innovation platform, such as wafer-scale directed self-assembly with improved fidelity, precision, and capability for nonregular structures—as well as the metrology tools to assess the quality of the structures, with a focus on making these techniques manufacturable.

The national labs should have continued investment in world-class user facilities as well. These can not only serve as some of the hub facilities mentioned above for various geographic areas, but also should focus on having large, one-of-a-kind tools that only need to be utilized for specialized measurements on a nonregular basis and/or are too expensive and impractical to maintain at multiple locations. The NIST neutron source and the various DOE light sources are good examples of these, but other examples include the kind of experimental cutting-edge metrology tools that NIST develops as part of its mission. Maintaining investment for these and for cost-effective access is vital for pushing the edge of nanoelectronics.

Finally, nanoelectronics continues to rely heavily on advanced modeling and simulation tools, so resources such as the nanoHub (http://nanohub.org/) are crucial. While the model for many of these has been an initial multiyear investment by NSF, after which they are expected to become self-sufficient, ongoing funding by NSF or another government source is a much better model to ensure that they have sufficient support to remain on the cutting edge and be accessible to all researchers.

5 R&D Investment and Implementation Strategies

Given the increased need to consider multiple aspects of a problem from the beginning, e.g., the need to have circuit and architecture experts actively involved in materials and device research early on, funding of large, sustainable, multidisciplinary teams at virtual centers that work together on the next big technological advances will become even more important to understanding how new phenomena can be utilized effectively. For example, involving material scientists, physicists, device designers, modelers, and circuit and system design specialists at the onset of a project will definitely provide a more realistic assessment of how a new device will really perform compared to CMOS. In order to be effective, these virtual centers must of course have a very clear mission—even for basic research—if the working groups at the centers are ever to function as teams rather than as just a collection of individual researchers. In building these teams, it is also important to balance getting the best researchers in any given field (which often requires geographically distributed, multi-institution teams) with enabling the most effective collaboration (which is often easier to obtain at a single institution). A good balance can be achieved by investing adequate funds to support at least 2–3 students or post-docs per principal investigator (PI), and to fund at least 2–3 PIs per university. This ensures that the center mission gets sufficient mindshare of each of the PIs, and that it has a critical mass of collaborators at each university. The total number of universities in the center will depend on the scope of the project and number of disciplines needed.

While large efforts will play an increasingly important role in encouraging innovations that cross boundaries, it is important to also continue robust funding for small, high-risk projects that consist of only one or two PIs in a new area. These often get overlooked, particularly in the peer-review process, but can result in breakthroughs that drive much larger research efforts in the future. It is important when funding these smaller projects to not burden them with the same level of additional non-research goals expected of larger centers. Education, outreach, and broad impact requirements are laudable contributions that the large centers should make as part of their missions, but smaller projects should be focused on the research itself for maximum value from a small group.

Regardless of the size of the research project, it is also crucial to increase the interaction between university and industrial researchers, even for basic research. At a minimum, it is important to increase the utilization of industrial scientists as evaluators for new proposals, particularly in the area of nanoelectronics and nanomagnetic devices, since often the academic community cannot evaluate them properly without knowing the actual needs or the current state of established technologies in production. There is also an inherent conflict in the purely academic peer review process, where the experts are also potential competitors for the funding. Heightened risk-aversion at the agencies due to tightened budgets can often lead to funding of less innovative, more evolutionary proposals. Industry can serve as a catalyst for pushing studies to areas that are more potentially game-changing, because they are less likely to choose evolutionary studies that they could potentially do more expediently themselves in-house. Beyond just proposal review, real collaboration should involve actual exchange of personnel, for example, graduate students spending time in industrial settings or industrial researchers on assignment at university labs. This facilitates not only the sharing of resources but more importantly the sharing of expertise. For example, working on devices in a university without knowing if something is actually manufacturable can waste both time and money; at the same time, the trial-and-error approach that is sometimes employed for expediency in industry can greatly benefit from input from an academic research perspective.

The best R&D partnerships involve government, university, and industry each playing to their own strengths. An example of this, which has been highlighted in various National Nanotechnology Initiative (NNI) strategic reports, is the Nanoelectronics Research Initiative (NRI). The NRI has set up four regional centers for doing research on post-CMOS devices across the country. Each center receives significant funding from the primary state—and in some cases, city—where it is headquartered, as well as from industry and NIST. In addition, NRI co-funds individual projects at existing NSF Nanoscale Science and Engineering Centers. While the industry funding itself is smaller than the government contributions (about 20% of the total), it is large enough to ensure that industry actively engages with all the centers, not only in the proposal selection process at the NRI centers, but also by sending assignees to the universities and NSF centers and bringing graduate students back to the member companies for internships and ultimately as permanent hires after graduation. Despite the long-term nature of the mission, this collaboration has created a goal-oriented, basic science research program that has already accelerated the pace of device-relevant scientific advancements and will ultimately result in accelerated technology transfer back to the companies for commercialization.

A new model to consider that would take this collaboration a step further would be a true government-university-industry (GUI) innovation center. In the United States, the GUI would likely be located at one of the DOE National Science Resources Centers (NSRCs; http://www.science.doe.gov/nano/), the NIST Nanofab (http://cnst.nist.gov/nanofab/nanofab.html), or at a mixed academic-industry research facility such as Albany Nanotech. Other countries have similar facilities that could be utilized. The GUI would be funded largely by a Federal agency, with some small cost-share by industry and facility support by the host lab. It would focus a small team of academic, industrial, and government researchers on a specific nanoelectronics "hard problem." This should be a challenge that addresses barriers to progress in technology for which no known solutions exist; requires fundamental advances in understanding, prediction, and control of processes and materials, leading to a proof-of-concept demonstration; is achievable with a multidisciplinary team of approximately 10–15 researchers working for 3–4 years; aligns with the national laboratories' mission to promote U.S. innovation and industrial competitiveness and with lab expertise and facility capabilities; and if successful, will have a significant potential economic impact by creating a new innovation platform to enable the industry to advance in an important product area. An example would be the creation of a low-power, room-temperature multiferroic material to allow the electrical manipulation of magnetic domains and/or spins, or the creation of a large-area graphene bilayer that demonstrates a room-temperature exciton condensate for ultralow energy device and interconnect applications.

The key for these kinds of partnerships is for each party to contribute based on its own assets: e.g., the Federal agencies contribute the bulk of the funding for the basic research; the state governments contribute infrastructure support both on-campus for research facilities as well as adjacent to campus for incubator labs and technology parks to encourage rapid uptake of (and economic development from) the technologies developed; and industry contributes sufficient funds to keep its attention, but more importantly, to foster the active involvement and guidance of its personnel in the research process and the technology transfer.

While the kinds of partnerships and multidisciplinary approaches to research mentioned here have always been important, they take on an increasing urgency in the coming decade as nanoelectronics converges with other application areas. For example, nanosensors will be deployed in increasing numbers and in a wider variety of environments, driving new requirements on the nanoelectronics that must be included in the research from the beginning: RFID tags must be flexible, robust, energy-scavenging, and dirt-cheap; chemical sensors to detect toxic gases (either in industrial settings or in Times Square) must be increasingly sensitive but also increasingly specific to only the target molecules; and sensors to be embedded into the human body must be able to withstand the chemical environment, measure, record and transmit the desired information, and be completely harmless to the biological system itself. Balancing this increasing number of (often conflicting) requirements drives the need for increased investment in multidisciplinary teams that are focused from the beginning on specific systems and application areas for the basic science research.

6 Conclusions and Priorities

In nanoelectronics and nanomagnetics, the focus must continue to be on increasing functionality per dollar, to both improve existing products and enable entirely new product areas. The major barriers to this are not necessarily in just making devices smaller, but in reducing the power of the devices to allow increased density and in controlling the variability of the structures to allow large-scale integration. With this in mind, the key priorities for funding in the next 10 years should focus on achieving the following six goals:

- Fabrication: Achieve 3D near-atomic-level control of reduced-dimensional materials.
- Fabrication: Combine lithography and self-assembly to pattern semi-arbitrary structures down to 1 nm precision.
- Devices: Discover devices for logic and memory that operate with greatly reduced energy dissipation.

- Devices: Exploit spin for memory, logic, and new functionality.
- Architecture: Integrate architecture and nanoscale device research for unique computation functionality.
- Architecture: Increase focus on emerging, non-IT applications.

When pursuing these priorities, it is important that a balance be struck between pursuing the fundamental science and materials research—vital to enabling the breakthroughs needed for advancing current technology and discovering new areas for nanoelectronic exploitation—and pursuing the ultimate goal of integrating these discoveries into real products and innovations in the future. Closer connections must be established between the research and the development communities, and between the academic and the industry labs, to be sure that issues related to large-scale manufacturing are included as a natural part of the research process, mitigating the chances of unpleasant surprises cropping up after several years of effort. The increasing diversity of potential applications for nanoelectronics—including the areas of medicine, energy, sensors, and the entire gamut of "smart" products and infrastructure for improving our lives and protecting the planet—requires better connections between research disciplines. In particular, it is important to think from the beginning at a systems level, rather than at the level of a single device or material, to let the application drive the research in the most effective directions. This cross-disciplinary, systems approach, as well as the closer relationship between research and manufacturing, should foster accelerated transfer of the technologies—including unforeseen spin-off technologies—out of the research labs and into the development and manufacturing world.

7 Broader Implications of Nanotechnology R&D on Society

Nanoelectronics has clearly been the major driver of the economy for arguably the last 50 years. It has been estimated that IT-producing and intensive IT-using industries currently account for over a quarter of the U.S. Gross Domestic Product (GDP) and drive 50% of U.S. economic growth [73]. This remarkable impact has largely been due to the power of scaling to increase the function/dollar of semiconductor chips each year, and hence there is an urgent need to find nanoelectronic devices that will continue to drive this economic engine. At the same time, future nanoelectronic semiconductors will be crucial for solving many of the other major challenges facing society today.

For example, finding a lower-energy device will help tackle many energy-related challenges facing the world. First, conversion to low-energy devices could significantly reduce energy consumption of IT-based systems, such as the growing number of data centers, which collectively accounted for 1.5% of electricity use in 2006 and are on track to double in 5 years [74]. Despite enormous efficiency advances made in the past, the switching energy of today's transistor is still 10,000 times the theoretical lower limit [27], indicating that significant efficiency gains are still to be made.

Second, semiconductor chips enable almost all of the energy-efficiency solutions envisioned today. They will be in advanced sensors that monitor and control everything from small appliances to "smart buildings" and will be the brains of the smart grid, tracking the intelligent and efficient generation, transmission, and use of energy across the country. A recent report by the American Council for an Energy-Efficient Economy [75] estimated that broadly implementing semiconductor-enabled technologies for energy efficiency could save the United States 1.2 trillion kWh of energy and reduce carbon dioxide emissions by 733 MMT by 2030. Lastly, continued scaling is crucial to high-performance computing (HPC), which is an invaluable tool for understanding materials and phenomena that are fundamental to energy solutions, including modeling complex nanostructured catalysts, fuel cell components, and transport of carbon during sequestration on geologic scales.

HPC has been behind nearly every major scientific advance and innovation in the past decade, not only in energy, but also in materials science, engineering, life sciences, climate and environment, and defense and security. In biology in particular, the sequencing of the human genome was arguably as much a triumph of computing technology as it was of medical science. And increased computational capability is crucial for advancing microbiology and chemistry, from the study of protein folding to new drug discovery.

In addition to improvements in computation, the bioengineering community continues to develop new instrumentation and sensors that can be implanted in the body for continuous health monitoring. Many of these sensors require sophisticated control electronics, power sources, and transmit/receive modules. Nanoelectronics technology promises more functionality in smaller devices, allowing ever-increasing complexity in the *in vivo* sensors. Targeted applications include new forms of drug delivery and DNA sequencing, tagging of bio-molecules with semiconductor quantum dots, and other nano-actuators and sensors at the cellular and molecular level. By enabling new diagnostic and treatment options for healthcare providers, nanoelectronics will undoubtedly have a profound effect on medical practice in the near future and will be a potent driver of growth in the biotechnology sector.

The increasing proliferation of mobile electronic devices and access to high-speed connections is changing the entire nature of how we interact as a society and a planet. In the next decade, it is foreseeable—perhaps inevitable—that the majority of the people in the world will have constant access to the world's information, and each other, through both the traditional web-based interfaces we see today and, increasingly, through forms of virtual and augmented reality. This will accelerate the move to more remote business interactions, more globalization of the workforce, and more remote delivery of entertainment, legal guidance, and education, as well as services once thought to be immune from "outsourcing," such as healthcare (including surgery). The ubiquitous access to information will continue to raise privacy concerns as individuals balance their desire to take full advantage of the online world with maintaining control over information they consider sensitive. This will inevitably lead to new social definitions of what is "private" and what is "public" and new cultural norms for what is acceptable interaction, made even more interesting by the increasing global, multicultural, multiethnic, multi-religious nature of online interaction.

Imagine how accelerated this global interaction will become if real-time, natural language translation becomes accessible to the everyday person, which it will, if not in the coming decade then in the next. All of these changes will inevitably present new challenges to the current business models of corporations, as well as to the current economic models of nations, as both struggle to take advantage of the vast new opportunities offered by truly globally accessible information, workforce, and supply chain, while maintaining their own competitive advantages and the well-being of their employees and citizens.

8 Examples of Achievements and Paradigm Shifts

8.1 Nanomagnetics

Contact person: S. Wolf, University of Virginia

Magnetic field-switched MRAM does not scale very well below about 90 nm because the writing of the bit requires producing a magnetic field using two perpendicular wires (word and bit line), and as the bit size shrinks, the magnetic anisotropy energy to maintain the information stored has to increase, causing the write current to increase—counter scaling (see Fig. 3). Thus, spin torque transfer switching as described above needs to be perfected. The new MRAM (STT-RAM or ST-MRAM, depending on who is developing it) will scale perfectly, provided the switching current densities can be reduced below 10^6 A/cm^2 (Fig. 3). In order to accomplish these

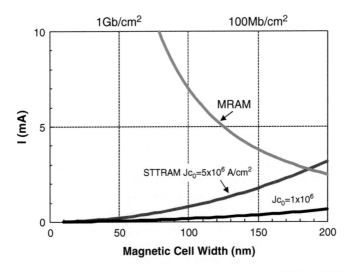

Fig. 3 Comparison of writing current scaling trends between *MRAM* and STTRAM (Wolf et al. [76])

Table 1 Memory technology performance comparison

	SRAM	DRAM	Flash (NOR)	Flash (NAND)	FeRAM	MRAM	PRAM	STT-RAM
Non-volatile	No	No	Yes	Yes	Yes	Yes	Yes	Yes
Cell size(F^2)	50–120	6–10	10	5	15–34	16–40	6–12	6–20
Read time(ns)	1–100	30	10	50	20–80	3–20	20–50	2–20
Write/Erase time(ns)	1–100	50/50	1 μs/10 ms	1 ms/0.1 ms	50/50	3–20	50/120	2–20
Endurance	10^{16}	10^6	10^5	10^5	10^{12}	$>10^{15}$	10^{10}	$>10^{15}$
Write power	Low	Low	Very high	Very high	Low	High	Low	Low
Other power consumption	Current leakage	Refresh current	None	None	None	None	None	None
High voltage required	No	2 V	6–8 V	16–20 V	2–3 V	3 V	1.5–3 V	<1.5 V
			Existing products				*Prototype*	

Source: Wolf et al. [76]

goals there needs to be considerable research in new magnetic materials and novel magnetic structures. However, if STT-RAM is successful, it can be a universal memory that can ultimately replace most or all of the existing semiconducting memory technologies (see Table 1). There are other unique ways to utilize the MTJ in novel memories, such as race track memory [21], which requires considerable research focused on understanding magnetic domain walls and domain wall motion. In fact, there is a type of MRAM that utilizes a movable domain wall to write the information to the free layer. This also involves spin torque, but the details of how domain walls are driven is still of significant scientific interest.

Another potentially fruitful goal for the next decade is the exploitation of spin torque nano-oscillators for signal generation and detection. These oscillators can be tuned over a very wide range (many octaves), but require large magnetic fields for the highest frequencies (approaching 100 GHz). Theoretically, these oscillators can utilize internal or exchange fields, but this has not yet been demonstrated. Further research on the potential and limitations of these oscillators will ultimately determine their utility.

Magnetic cellular automata have recently emerged as a potentially very-low-power logic paradigm [77]. These structures take advantage of the collective nature of magnetism and dipolar and exchange coupling between nanoscale magnets to provide low-power switching and information transfer. Some recently proposed structures utilize ferroelastic coupling at a ferroelectric/ferromagnetic interface to control the direction of the easy axis of the ferromagnet and provide a means for reconfiguring a magnetic cellular automata array on the fly (Wolf et al. [76]). The ability to reconfigure the array also allows the array to be very regular and amenable to self-assembly techniques (see Sect. 4, Infrastructure Needs).

If semiconductor spintronics is to succeed as a potential supplement/replacement for non-magnetic semiconductor classical electronics, robust high-Curie-temperature dilute magnetic semiconductors must be discovered or proven. In addition, single spins in semiconductors have emerged as one of the potential sources for qubits for quantum information processing. In particular, nitrogen-vacancy (NV) centers in diamond are a very promising potential spin qubit, but there still needs to be considerable research into controlling the location of and interactions between these "defects" [78].

8.2 Spin-Transfer Torques Yield New Nanoscale Magnetic Technologies

Contact person: Dan C. Ralph, Cornell University

The ability to fabricate nanometer-scale devices is enabling new magnetic technologies by allowing the use of a recently-discovered mechanism called spin-transfer torque to achieve efficient electrical control over the orientation of magnetic components. This new technique can apply torques to a magnet that are hundreds of times stronger per unit current compared to older methods that rely on magnetic fields. This innovative type of magnetic control is being developed rapidly to produce new magnetic memory technologies and also high-frequency devices for signal processing.

The idea behind spin-transfer torque relies on the fact that every electron possesses a built-in spin as well as electrical charge. In most existing electronics technologies this spin does not affect device function because when an electron moves through a metal or semiconductor the axis of the spin can be reoriented randomly, so that on average the spin orientation is not preserved over length scales longer than a few 100 nm. However, with recent advances that allow the production of devices smaller than 100 nm, it has now become possible to develop new technologies which utilize spin-polarized electrical currents that maintain a well-defined spin axis.

Spin-transfer torque is implemented using devices consisting of stacks of magnetic and non-magnetic materials with nanometer-scale thicknesses (see Fig. 4), with an electrical current flowing perpendicular to the layers in the stack. The first step in operation is to orient the spin axes of the flowing electrons so that on average their spins point in the same direction. This is done by having the electrons pass through a first layer of magnetic material, which acts as a filter to align electron spins in the same direction as its magnetization. These spin-polarized electrons then flow through a non-magnetic spacer material until they reach a second magnetic layer. As long as the thickness of the spacer layer is less than about 100 nm, the spin direction of the electrons is preserved through the spacer. When the spin-polarized electrons reach the second magnetic layer, they undergo a second stage of spin filtering and in the process they can transfer some of their spin to the second magnetic layer, thereby applying a torque that can be strong enough to reorient the magnetization of this layer. Depending on the device design, this torque can be used either to switch the magnetic orientation of the second layer between two different angles, providing the basis of a magnetic memory, or it can be used to excite the magnetic layer into a dynamical state in which its magnetization precesses at gigahertz-scale frequencies.

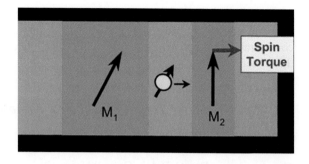

Fig. 4 Spin-polarized electrons produced by spin filtering from one magnetic layer can apply a strong torque to a second magnetic layer downstream (Courtesy of D. Ralph)

The first theoretical predictions of spin transfer torque were made in 1996 [79], and the first experiments which showed that the torque is strong enough to produce magnetic switching were performed in 2000 [21]. In the 10 years since, the science and technology of the effect have advanced very rapidly. Both large companies (Hitachi Ltd., Hynix, Intel, IBM, Micron, Qualcomm, Sony, Toshiba) and newer start-ups (Avalanche, Crocus, Everspin, Grandis, MagIC, Spin Transfer Technologies) have active development programs. Several have already demonstrated multi-megabit random access memory circuits in which spin torque is used to write and erase magnetic information. These memories can provide excellent performance in regard to speed, low cost, high density, and no wear-out, and have the added advantages that they are non-volatile (they retain information when power is off) and it should be possible to scale their device size to the same smallest dimensions possible for silicon processing. No silicon-based memory can combine all of these virtues (Fig. 5).

While the companies are somewhat secretive about recent progress, the first commercial spin-torque memory products are likely just 2–3 years from market. They are expected to be aimed initially at specialized applications, e.g., replacing battery-backed SRAM and providing radiation-hard nonvolatile memories for aerospace equipment. As companies grow more confident in the technology, the expectation is that spin-torque memories will spread to larger markets, with a particular focus on embedding memory in microprocessors to increase their performance and reduce power consumption. If successful in this area, spin-torque memories could acquire a several billion dollar per year share of the semiconductor memory market.

Other spin-torque-based devices may also enable entirely new types of technology. Based on the ability of spin torque to excite magnetic layers into dynamically-precessing states [80], it is possible to make microwave-signal sources that are unique in being only a few nanometers in size and frequency tunable. Spin-torque devices have also exhibited performance better than existing technologies for microwave detection and high-speed signal-processing functions. Research is underway to develop these capabilities to make spin-torque-based microwave sources and detectors for short-distance chip-to-chip communications and other applications.

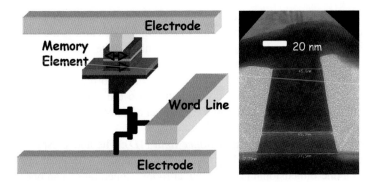

Fig. 5 Schematic of spin-torque memory circuit, and a cross section of an individual 40 nm wide memory element (Images courtesy of Daniel Worledge, IBM)

8.3 High-Performance Electronic Device Applications on Carbon Nanotubes

Contact person: J. Appenzeller, Purdue University

While many different applications have been explored based on carbon nanotubes (CNTs), this section will exclusively comment on major accomplishments, key applications and challenges in the area of high performance electronics which—in the author's opinion—utilizes the unique aspects of carbon nanotubes in the most promising way. The emphasis of the following sections is on high-speed switching at low-power consumption levels with carbon nanotubes as the key building block for future nanoelectronics.

8.3.1 Major Accomplishments

Ballistic transport: Electronic transport in condensed matter is typically impeded by scattering events that ultimately limit the switching speed of conventional field-effect transistor (FET) applications. Ideally, scattering free transport allows for ballistic transport conditions – the most desirable situation for high-performance device applications. Carbon nanotubes are to the best of my knowledge the only class of materials that have shown ballistic transport at room-temperature [81–84] as a result of the reduced phase space available for backscattering events. This property and the fact that CNTs are "natural" ultra-thin body channels are critical ingredients of the high-performance devices discussed in the next section.

Potential for low-power applications:
- *Band-to-band tunneling transistor*: The most effective way to reduce power consumption in electronic devices is to reduce the supply voltage. Ideally, this is

accomplished without sacrificing on-state performance. A device concept that is discussed by many groups worldwide to accomplish this task is the band-to-band tunneling transistor (T-FET). Carbon nanotubes were the first material that showed experimentally that the T-FET can indeed operate with an inverse sub-threshold slope well below 60 mV/dec [85]. Figure 6 illustrates the operation principle and the impact of the gate voltage on the band structure. Detailed simulation work further highlights the potential of carbon nanotubes for ultra-low power device applications [86, 87].

- *Operation in the quantum capacitance limit (1D)*: The above aspect is further supported by the critical finding that carbon nanotubes—an ideal implementation of a one-dimensional (1D) system at room temperature—can operate in the so-called quantum capacitance limit (QCL). Unlike conventional FETs, 1D systems allow controlling the band movement (i.e., the position of the conduction and valence band relative to the source and drain Fermi level) rather than the charge by means of the gate voltage [88] for thin but achievable gate dielectrics. This change in device operation has important implications for the power consumption of future generations of carbon nanotube devices. In fact, simulations [89] clearly show that while the gate delay scales with channel length as in classical devices the power delay product benefits substantially from the QCL regime. Since the quantum capacitance scales with the channel length L but is independent of the gate oxide thickness t_{ox}, shorter channel lengths translate into a smaller total capacitance – an effect that is normally compensated by the thinner gate oxide required to prevent short-channel effects in scaled FETs. This means that 1D CNTFETs (if properly scaled) will allow for a substantial reduction in power consumption if compared to their conventional counterparts.

Ring oscillator: In addition to the above device-related major accomplishments, the most critical breakthrough in terms of circuit implementations in the field of carbon nanotubes is certainly the experimental realization of a 5-stage ring oscillator on an individual carbon nanotube [90] (Fig. 7). The experimental demonstration not

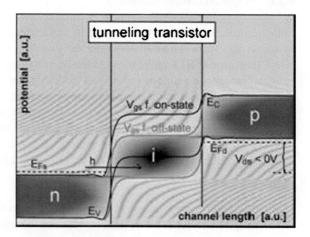

Fig. 6 Conduction and valence band for a p-type T-CNTFET in the on- and off-state. Holes are injected from the source for sufficiently negative gate voltages [85]

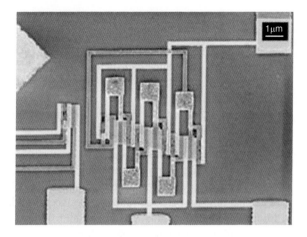

Fig. 7 Implementation of a CMOS-type 5-stage ring oscillator on a single carbon nanotube [90]

only shows that CMOS-type architectures can be combined with carbon nanotubes but also that work-function tuning of the metal gates is an option to adjust threshold voltages for CNTFETs.

8.3.2 Key Application Areas

Low-power electronics: The above discussed findings that carbon nanotubes are ballistic, one-dimensional conductors that can operate in the quantum capacitance limit and can be utilized to create novel devices—such as the band-to-band tunneling transistor—are the key for the author's conclusion about CNTs' particular usefulness for low-power device applications. Combining the intrinsic advantages of carbon nanotubes in future generations of devices and circuits is believed to be the most beneficial application realm.

Ultimate CMOS: Because of their ultra-small body *and* their excellent transport properties, CNTs also naturally lend themselves to ultimately scaled CMOS applications. While the demonstration of a CNT-based field-effect transistor with a channel length of < 10 nm is lacking today (2010), CNTs are believed to be scalable into this channel length regime while providing long channel-type device characteristics. Ideal electrostatic conditions translate into a superior suppression of short channel effects.

8.3.3 Key Challenges

Densely aligned nanotube arrays: For all of the above electronic applications to become reality, the use of individual CNTs is insufficient. Instead, it is mandatory to arrange large numbers of CNTs into parallel arrays with a pitch in the ~10 nm range.

A densely packed array of semiconducting CNTs will allow for the required current scaling that is typically accomplished in conventional circuits by adjusting the channel width of the individual devices. In order to recover this width-scaling without sacrificing performance by utilizing "too few" CNTs, a high packing density of CNTs is highly desirable and—despite great progress in this area—not yet state-of-the-art. Consequently, improvements and further invention will be needed to facilitate the formation of dense CNT arrays through the controlled synthesis of nanotubes of given diameter and chirality.

Low-power circuit implementation: To go beyond demonstrations on the device level and simple circuit implementations, there is the clear need to develop novel low-power circuits based on nanotubes' unique tunneling properties. Since it is to date unclear how far parasitics and interconnects will impact the total performance of future CNT-based circuits, it is important to study CNTs in this context in much greater detail to make future high-performance low-power electronic applications become a reality.

8.4 Graphene Electronics

Contact person: P. Avouris, IBM

The graphene field is barely 5 years old, and fundamental understanding of the science is still incomplete. For example, there are still uncertainties with respect to the nature of the scattering mechanisms that determine electrical transport in graphene. The nature of the interactions of graphene with insulating substrates and metal contacts is still not well understood.

A clear avenue for graphene applications is RF devices. The advantage lies not only in the achievement of extremely high operating frequencies that will enable new types of applications in communications and imaging, but also in the possibility of a simpler and cheaper fabrication technology than that required by present III-V semiconductor-based technologies. For this to be accomplished, a number of advances are needed, particularly in graphene synthesis. There is a need to develop graphene growth techniques on a large scale (for 8–12 in. wafers). Currently, the homogeneity of the synthesized graphene is not adequate, and strict control of layer thickness over large (wafer-scale) areas is not yet realized. Structural defects and impurities are not under full control, and their effect on the electrical properties of graphene is not well understood. Correspondingly, diagnostic techniques need to be developed for measuring homogeneity of growth, mobility, and sample purity on a large scale. This is likely to require integration of a variety of existing technologies, e.g., electron microscopy, interferometry, and Raman imaging in single tools.

Development of chemical vapor deposition (CVD) methods that could replace the expensive SiC approach is highly desirable. CVD on copper is particularly promising because copper does not dissolve carbon, and thus the graphene growth is self-limiting at the monolayer level. Dissolution of the copper substrate can be used to transfer graphene onto polymer substrates utilizing roll-to-roll technology, and the graphene can be used in applications requiring transparent conductive electrodes, such as photovoltaics and displays [91]. Further purification to remove dopants may enable the use of this approach in device applications.

While carriers in free graphene have very high mobilities, the deposition of insulating gate films on it—usually inorganic oxides such as SiO_2, HfO_2, and Al_2O_3—strongly degrades this mobility. Seeding layers between graphene and these oxides are needed to foster proper nucleation and prevent this degradation; research on such appropriate layers is needed [92].

Another area that needs attention is the thermal management of the devices. Since graphene is bonded to substrates by van der Waals forces, phonon flow and power dissipation are impeded; as a result, self-heating by current flow in graphene devices can be significant [93, 94].

Advances in device architecture are expected. Device scaling and minimization of parasitics would drastically improve performances, particularly power gain. High-breakdown-threshold, high-k dielectrics are needed. Such dielectrics would also allow the opening of larger bandgaps in bilayer graphene by applying higher E-fields. Such devices can potentially be used in digital devices and also as THz emitters and photodetectors.

For optoelectronic applications of graphene, increases in photoresponsivity are highly desirable. This may, for example, involve multilayer assemblies with separators to minimize current screening effects, integration with silicon waveguides, or plasmonic enhancement.

8.5 Heterogeneous Nanowire Devices

Contact person: T. Picraux, Center for Integrated Nanotechnologies (CINT), Sandia National Laboratories

Just as the major advances in planar-technology-based Si microelectronics over the last two decades have been largely based on the introduction of new materials, nanomaterials synthesis approaches are opening up new classes of electronic-quality heterogeneous materials for advanced device exploration. Bottom-up assembly of semiconducting nanowires opens the possibility to create new materials combinations and to achieve electronic and optoelectronic devices not possible by conventional planar processing. Examples include the CVD growth of Si, Ge, InAs, InGaP, AlGaN, ZnO, and other semiconductor nanowires based on the vapor-liquid-solid technique. Single crystalline combinations of these materials in both core-shell and axial heterostructured form (Fig. 8) can be synthesized to create new band offsets and strain-induced band modifications for bandgap-engineered device structures not previously imagined.

The lack of lateral constraint and strain sharing for these nanostructures means they can be fabricated without strain-relieving misfit dislocations, opening new

Fig. 8 Heterogeneous nanowire growth, showing structures with different materials introduced both vertically and radially during growth

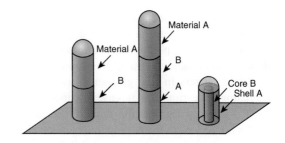

opportunities for high-performance electronic and optoelectronic devices. As an example, Si-Ge axial heterostructured tunneling field-effect transistors with better performance (higher drive currents and better subthreshold slopes) than possible with top-down homogeneous devices may be possible. Other possibilities include integrating optical emitters and detectors with logic and memory devices, and creating new spin-based devices. While the methods to grow such materials have been established, controlled device structures and electrical doping of new heterogeneous materials combinations and device possibilities remain to be explored.

A major challenge that nanoscience must address in order to exploit such new heterogeneous nanoscale devices is to develop methods to integrate these structures onto Si wafers, 3D flexible circuits, or other device platforms (see also Chapter "Synthesis, Processing, and manufacturing of Components, Devices, and Systems"). Methods to achieve high registry and reproducible structures for large arrays of heterogeneous devices will be essential for their eventual use. New methods to integrate bottom-up and top-down fabrication in large, highly reproducible arrays must be developed. Development of new approaches to device architectures and new combinations of processes for improved patterning and integration of the devices into overall structures is a grand challenge for nanoscience.

8.6 Biologically-Inspired Intelligent Physical Systems for Computing

Contact person: Kang Wang, University of California, Los Angeles (UCLA)

Nanoscience and nanoengineering offer the promise of new properties and improved materials. Yet with these new potentials comes challenges. The next wave of electronic systems beyond today's CMOS-dominated electronics needs to emerge from completely different concepts in order to further information systems beyond the limit of CMOS scaling. One possibility is intelligent physical systems using new nanoscale physical concepts and properties.

Our current understanding of the natural world lacks an effective description of the evolution of complexity evident in all living systems. The self-organized criticality (SOC) of nanoscale systems with an almost infinite correlation length (scale-invariant) may be explored to understand the emergence of intelligence, and it may be used to understand and solve some of the most highly interactive, complex problems. If realized, this approach may lead to an entirely new paradigm of information processing beyond today's computers. This new processing capability might be applied to understand complex problems as diverse as the complexity of living systems, the nature of human intelligence, and complex physical worlds and social issues. This capability would find broad utility in many applications, such as autonomous systems, decision making for complex problems (e.g., in finance), improved situational awareness and decision support in complex military engagements, internet traffic management, earth quake prediction, and other areas. New research directions should be directed to the study of SOC and its properties, as well as the utilization of SOC for information processing.

9 International Perspectives from Site Visits Abroad

9.1 United States-European Union Workshop (Hamburg, Germany)

Panel members/discussants

Clivia Sotomayor Torres (co-chair), Centre d'Investigació en Nanociencia i Nanotecnologia, Spain
Jeffrey Welser (co-chair), IBM, United States,
J-P Bourgouin Alternative Energies and Atomic Energy Commission (CEA), France
D. Dascalu, IMT Bucharest, Romania
Jozef Devreese, University of Antwerp, Belgium
M. Dragoman , IMT Bucharest, Romania
Mark Lundstrom, Purdue University, United States
A. Mueller, IMT Buchares, Romania
E. Nommiste, University of Tartu, Estonia
H. Pedersen, European Commission, Belgium
M. Penn, Future Horizons Ltd., UK
John Schmitz, NXP Semiconductors, Germany
Wolfgang Wenzel, Karlsruhe Institute for Technology, Germany

Overall, the workshop reinforced many of the same priorities as the previous U.S. workshop, but with additional emphasis being place on the use of nanoelectronics to augment current—the so-called "More than Moore" approach (Fig. 9). A more thorough review of the European nanoelectronics research strategy can be found in the proceedings from the EU Nanoelectronics Workshop 2009 (FP7 2009).

Listed below are the top ten research priorities—expanded from the six developed by the US workshop—that resulted from the meeting, with the EU additions in bold:

- Fabrication: Achieve 3D near-atomic-level control of reduced-dimensional materials.

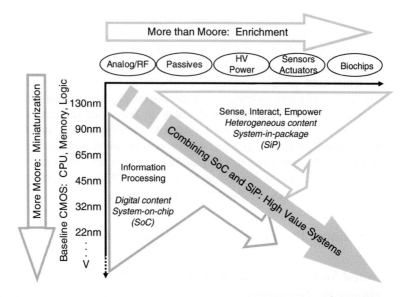

Fig. 9 Importance of pursuing alternate technologies to enable "More than Moore" (Courtesy of John Schmitz, "Reliability in the More than Moore Landscape", IEEE International Integrated Reliability Workshop; October 18, 2009; S. Lake Tahoe, California. http://www.iirw.org/09/IIRW_keynote_final_posting.pdf)

- Fabrication: Combine lithography and self-assembly to pattern semi-arbitrary structures with 1 nm precision, *possibly learning from and using biology and supramolecular chemistry.*
- Theory and Simulations: Concurrent development of multi-physics and multi-scale models with predictive capability for devices and circuits up to non-equilibrium systems.
- Devices: Discover devices for logic and memory that operate with greatly reduced energy dissipation, *exploring also sub-$k_B T$ information processes and stochastic processes.*
- Devices: Exploit spin for memory, logic and new functionality.
- Devices: Develop phonon physics and noise in physical systems applied to a single device and or function, towards reliable circuits and systems with lower switching energy and less energy demand for data transfer, for an overall superior thermal management.
- Devices: Exploit charge transfer and functions based on the Si-biology (Si-protein and Si-molecule) interfaces for heterogeneous integration at the nm-scale.
- Devices and systems: Incorporate analogue functionality, essential for sensors and actuator products. New architecture concepts to partition and integrate analogue and digital functions or blocks while meeting requirements concerning e.g. noise, linearity and energy.
- Architecture: Integrate architecture and nanoscale device research for unique computation functionality *and develop theory of information for computing.*
- Architecture: Increase focus on emerging, non-IT applications.

In addition to seeking new functionality through More than Moore research, these added priorities also indicate a strong focus on integrating fundamental science with the more technology-based work. There is a strong emphasis on learning from the biological / natural world to help advance not only fabrication technology, but also methods to manage noise and variability and even new methods of doing computation. There is also more emphasis placed on the need for improving analogue technology, both for current applications which do not scale well (e.g., RF transistors) and for interfacing to the non-digital world through sensors and actuators. All of this leads to an increased need for interdisciplinary teams that can link from materials & physics to technology & system design (Fig. 10), as well as modeling tools that can span all of these areas.

Lastly, there is increasing concern in the EU (shared in the United States) on how to ensure access to the facilities needed for this work. The increasing costs of maintaining a state-of-the-art research laboratory in nanoelectronics should be mitigated by a research infrastructure policy that includes access to both technology and simulation tools, thus bridging the gap between proof of concept and a potential viable technology. It is unlikely that there is a single type of infrastructure suitable for all research in Nanoelectronics. Coordination at regional and global levels is needed.

This cross-disciplinary, systems approach, as well as the closer relationship between research and manufacturing, should foster accelerated transfer of results—including unforeseen spin-off technology—out of the research labs and into the development and manufacturing world. Given the importance of nanoelectronics as not only the growth engine for the world economy over the past half century, but also its critical role in almost every major scientific and engineering discipline working to solve challenges in all parts of society, the US and EU researchers agreed it is crucial that both the research and resulting product innovations continue to progress.

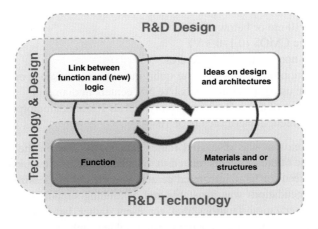

Fig. 10 Relationship between Design and Technology in Nanoelectronics (From EP FP7 project NANO-TEC, Coordinator CM Sotomayor Torres, http://www.fp7-nanotec.eu)

9.2 United States-Japan-Korea-Taiwan Workshop (Tokyo/Tsukuba, Japan)

Panel members/discussants

Ming-Huei Hong (co-chair), National Tsing Hua University (NTHU), Taiwan, ROC
Stu Wolf (co-chair), University of Virginia, United States
Masashi Kawasaki, Tohuku University, Japan
Yoshihige Suzuki, Osaka University, Japan
Kyounghoon Yang, KAIST, Korea
Naoki Yokoyama, Fujitsu Laboratories, Ltd., Japan
Iwao Ohdomari, JST, Japan
Tetsuji Yasuda, AIST, Japan
Shinji Yuasa, AIST, Japan

In many ways this workshop reinforced many of the same priorities of the U.S. workshop, including spintronics and spin torque devices for memory and logic, but like the European workshop there was a large focus on the "More than Moore" aspect of continuing the development of CMOS and CMOS related devices, moving ultimately to devices that are "beyond CMOS" like nano resonant tunneling diodes (nRTDs) for low power electronics utilizing non-Boolean multi-valued logic and neural networks. Some other areas that were discussed in some detail were directions for dissipationless electronics, including superconductivity and topological insulators, state variables beyond charge and spin using Mott transitions, i.e., Mottronics, and orbital degrees of freedom (Orbitronics). There was also more of an emphasis on quantum information than in the prior workshops, and this became one of the research goals (see below). There was also an emphasis on other areas of electronics beyond just logic and memory, and this included things such as photovoltaics and batteries, which were probably more thoroughly covered in other sessions.

There were five specific research goals for the next 5–10 years that were emphasized in this workshop:

- Achieving 10 nm or below CMOS.
- Low voltage CMOS (0.1–0.3 V).
- Logic with non-volatile architectures.
- Demonstration of a ten solid-state qubit quantum processor operating at room temperature using novel materials like NV centers in diamond.
- Demonstration of a non-Boolean, prototype RTD IC for neuromorphic processing with power reductions of factors of 10–100.

Another area that was discussed rather extensively was on the infrastructure needs for some of the problems related to the advancement of technologies as they mature. The need for large integrated centers was clearly emphasized. One of the major recommendations was for stable and continuing support for the large processing/characterization facilities that represent rather large investments. Also emphasized was the need for a continuing and stable supply of scientists, engineers, and technicians to support the research and the large facilities. Finally, it was recommended that an incentive should be provided to students from the lowest grade through high school to support the pursuit of nano endeavors and to encourage them to become scientists and engineers.

The final topic for discussion that related to overall strategies emphasized two key areas:

- Increase efforts to ensure the globalization of research, so that key results are properly disseminated throughout the community and that tools for education are also globalized.
- Assure that nanoelectronics remains a key technology area for research, because it provides a foundation for many other areas including biomedical, energy, information and communication technologies.

9.3 United States-Australia-China-India-Saudi Arabia-Singapore Workshop (Singapore)

Panel members/discussants

Andrew Wee (co-chair), National University of Singapore
Stu Wolf (co-chair), University of Virginia, United States
Michelle Simmons, University of New South Wales, Australia
Wei Huang, Nanjing University of Posts and Telecommunications, China
Yong Lim Foo, Institute of Materials Research & Engineering (IMRE), Singapore

The main additions to the Chicago workshop vision for the next 10 years were threefold:

- New approaches to molecular electronics.
- Increased integration of disparate materials, e.g., Ge and III-V's on Si and inorganic-organic hybrids.
- Increased focus on Quantum information including developing multi-qubit, scalable architectures and coupling optical, microwave and solid state systems.

The following chart (Fig. 11) was discussed in some detail despite the fact that it came from the International Technology Roadmap for Semiconductors (2009).

The goals for 2020 were discussed using the Chicago meeting as the template, and the goal of achieving 3D near-atomic level control of reduced dimensional material was validated, but with the following additions: One of the perceived barriers, in addition to the control and characterization of materials at the atomic scale, was the ability to integrate tools at different length scales for fast, cheap fabrication. For solutions, we added the use of controlled chemical synthesis of atomic scale structures, with the example of growth and integration of graphene nano-ribbons and inorganic nanowires. The fabrication and device goals from the Chicago meeting were discussed and validated, but there was an important goal that was added: exploiting the quantum state for computation and secure communication. The barriers to this included the lack of a current pathway to scale-up to a size that would outperform conventional computation, the inability to transport a quantum

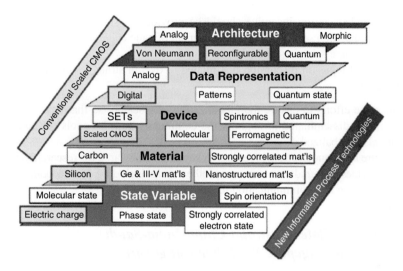

Fig. 11 International Roadmap for Semiconductors: a taxonomy for emerging research in nanoinformation processing devices (the technology entries are representative but not comprehensive) (From International Roadmap for Semiconductors 2009)

Development of quantum information processing

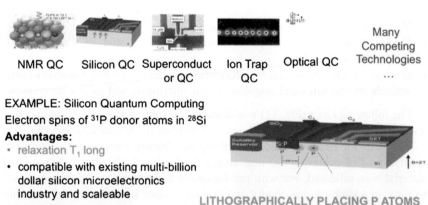

NMR QC   Silicon QC   Superconduct   Ion Trap   Optical QC   Many Competing Technologies
                        or QC   QC

EXAMPLE: Silicon Quantum Computing
Electron spins of ^{31}P donor atoms in ^{28}Si

Advantages:
- relaxation T_1 long
- compatible with existing multi-billion dollar silicon microelectronics industry and scaleable

Challenges:
- require the ability to dope Si with atomic precision aligned to nanometer sized surface gates in 3D
  → scanning probes and MBE
  → directed self assembly

LITHOGRAPHICALLY PLACING P ATOMS
Fine   Atomic

Fig. 12 Quantum information processing approaches (Courtesy of M. Simmons)

state, and a lack of multi-scale modeling and simulation. The solution, of course, would be to investigate different architectures, integrate solid state, optical and microwave approaches, and develop improved modeling and simulation tools.

Michelle Simmons presented a slide (Fig. 12) that depicted the development of a quantum information processor at the University of New South Wales that has been making significant progress over the last decade. This slide represents a very nice nanoscale tour de force.

A final addition to the Chicago meeting was the development of a new global R&D strategy that would emphasize the formation of international networks to leverage complementary strengths.

References

1. G.E. Moore, Cramming more components onto integrated circuits. Electronics **38**(8), 114–117 (1965)
2. R.H. Dennard, F.H. Gaensslen, H.-N. Yu, V.L. Rideout, E. Bassous, A.R. LeBlanc, Design for ion-implanted MOSFET's with very small physical dimensions. IEEE J. Solid State Circ. **SC-9**(5), 256–268 (1974)
3. D.J. Frank, Power-constrained device and technology design for the end of scaling. Int. Electron. Devices. Meet. (IEDM) 2002 Dig, 643–646 (2002). doi: 10.1109/IEDM.2002.1175921
4. B.H. Lee, S.C. Song, R. Choi, P. Kirsch, Metal electrode/high-k dielectric gate-stack technology for power management. IEEE Trans. Electron Devices **55**(1), 8–20 (2008). doi:10.1109/TED.2007.911044
5. S. Sivakumar, Lithography challenges for 32 nm technologies and beyond. Int. Electron. Device. Meet. **2006**, 1–4 (2006). doi:10.1109/IEDM.2006.346952
6. B. Wu, A. Kumar, Extreme ultraviolet lithography: a review. J. Vac. Sci. Technol. B Microelectron. Nanometer Struct. **25**(6), 1743–1761 (2007). doi:10.1116/1.2794048
7. H. Schift, Nanoimprint lithography: an old story in modern times? A review. J. Vac. Sci. Technol. B Microelectron. Nanometer Struct. **26**(2), 458–480 (2008). doi:10.1116/1.2890972
8. D. Pires, J.L. Hedrick, A. De Silva, J. Frommer, B. Gotsmann, H. Wolf, D. Michel, U. Duerig, A.W. Knoll, Nanoscale three-dimensional patterning of molecular resists by scanning probes. Science **328**, 732–735 (2010)
9. G.S. Craig, P.F. Nealey, Exploring the manufacturability of using block copolymers as resist materials in conjunction with advanced lithographic tools. J. Vac. Sci. Technol. B **25**(6), 1969–1975 (2007). doi:10.1116/1.2801888
10. W. Lu, A.M. Sastry, Self-assembly for semiconductor industry. IEEE Trans. Semicond. Manuf. **20**(4), 421–431 (2007). doi:10.1109/TSM.2007.907622
11. S. Tiwari, F. Rana, K. Chan, H. Hanafi, Wei Chan, D. Buchanan, Int. Electron. Devices. Meet., 521–524 (1995). doi: 10.1109/IEDM.1995.499252
12. K.-M. Chang, Silicon nanocrystal memory: technology and applications. Int Solid-State Integr Circ Technol (ICSICT '06), 25–728 (2006). doi: 10.1109/ICSICT.2006.306469
13. Freescale Semiconductor, Inc, Thin film storage (TFS) with flex memory technology (2010), Available at: http://www.freescale.com/webapp/sps/site/overview.jsp?nodeId=0ST287482188DB3. Accessed 14 May 2010
14. G.W. Burr, M.J. Breitwisch, M.F. Franceschini, D. Garetto, K. Gopalakrishnan, B. Jackson, B. Kurdi, C. Lam, L.A. Lastras, A. Padilla, B. Rajendran, S. Raoux, R.S. Shenoy, Phase change memory technology. J. Vac. Sci. Technol. B **28**, 223–262 (2010). doi:10.1116/1.3301579
15. L.O. Chua, Memristor: the missing circuit element. IEEE Trans. Circuit Theory **CT18**, 507–519 (1971). doi:0.1109/TCT.1971.1083337

16. D.B. Strukov, G.S. Snider, D.R. Stewart, R.S. Williams, The missing memristor found [Letter]. Nature **453**(7191), 80–83 (2008). doi:10.1038/nature06932
17. J.J. Yang, M.D. Pickett, X. Li, D.A.A. Ohlberg, D.R. Stewart, R.S. Williams, Memristive switching mechanism for metal-oxide-metal nanodevices. Nat. Nanotechnol. **3**, 429–433 (2008)
18. P. Vontobel, W. Robinett, J. Straznicky, P.J. Kuekes, R.S. Williams, Writing to and reading from a nano-scale crossbar memory based on memristors. Nanotechnology **20**, 425204 (2009)
19. J. Borghetti, Z. Li, J. Strasnicky, X. Li, D.A.A. Ohlberg, W. Wu, D.R. Stewart, R.S. Williams, A hybrid nanomemristor/transistor logic circuit capable of self-programming. Proc. Natl. Acad. Sci. U.S.A. **106**, 1699–1703 (2009)
20. J. Borghetti, G.S. Snider, P.J. Kuekes, J.J. Yang, D.R. Stewart, R.S. Williams, 'Memristive' switches enable 'stateful' logic operations via material implication. Nature **464**, 873–876 (2010). doi:10.1038/nature08940
21. J.A. Katine, F.J. Albert, R.A. Buhrman, E.B. Myers, D.C. Ralph, Current-driven magnetization reversal and spin-wave excitations in Co/Cu/Co pillars. Phys. Rev. Lett. **84**, 3149–3152 (2000)
22. M. Jurczak, N. Collaert, A. Veloso, T. Hoffmann, S. Biesemans, Review of FINFET technology. IEEE Int. SOI Conf, 1–4 (2009). doi:10.1109/SOI.2009.5318794
23. W. Lu, Nanowire based electronics: challenges and prospects. IEDM **2009**, 1–4 (2009). doi:10.1109/IEDM.2009.5424283
24. P. Avouris, J. Appenzeller, R. Martel, S. Wind, Carbon nanotube electronics. Proc. IEEE **91**(11), 1772–1784 (2003). doi:10.1109/JPROC.2003.818338
25. R.K. Cavin, V.V. Zhirnov, Silicon nanoelectronics and beyond: reflections from a semiconductor industry-government workshop. J. Nanopart. Res. **8**, 137–147 (2004)
26. R.K. Cavin, V.V. Zhirnov, G.I. Bourianoff, J.A. Hutchby, D.J.C. Herr, H.H. Hosack, W.H. Joyner, T.A. Wooldridge, A long-term view of research targets in nanoelectronics. J. Nanopart. Res. **7**, 573–586 (2005)
27. R.K. Cavin, V.V. Zhirnov, D.J.C. Herr, A. Avila, J. Hutchby, Research directions and challenges in nanoelectronics. J. Nanopart. Res. **8**, 841–858 (2006)
28. J.J. Welser, G.I. Bourianoff, V.V. Zhirnov, R.K. Cavin, The quest for the next information processing technology. J. Nanopart. Res. **10**(1), 1–10 (2008)
29. K.S. Novoselov, A.K. Geim, S.V. Morozov, D. Jiang, Y. Zhang, S.V. Dubonos, I.V. Grigorieva, A.A. Firsov, Electric field effect in atomically thin carbon films. Science **306**, 666–669 (2004). doi:10.1126/science.1102896
30. J. Wang, J.B. Neaton, H. Zheng, V. Nagarajan, S.B. Ogale, B. Liu, D. Viehland, V. Vaithyanathan, D.G. Schlom, U.V. Waghmare, N.A. Spaldin, K.M. Rabe, M. Wuttig, R. Ramesh, Epitaxial $BiFeO_3$ multiferroic thin film heterostructures. Science **299**, 1719 (2003)
31. M.I. D'yakonov, V.I. Perel', Possibility of orientating electron spins with current. JETP Lett. **13**, 467–469 (1971)
32. Y.K. Kato, R.C. Myers, A.C. Gossard, D.D. Awschalom, Observation of the spin Hall effect in semiconductors. Science **306**, 1910 (2004)
33. J. Wunderlich, B. Kaestner, J. Sinova, T. Jungwirth, Experimental observation of the spin-Hall effect in a two-dimensional spin-orbit coupled semiconductor system. Phys. Rev. Lett. **94**, 047204 (2005)
34. B.A. Bernevig, T.L. Hughes, S.-C. Zhang, Quantum spin hall effect and topological phase transition in HgTe quantum wells. Science **314**(5806), 1757 (2006). doi:10.1126/science.1133734
35. Y.L. Chen, J.G. Analytis, J.-H. Chu, Z.K. Liu, S.-K. Mo, X.L. Qi, H.J. Zhang, D.H. Lu, X. Dai, Z. Fang, S.C. Zhang, I.R. Fisher, Z. Hussain, Z.-X. Shen, Experimental realization of a three-dimensional topological insulator, Bi2Te3. Science **325**, 178–181 (2009). doi:10.1126/science.1173034
36. M. König, S. Wiedmann, C. Brüne, A. Roth, H. Buhmann, L.W. Molenkamp, X.-L. Qi, S.-C. Zhang, Quantum spin hall insulator state in HgTe quantum wells. Science **318**(5851), 766–770 (2007). doi:10.1126/science.1148047
37. T. Dietl, H. Ohno, F. Matsukura, J. Cibert, D. Ferrand, Zener model description of ferromagnetism in zinc-blend magnetic semiconductors. Science **287**, 1019 (2000)

38. H. Ohno, A. Shen, F. Matsukura, A. Oiwa, A. Endo, S. Katsumoto, Y. Iye, (Ga, Mn)As: a new diluted magnetic semiconductor based on GaAs. Appl. Phys. Lett. **69**, 363 (1996). doi:10.1063/1.118061
39. K. Terabe, T. Hasegawa, T. Nakayama, M. Aono, Quantized conductance atomic switch. Nature **433**, 47–49 (2005)
40. P.R. Wallace, The band theory of graphite. Phys. Rev. **71**, 622–634 (1947). doi:10.1103/PhysRev.71.622
41. J.W. May, Platinum surface LEED rings. Surf. Sci. **17**, 267–270 (1969). doi:10.1016/0039-6028(69)90227-1
42. A.J. Van Bommel, J.E. Crombeen, A. Van Tooren, LEED and Auger electron observations of the SiC(0001) surface. Surf. Sci. **48**, 463–472 (1975). doi:10.1016/0039-6028(75)90419-7
43. H.P. Boem, A. Clauss, G.O. Fischer, U. Hofmann, Thin carbon leaves. Z. Naturforsch. **17b**, 150–153 (1962)
44. K.S. Novoselov, A.K. Geim, S.V. Morozov, D. Jiang, M.I. Katsnelson, I.V. Grigorieva, S.V. Dubonos, A.A. Firsov, Two-dimensional gas of massless Dirac fermions in graphene. Nature **438**, 197–200 (2005). doi:10.1038/nature04233
45. Y. Zhang, Y.-W. Tan, H.L. Stormer, P. Kim, Experimental observation of the quantum Hall effect and Berry's phase in graphene. Nature **438**, 201–204 (2005)
46. A.K. Geim, A.K. Novoselov, The rise of graphene. Nat. Mater. **6**, 183–191 (2007). doi:10.1038/nmat1849
47. K.I. Bolotin, K.J. Sikes, Z. Jiang, G. Fundenberg, J. Hone, P. Kim, H.L. Stormer, Ultrahigh electron mobility in suspended graphene. Solid State Commun. **146**, 351–355 (2008). doi:10.1016/j.ssc.2008.02.024
48. A.K. Geim, Graphene: status and prospects. Science **324**, 1530–1534 (2009). doi:10.1126/science.1158877
49. E. McCann, Asymmetry gap in the electronic band structure of bilayer graphene. Phys. Rev. B **74**, 161403(R) (2006). doi:10.1103/PhysRevB.74.161403
50. J.B. Oostinga, H.B. Heersche, X. Liu, A.F. Morpurgo, L.M.K. Vandersypen, Gate-induced insulating state in bilayer graphene devices. Nat. Mater. **7**, 151–157 (2008). doi:10.1038/nmat2082
51. F. Xia, D.B. Farmer, Y.-M. Lin, P. Avouris, Graphene field-effect transistors with high on/off current ratio and large transport band gap at room temperature. Nano Lett. **10**, 715–718 (2010). doi:10.1021/nl9039636
52. C. Berger, Z. Song, T. Li, X. Li, A.Y. Ogbazghi, R. Feng, Z. Dai, A.N. Marchenkov, E.H. Conrad, P.N. First, W.A. de Heer, Ultrathin epitaxial graphite: 2D electron gas properties and a route toward graphene-based nanoelectronics. J. Phys. Chem. B **108**, 19912–19916 (2004). doi:10.1021/jp040650f
53. A. Reina, X. Jia, J. Ho, D. Nezich, H. Son, V. Bulovic, M.S. Dresselhaus, J. Kong, Large area, few-layer graphene films on arbitrary substrates by chemical vapor deposition. Nano Lett. **9**, 30–35 (2009). doi:10.1021/nl801827v
54. X. Li, W. Cai, J. An, S. Kim, J. Nah, D. Yang, R. Piner, A. Velamakanni, I. Jung, E. Tutuc, S.K. Banerjee, L. Colombo, R. Ruoff, Large-area synthesis of high-quality and uniform graphene films on copper foils. Science **324**, 1312–1314 (2009). doi:10.1126/science.1171245
55. Y.-M. Lin, H.-Y. Chiu, K.A. Jenkins, D.B. Farmer, P. Avouris, A. Valdes-Garcia, Dual-gate graphene FETs with of 50 GHz. IEEE Electron Device Lett. **31**, 68–70 (2010). doi:0.1109/LED.2009.2034876
56. J.S. Moon, D. Curtis, M. Hu, D. Wong, C. McGuire, P.M. Campbell, G. Jernigan, J.L. Tedesco, B. VanMil, R. Myers-Ward, C. Eddy, D.K. Gaskill, Epitaxial-graphene RF field-effect transistors on Si-face 6 H-SiC substrates. IEEE Electron Device Lett. **30**, 650–652 (2009). doi:10.1109/LED.2009.2020699
57. Y.-M. Lin, C. Dimitrakopoulos, K.A. Jenkins, D.B. Farmer, H.-Y. Chiu, A. Grill, Ph Avouris, 100-GHz transistors from wafer-scale epitaxial graphene. Science **327**(5966), 662 (2010). doi:10.1126/science.1184289

58. F. Xia, T. Mueller, Y.-M. Lin, A. Valdes-Garcia, Ph Avouris, Ultrafast graphene photodetector [Letter]. Nat. Nanotechnol. **4**, 839–843 (2009). doi:10.1038/nnano.2009.292
59. T. Mueller, F. Xia, P. Avouris, Graphene photodetectors for high-speed optical communications [Letter]. Nat. Photonics **4**, 297–301 (2010). doi:10.1038/nphoton.2010.40
60. V.V. Chelanov, V. Fal'ko, B. Altshuler, The focusing of electron flow and a Veselago lens in graphene p-n junctions. Science **315**, 1252–1255 (2007)
61. S.K. Banerjee, L.F. Register, E. Tutuc, D. Reddy, A.H. MacDonald, Bilayer pseudospin field-effect transistor (BiSFET): a proposed new logic device. IEEE Electron Devices. Lett. **30**(2), 158–160 (2009)
62. H.Min, G. Borghi, M. Polini, A.H. MacDonald, Pseudospin magnetism in graphene. Phys. Rev. B **77**(17 Jan), 041407–1 (2008). doi: 10.1103/PhysRevB.77.041407
63. J.-J. Su, A.H. MacDonald, How to make a bilayer exciton condensate flow. Nat. Phys. **4**, 799–802 (2008)
64. T. Miyazaki, N. Tezuka, Giant magnetic tunneling effect in Fe/Al_2O_3/Fe junction. J. Magn. Magn. Mater. **139**, L231–L234 (1995)
65. J.S. Moodera, L.R. Kinder, T.M. Wong, R. Meservey, Large magnetoresistance at room-temperature in ferromagnetic thin-film tunnel-junctions. Phys. Rev. Lett. **74**, 3273–3276 (1995)
66. S. Tehrani, J.M. Slaughter, E. Chen, M. Durlam, J. Shi, M. DeHerrera, Progress and outlook for MRAM technology. IEEE Trans. Magn. **35**, 2814–2819 (1999)
67. B.N. Engel, J. Akerman, B. Butcher, R.W. Dave, M. DeHerrera, M. Durlam, G. Grynkewich, J. Janesky, S.V. Pietambaram, N.D. Rizzo, J.M. Slaughter, K. Smith, J.J. Sun, S. Tehrani, A 4-Mb toggle MRAM based on a novel bit and switching method. IEEE Trans. Magn. **41**, 132–136 (2005)
68. W.H. Butler, X.G. Zhang, T.C. Shulthess, J.M. MacLaren, Spin-dependent tunneling conductance of Fe/Mg.OFe sandwiches. Phys. Rev. B **63**, 0544416 (2001)
69. S.S.P. Parkin, M. Hayashi, L. Thomas, Magnetic domain wall racetrack memory. Science **320**, 209–211 (2008)
70. S.A. Wolf, D.D. Awschalom, R.A. Buhrman, J.M. Daughton, S. Von Molnar, M.L. Roukes, A.Y. Chelkanova, D.M. Treger, Spintronics: a spin based electronics vision for the future. Science **294**, 1488–1495 (2001)
71. H.M. Saavedra, T.J. Mullen, P.P. Zhang, D.C. Dewey, S.A. Claridge, P.S. Weiss, Hybrid approaches in nanolithography. Rep. Prog. Phys. **73**, 036501 (2010). doi:10.1088/0034-4885/73/3/036501
72. W. Eerenstein, N.D. Mathur, J.F. Scott, Multiferroic and magnetoelectric materials. Nature **442**(17), 759–765 (2006). doi:10.1038/nature05023
73. D. Jorgenson, Moore's law and the emergence of the new economy. Semiconductor Industry Association 2005 annual report: 2020 is closer than you think, pp. 16–20 (2005), Available online: http://www.sia-online.org/galleries/annual_report/Annual%20Report%202005.pdf
74. U.S. Environmental Protection Agency (U.S. EPA). (August 2). Report to Congress on server and data center energy efficiency Public Law 109–431 (2007), Available online: http://www.energystar.gov/
75. American Council for an Energy-Efficient Economy, Semiconductor technologies: the potential to revolutionize U.S. energy productivity (ACEEE, Washington, D.C, 2009), Available online: http://www.aceee.org/pubs/e094.htm
76. S.A. Wolf, J. Lu, M. Stan, E. Chen, D.M. Treger, The promise of nanomagnetics and spintronics for future logic and universal memory. *Proc. IEEE.* **98**, 2155–2168 (2010), Available at: http://ieeexplore.ieee.org/stamp/stamp.jsp?arnumber=05640335. Accessed 12 Dec 2010
77. A. Orlov, A. Imre, G. Csaba, L. Ji, W. Porod, G.H. Bernstein, Magnetic quantum-dot cellular automata: recent developments and prospects. J. Nanoelectron. Optoelectron. **3**, 55–68 (2008)
78. R. Hanson, D.D. Awschalom, Coherent manipulation of single spins in semiconductors. Nature **453**, 1043–1049 (2008)
79. J.C. Slonczewski, Current-driven excitation of magnetic multilayers. J. Magn. Magn. Mater. **159**, L1–L7 (1996)

80. S.I. Kiselev, J.C. Sankey, I.N. Krivorotov, N.C. Emley, R.J. Schoelkopf, R.A. Buhrman, D.C. Ralph, Microwave oscillations of a nanomagnet driven by a spin-polarized current. Nature **425**, 380–383 (2003)
81. T. Durkop, S.A. Getty, E. Cobas, M.S. Fuhrer, Extraordinary mobility in semiconducting carbon nanotubes. Nano Lett. **4**, 35–39 (2004)
82. A.D. Franklin, G. Tulevski, J.B. Hannon, Z. Chen, Can carbon nanotube transistors be scaled without performance degradation? IEEE IEDM Tech. Dig. 561–564 (2009)
83. A. Javey, J. Guo, Q. Wang, M. Lundstrom, H. Dai, Ballistic carbon nanotube field-effect transistors. Nature **424**, 654–657 (2003)
84. S.J. Wind, Appenzeller, Ph Avouris, Lateral scaling in carbon nanotube field-effect transistors. Phys. Rev. Lett. **91**, 058301-1–058301-4 (2003)
85. J. Appenzeller, Y.-M. Lin, J. Knoch, Ph Avouris, Band-to-band tunneling in carbon nanotube field-effect transistors. Phys. Rev. Lett. **93**, 196805-1–196805-4 (2004)
86. J. Appenzeller, Y.-M. Lin, J. Knoch, Ph Avouris, Comparing carbon nanotube transistors – the ideal choice: a novel tunneling device design. IEEE Trans. Electron Devices **52**, 2568–2576 (2005)
87. S.O. Koswatta, D.E. Nikonov, M.S. Lundstrom, Computational study of carbon nanotube p-i-n tunneling FETs. IEDM Tech. Dig. **2005**, 518–521 (2004)
88. J. Knoch, J. Appenzeller, Carbon nanotube field-effect transistors—The importance of being small, in *AmIware, hardware technology drivers of ambient intelligence*, ed. by S. Mukherjee, E. Aarts, R. Roovers, F. Widdershoven, M. Ouwerkerk (Springer, New York, 2006), pp. 371–402
89. J. Knoch, W. Riess, J. Appenzeller, Outperforming the conventional scaling rules in the quantum capacitance limit. IEEE Electron Devices Lett. **29**, 372–374 (2008)
90. Z. Chen, J. Appenzeller, Y.-M. Lin, J.S. Oakley, A.G. Rinzler, J. Tang, S. Wind, P. Solomon, Ph Avouris, An integrated logic circuit assembled on a single carbon nanotube. Science **311**, 1735 (2006)
91. S. Bae, H.K. Kim, Y. Lee, X. Xu, J.-S. Park, Y. Zheng, J. Balakrishnan, D. Im, T. Lei, Y.I. Song, Y.J. Kim, K.S. Kim, B. Özyilmaz, J.-H. Ahn, B.H. Hong, S. Iijima, 30-inch roll-based production of high-quality graphene films for flexible transparent electrodes. Mater. Sci. (2010). doi:arXiv:0912.5485 [cond-mat.mtrl-sci]. Forthcoming
92. D.B. Farmer, R. Golizadeh-Mojarad, V. Perebeinos, Y.-M. Lin, G.S. Tulevski, J.C. Tsang, P. Avouris, Chemical doping and electron–hole conduction asymmetry in graphene devices. Nano Lett. **9**, 388–392 (2009). doi:10.1021/nl803214a
93. D.H. Chae, B. Krauss, K. von Klitzing, J.H. Smet, Hot phonons in an electrically biased graphene constriction. Nano Lett. **10**, 466–471 (2010). doi:10.1021/nl903167f
94. M. Freitag, M. Steiner, Y. Martin, V. Perebeinos, Z. Chen, J.C. Tsang, P. Avouris, Energy dissipation in graphene field-effect transistors. Nano Lett. **9**, 1883–1888 (2009). doi:10.1021/nl803883h
95. L. Berger, Emission of spin waves by a magnetic multilayer traversed by a current. Phys. Rev. B **54**, 9353–9358 (1996)

Applications: Nanophotonics and Plasmonics

Evelyn L. Hu, Mark Brongersma, and Adra Baca

Keywords Nanophotonics • Plasmonics • Micro- and Nanocavities • Circuits • International perspective

1 Vision for the Next Decade

Both nanophotonics and plasmonics concern investigations into building, manipulating, and characterizing optically active nanostructures with a view to creating new capabilities in instrumentation for the nanoscale, chemical and biomedical sensing, information and communications technologies, enhanced solar cells and lighting, disease treatment, environmental remediation, and many other applications. Photonics and plasmonics share the characteristic that at least some of their basic concepts have been known for 40–50 years, but they have come into their own only in the last 10 years, based on recent discoveries in nanoscience. Photonic materials and devices have played a pervasive role in communications, energy conversion, and sensing since the 1960s and 1970s. Photonics at the nanoscale, or *nanophotonics* might be defined as "the science and engineering of light-matter interactions that take place on wavelength and subwavelength scales where the physical, chemical, or structural nature of natural or artificial nanostructure matter controls the interactions" [1]. Broadly speaking, over the next 10 years

---

E.L. Hu (✉)
Harvard School of Engineering and Applied Sciences, 29 Oxford Street,
Cambridge, MA 02138, USA
e-mail: ehu@seas.harvard.edu

M. Brongersma
Stanford University, Durand Building, 496 Lomita Mall, Stanford, CA 94305-4034, USA

A. Baca
Corning, Inc., 1 Riverfront Plaza, Corning, NY 14831-0001, USA

nanophotonic structures and devices promise dramatic reductions in energies of device operation, densely integrated information systems with lower power dissipation, enhanced spatial resolution for imaging and patterning, and new sensors of increased sensitivity and specificity.

Plasmonics aims to exploit the unique optical properties of metallic nanostructures to enable routing and active manipulation of light at the nanoscale [2–4]. It has only been in the past 10 years that the young field of plasmonics has rapidly gained momentum, enabling exciting new fundamental science as well as groundbreaking real-life applications in the coming 10 years in terms of targeted medical therapy, ultrahigh-resolution imaging and patterning, and control of optical processes with extraordinary spatial and frequency precision. In addition, because plasmonics offers a natural integration compatibility with electronics and the speed of photonics, circuits and systems formed of plasmonic and electronic devices hold promise for next-generation systems that will incorporate the best qualities of both photonics and electronics for computation and communication at high speed, broad bandwidth, and low power dissipation

1.1 Changes in the Vision over the Last 10 Years

Despite their relevance to many key technologies going forward, neither nanophotonic nor plasmonics devices were explicitly considered in the 1999 report *Nanotechnology research directions: Vision for nanotechnology R&D in the next decade* [5]. In part, this may have resulted from the fact that one-dimensional nanoscale structures, such as antireflection coatings and distributed Bragg mirrors, have long been a part of optical design and engineering. The first quantum well laser, operating at room temperature with an active layer 20 nm thick, was reported in 1978 [6]; today quantum well lasers are the standard for room-temperature solid-state semiconductor devices; they have high efficiency and can be manufactured at relatively low cost.

Many of the critical concepts for today's rapidly growing and diverse field of nanophotonics were established in past decades, but recent progress in the ability to control materials at the nanoscale in multiple dimensions has allowed the validation of those concepts and the anticipation of yet more intriguing and powerful photonic behaviors. For example, an early paper by Yablonovitch [7] discussed dielectric materials with spatial variations in index of refraction on the order of a wavelength of light. He anticipated that these *photonic crystals* would have a dramatic effect on the spontaneous emission of light within these structures. Negative index materials, a component of a class of *metamaterials,* were anticipated as early as the 1960s by Veselago [8]. *Surface plasmons* have played an important role in surface-enhanced Raman spectroscopy (SERS), an active area of research since the late 1970s [9, 10]. Today's research in plasmonics encompasses an even broader range of structures and applications [11], as is the case for nanophotonics.

1.2 Vision for the Next 10 Years

Advances in the fabrication of optical structures at the nanoscale and improved control of materials properties have allowed researchers to demonstrate and realize the potential of nanophotonics and plasmonics, and they provide strong impetus for further investments in these fields. In the next 10 years, we envision that nanophotonics and plasmonics will have dramatic enabling capabilities for new medical therapies; low-power, high-bandwidth, and high-density computation and communications; high-spatial-resolution imaging and sensing with high spectral and spatial precision; efficient optical sources and detectors; and a host of profound scientific discoveries about the nature of light–matter interactions.

2 Advances in the Last 10 Years and Current Status

2.1 Nanophotonics

Over the last 5–10 years we have witnessed an enormous increase in the number and diversity of photonics applications. This has resulted from the significant advances in computational design tools and their accessibility, the emergence of new nanofabrication techniques, and the realization of new optical and structural characterization methods. At the forefront of these advances are the developments in the area of micro- and nano-photonic devices that have dimensions on the order of or below the wavelength of light.

Advances in electromagnetic and electronic device simulation tools have been tremendous over the last decade. Good commercial and freeware codes (e.g., finite difference time domain (FDTD), discrete dipole approximation (DDA), boundary element methods (BEM), or finite-difference frequency domain (FDFD)) codes are inexpensive and available to virtually everyone. Improvements in nanofabrication techniques, greater accessibility of high-resolution patterning (e.g., electron beam lithography), and pattern transfer processes (e.g., low-damage ion-assisted etching) have produced photonic crystal, microdisk, and ring resonator devices with exceptional performance.

Discovery in nanophotonics has been enabled by the accessibility of optical nanoscale characterization tools such as scanning near-field optical microscopy (SNOM). Accessibility results from the increasing number of commercial companies that sell such equipment and because the tools have become substantially more user-friendly.

Similarly, progress in nanophotonics has benefited from advances in structural characterization tools such as atomic force microscopy (AFM), nano-Auger, nano-secondary ion mass spectrometry (nano-SIMS), scanning electron microscopes (SEMs), and transmission electron microscopes (TEM). These instruments have had a major impact on the ability to correlate the size, atomic structure, and spatial arrangements of nanostructure to the observed optical properties.

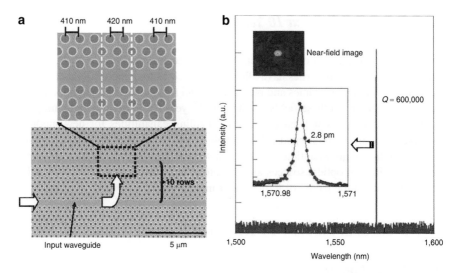

Fig. 1 (a) SEM of fabricated photonic crystal structure. (b) spectrum of the cavity, showing Q = 600,000 [12]

The progress in both simulation and fabrication of nanophotonic structures has resulted in the formation of ultrahigh-quality (Q, meaning low-optical-loss) optical structures [12] (Fig. 1). These structures, in turn, have allowed researchers to engineer distinctive optical states, localize and slow the velocity of light, and create efficient light-emitting sources and strongly coupled light–matter interactions, resulting in new quantum mechanical states.

In particular, for dielectric materials:

- *Slowing of light* has been observed in photonic crystal waveguides [13, 14] and in coupled ring resonators [15]. This is an achievement not only scientifically, but also for the implication in controlled delay and storage of light in compact, on-chip information processing (Fig. 2).
- *Strong coupling between dielectric nanocavities and quantum dots* has by now been observed by a number of research groups [17–19]. The exciton-photon (polariton) states that result can form the basis of quantum information schemes, or produce ultra-low threshold lasers.

2.2 Plasmonics

Although surface-enhanced Raman spectroscopy (SERS), one of the first "killer applications" of metallic nanostructures, was discovered in the 1970s [10, 20, 21], the young field of plasmonics only started to rapidly spread into new directions in the late 1990s and early 2000s. At that time it was demonstrated in rapid succession

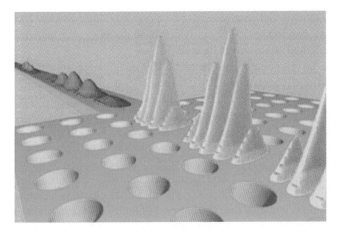

Fig. 2 Illustration of pulse compression and intensity increase of pulse after entering the slow-light regime [16]

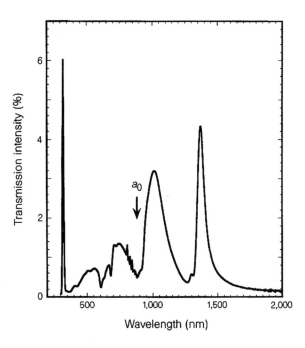

Fig. 3 Zero-order transmission spectrum for a square array of 150 nm holes with a period of 900 nm in a 200 nm thick Ag film. The maximum transmitted intensity occurs at 1,370 nm, nearly ten times the diameter of an individual hole in the array [23]

that metallic nanowires can guide light well below the seemingly unsurpassable diffraction limit [22], that metal films with nanoscale holes show extraordinarily high optical transmission [23] (Fig. 3), and that a simple thin film of metal can serve as an optical lens [24]. Plasmonic elements further gained importance as popular components of *metamaterials*, i.e., artificial optical materials with rationally designed geometries and arrangements of nanoscale building blocks. The burgeoning field of transformation optics elegantly demonstrates how such materials can facilitate unprecedented control over light [25].

As these novel phenomena captured the imagination of a broad audience, researchers in the lab started to also gain respect for some of the severe limitations of metals. The most important challenge that still persists today is that metals exhibit substantial resistive heating losses when they interact with light. For this reason, it will be valuable to explore ways to get around that issue. In some cases, local heat generation may be used to advantage, and in some cases heating losses can be neglected. In the coming decade, it will be worth exploring the use of new materials such as transparent conductive oxides and to engineer the band structure of metals [26].

A very diverse set of plasmonics applications has emerged in the last 10 years. Early applications included the development of high-performance near-field optical microscopy (NSOM) and biosensing methods. More recently, many new technologies have emerged in which the use of plasmonics seems promising, including thermally assisted magnetic recording [27], thermal cancer treatment [28], catalysis and nanostructure growth [29], solar cells [30, 31], and computer chips [32, 33]. High-dielectric-constant materials also can effectively be used as antennas, waveguides, and resonators, and their use deserves further exploration [34, 35]. Materials that exhibit strong optical resonances are particularly interesting, because they can exhibit high positive, negative, and near-zero magnitudes of the dielectric constant and/or permeability.

Several of the applications noted above capitalize on light-induced heating, which was originally considered a weakness of plasmonics. After researchers realized that long-distance information transport on chips with plasmonic waveguides would suffer too strongly from heating effects [36], it now has been established that modulators and detectors can be achieved that meet the stringent power, speed, and materials requirements necessary to incorporate plasmonics with CMOS technology. Plasmonic sources capable of efficiently coupling quantum emitters to a single, well-defined optical mode may first find applications in the field of quantum plasmonics and later in power-efficient chip-scale optical sources [37, 38]. In this respect, the recent prediction [39] and realization [40–42] of coherent nanometallic light sources constitutes an extremely important development (see Fig. 4).

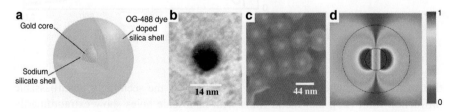

Fig. 4 Diagram of a "spaser": Non-linear optical effects from surface plasmons. (**a**) Schematic of the hybrid structure, (**b**) TEM of the Au core of the structure, (**c**) SEM of the hybrid components. (**d**) Field strengths of the spaser in operation (simulated) [41]

In terms of advances in theory and simulation related to plasmonics, photonics, and electronics, Nader Engheta [43] recently developed an elegant theoretical framework that treats nanostructured optical or "metactronic" circuits much akin to conventional electronic circuitry. In this framework, insulators are modeled as capacitors, metals as inductors, and energy dissipation (heating) can be accounted for by introducing resistors. The desired response of an optical circuit can now be realized simply through the optimization of an electronic circuit.

3 Goals, Barriers, and Solutions for the Next 5–10 Years

3.1 Nanophotonics

3.1.1 Achieve Integration with Electronic Circuits for Ultrasmall, Ultrahigh-Speed Information Communications Applications

Advances in nanophotonic device structures have proceeded rapidly in the past 10 years. Using these miniscule optical components, one could envision creating photonic circuits with scales of integration rivaling today's electronic integrated circuits. However, we have by no means fully capitalized on the enormous information capacity and high speed of photons that could enable novel optical information processing capabilities of almost unthinkable speed. Whereas optical fibers are already the preferred information link for long distance, it is now becoming eminently clear that photons are also going to play a number of important roles in future computers. Although on-chip and chip-to-chip optical interconnects have been explored for some time, the accelerated growth in microprocessor performance and the recent emergence of chip multiprocessors (CMP) have dramatically increased the need for the kind of communications infrastructure that nanophotonics can address. With the rapid advances in CMOS and CMOS-compatible processing techniques, it comes as no surprise that Si-based devices, such as the coupled resonator-waveguide shown in Fig. 5 have started to gain popularity, despite the indirect gap nature of Si [45, 55]. In addition, ultrasmall-dimensioned lasers utilizing semiconductor or metal materials can provide highly efficient optical sources that can be readily integrated on-chip [46].

3.1.2 Control Light Trapping and Device Integration for Applications in the Living World

Besides its increased importance for information technology, the photon is an essential source of energy for the living world and is finding use in a larger number of applications in chemistry, biology, and medicine. New light trapping technologies for solar cells, photocatalysis for clean fuel generation, thermal/heat management,

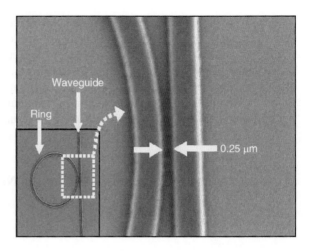

Fig. 5 SEM image of a Si-based ring resonator coupled to a waveguide. *Inset* shows the entire ring structure [44]

and other "green" photonics technologies are expected to gain tremendous momentum in the coming decade. Miniature integrated optoelectronic sensor platforms that are coupled to information networks may well revolutionize biology and healthcare. Because electronics and photonics can be integrated effectively on a common platform, a myriad of new, inexpensive applications can be realized that will benefit from the same economy of scales as Si-based electronics is capitalizing on.

Next to the visible and near-IR range, the mid-IR and THz frequency regimes will gain importance as more inexpensive, user-friendly infrastructure (i.e., sources, switches, detectors) becomes available and the number of applications in biology, homeland security, and defense grows. As photonic devices keep on shrinking to the nanoscale, new physics also emerges that could enable new ways to route and actively manipulate photons.

3.1.3 Use Light to Control Thermal and Mechanical Performance of Materials

There has been much recent excitement at the coupling of optical modes with mechanical modes: phenomena that occur at the right match of the mechanical and optical energies that apply to nanostructures (Fig. 6). This promises much scientific richness in the interplay of temperature, mechanical modes, and optics, and also promises new device applications.

3.2 Plasmonics

3.2.1 Achieve Control over the Flow of Light

Ultrasmall plasmonic or high index semiconductor cavities and waveguides (Fig. 6) are currently giving rise to new, unexpected opportunities. Plasmonic structures and metamaterials consisting of deep-subwavelength building blocks enable light to be

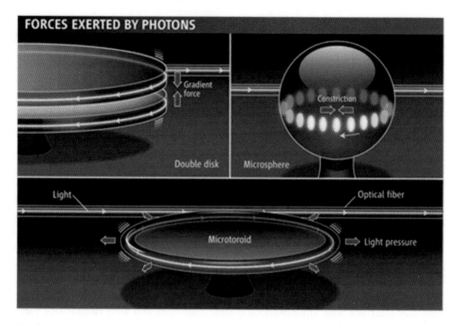

Fig. 6 Light can induce vibrations in a micron-size optical resonator by exerting pressure (*bottom*), creating pushes or pulls (*top left*), or causing the material to constrict (*top right*) [47]

concentrated and actively manipulated in new ways. Along with new mathematical frameworks, such as transformation optics and newly emerging simulation tools, recent advances in the laboratory are describing how we may attain ultimate control over the flow of light.

The most important advances in plasmonics seem to rely heavily on one key property of engineered metallic structures: they exhibit an unparalleled ability to concentrate light. Even a simple spherical metallic nanoparticle can serve as a tiny antenna capable of capturing and concentrating light waves, and its basic operation is quite similar to the ubiquitous radio frequency (RF) antennas at work in cell phones or radios. By squeezing light into nanoscale volumes, plasmonic elements also allow for fundamental studies on light–matter interactions at length scales that have been inaccessible in even the most advanced dielectric components such as photonic crystals. The relevance of confining light is most easily explained by considering a resonator or cavity with a (mode) volume V_M. Many physical processes are dramatically enhanced when light is forced to interact with materials confined to ultrasmall volumes. Of course, the strength of the interaction in plasmonic resonators is diminished by optical losses that can shorten the lifetime of a photon in such a cavity; this lifetime is often quantified by the optical quality factor, or Q, and the effectiveness of many optical processes tends to scale with the ratio of Q/V_m to some power. Despite the modest values of Q in nanoscale metallic cavities (typically between 10 and 1,000), metallic cavities can have such small V_M values that often they are capable of outperforming dielectric cavities with Qs that are orders of magnitude higher. The small size and low Q values of plasmonic structures come with the added benefit of an ultrahigh speed (<100 femtosecond (fs),[1] [48]) and broadband optical response.

3.2.2 Exploit Synergies Between Plasmonics, Photonics, and Electronics

Over the last decade, it has gradually become clear what role plasmonics can play in future device technologies and how it can complement electronics and conventional photonics. Each of these device technologies can perform unique functions that play to the strength of the key materials. The electrical properties of semiconductors enable the realization of truly nanoscale elements for computation and information storage; the high transparency of dielectrics (e.g., glass) facilitates information transport over long distances and at very high data rates. Unfortunately, semiconductor electronics is limited in speed by interconnect (RC) delay-time issues, and photonics is limited in size by the fundamental laws of diffraction. Plasmonics offers precisely what electronics and photonics do not have: the size of electronics and the speed of photonics (Fig. 7). Plasmonic devices might therefore naturally interface with similar-speed photonic devices and with similar-size electronic components, increasing the synergy between these technologies.

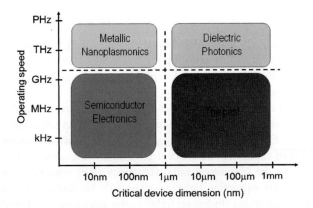

Fig. 7 Graph of the operating speeds and critical dimensions of different chip-scale device technologies. Different domains in terms of operating speed and device sizes are expected to heavily rely on the unique material properties of semiconductors (electronics), insulators (photonics), and metals (plasmonics). The dashed lines indicate physical limitations of different technologies; semiconductor electronics is limited in speed by interconnect delay time issues to about 10 GHz. Dielectric photonics is limited in size by the fundamental laws of diffraction. Plasmonics can serve as a bridge between similar-speed dielectric photonics and similar-size nano-electronics (Adapted from Brongersma and Shalaev [3])

---

[1] 10^{-15} of a second.

The semiconductor and photonics industries have continued to rapidly develop, and it will be exciting to see what the next decade will bring for plasmonics in these industries. In order to continue and even accelerate the advances in nanophotonics and plasmonics, a number of barriers will have to be overcome. The barriers relate to the enabling resources in the scientific and technological infrastructure, and thus they will be described in the next section.

4 Scientific and Technological Infrastructure Needs

4.1 Develop Simple Design Rules and Coupled Optical Simulations for Nanophotonics and Nanoplasmonics

Although advances in electromagnetic and electronic device simulation tools have been considerable, new challenges have also emerged. As the complexity of integrated photonic systems increases, the development of simple design rules for nanophotonics and plasmonics components is absolutely essential. The power of good design rules lies in their ability to hide much of the complexity within an individual device; the aim is to capture the essence of the device function and focus on its interactions with other devices. Such simplifications then enable the construction of system-level theories and simulators that can predict the behavior of larger circuits. Theoretical frameworks should continue to be developed to address functions of nanostructured optical, or metatronic, circuits.

To properly predict the behavior of optoelectronic circuits, simulation tools are required that can simultaneously capture the flow of electrons and photons and their impact on each other. Currently, these types of simulations tend to be performed separately for the electrons and photons and then combined in an *ad hoc* fashion. The coupling to other degrees of freedom (including mechanical, thermal, magnetic, biological) will become important as well. Coupling of optical simulations to quantum mechanical simulation tools is also needed for an increasing number of applications. In many optical simulation tools, the materials properties are taken into account by utilizing macroscopic properties, such as the real and imaginary part of the dielectric constant. This is inadequate for many nanostructured active and passive systems, and we will need to return to more basic theories.

4.2 Support New and Expanded Fabrication Tools and Facilities for Nanophotonic and Plasmonic Devices and Circuits

Over the last decade, the field of nanotechnology has provided a wide range of new nanofabrication and synthesis techniques. With the rapid advances in CMOS and CMOS-compatible processing techniques, it comes as no surprise that silicon-based devices have started to gain popularity (despite the indirect gap nature of Si). This trend is further augmented by the fact that more materials are becoming allowed in Si fabs, and new techniques such as wafer bonding enable convenient ways to add materials with more desirable optical properties for active devices (e.g., III-Vs). It will be of value to educate and train the next generation of optical engineers to take advantage of Si fabs and provide easy access to such facilities. It also will be important to build new infrastructure to facilitate the anticipated increase in demand.

Other important fabrication or prototyping techniques for nanophotonics/nanoplasmonic circuits and devices are state-of-the-art electron beam lithography and focused ion beam milling tools. These tools are extremely powerful and enable the realization of feature sizes below the 10 nm length scale. The cost of such tools will need to come down in order to provide access to an increasing number of users. The increasing divergence between capital equipment costs and equipments grants in the United States is making it harder and harder to provide critical infrastructure to research groups in academia, and industry is facing similar challenges. For this reason, it is also of the utmost importance to further develop less expensive printing technologies (e.g., nanoimprint lithography), patterning techniques (i.e., nanosphere lithography), and materials chemistry approaches (e.g., directed self-assembly, nanoparticles and nanowire growth, and catalysis). Many of these provide access to similar (10 nm) length-scales, although cleanliness, reproducibility, and versatility need to be further improved.

All nanophotonics fabrication technologies will require (1) higher precision tools to control structure size, shape and placement, (2) larger reproducibility, (3) improved control over the material's purity, and (4) new ways to integrate and interface with electronics and conventional optics. Chip-scale devices face additional challenges in terms of cleanliness requirements and materials compatibilities. We must also give thought to newly emerging health and safety issues associated with the use of photonic and plasmonic (and all) nanostructures.

4.3 Expand Optical and Structural Characterization Capabilities for Nanophotonic and Plasmonic Materials and Devices

While the research community has benefited greatly from access to optical nano-charaterization tools, quantitative interpretation of SNOM images is still a challenge and requires significant expertise, the development of better SNOM tips, and better software. There is an increasing need for characterization tools that can correlate the structural and materials properties at the atomic-scale to the optical properties of nanophotonic devices and nanostructured materials. Emerging optical characterization tools in the fields of nanophotonics and plasmonics include cathodoluminescence, optical scanning tunneling microscopy, and electron energy loss spectroscopy (EELS) in a transmission electron microscope; these techniques have already enabled studies of plasmonic modes with nanometer resolution (1–2 orders of magnitude better than SNOM). Equipment like this will help realize simulation tools that properly take into account the optical, electronic, and quantum mechanical effects that will play an increasingly important role in nanophotonic and nanoplasmonic devices.

Researchers will also continue to rely on the availability of and access to state-of-the-art structural characterization tools, as their performance parameters keep on improving every year.

4.4 Build New Educational Systems that Promote Diversity, Interdisciplinarity, and Collaboration

As the fields of nanophotonics and plasmonics expand rapidly into many applications, ranging from information transport to biology, it will also be important to restructure and further develop our education system in the right way. The interdisciplinary nature of these fields requires the development of a common language and new educational methods and tools. In such a rapidly-developing field that has widespread applications, the background training of participants in the field will rapidly diversify. For example, just 10 years ago a symposium on plasmonics was rare, and the audience would primarily consist of hardcore theoreticians and a few scattered experimentalists. Currently, there are well over 20 plasmonics symposia/conferences around the globe annually, filled with people from academia and industry with backgrounds in physics, chemistry, materials science, electrical engineering, biology, chemistry, and medicine. In order to take advantage of the many views and uses of optics, a diverse educational program should be constructed to reduce misunderstandings due to the differences in vocabulary used to identify common concepts and enhance new and valuable collaborations.

5 R&D Investment and Implementation Strategies

Over the past decade, there has been significant maturation of the nanotechnology facilities network supported by the Federal Government that is focused on fundamental understanding of nanotechnology. The National Nanotechnology Initiative has provided a framework for the development of numerous centers supported by both Federal and state agencies. Many of these centers have been developed as complementary centers involving coordinated Federal, university, and industry funding sources. Major investments have been made by the Department of Defense, Department of Energy, National Institutes of Health, and National Science Foundation, among other Federal agencies.

The center focus is important in bringing together the numerous critical building blocks and kinds of expertise that comprise research in nanophotonics, linking complex simulations, materials synthesis, state-of-the-art fabrication, and characterization. The breadth and diversity of these centers also helps to link scientific innovation with applications.

In order for nanophotonics to have the greatest possible opportunity for impact, innovative means should be sought for coordinating these research facilities and capabilities with industry:

- Decisionmakers at academic institutions should be educated about and sensitized to needs-driven research that is critical to industry.
- Means should be explored for expediting technology transfer from the laboratory to processing facilities and accelerating scale-up and manufacture of innovations in nanophotonics.

6 Conclusions and Priorities

The field of nanophotonics has advanced enormously in the past 10 years, but we have by no means fully capitalized on the potential of these technologies for high-speed information transmission, energy capture and storage, and sensing. Integrated advances in materials synthesis, high-resolution fabrication, computation and modeling of optical behavior, and nanoscale optical and structural characterization have given rise to exceptional demonstrations of photonic coherence, localization, and switching in compact structures with dimensions that are at or below the wavelength of light. Nanophotonic components appear to be making progress in true integration into mainstream electronic technologies.

Research priorities should involve bringing some of the substantial recent advances in this area to the next level of development in ways that may change the paradigm of imaging and information processing using photons. These areas include:

- The design and fabrication of optical cavities that truly allow full control of the lifetimes and interaction of photons. Among the many impacts that such cavities have had and could have in the future:
  - Recent achievements of ultralow-threshold lasers with 10 s of nanoWatts thresholds. We look in the future towards "threshold-less" lasers, where the efficiency of energy transfer between cavity and gain medium is so great that lasing can be initiated with miniscule power input to achieve exceptionally high power gains.
  - Recent achievement of "slowed light" in solid-state optical cavities. This presents the possibility, for the first time, of scalable, on-chip delay and storage of photons for photonic information processing, and provides a key enabler for all-optical processing with tremendous advantages in bandwidth and high-speed at greatly reduced power dissipation. We look in the future towards light slowed to the point where lossless storage of light can be achieved with storage times of milliseconds or longer.

- The rapid advances in the field of *plasmonics*, opening up tremendous new possibilities for the application of photonics in broad areas of application. Some of the many impacts that could be realized in this area include:
  - Recent advances in single-molecule imaging enabled by local-excitation and enhanced signal emission could ultimately result in the controlled and specific absorption and emission of light from single molecules.
  - Plasmonic structures have been used for the initial demonstrations of metamaterials at visible and near-infrared wavelengths. These are materials with negative index of refraction. Although there have been some initial demonstrations of enhanced imaging using these structures, we look ahead to the realization of true "superlenses" with spatial resolution substantially below the wavelength of light used.
  - There have been demonstrations of discrete plasmonic devices such as antennas and waveguides. The size scale and performance of these metallic components makes them ideally suited to be integrated within complex electronic circuits, leading to electro-optic circuits with increased bandwidth, with operation at higher frequency and lower dissipation.

The community of researchers that is key to capitalizing on the broad potentials of nanophotonics depends on advances in and accessibility of state-of-the art:

- Computational tools that can span different length scales, coupling optical simulations to quantum mechanical simulations and coupling optical modes with thermal, mechanical, magnetic, and electrical energy states.
- Optical and structural characterization tools that could help to correlate the structural properties of materials with their photonic performances.
- Nanofabrication techniques that allow high local precision (~1 nm), relative ease of use, and compatibility with different materials and technologies (e.g., electronics).

Continued investment in consortia and centers of excellence is also encouraged for this rapidly developing and multidisciplinary area, with a focus on new educational approaches and the ability to link fundamental science to the demonstration of compelling new applications.

7 Broader Implications for Society

As indicated earlier in this chapter, the applications of nanophotonics to societal benefit are profound and pervasive. Nanophotonic elements are already being integrated into the dominant silicon electronic platform to provide high-bandwidth, low-latency information transfer that challenges the scalability of future computing systems. Nanophotonics has an important role to play in more efficient energy-harvesting; in high-sensitivity, integrable sensing in biological, security, and other platforms; and in curing disease (e.g., see Fig. 8) and remediating environmental degradation.

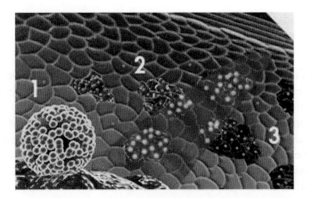

Fig. 8 Use of Au-coated plasmonic elements (nanoshells) for cancer treatment. (**a**) Each nanoshell is about 10,000 times smaller than a white blood cell. (**b**) Via antibodies that functionalize them, about 20 nanoshells cover a tumor cell. (**c**) Plasmonic responses concentrate externally delivered infrared radiation, selectively directed to and destroying the tumor cells [49]

In looking at the broad range of possible applications for nanophotonics, it is not simply the size of the components or systems that provides benefit, it is also the new physical mechanisms at play. The small (and ever smaller) size of nanophotonic device elements provides benefits for their integration and incorporation into heterogeneous material systems and into a variety of different technological platforms. The reduction in size of the nanophotonic components also gives rise to new physical behaviors, for example, control over optical states and frequencies, and the ability to localize light at sub-wavelength scales. As we gain further understanding and mastery of nanophotonic elements and systems, the benefits to society should increase considerably.

8 Examples of Achievements and Paradigm Shifts

8.1 Nanophotonics on a Chip

Contact person: Michal Lipson, Cornell University

Future electronics will rely on nanophotonics for transmitting information across the chip. This will enable the electronics industry to continue scaling in size and bandwidth without the existing limitations in power. In order to realize this vision, the optical elements are required to be compatible with silicon, the material of choice for electronics, or in other words, one requires *silicon nanophotonics* (Fig. 9). Devices that enabled this technology were proposed in 2004.

Until recently, silicon nanophotonics was viewed as highly limited due to the inability to amplify light and emit light efficiently in silicon. However, recent work by Cornell University researchers and others has shown that light amplification and

Applications: Nanophotonics and Plasmonics 433

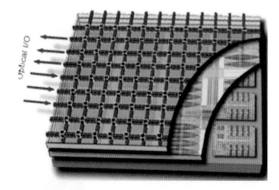

Fig. 9 IBM vision of silicon photonics for optical interconnects in future electronics

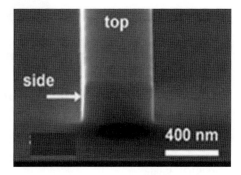

Fig. 10 Silicon waveguide

emission can be achieved by using silicon as a nonlinear optical material. One of the primary reasons for this is that the refractive index for silicon is very high, and thus extremely compact waveguides can be created that confine light very tightly (Fig. 10). This results in the ability of the devices to operate at very low power levels. While this is certainly an important property, these tightly confining waveguides offer the ability to use the waveguide dispersion to conserve momentum (i.e., phase match) of nonlinear processes, which greatly improves the efficiency.

This dispersion control has been applied to achieve amplification over extremely large bandwidths [50]. One could then envision using a single micron-size device for amplifying the entire telecommunications bandwidth. Based on such a device, researchers have also demonstrated an ultrasmall device that can emit light over very broad range of frequencies. The light emitted is laser-like and can be used as a source for lighting up silicon chips [51] (Fig. 11).

Silicon nanophotonics applications range from future computing to communication. Today, largely due to the advances described above, startups are already commercializing this technology. In addition, the computing industry (e.g., Intel, IBM), has significant efforts in silicon nanophotonics development that aim to embed this technology in future computing systems.

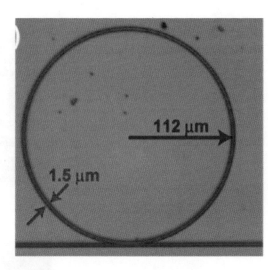

Fig. 11 Device used as a silicon optical oscillator

9 International Perspectives from Site Visits Abroad

9.1 U.S.–European Union Workshop (Hamburg, Germany)

Panel members/discussants

Fernando Briones Fernandez-Pola (co-chair), Spanish National Research Council (CSIC), Spain
Evelyn Hu (co-chair), Harvard University, Cambridge, Massachusetts, United States
M. Alterelli, European X-ray Free-Electron Laser (XFEL), Germany
Y. Bruynserade, Catholic University of Leuven, Belgium
J. C. Goldschmidt, Fraunhofer Institute for Solar Energy Systems (ISE), Germany
M. Kirm, Institute of Physics of the University of Tartu, Estonia
With thanks to C. Sotomayor-Torres, Centro d'Investigació en Nanociencia i Nanotecnologia (CINN), Spain

Distinguishing discussions in this workshop focused on the utilization and leveraging of innovation in photonics and plasmonics for the advancement of research in fields such as photovoltaic research, efficient catalytic processes, and improved medical imaging. As in many of the other workshops, substantial progress and promise was found in the rapidly growing area of plasmonics, but additional opportunities included:

- Rapid, reliable single-photon sources for quantum communications and computing
- Development of new, higher-spatial-resolution imaging and characterization tools, extending the characterization frequencies into the infrared and THz regimes
- Optical control of spin-based phenomena with higher spatial and time resolution

In addition, participants in this session saw the potential for better integration of nanophotonic and plasmonic components into existing nanoelectronic platforms.

Discussion also centered on the need for better utilization of new larger-scale facilities such as synchrotron radiation sources and free electron lasers, both for the characterization of nanoscale components that comprise the photonic elements and for potential roles in the fabrication of nanophotonic devices. Such high-power light sources have the capability to carry out 3D imaging of nanoscale structures with sub-picosecond time resolution.

In realizing the full benefits of innovation in photonics and plasmonics to the breadth of potential fields related to energy-efficient devices (e.g., photovoltaics) and biomedical research (e.g., imaging), participants pointed out the necessity of making broadly available a common infrastructure platform for materials synthesis, nanoscale fabrication, and characterization.

Discussions on R&D strategies that would promote rapid advances in this field brought forth ideas of collaborative research environments that in one sense was a common theme among all workshops in this area but also may have addressed issues unique to European research. These recommendations include:

- Promoting interaction between engineering and science by creating collaborative, critical-size R&D environments (cluster competencies) at the precompetitive research level
- Improving collaboration between universities and large research facilities
- Funding brilliant young researchers at substantial levels, modeled on existing European Research Council funding programs (http://erc.europa.eu)

Finally, although there was no specific discussion at the workshop, it should be noted that the European community has been active in developing strategic roadmaps in photonics and nanotechnology. The MONA (Merging Optics and Nanotechnologies) consortium, launched in June 2005, developed a roadmap for photonics and nanotechnologies in Europe by involving several hundred researchers in industry and academia, who participated in a series of workshops, symposia, and expert interviews over a period of 2 years [52] (Fig. 12). This roadmap was updated in January 2010 by *the Second Strategic Research Agenda in Photonics* [53] (Fig. 13).

As stated by Martin Goetzeler, CEO of OSRAM, in the foreword to that report,

> When the first agenda was published in 2006, photonics in Europe looked very different. Photonics21 had only just begun its task of building a community...Today more than 5,000 companies, most of them small and medium-size enterprises, manufacture photonics products in Europe. The sector employs almost 300 000 people directly and many more work for its suppliers. No fewer than 40 000 jobs have been created here in Europe within the last four years. Photonics innovations are key drivers for profitable growth. The world market for photonics products reached €270 billion in 2008, of which €55 billion was produced in Europe—a growth of nearly 30% since 2005. We are particularly strong in lighting, manufacturing technology, medical technology, defence [*sic*] photonics and optical components and systems.
>
> In September 2009 the European Commission designated photonics as one of five key enabling technologies for our future prosperity. This signifies not only the economic importance of photonics, but its potential to address what have been called the "grand challenges" of our time.

Fig. 12 Cover of roadmap report of merging optics and nanotechnologies (see MONA Consortium [52])

Fig. 13 Cover of second strategic research agenda in photonics (see Nanophotonics Europe Organization [53])

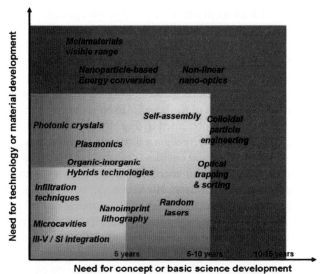

Fig. 14 Part of nanophotonics roadmap in the PhoReMoST document [54]

In addition, an "Emerging Nanophotonics Roadmap" (Fig. 14) has been formulated by the members of the EU Network of Excellence in Nano<u>Ph</u>otonics to <u>Re</u>alize <u>Mo</u>lecular <u>S</u>cale <u>T</u>echnologies (PhoReMoST). Further information can be found by accessing the website of the Nanophotonics Europe Association (http://www.nanophotonicseurope.org/).

9.2 U.S.–Japan–Korea–Taiwan Workshop (Tokyo/Tsukuba, Japan)

Panel members/discussants

Satashi Kawata (co-chair), Osaka University, Japan
Evelyn Hu (co-chair), Harvard University, Cambridge, Massachusetts, United States
Yasuhiko Arakawa, University of Tokyo, Japan
Susumu Noda, Kyoto University, Japan
Yong-Tak Lee, Gwangju Institute of Science and Technology, Korea
Chi-Kuang Sun, National Taiwan University and Academia Sinica, Taiwan

Professor Kawata began this session by a discussion of the long-term investments into photonics by Japan, Korea, and Taiwan. For example, as early as 1981, the Japanese government founded the Optoelectronics Joint Research Laboratory (OJL) to conduct basic research to enable fabrication of optoelectronic integrated circuits and transfer of the technology to its nine member companies at the end of the project.

Several topics were given more in-depth discussion at this workshop, representing the expertise and the sustained contributions made by the various workshop participants. Professor Kawata discussed the use of two-photon reduction to form 3-dimensional metallic nanostructures, of great potential interest for plasmonic applications. He also discussed "tip-enhanced" Raman spectroscopy, innovative instrumentation for imaging with super-resolution. Professor Arakawa, one of the leaders in the science and technology of semiconductor quantum dot photonic devices, discussed the advantages in speed and high-temperature stability offered by quantum dot lasers. His vision of the benefits to be gained from these nanostructured photonic devices can be represented by Fig. 15.

Professor Noda, internationally recognized for his contributions to the science and applications of photonic crystal cavities, discussed the progress and vision in the "ultimate control of photons" for energy-efficient sensors, optical sources, and next-generation information processing. His vision can be represented by Fig. 16.

There was a greater in-depth discussion of the applications of photonics and plasmonics to biomedicine, given by Professor Sun. His discussion ranged broadly and included the use of plasmonic structures and colloidal quantum dots as imaging agents for biomedical applications, as well as microbial diagnostics using novel photonics platforms. His vision for biomedical applications and platforms for nanophotonics is represented in Fig. 17.

Professor Lee discussed the importance of nanophotonics and plasmonics to information processing, providing on-chip integrated photonic sources, detectors, waveguides, and interconnects. His vision is represented in Fig. 18.

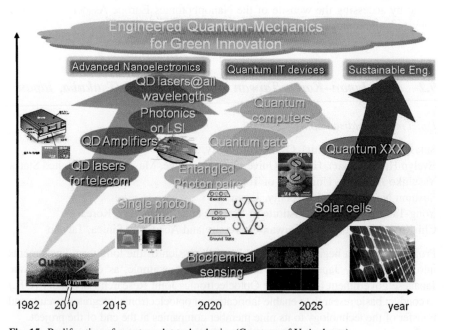

Fig. 15 Proliferation of quantum dot technologies (Courtesy of Y. Arakawa)

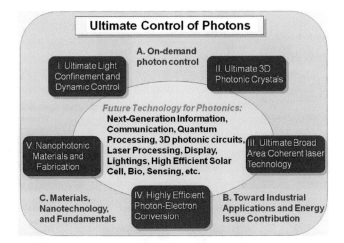

Fig. 16 Ultimate control of photons (Courtesy of S. Noda)

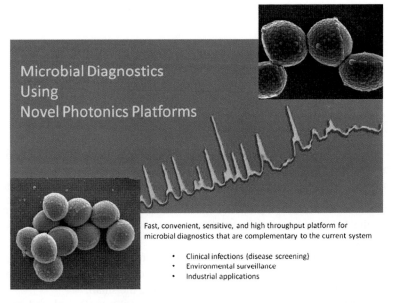

Fig. 17 Biomedical applications and platforms for nanophotonics (Courtesy of C.-K. Sun)

The workshop participants also discussed the need for long-term investment in people and ideas, working in collaborative environments, and providing new education/research centers that will allow exploration of new nanophotonic concepts that go beyond the purview of older, more conventional disciplines.

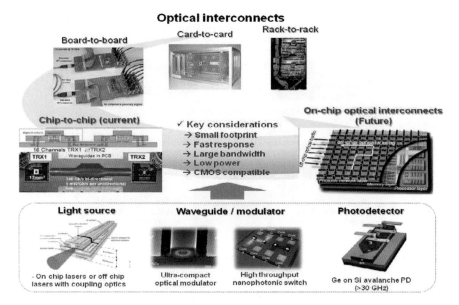

Fig. 18 Nanophotonics in integrated information systems (Courtesy of Y.T. Lee)

9.3 U.S.–Australia–China–India–Saudi Arabia–Singapore Workshop (Singapore)

Panel members/discussants

Paul Mulvaney (co-chair), University of Melbourne, Australia
Mark Lundstrom (co-chair), Purdue University, West Lafayette, Indiana, United States
Chennupati Jagadish, Australian National University, Australia
Chen Wang, National Center for Nanoscience and Technology, Beijing, China

The breakout session had input from Australian and Chinese participants, in addition to the U.S. moderators. Some differences in emphasis were apparent from these discussions.

Australia places a strong emphasis on environmental science at present. Nanophotonics is seen as an important platform for economical environmental sensors. There is a tremendous interest generally in "sensing the small."

Biosensors, bioassays, and other applications in the biological and medical fields are also seen as important drivers for nanophotonics. The highest-profile activities in this field are two consortia working towards the development of the bionic eye. The bionic ear was pioneered in Australia, and this technology for interfacing electronics to neural structures is seen as an important platform that makes the bionic eye look feasible.

Australia is ideally placed to exploit solar energy, given its high solar insolation, large area (about 85% of the area of the U.S. mainland), and low population density (national population is comparable to that of the State of California). Consequently, the potential of nanophotonics to provide solutions in photovoltaics and smart windows for efficient energy use are high-profile activities.

Less strong in Australia compared to the United States are the fields of interfacing to defense applications and using nanophotonics to provide advances in electronics industries. However, it is also apparent that Australian researchers see the integration of optical structures with conventional electronics as a very desirable long-term goal.

This breakout session identified important applications and goals for nanophotonics in the next 10 years:

- The all-optical chip
- Metamaterials operating at visible wavelengths
- Single-(bio)molecule detection
- Artificial photosynthetic systems for energy applications

Workshop participants in this session similarly discussed infrastructure needs:

- Collaboration (sharing of information, construction of joint databases)
- Networked fabrication, metrology, and characterization resources.
- Wider availability of open-source, multiscale computational tools (especially for designing nanophotonic circuits)

Here too, participants stressed the need for strategic, longer-term, perhaps "center-level" funding; precompetitive collaborations between universities and industry, and better training of a nanotechnologist generation. They also called for a "Nanophotonics Roadmap" (note the activities on "roadmapping" as carried out by the European Community, Sect. 9.2).

References

1. National Research Council of the National Academies (NRC), *Nanophotonics: Accessibility and Applicability* (National Academies Press, Washington, DC, 2008)
2. W.D. Barnes, A. Dereux, T.W. Ebbesen, Surface plasmon subwavelength optics. Nature **424**, 824–830 (2003)
3. M. Brongersma, V. Shalaev, The case for plasmonics. Science **328**, 440–441 (2010)
4. J.A. Schuller, E.S. Barnard, W. Cai, Y.C. Jun, J.S. White, M.L. Brongersma, Plasmonics for extreme light concentration and manipulation. Nat. Mater. **9**, 193–204 (2010). doi:10.1038/nmat2630
5. M. Roco, S. Williams, P. Alivisatos (eds.), *Nanotechnology Research Directions: Vision for Nanotechnology R&D in the Next Decade.* (NSTC/Springer, Washington, DC, 1999), previously Kluwer, 2000. Available online: http://www.nano.gov/html/res/pubs.html
6. R.D. Dupuis, P.D. Dapkus, N. Holonyak, E.A. Rezek, R. Chin, Room-temperature laser operation of quantum-well $Ga_{(1-x)}Al_{(x)}As$-GaAs laser diodes grown by metalorganic chemical vapor deposition. Appl. Phys. Lett. **32**(5), 295–299 (1978). doi:10.1063/1.90026
7. E. Yablonovitch, Inhibited spontaneous emission in solid-state physics and electronics. Phys. Rev. Lett. **58**(20), 2059–2062 (1987). Available online: http://www.ee.ucla.edu/labs/photon/pubs/ey1987prl5820.pdf
8. V.G. Veselago, Electrodynamics of substances with simultaneously negative values of sigma and mu. Sov. Phys. USPEKHI USSR **10**, 509–514 (1968)

9. M.C. Albrecht, J.A. Creighton, Anomalously intense Raman-spectra of pyridine at a silver electrode. J. Am. Chem. Soc. **99**, 5215–5217 (1977)
10. D.L. Jeanmaire, R.P. Van Duyne, Surface Raman spectroelectrochemistry, 1. Heterocyclic, aromatic, and aliphatic amines adsorbed on the anodized silver electrode. J. Electroanal. Chem. **82**(1), 1–20 (1977). doi:10.1016/S0022-0728(77)80224-6
11. M. Brongersma, Plasmonics: electromagnetic energy transfer and switching in nanoparticle chain-arrays below the diffraction limit. in *MRS. Symposium Proceedings H (Molecular Electronics)*, vol 582, Boston, 1999, p 502
12. B.-S.S. Song, T.A. Noda, Y. Akahane, Ultra-high-Q photonic double-heterostructure nanocavity. Nat. Mater. **4**, 207–210 (2005). doi:10.1038/nmat1320
13. H.K. Gersen, T.J. Karle, R.J.P. Engelen, W. Bogaerts, J.P. Korterik, N.F. van Hulst, T.F. Krauss, L. Kuipers, Real-space observation of ultraslow light in photonic crystal waveguides. Phys. Rev. Lett. **94**(7), 073903/1–4 (2005). doi:10.1103/PhysRevLett.94.073903
14. M.Y. Notomi, K. Yamada, A. Shinya, J. Takahashi, C. Takahashi, I. Yokohama, Extremely large group-velocity dispersion of line-defect waveguides in photonic crystal slabs. Phys. Rev. Lett. **87**, 253902/1–4 (2001). doi:10.1103/PhysRevLett.87.253902
15. Y.A. Vlasov, M. O'Boyle, H.F. Hamann, S.J. McNab, Active control of slow light on a chip with photonic crystal waveguides. Nature **438**, 65–69 (2005). doi:10.1038/nature04210
16. T. Krauss, Slow light in photonic crystal waveguides. J. Phys. D **40**(9), 2666–2670 (2007). doi:10.1088/0022-3727/40/9/S07
17. K.B. Hennessy, A. Badolato, M. Winger, D. Gerace, M. Atatüre, S. Gulde, S. Fält, E.L. Hu, A. Imamoglu, Quantum nature of a strongly coupled single quantum dot-cavity system. Nature **445**, 896–899 (2007). doi:10.1038/nature05586
18. J.S. Reithmaier, Strong coupling in a single quantum dot-semiconductor microcavity system. Nature **432**, 197–200 (2004)
19. T. Yoshie, A. Scherer, J. Hendrickson, G. Khitrova, H.M. Gibbs, G. Rupper, C. Ell, O.B. Shchekin, D.G. Deppe, Vacuum Rabi splitting with a single quantum dot in a photonic crystal cavity. Nature **432**, 200–203 (2004). doi:10.1038/nature03119
20. M.H. Fleischmann, P.J. Hendra, A.J. McQuillan, Raman spectra of pyridine adsorbed at a silver electrode. Chem. Phys. Lett. **26**, 163–166 (1974). doi:10.1016/0009-2614(74)85388-1
21. M. Moskovits, Surface-roughness and enhanced intensity of Raman-scattering by molecules adsorbed on metals. J. Chem. Phys. **69**, 4159–4162 (1978). doi:10.1063/1.437095
22. J. Takahara, S. Yamagishi, H. Taki, A. Morimoto, T. Kobayashi, Guiding of a one-dimensional optical beam with nanometer diameter. Opt. Lett. **22**(7), 475–477 (1997). doi:10.1364/OL.22.000475
23. T.L. Ebbesen, H.J. Lezec, H.F. Ghaemi, T. Thio, P.A. Wolff, Extraordinary optical transmission through sub-wavelength hole arrays. Nature **391**, 667–669 (1998). doi:10.1038/35570
24. J. Pendry, Negative refraction makes a perfect lens. Phys. Rev. Lett. **85**, 3966–3969 (2000). doi:10.1103/PhysRevLett.85.3966
25. V.M. Shalaev, Transforming light. Science **322**, 384–386 (2008). doi:10.1126/science.1166079
26. P.R. West, S. Ishii, G.V. Naik, N.K. Emani, V.M. Shalaev, A. Boltasseva, Searching for better plasmonic materials. Laser Photon. Rev. 1–13, (2010). doi:10.1002/lpor.200900055
27. W.P. Challener, C. Peng, A.V. Itagi, D. Karns, W. Peng, Y. Peng, X.M. Yang, X. Zhu, N.J. Gokemeijer, Y.-T. Hsia, G. Ju, R.E. Rottmayer, M.A. Seigler, E.C. Gage, Heat-assisted magnetic recording by a near-field transducer with efficient optical energy transfer. Nat. Photonics **3**, 220–224 (2009). doi:10.1038/nphoton.2009.2
28. L.S. Hirsch, R.J. Stafford, J.A. Bankson, S.R. Sershen, B. Rivera, R.E. Price, J.D. Hazle, N.J. Halas, J.L. West, Nanoshell-mediated near-infrared thermal therapy of tumors under magnetic resonance guidance. Proc. Natl. Acad. Sci. U.S.A. **100**(23), 13549–13554 (2003). doi:10.1073/pnas.2232479100
29. L.B. Cao, D.N. Barsic, A.R. Guichard, Plasmon-assisted local temperature control to pattern individual semiconductor nanowires and carbon nanotubes. Nano Lett. **7**(11), 3523–3527 (2007)
30. H.P. Atwater, A. Polman, Plasmonics for improved photovoltaic devices. Nat. Mater. **9**(3), 205–213 (2009)

31. R.W. Pala, J. White, E. Barnard, J. Liu, M.L. Brongersma, Design of plasmonic thin-film solar cells with broadband absorption enhancements. Adv. Mater. **21**, 3504–3509 (2009). doi:10.1002/adma.200900331
32. W.W. Cai, J.S. White, M.L. Brongersma, Compact, high-speed and power-efficient electrooptic plasmonic modulators. Nano Lett. **9**(12), 4403–4411 (2009)
33. L.S. Tang, E. Kocabas, S. Latif, A.K. Okyay, D.-S. Ly-Gagnon, K.C. Saraswat, D.A.B. Miller, Nanometre-scale germanium photodetector enhanced by a near-infrared dipole antenna. Nat. Photonics **2**, 226–229 (2008). doi:doi:10.1038/nphoton.2008.30
34. L.W. Cao, J.S. White, J.-S. Park, J.A. Schuller, B.M. Clemens, M.L. Brongersma, Engineering light absorption in semiconductor nanowire devices. Nat. Mater. **8**, 643–647 (2009). doi:10.1038/nmat2477
35. J.A. Schuller, T. Taubner, M.L. Brongersma, Optical antenna thermal emitters. Nat. Photonics **3**, 658–661 (2009). doi:10.1038/nphoton.2009.188
36. R. Zia, J.A. Schuller, A. Chandran, M.L. Brongersma, Plasmonics: the next chip-scale technology. Mater. Today **9**, 20–27 (2006)
37. A.M. Akimov, A. Mukherjee, C.L. Yu, D.E. Chang, A.S. Zibrov, P.R. Hemmer, H. Park, M.D. Lukin, Generation of single optical plasmons in metallic nanowired coupled to quantum dots. Nature **450**, 402–406 (2007)
38. A.J. Hryciw, Y.C. Jun, M.L. Brongersma, Electrifying plasmonics on silicon. Nat. Mater. **9**, 3–4 (2010). doi:10.1038/nmat2598
39. D.S. Bergman, M.I. Stockman, Surface plasmon amplification by stimulated emission of radiation. Phys. Rev. Lett. **90**, 027402 (2003)
40. M.T. Hill, Y.-S. Oei, B. Smalbrugge, Y. Zhu, T. de Vries, P.J. van Veldhoven, F.W.M. van Otten, T.J. Eijkemans, J.P. Turkiewicz, H. de Waardt, E.J. Geluk, S.-H. Kwon, Y.-H. Lee, R. Nötzel, M.K. Smit, Lasing in metallic-coated nanocavities. Nat. Photonics **1**, 589–594 (2007). doi:10.1038/nphoton.2007.171
41. M.Z. Noginov, G. Zhu, A.M. Belgrave, R. Bakker, V.M. Shalaev, E.E. Narimanov, S. Stout, E. Herz, T. Suteewong, U. Wiesner, Demonstration of a spaser-based nanolaser. Nature **460**, 1110–1112 (2009). doi:10.1038/nature08318
42. R.F. Oulton, V.J. Sorger, T. Zentgraf, R.-M. Ma, C. Gladden, L. Dai, G. Bartal, X. Zhang, Plasmon lasers at deep subwavelength scale. Nature **461**, 629–632 (2009). doi:10.1038/nature08364
43. N. Engheta, Circuits with light at nanoscales: optical nanocircuits inspired by metamaterials. Science **317**(5845), 1698–1702 (2007). doi:10.1126/science.1133268
44. V.R. Almeida, C.A. Barrios, R.R. Panepucci, M. Lipson, All-optical control of light on a silicon chip. Nature **431**, 1081–1084 (2004)
45. H. Hogan, Silicon photonics could save the computer industry. Photon. Spectra (Mar), 36 (2010), Available online: http://www.photonics.com/Article.aspx?AID=41611
46. R.F. Service, Ever-smaller lasers pave the way for data highways made of light. Science **328**, 810 (2010)
47. A. Cho, Putting light's light touch to work as optics meets mechanics. Science **328**(5980), 812 (2010). doi:10.1126/science.328.5980.812
48. M. Stockman, The spaser as a nanoscale quantum generator and ultrafast amplifier. J. Opt. **12**, 024004–024021 (2010). doi:10.1088/2040-8978/12/2/024004
49. K. Kelleher, Engineers light up cancer research. Emerging medicine: scientists design gold "nanoshells" that seek and destroy tumors. PopSci. (2003). Posted 6 Nov 2003. Retrieved 30 May 2010 from http://www.popsci.com/scitech/article/2003-11/engineers-light-cancer-research
50. M. Foster, R. Salem, D. Geraghty, A. Turner-Foster, M. Lipson, A. Gaeta, Silicon-chip-based ultrafast optical oscilloscope. Nature **456**, 81–84 (2008). doi:10.1038/nature07430
51. J. Levy, A. Gondarenko, M. Foster, A. Gaeta, M. Lipson, CMOS-compatible multiple-wavelength oscillator for on-chip optical interconnects. Nat. Photonics **4**, 37–40 (2009). doi:10.1038/nphoton.2009.259
52. MONA Consortium, Merging optics and nanotechnologies: a European roadmap for photonics and nanotechnologies (2008), Available online: http://www.ist-mona.org/

53. Nanophotonics Europe Organization, Lighting the way ahead. Photonics 21: second strategic research agenda in photonics. (European Technology Platform Photonics21, Dieseldorf, 2010), Available online: http://www.photonics21.org/download/SRA_2010.pdf
54. PhOREMOST Network of Excellence, *Emerging Nanophotonics* (PhOREMOST, Cork, 2008)
55. F. Xia, L. Sekaric, Y. Vlasov, Ultracompact optical buffers on a silicon chip. Nat. Photonics **1**, 65–71 (2007). doi:10.1038/nphoton.2006.42

Applications: Catalysis by Nanostructured Materials

Evelyn L. Hu, S. Mark Davis, Robert Davis, and Erik Scher

Keywords Catalysis • Nanostructured catalysts • Synthesis methods • Fuel cells • International perspective

1 Vision for the Next Decade

1.1 Changes in the Vision over the Last 10 Years

The 1999 *Nanotechnology Research Directions* report included nanoscale catalysis as one aspect of applications of nanotechnology to the energy and chemicals industries [1]. The vision centered on the recognition that "new properties intrinsic to nanostructures" could lead to breakthroughs in catalysis with high selectivity at high yield. An example cited in that report was the observation that, while bulk gold is largely unreactive, highly selective catalytic activity could be observed for gold nanoparticles smaller than about 3–5 nm in diameter [2]. Nanoparticles and nanostructured materials have traditionally played a critical role in the effectiveness of

E.L. Hu (✉)
Harvard School of Engineering and Applied Sciences, 29 Oxford Street,
Cambridge, MA 02138, USA
e-mail: ehu@seas.harvard.edu

S.M. Davis
ExxonMobil Chemical R&D, BTEC-East 2313, 5959 Las Colinas Boulevard,
Irving, TX 75039-2298, USA

R. Davis
Department of Chemical Engineering, University of Virginia, 102 Engineers' Way,
P.O. Box 400741, Charlottesville, VA 22904-4741, USA

E. Scher
Siluria, 2625 Hanover Street, Palo Alto, CA 94304, USA

industrial catalysts [3], but the past decade has witnessed significant advances in the control of nanoscale materials and the characterization and *in situ* probing of catalytic processes at the atomic, active site scale.

1.2 Vision for the Next 10 Years

The advances in nanostructured catalytic materials in the last 10 years support a new vision for nanoscience-inspired design, synthesis, and formulation of industrially important catalytic materials. The implications of further progress in "deterministic" nanocatalysis and the broad applications to energy and the environment underscore the importance of this area for future investment. As described in Sect. 2, a grand challenge and vision for catalysis at the nanoscale is providing catalytic materials that will more accurately and efficiently control reaction pathways through the precise control of the composition and structure of those catalysts, over length scales spanning 1 nm–1 μm.

Such control will enable the broad development and deployment of nanostructured catalysts that can more efficiently and selectively convert lower-grade hydrocarbons into higher-value fuels and chemical products, thereby reducing dependence on oil, providing more effective pollution abatement, and efficiently harnessing solar power, and redirecting energy selectively into driving thermodynamically uphill chemical processes. Science-based protocols to build and control composite particle structures should stimulate development of improved methods for catalyst scale-up and large scale manufacturing that will be needed to broadly allow penetration of nanoengineered materials into new commercial catalyst applications.

2 Scientific and Technological Advances in the Past 10 Years

A U.S. National Science Foundation (NSF) workshop, "Future Directions in Catalysis: Structures that Function at the Nanoscale," was held in 2003 at NSF headquarters in Arlington, VA [4]. Thirty-four distinguished participants, primarily from U.S. academic institutions, government agencies, national laboratories, and major companies assessed the state of the art in the field and provided vision statements on the future directions of catalysis research. The workshop was organized around three working groups focused on (1) synthesis, (2) characterization, and (3) theoretical modeling of catalysts. These topics formed the basic framework of a subsequent World Technology Evaluation Center (WTEC) assessment of the worldwide state of the art and research trends in catalysis by nanostructured materials; this study was conducted in 2009 by a panel of eight experts in the field [5].

Both studies (2003 and 2009) articulated the consensus view that synthesis of new nanoscopic catalysts will require fundamental understanding of molecular-scale self-assembly of complex, multicomponent, metastable systems, and that newly developed tools of nanotechnology will likely play a key role in the synthesis of new catalytic structures. Moreover, both studies reported on the extensive use of nanoscale characterization methods in catalysis and the need for new and improved *in situ* characterization methods that extend the limits of temperature, pressure, and spatial resolution to probe nanostructured catalysts in their working state. Finally, although significant advances in theoretical descriptions of complex reactions and models that span multiple time and length scales have been realized over the last decade, additional improvements are required to develop the predictive capabilities of computation, especially in liquid phase systems. A grand challenge that emerged from the 2003 workshop is "to control the composition and structure of catalytic materials over length scales from 1 nm–1 μm to provide catalytic materials that accurately and efficiently control reaction pathways" [4]. Although great strides have been made in fulfilling this challenge, significant hurdles remain.

2.1 Synthesis of Nanostructured Catalysts

The need to control surface structure and chemical reactivity at the nanometer length scale is paramount in catalysis research and engineering. Zeolites are archetypical high-surface-area crystalline solids with nanometer-size pores and cavities that allow for highly precise manipulation of catalytic components and diffusion paths of reagents and products. For example, research in the 1980s artfully demonstrated the exquisite control of metal particle sizes in zeolite pores for hydrocarbon reactions relevant to fuels production. In particular, Pt clusters composed of only a few atoms can be readily synthesized by chemical methods in the nanometer-sized channels of zeolite L to form a catalyst that is especially active and selective for the formation of benzene and toluene from linear alkanes. However, a common problem with nanoporous catalysts is the severe diffusional resistance encountered by reactants and products that have molecular dimensions similar to those of the pores. If side products or carbonaceous deposits block the entrance to these pores, the entire interior regions of the pores can be rendered inaccessible for catalysis.

To overcome these inherent diffusional limitations, various labs around the globe have synthesized microporous zeolites (i.e., zeolites with nanometer-sized pores) with an interconnected array of mesopores (i.e., pores of much larger dimension) to improve accessibility of the internal regions of the material to reactants and products. Multiple synthetic strategies for synthesizing these hybrid micro-mesoporous zeolites have evolved recently, to include the use of:

- Novel organic structure-directing agents such as quaternary ammonium ions to template the original zeolite synthesis as well as alkyl chains to promote aggregation of the structure directors into micellar units [6]
- Elemental carbon (in the form of carbon black, fibers, nanotubes, etc.) as a template for mesopores in the zeolite synthesis gel that can be eventually removed by complete oxidation at elevated temperature [7]

- Preformed zeolite nanoparticles as structural building blocks to form larger crystals in the presence of a surfactant [8]

In the above-mentioned synthetic methods, the resulting materials contain both micropores and mesopores within a single solid particle, which greatly facilitates transport of reactants and products to the interior regions where the active sites are located. Another approach to overcome diffusional limitations is to synthesize new zeolite structures with larger-pore systems, thus avoiding the creation of very small pore dimensions altogether [9]. Although this approach is appealing, there are significant challenges in the generalized synthesis of large pore materials as well as in their stability under catalytic reaction conditions.

Exciting advances have recently occurred in the synthesis of small metal and metal oxide nanoparticles with controlled size, shape, and specific surface orientations, such as Pd nanoparticles that expose the selective {100} plane for hydrogenation of dienes [10] and CeO_2 nanoparticles that expose the active {100} plane for CO oxidation [11]. These advances point toward a future in catalyst preparation that will enable practitioners to more rationally design and prepare catalysts with the specific surface structures that are required for high specificity, structure-sensitive catalytic reactions [12, 13].

2.2 Characterization of Nanostructured Catalysts

The rapid developments in *in situ* spectroscopic tools and atomic-resolution electron microscopy over the last decade have revolutionized the understanding of catalyst structures at the nanoscale. Modern catalysis laboratories now routinely utilize a broad suite of characterization methods such as adsorption, temperature programmed reactions, x-ray diffraction, scanning and transmission electron microscopy, and electron spectroscopies (x-ray, photoelectron, and Auger spectroscopies), as well as electron paramagnetic, and nuclear magnetic resonance, ultraviolet/visible, Raman, and infrared spectroscopies. Major progress has been made in characterizing catalysts in their working state, not simply prior to or after a catalytic reaction, with increased realization that catalysts are highly dynamic solids that often restructure under reaction conditions. Thus, new sample cells have been devised to interrogate materials by all of the photon-based spectroscopies as well as by EPR and NMR spectroscopy, while simultaneously measuring reaction kinetics and selectivities.

One recent development in characterization instrumentation involves the adaptation of electron-based methods such as x-ray photoelectron spectroscopy (XPS) and electron microscopy to conditions that more closely approach catalytic reaction conditions. For example, the high photon flux provided by a synchrotron radiation source as well as differential pumping of the reaction chamber allows acquisition of XPS data on catalytic surfaces up to millibar reaction pressures [14] (see Fig. 1). Likewise, new sample cells for transmission electron microscopy allow

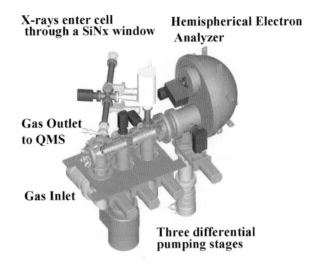

Fig. 1 Instrument for carrying out XPS at millibar pressures and capable of being operated at high flux (Courtesy of Fritz Haber Institute, Germany)

for the direct observation of reactive environments on metal nanoparticles at elevated temperature and mbar reaction pressure [15]. The use of aberration correction on high-resolution electron microscopes has recently enabled unprecedented resolution of features on catalyst samples so that observation of single metal atoms on supports is now possible.

Synchrotron radiation methods have continued to play a major role in characterization of catalysts, since hard x-rays can be used to probe solids at high temperatures and pressures or in the liquid phase without the need for the catalyst to be crystalline. Soft x-ray synchrotron spectroscopies can provide complementary information on local surface structure and composition. Therefore, the catalysis community worldwide has quickly implemented recent improvements in energy resolution, spatial resolution, and temporal resolution of synchrotron-based characterization methods.

It is now possible to simultaneously acquire small-angle and wide-angle x-ray scattering from catalysts while recording the x-ray absorption spectrum, all in a time-resolved manner. Indeed, some beamlines can acquire an EXAFS (extended x-ray absorption fine-structure spectroscopy) spectrum in less than 0.1 s, which is on the scale of a catalytic turnover event. Although commercial infrared (IR) and Raman microscopes have been available for some time and are used by some catalysis laboratories, the implementation of spatially-resolved x-ray absorption spectroscopy has taken longer to develop. Recent work in the Netherlands has achieved spatially resolved near-edge spectra on a single 600 nm iron oxide particle used for the Fischer-Tropsch synthesis reaction.

As catalyst research expands into new technology areas such as the conversion of biomass into liquid products, there will be a growing need for characterization methods that allow for interrogation of solid structures in liquid environments. Thus, the important role of national-level user facilities for electron microscopy, magnetic resonance, and synchrotron radiation studies in catalysis research will likely grow in the coming decade.

2.3 Theory and Simulation in Catalysis

Computational catalysis has reached the stage over the last decade in which it provides a necessary complement to experimental research in the field (Fig. 2). Advances in computer processor speeds, large scale implementation of parallel architectures, and development of more efficient theoretical and computational methods allow for complex simulations of catalytic reactions on solid surfaces that often match well to experimental results. As noted in the 2009 WTEC panel report, *An International Assessment of Research in Catalysis by Nanostructured Materials*, theory and simulation are now able to predict structures and properties of well-defined model catalysts, simulate spectra provided by a variety of experimental tools, elucidate catalytic reaction paths and begin guiding the search for new catalytic materials and compositions.

The computational tools available today can provide structural information such as bond lengths and bond angles of surface adsorbates to within about 0.005 nm and 2°, respectively, and spectroscopic features of the adsorbates within a few percent of experiment. Moreover, adsorption bond energies, activation barriers, and overall reaction energies are now routinely calculated within a couple of days at an accuracy of about 20 kJ mol^{-1} or better. Although the energies estimated computationally are beyond chemical accuracy for predicting absolute reaction rates, computational catalysis can now be used to effectively discriminate between competing reaction paths and predict overall trends involving catalyst composition and reaction conditions.

Although there has been rapid progress in computational catalysis regarding the description of catalytic reactions at the molecular level, there has been less progress

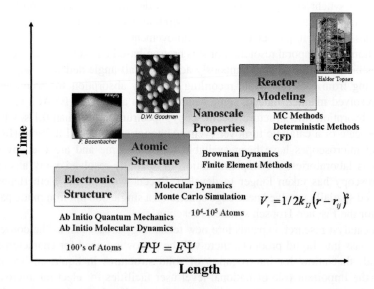

Fig. 2 Hierarchy of time and length scales in heterogeneous catalysis, and associated modeling methods [5]

on the use of theory and simulation to guide catalyst synthesis (for a recent review, see [16] and references therein). Catalyst preparation often involves the self-assembly of metastable structures that are guided by weak interactions with molecules within the system. These structures are often further modified by dynamic rearrangements during catalytic reactions. Thus, catalyst design and predictive synthesis presents a serious challenge to the computational community, since weak forces are not well-described by current theoretical methods, and non-equilibrium structures are difficult to find computationally.

2.4 Areas of Application and Economic Impacts

Catalyst manufacturing was one of the earliest commercial application areas for modern nanoscience methods. Important examples include sol-gel wash coat processing for the fabrication of automobile catalytic converters and use of zeolites with tailored framework structures for shape-selective chemical and fuel processing. As noted earlier, direct applications of nanotechnology in catalyst manufacturing have grown rapidly over the past decade as new synthesis and characterization methods have become available. It is estimated that synthetic methods derived from modern nanotechnology are now utilized in 30–40% or more of global catalyst products (Payne 2010; see also The Catalyst Group Resources 2010) see also [17]. Because these commercial methods are proprietary, relatively few application details have been disclosed in the open literature. Specific examples of recent commercial catalyst applications are given in Sect. 8. Further examples can be found in recent multiclient studies by the Catalyst Group [17, 18].

The global catalyst business represents an $18–20 billion-dollar per-year enterprise with applications primarily involving petroleum refining, chemicals processing, and environmental catalysis (used with permission, [19]. Most commercial catalysis innovations continue to be targeted at achieving significant improvements in product selectivity and energy efficiency for conversion and upgrading of about 80 million barrels per day of petroleum into transportation fuels and petrochemicals globally. With product values currently in the range of about $100 US dollars or more per barrel of feedstock, global economic impacts across the full value chain for these technologies is in the range of several trillion dollars annually. For this reason, it appears likely that catalysis currently remains the most successful commercial application area for modern nanotechnologies.

The 2009 WTEC catalysis report [5] found that conversion of nonpetroleum feedstocks such as coal, natural gas, and biomass to fuels, energy, and chemicals is a high priority in nearly all of the countries visited in that study. China, in particular, has a major emphasis on non-conventional energy applications, especially those involving the conversion of coal to liquid fuels and chemicals. Significant activities in photocatalysis, hydrogen generation, and fuel cells are carried out in many locations. There is a general recognition that energy carriers and chemicals should be produced, and ultimately used, with as little impact on the environment as possible; catalytic solutions are thus being pursued in this framework of environmental sustainability. Catalytic production of ultra-low-sulfur fuels, use of renewable carbon sources and sunlight, conversion of the greenhouse gas CO_2 to useful products, highly selective oxidation of hydrocarbons, and catalytic after-treatment of waste streams are all being pursued vigorously around the globe. A growing area of interest is the catalytic transformation of various renewable plant sources to energy-relevant compounds such as bio-oil (a liquid feedstock that can be upgraded in combination with refinery streams), biodiesel fuel, hydrogen, alcohols, and so forth.

2.5 Summary Statement

Catalysis by nanostructured materials is an active area of research around the globe. Its rate of growth appears to be increasing faster than that of all science, according to a detailed bibliometric analysis [5], presumably because of increasing concerns regarding future energy security and environmental sustainability. The European Union (EU) currently holds a dominant position in the world in terms of research paper output, but the rapid growth of research in Asia over the previous decade is challenging that position. Investment levels in catalysis research in the EU and Asia appear to be significantly greater than that in the United States over the last decade. A bright spot in the analysis confirms that recent U.S. publications are the most cited in the world, which suggests that research funds in the United States are distributed effectively to the highest-quality laboratories.

3 Goals, Barriers, and Solutions for the Next 5–10 Years

Over the last 30 years, the synthesis and characterization of novel nanoscale materials has advanced greatly in both academic and industrial research settings. In particular, over the last decade, programs such as the National Nanotechnology Initiative (NNI) have helped, directly and indirectly, push forward the tools and techniques required to understand and improve upon the properties of nanoscale materials. scanning transmission electron microscopy (STEM), atomic force microscopy/electric force microscopy (AFM/EFM), and focused ion beam (FIB) technique. In parallel, the development of high-throughput screening tools such as those developed by scientific equipment companies Symyx, Avantium, Hte, and others has allowed for a dramatic increase in the total amount of empirical data available. This mass of data has benefitted the experimental community directly, and it has as well provided additional pools of data for the theoretical community to develop and refine models of reaction mechanism models. The growth in national computing centers such as those at the Lawrence Livermore and, Argonne National Laboratories, LBNL, etc. has also allowed for the creation of more and more accurate mechanistic models over the last decades.

Given the great advances in synthetic methods, characterization, theory, informatics, and HTE tool development, the next challenge is to significantly broaden the practical application of this knowledge and capability. There remain many challenges in the energy and commodity chemicals industries that appear well suited to solutions based on new nanostructured catalyst systems. This work needs to take place in both academic and industrial labs, both on the fundamental research that has been enabled by nanomaterials, as well as the application of those nanomaterials towards solving the hardest problems in the field of catalysis. Some of the "Holy Grail" sought-after catalytic reactions include: selective oxidation of alkanes, oxidative dehydrogenation of alkanes to alkenes, carbon dioxide reforming, and—perhaps the most challenging case of selective oxidations—the oxidative coupling of methane to ethylene [20].

Many of these and other commercially valuable reactions have been pursued by industry and academic researchers for many years, but in most cases this was done prior to recent advances in nanomaterials development, and prior to the widespread use of high-throughput screening and improved modeling techniques. Two specific examples of reactions that would greatly benefit from the recent advances are described further below. If either of these is solved in the next 10 years, the country could greatly benefit greatly via better use of large projected reserves of natural gas.

3.1 Carbon Dioxide Reforming

Carbon dioxide reforming (CDR) of methane is a process for converting CO_2 in process streams or naturally occurring sources into the more valuable chemical product, syngas (a mixture of hydrogen and carbon monoxide). In CDR, a mixture of two potent greenhouse gases, methane and CO_2, is converted into syngas according to the following reaction: $CH_4 + CO_2 \rightarrow 2CO + 2H_2$. Syngas can then be converted into a wide range of hydrocarbon products through established commercial processes such as the Fischer-Tropsch reaction to form liquid fuels including methanol, ethanol, diesel, and gasoline. The result is a powerful technique to not only remove CO_2 from the atmosphere and reduce the emissions of methane (another greenhouse gas) but also to create a new alternative source for fuels that are not derived from petroleum. Reforming with CO_2, rather than the conventional steam methane reforming using H_2O could be attractive in areas where water is not readily available, and it yields syngas with a 1:1 H_2/CO ratio, which is a preferable feedstock for the Fischer–Tropsch synthesis of long-chain hydrocarbons.

Unfortunately, no established industrial technology for CDR exists today in spite of its tremendous potential value. A primary problem is due to side reactions from catalyst deactivation (i.e., coking) induced by carbon deposition via the Bouduard reaction ($2CO \rightarrow C + CO_2$) and/or methane cracking ($CH_4 \rightarrow C + 2H_2$) resulting from the high-temperature reaction conditions [21]. The coking effect is intimately related to the complex reaction mechanism, and the associated reaction kinetics of the catalysts employed in the reaction.

Improving the selectivity of the reaction towards the desired syngas product while simultaneously reducing the coking on the surface of the catalyst will require novel approaches to catalyst surface design and fabrication, most likely enabled by the advances in nanomaterials synthesis and characterization.

3.2 Oxidative Coupling of Methane to Ethylene

Ethylene is the largest chemical intermediate produced in the world and is utilized in a wide range of important industrial products, including plastics, surfactants, and pharmaceuticals. It has a worldwide annual production in excess of 140 million metric tons, of which over 25% is produced in the United States. Global demand growth for ethylene remains robust at about 4% per year [22]. Ethylene is primarily manufactured via high-temperature steam cracking of naphtha, with a smaller amount being made from ethane. The endothermic cracking reaction requires high temperatures (>900°C) and high energy input for both the reaction and product-separation separations processes. This results in steam cracking being one of the largest consumers of fuel as well as the largest CO_2 emitter of any commodity chemical product [23].

A potentially promising reaction for direct natural gas activation is the oxidative coupling of methane (OCM) to ethylene: $2CH_4 + O_2 = C_2H_4 + 2H_2O$. The reaction is exothermic ($\Delta H = -67$ kcal/mole) and to date has only been shown to occur at very high temperatures (>700°C). In the reaction, methane (CH_4) is activated on the catalyst surface, forming methyl radicals which then couple to ethane (C_2H_6), followed by dehydrogenation to ethylene (C_2H_4). Since the OCM reaction was first reported over 30 years ago, it has been the target of intense research interest, but conventional catalyst preparations have not yet afforded commercially attractive yields and selectivities. Numerous publications from industrial and academic labs have consistently demonstrated characteristic performance of high selectivity at low conversion of methane, or low selectivity at high conversion [24]. To break this cycle, a catalyst with a specific activity for CH_4 bond activation at lower temperatures (<700°C) must be developed. It has been over 15 years since the last large-scale efforts in industry and academia were focused on this reaction. During that time, the advent of bottoms-up nanomaterials synthesis, high-throughput screening, advanced characterization, and improved modeling have all made great advances. If these advances are now applied to this problem, it appears that significant strides could be made in pushing the selectivity higher at lower temperatures, thereby resulting in a commercially viable process which could simultaneously reduce our dependence on oil, and reduce our CO_2 footprint.

Breakthroughs in difficult, yet high-reward, reactions like OCM will likely require focused effort and broader partnerships between academia, the national labs, innovative small businesses, and the multinational chemical companies.

4 Scientific and Technological Infrastructure

In recent years, substantial progress has been made in the synthesis of new, nanoscopic catalysts. A critical contributor to that progress lies with advances in instrumentation that makes possible monitoring of catalysts in their working state. This allows researchers to close the loop between catalyst structure and function. Further progress requires a better fundamental understanding of the molecular-scale self-assembly of complex, multicomponent, metastable systems. This in turn will depend on critical advances in theory and simulation, as well as the development and accessibility of new *in situ* characterization methods that will provide the requisite spatial, energy, and time resolution, allowing the study of catalysts in their working state.

In particular, instrumentation must be developed that will allow probing of nanostructured catalysts under the appropriate conditions of pressure and temperature, while maintaining the highest spatial resolution possible. High-resolution microscopy in which dynamic reshaping of metal nanocrystals can be studied *in situ* would provide important insights into the mechanisms of catalytic action in nanostructures. Complementary, integrated means of carrying out high-throughput synthesis and screening of nanostructured catalysts would allow faster convergence to optimal structures and compositions.

As stated in Sect. 2, synchrotrons have played a major role in catalyst characterization, since the hard x-rays they generate can probe solids at high temperatures and pressures (catalytic "working conditions"), or in the liquid phase, without the requirement that the catalyst be crystalline. It is important that large-scale, complex, and expensive facilities such as synchrotrons, high resolution electron microscopes, and high field magnetic resonance equipment be made broadly available to the community of researchers in nanocatalysis.

While *computational techniques* related to theory and simulation of catalytic reactions have recently shown major progress, further improvements are needed to enhance the *accuracy* of computational methods, particularly for reacting systems, and to develop *predictive* capabilities. The accuracy of the methods is still an issue for reacting systems, and there needs to be a better connection between the simulation of adsorbates on surfaces and their rate of transformation to products. Thus, more work is needed to link *ab initio* methods used to describe the structure and energetics associated with adsorbed species to the atomistic simulations needed to describe their diffusion and reaction on the surface. More capability is also needed in the use of theory and simulation to guide *catalyst synthesis* at both the atomic scale and at larger length scales representative of catalyst support structures. Theoretical methods also do not handle very well the description of photoexcitation events that would occur during photocatalysis, an area that is likely to become more important in the next decade as researchers search for better ways to utilize solar energy.

5 R&D Investment and Implementation Strategies

The 2009 WTEC catalysis report [5] noted that "the overall level of investment in catalysis research in Europe appears to be higher than that in the United States." The report also noted that cooperation between universities and companies in this field is more common in Europe and Asia (related to the intellectual property environment), and also that "European and Asian countries have done an excellent job combining academic research with national laboratory activities." All of these observations provide some valuable benchmarks for future U.S. investments in this area. Major developments in computation and synthesis are required, and perhaps most importantly, the appropriate instrumentation must be developed and made available to the community of researchers. The ability to monitor catalysts with the appropriate resolution (spatial, temporal) in their working state is critical to understanding and thus control of nanostructured catalysts. Investments should be made to make more widely accessible both large-scale national facilities and specialized state-of-the-art instrumentation, and funding should also support the development of next-generation instrumentation. Nanostructured catalysis has the potential to profoundly affect our quality of life, and this should be reflected through investments in research and development in this area, commensurate with societal impacts in the areas of fuels, chemicals, and energy efficiency.

From a practical application standpoint, two of the more promising, but difficult, frontier areas for nanotechnology in catalysis are predictive design and fabrication of bulk catalyst particles with nanoengineered porosity over multiple length scales. This will require development of better, cost-effective structure-directing methods to further control the size, shape, and atomic surface structure of highly dispersed materials, as well as of methods to better produce well defined micro-and mesoporous support structures. Better synthetic methods to build and control composite particle architecture over multiple length scales should spur development of improved tools for catalyst scale-up and large-scale manufacturing that will be needed to broadly allow the penetration of nanoengineered materials into commercial applications.

There is also a strong need for improved reaction modeling capabilities that better couple catalyzed reaction pathways with molecular transport in micro- and mesoporous materials over multiple length scales, e.g., multiscale modeling integrating transport in multicomponent powders or pellets with fundamental surface reaction kinetics. Advances in this area should strongly facilitate more efficient design, development, and optimization of next generation commercial catalyst structures.

6 Conclusions and Priorities

Catalysis by nanostructured materials offers tremendous potential for more efficient, low-cost, and environmentally sustainable production of energy carriers and chemicals in manners consistent with environmental sustainability.

Although tremendous progress has been made in this area, substantial challenges remain. Research priorities should build on the substantial recent progress in this field to be able to dramatically affect some of the 'Holy Grail' reactions described in Sect.3 in Chapter "Applications: Nanophotonics and Plasmonics". In particular:

- There have been recent successes in beginning to characterize some catalytic processes "in the working state", under "real" conditions of elevated temperature, pressure and reactant flux. In addition monitoring of "single turnover" events, that is, single catalytic events, has been realized. The ultimate objective is a complete snapshot of a multi-step catalytic process under "working state" conditions.
- There has been steady progress in the ability to control the size, structure and crystalline composition of nano-sized catalysts. To fully exploit these structures, it is critical that progress be made in ensuring the robustness and stability of those nano-sized catalysts through the catalytic process itself.
- Ultimately, research priorities for nanostructured catalysts have the overall goal of realizing precise control over the composition and structure of catalysts over length scales spanning 1 nm–1μm, allowing the efficient control of reaction pathways.

In order to realize those researcher priorities, better understanding and control is required in the synthesis of molecular-scale, multicomponent, metastable systems and the assembly of these systems into larger, bulk composite particle structures. Advances in nanoscale characterization techniques are needed, especially those that will allow *in situ* monitoring under the conditions of temperature and pressure that characterize the "working state" of the catalysts. Finally, significant advances are required in theoretical descriptions of complex catalytic reactions that will provide predictive capabilities.

Given the critical importance of the highest-quality characterization tools, a priority should be in the investment of resources to develop instrumentation to facilitate *in situ* nanoscale characterization. Another priority should be to provide the means of making such instrumentation, including national resources such as synchrotron beamlines, more widely accessible to the research community. Given the importance of this area to meeting environmental, energy, and sustainability challenges, strong industry-academic collaborations should be encouraged and facilitated.

In addition to methods of studying catalytic reactions, significant effort should be dedicated to solving the most challenging and beneficial reactions discovered but not conquered over the last 50 years. In particular, developing catalysts that enable the successful use of natural gas feedstocks for commodity chemicals, ranging from the ubiquitous olefins (ethylene and propylene), to aromatics and gasoline, instead of the current predominant use as a heat source. If dedicated effort were given to one or two of these reactions, significant progress from R&D to commercial demonstration-scale plants could likely be achieved in the next 10 years.

7 Broader Implications for Society

More efficient and selective catalysts could have a profound impact on energy generation and efficiency, pollution abatement, the production of commodity and specialty chemicals and pharmaceuticals, and global economic health and development. Thus, R&D in the area of nanostructured catalysts will have a profound effect on society and its standard of living. Indeed, the WTEC catalysis report [5] noted that the key applications stimulating most catalysis research worldwide are related to energy and the environment. Not only will developments in catalysis directly impact society through new, improved, and cheaper, products, but larger-scale impacts similar to those of the Haber-Bosch process of the early 1900 s are possible. For example, enabling the use of natural gas as a source of commodity chemicals, instead of burning it will have many positive impacts simultaneously: e.g. decrease U.S. dependence on oil, increase the value of natural gas in North America, decrease CO_2 and total greenhouse gas emissions, and decrease global demand on petroleum.

8 Examples of Achievements and Paradigm Shifts

8.1 Direct Observation of Single Catalytic Events

Contact person: Robert R. Davis, University of Virginia

The tools of nanotechnology developed over the last decade enabled researchers to observe and analyze for the first time individual catalytic events on solid surfaces under realistic operating conditions. Whereas spectroscopic methods have been used for many years to probe the reactivity of catalytic surfaces and microscopic methods were able to reveal the atomic structures of the catalysts themselves and sometimes the organization of adsorbate overlayers on the catalytic surfaces, single-molecule resolution of catalytic turnover events has proven to be elusive. In 2006, M. Roeffaers and colleagues published a landmark paper in *Nature* in which they demonstrated how wide field microscopy can be used to map the spatial distribution of catalytic activity on a crystal surface by counting single turnover events [25]. While individual elementary reactions may occur on the subpicosecond time scale and therefore cannot be followed by this methodology, it is quite evident that the overall turnover of a catalytic cycle can be monitored with typical turnover periods in range of 10^{-2}–10^2 s.

The principle of the Roeffaers experiment is illustrated schematically in Fig. 3. The catalyst was a Li-Al layered double hydroxide (LDH) in the form of stacked sheets that assemble into prismatic crystals with large basal {0001} planes. The presence of both Li^+ and Al^{3+} in the gibbsite-type sheets results in a net positive charge that is balanced by anions located between the sheets. When those charge-balancing species

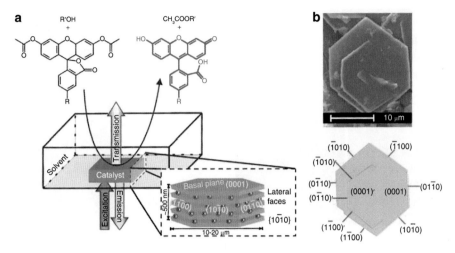

Fig. 3 Experimental set-up. (a) Schematic drawing of the experimental set-up: the LDH particle is exposed to fluorescein ester (R=–COOH for C-FDA; R=–H for FDA) in R′OH solution (R′=–H for hydrolysis; R′=–nC_4H_9 for transesterification). A wide field microscope with 488 nm excitation light was used. The inset shows the different crystallographic faces of a hexagonal LDH crystallite with indication of the Miller indices. (b) Scanning electron micrograph of a typical LDH crystal with assignment of the different crystal faces for the intergrown crystal (From Roeffaers et al. [25])

are OH^- anions, the solid LDH functions quite effectively as a Bronsted base and can therefore catalyze reactions such as hydrolysis, esterification and transesterification. Roeffaers et al. mounted a single crystal of the LDH catalyst onto a microscope slide with the basal {0001} plane parallel to the cover glass. The researchers then proceeded to use a non-fluorescent ester of fluorescein, 5-carboxyfluorescein diacetate (C-FDA) as a probe of the catalytically active base sites of the LDH because the reaction products from either hydrolysis of the C-FDA ester in water or transesterification of the ester with butanol are fluorescent. As shown in Fig. 3, irradiation of the hexagonally-shaped crystal of LDH with an appropriate wavelength allows for direct observation of the hydrolysis or transesterification of C-FDA in condensed media.

The bright spots in Fig. 4 correspond to the catalytic formation of single fluorescent product molecules formed by either transesterification with butanol (Fig. 4a–d) or hydrolysis of C-FDA (Fig. 4e–h). The authors were able to record the time evolution of the spot pattern (corresponding to catalytic turnover of individual active sites) and the movement of spots across the surface from diffusion of reaction products. The distributions of reaction rates as summarized in Fig. 4c, g enable direct correlation of individual site activities with spatial positions on the catalyst surface.

The exciting observation of catalytic turnover at a single-site level was soon followed by another study in which a redox catalytic reaction was monitored at the molecular level as it occurred on individual metal nanoparticles (Fig. 5). Xu et al. [26] discovered that spherical gold nanoparticles catalyze the reduction of non-fluorescent resazurin to fluorescent resorufin by NH_2OH, which provided the basis for single

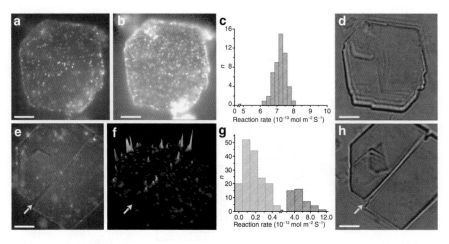

Fig. 4 Wide field images of catalytic reactions on individual LDH particles. Transesterification of C-FDA with 1-butanol at 40 nM (**a**) and 700 nM (**b**) ester concentration on the same LDH crystal. (**c**) Distribution of initial reaction rates for 1 µm² domains on the crystal faces ($n=50$). (**d**) Transmission image of the crystal. (**e-h**) Hydrolysis of 600 nM C-FDA on LDH crystal. (**e**) Fluorescence image, showing formation of single product molecules mainly at crystal edges (96 ms per image). (**f**) Accumulated spot intensity on the same crystal over 256 consecutive images. (**g**) Distribution of initial reaction rates for 1 µm² domains on the faces of the LDH crystal ($n=207$). The distribution clearly shows two statistically different subpopulations. The fast population corresponds to active domains located on the {1010} faces (*right graph*), whereas the {0001} faces host the slow population (*left graph*). (**h**) Transmission image. Scale bars, 5 µm. *Arrows* in e, f, h indicate the same viewing direction on the same crystal [25]

molecule observation, as depicted in Fig. 5a. [26]. The evolving catalytic reactions on a single 6 nm Au particle were recorded by monitoring the bursts of light at a fixed location (marked by the arrow in Fig. 5b) and plotting the light intensity with time, as shown in Fig. 5c.

The intensity of the fluorescence bursts was generally the same value, which indicates the turnover of one reactant resazurin molecule to one product resorufin prior to desorption from the gold particle. (In about 1% of the trajectories, the intensity of a spot was double the normal value, which suggested that two product molecules were co-adsorbed on the gold nanoparticle, as illustrated in Fig. 5d.) The time period in the center of Fig. 5c labeled as τ_{off}, in which there was no fluorescence from the Au particle, corresponds to the time in which the Au particle was either bare or had non-fluorescent species adsorbed on its surface. The time period labeled as τ_{on} corresponds to the time that product was adsorbed on the Au surface, prior to desorption into the liquid phase. From the intensity profile, a statistical measure of the overall catalytic turnover frequency of the reaction on *a single gold nanoparticle* was calculated.

The ability to observe single-molecule catalytic events on surfaces allows the field of heterogeneous catalysis to explore the influence of metal particle size on a variety of reaction kinetic parameters such as activity, selectivity, and deactivation, as well as to explore the effects of dynamic restructuring of catalytic particles on surface reactivity. A major challenge with translating this technique into general

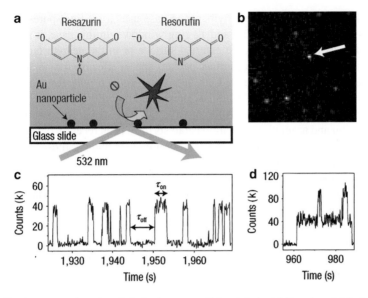

Fig. 5 Single-turnover detection of single-Au-nanoparticle catalysis. (**a**) Experimental design using total internal reflection fluorescence microscopy. (**b**) A typical image (~18×18 μm²) of fluorescent products during catalysis taken at 100 ms per frame. The pixel size is ~270 nm, which results in the pixilated fluorescence spots. (**c**) A segment of the fluorescence trajectory from the fluorescence spot marked by the *arrow* in *b* at 0.05 μM resazurin and 1 mM NH$_2$OH. (**d**) A segment of another fluorescence trajectory showing two on-levels at the same conditions (From Xu et al. [26])

catalysis practice involves the present requirement for a change in the fluorescent nature of a molecule upon reaction. This constraint prevents the study of the vast majority of commercially-relevant chemical transformations by this technique. Nevertheless, the fundamental concepts derived from these works and continued developments in microscopic methods will surely lead to a new era of nanotechnology used in the study of catalytic reactions on solid surfaces.

8.2 Representative Commercial Catalyst Applications

Contact person: S. Mark Davis, ExxonMobil Chemical

Two recent example applications for nanoengineered microporous materials into shape-selective ExxonMobil catalytic processes are (1) selective hydroisomerization catalysts for producing high-quality lube basestocks from paraffinic gas oils (process also known as MSDW™), and (2) catalysts for producing ethylbenzene from ethylene and benzene more selectively (also known as EBMax™) [27]. Normal paraffin hydroisomerization takes place inside the channels of a proprietary, medium-pore-size zeolite. By carefully tailoring the pore size and crystal morphology, it is possible to create large differences in molecular diffusion rates, so that normal paraffins are able to enter and preferentially react to produce methyl-branched isoparaffins, but larger or highly branched molecules are excluded from accessing the active sites.

A different strategy in the ethylbenzene case uses a medium-pore-size zeolite containing a mixture of 10-ring and 12-ring windows. In this case, it appears that most catalysis takes place on the external surface or in the near-surface region of the zeolite. Specifically, 12-ring surface pockets appear effective for single alkylation reactions that produce the desired ethylbenzene product, whereas molecules that diffuse inside into the zeolite cages have restricted mobility and are more prone to multiple alkylation reactions. Alternative catalyst systems for paraffin hydroisomerization and aromatics alkylation catalysts have been recently developed using similar principles in other industrial laboratories including those of Chevron, Axens, Sinopec, and UOP.

Haldor Topsoe Laboratories has recently developed several new catalyst families where the dispersed phase structure is strongly influenced by modern nanosynthesis methods. One innovation is the BRIM™ catalyst family for hydrodesulfurization (or HDS) of diesel and other distillate petroleum streams [28]. In these systems, special catalyst synthesis methods are used to produce highly dispersed, promoted molybdenum sulfide particles that are more active and are sulfided more efficiently during catalyst pretreatment. Recent, detailed characterization studies using high-resolution electron microscopy and complementary methods showed that the edges of the small metal sulfide particles show semi-metallic electronic states that contribute to high catalytic activity [29]. Haldor Topsoe has also recently commercialized new Fence™ catalysts for syngas conversion to methanol. In this case, small alumina particles with controlled size distribution are intimately formulated together with copper on zinc oxide, which is the active catalyst phase. The alumina acts as a structural promoter, which efficiently inhibits copper sintering and thereby improves long-term catalyst activity maintenance [30].

Headwaters NanoKinetix, Inc., has also reported application of nano-based synthetic methods into the formulation of new commercial catalyst systems. One example is the NxCat™ system for direct hydrogen peroxide synthesis from hydrogen and oxygen ([31]; see also [32, 33]). Proprietary techniques are used to produce uniform supported 4 nm platinum-palladium alloy particles with mostly (110) surface orientation that is needed for high selectivity. This new catalyst system received an EPA Green Chemistry Award from the U.S. Environmental Protection Agency (EPA) in 2007 and is now being commercialized through a joint venture between Headwaters and Evonik. Another Headwaters innovation includes the NxCat™ system for more selective reforming of naphtha to produce high-octane gasoline. Earlier surface science and catalysis studies by Somorjai and colleagues [34] showed that (111) platinum surface orientation is needed to more selectively catalyze paraffin dehydrocyclization to high-octane aromatics. Scientists at Headwaters identified new, structure-directing synthetic methods for producing high-area supported catalysts that appear to exhibit a higher degree of this platinum surface orientation ([31]; see also [35–39]). Headwaters has also reported application of nanoengineering in development of new hydroprocessing catalyst systems.

Modern nanotechnology has also had significant influence on the tools utilized to carry out industrial catalyst research and development. High-throughput, combinatorial methods for rapid robotic synthesis, characterization, and parallel screening at the laboratory scale have fundamentally changed the catalyst discovery process. Enabling technologies for these tools span several areas, including microsensors, micromachines, microelectronics, fast analytics, and data visualization and modeling. While much work remains to more directly extend these types of tools into catalyst development and commercial manufacture, there appears to be exciting potential for these types of developments over the next several years.

8.3 Fuel Cell Electrocatalyst R&D

Contact person: Alex Harris, Brookhaven National Laboratory[1]

Barriers to commercialization of fuel cells have been substantially reduced by research on electrocatalyst materials. Studies undertaken at BNL since 2002 led to the realization that the electrochemical activity of metals could be tuned by putting a single atomic layer of one metal over another, and more specifically by synthesis of tailored core-shell nanoparticles, with one or two layers of atoms of one active metal decorating a core particle comprised of a second metal. These core-shell nanocatalysts were further developed under funding by the fuel cell program in EERE in the period 2003–2010. The research has resulted in electrocatalysts for fuel cells that demonstrate a 4 to 20-fold reduction in the amount of expensive platinum necessary for good performance and show important improvements in long-term stability.

Based on these promising results and starting in 2005, BNL entered into cooperative research and development agreements (CRADAs) funded by industrial partners such as (GM, Toyota, UTC Fuel Cells, and Battelle to further progress in scale-up and to assess commercial potential. The commercial partners now believe that these core-shell nanocatalysts are currently the most promising electrocatalyst path for commercialization of low-temperature fuel cells for automotive and stationary applications.

This R&D program has generated numerous patents and patent applications, and the intellectual property portfolio has been an important factor in attracting CRADA and commercialization partners, enabling opportunities for a competitive market position. BNL's Laboratory's Technology Maturation Program (TMP) provides funding to advance the technological readiness for laboratory inventions, including scale-up or life-cycle demonstrations/testing, in order to improve commercial potential. In this case, the BNL TMP funding supported the scale-up of the synthesis from test quantities (milligrams or less) to 10 gram batches and the development of an in-house membrane electrolyte assembly (MEA) fabrication and fuel cell test station for testing larger-scale assembled fuel cells. This scale-up effort has been critical to moving the technology from the laboratory to deployment.

---

[1] Supported by the Department of Energy (DOE) Offices of Basic Energy Sciences (BES) and of Energy Efficiency and Renewable Energy (EERE).

9 International Perspectives from Site Visits Abroad

With the exception of the Chicago Workshop on the Long-Term Impacts and Future Opportunities for Nanoscale Science and Engineering held in Chicago, March 9–10, 2010, catalysis was not discussed within its own breakout session in the Hamburg, Tokyo/Tsukuba, or Singapore international workshops associated with this study, although the theme of catalysis did came up in sessions such as photonics and plasmonics and energy. Within these sessions there was a general recognition of the importance of more efficient and specific catalysts in a variety of energy-efficient processes. For example, nanophotonic devices might locally and specifically couple in light to photocatalysts, driving the catalytic process.

At the U.S.-Japan-Korea-Taiwan workshop held in Tsukuba, Professor Kazunari Domen of the University of Tokyo presented some work specifically focused on catalysis, including work being carried out by Professor M. Haruta of Tokyo Metropolitan University, and a long-time leader in this field. Fig. 6 represents

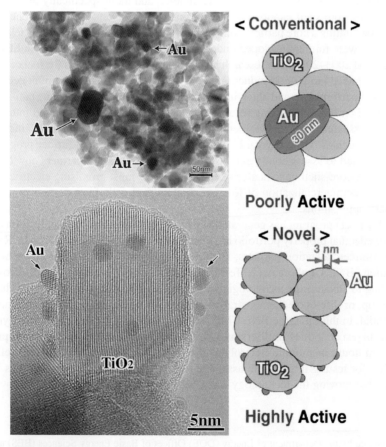

Fig. 6 Comparison of the nanostructure of poorly active and highly active Au-based catalysts (Courtesy of M. Haruta)

Haruta's work, shown at the workshop, and illustrates the differences in nanostructure of poorly active and highly active Au-based catalysts. Otherwise, no significant considerations were given to specific visions, opportunities, and challenges for nanoscale catalysis. It should be noted, however, that the 2009 WTEC study included a broad review of international activities, and the reader should refer to that report for a recent global perspective in nanotechnology research trends applied to catalysis [5].

References

1. M.C. Roco, S. Williams, P. Alivisatos, *Nanotechnology Research Directions: IWGN Workshop Report* (WTEC, Baltimore, 1999)
2. M. Haruta, Size- and support-dependency in the catalysis of gold. Catal. Today **36**(1), 153–160 (1997). doi:10.1016/S0920-5861(96)00208-8
3. A.Y. Bell, The impact of nanoscience on heterogeneous catalysis. Science **14**, 1688–1691 (2003). doi:10.1126/science.1083671
4. M. Davis, D. Tilley, *Future directions in catalysis* (National Science Foundation, Washington, 2003)
5. R. Davis, V.V. Guliants, G. Huber, R.F. Lobo, J.T. Miller, M. Neurock, R. Sharma, L. Thompson, An international assessment of research in catalysis by nanostructured materials (WTEC, Baltimore, 2009), Available online: http://www.wtec.org/catalysis/WTEC-CatalysisReport-6Feb2009-color-hi-res.pdf
6. M. Choi, H.S. Cho, R. Srivastava, C. Venkatesan, D.H. Choi, R. Ryoo, Amphiphilic organosiline-directed synthesis of crystalline zeolites with tunable mesoporosity. Nat. Mater. **5**(9), 718–723 (2006). doi:10.1038/nmat1705
7. C. Christensen, I. Schmidt, A. Carlsson, K. Johanssen, K. Herbst, Crystals in crystals. J. Am. Chem. Soc. **127**, 8098–8102 (2005)
8. D. Li, D. Su, J. Song, X. Guan, K. Hofmann, F.S. Xiao, Highly stream-stable mesoporous silica assembles from preformed zeolite precursors at high temperature. J. Mater. Chem. **15**, 5063–5069 (2005)
9. A. Corma, M. Diaz-Cabana, J. Jorda, C. Martinez, M. Moliner, High-throughput synthesis and catalytic properties of a molecular sieve with 18- and 10-member rings. Nature **443**, 842–845 (2006). doi:10.1038/nature05238
10. F. Berhault, L. Bisson, C. Thomazeau, C. Verdon, D. Uzio, Preparation of nanostructured Pd particles using a seeding synthesis approach. Appl. Catal. A **327**(1), 32–43 (2007). doi:10.1016/j.apcata.2007.04.028
11. E. Aneggi, J. Llorca, M. Boaro, A. Trovarelli, Surface-structure sensitivity of CO oxidation over polycrystalline ceria powders. J. Catal. **234**(1), 88–95 (2005). doi:10.1016/j.jcat.2005.06.008
12. I. Lee, F. Delbecq, F. Morales, M. Albiter, F. Zaera, Tuning selectivity in catalysis by controlling particle shape. Nat. Mater. **8**, 132–138 (2008). doi:10.1038/nmat2371
13. C. Witham, W. Huang, C. Tsung, J. Kuhn, G. Somorjai, F. Toste, Converting homogeneous to heterogeneous in electrophilic catalysis using monodisperse metal nanoparticles. Nat. Chem. **2**, 36–41 (2010). doi:10.1038/nchem.468
14. M. Salmeron, R. Schlogl, Ambient pressure photoelectron spectroscopy. Surf. Sci. Rep. **63**(4), 169–199 (2008). doi:10.1016/j.surfrep.2008.01.001
15. P. Hansen, J. Wagner, S. Helveg, J. Rostrop-Nielsen, B. Clausen, H. Topsoe, Atom-resolved imaging of dynamic shape changes in supported copper nanocrystals. Science **15**, 2053–2055 (2002). doi:10.1126/science.1069325
16. J. Norskov, T. Bligaard, J. Rossmeis, C. Christensen, Towards the computational design of solid catalysts. Nat. Chem. **1**, 37–46 (2009). doi:10.1038/nchem.121

17. The Catalyst Group Resources, *Understanding Nano-Scale Catalytic Effects*. CAP Client-Private Report (The Catalyst Group, Spring House, 2010)
18. The Catalyst Group Resources, *Advances in Nanocatalysts and Products*. CAP Client-Private Technical Report (The Catalyst Group, Spring House, 2002)
19. The Catalyst Group Resources, *Intelligence report: Business Shifts in the Global Catalytic Process Industries, 2007–2013*. CAP Client-Private Report (The Catalyst Group, Spring House, 2008)
20. J. Haggin, Chemists seek greater recognition for catalysis. Chem. Eng. News **71**(22), 23–27 (1993). May 31 http://www.osti.gov/energycitations/product.biblio.jsp?osti_id=6347886
21. V. Kroll, H. Swaan, C. Mirodatos, Methane reforming reaction with carbon dioxide over Ni/SiO_2 catalyst-I. Deactivation studies. J. Catal. **161**(1), 409–422 (1996). doi:10.1006/jcat.1996.0199
22. T. Ren, M. Patel, Basic petrochemicals from natural gas, coal and biomass. Resour. Conserv. Recycl. **53**(9), 513–528 (2009). doi:10.1016/j.resconrec.2009.04.005
23. M. Ruth, A. Amato, B. Davidsdottir, Carbon emissions from U.S. ethylene production under climate change policies. Environ. Sci. Technol. **36**(2), 119–124 (2002)
24. J. Labinger, Oxidative coupling of methane. Catal. Lett. **1**, 371–375 (1988). doi:10.1007/BF00766166
25. M. Roeffaers, B. Sels, H. Uji-i, F. DeSchryver, P. Jacobs, D. Devos, J. Hofkens, Spatially resolved observation of crystal-face-dependent catalysis by single turnover counting. Nature **439**, 572–575 (2006). doi:doi:10.1038/nature04502
26. W. Xu, J. Kong, Y.T. Yeh, P. Chen, Single-molecule nanocatalysis reveals heterogeneous reaction pathways and catalytic dynamics. Nat. Mater. **7**, 992–996 (2008). doi:10.1038/nmat2319
27. J. Santiesteban, T. Degnan, M. Daage, Advanced catalysts for the petroleum refining and petrochemical industries. Paper presented at the14th International Congress on Catalysis, Seoul, 2008
28. Haldor Topsoe, Corporate Web site (2010), http://www.topsoe.com/research/Research_at_Topsoe.aspx
29. J. Laurisen, J. Kibsgaard, S. Helveg, H. Topsoe, B. Clausen, E. Laegsgaard, F. Besenbacher, Size-dependent structure of MOS_2 nanocrystals. Nat. Nanotechnol. **2**(1), 53–58 (2007)
30. K. Svennerber, From science to proven technology – development of a new topsoe methanol synthesis catalyst MK-151. Paper presented at the 2009 World Methanol Conference, Miami, 2009
31. Headwaters NanoKinetix, Inc. n.d. Corporate Web site, http://www.htigrp.com/nano.asp
32. M. Rueter, B. Zhou, S. Parasher, Process for direct catalytic hydrogen peroxide production. U.S. Patent 7,144,565, 2006
33. B. Zhou, M. Rueter, S. Parasher, Supported catalysts having a controlled coordination structure and methods for preparing such catalysts. U.S. Patent 7,011,807, 2006
34. S.M. Davis, F. Zaera, G.F. Somorjai, Surface structure and temperature dependence of n-hexane skeletal rearrangement reactions. J. Catal. **85**(1), 206–223 (1984). doi:10.1016/0021-9517(84)90124-6
35. H. Trevino, Z. Zhou, Z. Wu, B. Zhou, Reforming nanocatalysts and methods of making and using such catalysts. U.S. Patent 7,541,309, 2009
36. B. Zhou, M. Rueter, Supported noble metal nanometer catalyst particles containing controlled (111) crystal face exposure. U.S. Patent 6,746,597, 2004
37. B. Zhou, H. Trevino, Z. Wu, Reforming catalysts having a controlled coordination structure and methods for preparing such compositions. U.S. Patent 7.655,137, 2010
38. B. Zhou, H. Trevino, Z. Wu, Z. Zhou, C. Liu, Reforming nanocatalysts and method of making and using such catalysts. U.S. Patent 7,569,508, 2009
39. Z. Zhou, Z. Wu, C. Zhang, B. Zhou, Methods for manufacturing bi-metallic catalysts having a controlled crystal face exposure. U.S. Patent 7,601,668, 2009

Applications: High-Performance Materials and Emerging Areas*

Mark Hersam and Paul S. Weiss

Keywords Nanocomposite • Nanofibers • Metamaterials • Carbon nanotubes • Nanomanufacturing • Combinatorial • Separation • Purification • Aeronautics • Nanofluidics • Nanostructured polymers • Wood products • International perspective

1 Vision for the Next Decade

1.1 Changes in the Vision over the Last 10 Years

The field encompassed by the term "nanomaterials" has changed dramatically over the past 10 years. While it was useful in 2000 to describe nanomaterials on the sole basis of our ability to understand and to control matter at the nanoscale where material properties possess a distinct size-dependence, the field has now grown well beyond that earlier definition. For example, it is evident in 2010 that additional factors beyond constituent nanoparticles (e.g., high surface/interface area, proximity, and novel chemical, physical, and/or biological moieties) are also playing major roles. New work envisioned at the turn of the twenty-first century [1, 2] on organic-inorganic composite materials, multifunctional materials, self-healing materials, and nanoscale sensors has become a reality in the past 10 years, while exploratory ideas about bio-abiotic heterogeneous systems, information-nano-bio integration,

---

*With contributions from: Richard Siegel, Phil Jones, Fereshteh Ebrahimi, Chris Murray, Sharon Glotzer, James Ruud, John Belk, Santokh Badesha, Adra Baca, David Knox.

M. Hersam (✉)
Department of Materials Science and Engineering, Northwestern University, 2220 Campus Drive, Evanston, IL 60208, USA
e-mail: m-hersam@northwestern.edu

P.S. Weiss
California NanoSystems Institute, University of California, 570 Westwood Plaza, Building 114, Los Angeles, CA 90095, USA
e-mail: stm@ucla.edu

electronic-neural interfaces, and nano-information-biological-cognitive systems [3] have received increased recognition in the scientific community and funding programs.

The past decade has seen an evolution in scientific understanding and capabilities, from working with isolated nanoparticles, nanotubes, and nanowires to working with hybrid nanocomposites whose properties are controlled not only by the constituent nanomaterials but also by their morphology, spatial anisotropy, and relative proximity with respect to one other and the host matrix. The worldwide scientific focus on nanostructured hybrid material systems based upon synthetic or natural polymers combined with metal, ceramic, carbon, or natural (e.g., clay) nanostructures represents a truly revolutionary change in our thinking and in our ability to create nanostructured materials and coatings to solve real problems to benefit society. Specific examples include:

- Biopolymer/inorganic-nanoparticle hybrid assemblies for targeted drug delivery or precise environmental sensors
- Synthetic polymer coatings containing dispersed enzymes
- Ceramic or metal nanoparticles for antiseptic or antifouling surfaces
- Diverse synthetic polymer/clay or nanoparticle dispersions for various industrial applications

1.2 Vision for the Next 10 Years

Over the next 10 years, R&D programs will focus on issues that will improve the performance, multifunctionality, integration, and sustainability of nanomaterials in a range of emerging and converging technologies. In particular, methods are required for nanomaterials and nanosystems by design; scaling up high-quality and monodisperse nanomaterial production; for rapidly measuring and characterizing quality and reproducibility of manufacturing processes incorporating nanomaterials; and for manufacturing nanostructures into bulk materials, coatings, and devices while retaining enhanced nanoscale properties. Through controlled assembly of nanoconstituents that have distinct properties, the next generation nanocomposite materials are expected to have the unique and powerful attribute of independent tunability of previously coupled properties. For example, bulk materials with high electrical conductivity typically possess high thermal conductivity, whereas these properties have the potential of being decoupled in next-generation nanocomposites. This specific example has broad implications for thermoelectric devices that convert waste heat into useful electricity. Similarly, the decoupling of electrical conductivity and optical reflectivity would enable a new class of transparent conductors that could serve as the basis of improved photovoltaic and display technologies. Ultimately, the rational assembly of nanomaterials into nanocomposites will yield high-performance materials with new combinations of properties, thus underpinning the development of previously unrealizable applications.

2 Advances in the Last 10 Years and Current Status

The main advances in nanostructured materials and coatings in the past 10 years have centered on hierarchical hybrid nanostructured material systems consisting of a potentially wide range of biological and non-biological building blocks; nanostructure-matrix interfaces in such nanocomposites; and the rapidly developing ability to create engineered interfaces with novel chemical, physical, and/or biological moieties for a variety of systems. While these high-performance nanomaterials have arguably touched and impacted nearly all fields of science and engineering, this section will provide a few specific examples for illustrative purposes.

2.1 Nanofibrous Media

During the past decade, nanotechnology has enabled significant expansion of consumer and industrial products by the use of nanofibrous media. Nanofibrous media are fabricated by techniques such as electrospinning and "islands-in-the sea" fiber spinning. While nanofibers have been known for some time, the concepts around the use of nanofibers have finally come into their own with nanotechnology allowing an understanding of their usage [4]. Specifically, companies such as Donaldson, E-Spin, United Air Specialists, and Argonide have built multimillion-dollar businesses using nanofibrous membranes. Initially, these membranes were developed for simple air filtration due to the achievement of high filtration efficiency at low operating pressure. More recently, these membranes have been incorporated into more sophisticated filtration systems with other functional effects such as bacteria removal using titanium dioxide fibrous webs, as depicted in Fig. 1 [5]. Recently, the National Aeronautics and Space Administration (NASA) recognized Argonide Corporation for its NanoCeram® water filter, which is capable of filtering >99.99% of hazardous particles from water [6].

While specific data regarding the growth of nanofiber-web-based materials are confidential to specific company product lines, it is estimated that nanofiber-web-based filtration media have grown by approximately $500 million in the previous

Fig. 1 Titanium dioxide fibrous web decorated with TiO_2 that is used for bacteria removal in advanced filtration systems [5]; false color SEM image courtesy of David Knox, MeadWestvaco Corporation

decade from a base of about $200 million. Early nanofiber-web-based filtration media were used almost exclusively in military applications; these media are now found in automotive and other applications. This expansion has been driven by improved understanding of nanofiber phenomena such as the origin of observed low pressure drops and particle interactions.

Beyond air filtration, the high degree of hydrophobicity in nanofibrous materials yields "instant clean" effects on everything from shirts to tablecloths to surgical and medical devices. For example, Nano-Tex has developed technology and products that incorporate nanowhiskers to allow for virtually stain-free surfaces. This application alone has already blossomed into an estimated $200 million per year market.

2.2 Nanocrystalline Metals

Ten years ago, it was known that nanostructured metals were strong, but it was believed that they were brittle in nature, which would limit their applications as structural materials. In the past decade, extensive progress has been made in the area of processing defect-free nanocrystalline metals using electrodeposition and plastic deformation techniques [7–9]. Consequently, it has been proven that nanocrystalline metals are not inherently brittle and that their tensile ductility befits their high strength [10, 11]. Figure 2 compares the tensile stress-strain curve of a nanocrystalline Ni-Fe alloy with a medium carbon quenched and tempered steel demonstrating the high strength (over 2 GPa) of the single-phase nanocrystalline metal and its reasonable ductility in comparison to a high-performance steel. The fact that nanocrystalline metals and alloys have reasonable ductility makes them candidates for multifunctional structural applications, where other properties such as corrosion resistance, magnetic properties, and optical properties are important in addition to

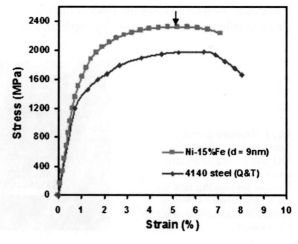

Fig. 2 A comparison of the tensile stress-strain curve of a nanocrystalline Ni-Fe alloy (average grain size of 9 nm) with a structural hardened steel (Courtesy of Fereshteh Ebrahimi, University of Florida)

high strength. Through the past decade, it has been demonstrated experimentally and by computational techniques that constraints imposed by the small grain size of nanocrystalline materials makes mechanisms that are relatively inconsequential in conventional polycrystalline metals to become dominant at the nanoscale. Examples include:

- Observation of mechanical twinning in face-centered cubic (fcc) metals with relatively high stacking-fault energy, such as aluminum and copper [12, 13]
- Motion of unpaired partial dislocations in fcc metals [14]
- Grain boundary-mediated deformation and grain rotation [15–17]
- Grain growth and coupled grain boundary migration [18–22]

Because of the dominance of the grain-boundary-mediated deformation in nanocrystalline materials, the structure of the boundaries plays a significant role in the plastic deformation of these materials. For instance, low-temperature annealing can relax the grain boundaries of as-processed materials considerably and makes the processes of dislocation nucleation from grain boundaries and grain boundary sliding more difficult [23]. To illustrate this point, Fig. 3 demonstrates that low-temperature annealing can result in an increase in strength and significant loss of ductility [24].

Simulation studies have elucidated that as the grain size is reduced below a critical level (10–20 nm, depending on the metal), the dominant deformation mechanisms change from dislocation activities within the crystals to grain-boundary-mediated deformation [14]. This transition causes a loss of strength, resulting in the so-called "inverse Hall-Petch" phenomenon. Detailed investigations using scattering techniques have revealed that because of the mechanistic differences in the plastic deformation, the overall deformation behavior of nanocrystalline metals is also different in comparison to conventional metals [25].

The experimental and simulation data suggest that during the plastic deformation of nanocrystalline materials, only a fraction of grains deform plastically [26–28]. Therefore, these materials can be envisioned as composite materials consisting of both

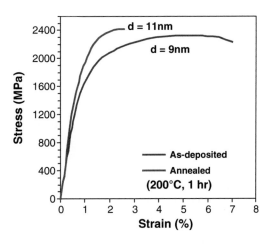

Fig. 3 Tensile stress-strain curves for as-deposited and low-temperature annealed Ni-15%Fe nanocrystalline materials (Courtesy of Fereshteh Ebrahimi, University of Florida)

elastic (smaller grains) and plastic (larger grains) components. More specifically, the large grains contribute to plastic deformation and the smaller elastic grains contribute to strain hardening. Consequently, strength and ductility of nanocrystalline metals depend strongly on the grain size distribution. Based on this idea, it has been demonstrated that metals with duplex grain size distributions can exhibit an optimum combination of strength and ductility [29]. Emerging results on the fracture of nanocrystalline metals suggest that in contrast to conventional fcc metals, whose fracture mode occurs by the plasticity-induced microvoid coalescence mechanism, nanostructured fcc metals appear to fracture by cleavage (i.e., by breaking atomic bonds) [30, 31].

It has been known for nearly 50 years that reducing the specimen size results in different stress-strain behavior [32], and thin samples such as whiskers exhibit strength levels close to the ideal strength [33]. In the past 10 years, more experimental results have confirmed that smaller samples are indeed stronger [34]. This increase in strength has been attributed to the lack of dislocation sources. One of the concerns regarding the application of nanocrystalline materials has been their stability or lack thereof. In the last decade, it has been established that alloying can significantly stabilize the grain boundaries through the solute drag mechanism [35, 36]. For example, Fig. 4 shows that the grain growth in Ni can be retarded significantly by the addition of Fe [35].

2.3 Cellulose-Based Nanomaterials

Over the past decade, the forest products industry has identified nanotechnology as a means to tap the enormous undeveloped potential of trees as photochemical "factories" that produce abundant sources of raw materials using sunlight and water. Forest biomass resources provide a key platform for sustainable production of renewable, recyclable, and environmentally preferable materials to meet the needs of society in the twenty-first century. Wood-based lignocellulosic materials (i.e., forest biomass) provide a vast material resource and are geographically dispersed.

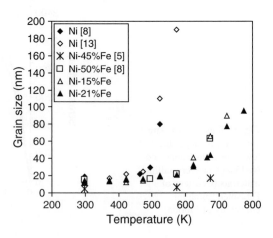

Fig. 4 Grain size evolution as a function of annealing temperature for pure Ni and Ni-Fe alloys (Courtesy of Fereshteh Ebrahimi, University of Florida)

Applications: High-Performance Materials and Emerging Areas

The forest products industry nanotechnology roadmap [37] identifies the industry vision as "sustainably meeting the needs of present and future generations for wood-based materials and products by applying nanotechnology science and engineering to efficiently and effectively capture the entire range of values that wood-based lignocellulosic materials are capable of providing." In addition, the forest products industry sees its inherent strengths as including stewardship of an abundant, renewable, and sustainable raw material base; supporting a manufacturing infrastructure that can process wood resources into a wide variety of consumer products; and being uniquely positioned to move into new, growth markets centered on bio-based environmentally preferable products.

At the nanoscale, wood is composed of elementary nanofibrils (whiskers) that have cross-sectional dimensions of about 3–5 nm and are composed of cellulose polymer chains arranged in ordered (crystalline) and less-ordered (amorphous) regions [37–40]. Wood is approximately 30-40% cellulose by weight, with about half of the cellulose in nanocrystalline form and half in amorphous form (Fig. 5). Nanocrystalline cellulose is relatively uniform in diameter and length; these dimensions vary with plant species. Cellulose is the most common organic polymer in the world, representing about 1.5×10^{12} tons of the total annual biomass production. Cellulose is expressed from enzyme rosettes as 3–5 nm–diameter fibrils that aggregate into larger fibrils up to 20 nm in diameter. These fibrils self-assemble in a

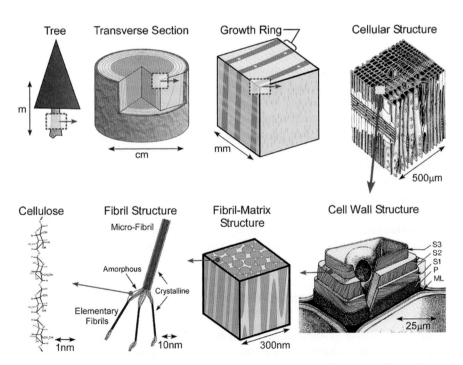

Fig. 5 Wood is a cellular hierarchical biocomposite made up of cellulose, hemicellulose, lignin, extractives, and trace elements. At the nanoscale level, wood is a cellulosic fibrillar composite (Courtesy of Phil Jones, Imerys)

manner similar to liquid crystals, leading to nanodimensional and larger structures seen in typical plant cell walls. The theoretical modulus of a cellulose molecule is ca. 250 GPa, but measurements for the stiffness of cellulose in the cell wall are ca. 130 GPa. These measurements imply that cellulose is a high-performance material comparable with the best fibers technology can produce.

Because of the hundreds of millions of tons of wood available for processing, commercial production is both sustainable and renewable and represents an industrially significant supply. High-value, renewable nano-enabled composites can be produced by identifying commercially attractive methods to liberate both nanofibrils and nanocrystalline cellulose and by establishing methods for characterization, stabilization, and blending of these wood-based nanomaterials with a variety of other nanomaterials.

3 Goals, Barriers, and Solutions for the Next 5–10 Years

Over the next 10 years, nanoscience researchers will focus on a range of issues to improve the performance, multifunctionality, integration, and sustainability of nanomaterials in a variety of emerging and converging technologies. Specific goals are described in the subsections below.

3.1 *Separation, Fractionation, and Purification*

The defining feature of a nanomaterial is that its properties depend not only on composition but also size and shape., Any polydispersity in size or shape leads to inhomogeneous properties that are generally undesirable in commercial technologies. Consequently, researchers will seek scalable and economical strategies for separating, fractionating, and purifying nanomaterials as a function of size and shape, thus yielding monodisperse nanoscale building blocks. The resulting uniform properties will enable reliable and reproducible performance in devices, technologies, and composite materials based on nanomaterial constituents. Furthermore, with monodisperse nanoscale building blocks in hand, controlled polydispersity can also be achieved and may be desirable where a controlled range of properties are required (e.g., photovoltaic technology requires conductive films with a range of optical transparencies matching the broadband solar spectrum).

3.2 *Hierarchical Metamaterials*

With monodisperse nanomaterials, it will become possible to assemble them into ordered crystalline structures where the constituent nanoparticles play the role of artificial atoms (Fig. 6). While some progress has been made along these lines in the field of quantum dots in the past decade, over the next decade, additional levels

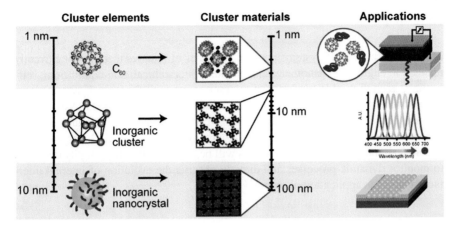

Fig. 6 Fullerenes, atomic clusters, and larger inorganic nanocrystals can be used as assembly elements for creating materials with tailored properties. Applications include photovoltaics (*top*), optical biosensors (*middle*), and electronics (*bottom*) ([41]; Courtesy of Paul S. Weiss, UCLA)

of control will be sought. Through controlled self-assembly and directed assembly of nanoconstituents of distinct properties, the next-generation nanocomposite material will have the unique and powerful attribute of independent tunability of previously coupled properties. For example, bulk materials with high electrical conductivity typically also possess high thermal conductivity; however, these properties have the potential of being decoupled in next-generation nanocomposites. This specific example has broad implications for thermoelectric devices that convert waste heat into useful electricity. Similarly, the decoupling of electrical conductivity and optical reflectivity would enable a new class of transparent conductors, which serve as the basis of photovoltaic and display technologies. Ultimately, the goal will be to rewrite the solid state physics textbooks regarding relationships between charge, energy, spin, phonons, excitons, and mass transport.

3.3 Nanomanufacturing

To bring nanomaterials to the mainstream marketplace, significant improvements in nanomanufacturing will be required. Problems surrounding scale-up, cost, sustainability, energy efficiency, process control, and quality control need to be overcome in order to realize mass production of microscale and macroscale nanocomposite materials that reliably retain the attributes of the nanoscale constituents. Fabrication of bulk products with retained nanoscale features will likely require scalable self-assembly techniques, durable surface-modification methods, and hierarchical processing and assembly. Ideally, this nanomanufacturing optimization will be informed by advanced computational capabilities that will allow improved design principles to be incorporated into the nanocomposite fabrication process.

3.4 Inspiration from Biology

Materials in biological systems possess a number of features that are not currently realized in engineered nanomaterials, including hierarchical, non-equilibrium, self-healing, reconfigurable, and defect-tolerant structures. Biological systems also possess optimized interactions between organic and inorganic media. Research in the next decade will seek to emulate this powerful combination of features in nanocomposite materials. To achieve these goals, improved control, characterization, and understanding of internal interfaces will be required. It will be highly desirable to monitor dynamic processes and directly observe the evolution of internal interfaces under the application of external stimuli.

3.5 Combinatorial and Computational Approaches

Nanocomposite materials represent an immense phase space that includes the size, shape, and composition of the nanoconstituents, surface functionalization, and matrix. This phase space cannot be efficiently explored via serial, empirical study. Consequently, alternative strategies, including massively parallel combinatorial approaches and multiscale modeling, will be pursued in the next decade. The former will require innovative experimental design and parallel characterization capabilities, while the latter will benefit from expected improvements in computational power and algorithm optimization. Efforts to improve the performance and durability of nanoparticle surface functionalization can likely be guided by previous and ongoing research in surface science. These hierarchical hybrid nanostructured material systems with appropriate functionalities will enable application to a much wider range of important societal problems in the areas of energy, environment, and health care. Towards these ends, researchers must not only continue to create the fundamental understanding and necessary tools to assemble and fully characterize these nanomaterials and systems, but also to develop reliable design capabilities to move from an era of "best-guess" nanomaterials to truly engineered systems.

3.6 Emerging and Converging Technologies

Realization of the aforementioned goals will create a suite of new nanocomposite materials with unprecedented properties and unique combinations of properties. In particular, the unique combinations of properties will allow previously disparate technologies to converge into single, multifunctional platforms. Specific examples include thermoelectrics, transparent conductors, combined supercapacitor and battery structures, integrated diagnostic and therapeutic devices, sensors/actuators, optoelectronics, and communication/computational systems.

4 Scientific and Technological Infrastructure Needs

To realize the goals that were delineated in the previous section, significant improvements in scientific and technological infrastructure will be required. For example, new instrumentation is needed to investigate the localized nature of interfaces in nanostructured material systems and to probe the fundamental properties of nanomaterials (e.g., mechanical, electrical, thermal, chemical, biological, optical, and magnetic). Reliable multiscale design techniques for hierarchical nanocomposites are also critical. Only with the widespread availability of these tools, both experimental and theoretical, will the field reach its future potential.

Beyond these infrastructure needs, it is imperative to have a highly trained, multidisciplinary workforce. Highly skilled scientists and engineers should be continually encouraged and greatly facilitated on a global scale. Within the United States, the educational system from K–12 through postgraduate has to be significantly improved, and greater encouragement must be given to all students, particularly those in under-represented groups (e.g., women and minorities) to develop their interests and capabilities in the Science, Technology, Engineering, and Mathematics (STEM) fields. A creative new system of national scholarships for students following STEM career paths might incentivize such an effort. Furthermore, improved educational materials are needed—ranging from conventional textbooks to online curricular materials—to promote better teaching and learning in the emerging field of nanomaterials.

Ongoing guidance to realize these scientific and technological infrastructure needs should be provided by a panel of experts. This panel should be heavily populated with people who have already achieved success and have proven leadership and vision (e.g., Nobel laureates, National Academy members, business leaders).

5 R&D Investment and Implementation Strategies

While the investment of taxpayer funds in the U.S. National Nanotechnology Initiative and similar government initiatives globally has been considerable, the return on investment has been even greater. Consequently, continued investment at even higher levels is needed to secure the gains already made in this rapidly advancing, but still developing field. For example, the U.S. National Science Foundation Nanoscale Science and Engineering Centers have been a resounding success. However, some of these centers are nearing their 10-year funding limit. A competitive mechanism should be instituted for national investment in these centers to be sustained and even increased in order to secure their research successes and to develop them into intellectual property that can directly benefit society. In addition, individual and small group grants warrant continued support to seed efforts that can eventually blossom into the basis of additional center-level initiatives. Toward that end, it would be desirable to incentivize the teaming of smaller nanoscience research efforts into networks in an effort to minimize redundant work. Interdisciplinary and international collaborative efforts are also likely to yield important advances in this field. As the nanomaterials field matures, future science and technology policy should not only identify areas of research and development that deserve support but also identify specific directions where support should be reduced or eliminated (i.e., winners *and* losers should be identified).

Investments by the Federal Government should be matched by state and local governments, as well as by the private sector (i.e., industry) in public-private partnerships. A global industry consortium would be particularly useful for funding efforts with commercial potential. Nanotechnology is rapidly progressing to the point where all potential funding bodies will reap the benefits of these investments in better national and local security, improved economies and quality of life, and significant job (and hence tax) creation. Such leveraged investments should become the norm for fields such as nanotechnology that have broad positive impact on society.

6 Conclusions and Priorities

The past decade has seen an evolution from nanomaterials that are isolated nanoparticles, nanotubes, and nanowires to nanomaterials that are hybrid nanocomposites where the properties are controlled not only by the constituent nanomaterials but also by their morphology, spatial anisotropy, and relative proximity with respect to each other and the host matrix. The worldwide focus on nanostructured hybrid material systems represents a true paradigm shift in our thinking and in our ability to create nanostructured materials and coatings to solve real problems that benefit society. The next 10 years will see a research focus on a range of issues that will improve the performance, multifunctionality, integration, and sustainability of nanomaterials in a range of emerging and converging technologies. Specific priorities include:

- Separation, fractionation, and purification in an effort to realize nanomaterials with monodispersity in composition, size, and shape
- Realization of hierarchical metamaterials with independent tunability of previously coupled properties
- Improvements in nanomanufacturing capabilities, including solving problems related to scale-up, cost, sustainability, energy efficiency, process control, and quality control
- Realization of nanomaterials with biologically inspired attributes, including non-equilibrium, self-healing, reconfigurable, and defect-tolerant structures in hybrid organic/inorganic media
- Combinatorial and computational approaches that enable efficient exploration of the vast phase space for nanocomposites, including the size, shape, and composition of the nanoconstituents, surface functionalization, and matrix
- Utilization of new nanocomposite materials with unprecedented properties and unique combinations of properties in emerging and converging technologies

7 Broader Implications for Society

Throughout history, advances in materials have fueled advances in technology. Undoubtedly, nanomaterials and nanocomposites are already having and will continue to have profound positive benefits for society. In particular, through controlled assembly of nanoconstituents of distinct properties, next-generation nanocomposite materials will have concurrent and independent tunability of distinct properties (e.g., mechanical, electrical, thermal, chemical, biological, optical, and magnetic). In this manner, next-generation nanocomposites are expected to decouple properties that are intimately intertwined in bulk materials. For example, the decoupling of electrical and thermal conductivity will have broad implications for thermoelectric devices that convert waste heat into useful electricity. Similarly, the decoupling of electrical conductivity and optical reflectivity will enable improved transparent conductors, thus leading to improved photovoltaic and display technologies. Ultimately, the rational assembly of nanomaterials into nanocomposites will yield high-performance materials with new combinations of properties, thus driving the development of previously unrealizable applications as schematically depicted in Fig. 7.

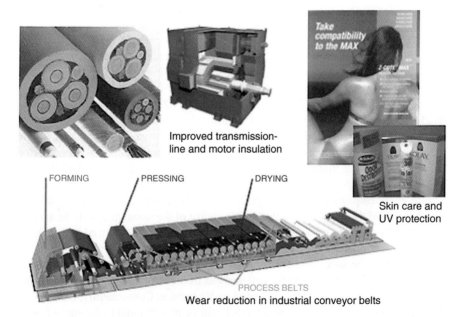

Fig. 7 Real-world examples of applications of benefit to society that are based on nanostructured hybrid materials (Courtesy of Richard Siegel, Rensselaer Polytechnic Institute)

8 Examples of Achievements and Paradigm Shifts

8.1 Monodisperse Single-Walled Carbon Nanotubes

Contact person: Mark Hersam, Northwestern University

Single-walled carbon nanotubes (SWNTs) are high-aspect-ratio cylinders of carbon ~1 nm in diameter whose walls are one atomic layer thick with an atomic arrangement analogous to graphite. The atomic structure of a SWNT is uniquely defined by a two-dimensional chiral vector whose components are typically specified by a pair of positive integers: (n,m). This so-called chirality of the SWNT dictates its resulting properties. Unfortunately, current synthetic methods for producing SWNTs lack control over the chirality, thus leading to significant polydispersity in the resulting properties of as-synthesized SWNTs. Consequently, while many applications have been proposed for SWNTs, their widespread use in high-performance technology such as electronics, photonics, and sensors has been limited to date by their inhomogeneity.

In an effort to realize the technological promise of SWNTs, many techniques have been developed to sort SWNTs by their physical and electronic structures. Leading examples are dielectrophoresis [42], chemical functionalization [43], selective etching [44], controlled electrical breakdown [45], anion exchange chromatography [46], and size-exclusion chromatography [47]. While each of these approaches has its own attributes and has been implemented for small-scale use in research laboratories, none of them have yet been adopted for industrial-scale separation of SWNTs.

In 2005, an alternative strategy was developed for sorting SWNTs, called density gradient ultracentrifugation (DGU), which combines several desirable attributes for large-scale production, including scalability, compatibility with a diverse range of raw materials, non-covalent and reversible functionalization chemistry, and iterative repeatability [48, 49]. Historically, DGU has been widely utilized in biochemistry and the pharmaceutical industry for separating subcellular components such as proteins and nucleic acids [50]. DGU works by exploiting subtle differences in buoyant density; in particular, the species of interest are loaded into an aqueous solution that possesses a known density gradient. Under the influence of a centripetal force introduced by an ultracentrifuge, the species will sediment toward their respective isopycnic points (i.e., the position where their density matches that of the gradient). With suitable choice of the initial gradient, the species will spatially separate by density, at which point they can be removed by a process known as fractionation.

For DGU to be successful for sorting SWNTs, the buoyant density of a SWNT must be directly related to its physical and electronic structure. Since a SWNT is a hollow cylinder, all of its mass is located on its surface. Consequently, the buoyant density (mass-to-volume ratio) of a SWNT will be proportional to the surface area-to-volume ratio for a cylinder, which is inversely proportional to diameter. If DGU occurred in vacuum, then the SWNT buoyant density would follow this simple inverse relationship with its diameter. However, since DGU occurs in aqueous solution and SWNTs are strongly hydrophobic, amphiphilic surfactants must be used to disperse the SWNTs. Consequently, the actual buoyant density of a SWNT in a DGU experiment will be a function both of the geometry of the SWNT and the thickness and hydration of the amphiphilic surfactant coating. When a surfactant is chosen that uniformly and identically encapsulates all of the SWNTs in solution, then the buoyant density will remain only a function of the SWNT diameter. On the other hand, if a surfactant or combination of surfactants is chosen that inequivalently encapsulate SWNTs as a function of their electronic structure (e.g., metal versus semiconducting), then DGU can be used to sort SWNTs by properties beyond simple geometrical parameters. Ultimately, the combination of clever surfactant chemistry and DGU enables wide tunability for sorting SWNTs.

Figure 8 outlines the DGU process for SWNTs produced by the CoMoCAT® growth strategy [51]. The CoMoCAT method produces SWNTs by carbon monoxide disproportionation using a proprietary cobalt/molybdenum catalyst. Even though CoMoCAT SWNTs possess a relatively narrow diameter distribution (0.7–1.1 nm), a number of distinct chiralities can be identified from optical absorption spectra. The first step in the DGU process is to disperse the SWNTs in aqueous solution using an amphiphilic surfactant such as sodium cholate. Although ultrasonication leads to a high yield of individually encapsulated SWNTs, some small bundles of SWNTs remain as indicated in Fig. 8a. Since the bundles possess a

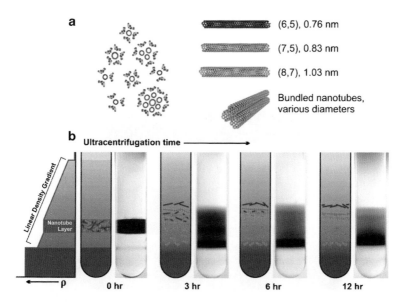

Fig. 8 (a) Cross-sectional schematic of surfactant-encapsulated SWNTs; three specific SWNT chiral vectors and respective diameters are identified. (b) Schematic representation and corresponding photographs of an ultracentrifuge tube at four different points in the DGU process (Courtesy of Mark Hersam, Northwestern University)

higher density than individually encapsulated SWNTs, they simply sediment to the bottom of the density gradient during DGU.

In the second step, the dispersed SWNT aqueous solution is injected into a linear density gradient that is formed from a solution of water and iodixanol. Iodixanol, $C_{35}H_{44}I_6N_6O_5$, is a water-soluble molecule that possesses a higher density than water. Therefore, by varying the concentration of iodixanol in the aqueous solution, a density gradient can be formed. Figure 8b schematically shows the initial density gradient profile and the starting position of SWNTs. By injecting the SWNTs near their isopycnic point in the gradient, the distance that they need to travel, and thus the ultracentrifugation time, are minimized.

Figure 8b also shows schematics and corresponding photographs of the ultracentrifuge tube at different points in the DGU process. After 3 h of DGU at a centripetal acceleration of 288,000 g, the SWNTs have begun to sediment but have not yet reached their equilibrium position in the density gradient. Then, after 6 h of DGU, layering of the SWNTs by their physical and electronic structure becomes apparent. Finally, after 12 h of DGU, the SWNTs have clearly layered to the point where fractionation can commence. Evidence for a successful DGU run includes the formation of visibly colored bands, as can be seen in Fig. 8b. Using one of several fractionation methods, these colored bands can then be removed from the centrifuge tube and collected sequentially in optical cuvettes. Optical purity of the resulting SWNT solutions is a direct indicator of monodispersity in the SWNT physical structure and electronic properties.

Figure 9 contains a photograph of five distinct, monodisperse SWNT fractions produced from one DGU run.

Fig. 9 Following DGU and subsequent fractionation, optically pure SWNT samples are isolated into distinct cuvettes. The color differences between these vials provide clear visual evidence for the success of the DGU process in sorting SWNTs by their physical and electronic structure (Courtesy of Mark Hersam, Northwestern University)

Applications: High-Performance Materials and Emerging Areas

The positive attributes of DGU are multifold. Easily controlled parameters such as surfactant chemistry, initial density gradient profile, and ultracentrifugation acceleration and time provide sufficient flexibility to accommodate a broad range of SWNT raw materials. In addition, the use of non-covalent and reversible surfactant chemistry implies that the encapsulating molecules can be easily removed via dialysis or copious rinsing, thus returning the SWNTs to their native state. The ability to remove the surfactants is particularly important for electronic applications where functionalization chemistry can compromise electrical contacts. Another advantage of DGU is that it can be iteratively repeated. In particular, following the first round of DGU, the best fraction can be removed and placed into a second gradient, at which point the DGU process can be repeated. In this manner, nearly arbitrary levels of purity can be achieved through multiple iterations of DGU. Finally, because DGU has already had widespread use in the pharmaceutical industry, the scalability and economic viability of DGU have already been demonstrated.

The future prospects are promising for SWNTs prepared by DGU. Commercialization of this approach has already been initiated by a start-up company called NanoIntegris (http://www.nanointegris.com/). In addition, DGU-prepared SWNT samples have successfully been exploited in field-effect transistors [52] and metallic coatings [53]. In research laboratories, the optical purity of DGU-prepared SWNT samples has also been exploited in time-resolved pump-probe laser spectroscopy studies of ultrafast carrier dynamics [54, 55]. Additional applications that will likely benefit from monodisperse SWNTs include transparent conductors, high-speed integrated circuits, biosensors, and nanocomposite materials.

8.2 Quantum Dot Application in Imaging

Contact person: James Murday, University of Southern California

In the early days of photography, film-based cameras were used to capture images (Fig. 10, left). In more recent years, charge-coupled device (CCD) cameras replaced film with silicon and ushered in digital photography. Then, as cameras became increasingly portable, the CMOS camera was developed (Fig. 10, center). The image sensor is produced by a CMOS process (hence is also known as a CMOS sensor), and it has emerged as an alternative to the CCD imager sensor. The CMOS active pixel sensor is most commonly used in cell phone cameras, web cameras, and in some digital single-lens reflex cameras; however, silicon-based image sensors only capture on average 25% of visible light.

QuantumFilm (Fig. 10, right) has been developed by InVisage (http://www.invisageinc.com/) under the guidance of Edward H. Sargent at the University of Toronto. The technology is based on semiconductor quantum dots [56] and integrates with standard CMOS manufacturing processes. QuantumFilm captures 90–95% of the light, enabling better pictures in even the most challenging lighting conditions. It works by capturing an imprint of a light image and then employing

Fig. 10 Evolution of photography from film to CMOS imaging to the quantum-dot-based technologies being developed by InVisage (Figures courtesy of James Murday, University of Southern California)

the silicon beneath it to read out the QuantumFilm. The first application—projected for 2011 [57]—will enable high pixel count and high performance in tiny form factors, breaking the inherent performance–resolution tradeoff of silicon.

Normally, cameras that are sensitive in the infrared and thus can image at night cannot be made using low-cost silicon processing, because, by virtue of its fixed bandgap, silicon is insensitive to wavelengths longer than 1.1 μm. InVisage spin-coats ~5 nm diameter PbS nanocrystals onto a chip, allowing exceptional device performance to be achieved in the shortwave infrared (SWIR) at a fraction of the cost of epitaxy-based, compound semiconductor IR sensors. InVisage is working to improve the sensitivity of the photodetector technology [58] and to integrate it with prefabricated silicon read-out integrated circuits to realize SWIR-sensitive focal arrays [59]. This technology has the potential to bring tremendous improvements to security systems and consumer electronics applications, because existing infrared imaging technologies are prohibitively expensive for high-volume civilian security markets. Other potential applications of quantum dot light absorption include medical imaging and solar energy conversion.

8.3 Nanotechnology-Based Paradigm Shifts in Aerospace

Contact person: Michael Meador, NASA

Nanotechnology has the potential to significantly impact future aircraft and space exploration missions. Use of nanostructured materials can enable the development of new aircraft and spacecraft that are significantly lighter than current vehicles and have enhanced performance, increased durability, and improved safety. Nanoelectronics can lead to new devices that are more radiation- and fault-tolerant and have built-in redundancies necessary for use in long-duration space exploration missions. Use of quantum dots and other nanostructures can enable the fabrication of lightweight, flexible, and durable photovoltaic devices to power future exploration missions. Nanoscale electrode materials can lead to new fuel cells and batteries with higher specific power and energy for use in both aircraft and spacecraft. A few examples of where nanotechnology developments are likely to impact future missions are given below.

8.3.1 Aircraft

Concerns about the environment will drive the design of new aircraft that have reduced fuel consumption, lower noise, and reduced emissions. Future aircraft designs, such as the Blended Wing Body concept (Fig. 11) currently being evaluated by the NASA, will be radical departures from the conventional "tube and wing" construction that has been used in aircraft for more than 100 years. Nanotechnology-derived materials will be used heavily in these vehicles. Carbon-nanotube-enhanced fibers [60] could enable the development of new lightweight composite materials that will reduce the weight of these vehicles by as much as 40%. In addition, these new nanocomposites will have higher electrical and thermal conductivity than conventional composites.

Lightning strike is a major concern in composite aircraft such as the Boeing 787. Typically, a thin copper or aluminum mesh is applied to the surface of these aircraft to improve their conductivity and provide for lightning strike protection. This mesh adds a significant amount of weight to the aircraft and requires additional labor (and expense) to apply. Use of carbon nanotube wire in place of copper wire in the aircraft power distribution system will lead to significant weight reductions. More than 4,000 lb. of wire is used in a conventional commercial aircraft such as the

Fig. 11 Nanotechnology will be used extensively in future aircraft designs including this blended wing body subsonic aircraft (Courtesy of Michael Meador, NASA)

Boeing 747. Carbon nanotube wires, such as those currently under development at Nanocomp Technologies, Inc., have higher current-carrying capability than copper at one-seventh the density, are more mechanically robust than copper, and are not susceptible to oxidation.

8.3.2 Space Exploration

Nanotechnology has the potential to positively impact space exploration missions. Carbons nanotube–enhanced fibers can enable significant reductions in spacecraft weight. Cryogenic propellant tanks can also benefit from nanotechnology. Cryotanks account for more than 50% of the dry weight (weight without propellant) of a spacecraft. Use of composites in place of metal alloys in these tanks can lead to weight reductions on the order of 30%. However, composites have had mixed success in cryotank applications as a result of the inherent permeability of polymers and composites to low molecular weight gases (e.g., hydrogen), the incompatibility of organic materials with liquid oxygen, and the propensity of composites to microcrack due to thermal cycling during use. Typical approaches to mitigate these issues involve the use of a metal liner on the inner wall of the tank, but this strategy increases weight, complexity, and cost of tank manufacturing, and the liner can delaminate from the tank wall due to the mismatch in coefficients of thermal expansion between the metal liner and composite tank substrate. Recent work has shown that the addition of organically modified clays to toughened epoxy resins leads to a 60% reduction in hydrogen permeability, enhanced compatibility with liquid oxygen, and improved resistance to microcracking [61]. Use of these nanocomposites could enable the fabrication of linerless nanocomposite cryotanks.

Increasing the capability of future robotic exploration missions will require instrumentation that is lighter in weight, more compact, and lower power than currently available. Carbon nanotube emitters have been utilized to develop compact, low-power mass spectrometers for use in exploration missions [62].

8.4 Developments in Nanofluidics

Contact person: John Rogers, University of Illinois at Urbana-Champaign

The field of nanofluidics, like other areas of nanoscience, derives its driving force from four principal features: (a) accessing new physical phenomena on the nanoscale; (b) enormous increases in the importance of surfaces; (c) the elevation of diffusion to a practical method of mass transport; and (d) the ability to build structures that are commensurate in size with molecular assemblies and even single molecules. Important developments in the past decade have exploited all of these.

8.4.1 New Phenomena

Concentration polarization. An interesting phenomenon that has been reported when ion-selective nanofluidic channels are connected to microfluidic channels is enrichment/depletion of ions at micro/nanofluidic junctions. Referred to as concentration polarization, the phenomenon has been exploited for applications such as water desalination.

Fluidic diodes/active elements. By controlling the spatial distribution of surface charge density on the interior of a nanopore and using a gate electrode, nanopores have been shown to function as diodes, bipolar, and metal oxide semiconductor (MOS)-like devices, opening up opportunities for nanofluidics-based logic circuitry.

3D fluidic switches. Nanocapillary array membranes, consisting of arrays of high-aspect-ratio cylindrical nanopores connecting the opposing surfaces of micrometer thick membranes, can function as electrically addressable fluidic switches supporting 3D integration of microfluidics with no moving parts (Fig. 12).

Electroosmotic flow (EOF) of the 2nd kind/fluidic vortices/convection. When nanopores are interfaced with microchannels, the net space charge at the micro/nanofluidic junction has been shown to give rise to nonlinear electrokinetic transport due to induced pressure, induced electroosmotic flow of the second kind, and complex flow circulations. These mechanisms can be exploited for numerous applications, including rapid mixing of species, which can enable the analog of stopped-flow reactors on the femtoliter-picoliter volume scale.

Rotation-translation coupling. When the rotational motion of a molecule, e.g., of water, is severely restricted, such as in single-file water in a nanopore, the frustrated rotational motions have been shown to be coupled to translational motions, leading to a preference for water molecules to move in the direction of the dipole. This phenomenon has been exploited for enhanced molecular transport.

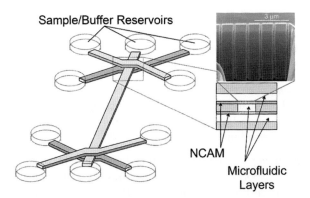

Fig. 12 Schematic diagram of a nanofluidically enabled integrated microfluidic circuit. Inset (*right*) shows scanning electron microscope (SEM) image of a focused ion beam (FIB)–milled nanocapillary array membrane (*NCAM*), which acts as a digital nanofluidic switch to control transport in the three-dimensional circuit (courtesy of John Rogers, University of Illinois at Urbana-Champaign).

8.4.2 Surfaces

Surface conduction. A much greater understanding of the role of surface conduction in the overall current conduction process in nanopores, especially at low ionic strength, has been obtained through various quantitative experiments.

Structured water. Water confined in nanopores of a critical size under ambient conditions has been shown to undergo a transition into a state having ice-like mobility with an amount of hydrogen bonding similar to that in liquid water. Thus, nanopores can provide an environment in which the dynamics of phase changes may be studied directly.

8.4.3 Diffusive Transport

Reactive mixing. Entrained flow microfluidic geometry has been used to illustrate the possibilities for reactive mixing using only diffusion for mass transport at sufficiently small distances.

Nanofluidic reactors. Nanofluidic channels have been exploited to confine species for enhanced reactivity. Homogeneous confined DNA has been utilized for restriction enzyme mapping at high efficiency, and heterogeneous molecular recognition (anti-insulin) and enzymatic (oxidoreductases) species exhibit enhanced reactivity in nanofluidic channels.

8.4.4 Commensurate Molecular Structures

Entropic DNA separations. The microfluidic-nanofluidic interface constitutes an entropic barrier that has been elegantly exploited for DNA separations. Confinement-induced entropic forces have been realized in a number of different nanofluidic geometries.

Resistive pulse sensing. The generic reduction in current in a nanopore can be exploited to observe the passage of chemical species that either occupy a significant fraction of the pore volume or alter the pore conductivity in other ways.

Stochastic sensing. The resistive pulse idea has been coupled to the presence of single biological ion channels with stochastic blockage elements that respond to the binding/unbinding of analytes. Monitoring the on/off statistics can then be used to assay for the concentration of analytes.

Nanopore sequencing. A promising methodology for DNA sequencing is to electrophoretically transport DNA through nanopores. Recent results have shown that nanopores have been used for sequencing DNA, identify defective DNA structures, and separate single-stranded DNA from double-stranded DNA structures.

Single-molecule studies/zero-mode-waveguides. Ultrasmall-volume (zL) pores in thin opaque metal films can be irradiated in order to produce non-propagating (cut-off) optical modes that can be used to interrogate chemistry within the pores. The ultrasmall volumes access single-molecule dynamics for macromolecules (e.g., enzymes) that have µM to mM dissociation constant values.

8.4.5 Technical Advances

Concentration polarization desalting. By placing a nanofluidic interface in one of the arms of a Y-shaped microfluidic element, concentration polarization has been exploited to divide a stream of sea water into desalted and concentrated-salted streams. This approach can be effective for small-scale to medium-scale desalinization systems with battery powered operation.

Multidimensional chemical analysis. Using nanofluidic elements to control flow in multilevel microfluidic architectures, multidimensional chemical analysis (e.g., two-dimensional separations coupling electrophoresis and micellar electrokinetic chromatography) has been accomplished.

8.5 Polymer Nanocomposites

Contact person: Richard Siegel, Rensselaer Polytechnic Institute

During the past decade, significant advances have been made in both creating and understanding a wide range of polymer nanocomposites with greatly improved properties [63, 64]. These include a variety of synthetic and natural polymer matrices combined with a host of nanoscale building blocks (e.g., nanoparticles, nanotubes, or nanolayers). Such nanostructured fillers, with their extremely high specific surface areas, are able to very strongly influence, and even dominate, the bulk properties of these nanocomposites by means of their interaction with the polymer chains of the matrix. Learning to specify and control these interactions has led over the past 10 years to the capability to now create polymer nanocomposites with individual properties, and even multifunctional sets of properties, that are desirable for many commercial applications.

For example, as shown in Fig. 13, it has recently been possible to tailor the surfaces of SiO_2 nanoparticles by means of reversible addition-fragmentation chain transfer (RAFT) polymerization and click chemistry [65], attaching an inner functional polymer layer and an outer matrix compatible layer to the nanoparticles, to greatly improve the properties of epoxy insulation material (J. Gao, S. Zhao, L.S. Schadler, and H. Hillborg, private communications 2010).

It has also become possible to theoretically model [66, 67] the organization of these nanoparticle-polymer systems in terms of the number density and length of such grafted, brush-like polymers and then to experimentally demonstrate the predictive nature of these models [68]. Such capabilities for structure and property control, as they are developed and expanded, will enable broad future applications of polymer nanocomposites.

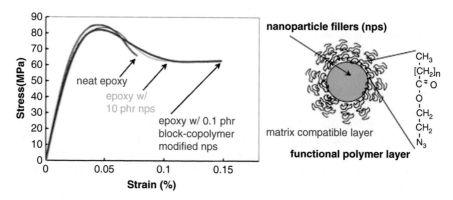

Fig. 13 Improved epoxy nanocomposites for electrical insulation have increased strain-to-failure, fatigue resistance, electrical breakdown strength (30%), thermal conductivity, and endurance strength (×10), and lower constant of thermal expansion (CTE)

9 International Perspectives from Site Visits Abroad

9.1 United States-European Union Workshop (Hamburg, Germany)

Panel members/discussants

H. Peter Degischer (co-chair), University of Technology, Wien (Vienna), Austria
Mark Hersam (co-chair), Northwestern University, United States
Costas Charitidis, National Technical University of Athens, Greece
Michael Moseler, Fraunhofer Institute for Mechanics of Materials (IWM), Germany
Inge Genné, VITO NV (Flemish Institute for Technological Research), Belgium

The field encompassed by the term "nanomaterials" has changed dramatically over the past 10 years. The vision from 2000 to produce cost-efficient structural materials for automotive, consumer appliance, tooling, and container industries has been transferred in 2010 to advanced high-tech niche products. The opportunities for significant improvements in structural materials with nanoscopic reinforcements of extraordinary properties and the exploitation of nanoscale grain size in inorganic materials are seen more specifically to have the potential for multifunctionality. Enabling impacts were expected of high surface and interface areas at the nanoscale, novel chemical reactivity and/or physical interactions, and biochemical properties exploitable in medical implants. Actually, engineered materials are required that offer property profiles that cannot be fulfilled by existing materials, e.g., the combination of high strength/stiffness and toughness at elevated temperatures, wear and corrosion resistance provided by lightweight materials, etc. Such challenging mechanical performance is mostly required in combination with physical properties such as high thermal conductivity and stability, and/or optical transparency.

Extensive research has been dedicated to processing of nanocomposites and nano-grained materials. Laboratory-scale processing of nanoparticles, nanotubes, and nanowires is available. Dispersion-strengthened metals and polymers with sub-micrometer-scale fillers existed already but have been further improved in an incremental manner. The distribution of the nanoscopic reinforcement and the consolidation with polymer, metal, or ceramic matrices, combined with the design of interfaces between the constituents to transfer properties, remain fertile areas of research. Hybrid nanocomposites of different materials at various length scales are considered for the design of unique property profiles. Interconnected reinforcements and interpenetrating nanocomposites provide significant thermal stability.

Sub-μm aluminum grains with natural nanoscopic oxide skin compacted to a network of flexible ceramic closed cells filled with ductile aluminum exhibit extraordinary elevated temperature strength and toughness. Inorganic particles in biopolymers have been developed for nanocontainments for targeted drug delivery. Methods to render self-assembling molecules creating periodic polymer structures have been developed. Nanoporous materials have been produced for physico-chemical applications. Embedding of conducting nanoparticles into polymer matrices has been achieved to provide transparent conducting foils of considerable strength. Sub-μm coatings of all material categories are being developed by embedding nanoscale constituents to improve physical, chemical, and mechanical surface performance. Nano-grained metals produced by various processing methods of severe plastic deformation achieve macroscopic components surpassing the exploitation of the Hall-Petch relation for strength by maintaining the disordered crystalline state up to appreciable temperatures.

In addition, significant advances have occurred in the field of microstructural characterization of size, morphology, and spatial distribution of constituents of heterogeneous materials, including three-dimensional methods at the sub-μm scale (e.g., focused ion beam sectioning, 3D high-resolution TEM, and third-generation focused synchrotron radiation sources). The reliable determination of properties at the nanoscale with high accuracy deserves further improvement and standardization. The incorporation of microsensors made of functional nanomaterials into extremely loaded structural components could be used for health monitoring.

The number of publications on the afore mentioned topics is still increasing. Solutions are sought for uniform dispersion of significant volume fractions, increased interface bonding, and reduction of porosities. It is evident that the interfaces in nanocomposites play a dominant role in the achievable property profiles. The unsatisfactory distribution and the limited transferability of the attractive properties of nanotubes to the matrix represent the main limitation for the exploitation of such nanoscopic reinforcements.

Enabling features of nanocomposites are not yet found, and conventional composites with matrices of polymers, metals, or ceramics can be identified that outperform the mechanical properties of nanocomposites. More emphasis is required to predict the potential of achieving previously unrealizable property profiles.

Based on the acquired knowhow, modeling and simulation can be applied to assess the multitude of material combinations from producible architectures of nanoconstituents.

Processing research should aim to give realistic visions of the variability of technologies to achieve structures controlling size-dependent interactions of the nanocomposite constituents by interface design strategies. Assemblies with tunable performance are demanded. Estimations of the required efforts to produce nanomaterials are required to assess the competitiveness of "nano" solutions. Process developments are to be combined with quality control methods applicable to successful up-scaling of the technology. Opportunities are envisaged in self-assembling methods, *in situ* methods to create hierarchical structures, and in building-block architectures.

Methods for structural characterization are available, but standardized property definitions and methods of determination would be required in each production phase. Therefore, research is required to correlate multiscale structural features with properties. Advancements should be pursued in the combination of properties for structural applications that are not achieved by existing materials.

Scientific knowledge-based calls should be sent out asking for interdisciplinary research proposals combining theory, modeling, simulation, processing, characterization, and experimental verification of a hypothesis submitted to critical evaluations. Boards consisting of scientists, industrial researchers, market experts, and funding agencies should work on the analysis of the strengths, weaknesses, and opportunities of a proposed solution as well as on the market tendencies (SWOT analysis) of the project objectives. Funding conditions should attract the best researchers in the field and allow them to work efficiently with as little bureaucratic burden as possible. Critical scientific reviewing of publications by experts needs to be enhanced to avoid diversion down unproductive tracks.

Business models are required to assess the market tendencies as well as the benefits expectable by applying nanomaterials. The up-scaling from successful laboratory samples to industrially reproducible products requires a promising profitable prognosis and reasonable access to venture [and other] capital resources, particularly for the development of small-scale niche products. Environmental, health, and safety (EHS) concerns must be considered in the context of life-cycle assessments for promising products. Regulations shall be globally agreed upon to replace the existing uncertainty regarding health and safety concerns.

Over the next 10 years, researchers will focus on a range of issues to improve the performance, multifunctionality, integration, and sustainability of nanostructured and hybrid material systems for structural applications. Functional nanomaterials also require certain mechanical property profiles, where research on structural materials can contribute to the development of reliable products. In particular, the field requires technologies for scaling-up high-quality nanomaterial production, designed to achieve independent tunability of previously coupled properties. Examples of desirable properties of new nanostructured materials include high stiffness and strength combined with toughness and low weight, complemented with elevated temperature and fatigue resistance, possibly combined with conductivity.

9.2 United States-Japan-Korea-Taiwan Workshop (Tsukuba, Japan)

Panel members/discussants

Sang-Hee Suh (co-chair), Korea Institute of Science & Technology
Mark Hersam (co-chair), Northwestern University, United States
Takuzo Aida, University of Tokyo, Japan
Hideo Hosono, Tokyo Institute of Technology, Japan
Soo Ho Kim, Korea Institute of Materials Science
Li-Chyong Chen, National Taiwan University
Hidenori Takagi, University of Tokyo, Japan

In the last 10 years, major advances with broad technological impact have been achieved in almost all categories of materials, including the following:

- *Polymer materials*, e.g., controlled radical polymerization, block copolymers, polymer brushes, dendrimers, supramolecular one-dimensional polymers, and metal-oxide frameworks
- *Structural materials*, e.g., high-strength, high-functionality, lightweight, automotive materials (achieved but not economically feasible); wear-resistant materials (3–4x improvement); and corrosion-resistant coatings for cutting tools and semiconductor processing equipment
- *Electronic materials*, e.g., electronic, optoelectronic, spintronic, energy, phase change materials, correlated electrons for thermoelectrics; nanostructured silicon for light-harvesting (anti-reflection improved by 10x, also broadband); GaN nanomaterials with 1000x improvement in photoconductivity; transparent amorphous oxide semiconductors for high-mobility electronics (LCD/OLED back plane)
- *Carbon-fiber-based materials*, e.g., aircraft/aerospace (a $1 billion industry); carbon-fiber-based nanomaterials used in the Chevy Corvette (high-performance, high-cost); carbon-based nanomaterials (e.g., graphene), which present promise for transparent conductors
- *Catalysts*, e.g., Pt nanoparticle/CNT/graphene nanocomposite for fuel cells requires 1/10 Pt level with better performance; self-cleaning materials represent a $1 billion market (mainly titanium oxide, which is also present in cosmetics, textiles, and paints)

A wide range of goals have been identified for the next 5–10 years in the area of high-performance materials:

- Low-cost manufacturing processes are required to bring high-performance materials to the marketplace.
- New functionality needs to be realized from materials based on earth-abundant elements.
- Renewable resources should be used as raw materials (e.g., biofibers, plants, wood, etc.).

- Specific focus should be devoted to green nanotechnology (e.g., new catalysts, energy conversion/transport/storage (entropic materials), and water-based plastics [aquamaterials]).
- Computational design will become increasingly important to save time and money in developing nanomaterials.
- Self-healing materials, biomimetic materials, and bionanomaterials have been demonstrated but need to be transitioned to real applications (e.g., paints, implants, regenerative medicine).

In addition to technological drivers, a range of fundamental issues remain, including the needs for:

- Improved understanding and control of surfaces/interfaces (especially organic/inorganic interfaces)
- Development of computational approaches for electron correlation
- Realization of properties/materials by design using combined computational/combinatorial approaches

These goals require improvements in the scientific and technological infrastructure and innovative R&D investment and implementation strategies. From the infrastructure perspective, a variety of specific items are needed, including:

- Characterization/observation methods for interfaces/surfaces in the operating state (e.g., next-generation neutron and synchrotron sources)
- Nanofabrication user facilities (including ongoing support)
- Human resource training

R&D investment and implementation recommendations include:

- Long-term funding (5 years guaranteed + 5 year renewal)
- International funding
- Continuation of fundamental research, funded by government
- Government-funded R&D focused on social issues (energy, environment, water, food, health)
- Cross-cutting, multidisciplinary funding opportunities
- Applications/commercialization funded with contributions from industry and private investors

9.3 United States-Australia-China-India-Saudi Arabia-Singapore Workshop (Singapore)

Panel members/discussants

Jan Ma (co-chair), Nanyang Technological University, Singapore
Mark Hersam (co-chair), Northwestern University, United States
Rose Amal, University of New South Wales, Australia
Julian Gale, Curtin University of Technology, Australia

Zhongfan Liu, Peking University, China
Koon Gee Neoh, National University of Singapore
Yee Yan Tay, Nanyang Technological University, Singapore

Over the past decade, the idea of manipulating materials at the nanoscale has been intriguing researchers globally. In 2000, researchers were focused on novel methodologies to control the size of different materials while attempting to explore many unique and distinctive size-dependent properties. The intensive study of nanoparticles, nanowires, and nanotubes is a good example. In particular, this early work attempted to develop an understanding of the chemical and physical properties of individual nanoconstituents. The nanomaterials were also effectively incorporated into coatings and bulk nanocomposites with clearly identifiable applications making use of the spatial anisotropy, morphology, and the relative proximity of nanoconstituents with respect to each other and the host matrix. Specific examples include nano-bio interfacing in the field of biotechnology, where nanomaterials are interfaced with a biological component for biosensing and drug delivery applications. Photocatalytic features of materials at the nanoscale dimensions are useful for self-cleaning surface applications, which have been successfully commercialized in the paint industry. In addition, metal-organic framework emerged for many state-of-the-art applications such as hydrogen storage.

Over the next 10 years, we anticipate new directions for various aspects of nanomaterials, built on these early achievements. For example, the study of two-dimensional and three-dimensional open structure architectures at the nanometer scale will attract significant attention. A specific aspect of this work in the next 10 years will be the development of nanoporous membranes with ballistic selective water transport, which could impact technologies such as water purification. While various unique nanostructures have been emphasized, it is also necessary to address the physical properties of nanocomposites, where interfaces are expected to play a dominant role. In particular, the optimization of internal interfaces will likely allow fluxes such as charge, energy, spin, etc., to be controlled independently. The introduction of computational guidance on realistic nanomaterials will also provide many useful predictive properties that are essential for the design of nanoscale devices.

The realization of these goals will require essential supporting aspects. For example, it is necessary to develop advanced characterization techniques that are capable of *in situ* and *in vivo* studies with concurrent time and spatial resolution. This type of instrumentation will facilitate the development of fundamental concepts, thus inviting new views of complex systems at short time and length scales. It is also important to establish affordable shared facilities with stable funding for recurring costs. While it has been highlighted previously that methodologies for scaling-up should be emphasized, it is also critical that the requisite infrastructure be in place to facilitate cost reduction in early-stage manufacturing. Multidisciplinary research will ultimately drive future innovation; thus, funding agencies need to provide ample proposal opportunities along these lines.

References

1. M.C. Roco, R.S. Williams P. Alivisatos, IWGN workshop report: nanotechnology research directions (Kluwer, Dordrecht, 2000), Available online: http://www.nano.gov/html/res/pubs.html
2. R.W. Siegel, E. Hu, M.C. Roco (eds.), *Nanostructure Science and Technology. A Worldwide Study: WTEC Panel Report on R&D Status and Trends in Nanoparticles, Nanostructured Materials, and Nanodevices* (Kluwer Academic Publishers, Dordrecht, 1999)
3. M.C. Roco, W.S. Bainbridge (eds.), *Converging Technologies for Improving Human Performance: Nanotechnology, Biotechnology, Information Technology and Cognitive Science* (Springer, Dordrecht, 2003), Available online: http://www.wtec.org/ConvergingTechnologies/Report/NBIC_report.pdf
4. T. Grafe, M. Gogins, M. Barris, J. Schaefer, R. Canepa, Nanofibers in filtration applications in transportation. Presented at *Filtration 2001 International Conference and Exposition of the INDA* (Association of the Nonwoven Fabrics Industry), Chicago, 3–5 Dec 2001, Available online: http://www.asia.donaldson.com/en/filtermedia/support/datalibrary/050272.pdf
5. T. Yuranova, R. Mosteo, J. Bandata, D. Laub, J. Kiwi, Self-cleaning cotton textiles surfaces modified by photoactive SiO_2/TiO_2 coating. J. Mol. Catal. A Chem. **244**, 160 (2006)
6. National Aeronautics and Space Administration (NASA), "Spinoff" (brochure), (2009), Available online: http://www.sti.nasa.gov/tto/Spinoff2009/pdf/Brochure_09_web.pdf
7. A.A. Karimpoor, U. Erb, K.T. Aust, G. Palumbo, High strength nanocrystalline cobalt with high tensile ductility. Scr. Mater. **49**, 651–656 (2003)
8. H. Li, F. Ebrahimi, Transition of deformation and fracture behaviors in nanostructured face-centered-cubic metals. Appl. Phys. Lett. **84**, 4037–4039 (2004). doi:10.1063/1.1756198
9. T.C. Lowe, R.Z. Valiev, The use of severe plastic deformation techniques in grain refinement. J. Miner. Met. Mater. Soc. **56**(10), 64–68 (2004)
10. C.C. Koch, K.M. Youssef, R.O. Scattergood, K.L. Murty, Breakthroughs in optimization of mechanical properties of nanostructured metals and alloys. Adv. Eng. Mater. **7**(9), 787–794 (2005). doi:10.1002/adem.200500094
11. H. Li, F. Ebrahimi, Tensile behavior of a nanocrystalline Ni–Fe alloy. Acta Mater. **54**(10), 2877–2886 (2006). doi:10.1016/j.actamat.2006.02.033
12. M. Chen, E. Ma, K.J. Hemker, H. Sheng, Y. Wang, X. Cheng, Deformation twinning in nanocrystalline aluminum. Science **300**(5623), 275–1277 (2003). doi:10.1126/science.1083727
13. X. Wu, Y.T. Zhu, M.W. Chen, E. Ma, Twinning and stacking fault formation during tensile deformation of nanocrystalline Ni. Scr. Mater. **54**(9), 1685–1690 (2006). doi:10.1016/j.scriptamat.2005.12.045
14. D. Wolf, V. Yamakov, S.R. Phillpot, A.K. Mukherjee, Deformation mechanism and inverse Hall-Petch behavior in nanocrystalline materials. Z. Metallkd. **94**, 1091–1097 (2003)
15. Z. Shan, E.A. Stach, J.M.K. Wiezorek, J.A. Knapp, D.M. Follstaedt, S.X. Mao, Grain boundary mediated plasticity in nanocrystalline nickel. Science **305**(5684), 654–657 (2004)
16. H. Van Swygenhoven, P.A. Derlet, Grain-boundary sliding in nanocrystalline fcc metals. Phys. Rev. **64**(22), 1–7 (2001)
17. H. Van Swygenhoven, P.M. Derlet, A.G. Froseth, Nucleation and propagation of dislocations in nanocrystalline fcc metals. Acta Mater. **54**(7), 1975–1983 (2006)
18. J. Cahn, Y. Mishinb, A. Suzukib, Coupling grain boundary motion to shear formation. Acta Mater. **54**(19), 4953–4975 (2006)
19. A.J. Haslam, D. Moldovan, V. Yamakov, D. Wolf, S.R. Phillpot, H. Gleiter, Stress-enhanced grain growth in a nanocrystalline material by molecular-dynamics simulation. Acta Mater. **51**, 2097–2112 (2003)
20. M. Jin, A.M. Minor, E.A. Stach, J.W. Morris, Direct observation of deformation-induced grain growth during the nanoindentation of ultrafine-grained Al at room temperature. Acta Mater. **52**(18), 5381–5387 (2004)

21. F. Mompiou, D. Caillard, M. Legros, Grain boundary shear–migration coupling – I. In situ TEM straining experiments in Al polycrystals. Adv. Mater. **57**(7), 2198–2209 (2009)
22. K. Zhang, J.R. Weertman, J.A. Eastman, Rapid stress-driven grain coarsening in nanocrystalline Cu at ambient and cryogenic temperatures. Appl. Phys. Lett. **87**(6), 19–21 (2005). doi:10.1063/1.2008377
23. A. Hasnaoui, H. Van Swygenhoven, P.M. Derlet, On non-equilibrium grain boundaries and their effect on thermal and mechanical behaviour: a molecular dynamics computer simulation. Acta Mater. **50**(15), 3927–3939 (2002)
24. F. Ebrahimi, H. Li, The effect of annealing on deformation and fracture of a nanocrystalline fcc metal. J. Mater. Sci. **42**(5), 1444–1454 (2007). doi:10.1007/s10853-006-0969-8
25. H. Li, H. Choo, Y. Ren, T. Saleh, U. Lienert, P. Liaw, F. Ebrahimi, Strain-dependent deformation behavior in nanocrystalline metals. Phys. Rev. Lett. **101**(1), 015502–015506 (2008). doi:10.1103/PhysRevLett.101.015502
26. E. Biztek, P.M. Derlet, P.M. Anderson, H. Van Swygenhoven, The stress-strain response of nanocrystalline metals: a statistical analysis of atomistic simulation. Acta Mater. **56**(17), 4846–4857 (2008)
27. F. Ebrahimi, Z. Ahmed, K.L. Morgan, Effect of grain size distribution on tensile properties of electrodeposited nanocrystalline nickel. Proc. Mater. Res. Soc. **634**, B2.7.1 2001
28. B. Raeisinia, C. Sinclar, W. Poole, C. Tomé, On the impact of grain size distribution on the plastic behavior of polycrystalline metals. Modell. Simul. Mater. Sci. Eng. **16**, 025001 (2008). doi:10.1088/0965-0393/16/2/025001
29. Y. Zhao, X. Liao, S. Cheng, E. Ma, Y. Zhu, Simultaneously increasing the ductility and strength of nanostructured alloys. Adv. Mater. **18**(17), 2280–2283 (2006). doi:10.1002/adma.200600310
30. F. Ebrahimi, A. Liscano, D. Kong, Q. Zhai, H. Li, Fracture of bulk face centered cubic metallic nanostructures. Rev. Adv. Mater. Sci. **13**, 33–40 (2006), Available online: http://www.ipme.ru/e-journals/RAMS/no_11306/ebrahimi.pdf
31. H. Li, F. Ebrahimi, Ductile-to-brittle transition in nanocrystalline metals. Adv. Mater. **17**(16), 1969–1972 (2005). doi:10.1002/adma.200500436
32. R.W.K. Honeycombe, *Plastic Deformation of Metals* (E. Arnold Publisher, London, 1984)
33. A. Kelly, N.H. MacMillan, *Strong Solids* (Clarendon, Oxford, 1986)
34. W.D. Nix, Exploiting new opportunities in materials research by remembering and applying old lessons. MRS Bull. **34**(2), 82–91 (2009)
35. F. Ebrahimi, H. Li, Grain growth in electrodeposited nanocrystalline fcc Ni-Fe alloys. Scr. Mater. **55**(3), 263–266 (2006). doi:10.1016/j.scriptamat.2006.03.05
36. C. Koch, R. Scattergood, K. Darling, J. Semones, Stabilization of nanocrystalline grain sizes by solute additions. J. Mater. Sci. **43**(23–24), 7264–7272 (2008)
37. American Forest and Paper Association (AF&PA) and Energetics, Inc., *Nanotechnology for the Forest Products Industry: Vision and Technology Roadmap* (TAPPI Press, Atlanta, 2005), Available online: http://www.agenda2020.org/Tech/vision.htm
38. J.P.E. Jones, T. Wegner, Wood and paper as materials for the 21st century. Proc. Mater. Res. Soc., 1187-KK04-06, (2009). doi:10.1557/PROC-1187-KK04-06
39. M.T. Postek, A. Vladar, J. Dagata, N. Farkas, B. Ming, R. Sabo, T.H. Wegner, J. Beecher, Cellulose nanocrystals the next big nano-thing? Proc. SPIE **7042**, 1–11 (2008). doi:10.1117/12.797575
40. C. Tamerler, M. Sarikaya, Molecular biomimetics: genetic synthesis, assembly, and formation of materials using peptides. MRS Bull. **33**, 504–512 (2008), Available online: http://compbio.washington.edu/publications/samudrala_2008d.introduction.pdf
41. S.A. Claridge, A.W. Castleman, S.N. Khanna, C.B. Murray, A. Sen, P.S. Weiss, Cluster-assembled materials. ACS Nano **3**, 244–255 (2009). doi:10.1021/nn800820e
42. R. Krupke, F. Hennrich, H. Lonheysen, M.M. Kappes, Separation of metallic from semiconducting single-walled carbon nanotubes. Science **301**(5631), 344–347 (2003). doi:10.1126/science.300.5628.2018

43. M.S. Strano, C.A. Dyke, M.L. Ursey, P.W. Barone, M.J. Allen, H. Shan, C. Kittrell, R.H. Hauge, J.M. Tour, R.E. Smalley, Electronic structure control of single-walled carbon nanotube functionalization. Science **301**(5639), 1519–1522 (2003). doi:10.1126/science.1087691
44. G. Zhang, P. Qi, X. Wang, Y. Lu, X. Li, R. Tu, S. Bangsaruntip, D. Mann, L. Zhang, H. Dai, Selective etching of metallic carbon nanotubes by gas-phase reaction. Science **314**(5801), 974–977 (2006). doi:10.1126/science.1133781
45. P.G. Collins, M. Arnold, P. Avouris, Engineering carbon nanotubes and nanotube circuits using electrical breakdown. Science **292**(5517), 706–709 (2001). doi:10.1126/science.1058782
46. M. Zheng, A. Jagota, M. Strano, A. Santos, P. Barone, S. Chou, B. Diner, M.S. Dresselhaus, R. McLean, G. Onoa, G. Samsonidze, E. Semke, M. Usrey, D. Walls, Structure-based carbon nanotube sorting by sequence-dependent DNA assembly. Science **302**(5650), 1545–1548 (2003). doi:10.1126/science.1091911
47. M. Zheng, E. Semke, Enrichment of single chirality carbon nanotubes. J. Am. Chem. Soc. **129**(19), 6084–6085 (2007). doi:10.1021/ja071577k
48. M. Arnold, A. Green, J. Hulvat, S. Stupp, M. Hersam, Sorting carbon nanotubes by electronic structure using density differentiation. Nat. Nanotechnol. **1**, 60–65 (2006). doi:10.1038/nnano.2006.52, Available online: http://www.nature.com/nnano/journal/v1/n1/full/nnano.2006.52.html
49. M. Arnold, S. Stupp, M. Hersam, Enrichment of single-walled carbon nanotubes by diameter in density gradients. Nano Lett. **5**(4), 713–718 (2005)
50. J.M. Graham, *Biological Centrifugation* (BIOS Scientific, Oxford, 2001)
51. B. Kitiyanan, W. Alvarez, J. Harwell, D. Resasco, Controlled production of single-wall carbon nanotubes by catalytic decomposition of CO on bimetallic CoMo catalysts, Chem. Phys. Lett. **317**, 497–503 (2000), Available online: http://www.ou.edu/engineering/nanotube/pubs/2000-1.pdf
52. M. Engel, J. Small, M. Steiner, M. Freitag, A. Green, M. Hersam, P. Avouris, Thin film nanotube transistors based on self-assembled, aligned, semiconducting carbon nanotube arrays. ACS Nano **2**(12), 2445–2452 (2008). doi:10.1021/nn800708w
53. A. Green, M. Hersam, Colored semitransparent conductive coatings consisting of monodisperse metallic single-walled carbon nanotubes. Nano Lett. **8**(5), 1417–1422 (2008). doi:10.1021/nl080302f
54. J. Crochet, M. Clemens, T. Hertel, Quantum yield heterogeneities of aqueous single-wall carbon nanotube suspensions. J. Am. Chem. Soc. **129**(26), 8058–8059 (2007). doi:10.1021/ja071553d
55. Z. Zhu, J. Crochet, M. Arnold, M. Hersam, H. Ulbricht, D. Resasco, T. Hertel, Pump-probe spectroscopy of exciton dynamics in (6,5) carbon nanotubes. J. Phys. Chem. C **111**, 3831–3835 (2007)
56. J. Tang, L. Brzozowski, D. Aaron, R. Barkhouse, X. Wang, R. Debnath, R. Wolowiec, E. Palmiano, L. Levina, A. Pattantyus-Abraham, D. Jamakosmanovic, E.H. Sargent, Quantum dot photovoltaics in the extreme quantum confinement regime: the surface-chemical origins of exceptional air- and light-stability. ACS Nano **4**, 869–878 (2010). doi:10.1021/nn901564q
57. K. Greene, Quantum dot camera phones. MITS Technol. Rev., 22 Mar 2010, Available online: http://www.technologyreview.com/communications/24840/page1/
58. G. Konstantatos, L. Levina, A. Fischer, E.H. Sargent, Engineering the temporal response of photonconductive photodetectors via selective introduction of surface trap states. Nano Lett. **8**, 1446–1450 (2008)
59. J.P. Clifford, G. Konstantatos, K.W. Johnston, S. Hoogland, L. Levina, E.H. Sargent, Fast, sensitive and spectrally tuneable colloidal-quantum-dot photodetectors. Nat. Nanotechnol. **4**, 40–44 (2008)
60. H.G. Chae, S. Kumar, Materials science: making strong fibers. Science **319**, 908–909 (2008). doi:10.1126/science.1153911
61. S.G. Miller, M.A. Meador, Polymer-layered silicate nanocomposites for cryotank applications, in *Proceedings of the 48th AIAA/ASME/ASCE/AHS/ASC Structures, Structural Dynamics, and Materials Conference*, Honolulu, 2007

62. S.A. Getty, T.T. King, R.A. Bis, H.H. Jones, F. Herrero, B.A. Lynch, P. Roman, P.R. Mahaffy, Performance of a carbon nanotube field emission electron gun. Proc. SPIE **6556**, 18 (2007). doi:10.1117/12.720995
63. R.A. Vaia, J.F. Maguire, Polymer nanocomposites with prescribed morphology: going beyond nanoparticle-filled polymers. Chem. Mater. **19**, 2736 (2007)
64. K. Winey, R. Vaia (eds.), Polymer nanocomposites. MRS Bull. **32**, 314–322 (2007), Available online: http://www.mrs.org/s_mrs/bin.asp?CID=12527&DID=208635
65. Y. Li, B.C. Benicewicz, Functionalization of silica nanoparticles via the combination of surface-initiated RAFT polymerization and click reactions. Macromolecules **41**, 7986–7992 (2008)
66. A. Jayaraman, K.S. Schweizer, Effect of the number and placement of polymer tethers on the structure of concentrated solutions and melts of hybrid nanoparticles. Langmuir **24**, 11119 (2008a)
67. A. Jayaraman, K.S. Schweizer, Structure and assembly of dense solutions and melts of single tethered nanoparticles. J. Chem. Phys. **128**, 164904 (2008b)
68. P. Acora, H. Liu, S.K. Kumar, J. Moll, Y. Li, B.C. Benicewicz, L.S. Schadler, D. Acehin, A.Z. Panagiotopoulos, A.V. Pryamitsyn, V. Ganesan, J. Ilavsky, P. Thiyagarajan, R.H. Colby, J.F. Douglas, Anisotropic self-assembly of spherical polymer-grafted nanoparticles. Nat. Mater. **8**, 354 (2009)

Developing the Human and Physical Infrastructure for Nanoscale Science and Engineering

James Murday, Mark Hersam, Robert Chang, Steve Fonash, and Larry Bell

Keywords Education and training • Infrastructure • Facilities • Informal education workforce • STEM • Cyberlearning • Partnerships • International perspective

1 Vision for the Next Decade

1.1 Changes of the Vision over the Last 10 Years

In 2000, the National Nanotechnology Initiative (NNI) Implementation Plan [1] recognized that nanoscale science and engineering education is vital to U.S. economic development, public welfare, and quality of life. But the Nanotechnology Research Directions report's vision for education and physical infrastructure only

---

J. Murday (✉)
University of Southern California, Office of Research Advancement,
701 Pennsylvania Avenue NW, Suite 540, Washington, DC 20004, USA
e-mail: murday@usc.edu

M. Hersam
Department of Materials Science and Engineering, Northwestern University,
2220 Campus Drive, Evanston, IL 60208, USA
e-mail: m-hersam@northwestern.edu

R. Chang
Department of Materials Science and Engineering, Northwestern University,
2220 Campus Drive, Evanston, IL 60208-3108, USA

S. Fonash
Nanotechnology Applications and Career Knowledge Center, 112 Lubert Building,
101 Innovation Boulevard, Suite 112, University Park, PA 16802, USA

L. Bell
Boston Museum of Science, Nanoscale Informal Science Education Network,
1 Science Park, Boston, MA 02114, USA

addressed (a) the need for a multidisciplinary university community based on the unitary concepts in nanoscale science and engineering, and (b) the need to enhance access to fabrication, processing, and characterization equipment ([2], p. 153). As described in Sect. 2, considerable progress has been made toward – and beyond – that early vision.

The importance of improvement in U.S. science, technology, engineering, and mathematics (STEM) education has been highlighted by three recent reports [3–5]. In 2000 the NSF initially focused attention on graduate programs, with goals described as follows: "Educational programs need to be refocused from microanalysis to nanoscale understanding and creative manipulation of matter at the nanoscale – A 5-year goal – is to ensure that 50% of research institutions' faculty and students have access to a full range of nanoscale research facilities, and student access to education in nanoscale science and engineering is enabled in at least 25% of the research universities" [6]. As the decade progressed, and the growing, pervasive impact of NSE became more compelling, the NSF vision for NSE education evolved (see Fig. 1) to include undergraduate, community college, K–12, and informal venues [8].

NSF and Department of Commerce (DOC) also recognized early on the importance of convergence among the emerging fields of nanotechnology, biotechnology, information technology, and cognition, NBIC [9], and the need to improve education and training to adapt to advances in emerging new technologies.

Nanotechnology is now seen as a driving force for all major industries worldwide and as playing an essential role in solving challenges in areas such as energy, water, environment, health, information management, and security. To compete effectively in world markets, it is now recognized there must be continued attention to NSE discoveries emanating from graduate education, motivated and skilled entrepreneurs who can transition discovery into innovative technologies, state-of-the-art equipment for fabrication and characterization, well-trained workers for the industrial communities, and well-informed, nanotechnology-literate citizens to sustain the workforce pipeline and public support. Attention must be given to expand and sharpen the education efforts in all of these venues.

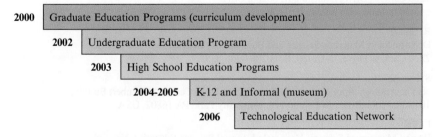

Fig. 1 Schematic of NSF investment in nanoscale science and engineering education, moving over time to broader and earlier education and training (After Murday [7])

1.2 The Vision for the Next 10 Years

The vision for nanoscale education in the coming decade is to build out the nation's human and technical infrastructure better enabling rapid, effective introduction of innovative nanotechnology-enabled products and processes that can address our shared social needs.

To realize this vision it will be necessary to:

- Continue the remarkable progress in science and engineering discovery facilitated by the marriage of research and education
- Sustain and update state-of-the-art fabrication and characterization facilities
- Incorporate new science and engineering knowledge, as appropriate, into the curriculum at all educational levels
- Ensure that the transdisciplinary nature of nanoscale science and engineering is not inhibited by traditional academic disciplinary constraints, but rather enriches and transcends them
- Facilitate the promises engendered by the convergence of nanotechnology, information technology, biotechnology, cognitive and other sciences (NBIC)
- Facilitate nanoscience-enabled innovative solutions to pressing global problems such as energy, water, environment, health, and information processes
- Provide first-rate training and recognition to teachers of K–12 science, technology, engineering, and math (STEM) classes
- Entice students into science and engineering careers
- Develop school-based nanotechnology career clusters that help prepare high school students for careers in nanotechnology
- Develop an informed, skilled workforce
- Include nanotechnology in retraining programs, such as those sponsored by the U.S. Department of Labor
- Enable workers and members of the general public to be sufficiently knowledgeable to understand the benefits and risks of nano-enabled technologies
- Provide adequate nanotechnology student and worker safety training
- Incorporate safety, social, and ethical issues into nanotechnology courses
- Institutionalize proven nanotechnology education programs

In addition, the state-of-the-art laboratory facilities that have been developed over the past 20 years – so critical not only to education, but also to research and technology development – must be provided with adequate operation and maintenance funding to ensure effective access by user communities (including students, small and medium enterprises, and larger-scale industries), be continually updated to sustain state-of-the-art capabilities, and coordinated nationally and globally in cost-sharing partnerships. Most important, those facilities must also be expanded to provide prototyping capabilities that will accelerate the transition of research discoveries into innovative technologies.

2 Advances in the Last 10 Years and Current Status

A main goal of the NNI has been to develop the educational resources, skilled workforce, and supporting infrastructure and tools to advance nanotechnology. A solid start has been made towards attaining this goal, but much remains to be done.

2.1 Physical Infrastructure

Nanoscale R&D user facilities have experienced significant advances in the past decade, both in the United States and around the world. Since 2000, about 100 national centers and networks and about 50 other research organizations focused on advanced R&D have been built or repurposed, together constituting a strong nanotechnology experimental infrastructure in the United States. The National Nanofabrication User Network (NNUN, www.nnun.org), a consortium of five leading university facilities providing nanofabrication user services to the R&D community, was established in 1994 with a focus on extending microelectronics; in 2003, NNUN was effectively morphed and expanded into the National Nanotechnology Infrastructure Network (NNIN; http://www.nnin.org/) with 14 participating sites distributed across the country. It has become an international model. The NNIN provides extensive support in nanoscale fabrication, synthesis, characterization, modeling, design, computation and hands-on training in an open, hands-on environment, available to all qualified users (see Sect. 8.8). The Network for Computational Nanotechnology, with Purdue as lead university, was created in 2002 to design, construct, deploy, and operate a cyber-resource for nanotechnology theory, modeling, and simulation (see Sect. 8.4). In the 2007–2008 time frame, large instrumentation user facilities were opened by the Department of Energy (DOE; http://www.er.doe.gov/bes/user_facilities/dsuf/nanocenters.htm) and the National Institute of Standards and Technology (NIST; http://www.cnst.nist.gov/nanofab/about_nanofab.html) (see Sects. 8.6 and 8.7, respectively). Appendix D has an extensive listing of the user facilities and other various university nanotechnology R&D centers and institutes that also have extensive instrumentation capabilities, most established in the last decade.

2.2 Education Drivers: New Fundamental Knowledge and Nano-Enabled Technology Innovation

Stimulated by the various nanotechnology programs around the globe [46], progress in nanoscale science and engineering over the past decade has been phenomenal. There were seven professional science/engineering journals with a nanotechnology focus in 2000; there are now over 90. The knowledge base has been growing exponentially, with the annual number of science/engineering professional publications increasing from 600 in 1990, to 13,700 in 2000, and to 68,000 in 2009.[1] The discoveries described in those many publications provide new knowledge that must be incorporated into the educational corpus.

Some examples follow of government and/or private programs created in response to the growing recognition in the past decade that nano-enabled technologies can be key contributors to solving problems of national priority:

- Information technology devices became nanoscale in three dimensions in about 2003 when the 90 nm semiconductor process node was realized; that became 45 nm in 2007, with 22 nm expected by 2012. Early in the 2000s, as the semiconductor industry looked forward, it realized the need for alternatives to complementary metal-oxide-semiconductor (CMOS) electronic devices, which face problems with continuing miniaturization, including quantum complications and heat dissipation. To accelerate progress in finding alternatives, the Semiconductor Research Corporation entered into a partnership with the National Science Foundation to establish a Nanoelectronics Research Initiative (NRI; http://nri.src.org/). SRC created four multidisciplinary NRI centers, with additional industry funding at selected NSF centers. NSF instituted a university program on "Science and Engineering Beyond Moore's Law" in 2009 (SEBML; http://www.eurekalert.org/pub_releases/2010-02/nsf-nsi020110.php) and a joint NSF-NRI solicitation "Nanoelectronics for 2020 and Beyond" in 2010 (NEB, NSF 10–614, with $18 M NSF and $2 M NRI funding) as part of the SEBML.
- There is growing awareness of the nanoscale role in medicine and health [10–12]. The annual investment by NIH in the NNI has grown from $32 million in 2000 to $360 million in 2010. A search on professional literature publications addressing the nanoscale in medicine and biology shows exponential growth beginning in roughly 2000: there were about 1,000 papers in 2000 compared with 11,000 in 2009.[2] A number of multidisciplinary centers for R&D in nanotechnology in medicine have been created, especially by the National Cancer Institute (NCI; http://nano.cancer.gov/; see Appendix D). A number of clinical trials utilizing nano-enabled technologies are in progress (see Sect. 8.10); MagForce Nanotechnologies AG received European regulatory approval for its Nano Cancer therapy in 2010.[3]
- DOE annual investment in the NNI has grown from $58 million in 2000 to $373 million in 2010, a reflection of the growing awareness of the role of nanotechnology in renewable energy and energy conservation (see http://www.nano.gov/html/about/symposia.html and http://www.energy.gov/sciencetech/nanotechnology.htm). Many, if not most, of the awardees of the DOE Energy

---

[1] These numbers are based on a simple keyword "nano*" search of the Thompson ISI Web of Science database; while easy to perform, this search generally undercounts the number of pertinent publications.
[2] This count is based on a keyword "nano* AND medic* OR nano* AND bio*" search of the Thompson ISI Web of Science database.
[3] http://www.magforce.de/english/home1.html

Frontier Research Center (EFRC) competition in 2009 had major nanoscale components in their research program (see Appendix D).
- The U.S. EPA has fostered nanoscale approaches to green manufacturing [13, 14]. Nanostructures are being exploited for environmental remediation, with careful attention to environmental, health, and safety issues [15].
- The nanoscale is expected to enable many innovative technologies that will lead to new jobs. Small businesses are a major contributor of new jobs. For several years Federal SBIR/STTR funds have been devoted to the transition of nanoscale research discoveries into innovative technologies, with ~$100 million awarded in 2008 alone [16]. The NIST Technology Insertion Program (TIP) has also funded nanotechnology efforts.

2.3 Education: Public/Informal, College/University, Community College, K–12

Education must evolve quickly to reflect the growing knowledge of the nanoscale and its application to new technologies. A perspective on the present status and challenges in NSE education can be found in several recent major reports and in the workshop report *Partnership for Nanotechnology Education* [7]. Education is a long-term investment; returns take at least one generation to become evident. Ten years ago, the NSF had the foresight to begin funding nanoscale education (formal and informal), together with research on the societal impact of nanotechnology [17]. The NNI agencies as a group have taken a multipronged approach to funding nanoscale education. The significant progress over the past decade in nanoscale science and engineering education can be reviewed in several workshop reports of the National Center for Learning and Teaching in Nanoscale Science and Engineering Education [18–20]. Table 1 lists a number of websites with educational materials addressing the nanoscale; in addition, various nanoscale research center efforts include educational programs.

2.3.1 Status of Public/Informal NSE Education

The Nanoscale Informal Science Education Network (NISE Net) was established in 2005 for the purpose of creating a national infrastructure of informal science education institutions, in partnership with nanoscale science research centers. It was aimed at raising public awareness, understanding, and engagement with nanoscale science, engineering, and technology. In scanning the 2005 environment in which the NISE Net set out to achieve those impacts, [47] identified four major challenges:

- The content and pedagogy of nanoscale education was just then emerging.
- The field was just learning how to design informal education resources that will effectively communicate the nanoscale to public audiences in informal science education settings.

Developing the Human and Physical Infrastructure

Table 1 Websites with nanoscale science and engineering educational content

Organization/institution	URL
Access nano (Australia)	http://www.accessnano.org/
American Chemical Society	http://community.acs.org/nanotation/
European Nanotechnology Gateway	http://www.nanoforum.org
Institute of Nanotechnology	http://www.nano.org.uk/CareersEducation/education.htm
Intro to nanotechnology	http://www.nanowerk.com/news/newsid=16048.php
McREL classroom resources	http://www.mcrel.org/NanoLeap/
Multimedia educ. & courses in nanotech (largely European)	http://www.nanopolis.net
NanoEd resource portal	http://www.nanoed.org
NanoHub	http://nanohub.org/
Nanoscale Informal Science Education Network	http://www.nisenet.org
NanoSchool box (Germany)	http://www.nanobionet.de/index.php?id=139&L=2
Nanotech KIDS	http://www.nanonet.go.jp/english/kids/
Nanotechnology Applications and Career Knowledge (NACK) Center	http://www.nano4me.org/
Nanotechnology news, people, events	http://www.nano-technology-systems.com/nanotechnologyeducation/
NanoTecNexus	http://www.Nanotecnexus.org
NanoYou (European Union)	http://nanoyou.eu/
Nanozone	http://nanozone.org/
NASA Quest	http://quest.nasa.gov/projects/nanotechnology/resources.html
National S&T Education Partnership	http://nationalstep.org/default.asp
NNI Education Center	http://www.nano.gov/html/edu/home_edu.html
NNIN Education Portal	http://www.nnin.org/nnin_edu.html
NSF Nanoscience Classroom Resources	http://www.nsf.gov/news/classroom/nano.jsp
PBS – DragonflyTV	http://pbskids.org/dragonflytv/nano/
Taiwan NanoEducation	http://www.nano.edu.tw/en_US/
	http://www.iat.ac.ae/downloads/NTech/UAE_Workshop_Pamphlet2.pdf
The Nanotechnology Group, Inc.	http://www.tntg.org
Wikipedia	http://en.wikipedia.org/wiki/Nanotechnology_education

- At the informal science education (ISE) institutional level, there was little expertise, experience and incentive to do nanoscale education for the public.
- At the field level, there was limited experience in developing and working within a national supportive network.
- Between 2005 and 2009, the NISE Net made significant progress in addressing all four of these challenges, with the result that by 2009, it could be said that: "Overall, NISE Net has created a large-scale functioning network that is capable of promoting nanoscience education for the public across the nation. In the past 4 years NISE Net has developed strong and distributed leadership, built relationships

with hundreds of individuals and institutions, created functional organizational and communication structures, and developed an initial collection of programs and resources that are flexible and valued by the field. In these ways NISE Net has established itself as a knowledgeable and valuable resource for both informal science educators and nanoscience researchers. (Inverness Research Associates 2009)"

- While the concepts of nanotechnology were virtually unknown to the public 10 years ago, in a 2009 national survey of adults, 62% said they had heard at least a little about nanotechnology [21]. While this increase cannot be attributed specifically to informal science education, the capacity to engage the public in a wide range of topics related to nanoscale science and technology has grown significantly in the last decade. Nanoscale science research centers, often in partnership with science museums, have engaged in a wide range of educational outreach activities, some focused on reaching public audiences and some on reaching K–12 school audiences. These collaborations have produced hands-on activities for classrooms and science museums; exhibits for science museums and even the EPCOT Center at Disneyworld; activities for the museum floor and for out-of-school programs; and various kinds of media content for magazines, websites, and even large-screen theaters.

Data suggests that the public in the United States is generally supportive of research and development in the area of nanotechnology, even in the absence of detailed knowledge about it.[4] Those with knowledge and awareness generally seem to support it more than those who do not. Gaining knowledge about nanotechnology R&D, however, does not necessarily translate to increased support; it may translate to increased awareness of potential risks, which may increase uncertainty about whether benefits will exceed risks. On the other hand, awareness of risks need not translate into decreased support [22]. Public views on risks versus benefits of nanotechnology show a significant neutral response: about one half of respondents either gave a neutral response (benefits equal risks) or they said they did not know [48], which suggests that these opinions are open to change as Americans become more familiar with nanotechnology. A more recent study showed nanotechnology ranked 19th out of 24 in terms of overall risk when compared with other selected hazards [23].

Education efforts are likely to help Americans become familiar with nanotechnology slowly over time, but a significant newsworthy event is likely to have a larger and more immediate impact. Will the future events be a great breakthrough in something the public cares about dearly, or will it be some kind of accident or disaster? Either of these scenarios would wake public attention to nanotechnology and provoke a dramatic change in awareness. Which will come first? How do we prepare to handle public information, questions, and concerns if the future event is an accident of some kind?

---

[4]Flagg, B. Nanotechnology and the Public, Part 1 of Front-End Analysis in Support of Nanoscale Informal Science Education Network, 2005, http://informalscience.org/evaluations/report_149.pdf.

2.3.2 Status of College/University NSE Education

Most research-intensive universities now have nanotechnology-related science and engineering courses, many have centers or institutes focused on the nanoscale, and several have nanotechnology-oriented departments and colleges. As an example of the latter, the new campus of the College of Nanoscience and Engineering at the State University of New York University at Albany (http://cnse.albany.edu/about_cnse.html; see also Sect. 8.5) is seamlessly integrated with collaborating micro- and nanoelectronics companies.

There has been significant progress in the incorporation of NSE concepts into curricula and textbooks at the college- and graduate-school levels [7]. Nanoscience-based courses are being rapidly introduced at 2-year colleges, 4-year colleges, and universities (see listings at NanoEd Resource, http://www.nanoed.org/ and [7]). A number of 4-year-degree-granting institutions in the United States have created minors or concentrations in nanotechnology. These are generally comprised of 18 credits of nanotechnology courses combined with courses fundamental to nanotechnology such as quantum physics and physical chemistry. These minors allow students to graduate from engineering (e.g., electrical engineering, chemical engineering) and science programs (e.g., physics, chemistry, biology) with a minor (or concentration) in nanotechnology. An example of this approach is the nanotechnology minor at The Pennsylvania State University (see Sect. 8.3 and Appendix D in Murday [7]). The NSF Research Experience for Undergraduates (REU) program is an effective vehicle for introducing undergraduates to nanoscale research projects. The NUNN/NNIN REU program alone has given over 600 undergraduate students the opportunity to conduct nanotechnology research.

Several approaches to introducing NSE concepts into undergraduate education have been utilized in the past decade:

- *Supplemental approach*: The practice of inserting fundamental NSE concepts into existing courses has been successful around the country.
- *Inquiry and design principles approach*: Concept modules using the principles of inquiry and design were pioneered by the NSF-funded Northwestern University program, "Materials World Modules" (MWM; http://www.materialsworldmodules.org/). Discovery comes with asking the "right" question. Effective design requires a sound knowledge of the nanoscale concept to be applied. Each module includes a series of inquiry-based activities that culminate in a team-based design project. These principles have been successful in engaging students at the undergraduate level.
- *Cascade approach*: This method allows older students to learn and reinforce concepts in the process of teaching them to younger students, resulting in a "cascade" of learning from one student group to another. Older students gain a deeper understanding of the concepts studied, experience in creative thinking and design, and improved inquiry and communication skills. Younger students gain more exciting, age-appropriate instructional materials and are inspired by their slightly older peers' interest in science and engineering. In both cases, participating students

have designed some amazingly innovative activities, which include card games, web-based games, and simulations of nanoscale phenomena. The "Teach for America" program for students to take time off to teach demonstrates as that peers of comparable age (in this case, undergraduate college students) can communicate very effectively with younger (middle school) students [24].

An exciting trend is that more and more young professors are interested in science education, including nanoscale education, and are beginning to work with precollege teachers and their students. Graduate programs such as those offered by NSF science center programs, Graduate STEM Fellows in K–12 Education [25], Integrative Graduate Education and Research Traineeship [26], and others, provide opportunities for graduate students to become involved in teaching their junior peers.

2.3.3 Status of Community College NSE Education

Forty percent of U.S. college and university students overall begin their education in community colleges [27]; the statistics are higher for minority populations. Since it is estimated that present minority populations will be in the majority of schoolchildren by 2023 [28], the role of community colleges in teaching nanoscale science and technology to American students and workers may become even more significant than it is today.

Economic pressures can arise when community and technical colleges feel that they must create four semesters of new courses for 2-year degree programs in nanotechnology. The high costs can contribute to the reluctance of administrators to involve their institutions in nanotechnology education. Four-semester nanotechnology programs also create student enrollment pressures: there must be a "critical-mass" of student enrollment generated each semester to justify running such programs. Faculty, staff, and facilities resource issues arise with the realization that a meaningful nanotechnology education must expose students to state-of-the-art tools of nanotechnology fabrication and, most importantly, characterization. Often, 2-year institutions do not have the resources required to provide this exposure. Further, in areas of the United States, 2-year institutions find themselves relatively geographically isolated, precluding interactions with industry or research universities with nanotechnology facilities resources.

Several NSF-supported nanoscience-oriented Advanced Technological Education (ATE) programs (see National Science Foundation (NSF) [4], Appendix D, and Sect. 8.3) focus on training future nanoscale technicians. Students participating in these programs acquire classroom knowledge and high-tech industrial laboratory experiences. The National ATE Center for Nanotechnology Applications and Career Knowledge (NACK), created in 2008 by NSF, is charged with augmenting and further developing nanotechnology education at 2-year degree institutions across the United States. NACK offers nanotechnology courses that may be attended or viewed on the web, course units on the web, state-of-the art equipment

Fig. 2 *Darker shading* indicates states that have 2-year degree institution nanotechnology programs associated with NACK (from http://www.nano4me.org)

utilization experiences that may be attended or utilized on the web, web access to characterization equipment, and workshops on teaching nanotechnology. All information and web access is through the NACK website http://www.nano4me.org. Figure 2 is a map of regions that have nanotechnology programs at 2-year degree institutions.

2.3.4 Status of K–12 NSE Education

The pre-college level is where educational systems begin preparing nano-literate citizens, training future nano-technicians and engineers, and using nanoscale concepts to pique student interest in STEM. The lack of student interest in science/engineering is endemic in economically advanced countries [29] and constitutes a severe mid- to long-term threat for future economic and quality-of-life growth if academia and industry should fail to find the needed science/engineering workforce.

In the United States, impediments to achieving these education needs include the fact that each state has its own set of standards and learning goals, and there are significant disparities in standards between states and between individual school districts. Educational standards are of urgent importance to teachers who must prepare their students to perform well on standardized tests. The Council of Chief State School Officers (CCSSO) and the National Governors Association (NGA) are leading a common core learning standards effort (http://www.corestandards.org/) based on international benchmarking. The Mathematics and the English Language

Arts standards were released (June 2010) and have been adopted by 35 States by October 2010. The sciences will be the next topic to be addressed. The National Research Council (NRC) Board on Science Education (BOSE; http://www7.nationalacademies.org/bose/BOSE_Projects.html) has been working on a conceptual framework for new science education standards. A draft incorporating core disciplines of life sciences, earth and space sciences, physical sciences, and engineering and technology was released for public comment in August 2010; the revised version is due for release early in 2011.

NSE educational concepts are being linked to existing STEM standards and learning goals at the state and national levels. A series of fundamental NSE concepts (e.g., surface-area-to-volume ratio, size and scale, size-dependent properties, dominant forces, quantum phenomena, self-assembly, tools for nanoscale characterization, and others), have been mapped to STEM concepts that already are being taught in U.S. classrooms. These concepts are also being aligned with the national science standards developed by American Association for the Advancement of Science (AAAS), the National Science Education Standards [30], the national math standards developed by the National Council of Teachers of Mathematics (NCTM), and various state standards. This work is paving the way for nanoscale content to be inserted into a broad range of STEM courses, including Biology, Chemistry, Mathematics, Physics, Technology, Engineering, and General Science.

NSE does have a "wow" factor that gets students excited about STEM and motivates them to learn. Experience in U.S. classrooms demonstrates that students find the nanoscale both exciting and intriguing, especially when they are applied to new technologies and real-world challenges. Given that lack of student interest is a critical challenge facing STEM education nationally, and in many developed countries, this effect must not be undervalued. Nanoscale science concepts and their applications help students connect STEM to their everyday lives, and this motivates them to learn. In the words of one Nashville teacher, "We all want to be wowed, amazed, and captivated. If education can do this and teach important concepts, students will beat the doors of the school down demanding more. Nanotechnology fills the bill."

Many programs already have a strong nanotechnology education and outreach component:

- A number of NSE education research centers with funding from NSF, DOE, and other agencies are charged with transferring these concepts into U.S. classrooms. Examples are the NSF Nanoscale Science and Engineering Centers (NSEC) and the DOE Nanoscale Science Research Centers (NSRC).
- Other programs such as the NSF Materials Research Science and Engineering Centers (MRSEC), Engineering Research Centers (ERC), and Science and Technology Centers (STC) have also developed NSE education components. Activities vary among centers but may include classroom outreach, content development, and teacher professional development, often with funding from the NSF Research Experiences for Teachers (RET) program.
- Other new nano-specific education projects have been funded under existing programs. For example, the NSF has funded nano-specific learning research by BSCS, Nanoteach, and McREL (see Table 1).

Challenges abound to engaging students in nanoscale science and technology. A large percentage of teachers are not prepared to teach mathematics and science. Further, their curricula are already so crowded that it is difficult to see where new topics can be added. And once these challenges have been considered, how can we teach students about what they cannot directly see or feel? How can we give them a real "nano" experience without bringing expensive, high-tech equipment into their classrooms? How can we make clear connections to real-world applications and future careers? Horizontal and vertical integration of NSE education programs is viewed as necessary to achieve the goals of educating the workforce, changing formal and informal education, and having an informed public.

Horizontal integration: In most schools, STEM is taught from distinct (traditional) disciplinary perspectives. While these perspectives can be useful, they can also result in student understanding that is disjointed and isolated from real-world systems and applications. Because NSE education concepts describe phenomena that occur across all STEM disciplines (e.g., see Fig. 3), these concepts transcend traditional disciplinary boundaries. In particular, Chemistry, Physics, and Biology merge at the nanoscale, addressing similar problems using similar techniques. In addition, at the nanoscale, unique materials properties and complex fabrication techniques require a merging of science and engineering. It is therefore quite natural to take an integrated "systems" approach to learning and teaching nanotechnology. Moreover, compelling connections can be made to social studies, physical education,

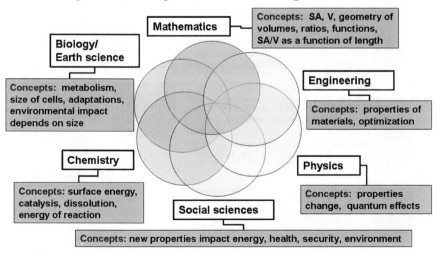

Fig. 3 Example of the horizontal integration of nanotechnology knowledge of surface-area-to-volume (SA/V) ratio, leading to repeated exposures of this concept to students in STEM and other subjects (Courtesy of Nanotechnology Center for Learning and Teaching)

and the arts that increase the relevance of nanoscale science and technology to society and the economy. Taking a horizontal, transdisciplinary approach provides the opportunity to teach STEM in an integrated way and enables students to appreciate the natural overlap of these disciplines. Moreover, it serves to unify STEM curricula by allowing students to experience a nanoscience concept repeatedly from different angles and in different contexts. Public school districts in Los Angeles and Nashville are piloting integrated science curricula that use nanoscale science as a core (or unifying) discipline and link NSE concepts to both STEM and non-STEM courses.

Vertical integration: Our current educational structure is often characterized by large achievement and learning gaps between levels. Incorporating nanoscale education into grades 7–16 affords an opportunity to study how students learn fundamental nanoscience concepts at each level, to monitor student longitudinal progressions, and to develop learning trajectories to guide effective teaching across levels. The learning research suggests that teaching the same fundamental NSE concept (such as size and scale or SA/V ratio; see Fig. 4) at multiple grade levels can result in deeper understanding of STEM concepts and in more efficient transitions from one grade level to another. This vertical integration is both time-saving and effective.

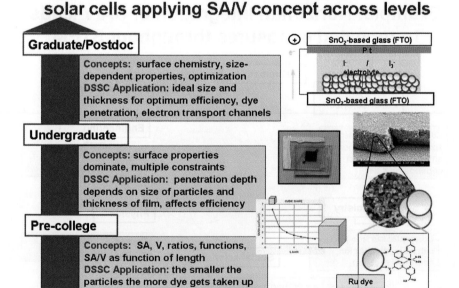

Fig. 4 Example of vertical integration of nanotechnology knowledge for dye-sensitized solar cells, applying the concept of SA/V ratios at higher conceptual levels over time (Courtesy of Nanotechnology Center for Learning and Teaching)

2.3.5 International Integration

The U.S. K–12 STEM education does not compare well with that of many countries around the world, especially those with whom U.S. businesses must compete for new markets. A recent report by the National Governors Association and Council of Chief State School Officers emphasizes the need for international benchmarking as a means toward improvement of educational standards [28]. Since nanotechnology research, development, and deployment are growing globally, all countries (but especially the United States) can benefit from learning what works best in the incorporation of the nanoscale into STEM curricula in other countries of the world.

3 Goals, Barriers, and Solutions for the Next 5–10 Years

The remarkable progress in nanoscale research discovery, coupled with the growing opportunities in innovative nano-enabled technology, presents new infrastructure challenges. In the next 10 years, while sustaining the discovery progress, how might the transition from discovery into technology be accelerated? Further, by virtue of its inherent transdisciplinarity, its fascinating real-world applications, and its immediate relevance to future career opportunities, nanoscale science education is an ideal means of increasing student interest and achievement in STEM disciplines. A national strategy should be developed to expand the current efforts in STEM/NSE education, to help ensure that citizens around the world can reap the economic, social, and educational benefits of nanotechnology.

3.1 Physical Infrastructure

There has been truly impressive growth in nanoscience/engineering facilities over the past decade. A prime goal is to fully exploit those facilities for science and engineering, while continually adding new capabilities as they are developed. It is vital to critically assess their utilization and effectiveness, followed by corrective actions, if needed to gain higher efficacy.

While new analytical and fabrication tool development over the past 20 years have provided many impressive new capabilities (see in chapter "Investigative Tools: Theory, Modeling, and Simulation and Sect. 8.13 for examples), a number of NNI workshops have pointed toward the importance of persisting in advanced tools development.[5] When these tools are either expensive and/or difficult to operate,

---

[5] Nanomaterials and Human Health & Instrumentation, Metrology, and Analytical Methods (November 17–18, 2009); Nanomaterials and the Environment & Instrumentation, Metrology, and Analytical Methods (October 6–7, 2009); Nanotechnology-Enabled Sensing (May 5–7, 2009); X-Rays and Neutrons: Essential Tools for Nanoscience Research (June 16–18, 2005); Nanotechnology Instrumentation and Metrology (January 27–29, 2004). See http://www.nano.gov/html/res/pubs.html for associated publications.

they must be added into the national user facilities to make them accessible to a wide variety of users, including teachers and students.

The international workshops all pointed to a new R&D facility need. The success of centers addressing fabrication and characterization for discovery needs to be emulated in a new suite of centers whose focus will be on the transition (or translation) of discovery into innovative technology.

3.2 Informal/Public Outreach in NSE

Having successfully built a network focused on, and capable of, providing nanoscale informal educational experiences to the public, the goal of NISE Net for the next 5 years should be to use that network to expand the number of people it reaches to the millions. To accomplish this goal, objectives include the following:

- Perform a SWOT analysis of the public outreach to determine the value to current activities, and use this benchmark data to design efficient and effective public education models
- Deepen the involvement in nanoscale ISE of over 100 institutions (universities, museums, public radio, etc.) and expand their nanoscience and technology offerings by infusing "nano" content throughout their educational programming
- Leverage resources through connections with other networks of informal educators and researchers to further spread the reach of NSE education for audiences in both informal and formal educational settings
- Develop sustainable partnerships between informal educational institutions and research institutions that can provide ongoing benefits to both when NISE Net funding expires
- Explore electronic media for innovative approaches to reaching youth who are adopting these new communications mechanisms

More broadly, two interrelated strands of educational activities should continue: preparation of the future workforce and preparation of consumers and citizens.

3.2.1 Preparation of the Future Workforce

Preparation of the future workforce is strongly associated with formal education but is supported by informal science education as well. From an informal education perspective, aspects of future workforce preparation should take place in four stages.

- *Stage 1 – Primary (grades K–5)*: The goal for ISE education for young children is to grow their interest in science, technology, engineering, mathematics – and society in general (STEMS subjects). This can include fostering a fascination with unusual properties of materials, size-dependent phenomena, and other topics related to the nanoscale, but it is not yet important that students recognize

nanotechnology as a separate research field. Development of materials and educators' professional qualifications within informal science education should help support the work of this stage to build general interest in STEMS and to add nanoscale-related phenomena to existing science-based activities.
- *Stage 2 – Secondary (grades 6–12)*: The goal for ISE for middle school and high school youth is to build their underlying knowledge, awareness, and skills in STEMS while maintaining and growing their interest. Informal education should also help inform parents about areas of STEMS that their children might find of interest as potential careers. Highlighting applications and societal implications can also be a motivating factor.
- *Stage 3 – College*: The goal for ISE for undergraduate and graduate students is to give them the opportunity to contribute to public and K–12 educational activities by presenting their research and talking with the public about it. Developing this ability among future scientists is key to the education of consumers and citizens and can help students think about the big issues in their research and their careers.
- *Stage 4 – Workplace*: The goal for ISE in the workplace is to increase the number of scientists, engineers, and technicians who can work in nanotechnology by retraining; for their education as consumer and citizens; and can contribute their knowledge in various informal educational environments to public and K–12 educational activities.

3.2.2 Preparation of Consumers and Citizens

In the absence of either a noteworthy breakthrough or a severe accident involving nanotechnology, widespread public education activities can and should build public awareness of nano-enabled technologies. An emphasis on the nanoscale as an essential pathway to resolving societal problems, such as renewable energy, health/medicine, and environment, might frame the nanoscale in terms more conducive to public interest. The public presently does not believe nanoparticles are particularly risky per se (Berube et al. submitted). In the absence of an immediate reason for the public to care personally about nanotechnology, this is likely to be incremental and slow and needs to be sustained to have long-term impact. This requires continually paying attention to increasing public awareness and knowledge at a measured pace while building the capacity of informal educators and other science communicators to inform the public as future events raise the level of public awareness and interest. It is also necessary to develop the capacity for the public to assess the benefits and risks of nanotechnology and to gain a genuine capacity to influence policy. As Lee and Scheufele [31] have noted:

> Public outreach is not a matter of promoting pro-science views among the general public or simply improving literacy. Public outreach involves effective communication with all stakeholders (scientists, citizens, policy makers, etc.) Currently, public debate about issues such as stem cell research is dominated by ardent interest groups and partisan players. Scientific views are not heard.

For the public, key factors in dealing with uncertain risk are trust and control. Data shows that the public wants both information and government regulation. The public feels that university scientists are a trustworthy source of information about nanotechnology. The public also has a high level of trust in science museums. The government and news media are least trusted "to tell the truth about the risks and benefits of nanotechnology". According to research conducted in connection with the Center for Nanotechnology in Society, "university researchers and scientists working for nano businesses are among the most trusted sources for information about nanotechnology" (Scheufele 2007). Unpublished results from the same study, however, show that government and news media are least trusted to tell the truth about the risks and benefits of nanotechnology.

All of this suggests a strategy for public education and engagement that overlaps with governance – that is, creating a science/technology governance structure that better involves the public. There is a need to build two-way communication between scientists and public and to include assessment of risks and benefits in the dialogue. Far exceeding the number of informal science educators, the scientists, engineers, and technical staff in the United States can be a valuable resource in talking with the public, but they need to be given the tools to do so effectively. Mechanisms need to be created to effectively engage the public in long-term thinking about societal implications, risks, and benefits, and about issues regarding government controls, regulation, and priorities in nanotechnology research and development.

In order to achieve this proactive kind of governance and public engagement model, collaborations should be fostered among science research institutions, social science institutions, and informal science education institutions, with a specific focus on developing:

- Easy-to-communicate notions about the future promise of research and ways to equip informal educators and scientists to communicate with the public about these ideas
- Broad knowledge about societal implications of future technologies, ways to equip informal educators and scientists to consider such issues in connection with their work, and ways to communicate with the public about them
- Mechanisms for public participation in technology assessment that fosters dialogue between scientific and public participants, leading to informed public involvement in policymaking

3.3 College/University NSE Education

Nanoscale science and engineering discoveries will continue to develop at a rapid pace; innovative, nano-enabled technologies already are penetrating the marketplace. The implications for undergraduate and graduate education are likely to be profound, but hard to anticipate. There needs to be continuing dialog and ongoing experimentation on curricula and degree requirements, working with industry to assure relevance to the workforce. Since the research into NSE is truly international in scope, an important goal is to provide experience in international venues as part of the degree process. A year "abroad" has long been valued as part of a liberal arts education; some variant on this idea should also be considered for nanoscale science/engineering education. Post-doctorate work abroad should be encouraged for scientists, but so should opportunities for other students, so as to serve not only those science students looking toward university research careers but also those students looking towards careers in business and industry.

Although no mandated teaching and learning standards exist at the college level (beyond professional accreditation), strong disciplinary boundaries are reinforced both by the accreditation process and by competition of departments for funding and resources. These disciplinary boundaries are often more rigid than those at the pre-college level. Priorities for improving NSE education at the college level include developing innovative nanoscale courses and degree programs, fostering cross-disciplinarity, and supporting NSE education community-building through faculty workshops, content development workshops, online content sharing, and networking events. Global partnering to share content and best practices will be essential in the new "flat world."

The weakest NSE elements in university education involve nanoscience education for students who are not majoring in science or engineering. Incorporating science-related societal issues as a meaningful component of liberal arts courses, and other creative interfaces for students in science and education, would offer opportunities to increase science/technology literacy and advocacy.

3.4 Community College NSE Education

While U.S. research universities are acknowledged as world leaders in science and engineering, competition is growing globally – especially at the nanoscale. To meet national needs for researchers in the near future, it will become more important to nurture U.S.-born students' interest in STEM. Because 40% of college students get their start in community colleges, the Opportunity Equation report [27] recommends closer interaction between community colleges and the universities. NSF, the Department of Education (DoEd), and other agencies with relevant missions must have a goal to foster nanoscale curricula development and evaluation that is appropriate for community colleges and to ensure meaningful collaborations between community colleges and the national nanocenters and networks. Nanotechnology should be included in DoEd's Department of College and Career Transitions program.

3.5 K–12 NSE Education

A primary goal in education is the insertion of NSE into national and state K–12 standards of learning. Incorporation of NSE into K–12 curricula and textbooks has lagged the rapid development of new scientific knowledge about the nanoscale.

If this is to be rectified, it will be essential to introduce nanoscale science/engineering into the various state learning standards that effectively determine what is taught in primary and secondary school classrooms. Otherwise curriculum changes at the K–12 levels of education will be minimal, and increasing nanoscience literacy toward a productive workforce will not be maximized [7]. A key objective is to identify the fundamental nanoscale concepts and to map them onto existing state and national STEM learning standards such as the American Association for the Advancement of Sciences Benchmarks for Science Literacy (http://www.project2061.org/publications/bsl/default.htm), the National Science Education Standards [30], and the National Council of Teachers of Mathematics Standards and Focal Points (http://www.nctm.org/standards/default.aspx?id=58). Nanoscience educators should also work with these organizations to integrate specific nanoscience concepts into the standards and cooperate with organizations now working to nationalize learning standards. Increasing the connections between NSE/STEM and non-STEM disciplines such as social sciences, business sciences, language arts, sports, and fine arts also will help students understand the relevance of NSE and STEM to their everyday lives and to solving global challenges in energy, environment, health, communications, and others.

In the U.S. the National Governors Association has approached Achieve, Inc., with the task of preparing internationally benchmarked, common core learning standards in the sciences that might be adapted by each state for its own academic content standards. The National Science and Technology Council's National Science, Engineering and Technology Subcommittee (NSET), which coordinates NNI endeavors, has initiated contact with the National Governors Association, the Council of Chief State School Officers, and Achieve, Inc., to support their efforts to appropriately introduce the nanoscale into common core educational standards for developmental progression in the sciences (life, physical, earth/space) and engineering/technology. Assistance should be provided to those efforts. Because each state will determine its own process, participants in the many Federal centers for nanotechnology R&D should begin working with their own state departments of education toward forward-looking revision of their science learning standards.

3.5.1 K–12 Curricula

Standards provide the target; curricula provide the means. To regain prominence in science, technology, and engineering, the United States must stop having a haphazard approach to curriculum development. Funding is needed to allow for the design, development, testing, and implementation of coherent NSE curriculum models that will allow 7- to 16-year-old students to develop an integrated understanding of core science ideas that underpin nanoscale science and engineering. Such curricula should focus on helping students develop progressively deeper understanding of the core ideas.

Developing the Human and Physical Infrastructure 521

3.5.2 K–12 Teaching Aids

The sciences (biology, chemistry, earth/space, and physics) and engineering require hands-on experiences as part of the learning process. While inclusion of laboratory work at local schools is a necessity, some laboratory learning may be beyond the capability and/or budget of local schools and personnel. The NNIN, NSEC, and DOE nanocenters, and the NIST Center for Nanoscale Science and Technology should have a goal to work with the National Science Teachers Association and the DoEd toward the preparation of on-site and/or remote access to high-end laboratory facilities that can contribute to the K–12 NSE education process.

3.5.3 Teacher Education and Training (Professional Development)

There will be growing inclusion of nanoscale science, engineering, and technology into standards of learning. There are also growing learning resources that address nanoscale science, engineering, and technology. Teachers will need to be trained to use these resources, and there must be a goal to provide that training. The nation's nanocenters can be a vital resource to provide materials, training, and information. They should be encouraged to be more proactive toward K–12 teacher training.

Person-to-person contact remains the most effective mode of teaching and learning. The various university-based nanocenters should mobilize their undergraduate and graduate students to engage in K–12 education at the nanoscale. Federal funding agencies must provide an adequate budget allowance for this work. In tenure and promotion decisions, universities must recognize the efforts of their faculty to supervise activities of their students who are engaged in teaching nanoscience concepts to younger students.

3.6 Across NSE Educational Levels: Global Partnerships and Cyberlearning

3.6.1 Global Partnerships

The challenges in NSE education are global in character. Cooperation with global partners was an explicit topic at the Global Workshop on Nanoscale Science and Engineering Education in November 2008 National Center for Learning and Teaching in Nanoscale Science and Engineering Education (NCLT) [20]. The primary finding at this workshop was that there is a need for an expansive NSE education community, both in real space and in virtual space, to support the development and exchange of resources and build excitement around nanoscale science and engineering education. The NSF Office of International Science and Engineering can potentially play a significant role in helping make this possible.

3.6.2 Cyberinfrastructure/Education Through Social Media

Nano-enabled technologies can improve standards of living across the world. With the decline in the last few years in the number of science journalists, there is an opportunity for the NNI agencies, university and industrial programs, and other stakeholder groups to develop a continuing stream of information that can keep the public informed about benefits and risks emanating from nanotechnology-based research and commercialization. The rapid growth in information technologies is creating new interaction paradigms that should be exploited (e.g., Wikipedia, FaceBook, Second Life, YouTube, and Kindle) to reach young and IT-literate learners who use those new media tools [32, 49, 50]. Cyber-education should be included in the suite of learning paradigms to engage students. NSF, with its interest in cyberlearning, should take the initiative, but the DoEd must also be engaged to ensure a continuing effort.

There are a number of websites (such as those listed in Table 1) with materials that address curricula supplements, teaching aids, and science fair projects. The NACK website (http://www.nano4me.org/) is focused on supporting nanotechnology education at 2-year degree granting institutions. In addition, NSF's university-based Nanoscale Science and Engineering Centers (NSECs) have been very productive in terms of developing innovative approaches to NSE education. However, web-based NSE educational materials are widely dispersed, are of non-uniform format, and have varying degrees of refinement. The DoEd, working closely with the National Science Teachers Association (NSTA) and cyber-oriented curriculum developers (such as those listed in Murray [7], Table 6), should create a central website to disseminate the information. The NSTA should serve as the evaluator for quality control to ensure NSE-focused educational website materials are of high quality, are in a format readily utilized by K–12 teachers, are carefully indexed to the various state learning standards, and can be readily accessed from the NSTA website. Additional well-designed, highly interactive, media-rich, online learning tools should continue to be developed.

NanoHUB, the National Center for Learning and Teaching (NCLT), NNIN, NACK, and other cyberinfrastructure resources focused on nanotechnology – all highly beneficial – need to be better publicized regarding accessibility, targeted user levels, customizability (in terms of both targeted audiences and user interface), interoperability with other systems, and service and training offerings. Consideration should be given to the research and development of an overall mechanism for efficient search, access, and use of cyberinfrastructure resources focused on nanoscale science and technology with potential relevance to education at all levels. Such a mechanism would likely entail creating an inventoried, analyzed, tagged registry. Also worth considering are mechanisms to enable greater interoperability with emerging cyber-nanotechnology resources such as remote access to and control of state-of-the-art nanotechnology characterization equipment; properties databases; applications of NSE; environmental, health, and safety implications; and general educational paradigms such as virtual and immersive environments, simulations, and games. Such integration and knowledge sharing offer great promise to accelerate users' discovery of and learning about nanoscale science and technology.

Wikipedia is becoming the de facto encyclopedia of the twenty-first Century. The Wikipedia entries on nanotechnology should be routinely updated and expanded. This task might best be accomplished by mobilizing the impressive variety of talent and expertise at the various nanocenters. K–12 science teachers should be involved to ensure the information is structured in ways that can be readily absorbed by students at various grade levels.

4 Scientific and Technological Infrastructure Needs

4.1 Facility Infrastructure

User facilities' budgets must be adequate to provide operating funds and local experts who can assist the novice users in effective use of the state-of-art equipment. There would be value for vetted intellectual property (IP) model agreements that might be used as a trusted basis for companies seeking access to the facilities.

While new analytical and fabrication tool development over the past 20 years has provided many impressive new capabilities (see Sect. 8.13 for recent examples), a number of workshops[6] have pointed out the importance of continuing to do cutting-edge R&D. When these tools are either expensive and/or difficult to operate, they must be added into the user facilities.

As nano-enabled technologies become ever more sophisticated, the equipment needed to make/measure/manipulate will also grow in complexity and cost. International and industrial partnerships need to be fostered as a means of sharing the cost of these new capabilities.

The international workshops all pointed to a new R&D facility need. The success of centers addressing fabrication and characterization for discovery needs to be emulated in a new suite of centers whose focus will be on the transition (or translation) of discovery into innovative technology. This will require user facilities with capability in manufacturing and prototyping. Since there will be a wide diversity in manufacturing/prototyping requirements, a suite of user facilities will be necessary, each addressing a different suite of technologies and processes.

4.2 Workforce Development (Industry) NSE Education

Preparation for employment is an important aspect of the educational process. In our rapidly evolving world, the needs of industry are fluid, due to changing technologies and growing global competition. Nanoscale science will be instrumental

---

[6] NIST Workshop on Instrumentation, Metrology, and Standards for Nanomanufacturing (Oct. 17–19, 2006); NNI Workshop on Instrumentation and Metrology for Nanotechnology (Jan. 27–29, 2004); NNI Workshop, Research Directions II (Sept. 8–10, 2004); Chemical Industry R&D Roadmap for Nanomaterials by Design (30 Sept.–2 Oct. 2002).

in technological change. Many countries have followed the lead of the United States and established a nanotechnology initiative [33]. Those initiatives tend to be more focused on targeted technology development than is the United States; consequently, there will likely be strong global competition for people trained in nanoscience and technology at various levels. The Department of Labor needs to work with industry groups and with professional science and engineering societies to develop accurate assessments of domestic nanotechnology-based workforce needs, including the effects of growing education and job opportunities in other countries. These needs must be factored into the educational system.

4.3 Informal/Public NSE Education

A key infrastructure need is educators who are interested in, familiar with, knowledgeable about, and comfortable with teaching nanoscale science, engineering, and technology. This is true in both formal and informal education, but the focus here is informal. This includes educators in science museums, children's museums, university outreach programs, after-school and out-of-school programs, libraries, educational television, radio, and the Internet.

A second infrastructure need is an easily accessible collection of educational materials and activities that can engage the public of various ages with a wide range of nanoscale science, engineering, and technology topics, applications, and societal implications – and that are well suited to the wide range of informal learning environments in which people learn outside of school. In addition to the NiseNET online resources (http://www.nisenet.org/), a superb example can be found at the Munich Deutsches Museum (http://www.deutsches-museum.de/ausstellungen/neue-technologien/), which has a highly interactive exhibit featuring nano- and biotechnologies.

Especially given the decline in the number of science journalists, there is a need and opportunity for the NNI member agencies and the National Nanotechnology Coordination Office (NNCO), university and industrial nanotechnology programs, and other stakeholder groups to develop a continuing stream of information to inform the public of the benefits and risks stemming from scientific and technological progress at the nanoscale. The rapid growth in information technologies is creating new interaction paradigms that might be exploited using electronic media such as Wikipedia, FaceBook, Second Life, YouTube, and Kindle [32].

4.4 College/University NSE Education

The multidisciplinary/transdisciplinary nature of nanoscale science and engineering will continue to require center/institute type activity at colleges and universities. These centers/institutes should be widely dispersed geographically such that their facilities and capabilities are accessible to others without the need for extensive travel time. Further, introducing web-controlled instrumentation for remote access of selected instruments can potentially provide laboratory experiences for K–12 and community college students whose schools otherwise could not provide them.

Nanoscale science and engineering R&D will create many opportunities for innovative, disruptive technologies. Motivated, skilled entrepreneurs will be essential to carry the research discoveries into marketable technologies. College/university educators must find ways to nurture budding entrepreneurs. Models include the MIT Institute of Nanotechnologies "Soldier Design Competition" (http://web.mit.edu/isn/newsandevents/designcomp/finals.html), the University of South Carolina's "Ivory Tower to Marketplace: Entrepreneur Laboratory" (http://www.nano.sc.edu/news.aspx?article_id=27) and the University of Southern California Steven's Institute for Innovation's "Student Innovator Showcase and Competition" (http://stevens.usc.edu/studentinnovatorshowcase.php).

4.5 Community College NSE Education

Many community colleges do not have the faculty, staff, and facilities resources to offer a meaningful nanotechnology education that includes exposure to state-of-the-art nanotechnology fabrication tools and, more importantly, to nanotechnology characterization tools. The model developed and supported by the NACK Center (see Sect. 8.3) uses resource-sharing among community colleges, research universities, and NACK to bring a meaningful nanotechnology education experience to 2-year degree students. This experience can include attending nanotechnology courses at the research university partner, using NACK web-available course materials, and using NACK equipment accessed via the web. The community colleges benefit by being able to give a state-of-the-art nanotechnology education; the research universities benefit by gaining access to a new source of potential students for 4-year degree programs.

4.6 K–12 NSE Education

The growing knowledge base at the nanoscale, coupled with the importance of nano-enabled technology solutions to high-priority global challenges in renewable energy, energy conservation, potable water, heath, and environment, compel introduction of NSE into K–12 curricula. Funding is needed to allow for the design, development, testing, and implementation of a coherent curriculum that would allow 7- to 18-year-old students to develop an integrated understanding of core science ideas that underpin nanoscale science and engineering. Such a curriculum would focus on helping students develop progressively deeper understanding of core ideas. Such a process calls for changes in the standards that focus on teaching big ideas, with a focus on developing a deeper understanding of these ideas. In addition to the standards and curricula requirements, web-based materials addressing the nanoscale, vetted for accuracy and effectiveness, need to be created and presented in formats convenient to K–12 teachers and students. These materials must be correlated with United States and internationally benchmarked common core standards.

4.7 Global Partnerships in NSE Education

There are numerous groups around the world addressing STEM education, nanoscale education, nanoscale science and engineering research, and nano-enabled technologies. There is an immediate challenge to integrate these various communities.

5 R&D Investment and Implementation Strategies

5.1 Facilities

The impressive array of existing nanoscience R&D user facilities faces two key challenges: to provide sustained operating funds to support local expertise available to assist itinerant users, and to continually maintain/update the instrumentation. Nanocenters with unique, expensive instruments have been developed by many nations. As nanoscale science and engineering require more sophisticated systems, it is inevitable that new instrumentation will be expensive. Furthermore, the sensitivity of those instruments will require buildings engineered to adequately manage "noise" that can affect instrument results, i.e., fluctuations in temperature, pressure, cleanliness, electromagnetic radiation, vibration, acoustics, electrical power quality, etc.; those buildings will be expensive to build and operate. International and industrial partnerships will become more important as an approach to share the cost burden.[7] One example is the government-university-industry partnership that has created the College of Nanoscale Science and Engineering facility at the University of Albany, New York (see Sect. 8.5).

The NSF National Nanotechnology Infrastructure Network (NNIN), NSEC, DOE nanocenters, and the NIST Center for Nanoscale Science and Technology are working with the NSTA and the DoEd toward preparation of on-site and/or remote access to high-end user facilities that can contribute to the education process at various levels.

Some of the present centers, especially those with a manufacturing focus, could be morphed into user facilities that provide prototyping capabilities. However, to fully represent the breadth of need capability, it is likely additional user-facilities will need to be created.

---

[7] Buildings for Advanced Technology", pending report from the NNI workshops in 2003, 2004 and 2006.

5.2 Federal Agency Investment in NSE Education

As has been noted in several places above, nanoscale science and engineering present numerous challenges to the K–Gray (i.e., continuing education for preschool through the senior years) education system. In the United States, the National Science Foundation has borne the major cost of developing nanoscale science and engineering education. As NSE leads to pervasive nano-enabled technologies, other agencies also must contribute toward the educational process. Table 2 illustrates education/training programs across the U.S. Federal government that should increase their participation in NSE educational efforts. That being said, as the prime agency responsible for STEM education, NSF must continue and expand its efforts as well.

The NSET, which has representation from 25 participating Federal agencies, should create a nanotechnology education and workforce working group to support agency efforts in addressing education and workforce issues. The working group should be guided by an education- and workforce-focused consultative board comprising the various principal stakeholders.

NSF's investments in cyberinfrastructure (e.g., [34]), along with those of other agencies, have resulted in a broad range of state-of-the-art, distributed digital resources, some for nanotechnology and science research, some for nanotechnology learning and education, and some for nanotechnology events and news. These cyber-nanotechnology resources vary in terms of their target audiences (education level, country of origin, original purpose), quality, level of integration with other

Table 2 Federal education programs with potential for NSE content

Agency	Program	URL
DOD	National defense education program	http://www.ndep.us/
DOE	Energy education	http://www1.eere.energy.gov/education/
	National labs	http://www.energy.gov/morekidspages.htm
DoEd		http://www.ed.gov/index.jhtml
DOL	Training, continuing education	http://www.doleta.gov/
DOT	Education and research	http://www.dot.gov/citizen_services/education_research/index.html
EPA	Teaching center	http://www.epa.gov/teachers/
NASA	Education program	http://www.nasa.gov/offices/education/programs/index.html
NIH	Office of science education	http://science.education.nih.gov/home2.nsf/feature/index.htm
NIST	Educational activities	http://www.nist.gov/public_affairs/edguide.cfm
NOAA	Education resources	http://www.education.noaa.gov/
NSF	Education & human resources	http://www.nsf.gov/funding/pgm_list.jsp?org=ehr
USDA	NRCS	http://soils.usda.gov/education/resources/k_12/
	AFSIC	http://www.nal.usda.gov/afsic/AFSIC_pubs/K–12.htm
	NIFA (formerly CSREES)	http://www.agclassroom.org/

cyberinfrastructure resources, usage, and usage reporting. Most lack inclusion of meta-information, which may limit knowledge transfer across user communities in terms of discoverability, search ability, and adaptability. The emerging nanoscale education community must be able to exploit existing cyberinfrastructure resource investments more effectively.

5.3 Informal/Public NSE Education

In the 2010–2015 time frame, it will be important to continue investment in the Nanoscale Informal Science Education Network (NISE Net) to promote informal public understanding of NSE. Goals should include achieving deeper sustainable impact on 100 institutions of informal education, broader reach across the range of organizations that engage in informal education, stronger connections to K–12 formal educational efforts, and greater inclusion of applications and societal perspectives in educator professional development and in the collection of educational materials.

Under the umbrella of broader impact requirements for center grants, there should continue to be strong encouragement by funding agencies of socially relevant educational outreach and public engagement activities, with encouragement to partner with informal science education organizations to generate these activities. Funding of research centers, and not only of small, individual research projects, is essential to realizing the critical mass necessary to make such collaborations productive. Research centers or research networks should be created and funded to address the grand challenges to society, span research at the nanoscale and at other scales of matter, provide specific funding for public participation in technology assessment and governance, and build an R&D agenda shared by the public, scientists, and policymakers. There will be an ongoing need to invest in educator training and materials development, with the goals of finding new ways to effectively communicate critical ideas about nanotechnology to the public, to keep abreast of new developments, and ensure that the educational momentum developed in the initial years of the NNI is well positioned to capitalize on public interest generated by future developments, whether they be positive breakthrough applications or events that raise concern.

The NNCO should take a more formative and supportive role in organizational and public relations related to nanotechnology developments, not to manage perceptions but to provide accurate and objective information in multiple public venues, using multiple platforms. Web communication must shift from passive web presences to active Web 2.0 efforts to engage highly relevant publics with nanoscience. Platforms like Facebook and Twitter are important mechanisms to communicate with the public in ways they are familiar with.

To date, the NISE Net has largely examined how "nano" fits into the size scale of materials, but a change of emphasis may be needed. Several studies have suggested that students will respond best to STEM in terms of addressing societal problems.

Now that nano-enabled technologies are beginning to proliferate, it would be timely to develop exhibits and programs associated with the impacts of those nano-enabled technologies.

NSF, which is the principal funding source for new nanoscale science and engineering projects, should take the lead in establishing links between museums and the national and international research communities for new exhibit and program development. Other stakeholders such as Federal funding agencies and industry representatives must also be contributors, since they will be engaged in the translational efforts that lead to technology impact.

5.4 College/University NSE Education

In the second decade of the U.S. National Nanotechnology Initiative there will be growing emphasis on transitioning science discovery into innovative technology. Industry involvement will be crucial. The NSF Engineering Research Centers already require partnerships with industry; any new NSF Center competitions focused on the nanoscale should incorporate industrial partners to foster innovation and workforce development. The NSF Grant Opportunities for Academic Liaison with Industry (GOALI) and Industry & University Cooperative Research Centers (I/UCRC; http://www.nsf.gov/eng/iip/iucrc/) programs should be exploited to a greater degree. Further, industry can provide its own incentives to colleges/universities through partnerships such as the Semiconductor Research Council Education Alliance (http://www.src.org/alliance/), Hewlett Packard Lab's Open Innovation Office (http://www.hpl.hp.com/open_innovation/), IBM's University Research & Collaboration (https://www.ibm.com/developerworks/university/research/), and the ASEE/NSF Corporate Research Postdoctoral Fellowship for Engineers (https://aseensfip.asee.org/jobs/57).

5.5 Community College NSE Education

NSF, DoEd, and other agencies with relevant missions should foster development and evaluation of nanotechnology curricula appropriate for community colleges, and ensure meaningful collaborations between the community colleges and the nanocenters. The DoEd's Department of College and Career Transitions program should ensure nanotechnology is included in that program.

To address the facilities issue, state governments, working in concert with the DoEd and local industries, need to develop mechanisms to enable interactions between the research universities, national laboratory facilities, and community colleges. Some states are taking steps and making good progress toward that integration: Texas has established the Texas Nanotechnology Workforce Development Initiative (http://nanotechworkforce.com/resources/workforce.php), and Pennsylvania has established the Center for Nanotechnology Education and Utilization (http://www.cneu.psu.edu/abHomeOf.html). One solution is integrating community college courses with nearby research universities, where they are available.

5.6 K–12 NSE Education

Without incorporating the current understanding of nanoscale science and engineering into science and engineering learning standards in each of the states, action at the K–12 levels of education will be minimal, and increasing nanoscience literacy toward a productive workforce will be inadequate. The NSE community, encouraged and funded by NSF and DoEd, must work with the National Governors Association and the Council of Chief State School Officers on the internationally benchmarked common core standards.

There will be growing inclusion of nanoscale science, engineering, and technology into standards of learning. The NSE community, encouraged and funded by NSF and DoEd, must work with the NSTA, professional science and engineering societies, and other pertinent organizations (including ones with international perspectives) to develop curricula, learning resources, and teacher training.

5.7 Global Partnerships in NSE Education

A focal international activity is needed to identify, validate, and integrate the many nanoscale education capabilities that presently exist around the world and to assess what is still needed.

6 Conclusions and Priorities

6.1 Facilities

- The nanoscale research facilities at centers and user-facilities currently are well equipped, but nanoscale science, engineering and technology is changing rapidly; it will be a challenge to keep those facilities to-to-date. There must be continued attention paid to facility infrastructure so as to sustain operation costs and the availability of local expertise. Periodic updates of instrumentation will be necessary, with careful consideration to most effective use of funds (i.e., no unwarranted duplication). Partnerships between countries, industries, and universities will be needed to sustain the funding for needed capabilities. Continued

NSF support for centers addressing the nanoscale is essential, since the funding necessary to acquire and sustain measurement/fabrication facilities can only be obtained through center-level efforts.

In addition, some of the present centers, especially those with a manufacturing focus, should be morphed into user facilities that provide prototyping capabilities. To fully represent the breadth of need manufacturing/prototyping capability, it will likely be necessary to create some additional user-facilities.

The U.S. investment in nanotechnology education has spawned a new paradigm for STEM education, which if broadly implemented, will allow the United States to maintain its global leadership in science and technology. This new paradigm is based on three types of integration – horizontal, vertical, and system – that can drastically improve student learning and significantly save education costs:

- *Horizontal integration* proceeds from the unique fusion of physical and biological properties of matter that exists at the nanoscale together with engineering design. Teaching this fusion has shown that eliminating traditional distinctions between STEM disciplines can produce deeper and broader student understanding. Exposure to cross-cutting ideas helps students learn to synthesize concepts from diverse sources and apply them to new situations. Integration with social and business sciences offers social relevance.
- *Vertical integration* means providing sustained, high-quality educational opportunities from precollege to graduate levels. Early exposure and consistent reinforcement saves time and minimizes redundancy and misconceptions. New discoveries can be rapidly transferred from the laboratory to the classroom through the participation of university researchers in developing precollege course content and in training teachers.
- *System integration* bridges the gap between basic science and engineering concepts and their applications. Hands-on design challenges build student confidence and emphasize critical thinking and innovation – key skills for a globally competitive workforce.

Although these principles will vastly improve the overall quality and effectiveness of national STEM education, it is important that NSE education should also be a focus. Discoveries and inventions at the nanoscale are driving technological progress and have become essential to economic development and global welfare. Our students must be prepared as nano-literate workers, consumers, and citizens

Partnership is emphasized for these activities because those interested in NSE education believe it is time to: (a) broaden the education efforts to explicitly include the many stakeholder communities; (b) establish a more enduring infrastructure than just the periodic Nanoscale Science and Engineering Education workshops (held in 2006, 2007, and 2008); and (c) develop partnerships to meet the challenges and identify the opportunities provided by the global advances at the nanoscale. This is consistent with the Carnegie Opportunity Equation report [27] that calls for a national mobilization in education exploiting partnerships.

Specific items for attention include:

- Internationally benchmarked standards must be created for the inclusion of nanoscale science and engineering in K–12 education. Since NSE is still rapidly evolving, those standards must be periodically vetted for currency.
- Curricula incorporating nanoscale science and engineering, and correlated to the standards, must be created, vetted by the teaching communities, and made readily available. Web-based and other electronic media must be utilized to better connect with new students who are growing up with those information media.
- Establish a network of regional hub sites – the Nanotechnology Education Hub Network – as a sustainable infrastructure for accelerating nanotechnology education. The hub sites would allow for efficient regional and national field-testing of new content and methodologies.
- There are numerous groups around the world addressing STEM education, NSE education, nanoscale science and engineering research, and nano-enabled technologies. There is an immediate challenge to integrate these various communities. There is a need to identify, validate, and integrate the many NSE educational capabilities that presently exist and to assess what is still needed.
- The NSET should create a working group on nanotechnology education and workforce to support agency efforts to address education and workforce issues. An education and workforce-focused consultative board to the NSET should also be created, comprising the various principal stakeholders. NNCO funds (or other contributions from the various Federal NNI agencies) should be used for this effort
- Consideration should be given to the research and development of an overall mechanism for efficient search, access, and use of cyber-infrastructure resources focused on nanoscience and nanotechnology; this has potential relevance to education at all levels. Such a mechanism would likely entail creating an inventoried, analyzed, tagged registry. Also worth considering are mechanisms for enabling greater interoperability with emerging cyber-nanotechnology resources such as remote access to and control of state-of-the-art nanotechnology characterization equipment; databases for properties and applications; environmental, health, and safety implications and best practices; and general educational paradigms, such as virtual and immersive environments, simulations, and games. Such integration and knowledge-sharing offer great promise for accelerated discovery and learning.
- The DoEd, working closely with the NSTA and cyber-oriented curriculum developers (such as those listed in Murday [7], Table 6), should create a core resource website for information and activities in support of NSE education. The NSTA should serve as the evaluator or resource for quality control to ensure the materials of NSE-focused educational websites are of high quality, are in a format readily utilized by K–12 teachers, are carefully indexed to the various state learning standards, and can be readily accessed from the NSTA website.

Developing the Human and Physical Infrastructure 533

Additional well-designed, highly interactive, media-rich, online learning tools should continue to be developed.
- There must be a close and continuing dialogue between industry and the education communities so that education standards and curricula adequately reflect workforce needs.
- Public concern over environmental safety and health issues at the nanoscale compels attention to both the formal and informal education processes. Workers and members of the general public may be in contact with nanomaterials in various forms during manufacture or in products and should be sufficiently knowledgeable to understand both benefits and risks.
- NSF, DoEd, and other agencies with relevant missions should foster nanotechnology curricula development and evaluation that is appropriate for community colleges and ensure meaningful collaborations between the community colleges and the nanocenters. The DoEd's Department of College and Career Transitions program addressing articulation should ensure nanotechnology is included in that program.
- The NSF National Nanotechnology Infrastructure Network (NNIN), NSEC, DOE nanocenters, and the NIST Center for Nanoscale Science and Technology should work with the NSTA and the DoEd to facilitate on-site and/or remote access to the higher-end user facilities that might contribute to K–12 STEM education.
- Wikipedia is becoming the de facto global encyclopedia. Current entries addressing nanoscale science and engineering in Wikipedia are woefully inadequate and must be routinely updated, expanded, vetted, and sustained.

While all of the items above have merit, higher priority should be given to:

- Incorporation of nanoscale science and engineering knowledge into education at all levels, but especially in the K-12 grades where all citizenry can be informed of the growing knowledge and its technological implications. The development of internationally benchmarked standards for the role of the nanoscale in K-12 education is critical; without those standards other priorities may preclude adequate attention to the nanoscale.
- The development of user facilities to provide prototyping capabilities. Without facilitating the transition of science/engineering discovery into innovative technologies, continued support for nanoscale science and engineering could be jeopardized.
- Establish a sustainable National Nanotechnology Education Hub Network for accelerating NSE/STEM education at all levels through content/learning tool development, teacher/workforce training, and informal education programs.
- Reflecting the growing use of electronic media, both formal and informal education in nanoscale science, engineering and technology needs to better exploit high-quality, web-accessible content. Further, user-friendly web-based remote control of facility analytical and fabrication tools are needed to make those capabilities more broadly available.

7 Broader Implications for Society

Education and physical infrastructure development are enabling conditions for realizing the promise of nano-enabled technology solutions expected to impact many of the pressing societal problems. Further, the excitement associated with addressing those problems may redress the lack of interest in science and engineering careers for native students in developed countries.

8 Examples of Achievements and Paradigm Shifts

8.1 NCLT: Establishing a Nanotechnology Academy as a Small Learning Community in a Los Angeles Unified School District High School (http://www.nclt.us/)

Contact person: Robert Chang, National Center for Learning and Teaching in Nanoscale Science and Engineering (NCLT)

NCLT has spearheaded the development of a model to systematically introduce nanoscience curriculum into secondary education in the United States. The goal is to create a more concentrated learning environment with a focus on nanotechnology by adopting the Small Learning Community (SLC) program already existing in some high schools. The SLC concept was developed by DoEd in response to today's "knowledge-based" economy as a means to increase students' literacy, analytic, and mathematical skills so they can be more successful in postsecondary education or the workforce. These SLCs are, in essence, schools within schools. Each forms a smaller learning cluster that is separate, autonomous, and distinct as a sub-school unit. These smaller school units often have advantages over the larger schools, especially because they make it easier to focus on more personalized teaching and learning, such as on the theme of nanotechnology.

NCLT has partner with Los Angeles Unified School District's (LAUSD) Taft High School to create a pilot nanotechnology SLC during the 2009–2010 school year. The goal is to help students in the SLC to obtain a strong set of core skills in nanoscience and nanotechnology. NCLT created a 15-min promotional DVD for Taft to be able to "market" its newly created Nanotechnology Academy during spring student recruitment. Taft teachers held orientation meetings with students and parents at feeder schools, where they distributed Nano Academy brochures and discussed Academy goals. To further heighten student interest in nanotechnology, NCLT provided a Nano-Tex™ lab coat to one of the science teachers. During a lunch recess, the teacher demonstrated the fabric's nanostructure-based nonstaining properties by allowing students to throw staining liquids on the pristine white lab coat; they were amazed at how nanotechnology could prevent staining on the coat. In another promotional activity, NCLT designed a contest for several science classes to use nanotechnology concepts to maximize a geyser erupting from a carbonated drink. After several weeks of classroom experimentation, the contest was promoted as a school-wide event. The recruiting goal for the Nano Academy is to serve 120 ninth and tenth grade students in the first year and continue to grow in succeeding years to include grades 9 through 12 and eventually to serve about 350 students per year.

Working closely with target Academy teachers who participated in a 2-week nanotechnology summer workshop at Northwestern, NCLT developed a course syllabus for the Academy's introductory courses on nanotechnology during the 2009–2010 implementation. The Academy plans to offer two classes in the first year: Intro to Nano (for grades 9–10) and Nano Tech (for grades 10–12). NCLT continues to work closely with Los Angeles high school teachers to:

- Develop a 4-year nanotechnology-based curriculum
- Develop cross-cutting nanotechnology projects
- Develop essential and guiding questions for use in English, Social Sciences, and other SLC classes, leading to fully interdisciplinary curricula and resource materials
- Carry out professional development for SLC teachers and train a cadre of lead SLC teachers
- Assist the SLCs in building strong partnerships with area universities (e.g., University of California Los Angeles, University of Southern California, and California State University Northridge), industry (Boeing's JRL Laboratories and General Motors), technical societies, etc.
- Collaborate with LAUSD to develop an evaluation plan to assess student achievement levels

8.2 Informal Education: NISE Net's NanoDays (http://www.nisenet.org/nanodays)

Contact person: Larry Bell, Museum of Science, Boston

Organized by the Nanoscale Informal Science Education Network (NISE Net), science museums and research centers have worked together to present NanoDays activities at over 200 sites across the nation during a week each spring from 2008 to 2010 to help inform the public about nanoscale science, engineering, and technology in exciting, creative, hands-on ways. NISE Net is funded by a cooperative agreement with the National Science Foundation's Division of Research on Learning in Formal and Informal Settings and led by the Museum of Science in Boston, the Science Museum of Minnesota in St. Paul, and the Exploratorium in San Francisco.

Informal educational activities about nanoscale science and engineering were virtually nonexistent 10 years ago, and even fundamental ideas about the behavior of matter at the nanoscale were mostly absent from informal science education.

For the most part, informal science educators felt that content of this sort was too difficult for the public to understand, not really exhibitable, and of little interest. As a result, prior to 2005, nanotechnology was covered in only a small number of institutions of informal science education. Inverness Research Associates (2009) reported that at that time there was little expertise, experience, or incentive to do nanoscale science education for the public. Neither science museums nor science research institutions had all the requisite capacities to carry out high-quality nanoscale science education, and they found little incentive to develop such capacity, as it was unclear that their audiences had a driving interest in the topic. Today, hundreds of informal educational institutions, many working in collaboration with university research centers, have introduced activities about nanoscale science and technology into their educational programs. Inverness has found (2009) that national NISE Net efforts in the last part of the decade, in particular NanoDays, have been catalytic in engaging new informal science education institutions and scientists to enter into nanoscale science education. The first NanoDays event in April 2008 had approximately 100 institutional participants; the second and third events in April 2009 and April 2010 each had approximately 200 institutional participants (Fig. 5).

NanoDays has not only introduced science museums to hands-on activities related to nanotechnology but also has created collaborations between science museums and nanoscale research centers. The public has benefited from the knowledge and expertise shared by researchers and graduate students, and students have gained valuable experience in communicating with the public about their areas of research.

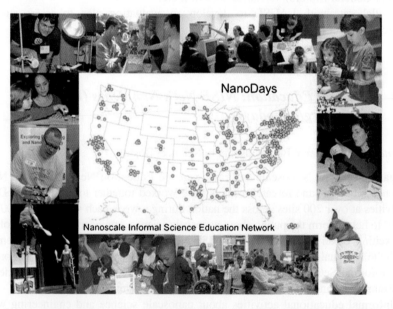

Fig. 5 NanoDays map showing the distribution of NanoDays kits to over 200 sites in all 50 states, Washington, DC, and Puerto Rico, with photographs of some of the wide range of activities that have taken place across the nation (Courtesy of L. Bell)

8.3 Pennsylvania State University Center for Nanotechnology Education and Utilization (http://www.nano4me.org/index.html)

Contact person: Steve Fonash, Pennsylvania State University

The Pennsylvania State University (Penn State) Center for Nanotechnology Education and Utilization is the research university partner to the Pennsylvania community and technical colleges. Through a resource-sharing partnership entitled the Pennsylvania Nanofabrication Manufacturing Technology (NMT) program, 27 academic institutions in Pennsylvania are able to offer a total of 54 two-year and four-year nanotechnology degrees [35]. The students in all of these programs must spend one semester at Penn State attending the hands-on nanotechnology fabrication, synthesis, and characterization immersion provided by the Capstone Semester, a six-course hands-on experience exposing students to state-of-the art equipment and clean room facilities (Fig. 6). The 18 credits of coursework can be used toward an associate or baccalaureate degree, used to earn an NMT Certificate, or both, depending on the specific program of the student's home institution. Refinement of the capstone semester is carried out in close consultation with the industry members of the NMT program's advisory board.

NSF, with contributions from the Penn State University Center for Nanotechnology Education and Utilization, supports the National Nanotechnology Applications and Career Knowledge Center. NACK addresses the widespread need for a workforce possessing strong nano- and microtechnology 2-year degrees [36]. NACK leadership also believes that bringing a nanotechnology education to 2-year degree institutions and exposing students attending these institutions to this exciting field is a motivating force that can bring a new demographic to 4-year STEM degree programs and beyond.

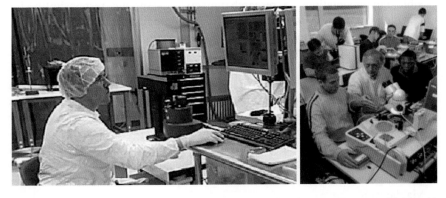

Fig. 6 NACK's laboratory practice as part of the NMT Capstone Semester (Courtesy of S. Fonash)

NACK has introduced a number of paradigm shifts designed to give the United States a well-trained nano- and micro-technology workforce. These shifts address four key issues faced by many community and technical colleges as they consider developing nanotechnology programs:

- *Economic pressures.* To alleviate the economic burden of creating and sustaining four semesters of new courses, NACK has introduced the concept of a capstone semester (as described above for the NMT program). The capstone semester consists of a suite of courses designed to give students from various science and technical programs (e.g., biology, engineering technology, chemistry, and physics) an immersion experience in nanotechnology. There is a skill set requirement rather than a course set requirement for entry into the capstone semester. The entry skill set requirements can be met by traditional biology, chemistry, engineering technology, math, materials science, and/or physics courses available at most 2-year institutions. Institutions thus do not need to develop four semesters of new courses. Students emerge from the NMT capstone semester with an exit skill set developed by the NACK Industry Advisory Board.
- *Student enrollment pressures.* The nanoscience capstone semester approach eliminates the pressure to maintain a baseline student enrollment in a high-tech program. Students move from traditional programs into the nanotechnology immersion capstone semester provided by a university. The critical mass of students needed to economically maintain a nanotechnology education experience only must be attained by the university for the capstone semester.
- *Pressures on faculty, staff, and facilities resources.* The NACK approach to the faculty, staff, and facilities issues faced by 2-year degree-granting institutions is one based on resource sharing. It entails several components: (1) sharing facilities and (2) sharing courses. Sharing facilities means 2-year-degree students using the facilities at a research university to obtain hands-on nanotechnology exposure, or it means community colleges themselves setting-up a teaching clean room facility to be shared among institutions in a given area. In the NACK approach, sharing courses has the following possible implementations: a research university assuming responsibility for teaching the capstone semester for students attending from community colleges, community college students using web-accessible courses provided by NACK, and community college faculty using units from NACK's web-accessible courses.
- *Geographic isolation of some 2-year degree institutions.* The NACK approach to overcoming the drawbacks of geographic isolation in teaching students interested in nanotechnology is two-fold: offering the capstone semester courses online for downloading, and providing web-based access to equipment. The NACK philosophy is that it is best to operate a tool (field-emission scanning electron microscope, scanning probe microscope, etc.) using a computer right next to it, but second-best is to operate the tool with a computer via the web.

8.4 Network for Computational Nanotechnology (http://www.ncn.purdue.edu/home/; http://nanohub.org/)

Contact person: George Adams, Network for Computational Nanotechnology

The Network for Computational Nanotechnology (NCN) supports the National Nanotechnology Initiative by designing, constructing, deploying, and operating the nanoHUB.org national cyber-infrastructure for nanotechnology theory, modeling, and simulation. NCN was established in September 2002 and is funded by the National Science Foundation to support the NNI.

Powered by the HUBzero.org platform created at Purdue University and released as open source software on April 14, 2010, nanoHUB.org is a science gateway where users can run any of over 170 nanotechnology simulation programs using their web browsers with just the click of a button. In the 12 months prior to May 2010, nanoHUB users ran 340,000 such simulations (see map of U.S. users, Fig. 7). They also learned about nanotechnology from 2,100 educational resources, including state-of-the-art seminars and complete courses authored by over 660 members of the nanotechnology research and education community.

The nanoHUB software has already had a strong impact on U.S. NSE education. The simulations and resources of nanoHUB have directly supported 575 research papers in the nanotechnology literature to date. In 2009, faculty at 76 universities used nanoHUB in 116 classes, including all top 50 U.S. engineering schools and 88% of the top 33 physics and chemistry schools. It should be noted that nanoHUB is reaching students at all academic levels, and it has assumed a strong role in the science education of minority and nontraditional students. It is used at 25% of the

Fig. 7 Users of the National Nanotechnology Network in the United States

256 U.S. minority-serving institutions with STEM discipline degree programs, at 32% of Historically Black Colleges and Universities, and at 39% of institutions with high Hispanic enrollment. Bruce Barker, President of Chippewa Valley Technical College, Eau Claire, Wisconsin, has pointed to nanoHUB's value to his students, "We have a high percentage of non-traditional students, many of whom are older and starting new careers, or who are coming from disadvantaged families; nanoHUB provides them with a toe-hold to a wider academic world." (from telephone interview Jan. 2010; courtesy of G.B. Adams).

NCN is leading an effort to rethink electronic devices from the nanoscale perspective. With support from Intel Foundation, NCN has created "Electronics from the Bottom Up," courseware that may reshape the teaching of nanoelectronic technology and will train a new generation of engineers to lead the twenty-first Century semiconductor industry. Using these new concepts, NCN is building an electronic device simulation platform that powers several tools on nanoHUB and runs efficiently on the largest computers on the national grid. Recent achievements include modeling the effect of a single donor atom in the active region of a FinFET nanotransistor [37]. Dr. Dmitri E. Nikonov, Components Research, Technology, and Manufacturing Group, Intel Corporation, who is charged with using simulation to evaluate beyond-CMOS electronic devices (the next generation of transistors), asserts that, "nanoHUB tools are indispensible to the mission of my department."

8.5 University of Albany, College of Nanoscale Science and Engineering (http://cnse.albany.edu/)

Contact person: Alain Diebold, University of Albany, State University of New York (SUNY)

The College of Nanoscale Science and Engineering (CNSE) of the University at Albany, State University of New York (SUNY), is the first college in the world dedicated to education, research, development, and deployment in the emerging disciplines of nanoscience, nanoengineering, nanobioscience, and nanoeconomics. At CNSE, academia, industry, and government have joined forces to advance atomic-scale knowledge, educate the next-generation workforce, and spearhead economic development. The result is an academic and corporate complex that's home to world-class intellectual capital, unmatched physical resources, and limitless opportunities.

CNSE has reshaped the traditional "silo" type college departmental structure into four cross-disciplinary constellations of scholarly excellence in nanoscience, nanoengineering, nanobioscience, and nanoeconomics. Through this game-changing paradigm, students engage in unique hands-on education, research, and training in the design, fabrication, and integration of nanoscale devices, structures, and systems to enable a wide range of emerging nanotechnologies. Students are supported by internships, fellowships, and scholarships provided by CNSE and its array of global corporate partners. CNSE complements its groundbreaking bachelor's, master's and PhD programs in nanoscale science and engineering with educational outreach to elementary, middle, and high schools; partnerships with community colleges and academic institutions around the world; and certificate-level technical training. This unprecedented effort is designed to educate the next generation of nanotechnology-savvy professionals and build the foundations of a skilled nanotechnology workforce at every level.

The setting for this pioneering research and educational blueprint is CNSE's world-class Albany NanoTech Complex (Fig. 8), the most advanced research enterprise of its kind anywhere in the world. This 800,000-square-foot megaplex has attracted more than $6 billion in public and private investment and houses the only fully-integrated, 300 mm wafer, computer chip pilot prototyping and demonstration line within 80,000 square feet of Class 1–capable clean rooms. Over 2,500 scientists, researchers, engineers, students, and faculty work at CNSE's Albany NanoTech Complex, which serves as a primary research and development location for a host of leading global high-tech corporations co-located on-site, including IBM, AMD, GlobalFoundries, International SEMATECH, Tokyo Electron, Applied Materials, ASML, Novellus Systems, Vistec Lithography, Atotech, and many more. An expansion currently in the planning stages is projected to increase the size of the CNSE Albany NanoTech Complex to over 1,250,000 square feet of state-of-the-art infrastructure, housing over 105,000 square feet of Class 1–capable clean rooms. Once completed, the expanded CNSE Albany NanoTech Complex is expected to house over 3,750 scientists, researchers, and engineers from CNSE and global corporations.

Buoyed by its unparalleled combination of intellectual know-how and leading-edge technological infrastructure, CNSE's Albany NanoTech Complex is the site of some of the world's most advanced nanoscale research, development, and commercialization activities. Here, academic and corporate scientists engage in innovative research in a variety of fields, including clean energy and advanced sensor and environmental technologies; advanced CMOS and post-CMOS nanoelectronics; 3D integrated circuits and advanced chip packaging; ultra-high-resolution optical, electron, and EUV lithography; and nanobioscience and nanomedicine. The result is a vibrant and powerful entity that is driving critical innovations to address the most important challenge areas facing society, from energy, the environment, and health care to military, telecommunications, information technology, and transportation, among many others.

Fig. 8 Photos of the Albany NanoTech Complex and labs (Courtesy of A. Diebold)

8.6 DOE Nanoscale Science Research Centers (http://www.nano.energy.gov)

Contact person: Altaf Carim, Department of Energy Office of Basic Energy Sciences

There are five DOE Office of Science Nanoscale Science Research Centers (NSRCs), as follows:

- Molecular Foundry – Lawrence Berkeley National Laboratory (California)
- Center for Nanoscale Materials – Argonne National Laboratory (Illinois)
- Center for Functional Nanomaterials – Brookhaven National Laboratory (New York)
- Center for Integrated Nanotechnologies – Los Alamos National Laboratory and Sandia National Laboratory (New Mexico)
- Center for Nanophase Materials Sciences – Oak Ridge National Laboratory (Tennessee)

Taken together, these centers provide resources unmatched in the world. The new nanocenter buildings contain clean rooms, laboratories for nanofabrication, one-of-a-kind signature instruments, and other instruments (such as nanopatterning tools and proximal probe microscopes) not generally available except at major scientific user facilities. These facilities are designed to be the nation's premier user centers for interdisciplinary research at the nanoscale, serving as the basis for a national/international program that encompasses new science, new tools, and new computing capabilities.

Each center is housed in a new laboratory building near one or more existing DOE facilities for x-ray, neutron, or electron scattering to take advantage of their complementary capabilities; those facilities include the Spallation Neutron Source at Oak Ridge; the synchrotron light sources at Argonne, Brookhaven, and Lawrence Berkeley; and semiconductor, microelectronics and combustion research facilities at Sandia and Los Alamos. Each center has particular strengths in different areas of nanoscale research, such as materials derived from or inspired by nature; hard and crystalline materials, including the structure of macromolecules; magnetic and soft materials, including polymers and ordered structures in fluids; and nanotechnology integration. User access to center facilities is through submission of proposals that are reviewed by independent proposal evaluation boards.

8.7 NIST Center for Nanoscale Science and Technology (http://www.nist.gov/cnst/)

Contact person: Robert Celotta, NIST CNST

The Center for Nanoscale Science and Technology (CNST) at the National Institute of Standards and Technology was established in May 2007 to accelerate innovation in nanotechnology-based commerce. It supports the development of nanotechnology through research on measurement and fabrication methods and technology. The CNST has a unique design that supports the U.S. nanotechnology enterprise through the readily available, shared-use NanoFab facility, as well as by providing opportunities for collaboration in multidisciplinary research on new nanoscale measurement methods and instruments, which are housed in buildings designed to provide stringent environmental controls on particulate matter, temperature, humidity, vibration, and electromagnetic interferences. It also serves as a hub linking the international nanotechnology community to the comprehensive measurement expertise available throughout NIST.

The NanoFab is accessible to industry, academia, and government agencies on a cost-reimbursable basis, providing researchers economical access to and training on the advanced tool set required for cutting-edge nanotechnology development. The NanoFab provides a comprehensive suite of tools and processes for nanofabrication and measurement. It includes a large, dedicated clean room – with all the tools operated within an 8,000 Sq.Ft (750 m^2) Class-100 space – and additional tools in adjacent, high-quality laboratory space. Over 65 major tools are available for e-beam lithography, photolithography, nanoimprint lithography, metal deposition, plasma etching, atomic layer deposition, chemical vapor diffusion, wet chemistry, and silicon micro- and nanomachining. The facility is accessible through a straightforward application process designed to get users into the clean room in a few weeks.

CNST research is creating the next generation of nanoscale measurement instruments and fabrication methods, which are made available through collaboration with its multidisciplinary scientists and engineers. This research is agile and highly interactive by design, with significant contributions from a rotating cadre of postdoctoral researchers, and with many collaborative projects involving both NIST scientists and others in the United States and abroad. The CNST is currently giving priority to the following three research areas:

- *Future electronics.* In support of continued growth in the electronics industry beyond CMOS technology, the CNST is developing new methods to create and characterize devices, architectures, and interconnects for graphene, nanophotonic, nanoplasmonic, spintronic, and other future electronic devices.
- *Nanofabrication and nanomanufacturing.* The center is advancing the state of the art in nanomanufacturing by developing measurement and fabrication tools for both lithographic ("top-down") and directed assembly ("bottom-up") approaches.
- *Energy storage, transport, and conversion.* This research is focused on creating new methods for elucidating light–matter interaction, charge and energy transfer processes, catalytic activity, and interfacial structure at the nanoscale in energy-related devices.

8.8 National Nanotechnology Infrastructure Network (http://www.nnin.org/)

Contact person: Sandip Tiwari, National Nanotechnology Infrastructure Network

The National Nanotechnology Infrastructure Network (NNIN) is a collective of 14 university-based facilities with the mission to enable rapid advancements in nanoscale science, technology, and engineering through open and efficient access for fabrication purposes. Its facilities-based infrastructure resources are openly accessible to the nation's students, scientists, and engineers who come from academia, small and large companies, national laboratories, etc., providing the capacity to translate ideas to practice across disciplines and new frontier areas. NNIN also supports experimental efforts and an independent interdisciplinary theoretical effort through computational resources, where the emphasis is on modeling and simulation of advanced scientific problems of the nanoscale via open and tested software, hardware, and basis information. NNIN leverages its infrastructure resources and geographic and institutional diversity to conduct other activities with broader impact: in education, in enhancing diversity in technical disciplines, in societal and ethical implications of nanotechnology, and in health and environment efforts. Figure 9 shows the member institutions of NNIN.

NNIN serves the world's largest community of experimental graduate students, engineers, and scientists under one umbrella. During 2009–2010, NNIN resources were used by over 5,300 unique users for a significant part of their experimental work. Of these, over 4,000 were graduate students, about 800 were industrial users, and the rest were from U.S. state and Federal laboratories and foreign institutions. More than 300 small companies used NNIN facilities. Nearly 3,200 publications, several of them the significant scientific and engineering highlights of the year, resulted from the work of the user community. Nearly a quarter of PhD awards in "nano"-related disciplines utilized NNIN resources. Over 10% of the professionals of small companies supported through SBIR grants also utilized NNIN resources.

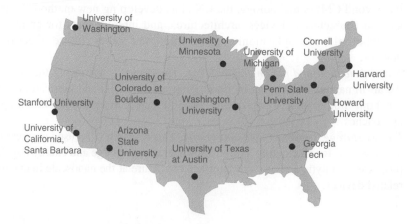

Fig. 9 NNIN sites (Source: http://www.nnin.org)

Developing the Human and Physical Infrastructure   545

Taken together, these figures show that NNIN is a major national force in human resource development and in R&D and commercialization of nanotechnology.

NNIN also has a broad portfolio in education and outreach. Its programs include (a) Research Experience for Undergraduates (REU); (b) International Research Experience for Undergraduates (iREU); (c) Laboratory Experience for Faculty (LEF); (c) Showcase for Students; (d) international Winter School for Graduate Students (iWSG); (e) Symposia; (f) *Nanooze* (a magazine for elementary and middle school students); (f) Open Textbook; and (g) technical workshops & symposia. Local activities from NNIN include day and longer camps for middle and high school students, local outreach through workshops for teachers, school and community-connected activities, and from Howard University, a laboratory on wheels that brings nanotechnology activities to high schools in the eastern part of the country.

NNIN's research efforts also include examination and development of (a) understanding of interdisciplinary collaborations and their impact on research, (b) the impact on competitiveness and the process of technology transfer and industrial innovation and innovation by industry through NNIN, (c) impact of government-funded faculty research and the faculty's interaction with industry in technology transfer, (d) impact of intellectual exchange, openness, and sharing such as in a network in conduct and impact of research, and (e) the ethical issues related to nanotechnology and fostering of ethical conduct.

Throughout the breadth of its network activities, NNIN seeks the integration and development of consciousness about social and ethical issues (SEI) related to nanoscience while being the nation's leading open-access scientific research facility. SEI efforts within NNIN therefore embody both the network's research and educational pursuits. NNIN has organized its SEI efforts to take advantage of the network's unique strengths as a national resource with geographic diversity, technical breadth, and community interests. Within its user community, NNIN provides training and educational opportunities through SEI modules and teaching materials employed in training and ongoing educational programs, selective incorporation of these in the network educational activities (REU, RET, workshops, symposia, etc.), and in broader outreach activities. NNIN provides opportunities for national researchers via competitive travel and seed grants to support related SEI investigations. NNIN stimulates and facilitates SEI research on the network's unique, world-leading strength – the largest collection of nanotechnology users (students and professionals), communities (academe and industry), and relevant technologies.

8.9 Nanotechnology Characterization Laboratory (http://ncl.cancer.gov/)

Contact person: Scott McNeil, NCL

The Nanotechnology Characterization Laboratory (NCL) was established as an interagency collaboration among the National Cancer Institute (NCI), the National Institute of Standards and Technology (NIST), and the U.S. Food and Drug

Administration (FDA) to evaluate the safety and efficacy of biomedical nanomaterials. The NCL is a national resource for all nanotechnology researchers and helps facilitate the regulatory review process by conducting preclinical characterization using standard methods. By providing critical infrastructure and characterization services to nanomaterial providers, the NCL accelerates the transition of nanoscale particles and devices into clinical applications. The NCL's activities are markedly speeding the development of nanotechnology-based products, reducing associated risks, and encouraging private sector investment in this promising area of technology development.

Candidates for NCL characterization are selected through an application process, and characterization of accepted nanomaterials is provided at no cost to the submitting investigator. As part of its standardized assay cascade, the NCL characterizes the nanomaterial's physicochemical characteristics, its in vitro immunological and toxicological properties, and its in vivo compatibility using animal models. The time required to characterize a nanomaterial from receipt through the in vivo phase is 1 year or more. Data derived from the NCL assay cascade are intended to be included in an investigator-led filing of an Investigational New Drug (IND) or Investigational Device Exemption (IDE) with the FDA but also can be used in scientific publications or for promotion purposes (e.g., to garner investment). The NCL website provides more information on the NCL application process or the NCL Assay Cascade.

In addition to characterization of applicant-provided nanomaterials, the NCL also collaborates with several other government agencies, including the FDA's National Center for Toxicological Research (NCTR), the National Institute of Allergy and Infectious Disease (NIAID), and the National Institute of Environmental Health Sciences (NIEHS) to characterize nanoparticles for other applications, such as evaluation of environmental, health, and safety concerns. These agencies have come together to promote knowledge and data sharing and to coordinate research efforts on the potential risks of nanotechnology in medicine and the environment.

8.10 Three National Institutes of Health (NIH) Nanotechnology-Related Networks

Contact person: Jeff Schloss, NIH, National Human Genome Research Institute

8.10.1 National Cancer Institute (NCI), Alliance for Nanotechnology in Cancer (http://nano.cancer.gov/)

Contact person: Piotr Grodzinski, NIH/NCI

The National Cancer Institute launched its Alliance for Nanotechnology in Cancer in 2004 to fund and to coordinate research that seeks to apply advances in nanotechnology to the detection, diagnosis, and treatment of cancer. The program is highly translational and focuses on techniques capable of producing clinically useful procedures. The Alliance builds on the premise of multidisciplinary research, engaging technology developers: chemists, engineers, and physicists, as well as biologists and clinicians – the community capable of identifying the most pressing needs of clinical oncology that are not met with currently available approaches. The Alliance operates as an integrated constellation of Centers for Cancer Nanotechnology Excellence (CCNEs) and smaller collaborative Cancer Nanotechnology Platform Partnerships (CNPPs), together with the Multidisciplinary Research Training awards and the National Characterization Laboratory (NCL). NCL, which is a collaborative effort with NIST and FDA, has become a national resource for nanomaterials characterization and developed an extensive cascade of assays to evaluate physical properties of nanostructures and their behavior in in vitro and in vivo environments.

Three challenge areas for the implementation of nanotechnologies into cancer research and oncology are:

- *Early diagnosis using in vitro assays and devices or in vivo imaging techniques.* Novel nanotechnologies can complement and augment existing genomic and proteomic techniques to analyze variations across different tumor types, thus offering the potential to recognize early the onset of the disease with sensitivity and specificity, which is not currently possible. Sensitive biosensors constructed of nanoscale components (e.g., nanocantilevers, nanowires, and nanochannels) can recognize genetic and molecular events and have reporting capabilities. The imaging contrast agents based on nanotechnologies (e.g., optical, magnetic resonance, ultrasound) are expected to be capable of identifying tumors that are significantly smaller than those detected with current technologies.
- *Multifunctional nano-therapeutics and post-therapy monitoring tools.* Because of their multifunctional capabilities, nanoscale devices can contain both targeting agents and therapeutic payloads – features useful in drug delivery to areas of the body that are difficult to access because of a variety of biological barriers, including those developed by tumors. Thus, multifunctional nanoscale devices offer the opportunity to utilize new approaches to therapy – "smart" nanotherapeutics may provide clinicians with the ability to locally deliver the drug at lower doses, while maintaining high efficacy or time the release of an anticancer drug or deliver multiple drugs sequentially in a timed manner or at several locations in the body.
- *Devices and techniques for cancer prevention and control.* Nanotechnology can play a vital role in establishing novel approaches to the disease's prevention. For example, nanoscale devices may prove valuable in delivering or mimicking polyepitope cancer vaccines that engage the immune system or cancer-preventing nutraceuticals or other chemopreventive agents in a sustained, timed-release, and targeted manner. Nanotechnology can also enable techniques allowing for effective disease management and elimination of cancer spread to other organs.

The Alliance has generated very strong scientific output, which includes over 1,000 peer-reviewed publications in highly regarded scientific journals (an average impact factor ~7) and more than 200 patent disclosures/applications. In addition to scientific advances, the association of the program with nearly 50 industrial entities (ranging from PI-initiated start-up companies to collaborations with large multi-national firms) has established a vital commercial outlet for produced technologies. Currently, these companies along with the investigators from the ANC, are engaged in several nanotherapy and imaging clinical trials. Several additional companies have a nanotechnology application in advanced, pre-IND stage of technology development.

Due to the success of the ANC program, the National Cancer Institute approved reissuance. Phase II will start in September 2010 and will be funded for another 5 years. Phase II of the program will consists of Centers for Cancer Nanotechnology Excellence (CCNEs), Cancer Nanotechnology Platform Partnerships (CNPPs) and National Characterization Laboratory (NCL), similar to Phase I. In addition, Cancer Nanotechnology Training Centers (CNTCs) and Path to Independence awards will be also included to strengthen training and education facet of the program.

8.10.2 NHLBI Programs of Excellence in Nanotechnology (http://www.nhlbi-pen.net/)

Contact person: Denis Buxton, NIH/National Heart, Lung, and Blood Institute

The goal of National Heart, Lung, and Blood Institute (NHLBI) Programs of Excellence in Nanotechnology is to develop nanotechnology-based tools for the diagnosis and treatment of heart, lung, and blood diseases, and to move the translation of these technologies towards clinical application. Initially funded in 2004 with awards to four centers, the program brings together multidisciplinary teams from the biological, physical and clinical sciences for the focused development and testing of nanoscale devices or devices with nanoscale components, and applies them to cardiovascular, hematopoietic and pulmonary diseases. The program also develops investigators with the interdisciplinary skills to apply nanotechnology to heart, lung, and blood disease problems. The program is related to NHLBI's Strategic Plan Goal II: To improve understanding of the clinical mechanisms of disease and thereby enable better prevention, diagnosis, and treatment. Program highlights are available in a series of newsletters (http://www.nhlbi-pen.net/default.php?pag=news).

8.10.3 NIH Nanomedicine Common Fund Initiative (http://nihroadmap.nih.gov/nanomedicine)

Contact person: Richard Fisher, NIH/National Eye Institute

The Nanomedicine Initiative is a 10-year program that started in Fiscal Year 2005 as part of the NIH Roadmap for Medical Research. Currently, it operates under the auspices of the Common Fund which was established by Congress in the 2006 NIH Reauthorization Act as a central funding authority to support programs that are novel, unique, experimental, and relevant across all components of the NIH. The overarching goal of the Nanomedicine Initiative is to move basic science studies toward translational endpoints. In particular, the NIH Nanomedicine Initiative initially funded eight centers that were challenged to use quantitative approaches to understand, from an engineering perspective, the design of biomolecular structural and functional pathways in cells, and to use that information to design and build functional biocompatible molecular tools to return the dysfunctional structures or systems back into "normal" operating ranges after function has been perturbed by disease. The multidisciplinary teams carrying out this initiative consist of researchers with deep knowledge of biology and physiology, physics, chemistry, math and computation, engineering, and clinical medicine. During the first few years, the work emphasized basic biological studies, while keeping in mind that the choice and design of experimental approaches are directed by the need to solve clinical problems. More recently the focus has moved toward application of the basic biological information to specific clinical problems. Work at the centers has been evolving on a trajectory toward preclinical models that test new solutions to specific diseases. The program involves over 300 investigators including dozens of post-doctoral research associates and graduate students at over 30 educational institutions located in 12 states, and internationally in five countries.

8.11 Dragonfly TV: Nanosphere (http://pbskids.org/dragonflytv/nano)

Contact person: Lisa Regalla, Twin Cities Public Television, Inc

DragonflyTV is an Emmy Award–winning children's multimedia science education program combining television, community outreach, print materials and science kits, and web-based information and activities. It is produced by Twin Cities Public Television in St. Paul, MN, and funded by the National Science Foundation. In its seventh season (2008), *DragonflyTV* teamed up with museums and research institutions nationwide to produce six, half-hour episodes on nano-scale science and engineering. The episodes are geared towards children ages 8–12 and follow a scope and sequence covering topics such as: size and scale, the structure of matter, size-dependent properties, nanoscale forces, applications, and societal implications.

Each episode includes two inquiry-based investigations, driven by kids, which reinforce the featured science concepts. In addition, each show contains a "Scientist Profile" that introduces role models and future careers in nanoscale science and engineering, and a segment called "Hey…Wait a Nanosecond," featuring real kids' opinions on societal implications.

Beyond broadcast, *DragonflyTV Nanosphere* resources include online videos, games, and activities; Educator Guides featuring inquiry-based activities for formal or informal use; a kids' nano "zine;" and the NanoBlast board game. These materials are distributed freely to educators through the National Science Teachers Association (NSTA), Association of Science Technology Centers (ASTC), and at NanoDays museum outreach events across the country.

8.12 Instrumentation for Education: NanoProfessor and NanoEducator (http://www.nanoprofessor.net/; http://www.ntmdt.com/platform/nanoeducator)

Contact person: James Murday, University of Southern California Office of Research Adv ancement

Two public-private partnerships, NanoInk and NT-MTD, are seeking to make twenty-first-Century education and workforce development in nanotechnology accessible to small 2- and 4-year colleges, by providing relatively low-cost instruments and curricula.

The NanoProfessor project has three components. The first is an accessible desktop nanofabrication machine (Fig. 10) simple enough for general students to operate at the nanoscale level. The second critical element is a worthwhile curriculum grounded in fundamental science and engineering concepts; it is an interdisciplinary curriculum designed to engage students in basic science learning through hands-on manufacturing and experiments with cutting-edge technology at the nanoscale level. The curriculum is being developed by a team of teachers, NanoInk professionals, and experts in instructional design.

Each unit and the overall course will be evaluated during development and throughout implementation. The third element is the active participation of educational institutions committed to the advancement of science, technology, engineering, and mathematics education. The educational partners will host the project, receive training for faculty members, and cooperate in the evaluation and dissemination of project outcomes.

The NanoEducator platform is a student-oriented scanning probe microscope (SPM) that was developed for use by even first-time microscope users; it can navigate through step-by-step operations. This device is designed to capture student interest in science at secondary and post-secondary-levels and to train future nanotechnologists in using both atomic force microscopy (AFM) and scanning tunneling microscopy (STM) techniques. Robust and foolproof, NanoEducator can help provide a broad, interdisciplinary understanding of different fields of nanoscience, allowing investigation of cells, viruses, bacteria, metals, semiconductors, dielectrics, polymers, etc. It is designed it to be cost-effective enough to equip a classroom with SPMs and comes complete with e-teach software, training literature, handbooks, and descriptive laboratory exercises.

Fig. 10 NanoInk desktop nanofabrication system

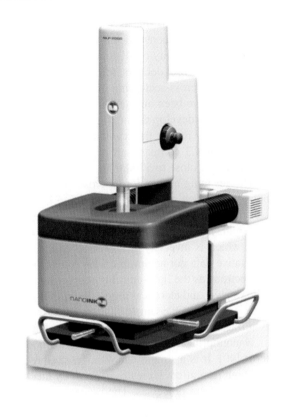

8.13 Three Illustrations of Newly Developed Innovative Nanoscale Instrumentation

Contact person: James Murday, University of Southern California Office of Research Advancement

The development of scanning tunneling microscopy/spectroscopy and the various forms of atomic force microscopy in the 1980s led to the rapid explosion of science and engineering at the nanoscale. But to meet the growing complexity and sophistication of nano-enabled technologies, additional developments are needed for better (i.e., faster, more precise, 3D, etc.) measurement. There have been many contributions during the first decade of the nano-initiatives. Three examples that have reached the commercialization stage are CT (computed tomography) imaging, scanning helium ion microscopy, and imprint lithography.

The nanoXCT-100 is a lab-based ultra-high-resolution CT scanner for 3D visualization of microscopic sample volumes [38]. Precision X-ray focusing optics deliver a resolution as fine as 50 nm, seamlessly extending the capabilities of X-ray CT beyond those of conventional scanners. Integrated Zernike phase contrast imaging enhances the visibility of all edges and interfaces when absorption contrast is low. The nanoXCT-100 delivers reliable 3D volumetric information otherwise only accessible by cross-sectioning or other destructive methods. Dr. Ge Wang and his colleagues have been awarded more than $1.3 million from the National Science Foundation to develop the next-generation nano-CT imaging system, which promises to greatly reduce the required dose of radiation. Virginia Tech and Xradia, a leading nano-CT company, are also collaborating on the project with a cost-sharing investment of close to $800,000.

The scanning helium ion microscope (SHIM or HeIM) is a new imaging technology based on a scanning helium ion beam [39]. This technology has several advantages over the traditional scanning electron microscope (SEM). Due to the very high source brightness and the short De Broglie wavelength of the helium ions, which is inversely proportional to their momentum, it is possible to obtain qualitative data not achievable with conventional microscopes, which use photons or electrons as the light-emitting source. Images offer topographic, material, crystallographic, and electrical properties of the sample. In contrast to other ion beams, there is no discernible sample damage, due to the relatively light mass of the helium ion. A surface resolution of 0.24 nm has been demonstrated.

Nanoimprint lithography (NIL) is a method of fabricating nanometer-scale patterns [40]. It is a simple nanolithography process with low cost, high throughput and high resolution. It creates patterns by mechanical deformation of imprint resist and subsequent processes. The process has been commercialized by at least three companies: Nanonex (http://www.nanonex.com), NIL Technology (http://www.nilt.com), and Molecular Imprints (http://www.molecularimprints.com). NIL has been incorporated into the 32 and 22 nm nodes of the International Technology Roadmap for Semiconductors.

9 International Perspectives from Site Visits Abroad

9.1 United States-European Union Workshop (Hamburg, Germany)

Panel members/discussants

Costas Charitidis (co-chair), Technical University of Athens, Greece
James Murday(co-chair), University of Southern California, United States
Nira Shimoni-Eyal, Hebrew University of Jerusalem, Israel
Helmuth Dosch, DESY, Germany
Massimo Altarelli, European XFEL GmbH, Germany
Dan Dascalu, University Politechnica Bucharest, Romania
Yvan Bruynseraede, Katholieke Universiteit Leuven, Belgium

Elaborating on European-wide attention to nanoscale science/engineering (e.g., see the European Nanotechnology Gateway: http://www.nanoforum.org, [41, 42]), there was an extensive European-wide study of the needs and opportunities for coordinating future research and development in nanomaterials science and nanotechnology for the advancement of technologies ranging from communication and information, health and medicine, future energy, environment and climate change to transport and cultural heritage. The result was the Gennesys report [43] that included challenges/opportunities for facilities and education. It provided the basis for this workshop report.

9.1.1 Facilities

The absence of central facilities for full parameter characterization of nanoparticles/materials is a bottleneck to advances in this critically important field; it is strongly encouraged a European Center of Excellence in nano-standardization and nanometrology be created. This would provide a scientific and industrial infrastructure in Europe for dimensional nanometrology and nanostandards specification (nanochemical, nanomechanical, nanobioproperties, nanoelectronics, etc.), reference materials, and standard measuring methods.

There are great advantages of nanomaterials, but they need to be carefully analyzed to identify any toxicology and environmental problems. Since the interactions of nanostructured materials with biological systems will be complex and levels of characterization different from the norm, a special center such as the U.S. National Institutes of Health Nanotechnology Characterization Laboratory should be established.

The strategic knowledge-based development of new nanomaterials, which are needed to solve urgent problems of society, requires creation of analytical science centers that cooperate in new ways with European research and have direct contact with neutron and accelerator-based x-ray facilities. These centers would address organizational aspects (availability, ease of access), educational aspects, and technical aspects (e.g., extreme focusing, nanobeam stations, small neutron beams, and transferring scientific results from laboratory conditions to realistic industrial ones).

On the one hand there is the European nanotechnology industry that needs to benefit from the best scientific data on materials it develops, on the other hand large facilities can provide such data but are not yet adapted to the industrial use. Hence the need to create an interface structure, EU-industry centers of excellence, whose mission is to bridge the gap between nanotechnology companies and large scientific facilities. This effort will seek to develop "pocket" facilities in order to democratize their use by making it accessible when it cannot be practiced on site, for example, control activities of production or treatment in hospital. Each EU-Industry Center of Excellence, would be focused around an urgent topic; the GENNESYS document proposes Softmatter Materials, Food, Science, Structural Nanomaterials, Nano-materials for Energy, and Nanomaterials for Cultural Heritage.

9.1.2 Education

A European action plan in nanomaterials education has to be worked out urgently to underpin a sustainable nanomaterials research strategy. Strong efforts must be undertaken to improve integration of nanomaterials education and research, particularly at the boundaries of disciplines and to prepare flexible and adaptable nanomaterials scientists and engineers for the future. An International Institute for Nanomaterials Education should be coordinated by the EC and/or other relevant agencies.

A new framework of cooperation between universities, national research institutes and industry needs to be developed. A Nanomaterials European College would ideally meet the requirements for the training of future materials scientists and engineers. It would be a central institute with satellite schools at the "Centers of Excellence" – and with close associations to recognized research universities and corporate research in industry throughout Europe. The International Institute for Nanomaterials would ideally complement the EIT (European Institute for Innovation and Technology). Joining both institutions would become the motor for "Innovative Europe" in nanomaterials science and technology.

In order to face up to the massive investments of major or emerging countries targeting world leadership in some economic areas, Europe and partner institutions must overcome the fragmentation of the human and materials resources of European research. This can be done by gathering the best teams to share these resources through integrating them into new European organizations such as the proposed GENNESYS European College of Excellence. The GENNESYS initiative represents a unique and attractive opportunity to gather and integrate geographically dispersed human resources and effectively scientific facilities for training and promoting activities.

Many companies throughout Europe and the world report problems in recruiting the types of graduates they need. For Europe to continue to compete alongside prestigious international institutions and programs on nanomaterials, it is important to create a "Europe Elite College" that provides a top-level education and the relevant skills mix. This should be a new institution, involving new "satellites" of leading universities and other institutions throughout Europe. Such a college should cover education, training, sciences and technologies for research, and have strong involvement by European industry. The elements for such a high level education are: multidisciplinary skills; top expertise in nanomaterials science and engineering; literacy in complementary fields; exposure to advanced research projects; literacy in key technological aspects; exposure to real technological problems; basic knowledge in social sciences, management, ethics, foreign languages; literacy in neighboring disciplines (international business, law, etc.); and interlinkages among education, research and industrial innovation. Students will be ready for that which research and development will provide. Sharing of post-docs, masters; and PhD students to foster the mobility of permanent researchers and professors between different institutions is needed to create "team spirit." The European College should have strong links with universities of excellence in Europe, nanomaterials research institutes, the research infrastructure, the centers of excellence, and industry.

Attention to the education/training of technicians is also necessary. Since experience with state-of-art fabrication/characterization/processing tools is essential, but difficult to sustain in a rapidly moving field such as nanoscale science and engineering, it will be necessary to work closely with the centers of excellence. Also needed is capability for remote access to that instrumentation.

Attention must also be devoted to primary and secondary education; in particular these education levels are important to stimulate interest in science/engineering careers. The EU has initiated the Know You project; during the 2009–2010 academic year 25 pilot schools have been teaching nanotechnology in their classrooms with a wide range of materials, including videos, online animations, games, workshops, virtual dialogues, and virtual experiments based on current research. Efforts such as this should be expanded and linked to international education efforts.

Consideration must be given to the research and development of an overall mechanism for efficient search, access, and use of cyber-infrastructure resources focused on nanoscience and technology with potential relevance to education at all levels. In particular, Wikipedia is becoming the *de facto* global encyclopedia; current Wikipedia entries addressing nanoscale science and engineering are woefully inadequate and must be updated, expanded, vetted, and sustained.

9.2 United States-Japan-Korea-Taiwan Workshop (Tokyo/Tsukuba, Japan)

Panel members/discussants

Hiroyuki Akinaga (co-chair), AIST, Japan
Mark Hersam (co-chair), Northwestern University, United States
Ryoji Doi, Ministry of Economy, Trade and Industry, Japan
Isao Inoue, University of Tsukuba, Japan
Chul-Gi Ko, Korea Institute of Materials Science, Korea
F. S. Shieu, National Chung Hsing University, Taiwan
Masahiro Takemura, National Institute for Materials Science, Japan
Taku Hon-iden, Tsukuba City, Japan
Iwao Ohdomari, Japan Science and Technology Agency, Japan

9.2.1 Facilities

In addition to the points made in the workshop held in Chicago [44], the need for international collaborations was emphasized. As nanoscale science matures into technology innovations, those collaborations will have to address export-import controls. There is need for joint research contract templates to minimize that difficulty. Further, while semiconductor technologies presently lead R&D, this is beginning to change rapidly. The user facility infrastructures are not keeping pace with

YEAR	STAGE OF EVOLUTION
~2000	Research by individual groups
~2005	User facility and network of user facilities
~2010	Problem-solving user facilities, and networking
~2015	User facility as a center of S&T formation
~2020	User facility in a society, as demonstrative test area and for outreach activities

Fig. 11 Evolution of user facilities from science and technology (S&T) to societal centers (From information provided by H. Akinaga)

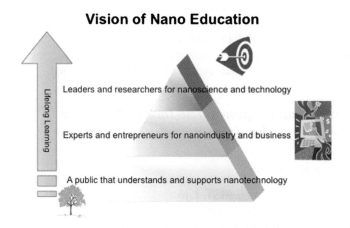

Fig. 12 Taiwan vision for nano-education (Courtesy of Prof. Fuh-Sheng Shieu, National Chung Hsing University, Taiwan)

the new needs. Finally, with the expected huge impact of nano-enabled technology toward the solution of societal needs, the user facilities need to evolve toward technology demonstration and public outreach activities (see Fig. 11).

9.2.2 Education

There was general agreement with the vision (see Fig. 12) presented by Prof. Fuh-Sheng Shieu, National Chung Hsing University, Taiwan. Taiwan, which has one of the more complete programs addressing education at the nanoscale [45], has begun a Phase II framework program that includes nanoeducation fundamental research, development of materials and curricula for elementary school through college, and popular science education.

There are education challenges and needs that must be addressed, including: multi-level collaboration amongst Universities, R&D Institutes, and Industry; communication with politicians, especially the next generation of policy makers – a challenge that suggests the incorporation of nanotechnology issues in business, law and medical education; reaching out to local governments which generally control K–12 education; and developing outreach materials that utilize nanotechnology as a new topic to interest students. For the latter there may be copyright issues associated with those materials impeding the rate of progress.

9.3 United States-Australia-China-India-Saudi Arabia-Singapore Workshop (Singapore)

Panel members/discussants

Hans Griesser (co-chair), University of South Australia
Mark Hersam (co-chair), Northwestern University, United States
Chennupati Jagadish, Australian National University, Australia
Chen Wang, National Center for Nanoscience and Technology, China
Wei Huang, Nanjing University of Posts and Telecommunications, China
Jayesh Bellare, Indian Institute of Technology, Bombay, India,
Salman Alrokayan, King Abdullah Institute for Nanotechnology, Saudi Arabia
Andrew Wee, National University of Singapore, Singapore
Jackie Ying, Institute of Bioengineering and Nanotechnology, Singapore
Freddy Boey, Nanyang Technological University, Singapore

9.3.1 Facilities

The capital equipment awards made to facilities must include explicit funding for ongoing operational costs (e.g., staffing/technicians, maintenance). Further there is need to attract, train and retain technicians and PhD-level staff to make the best use of the facility equipment. To continue the rapid progress in nanoscale science/engineering/technology it is necessary to invest in new nanofabrication, characterization, and computational tools and infrastructure, not only for research but also for educational needs. There must also be attention to standardized nanosafety protocols and guidelines.

9.3.2 Education

For K–12, the nanoscale can provide illustrations of multidisciplinary approaches toward the solution of social needs, a useful approach to avoid over specialization in students at their formative stage. Authenticated, peer-reviewed education aids should be made available, including the use of social media tools. University undergraduate and graduate students should be encouraged to visit K–12 classes.

At the University level, for students to be proficient at the nanoscale they need a working knowledge in physics, chemistry, engineering and biology – a challenge in our discipline centric education systems. At the undergraduate level, elective courses (including laboratories) are recommended, but they should have high scientific rigor (superficial courses would not be recommended). A web-based portal for authenticated, peer-reviewed teaching materials would have high value. Both joint nanoscience-journalism/communication programs and continuing education may help coping with the dwindling number of science journalists and teachers.

To better inform the public it will be necessary to be proactive in popular media, i.e. do not simply react to negative events. Nongovernmental organizations and activist groups need to be engaged in the dialog on nanosafety, and this may require different approaches in different countries based on local cultural mores. Informal science education (e.g., museums, mobile displays) is an important means for disseminating the successes of nanoscience to the general public.

References

1. NSTC/NSET, *The National Nanotechnology Initiative: The Initiative and its Implementation Plan*. (National Science and Technology Council, Washington, DC, July 2000). Available online: http://www.nano.gov/node/243
2. M.C. Roco, R.S. Williams, P. Alivisatos (eds.), Nanotechnology research directions: Vision for nanotechnology R&D in the next decade (1999). Available online: http://www.wtec.org/loyola/nano/IWGN.Research.Directions/
3. Members of the 2005 "Rising Above the Gathering Storm" Committee, *Rising Above the Gathering Storm, Revisited: Rapidly Approaching Category 5* (National Academies Press, Washington, DC, 2010). ISBN 10: 0-309-16097-9
4. National Science Board (2010), Preparing the Next Generation of STEM Innovators: Identifying and Developing Our Nation's Human Capital, Arlington VA, National Science Foundation (NSB 10–33).
5. President's Council of Advisors on Science and Technology (PCAST), Report to the President and Congress on the third assessment of the National Nanotechnology Initiative (Executive Office of the President, Washington, DC, 2010). Available online: http://www.nano.gov/html/res/otherpubs.html
6. M.C. Roco, Nanotechnology: A frontier for engineering education. Int. J. Eng. Ed. **18**(5), 488–497 (2002), Available online: http://www.ijee.dit.ie/articles/Vol18-5/IJEE1316.pdf
7. J. Murday. NSF workshop report: Partnership in nanotechnology education, Los Angeles, 26–28 April 2009, Available online: http://www.nsf.gov/crssprgm/nano/reports/educ09_murdyworkshop.pdf
8. M.C. Roco, Nanoscale Science and Engineering Education. Keynote presentation at the workshop Partnership for Nanotechnology Education, Los Angeles, 26 April 2009, Available online: http://www.nsf.gov/crssprgm/nano/reports/nni_09_0426_nanoeduc_usc_35sl.pdf
9. M.C. Roco, W.S. Bainbridge (eds.), *Converging Technologies for Improving Human Performance: Nanotechnology, Biotechnology, Information Technology and Cognitive Science* (Kluwer, Dordrecht, 2003). Available on line: http://www.wtec.org/ConvergingTechnologies/Report/NBIC_report.pdf
10. J.S. Murday, R.W. Siegel, J. Stein, J.F. Wright, Translational nanomedicine: Status assessment and opportunities. Nanomedicine **5**, 251–273 (2009)
11. Project NanoRoadSME (EU Sixth Framework Programme), Roadmap report concerning the use of nanomaterials in the medical and health sector (2006). Available online: http://www.nanoroad.net/index.php?topic=indapp#hm

12. K. Riehmann, S.W. Schneider, T.A. Luger, B. Godin, M. Ferrari, H. Fuchs, Nanomedicine—challenge and perspectives. Angew. Chem. Int. Ed. Engl. **48**, 872–897 (2009)
13. NSTC/NSET, *Nanotechnology and the environment*. Report of the NNI Workshop, 8–9 May 2003, Arlington. Available online: http://www.nano.gov/NNI_Nanotechnology_and_the_Environment.pdf
14. U.S. Environmental Protection Agency (USEPA), Research advancing green manufacturing of nanotechnology products (2009), Available online: http://www.epa.gov/nanoscience/quickfinder/green.htm
15. B. Karn, T. Kuiken, M. Otto. Nanotechnology and *in situ* remediation: A review of the benefits and potential risks. Environ. Health Perspect. **117**(12), 1823–1831 (2009), Available online: http://www.nano.gov/html/about/symposia.html
16. Nanotechnology, Science, and Engineering Subcommittee of the Committee on Technology of the National Science and Technology Council (NSET), National Nanotechnology Initiative: Supplement to the President's FY 2011 Budget. (NSET, Washington, DC, 2010), Available online: http://www.nano.gov/html/res/pubs.html
17. M.C. Roco, W.S. Bainbridge (eds.), *Nanotechnology: Societal implications* (Springer, New York, 2007)
18. National Center for Learning and Teaching in Nanoscale Science and Engineering Education (NCLT) (2006) Nanoconcepts in higher education. Presentations and findings of the second annual NCLT faculty workshop, 6–9 August 2006, California Polytechnic State University, San Luis Obispo, Available online: http://www.nclt.us/workshop/ws-faculty-aug06.shtml
19. National Center for Learning and Teaching in Nanoscale Science and Engineering Education (NCLT) (2008a). Best practices in nano-education. Presentations of the third annual NCLT faculty workshop, 26–29 March 2008, Alabama A&M University, Huntsville, Available online: http://www.nclt.us/workshop/ws-faculty-mar08.shtml
20. National Center for Learning and Teaching in Nanoscale Science and Engineering Education (NCLT), (2008b) Report of the global nanoscale science and engineering education workshop, Washington, DC, 13–14 Nov 2008, Available online: http://www.nclt.us/gnseews2008/
21. Hart Research Associates. Nanotechnology, synthetic biology, & public opinion: A report of findings based on a national survey of adults (2009), Available online: http://www.nanotechproject.org/publications/archive/8286/
22. B. Flagg, V. Knight-Williams. Summative evaluation of NISE Network's public forum: Nanotechnology in health care (2008), Available online: http://www.nisenet.org/catalog/evaluation/summative_evaluation_nise_networks_public_forum_nanotechnology_health_care
23. D. Berube, C. Cummings, J. Frith, A. Binder, R. Oldendick. Contextualizing nanoparticle risk perceptions. J. Nanopart. Res. (2010 Forthcoming)
24. S.W. Dugan, R.P.H. Chang, Cascade approach to learning nano science and engineering: A project of the National Center for Learning and Teaching in Nanoscale Science and Engineering. J. Mater. Educ. **32**(1–2), 21–28 (2010)
25. National Science Foundation (NSF). GK-12: Graduate STEM Fellows in K-12 Education (2010a), http://www.nsfgk12.org/
26. National Science Foundation (NSF), IGERT: Integrative Graduate Education and Research Traineeship (2010b). Available online: http://www.igert.org/
27. Carnegie Corporation of New York and Institute for Advanced Study (Carnegie-IAS), *The Opportunity Equation: Transforming Mathematics and Science Education for Citizenship and the Global Economy*. (Carnegie Corporation of New York-Institute for Advanced Study Commission on Mathematics and Science Education, New York, 2009), Available online: http://www.OpportunityEquation.org
28. National Governors Association (NGA), Council of Chief State School Officers (CCSSO), and Achieve, Inc., Benchmarking for success: Ensuring U.S. students receive a world-class education (2008), Available online: http://www.nga.org/Files/pdf/0812benchmarking.pdf
29. S. Sjoberg, C. Schreiner, The next generation of citizens: Attitudes to science among youngsters, in *The Culture of Science—How Does the Public Relate to Science Across the Globe?* ed. by M. Bauer, R. Shukl (Routledge, New York, 2010), Available online: http://www.ils.uio.no/english/rose/.
30. National Research Council (NRC) National Committee on Science Education Standards and Assessment, *National Science Education Standards* (National Academies Press, Washington, DC, 1996)

31. C.J. Lee, D.A. Scheufele, The influence of knowledge and deference toward scientific authority: A media effects model for public attitudes toward nanotechnology. J. Mass Commun. Q. **83**(4), 819–834 (2006)
32. D. Tapscott, *Grown up digital: How the net generation is changing your world* (McGraw Hill, Columbus, 2009)
33. X. Li, H. Chen, Y. Dang, Y. Lin, C.A. Larson, M.C. Roco, A longitudinal analysis of nanotechnology literature, 1976–2004. J. Nanopart. Res. **10**, 3–22 (2008)
34. National Science Foundation (NSF). NSF's Cyberinfrastructure vision for 21st century discover (2008), Available online: http://www.nsf.gov/pubs/2007/nsf0728/index.jsp
35. P.M. Hallacher, S.J. Fonash, D.E. Fenwick, The Pennsylvania Nanofabrication Manufacturing Technology (NMT) partnership: Resource sharing for nanotechnology workforce development. Int. J. Eng. Ed. **18**(5), 526–531 (2002)
36. S.J. Fonash, Nanotechnology and economic resiliency. Nanotoday **4**(4), 290–291 (2009)
37. G.P. Lansbergen, R. Rahman, C.J. Wellard, I. Woo, J. Caro, N. Collaert, S. Biesemans, G. Klimeck, L.C.L. Hollenberg, S. Rogge, Gate-induced quantum-confinement transition of a single dopant atom in a silicon FinFET. Nat. Phys. **4**, 656–661 (2008). doi:10.1038/nphys994
38. S. Wang, S.H. Lau, A. Tkachuk, F. Druewer, H. Chang, M. Feser, W. Yun, Nano-destructive 3D imaging of nano-structures with multi-scale X-ray microscopy, in *Nanotechnology 2008: Materials, Fabrication, Particles, and Characterization. Technical proceedings of the 2008 NSTI nanotechnology conference and trade show*, Vol. 1 (Nano Science and Technology Institute, Cambridge, 2008), pp. 822–825, Available online: http://www.nsti.org/procs/Nanotech2008v1
39. M.T. Postek, A.E. Vladar, Helium ion microscopy and its application to nanotechnology and nanometrology. Scanning **30**(6), 457–462 (2008)
40. S.V. Sreenivasan, Nanoscale manufacturing enabled by imprint lithography. MRS Bull. **33**(9), 854–863 (2008)
41. A. Hullmann. A European strategy for nanotechnology. Paper presented at the conference on nanotechnology in science, economy, and society, Marburg, 13–15 Jan 2005
42. M. Morrison (ed.). Sixth nanoforum report: European nanotechnology infrastructure and networks (2005), Available online: http://www.nanoforum.org/dateien/temp/European%20Nanotechnology%20Infrastructures%20and%20Networks%20July%202005.pdf?05082005163735
43. H. Dosch, M. Van de Voorde, *Gennesys White Paper: A New European Partnership Between Nanomaterials Science and Nanotechnology and Synchrotron Radiation and Neutron Facilities.* (MPI Inst. Metals Research, Stuttgart, 2009), Available online: http://mf.mpg.de/mpg/websiteMetallforschung/english/veroeffentlichungen/GENNESYS/
44. World Technology Evaluation Center (WTEC) (2010), Long-term impacts and future opportunities for nanoscale science and engineering nanotechnology. Workshop held on 9–10 March 2010, Evanston
45. C.-K. Lee, T.-T. Wu, P.-L. Liu, A. Hsu, Establishing a K-12 nanotechnology program for teacher professional development. IEEE Trans. Educ. **49**(1), 141–146 (2009)
46. M.C. Roco, International perspective on government nanotechnology funding in 2005. J. Nanopart. Res. **7**, 707–712 (2005)
47. J. Hirabayashi, L. Lopez, M. Phillips, The development of the NISE network A summary report (Part I of the summative report of the NISE Network evaluation). Inverness Research, May 2009, Available online: http://www.nisenet.org/catalog/assets/documents/overviewnise-network-evaluation
48. National Science Board. Science and Engineering Indicators (2010), Arlington, VA, National Science Foundation (NSB10–01)
49. Pew, Ideological News Sources: Who Watches and Why. Americans Spending More Time Following the News (2010). http://pewresearch.org/pubs/1725/where-people-get-news-print-online-readership-cable-news-viewers?src=prc-latest&proj=forum
50. Pew, Audience Segments in a Changing News Environment Key News Audiences Now Blend Online and Traditional Sources. Pew Research Center Biennial News Consumption Survey, http://www.pewtrusts.org/our_work_report_detail.aspx?id=42644

Innovative and Responsible Governance of Nanotechnology for Societal Development*

Mihail C. Roco, Barbara Harthorn, David Guston, and Philip Shapira

Keywords Nanotechnology innovation and commercialization • Responsible development • Global governance • Emerging technologies • Societal implications • Ethical and legal aspects • Nanotechnology market • Public participation • International perspective

1 Vision for the Next Decade

1.1 Changes in the Vision over the Last 10 Years

Nanotechnology has been defined as "a multidisciplinary field in support of a broad-based technology to reach mass use by 2020, offering a new approach for education, innovation, learning, and governance" [1]. The governance of nanotechnology development for societal benefit is a challenge with many facets ranging from fostering research and innovation to addressing ethical concerns and long-term human development aspects. The U.S. nanotechnology governance approach

---

*With contributions from: Skip Rung, Sean Murdock, Jeff Morris, Nora Savage, David Berube, Larry Bell, Jurron Bradley, Vijay Arora.

M.C. Roco (✉)
National Science Foundation, 4201 Wilson Boulevard, Arlington, VA 22230, USA
e-mail: mroco@nsf.gov

B. Harthorn
Center for Nanotechnology in Society, University of California,
Santa Barbara, CA 93106-2150, USA

D. Guston
College of Liberal Arts and Sciences, Arizona State University, P.O. Box 875603,
Tempe, AZ 85287-4401, USA

P. Shapira
Georgia Institute of Technology, D. M. Smith Building, Room 107, 685 Cherry Street,
Atlanta, GA 30332-0345, USA

has aimed to be "transformational, responsible, and inclusive, and [to] allow visionary development" [2]. Both domestically and globally, the approach to nanotechnology governance has evolved considerably in the last 10 years:

- The viability and societal importance of nanotechnology applications has been confirmed, while extreme predictions, both pro and con, have receded.
- An international community of professionals and organizations engaged in research, education, production, and societal assessment of nanotechnology has been established.
- From a science-driven governance focus in 2001, there is in 2010 an increased governance focus on economic and societal outcomes and preparation for new generations of commercial nanotechnology products.
- There is greater recognition and specificity given in governance discussions to environmental, health, and safety (EHS) aspects (see in chapter "Nanotechnology Environmental, Health, and Safety Issues") and ethical, legal, and social implications (ELSI) of nanotechnology. Considerable attention is being paid now to regulatory challenges, governance under conditions of uncertainty and knowledge gaps, use of voluntary codes, and modes of public participation in decision making. Overall, there is an increasing focus on "anticipatory governance."
- The vision of international and multinational collaboration and competition [3] has become a reality and intensified since the first International Dialogue on Responsible Development of Nanotechnology was held in 2004 [4].[1]

Through its long-term planning, R&D investment policies, partnerships, deliberate activities to promote public engagement, anticipate the social consequences of scientific practices, and integrate the social and physical sciences, nanotechnology is becoming a model for addressing the societal implications and governance issues of emerging technologies generally [5]. The commercialized nanotechnology innovation that accomplishes economic value for the nations that funded the research requires a supportive investment and workforce environment for manufacturing. Such environment has changed significantly in the last 10 years by transfer of manufacturing capabilities from "West" to "East", and places risk in taking the nanotechnology benefits in the United States and Europe as compared to Asia.

1.2 Vision for the Next 10 Years

Nanotechnology is expected to reach mass applications in products and processes by 2020, significantly guided by societal needs-driven governance. The shift to more complex generations of nanotechnology products, and the need to responsibly address broad societal challenges such as sustainability and health, are prominent.

---

[1] Also see reports of the Japan and Brussels dialogues in 2006 and 2008: http://unit.aist.go.jp/nri/ci/nanotech_society/Si_portal_j/doc/doc_report/report.pdf and http://cordis.europa.eu/nanotechnology/src/intldialogue.htm.

The transition in scientific capability to complex nanosystems and molecular bottom-up nanotechnology-based components will multiply the potential for societal benefits and concerns and will require enhanced approaches to building accountable, anticipatory, and participatory governance with real-time technology assessment:

- *Emphasis is expected to increase on innovation and commercialization* for societal "returns on investment" of nanotechnology in economic development and job creation, with measures to ensure safety and public participation. An innovation ecosystem will be further developed for applications of nanotechnology, including support for multidisciplinary participation, multiple sectors of application, entrepreneurial training, multi-stakeholder-focused research, continuing science to technology integration, regional hubs, private-public partnerships, gap funding, global commercialization, and legal and tax incentives. The balance between competitive benefits and safety concerns needs to be addressed in each economy by considering international context.
- *Nanotechnology will become a general-purpose enabling technology*, which – as with such prior technologies as electricity or computing – is likely to have widespread and pervasive applications across many sectors, combining incremental improvements with breakthrough solutions. Nanotechnology will become critical to commercial competitiveness in sectors such as advanced materials, electronics, and pharmaceuticals. Precompetitive nanoscale science and engineering platforms will provide the foundation for new activities in diverse industry sectors. Multidisciplinary horizontal, research-to-application vertical, regional hubs and system-integrated infrastructure will be developed. As nanotechnology grows in a broader context, it will further enable synthetic biology, quantum information systems, neuromorphic engineering, geoengineering, and other emerging and converging technologies.
- It will become imperative over the next decade to focus not only on how nanotechnology can generate economic and medical value ("material progress"), but also on how *nanotechnology can create cognitive, social, and environmental value ("moral progress")*.
- *Nanotechnology governance will become institutionalized* in research, education, manufacturing, and medicine, for optimum societal benefits.
- *Global coordination will be needed* for international standards and nomenclature, nano-EHS (such as toxicity testing and risk assessment and mitigation) and ELSI (such as public participation in achieving both benefits and safety, and reducing the gap between developing and developed countries). An international co-funding mechanism is envisioned.

2 Advances in the Last 10 Years and Current Status

Just a decade ago, governments, academia, and industry – in the United States and elsewhere in the world – commissioned a massive expansion of research and development in nanotechnology based on a long-term science and engineering vision. Systematic investment in research on societal dimensions of nanotechnology has been undertaken in the United States since 2001, in the European Union (EU) since 2003, in Japan since 2006, and in other countries as well as by international organizations (e.g., the Organization for Economic Co-operation and Development, International Organisation for Standardization, and International Risk Governance Council) since at least 2005. Societal dimensions were included as an essential part of the vision from the beginning of the U.S. National Nanotechnology Initiative (NNI) [6]. Nanotechnology has proven it has essential implications for how we comprehend nature, increase productivity, improve health, and extend the limits of sustainable development, among other vital topics.

2.1 Governance of Nanotechnology

Key challenges to nanotechnology governance have been recognized and implemented. These include developing the multidisciplinary knowledge foundation; establishing the innovation chain from discovery to societal use; establishing an international common language in nomenclature and patents; addressing broader implications for society; and developing the tools, people, and organizations to responsibly take advantage of the benefits of the new technology. To address those challenges, four simultaneous characteristics of effective nanotechnology governance were proposed and have been applied since 2001 [2]. Nanotechnology governance needs to be:

- Transformative (including a results or projects-oriented focus on advancing multi-disciplinary and multisector innovation)
- Responsible (including EHS and equitable access and benefits)
- Inclusive (participation of all agencies and stakeholders)
- Visionary (including long-term planning and anticipatory, adaptive measures)

These characteristics of nanotechnology governance continue to be important and applicable. United States examples of these four governance functions are presented in Table 1.

There is now an international community of scholars addressing not only research and education but also health and safety, ethics, and societal dimensions of nanotechnology. Examples of mechanisms and outputs include the National Science Foundation's "Nanotechnology in Society" network (begun in 2005), journals and publications (e.g., *Nanotechnology Law and Business* and *NanoEthics* journals, *Encyclopedia of Nanoscience and Society* [12]; and editorials in general, research-oriented journals such as *Nature Nanotechnology* and *Journal of Nanoparticle Research*), and the founding of the academic society, the Society for the Study of Nanoscience and Emerging Technologies (S.NET; http://www.the Snet.net) in 2009. From a position in 2000 where "science leaps ahead, ethics lags behind" [13], we are in 2010 in the process of achieving a more appropriate balance between science and ethics. A European Community (EC) "Code of Conduct for Research" has been proposed, but a common terminology and levels of national commitments have still to be reached internationally.

Table 1 Examples of U.S. applications of nanotechnology governance functions (2001–2010)

Nanotechnology governance aspect	Example 1	Example 2
Transformative function		
Investment policies	Support a balanced and integrated R&D infrastructure (NNI Budget requests, 2001–2010; about 100 new centers and networks)	Priority support for fundamental research, nanomanufacturing, healthcare (NIH/NCI cancer research), and other areas
Science, technology, and business policies	Support competitive peer-reviewed, multi-disciplinary R&D programs in NNI agencies	Support for innovation in converging technologies (nano-bio-info-others) at NSF, DOD, NASA
Education and training	Introduce earlier nanotechnology education (e.g., NSF's Nanoscale Center of Learning and Teaching 2005–, Nanotechnology Undergraduate Education 2002–, and K-16 programs)	Nanotechnology informal education extended to museums and Internet (e.g., NSF's Nanoscale Informal Science and Engineering network, 2005–)
Technology and economic trans-formation tools	Support integrative nanotechnology cross-sector platforms (e.g., Nanoelectronics Research Initiative 2004–)	Establish Nanomanufacturing R&D program at NSF in 2002; NSET Nanomanufacturing, Industry Liaison & Innovation working group (NILI), 2005–
Responsible function		
Environmental, health, and safety (EHS) implications	U.S. Congress: Nanotechnology R&D Act of December 2003 includes EHS guidance; OSTP, PCAST, and NRC make EHS recommendations; NNI publishes national strategy for nano-EHS, 2008	Program announcements since 2001 (NSF), 2003 (EPA), 2004 (NIH); NSET Nanotechnology Environmental and Health Implications working group (NEHI), 2005–
Ethical, legal, and social issues and other issues (ELSI+)	Ethics of nanotechnology addressed in publications (Roco and Bainbridge [6] and [7]; NGOs and UNESCO reports, e.g., UNESCO [8])	Program announcements for nano-ELSI (NSF, 2004–); Equitable benefits for developing countries (ETC – Canada 2005; CNS-UCSB [9])
Methods for risk governance	Risk analysis, including the social context, supported by NSF and EPA; applied in EPA, FDA, and OSHA policies	Multilevel risk nanotechnology governance in global ecological system [10]
Regulations and reinforcement	Nanotechnology-focused regulatory groups created at EPA, FDA, and NIOSH	Voluntary measures for nano-EHS at EPA, 2008
Communication and participation	Increased interactions among experts, users, and public at large via public hearings	Public and professional society participation in the legislative process for NNI funding

(continued)

Table 1 (continued)

Nanotechnology governance aspect	Example 1	Example 2
Inclusiveness function		
Partnerships to build national capacity	Foster interagency partnerships (25 agencies); industry-academe-state-Federal government partnerships (NNI support for three regional-local-state workshops)	Partnering among research funding and regulatory agencies for dealing with nanotechnology implications in the NSET Subcommittee and NEHI Working Group
Global capacity	International Dialogue Series on Responsible Nanotechnology (2004, 2006, 2008) initiating new activities; Follow-up on OECD, ISO, UNESCO	International Risk Governance Council reports on all nanotechnology and on food and cosmetics [10]
Public participation	Public input into R&D planning for nanotechnology EHS and ELSI after 2005	Combined public and expert surveys; public deliberations; informal science education (e.g., NSF)
Visionary function		
Long-term, global view	*Nanotechnology Research Directions* books (1999 & 2010); these inform the strategy of the U.S., EU, Japan, Korea, China, and other countries	Long-term effect of technology on human development (*Humanity and the Biosphere*, FFF and UNESCO [11]
Support human development, incl. sustainability	Research on energy and water resources using nanotechnology (DOE, NSF, EPA, others)	Research connecting nervous system, nanoscale physico-chemical mechanisms, brain functions, and education (NSF, NIH)
Long-term planning	Ten-year vision statements published for 2001–2010 (published in 2000) and 2011–2020 (this report, 2010)	NNI strategic plans every 3 years (last three in 2004, 2007, and 2010), followed by PCAST and NRC evaluations

On EHS-related issues (in chapter "Nanotechnology Environmental, Health, and Safety Issues"), the international research community has been implementing integrative work that brings together physical sciences and social sciences. Voluntary reporting schemes have been introduced, albeit with limited impact (e.g., via the U.S. Environmental Protection Agency in the United States, the California Department of Toxic Substance Control, and the Department of Farming and Rural Affairs in the UK). Standardization and metrology progress is taking place (see in chapter "Enabling and Investigative Tools: Measuring Methods, Instruments, and Metrology"). However, innovation is moving ahead of regulation, in part because regulatory bodies are waiting for standards (nomenclature, traceability methods, etc.). Two approaches are being developed in parallel in regulation of nanotechnology:

- Probing the extendibility of regulatory schemes like the Toxic Substances Control Act (TSCA) in the United States and the Registration, Evaluation and Authorization CHemicals Regulation (REACH) Act in the EU (both following a "developing the science" approach)
- Exploring (soft) regulatory and governance models that work despite insufficient knowledge for full risk assessment, including as ELSI research, voluntary codes, public engagement, observatories, public attitude surveys, and other instruments

Overall, the governance of nanotechnology has been focused on the first generation of nanotechnology products (passive nanostructures), with research and studies commencing on the next generations (see their descriptions in the chapter on Long View). Local governance innovations in places like Berkeley (CA), Cambridge (MA), Albany (NY), and in states like New York, California, Oklahoma, and Oregon, have provided "laboratories" for governance, including for regulatory and voluntary approaches. Their ideas have been modeled internationally and offer a perspective for future regional "innovation hubs" recommended later in this chapter.

2.2 Growth of Research and Outreach on Nanotechnology's Impact on Society

The report *Societal Implications of Nanoscience and Nanotechnology* [6] called for the involvement of social scientists from the beginning of the nanotechnology enterprise in large nanotechnology programs, centers, and projects. In 2000 there was very little attention paid to nanotechnology among the community of scholars that studies science and technology from a societal perspective [14]. Research, education, and professional activities in the societal aspects of nanotechnology, supported by the NNI agencies, have made significant progress in a short period of time. Nearly half of all articles on societal dimensions of nanotechnology today have at least one author from a U.S. institution, whereas only about one-quarter of all nanotechnology articles published from 2005 to 2007 had at least one U.S. author.

An early report on converging technologies ([7], xii) recommended that "Ethical, legal, moral, economic, environmental, workforce development, and other societal implications must be addressed from the beginning, involving leading ... scientists and engineers, social scientists, and a broad coalition of professional and civic organizations." There is now widespread agreement that it is better to address early the long-term EHS and ELSI issues related to converging and emerging technologies in a responsible government-sponsored framework but with broad stakeholder input, rather than having to adjust and respond to developments after the fact.

Research on societal implications of nanotechnology has been sponsored by the National Science Foundation (NSF) and other agencies involved in the National Nanotechnology Initiative (NNI) since September 2000, reaffirmed and strengthened by Congress (e.g., in the twenty-first Century Nanotechnology R&D Act of 2003) and National Research Council reports in 2002, 2006, and 2009. The second report by the President's Council of Advisors on Science and Technology on nanotechnology ([15], 38) exhorted NNI agencies to "engage scholars who represent disciplines that might not have been previously engaged in nanotechnology-related research… [and ensure that] …these efforts should be integrated with conventional scientific and engineering research programs." The development of general areas of attention was impacted by NNI funding, particularly funding through the NSF Nanoscale Interdisciplinary Research Team (NIRT) projects since 2001. The two Centers for Nanotechnology in Society (CNS) at Arizona State University (ASU) and the University of California, Santa Barbara (UCSB), founded by NSF in fall 2005, together with the NIRTs at the University of South Carolina-Columbia and Harvard, constitute a network for nanotechnology in society. Table 4 in Sect. 8.2 illustrates the considerable NSF investment in research and outreach on nanotechnology's impact on society. In March 2010, the NNI sponsored an EHS "Capstone" workshop that incorporated ELSI into discussions of how to shape the Federal investment in research on the environmental implications of nanotechnology.

2.3 Nanotechnology Innovation and Commercialization

New forms of organization and business models may originate with nanotechnology, in support of innovation. Innovation in nanotechnology generally involves a complex value chain, including large and small companies, research organizations, equipment suppliers, intermediaries, finance and insurance, ends users (who may be in the private and public sectors), regulators, and other stakeholder groups in a highly distributed global economy [16–18]. Most nanotechnology components are incorporated into existing industrial products to improve their performance.

Between 1990 and 2008, about 17,600 companies worldwide, of which 5,440 were U.S. companies, published about 52,100 scientific articles and applied for about 45,050 patents in the nanotechnology domain [19]. The growth in the number of patents and publications worldwide by private and public organizations has had a quasi-exponential trend since 2000 [20]. The ratio of corporate nanotechnology patent applications to corporate nanotechnology publications increased noticeably from about 0.23 in 1999 to over 1.2 in 2008; this changing ratio indicates a shift in corporate interest from discovery to applications. While most patents in nanotechnology are filed by large companies, small and medium-sized enterprises (SMEs) have increased their patent filings. For example, the proportion of World Intellectual Property Office Patent Cooperation Treaty patents in nanotechnology filed by U.S. SMEs compared with U.S. large companies increased from about 20% in the late 1990s to about 35% by 2006 [21].

The nanoscale science and engineering (NSE) patents authored by NSF grantees receiving support for fundamental research have a significantly higher citation index than all NSE patents [22]. This underlines the importance of fundamental research in the overall portfolio. Wang and Shapira [23] identified about 230 new nanotechnology-based venture start-ups formed in the United States through to 2005, about one-half being companies that had spun off from universities.

The broad nature of nanotechnology indicates that many geographical regions will have opportunities to engage in the development of nanotechnology. For example, while leading high-technology regions in the United States (such as the areas of San Francisco-Palo Alto and Boston) are at the forefront of nanotechnology innovation, other U.S. cities and regions also have clusters of corporations engaged in nanotechnology innovation. There is an extensive corridor of corporate nanotechnology activity along the East Coast, and there are multiple companies engaged in nanotechnology innovation in other traditional industrial areas of the Northeast and Midwest. Southern California also has prominent clusters of corporate nanotechnology activities, with emergent clusters also developing in the U.S. South (Fig. 1a).

In the period 1990–2009, 20 leading countries accounted for 93.8% of the 17,133 corporate publication/patent entries from 87 countries (Fig. 1b). The countries of the Organisation for Economic Co-operation and Development (OECD) together accounted for the major share of the world's corporate activity in nanotechnology publications and patents during that period. All of the OECD had 14,087 entries, of which 4,330 were from European OECD members. (All of the European Union countries combined had 4,390 entries.) Of the non-OECD countries, Japan and China dominated, with Taiwan, Russia, Brazil, and India also making distinguishable contributions to the total. The United States had 5,328 entries, Japan had 2,029 entries, and China had 1,989 entries.

A key factor for commercialized innovation and economic development is the nanotechnology development and "general technology development strength" of each nation [24]. The nations were ranked after those criteria. In nanotechnology development, the U.S. is the largest contributor followed by Japan and Germany. After the "general technology development strength", Korea, Japan, and Taiwan are best positioned, while the U.S. is close to the middle of 19 surveyed countries.

The balance between competitive benefits and safety concerns needs to be addressed in each country by considering international context. There is a risk to innovation-based prosperity and this has to be evaluated by considering the ensemble of societal effects.

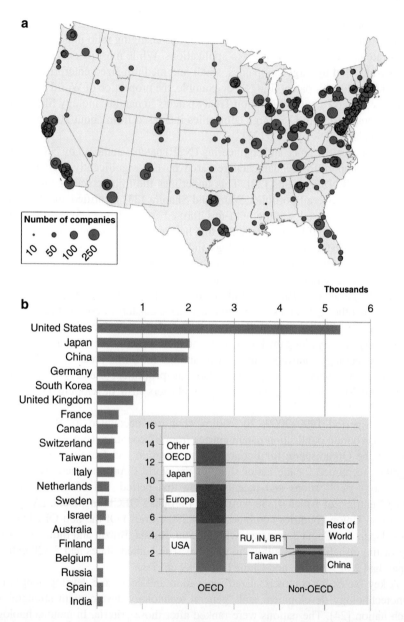

Fig. 1 Distribution of corporate entries into nanotechnology in the United States and other leading countries, 1990–2009. Analysis of companies reporting nanotechnology publications and/or patent records (applications or grants, all patent offices, 1990–July 2008), based on Georgia Tech global database of nanotechnology publications and patents. Cities with ten or more companies with entry into nanotechnology are mapped: (**a**) United States; (**b**) Leading countries and blocs; OECD indicates the 33 member countries of the Organisation for Economic Cooperation and Development; Europe = 20 European members of OECD with nanotechnology corporate entries; RU, IN, BR = Russia, India, and Brazil. (courtesy of Philip Shapira, Jan Youtie, and Luciano Kay)

Other key factors for innovation and corporate decision making in nanotechnology are recognizing consumers' values, their perceptions of the acceptability of products, and their responses to labeling. Taiwan's "nanoMark" approach recognizes legitimate applications of nanotechnology, and the labeling proposal under consideration by the EU, is focused on protecting the public against potential negative health effects. Consumer perceptions are affected by awareness education and access to information.

2.4 Public Perceptions of Nanotechnology

Surveys show that nanotechnology, when compared to other technologies, is not at the extreme, but close to biotechnology in terms of public perceptions about relative benefits and risks (Fig. 2). A meta-analysis of 22 public surveys conducted from 2002 to 2009 in the United States, Canada, Europe, and Japan found ongoing low levels of public familiarity with nanotechnology, with benefits viewed as outweighing risks by 3 to 1, but also a large (44%) minority who had not yet made up their minds about benefits or risks [25].

Public participation has been a central focus of an increasing amount of research. Upstream risk perception research [26, 27], small-scale informal science education activities with some engagement aspects such as science cafés, and U.S. public engagement activities around nanotechnology, such as Arizona State University's National Citizens' Technology Forum (NCTF) [28] and the comparative U.S.-UK and gender–focused deliberations at University of California–Santa Barbara, have been undertaken. In addition, there has been increasing use of scenarios and other foresight tools (including roadmaps, Delphi studies, etc.) in the last 10 years.

2.5 Prospects for Legislation

Social sciences scholars have scrutinized extant and prospective options for environmental health and safety regulation at the national level (e.g., [29–31]) and in the scientific [32] and industrial workplaces ([33]; the CNS-UCSB Nanotechnology and Occupational Health and Safety Conference 2007; Center for Environmental Implications of Nanotechnology industry survey 2009–2010). Davies [34] prepared a report on legislative aspects related to new generations of nanotechnology products and processes. The Chemical Heritage Foundation commissioned a study of nanomaterials' regulatory challenges across the product life cycle, an important direction for new research [35]. New legislative and regulatory initiatives are likely to focus on nanotechnology's environmental, health, and safety implications, as well as on the new generations of nanotechnology products. Such initiatives will be able to draw on this growing body of research.

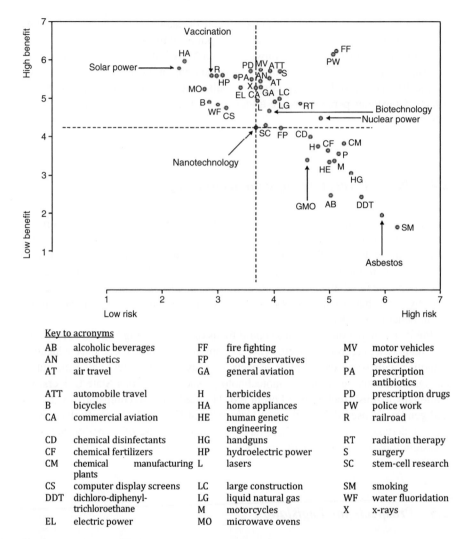

Fig. 2 Survey of public perceptions of nanotechnology products by CBEN (After Currall et al. 2006)

2.6 Addressing Grand Challenges for Societal Development

Nanotechnology may allow us to build a sustainable, society-focused technology through up-front design rather than retroactive problem solving (see in chapters "Nanotechnology for Sustainability: Environment, Water, Food, Minerals, and Climate" and "Nanotechnology for Sustainability: Energy Conversion, Storage, and Conservation"), use of molecular medicine and personalized health treatment (in chapter "Applications: Nanobiosystems, Medicine, and Health"), increased productivity (in Chapters "Synthesis, Processing, and Manufacturing of Components, Devices, and Systems" and "Applications: High-Performance Materials and Emerging Areas"), and early and continuing emphasis on multidisciplinary education (in chapter "Developing the Human and Physical Infrastructure for Nanoscale Science and Engineering").

Evaluating nanotechnology in the context of other emerging technologies is essential for overall development of societal benefit. For example, synthetic biology as an object of social study is perhaps as ill-defined today as nanotechnology was a decade ago. Research on nanotechnology applications has created many opportunities for social engagement in the process, from developing a strong understanding of the dynamics of emergent public perception and public opinion around nanotechnology to proposing new institutional modes of governance of nanotechnology development.

2.7 International Interactions and ELSI

A strategy was proposed in 2000 to create an international scientific R&D community driven by broad human development goals [3]. Many of those goals are still valid in 2010. Several different formats for international dialogue have emerged, each with strengths and limitations. Those formats include the International Dialogues on Responsible Research and Development of Nanotechnology (2004, 2006, 2008), and the Organisation for Economic Co-operation and Development (OECD). The first International Dialogue on Responsible Nanotechnology R&D, held in 2004 (http://www.nsf.gov/crssprgm/nano/activities/dialog.jsp) in Virginia (United States), was the first truly international meeting focused on a long-term view in nanotechnology; it was followed by similar meetings in 2006 in Tokyo (Japan) and in 2008 in Brussels (EU). The 2004 meeting inspired a series of loosely coordinated activities:

- October 2004–October 2005, Occupational Safety Group (UK, United States)
- November 2004, OECD/EHS group on nanotechnology begins
- December 2004, Meridian study for developing countries [36]
- December 2004, Nomenclature and standards (ISO, ANSI)
- February 2005, North–south Dialogue on Nanotechnology (UNIDO)
- May 2005, International Risk Governance Council (IRGC)
- May 2005, "Nano-world," Materials Research Society (materials, education)
- July 2005, Interim International Dialogue (host: EC)
- October 2005, OECD Working Party on Nanotechnology in the Committee for Scientific and Technological Policy (CSTP)
- June 2006, 2nd International Dialogue (host: Japan)
- 2006–2010, Growing international awareness in other national and international organizations of EHS, public participation, education for nanotechnology

Table 2 Websites with ELSI content

CNS at ASU	http://cns.asu.edu
CNS at UCSB	http://cns.ucsb.edu/
NSEC network (Nanoscale Science and Engineering Centers)	http://www.nsecnetworks.org/index.php
American Chemical Society	http://community.acs.org/nanotation/
European Nanotechnology Gateway	http://www.nanoforum.org
Institute of Nanotechnology	http://www.nano.org.uk/
NanoHub	http://nanohub.org/
Nanoscale Informal Science Education Network (NISEnet)	http://www.nisenet.org
NNI Education Center	http://www.nano.gov/html/edu/home_edu.html
National Nanotechnology Infrastructure Network (NNIN) ELSI Portal	http://www.nnin.org/nnin_edu.html
ICON (especially the Good Wiki project), Rice University	http://icon.rice.edu/about.cfm

Differences are noticeable today in the application of nanotechnology on a global scale [37]. Open-source "humanitarian" technology development increasingly is seen as key to nanotechnology applications in the developing world in vital, life-sustaining fields like water, energy, health, and food security (http://nanoequity2009.cns.ucsb.edu/).

The U.S. NNI agencies, followed by the EU, Japan, and Korea, have taken a multipronged approach to funding ELSI projects, which has yielded significant progress over the past decade. International perspectives reflecting opinions from over 40 countries are presented in Sect. 9 of this chapter. Table 2 lists a number of reference websites with ELSI materials addressing the nanoscale; in addition, various nanoscale center efforts listed in chapter "Developing the Human and Physical Infrastructure for Nanoscale Science and Engineering" include ELSI projects.

3 Goals, Barriers, and Solutions for the Next 5–10 Years

3.1 Prepare for Mass Use of Nanotechnology

We are advancing rapidly, but time is needed to grow ideas, people, infrastructure, and societal acceptance for mass application of nanotechnology; we still have only an early understanding of the full range of nanotechnology applications. Significantly, questions about the viability of nanotechnology applications are shifting to questions about how nanotechnology can address broad societal challenges in responsible ways. Global conditions that might be addressed by mass use of nanotechnology include population increase and aging; constraints on using common resources such as water, food, and energy; the competitive challenges and opportunities created by the growth of emerging countries such as Brazil, Russia, India, and China; and convergence with other emerging technologies such as modern biology, digital information technologies, cognitive technologies, and human-centric services. Such scientific, technological, and global societal changes require deep and cross-cutting actions over the next 10 years, creating the need for:

- An ecology of innovation specific to nanotechnology development
- Partnerships across disciplines, application sectors, and between and within regions
- A clear regulatory environment
- An international cross-domain informational system
- International organizations to promote common development aspects of nanotechnology R&D
- Greater cultural and political openness and commitment to international collaboration

3.2 Address Deficits in Risk Governance for the Next Generation of "Nanoproducts" as a Function of the Generation of the Product

In the next 10 years, we may see the emergence of early third- and fourth-generation nanotechnology-based devices and systems [38] (see also chapter on Long View). We have already seen the transition from first-generation passive nanotechnology products to second-generation active nanotechnology applications [39]. These shifts will present different and increased opportunities for societal impacts. They also will require enhanced approaches for governance and risk assessment and the further integration of anticipation, accountability, and open governance into R&D and innovation policies and programs. The main risk-governance deficits for the second to fourth generations of nanoproducts (including active nanodevices, nano-bio applications, and nanosystems) are the uncertain and/or unknown implications of the evolution of nanotechnology and its potential effects on people (e.g., human health, changes at birth, understanding of brain and cognitive issues, and human evolution); environmental effects across nanomaterial life cycles; and the lack of frameworks through which organizations and policies can address such uncertainties.

Governance approaches will need to evolve for new generations of nanotechnology products and productive processes, reflecting the increases in complexity and dynamics of nanostructured materials, devices, and systems (Fig. 3). Each product generation has its own unique characteristics: passive nanostructures, active nanostructures, complex nanosystems, and molecular nanosystems. Likewise, the four levels of risk-related knowledge shown in Fig. 3 and the associated technologies lead to the involvement of different types of actors and anticipate particular types of discourses.

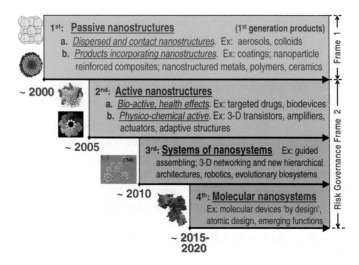

Fig. 3 Timeline for the beginning of industrial prototyping and commercialization of nanotechnology: Four generations of products and production processes [40]

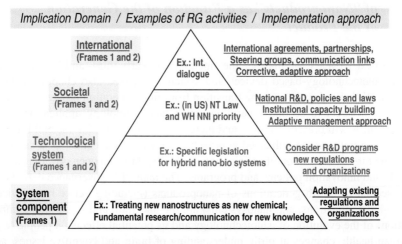

Fig. 4 Schematic for multilevel structure of risk governance for nanotechnology (NT stands for nanotechnology, WH for White House; after [2])

Between the first generation of nanoscale products and associated processes (referred to in Fig. 3 as Risk Governance "Frame 1") and the following three generations ("Frame 2"), there is a natural division in the level of risk. Knowledge of nanostructure behavior is better established for Frame 1, and the potential social and ethical consequences are expected to be more transformative for Frame 2 [40].

Figure 4 presents an attempt to categorize the levels of governance for the responsible function, mapping them to relevant risk-governance activities. Issues related to changes within nanoscale components of larger systems used in applications (such as nanoparticles in automobile paint) typically can be addressed by adapting existing regulations and organizations to the respective systems. Issues related to changes in a technological system (such as a new family of nanobiodevices and active nanostructures) can be best addressed by creating new R&D programs, setting new regulatory measures, and establishing suitable new organizations.

At the national level, typical risk governance actions include formulation of policies and enactment of legislation, which may be considered as we advance to nanosystems. At the international level, typical actions are international agreements, collaborative projects, and multi-stakeholder partnerships, which are needed as we advance to the third and forth generations of nanotechnology-based products, systems, and processes.

Specific risk deficits are associated with the second to fourth generations (Frame 2), due to their expected complex and/or evolving behavior [41]:

- There are uncertain or unknown implications, mostly because the products are not yet fabricated.
- There is limited knowledge on hazards and exposures and specific metrology.
- The institutional deficits (societal infrastructure, political system) are related to fragmented structures in government institutions and weak coordination among key actors.
- Risk communication deficits, i.e., significant gaps exist between distinct science communities and between science communities and manufacturers, industries, regulators, NGOs, the media, and the public.

The risks in Frame 2 are primarily related to assessment of the more complex behaviors of nanomaterials and prioritization of stakeholder concerns, which rest in part on value judgments:

- Risks to human biological and societal development
- Risks due to social structures: risks may be dampened but also induced and amplified by the effects of social and cultural norms, structures, and processes
- Public perception risks
- Trans-boundary risks: the risks faced by any individual, company, region, or country, which depend not only on their own choices but also on those of others

Risk-related knowledge may be simple risk, component complexity, system uncertainty, and/or ambiguity as a function of nanotechnology generation. Roco and Renn [41] proposed a risk management escalator (Fig. 5) as a function of the nanotechnology product generation. This gives a broad overview of the challenges and potential solutions to risk management and governance in the coming 10 years.

3.3 Create New Models for Innovation in Nanotechnology

Proposals by industry and NGOs for policy changes to facilitate innovation in the United States in nanotechnology include: increasing R&D tax credits, increasing

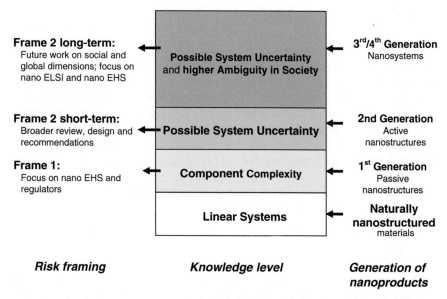

Fig. 5 Strategies as a function of the generation of nanoscale products (Fig. 3): Application to risk governance Frame 1 and Frame 2 [41]

support for precompetitive R&D, measures to provide capital for nanotechnology businesses, and changes in visa regulations to ensure access to highly-skilled technical talent [42, 43]. However, by themselves, such policies are unlikely to have major effects on the trajectories of nanotechnology innovation or to ensure that nanotechnology innovation addresses societal as well as economic objectives. To reach nanotechnology's full potential over the next decade, it is vital to combine economic support with meaningful incentives and frameworks to ensure responsible development that, besides technological and business goals, also addresses societal goals.

One promising model is development of regional multidisciplinary translational nanotechnology innovation hubs. These would undertake activities and develop networks to combine corporate and public sector users, researchers, EHS experts, and other stakeholders in strategies to stimulate, qualify, and diffuse nanotechnology innovation to meet societal goals. These hubs should also exploit complementary opportunities to engage traditional industries in nanotechnology-enabled innovation strategies, also involving manufacturing extension centers, universities, and other technology deployment capabilities. There may be regional opportunities to integrate translational nanotechnology innovation hubs with efforts to foster "nanoclusters" and "nanodistricts"; take an urban and regional systems approach to facilitate responsible innovation; and foster workforce training and development. There will be needs for informed decision making, clarity, anticipation, and coordination in regulatory processes to reduce uncertainty that will constrain nanotechnology innovation, yet also to ensure responsible and prudent development if those applications that may raise EHS concerns. At the same time, there will be opportunities for international and transnational collaboration to harmonize standards that will be helpful to the development of international markets for nanotechnology applications.

It is also vital to support the development of regional and state models for nanotechnology innovation. Since the establishment of the NNI in 2001, numerous state, regional, and local partnerships have been established, solely or as part of broader initiatives, to support and advance nanotechnology innovation and commercialization. (Seven major categories of partnerships, with representative examples, are noted in Sect. 8.1) In addition, there are some cross-state consortia backed by both academia and industry that are focused on advancing specific nanotechnology applications, such as the Western Institute of NanoElectronics. During the next 10 years, there will be a much greater emphasis on developing new models to support nanotechnology-based innovation and commercialization, on the societal returns to investment in nanotechnology, as well as on new measures to ensure safety. Establishment of public-private partnerships not only provides support for technical and commercial advances but also provides new means to engage the public in development of applications that are fundamentally translational; these emerging models may solve many problems of communicating with the public.

There will be multiple pathways through which nanotechnology innovation will be deployed and have impacts in and for industry between now (2010) and 2020. Nanotechnology is emerging as a general purpose technology, as initially proposed in the 1999 *Nanotechnology Research Directions* report and confirmed by later developments [18]. Early forecasts held that nanotechnology would contribute to approximately 10% of global manufacturing output by 2015 [6, 44]. The 2008–2010 global financial crisis and economic slowdown is temporarily dampening the current pace of nanotechnology's growth [45], but is not changing the underlying trajectory of development. In the near term, many of the innovations induced by nanotechnology are leading to both incremental improvements of existing products and, increasingly over time as we get closer to 2020, they are expected to lead to revolutionary architectures and functions.

3.4 Prepare Workers and the Public at Large for Nanotechnology Development

As the scale and scope of nanotechnology innovations picks up over the coming decade, there will be significant implications for employment and training (addressed in detail in chapter "Developing the Human and Physical Infrastructure for Nanoscale Science and Engineering"). The pervasive, general-purpose nature of nanotechnology means that impacts will be seen across all industry sectors. Whether in mature sectors such as plastics or packaging or in leading-edge industries such as electronics or aerospace, companies that lag in awareness, understanding, and applications of nanoscale materials, processes, and devices to their current and future lines of products and services are apt to be at a competitive disadvantage, with consequent risks to business survival and employment. At the same time, new jobs are likely to be created in enterprises of any size that can best identify and exploit the commercial opportunities that nanotechnology presents.

In this context, access to workers who have the skills to develop, acquire, produce, and manage nanotechnology-enabled innovations will be vitally important. It is important to ensure that those who will develop, apply, manage, and oversee innovations in nanotechnology are not only technically well-trained but also well-prepared to anticipate and address broader implications. Employees in corporate public, legal, and regulatory affairs and areas other than R&D will need increased knowledge of nanotechnologies as well.

3.5 Advance R&D Related to Ethics and Understanding of Societal Dimensions of Nanotechnology

The principal needs in the next 5–10 years relating to ethics and understanding of societal dimensions of nanotechnology are:

- A comprehensive understanding of nanotechnology in society, investigated by including "what goes into nanotechnology" (economic and social drivers, public expectations, cultural values, aspirations, etc.), in addition to "what comes out of nanotechnology" (applications and their effects)
- Integration of nanotechnology ELSI considerations into educational processes, including in-depth school curricula for interested students and establishing ELSI relationships to the processes of innovation and assessment (safety by design, responsible innovation)
- Global harmonization of traceability of measurement methods in standards and metrology; coordination of regulatory standards
- Integration of "life-cycle approaches" to materials testing (based on pre- and post-market product-testing, rather than predominantly on pre-production testing) (in chapter "Nanotechnology Environmental, Health, and Safety Issues")
- Implementation of "principles of green nanotechnology" – design principles for sustainability in light of life-cycle considerations (in chapter "Nanotechnology for sustainability: Environment, Water, Food, Minerals, and Climate")

3.6 Integrate Research for Applications and Implications of Nanotechnology

The approaches to nano-EHS and nano-life (such as biology, medicine, technology) science research need to be unified under the single objective of obtaining a rich understanding of the interactions of well-characterized engineered nanomaterials with biological systems. Integration of transformative and responsible aspects of nanotechnology in a unified R&D program is a priority.

3.7 Make Moral Progress

In the future, technological and economics decision making should consider larger issues of "moral progress."[2] Research on ethical, legal, and social issues is vital to understanding how to create social and environmental value in the development of science and technology (e.g., see [25–28, 46, 47]), which includes development of processes to address the diversity of views across different publics.

3.8 Build a Network for Anticipatory, Participatory, and Adaptive Technology Assessments

Aspects of technology assessment have been initiated since 2000, under the long-term planning and implementation of the NNI and open to the participation of major stakeholders. That long-term vision has been credited for the national and then global focus on nanotechnology R&D.

Participatory technology assessment is essential to responsible nanotechnology development. It has been proposed to establish a network to conduct participatory technology assessment activities that:

- Harness education, deliberation, and reflection to give a voice to everyday citizens who otherwise have minimal representation in the politics of science and technology
- Enable decision makers to take into account the informed views of their constituents regarding emerging developments in science and technology

A participatory nanotechnology assessment network would work with decision makers to identify timely and relevant topics for assessment, engage experts and the public nation-wide, facilitate in-depth learning and deliberative processes for thousands of participants, and disseminate the results to a general public audience of millions and to key decision makers. The home for this network could be a nonpartisan, policy research institution that can serve as an institutional link to government, eliciting input on technology assessment topics and functioning as a venue for disseminating results. The network would incorporate university participants who bring strengths in conceptual and methodological development in technology assessment methods, contributing to technical and social analysis, organizing participatory technology assessment

---

[2]Term coined by Susan Neiman, as quoted in "Why is the modern view of progress so impoverished?" (Onwards and Upwards section), *The Economist*, 19 Dec 2009.

exercises, and evaluating technology assessment projects. The network should also incorporate organizations (including science museums, science cafes, and citizen groups) that have capabilities in citizen engagement, collaboration with schools, and broad public education concerning science, technology, and society issues.

4 Scientific and Technological Infrastructure Needs

Nanotechnology infrastructure needs will change as a function of external conditions such as developments in other emerging technologies, increased requirements for sustainable development in a more crowded world, health and particularly aging, and globalization. A single top-down centralized investment approach may not be able to address such complexity. Several new infrastructures are needed, for both producers and users of nanotechnology, to enhance participation of the general public in decision making, inform policies, and expand international context. It will be necessary to assess business-to-consumer and business-to-business public nanotechnology product inventories, and create and test models of stakeholder engagement using emerging alternatives to the newspaper such as social media and Web 2.0 platforms. Other needs include:

- Horizontal integration of institutes and laboratories in safety, metrology, and societal implications research
- Establishment of platforms for international exchange on best practices, such as formal international traceability of measurement infrastructure, including an accreditation system
- Support for activities and infrastructure to connect the Global South to nanotechnology advancements to create better economic, health, and living conditions for the world's poor

5 R&D Investment and Implementation Strategies

Changing the focus of nanotechnology research from the Bohr and Edison quadrants in 2001–2010 to the Pasteur quadrant after 2010 (quadrants defined by Stokes [48]) has direct implications for R&D strategies:

- Platforms for R&D and innovation in nanotechnology need to be strengthened via:
    - Short and long-term framework policies and strategies to address manufacturing, healthcare, sustainable development, communication, and other societal needs
    - Regional capabilities and opportunities bringing together different stakeholders

- Linking innovation with society and equity in access and distribution of benefits
  - Cross-discipline, cross-sector information system on research, innovation and production
- Infrastructure for commercialization needs to be strengthened via:
  - Federal Government and state R&D investment and coordination
  - Regional partnerships
  - Public-private partnership platforms for precompetitive R&D and innovation in nanotechnology
- Continuity of investment for fundamental and applied research in this long-term initiative, and institutionalizing the R&D programs and funding mechanisms for nanotechnology
- Increased international exchanges, based on mutual benefit, to address opportunities for global R&D collaboration and competition
- Shift of ELSI work in the direction of probing citizens' expectations of the prosperity enabled by innovation contrasted with fears and objections to the means (land use, factories, tax/regulatory policy, someone might get rich) of accomplishing the necessary "economic value capture" from that innovation
- Development of improved assessment metrics

Table 3 gives several suggested strategies for R&D investment and implementation strategies, grouped by the four basic governance functions discussed earlier (e.g., see Table 1).

6 Conclusions and Priorities

A strong focus is needed in the next 10 years on improving anticipatory and participatory governance for nanotechnology that integrates the four basic functions of being transformative, responsible, inclusive, and visionary.

Improving open-innovation environments and creating better innovation mechanisms for nanotechnology has to be addressed with priority in the next decade as nanoscale science and engineering have established stronger foundations and expectations of societal outcome increase:

- Strengthening an *innovation ecosystem conductive to economic and safe application of nanotechnology*. This includes support for multidisciplinary participation, access to a diverse manufacturing base and multiple sectors of application, encouraging private-public partnerships and integration of capabilities, entrepreneurial training, multi-stakeholder-focused research, R&D platforms with continuing integration from research to technology application, regional hubs, research to commercialization gap funding, facilities for global commercialization, an outcome-drive culture encouraging creativity and innovation, and legal and

Table 3 Suggested function increases for future nanotechnology governance

Transformative function	Emphasize policies to develop focused, system-oriented R&D programs in nanomanufacturing, sustainability, and other priority areas
	Enhance the nanotechnology tools and facilitate the innovation cycle from discovery to invention to business models and to societal needs
	Strengthen priority investment in nanotechnology for human health, regenerating the human body, and maintaining working capacity while aging
	Investigate nanotechnology for sustainable natural resources (water, energy, food, clean environment)
	Develop new organizational and business models, including support for nanoinformatics
	Expand university and community college curricula supporting nanotechnology and converging emerging technologies (e.g., NSF's Nanoscale Center for Learning and Teaching)
	Foster nanotechnology research, education, and production clusters and regional hubs for various application areas to reduce the delay between inventions, technological development, and societal response
	Construct horizontally, vertically, and system-wide integrated infrastructure with open access
	Improve the metrics applicable to all projects and agencies in the United States
	Enhance international information systems to provide all researchers timely information
	Develop and implement informatics tools for nanomaterials, devices, and systems
	Create accreditation boards for traceability (reference materials, laboratories)
Responsible function	Establish research and regulations for the new (third and fourth) nanotechnology generations
	Implement/complete a predictive approach for toxicity of nanomaterials; establish user facilities to implement it
	Build a sustainable nanotechnology through up-front design rather than retro corrections
	Develop new systemic knowledge for a life-cycle approach to nanotechnology products
	Integrate nano-EHS and -ELSI considerations into the research process
	Develop an integrated, validated scientific platform for hazard, exposure, and risk assessment at a scale commensurate with technology growth (see in chapter "Synthesis, Processing, and Manufacturing of Components, Devices, and Systems")
	Sustain and expand the NSF's Nanotechnology in Society Network and create additional infrastructure within other NNI lead agencies
	Develop new methods, such as multicriteria decision analysis (e.g., Linkov et al. [49]; Tervonen et al. [50])
	Investigate nanotechnology for the poor [36]
	Institutionalize coordination of regulatory agencies and research organizations
	Use social science, history, philosophy, and ethics knowledge-base to research nano-ELSI rather than support actions subsidiary to outreach goals, e.g., draw on available theories & analysis of ongoing innovation trajectories

Inclusiveness function	Create public-private partnerships among Federal government, states, industry sectors, academe, and research foundations
	Address social issues of interest to many stakeholders, such as workforce displacement
	Develop a common information exchange domain for industry, researchers, regulators, consumers, general public
	Continuous contribution to OECD working groups on nanotechnology and related emerging technologies
	Global, cross-sector, and open source collaboration in the area of nano-EHS will be essential factors in the introduction of nanotechnology as a general purpose technology.
	To enhance participatory governance, increase the use of public and expert surveys and of emerging platforms of communication such as social media and Web 2.0 platforms
	Fund evidence-based nanotechnology risk communication based on public and expert mental models and risk perception research, media studies, and multi-pathway decision risk analysis
Visionary function	Study changing societal interactions due to converging and emerging technologies
	Develop operational aspects of anticipatory and participatory governance (e.g., [2, 25, 51, 52])
	Forecast long-term potential effects of nanotechnology on global warming; the next 1,000 years [11]
	Prioritize development of nanotechnologies for renewable energy, clean water, public health infrastructure, urban sustainability, and agricultural systems
	Prepare 10 year vision (2011–2020) (this report)
	Transition from a research-centric to a demand/user/application-centric focus

tax incentives. The balance between competitive benefits and safety concerns needs to be addressed in each country by considering international context.
- Create and sustain *mechanisms of innovation* for establishing nanotechnology infrastructure, economic development, job creations, quality of life, and national security. Several examples are:
  - Programs for public-private funding of *industry inspired fundamental and precompetitive research.* Previous examples in the U.S. are the NSF's Industry-University Cooperative Research Centers (IUCRC, since 2001), the Nanoelectronics Research Initiative (NRI, since 2004) partnering the Semiconductor Research Corporation (with NSF and more recently NIST), and NSF and Industrial Research Institute (IRI, in 2010–) program
  - *Focused research programs* where interdisciplinarity and partnering with industry is required (e.g., Nanotechnology Signature Initiatives, NNI, 2011–). Coordinate such programs across the breath and expertise of multiple agencies, through a variety of complementary funding mechanisms
  - *Funding innovation opportunities supplements* to research projects based on the research results obtained in the first half of the respective projects. A previous example in the U.S. is the NSF program solicitation "Grant Opportunities for Academic Liaison with Industry" (GOALI) combined with supplements offered by "Accelerating Innovation Research" (AIR) in 2010.
  - Creation and sustaining of regional public-private partnerships such as university-industry – government-local organizations research centers. Regional partnership models in U.S. are listed in Sect. 8.1
  - Support R&D multidisciplinary/multi-sector platforms with a long-term vision and planning (such as technology roadmaps). For example, in the U.S. the electronic, chemical industry and wood and paper industries have their own nanotechnology roadmaps
  - Support and maintain nanomanufacturing user facilities and education programs. Examples in the US are National Nanotechnology Infrastructure network (NNIN) and Sandia National Laboratory (SNL), and National Nanomanufacturing Network (NNN)
  - "High Tech Extension" is the direct connection of nanotechnology infrastructure to existing businesses, helping them improve existing products, develop new products, and expand employment (Sect. 8.1)
  - "Gap Funding," is accelerated commercialization assistance to entrepreneurial ventures (e.g., SMEs, university and/or corporate spinouts) in the form of technology transfer and early-stage funding on favorable terms (Sect. 8.1)
  - Provide nano-EHS regulatory assistance to companies, especially small and medium size
  - Support access of industry to data bases, research projects, user facilities and international collaboration
  - Provide education and supporting tolls for the introduction of nanotechnology for economical benefit and better paying jobs, to increase penetration of nanotechnology in both emerging and traditional industries

Priority actions in nano-EHS and ELSI for the next decade include the following:

- Integrate social science and humanities work with NSE research.
- Enhance public participation via ongoing, two-way/multi-way dialogues between nanotechnology community and organizations and civic organizations and lay publics. Articulate a new public engagement strategy, including reaching those least educated and those most dependent on Internet sources of information. Organize integrative activities for a broad set of NSE and societal dimensions researchers as well as various publics, including but not limited to scenario development workshops and informal science education. Make NSE experts accessible to policymakers for input.
- Provide more support for co-education of NSE and social science graduate students to develop interdisciplinary institutional cultures and national exchange networks; provide more opportunities to institutionalize and disseminate such practices.
- Develop structured (institutionalized) contexts for two-way communication between the public and researchers, as an important step in educating scientists and engineers about the legitimate bases for public concerns (and ongoing public support for science), as well as in educating the public about science and engineering and nanotechnology.
- Support research on the projected future "nano" workforce and on demographics for key nodes of nanotechnology-based industry development in United States and abroad.
- Give priority to evidence-based nanotechnology risk communication based on public and expert mental models and risk perception research, media studies, and multi-pathway decision risk analysis.
- Adopt an anticipatory, participatory, real-time technology assessment and adaptive governance model for nanotechnology so as to prepare the people, tools, and organizations for responsible development of nanotechnology. Evaluate how well social actors and regulatory institutions are prepared to deal with challenges from nanotechnology developments, e.g., new generation of products, dealing with knowledge gaps, and assignment of drug/device classifications.

Several overall possibilities for improving the governance of nanotechnology in the global self-regulating ecosystem are recommended (refer also to the examples in Table 3):

- Use open-source and incentive-based models
- Build a global, sustainable nanotechnology through up-front design rather than corrective actions
- Empowering stakeholders and promoting partnerships among them
- Implement long-term planning that includes international perspectives
- Institutionalize nanotechnology in research, education, and production processes

- Combine science-based voluntary and regulatory measures for nanotechnology governance and in particular for risk management [53, 54]
- Support an international co-funding mechanism for maintaining databases, nomenclature, standards and patents

7 Broader Implications for Society

This chapter already covers this topic in its main sections. One may underline that governance of nanotechnology is essential in realizing the benefits of the new technology, limiting its negative implications, and enhancing global collaboration. Further, nanotechnology development is interdependent and synergistic with other emerging technologies. Besides its key transformative effects in discovery, innovation, and specific applications, nanotechnology governance affects society at large and international interactions.

8 Examples of Achievements and Paradigm Shifts

8.1 Regional Partnerships in Nanotechnology

Contact person: Skip Rung, Oregon Nanoscience and Microtechnologies Institute (ONAMI)

Since the establishment of the NNI in 2001, numerous state, regional, and local partnerships have arisen, dedicated completely or in part to the advancement of nanotechnology. These partnerships may be grouped into seven major categories:

- State-backed organizations to enhance nanotechnology research capacity and state-funded programs to grow startup companies, with significant, but not exclusive, focus on nanotechnology (e.g., ONAMI and the Oklahoma Nanotechnology Initiative)
- State-funded programs to grow startup companies, some exclusive (e.g., Albany Nanotech) and other with significant, but not exclusive, focus on nanotechnology (e.g., Ben Franklin Technology Partners)
- Academically oriented infrastructure investments by states, including cost-share support from private sources (e.g., California NanoSystems Institute)
- Member-funded state/local trade associations (e.g., Colorado Nanotechnology Alliance)
- Member-funded national/international nanotechnology trade associations (e.g., NanoBusiness Alliance and the Silver Nanotechnology Working Group)
- Industry-sponsored academic-industry consortia (e.g., Western Institute of NanoElectronics)

- Industry-inspired fundamental research for an industry sector (e.g., Nanoelectronics Research Initiative involving NSF since October 2003 and NIST since 2007)

Funding, sustainability, and operational success for these kinds of partnerships can only occur in strong alignment with important stakeholder objectives that are able to out-compete other initiatives seeking public or voluntary private support. In the case of state investment (the majority of cases), the sole motive is economic development, requiring credible results in terms of jobs (ideally) or at least financial leverage. There is increasing pressure for such initiatives to become "self-supporting" (although with private and Federal funds), even in the case of activities for which the state economy is the primary beneficiary.

In the next 10 years, as the NNI increases its emphasis on commercialization, two regional/state initiative activities can be expected to grow in importance. The first activity, "High Tech Extension" (Fig. 6) is the direct connection of nanotechnology infrastructure to existing businesses, helping them improve existing products, develop new products, and expand employment. Easy and economical access to resources such as nanoscale materials characterization can expand the impact of nanoscience to a broader swath of the economy.

The second activity, known as "Gap Funding," is accelerated commercialization assistance to entrepreneurial ventures (e.g., SMEs, university and/or corporate spinouts) in the form of technology transfer and early-stage funding on favorable terms. While SBIR and STTR awards are vital tools in this regard, locally managed capital with an emphasis on launching growth companies is a necessary addition to the portfolio of commercialization programs, and one which lends itself well to Federal partnerships with state/regional initiatives. Federal and state partnerships for the "gap funding" of new ventures that commercialize NNI-funded technology R&D could accelerate commercialization by 2–4 years and ensure a focus on economic

The "High Tech Extension" Concept

Fig. 6 Nanoscience facilities and equipment can best benefit technology development when they are conveniently located and easy to use by businesses. Such access is especially important to the small and medium size enterprises that are critical for early-stage commercialization. State and regional economic development field staff can serve as "high-tech extension" agents

returns and job creation. The "gap" to be traversed with proposed short-term funding assistance is also known as the "valley of death" between business startup and commercial profitability, a particularly risky interim phase for advanced-technology businesses.

8.2 Examples of Research Projects on Societal Implications Established by NSF

Contact person: Mihail C. Roco, National Science Foundation

Table 4 lists the many projects established by the National Science Foundation through 2010 to support research on societal implications of nanotechnology research, development, and commercialization. (A number of these projects also support outreach to inform the American public regarding nanotechnology issues and involve them in governance discussions.)

8.3 Center for Nanotechnology in Society at ASU

Contact person: David Guston, Arizona State University

The Nanoscale Science and Engineering Center/Center for Nanotechnology in Society at Arizona State University (NSEC/CNS-ASU; http://cns.asu.edu) was established on October 1, 2005, with funding from the National Science Foundation. CNS-ASU combines research, training, and engagement to develop a new approach to governing emerging nanotechnology. The center uses the research methods of "real-time technology assessment" (RTTA) and guides them by a strategic vision of anticipatory governance. The anticipatory governance approach consists of enhanced foresight capabilities, engagement with lay publics, and integration of social science and humanistic work with nanoscale science and engineering research and education [55, 56]. Although based in Tempe, Arizona, CNS-ASU has major partnerships with the University of Wisconsin–Madison and the Georgia Institute of Technology, plus a network of other collaborators in the United States and abroad.

CNS-ASU has two types of integrated research programs, as well as educational and outreach activities (which are themselves integrated with research). Its two thematic research clusters, which pursue fundamental knowledge and create linkages across the RTTAs, are "Equity, Equality and Responsibility" and "Urban Design, Materials, and the Built Environment." The Center's four RTTA programs are:

- Research and Innovation Systems Assessment, which uses bibliometric and patent analyses to understand the evolving dynamics of the NSE enterprise
- Public Opinion and Values, which uses surveys and quasi-experimental media studies to understand changing public and scientists' perspectives on NSE

Table 4 Examples of NSF-sponsored projects supporting social implications inquiry, 2001–2010

Project[a]	Institution
Nanotechnology and its Publics	Pennsylvania State University
Public Information and Deliberation in Nanoscience and Nanotechnology Policy (SGER)	North Carolina State University
Social and Ethical Research and Education in Agrifood Nanotechnology (NIRT)	Michigan State University
From Laboratory to Society: Developing an Informed Approach to NSE (NIRT)	University of South Carolina
Intuitive Toxicology and Public Engagement (NIRT)	North Carolina State University
Data base and innovation timeline for nanotechnology	University of California Los Angeles
Social and ethical dimensions of nanotechnology	University of Virginia
Undergraduate Exploration of Nanoscience, Applications and Societal Implications (NUE)	Michigan Technological University
Ethics and belief inside the development of nanotechnology (CAREER)	University of Virginia
All NNIN and NCN centers have societal implications components	All 28 NSF nanotechnology centers and networks
NSEC: Center for Nanotechnology in Society at Arizona State University	Arizona State University
NSEC: Center for Nanotechnology in Society at University of California, Santa Barbara	University of California, Santa Barbara
NSEC: Nanotechnology in Society Project, Nano Connection to Society	Harvard University
NSEC: Center for Nanotechnology in Society: Constructive Interactions for Socially Responsible Nanotechnologies	University of South Carolina
CEIN: Predictive Toxicology Assessment and Safe Implementation of Nanotechnology in the Environment	University of California Los Angeles
CEIN: Center for Environmental Implications of Nanotechnology	Duke University
NNIN: National Nanotechnology Infrastructure Network (10%)	Cornell University
NIRT; Nanotechnology in the Public Interest: Regulatory Challenges, Capacity and Policy Recommendations	Northeastern University
Collaborative Grant: Bringing Nanotechnology and Society Courses to California Community Colleges	University of California, Santa Barbara

[a] Key to abbreviations of project types (in order of appearance): *SGER* Small Grant for Exploratory Research; *NIRT* Nanoscale Interdisciplinary Research Team *NUE* Nanotechnology Undergraduate Education in Engineering; *CAREER* Faculty Early Career Development Award; *NNIN* National Nanotechnology Infrastructure Network; *NCN* Network for Computational Nanotechnology; *NSEC* Nanoscale Science and Engineering Center; *CEIN* Center for the Environmental Implications of Nano-technology

- Anticipation and Deliberation, which uses scenario development and other techniques to foster deliberation on plausible NSE applications
- Reflexivity and Integration, which uses participant-observation and other techniques to assess the center's influence on reflexivity among NSE collaborators

Fig. 7 Participants in the first National Citizens' Technology Forum on Nanotechnology and Human Enhancement, conducted by CNS-ASU in March 2008 (Courtesy of David Guston)

The center's major conceptual-level achievement has been validating anticipatory governance as a richly generative strategic vision. Its three major operations-level achievements are: (1) completing the "end-to-end" assessment to create novel insights in a study of nanotechnology and the brain; (2) deepening the integration of NSE researchers into CNS-ASU; and (3) building collaborations for informal science education (ISE) on the societal aspects of NSE. Programmatic achievements include establishing an internationally adopted definition of nanotechnology to assemble and mine bibliographic and patent databases; conducting two national public opinion polls and a poll of leading nano-scientists; conducting the first National Citizens' Technology Forum on nanotechnology for human enhancement (Fig. 7); demonstrating that interactions between NSE researchers and social scientists can generate more reflexive decisions; sustaining an international research program on NSE and equity; and laying the foundations for a new research program in urban design, materials, and the built environment.

The center's principal intellectual merit derives from the large-scale, interdisciplinary ensemble that underpins it. The ability to embrace and facilitate interactions among disparate approaches to understanding nanotechnology, and to build complementary capacities to tap that knowledge for governance, is the critical intellectual contribution to which CNS-ASU aspires. Both in terms of publications and citations, the center's work has a substantial impact on scholarship. For broader impact, the center has coupled research, education, and outreach activities exceptionally well by training significant numbers of new scholars from the social sciences and nanoscience-based physical sciences, incorporating forefront research in new courses and ISE opportunities, and returning lessons learned and techniques developed for outreach back to the classroom. CNS-ASU has broadened the participation of under-represented groups by cultivating junior scholarship and raising issues of equity, gender, and disability as objects of programmatic study. The center has enhanced the infrastructure for research and education by organizing community-defining conferences, producing community-defining sources of knowledge, serving as an international hub for dozens of scholars, sharing data and instruments widely, and disseminating its results aggressively to its academic peers as well as to public, scientific, industry, and policy audiences.

8.4 Center for Nanotechnology in Society at UCSB

Contact person: Barbara Harthorn, University of California, Santa Barbara

The Center for Nanotechnology in Society at the University of California, Santa Barbara (CNS-UCSB), promotes the study of societal issues connected with emerging nanotechnology in the United States and around the globe. It serves as a national research and education center, a network hub among researchers and educators concerned with innovation and responsible development of nanotechnology, and a resource base for studying these issues in the United States and abroad. The work of the CNS-UCSB is intended to include multiple stakeholders in the analysis of nanotechnology in society and in discussion through outreach and education programs that extend to industry, community, and environmental organizations, policymakers, and diverse publics.

The intellectual aims of CNS-UCSB are twofold: to examine the emergence and societal implications of nanotechnology with a focus on the global human condition in a time of sustained technological innovation; and to apply empirical knowledge of human behavior, social systems, and history to promote the socially and environmentally sustainable development of nanotechnology in the United States and globally. These aims motivate research from many theoretical and methodological perspectives, provide the basis for industry-labor-government-academic-NGO dialogues, and organize the mentoring of graduate, undergraduate, and community college students and postdoctoral researchers.

CNS-UCSB researchers address a linked set of social and environmental issues regarding the domestic U.S. and comparative global creation, development, commercialization, consumption, and regulation of specific nano-enabled technologies for energy, water, environment, food, health, and information technology. The center addresses questions of nanotechnology-related societal change through research that encompasses three linked areas:

- Historical context of nanotechnology
- Nanotechnology and globalization, with an emphasis on East and South Asia
- Nanotechnology risk perception and social response studies among experts and publics; media framing of nanotechnology risks; and methods for engaging diverse U.S. publics in upstream deliberation about new technologies

CNS-UCSB has close ties with the internationally prominent nanoscience researchers at UCSB who are connected with the university's California NanoSystems Institute, Materials Research Laboratory, and National Nanotechnology Infrastructure Network; with ecotoxicology researchers in the UC Center for Environmental Implications of Nanotechnology (UC CEIN); and with social science research centers focused on relations among technology, culture, and society. These ties are enhanced by wider collaborations in the United States and abroad. U.S. collaborators are based at UC Berkeley, Chemical Heritage Foundation, Duke University, Quinnipiac University, Rice University, State University of New York (SUNY) Levin Institute, SUNY New Paltz, University of Washington, and

University of Wisconsin. Collaborators abroad are based at Beijing Institute of Technology, Cardiff University, Centre National de la Recherché Scientifique, University of British Columbia, University of East Anglia, University of Edinburgh, and Venice International University.

CNS-UCSB's novel graduate educational program co-educates societal implications and nanoscale science and engineering students. UCSB graduates in nanoscale science and engineering participate in CNS-UCSB research on, for example, science policy analysis, media coverage analysis, public deliberation, expert interviews on risk and innovation, Chinese patent analysis, and comparative state R&D policies.

8.5 Governance Toward Sustainable Nanotechnology

Contact person: Jeff Morris, U.S. Environmental Protection Agency

One objective of U.S. EPA's Nanomaterial Research Program is to shift thinking and behavior from managing risk to preventing pollution. Preventing pollution is one of main themes in the EPA *Nanomaterial Research Strategy* (http://www.epa.gov/nanoscience), while other themes directly support EPA research to understand what properties of different nanoscale materials may cause them to be, among other things, mobile, persistent, and/or bioavailable. This and other exposure-related information, together with research on what specific nanomaterial properties may influence toxicity, can inform the use of green chemistry and other approaches to foster the responsible design, development, and use of nanomaterials, including nanotechnology uses that directly or indirectly advance environmental protection. In addition to ensuring that existing nanomaterials are environmentally sustainable, EPA also needs to look for creative ways to develop nanomaterials in a sustainable manner.

The environmentally friendly research by EPA seeks to demonstrate how toxic chemicals can be avoided while producing nanoparticles and has been applied to one promising application: technology for cleaning up pollution that uses nanoscale zero valent iron (NZVI) to promote the breakdown of contaminants in ground water. The EPA team began by making NZVI by mixing tea with ferric nitrate. This process did not use any hazardous chemicals, such as sodium borohydride, which is commonly used to make nanoparticles. Not only did the process eliminate the use of hazardous chemicals, but the nanoparticles showed no significant signs of dermal toxicity. The researchers next used grape extract to make high-quality nanocrystals of gold, silver, palladium, and platinum [57]. The message behind this example is that moving toward sustainable nanotechnology means incorporating new thinking into materials research and development. The EPA research may or may not lead to "green nano" materials that can be commercialized. Nevertheless, it demonstrates that it is feasible to synthesize nanoparticles using nontoxic inputs, and that the real limits to the development and application of green chemistry approaches for nanotechnology lie in our own ingenuity.

8.6 Public Participation in Nanotechnology Debate in the United States

Contact person: David Berube, North Carolina State University

Public participation in science and technology debate has been convincingly shown to matter for normative, instrumental, and substantive purposes, and indeed this "participatory turn" is now evident in many countries [58]. In particular, effective public participation can serve a vital instrumental role in development of trust – essential in the nanotechnology case given the uncertainties about safety, extent of benefits, and longer term social risks. The NNI, through the NSF, has supported a number of efforts to include the public in science and technology policy decision making through a number of different formats and programs (see [5, 12]). Activities range from informal science outreach at museums (NISEnet), to science café – type informal community discussions at a number of sites, to longer-term informal "citizen schools" (e.g., at the University of South Carolina), and to multi-sited national engagement consensus conferences (CNS-ASU) and comparative cross-national public deliberations (CNS-UCSB). CNS's Public Communication of Science and Technology is conducting engagement activities on public perception of risks of nanoscience and on nanotechnology and food.

CNS-ASU's National Citizens' Technology Forum was modeled after Danish consensus conference but distributed across six U.S. locales. The NCTF on "nanotechnology and human enhancement" demonstrated that a high-quality deliberative activity can be organized at a national scale in the United States, and that a representative selection of lay citizens can come to discerning judgments about nanotechnology developments while they are still emergent [28]. CNS-UCSB's 2007 comparative U.S.-UK public deliberations were modeled on UK upstream deliberation efforts and included a between-groups design to compare deliberations on nanotechnology applications for energy and for health in the two countries [26, 27]. More recently CNS-UCSB in 2009 conducted an additional set of workshops, in deliberative groups, to examine more closely the role of gender differences, a consistent factor in diverging public views on risks.

About 53% of the public in the United States perceives little to no risk from nanotechnology [59]. The only nanotechnology applications to which the public regularly applies high negative EHS footprints are food-related. Important variables determining public perceptions of risk seem to be educational levels and socioeconomic categories more than cultural or religious identifiers, though culture and religion can be correlated to education and socioeconomic status.

There is a growing population of "newsless" Americans who do not seek out news from either traditional sources or digital media sources. Also, there is a growing body of Americans known as "net-newsers" who get most of their news information from Internet resources [60]. While some net-newsers clearly draw from traditional news that has migrated to the web, a growing number are turning to resources associated with the term "Web 2.0." These two phenomena pose special challenges for engaging the public in effective nanotechnology governance discussions. We must find new and creative ways to reach the newsless, and we must find creative ways to use social media engagement platforms to reach those individuals who are net-newsers. The swing toward net-newsing also means that much of what social science knows about the amplification of risk, which traditionally has been drawn from newspapers and television, will likely need to be reexamined.

8.7 Scenarios Approach: The NanoFutures Project

Contact person: Cynthia Selin, Arizona State University

The future of nanotechnology is not preordained and can therefore not be predicted. There are critical uncertainties surrounding both the technological pathways and the societal implications of discoveries on the nanoscale. The development of nanotechnology depends on choices made today, choices that occur throughout society in the boardroom, within the laboratory, in the legislature, and in shopping malls. There are numerous complex, interrelated variables that impinge upon what nanotechnology will ultimately look like in 10 years' time.

Future-oriented methods like scenario planning provide a means to structure key uncertainties driving the coevolution of nanotechnology and society [61]. These critical uncertainties range from the health of the U.S. economy, to regulatory frameworks, to public opinion, to the actual technical performance of many of nanotechnology's projected products. Anticipation and foresight, as opposed to predictive science, provide means to appreciate and analyze uncertainty in such a way as to maximize the positive outcomes and minimize the negative outcomes of nanotechnology [18, 51]. The value of scenario development in particular is to rehearse potential futures to identify untapped markets, unintended consequences, and unforeseen opportunities.

Three application areas are important to assess the prospective benefits and risks of nanotechnology:

- *Health and medicine*: Nanotechnology promises many breakthroughs in cancer treatment, drug delivery, and personalized medicine. The CNS has looked systematically at emerging diagnostic technologies and determined that critical choices revolve around the reliability and security of the data produced by the device and how well the device is managed and integrated within the larger medical system. If portable, fast, and reliable medical diagnostics are to yield positive societal benefits, questions regarding access must be adequately addressed.
- *Climate and natural resources*: Nanotechnology's development can be directed towards overcoming many of the planet's most urgent ills by generating products and processes that focus on conserving, protecting, and extending natural resources. One CNS-ASU scenario focused on generating drinkable water from air, which could enable off-the-grid survival and begin to address global demands for clean water.

- *Energy and equity*: Nanotechnology has much to offer towards producing greater efficiencies and cost savings in the energy domain. One particular scenario examined using nanotechnology-enhanced coolants to boost nuclear power generation. Describing such a future technology as a scenario provides a means to assess the broader barriers to and carriers of the innovation.

These anticipation and foresight approaches may take a variety of forms from traditional scenario planning to experiments with virtual gaming, simulation modeling, deliberative prototypes, and training modules. Such tools enable the scientific enterprise to become more responsive to shifting societal, political, and economic demands to produce more robust and relevant discoveries that address contemporary and future needs proactively.

8.8 Large Nanotechnology Firms as the Primary Source of Innovation and Under-Commercialization

Contact person: N. Horne, University of California, Berkeley

A small number of large multinational firms are responsible for a significant portion of nanotechnology patenting activity, yet competitive strategies artificially reduce their ability to commercialize products. New policies can change this trend.

Since 2000, nanotechnology discovery and innovation have flourished; nanotechnology has now reached the broad diffusion point of a general-purpose technology [62]. Large multinational enterprises (LMEs) remain the locus of most nanotechnology innovation relative to small and medium enterprises (SMEs) and universities, with moderate relative change over time (Table 5). Innovation occurs within LMEs due to the clustering of capital, including equipment and technically proficient labor, combined with deep market knowledge that maximizes application development.

Patenting is more concentrated in 2010 as compared to 2000, with over a quarter of all U.S. nanotechnology patents issued held by only 20 entities. And as of 2008, private R&D investment is now larger than public R&D investment. Moreover, LMEs now represent the largest source of capital annually, with less than 5% of total funding coming from the generally recognized source of innovation, venture capital. While this balance of relatively higher private funding is desirable, it further underscores the dominance of LMEs and the importance of ensuring high commercialization efficiencies for broader economic good.

Private firms are both effective commercialization drivers and a significant source of commercialization inefficiency. In all technology areas, at least one-third of technology products fully vetted through technical and market testing are not launched to market. Consistent findings of significant suppression rates emerge from empirical data across multiple applied nanotechnology market sectors sharing similar characteristics in the overall nanotechnology market, including longer exit periods and high initial capital investment requirements. The percentage of technically

Table 5 Top nanotechnology patent holders

	2004				2010		
Rank	Entity	Type	# U.S. nano patents		Entity	Type	# U.S. nano patents
1	IBM	LME	171		IBM	LME	257
2	UC regents	Univ.	123		Canon	LME	164
3	U.S. Navy	Govt.	82		Samsung	LME	137
4	Kodak	LME	72		UC Regents	Univ.	112
5	Minnesota mining	LME	59		HP	LME	112
6	MIT	Univ.	56		Hitachi	LME	78
7	Xerox	LME	56		Seiko	LME	80
8	Micron	LME	53		Olympus	LME	71
9	Matsushita	LME	45		Rice U.	Univ.	70
10	L'Oreal	LME	44		Nantero	SME	68
Total patents, top 10			761				1149
Percentage of total U.S. nanotechnology patents held by top 10 nanopatent assignees			14%				19%
Total patents, 2nd 10			309				496
Percentage of total U.S. nanotechnology patents held by next 10 nanopatent assignees			6%				8%
Percentage of total U.S. nanotechnology patents held by top 20 patent assignees			20%				27%

From Li et al. [63] and Graham [62]; the table cites data as originally published

and market-ready products not released to the market is on average between 40% and 50% (for technology products, see [64]; for pharmaceutical products, see [65]). The impact of regulatory review on pharmaceutical suppression is higher, of course, than for technology products. Policies to drive out sleeping patents are common in many industrialized nations via compulsory licensing and March-in clauses. These policies have been shown empirically to be ineffective due to significant underuse; firms do not use licenses because first-moving firms bear the costs, whereas subsequent firms would benefit financially [66].

The implications for 2020 are significant. Under current trends, continued government investment in basic and applied R&D combined with general economic recovery will create continued patenting and spin-out growth over the mid-term, despite a short-term shortage of venture capital funding. At the same time, a significant number of nanotechnology patents will be concentrated to a smaller set of actors. As a result, a limited number of large firms will continue to serve as both a significant source of intellectual property and under-commercialization in the near- and mid-terms. New policies to effectively drive out sleeping patents can increase nanotechnology's broader economic impact. Specifically, auctions across multiple-sector firms will offset the underuse of compulsory licensing; auctions should be carefully constructed to avoid distortions.

The goal of nanotechnology patent auctioning is to incentivize firms to release unused intellectual property (IP) by providing short- and mid-term profit for patents. With compulsory licensing, the number of potential bidders, and therefore the short-term valuation of intellectual property, are lower as compared to an open-auction market. Auctioning eliminates the weakness of compulsory licensing, as first-moving firms assume both the costs and the financial rewards of IP reassignment. Two factors determine the type of auction that would create the greatest efficiency: *private value*, in which bidding firms may have relevant IP that would significantly increase the value of an auctioned IP, and *information asymmetry*, in which bidding firms may have knowledge of the auctioned IP that would affect valuation. Given that nanotechnology products generally require many patents to create a final product, the withholding of a single patent critical to the success of a product could produce artificially high bids relative to the real value of the patent, simply due to timing. Concurrent rather than subsequent auctioning would prevent the overvaluation of such critical patent technology. Therefore, a uniform-price auction, otherwise known as a second-price sealed bid or Vickrey auction of multiple nanotechnology patents, would produce the most efficient reallocation of patents.

8.9 Decision Making with Uncertain Data

Contact person: Jeff Morris, U.S. Environmental Protection Agency

The history of regulation of industrial chemicals shows that regulatory agencies such as EPA have been unable to keep pace, in terms of acquiring and evaluating risk-related information, with the introduction of chemicals into society.[3] Yet it seems to be accepted by many government, industry, and NGO stakeholders that the appropriate path for nanotechnology governance is to follow the regulatory science model that has been used for decades for industrial chemicals.[4] This acceptance has important implications for the U.S. regulatory agencies under whose mandates nanotechnology risk issues fall. Christopher Bosso [29] has identified *institutional capacity* as a major issue arising from nanotechnology stakeholders'

---

[3] There are more than 84,000 chemical substances on the TSCA Chemical Substances Inventory; for only a small fraction of those has EPA received sufficient data to make risk determinations in accord with EPA's own risk assessment guidelines. On average, about 700 new substances are added every year. Information on the TSCA inventory may be found at http://www.epa.gov/oppt/newchems/pubs/invntory.htm. Also see [67].

[4] For discussion on regulatory science and its use in environmental decision making, see [68].

agreements that large amounts of data will be needed to inform decisions related to nanotechnology's environmental implications. Given the inability of regulatory agencies to adequately address the assessment needs of traditional industrial chemicals, it seems unlikely that regulators will have the capacity to keep up with nanotechnology's regulatory demands unless they adopt new approaches to governing the introduction of new substances, including but not limited to nanoscale materials, into society.

Related to institutional capacity is another issue raised by Bosso [29], the trade-off between taking action to anticipate risks and acquiring sufficient information to make defensible decisions about risks. Regulatory agencies traditionally have needed a large body of evidence to make decisions on chemical risks. It will take years, if not decades, to develop hazard and exposure databases as large as currently exist for such substances as asbestos.[5] The dilemma, therefore, is *how to instill anticipatory, risk-preventative behavior in governance institutions when little regulatory science data exist.* If those responsible for environmental decision making embrace the existing chemical assessment model as the principal approach to nanotechnology governance, the balance between being anticipatory and generating robust risk-information databases likely will become increasingly difficult and contentious.

The idea of *anticipatory technology evaluation* for nanomaterials fits within a larger national and global movement toward sustainable chemical, material, and product development and use. The people who invent, design, synthesize, fabricate, incorporate into products, use, regulate, and dispose of or recycle chemicals and other materials – including nanoscale materials – in many cases do not have adequate information (including but not limited to physical-chemical and/or material properties, life cycle, hazard, fate, exposure) to make decisions that lead to those chemicals or materials being designed, created, and managed in an environmentally sustainable manner. Nor do they often have information on the inputs (e.g., energy, starting materials) that go into, and the emissions that are released from, the fabrication of these substances. Without such information, environmental decision makers will not be able to overcome the current backlog of unassessed chemicals (including, increasingly, nanomaterials), let alone address the impacts of new materials from emerging technologies, such as nanoscale materials. The recent introduction of a TSCA reform bill in the United States, together with the European Community's progress toward implementing REACH, adds impetus to the need for innovative solutions to assessment approaches oriented toward the green design of chemicals, materials, and products.

---

[5] EPA's 1989 attempt to ban asbestos from products was overturned in 1991 by the Fifth Circuit Court of Appeals because, in essence, the court determined that EPA had not provided a sufficient regulatory science justification for the ban. See http://www.epa.gov/asbestos/pubs/ban.html. For a concise summary of the issue, see Environmental Working Group, "The Failed EPA Asbestos Ban," http://www.ewg.org/sites/asbestos/facts/fact5.php.

8.10 Penetration of Nanotechnology in Therapeutics and Diagnostics

Contact person: Mostafa Analoui, The Livingston Group, New York, NY

The past decade has witnessed a strong surge in research and product development around utilization of nanotechnology in life sciences (see in chapter "Applications: Nanobiosystems, Medicine, and Health"). During 2000–2010, nanotechnology publications and patents have shown a steady growth, while for nanobiotechnology the trend is showing a much faster growth, reflecting additional scientific investment both by public and private sectors [69]. This steady increase in scientific output and creation of intellectual properties, however, has not been matched with a similar pattern in investment, product development and commercialization [70]. This discrepancy in evolution of knowledge and market introduction is a common characteristic of innovative and emerging technologies.

An overwhelming level of investment is currently focused on reformulation and novel delivery of existing chemical and molecular entities. Consistently, more than 60% of nanomedicine R&D is allocated to this segment. There are several outstanding and successful developments. Perhaps the hallmark of such activities can be summarized in the journey that Abraxis took for development of nano-albumin formulated of paclitaxol (product known as Paclitaxel), one of the most cytotoxic agents. Abraxane has promised a safe therapy at much higher doses. Abraxane received FDA clearance for metastatic breast cancer in January of 2005. Since then, Abraxane has been prescribed to an increasing number of patients, with expanding indications. This product had more than $350 million sales in 2009 and was cornerstone for acquisition of Abraxis by Celgene for $2.9 billion. This is the largest merger and acquisitions deal to date in the nanomedicine field.

Examples of nano-formulated drugs approved and in the market are listed in Table 6, showing a market size of more than $2.6 billion in nanotechnology-based therapeutics in 2009, with no product in the market in 2000.

With more than $120 billion pharmaceutical products losing their patent protection between 2009 and 2014, this has started an avalanche of R&D and investment, which should come to fruition for patients and investors during 2010–2020. Perhaps the most promising products yet to come or new chemical/molecular entities based on a rational nanoscale-design addressing major chronic diseases such as Alzheimer's disease (AD), osteoarthritis and rheumatoid arthritis (OA/RA) and major improvement therapeutics for ophthalmic diseases such as Age-related Macular Degeneration (AMD) and Diabetic Macular Edema (DME). With current pipeline and increased R&D investment, some landscape-shifting management of such diseases via nanomedicine products is anticipated.

Nanotechnology-based diagnostics has gone through a significant landscape shift since 2000, when key promising areas (as a combination of ongoing research and blue-sky thinking) included nano-based contrast agents, nano-arrays for label-free sequencing, highly sensitive and specific assays and passive sensors. Quantum

Table 6 Selected nano-based therapeutics and their 2009 sales

Product	Particle type	Drug/application	Technology by/ licensed to	Status	2009 sales ($M)
TriCor	Nanocrystal	Fenofibrate	Elan/Abbott	Marketed	1,125.0
Rapamune	Nanocrystal	Sirolimus	Elan/Wyeth	Marketed	343.0
Ambisome	Liposomal	Amphotericin B	Gilead Sciences	Marketed	258.6
Abraxane (since 2005)	Nanoparticle	Paclitaxel	American Bioscience	Marketed	350
Doxil[a]	Liposomal	Doxorubicin	ALZA	Marketed	227.0
Emend	Nanocrystal	Aprepitant	Elan/Merck	Marketed	313.1
Abelcet	Liposomal	Amphotericin B	Elan	Marketed	22.6
Triglide	Nanocrystal	Fenofibrate	SkyePharma Pharmaceuticals	Marketed	28.0
Amphotec[a]	Liposomal	Amphotericin B	ALZA/Three Rivers Pharmaceuticals	Marketed	3.7
Total					$2,671 M

[a]Represents 2008 sales

dots (QDs) received broad attention as a promising optical contrast agent for in vitro and in vivo biological imaging. Despite significant progress in R&D on QDs, concerns with toxicity have prevented utilization of this product for human imaging. Nevertheless, there has been a significant program in enhancing several in vivo contrast agents (for CT and MR imaging), as well as in the introduction and validation of new class of agents that is expected to find their ways in clinical practice in next decade. Additionally, nano-based arrays and assays are gradually coming out of research laboratories into clinical markets. More than 50 companies are developing nanoparticle-based medicines for treating, imaging and diagnosing cancer in 2010 in the U.S. alone [71].

An example of such development is ultrasensitive detection of protein targets, using nanoparticle probe technology developed by Nanosphere, Inc. Nanosphere is using its patented gold nanoparticle probe technology to develop rapid, multiplexed clinical tests for some of the most common inherited genetic disorders, including certain types of thrombophilia, alterations of folate metabolism, cystic fibrosis, and hereditary hemochromatosis. Also, it must be noted that Nanosphere is a recent, pure-play nanodiagnostic company, which went public through IPO in 2007.

Currently nanodiagnostics concepts focus around utilization of nanoscale properties for:

- Ultrasensitive biomarker development/measurement
- Multi-assay for real-time in vitro assessment
- Clinical nano-tracers and contrast agents for establishing disease stage, drug PK/PD and monitoring therapy

Successful development of such ensembles of therapeutics and diagnostics for drug development will eventually lead to more effective utilization in clinical practice,

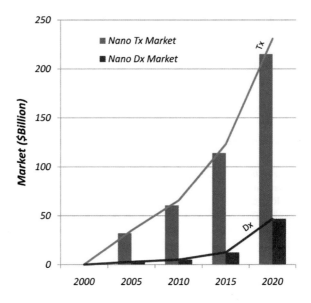

Fig. 8 Historical and projected markets for nanotherapeutics (Tx) and nanodiagnostics (Dx) (Baseline data and compounded annual growth rates are based on [72])

with the promise of moving toward "personalized medicine." Figure 8 compares historical and future market size for therapeutics and diagnostics products.

While we are not at a stage to claim availability of "personalized medicine" today (although depending on a chosen definition, one may claim this has been practiced in medicine for quite some time), we have certainly come a long way since 2000. In the next 10 years, nanotechnology is projected to make even greater contributions compared to the past 10 years (Table 7). Convergence of nanodiagnostics and nanotherapeutics, along with better understanding of the etiology of diseases, should provide game-changing solutions for prevention of disease, more effective patient management, and enhancing quality of life globally.

8.11 Products Enabled with Nanotechnology Generated $254 Billion in 2009

Contact person: Jurron Bradley, Lux Research

Since the U.S. National Nanotechnology Initiative sparked a boom of interest in the early 2000s, nanotechnology has enticed entrepreneurs, financiers, and corporate leaders with its potential to create value in a wide range of products and industries. For example, in 2009 businesses generated $254 billion in revenue from products touched by emerging nanotechnology, which is defined as the purposeful engineering of matter at scales of less than 100 nm to achieve size-dependent properties and functions.

Table 7 Major trends and projection in nanotherapeutics and nanodiagnostics 2000–2020

	2000	2010	2020
Therapeutics			
Reformulation	Academic Research	Several products approved and in the market	Fully developed market & deep pipeline of compounds with recent patent expiration
Novel Delivery	None	Several compounds in clinical trials	Multiple products in the market
Nano-based drug	None	Early stage R&D	Nano "blockbusters" addressing AD, OA/RA, CVD, DME/AMD
Diagnostics			
Assays and Reagents	None	Initial market entry	Main stream marketed products
In vitro Dx	None	A few approved/marketed products, more under development	Fully developed market. Multi-assay and hyper-sensitive solutions requiring minimal biological sample.
In vivo Dx	None	In vivo contrast agents under clinical trials	A few marketed products and deep pipeline
Theranostics (Tx+Dx)	None	Early stage R&D	A few game-changers paving the way toward personalized medicine. Significant steps toward nanobiosystem medicine.

There are three stages of the nanotech value chain, including nanomaterials (raw materials that make up the base of the nanotechnology value chain), nanointermediates (intermediate products – neither the first nor the last step in the value chain – that either incorporate nanomaterials or have been constructed from other materials to have nanoscale features) and nano-enabled products (finished goods at the end of the value chain that incorporate nanomaterials or nanointermediates). About 88% of 2009 revenue came from nano-enabled products, which are in big ticket markets like automobiles and construction (Fig. 9). The nanomaterials and nanointermediates portion *of the value chain supplied the other 12%, namely nanomaterials like zinc oxide, silver, and carbon nanotubes and nanointermediates like coatings and composites.*

In terms of sector, the manufacturing and materials sector – which includes industries like chemicals, automotive, and construction – accounted for 55% of the revenue in 2009, and the electronics and IT sector – which is dominated by computer and consumer electronics – contributed 30%. The healthcare and life sciences sector – primarily made up of pharmaceuticals, drug delivery, and medical devices – and the energy and environment sectors – comprised of energy applications like solar cells and alternative batteries – contributed 13% and 2%, respectively. In terms of region, the U.S. and Europe provided 67% of the revenue, followed by 37% from Asia and the remainder from the rest of the world (Fig. 9).

Innovative and Responsible Governance of Nanotechnology

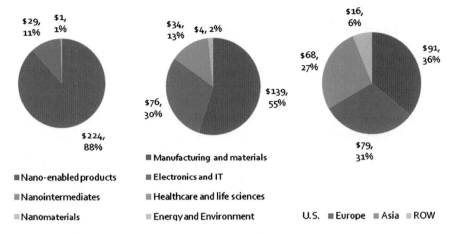

Fig. 9 Products touched by nanotechnology generated $254 billion in 2009

Table 8 Venture capital funding for nanotech totaled $792 million in 2009

(US$ million)	2000	2001	2002	2003	2004	2005	2006	2007	2008	2009
U.S.	$171	$145	$318	$301	$366	$566	$654	$683	$1,159	$668
Europe	$23	$34	$37	$25	$78	$69	$73	$54	$144	$108
Asia	$–	$48	$–	$–	$16	$6	$10	$2	$–	$5
Rest of world	$12	$27	$11	$44	$16	$19	$50	$35	$58	$12
Total	$206	$254	$366	$371	$476	$659	$787	$774	$1,360	$792

Venture capital funding increased steadily until 2008, but it experienced a significant decline during the 2009 economic crisis (Table 8).

9 International Perspectives from Site Visits Abroad

The following are summaries from the international WTEC "Nano2" workshops held in Germany, Japan, and Singapore, with a focus on international convergence in governance.

9.1 United States-European Union Workshop (Hamburg, Germany)

Panel members/discussants

Alfred Nordmann (co-chair), Technical University of Darmstadt, Germany
Mike Roco (co-chair), U.S. National Science Foundation

Rob Aitken, Institute of Occupational Medicine; SAFENANO, Edinburgh, UK
Richard Leach, National Physical Laboratory, UK
Ilmari Pyykkö, University of Tampere, Finland
Nira Shimoni-Eyal, Israel
Georgios Katalagarianakis, EU support, Greece
Christos Tokamanis, EU support

It was noted in this session that nanotechnology research amounts to a sociopolitical project. In the spirit of the "Nano2" study, this formulation underscores the desire to look beyond nanoparticles and other advances in nanomaterials to some of the more long-term prospects and ambitions of nanotechnology. The topics for the group included regulation, standardization, ethical, and societal dimensions. For each of these areas of inquiry, there is something different to report about international convergence.

From the point of view of regulation and the knowledge that is required to establish regulatory thresholds and procedures, progress in regard to nomenclature, measurement, characterization, standardization, and testing procedures appears painfully slow and lags behind the speed of commercial development and the introduction of products into the marketplace. While such a lag is not unusual in and of itself, there looms in this case the question whether the lack of progress owes to formidable systematic difficulties and the level of complexity. If so, this might prove to be a major obstacle for extending available regulatory methodologies in the near, medium, or even long term even to "first generation" nanomaterials.[6] The last 10 years saw the emergence of an at least two-pronged approach, and the next 10 years will see its further development:

- On the one hand, there is close attention being paid to the requirements for an adequate extension of existing regulatory frameworks, such as the need for the development of internationally standardized traceability methods. Greater effectiveness might be achieved by a greater cohesion among international funding schemes.
- On the other hand, numerous analyses and institutional innovations are focusing on the development of expanded soft-law regulatory schemes that can serve a stop-gap role in the absence of proper risk-assessment and classical regulatory monitoring. These institutional innovations comprise soft-law codes of conduct or certifications, observatories, public engagement exercises, and consumer conferences, ELSI research, and platforms for the exchange of best practices.[7] All of these largely informal institutions serve to observe what social scientists have analyzed as a collective experiment with emerging technologies. Here,

---

[6] It is the case, of course, that nanomaterials are already covered by, for example, the REACH regulatory framework—at the cost of either not considering sufficiently or *de facto* exempting the specificities arising from their nanoparticulate or nanostructured character.

[7] In this regard, the NanoCap project suggested the introduction of safety notes as a standard element of research publications, alongside the methods section. The note would merely describe what safety measures were actually taken in the laboratory and would thus contribute to best practices and the evolution of shared standards.

another avenue of research would integrate EHS and materials researchers more strongly with social scientists or regulators in order to explore together how far the notions of "safety by design," "precautionary science," "green nanotechnology," or "responsible innovation" can be developed.[8] Finally, it is in the arena of the collective experiment that epidemiology and a pre- and post-market product-testing approach receive greater attention than the ambition to determine pre-market and pre-production the toxicological properties of more or less generic nanomaterials. This includes increased emphasis on life-cycle analysis/assessments (LCA) and the development and improvement of LCA methodologies.

The call for international coordination and harmonization is loud and clear, and there are international working parties in a variety of venues. If international standards and the harmonization of traceability methods are not forthcoming, this is due partly to the duplication of research efforts and partly to the intractabilities of the problems at hand. The question of international convergence looks different in regard to ethical and societal dimensions. Here there is an initial emphasis on specific cultural values and citizens' attitudes at the national and European Union levels. Recognition of these differences is an important prerequisite for the international diffusion of nanotechnological products and processes.[9]

One can speak of a two-pronged approach related to ELSI:

- On the one hand, there is a proactive and anticipatory approach that consists of first imagining potential or likely future applications of nanotechnology in society and then to appreciate and evaluate their impacts. Here, prospects of human enhancement through use of nanotechnology are currently proving to be divisive.
- On the other hand, there are attempts to understand nanotechnology as a socio-political project – in other words, to see what societal and technological trajectories are continued and intensified by nanotechnology, to appreciate and assess the visionary dreams and societal expectations that drive nanotechnology research, and to seek out just where currently funded nanotechnology research is proving to be disruptive.[10] In the context of this approach, there remains much to be questioned that is now taken for granted.

The perceived division of moral labor ("ethical considerations are important but they should be delegated to advisory committees") has been and will continue to be

---

[8] Integrated approaches are visible in exemplary studies like Lawton, J. (ed.). 2008. *Novel materials in the environment: The case of nanotechnology*. London: Royal Commission on Environmental Pollution, also in studies of the International Risk Governance Council (IRGC).

[9] Here, internationalization of the debate is moved forward by academics through venues like the S.NET society or the Springer journal *NanoEthics*.

[10] This does not necessarily involve a consideration of long-term nanotechnological developments. Nanoparticles are already proving disruptive because they are so hard to classify and therefore do not fit classical assessment schemes. The use of biological properties in the construction of nanomaterials (virus-like structures as nanotechnological building blocks) may well prove even more disruptive.

challenged, as for example, by the Code of Conduct for Nanotechnology Research that has been proposed by the European Commission.

A society that observes itself in an experimental mode must repeatedly ask itself, "How are we doing?" Since answering this question involves judgments, interventions, and calls for action, this kind of assessment of how we are doing regarding nanotechnology goes beyond the role of nanotechnology observatories as they are currently conceived, and it will open up in the next decade new requirements for the inclusion of social science and humanities scholarship.

9.2 United States-Japan-Korea-Taiwan Workshop (Tokyo/Tsukuba, Japan)

Panel members/discussants

Tsung-Tsan Su (co-chair), Industrial Technology Research Institute, Taiwan
Mike Roco (co-chair), U.S. National Science Foundation
Yoshio Bandou, National Institute for Materials Science, Japan
Toshiyuki Fujimoto, National Institute of Advanced Industrial Science & Technology (AIST), Japan
Ivo Kwon, Ewha Women's University, Korea
Mizuki Sekiya, AIST, Japan

9.2.1 The Vision Has Changed in the Last 10 Years

- The initial focus was only on technical issues; now we also are addressing broader societal implications issues from economical outcomes and innovation to regulatory aspects.
- There has been a partial transition from science-driven policies to user-driven policies, e.g., applications-driven R&D. Some countries have always had a strong emphasis on applications-driven research. There is an increased emphasis now on "return on investment" – how many jobs can you create?
- Both EHS and ELSI are now addressed more realistically and with specificity.
- There is more emphasis now on a common international vision; more communication and acceptance of common approaches and goals, addressing global issues like lowering CO_2, energy, and the environment.
- Both negative and positive hype experienced initially have receded; extreme negative predictions of the early 2000s have not come to pass.

9.2.2 The Vision for the Next 10 Years

- Nanotechnology will be built into systems, e.g., nanotechnology to solve photovoltaic systems issues; use in transportation systems such as electric cars; biological applications such as in drug delivery, food, and agriculture, etc.; these will enable ubiquitous computing, communication, and sensing systems.

- Look to mass use of nanotechnology; many new products will emerge.
- Nanoscale science and engineering will be included in standards of learning by 2020.
- Nanotechnology will enable sustainable development.
- Nanotechnology may help to solve the world's problems, but there is concern of increasing the technological gap between developed and developing countries.
- Development of international activities will be institutionalized regarding nomenclature, standards, and patents related to nanotechnology, as well as in developing a common lexicon and improved toxicity evaluation, risk assessment, and mitigation.

9.2.3 The Main Goals for 2020

- Clear regulatory environment to enable commercialization, protect consumers and general public; this should include internationally acceptable, harmonized regulations.
- Promotion by international organizations (e.g., ISO, *International Electrotechnical Commission*) of professionalism in nanotechnology R&D; easy communication internationally will accelerate exchanges of opinions among people in different countries concerning nanotechnology and support for common standards for valid research methodologies.
- Shift to a new generation of nanotechnology-enabled products, including preparation of enabling manufacturing and monitoring tools and regulations.
- Reduction in cultural, political barriers (e.g., to international collaboration, acceptance of nanotechnology-enabled products).
- Creation of an international information system; databases, information sources for broad access from researchers, industry, regulators, political system, including different categories and lists of funded research projects.
- Training of young scientists internationally to understand societal implications of nanotechnology.

9.2.4 Main Infrastructure Needs

- Institutional mechanisms for international collaboration
- Ongoing support for ISO/TC229, IEC/TC113, OECD WPN; currently these are not permanent activities and serve only in advisory capacities.
- Fill the technology gap between developed and developing countries through international collaboration, e.g., the United Nations Environment Programme, United Nations Ethics Programme [8], Asia Nanotechnology Forum (ANF).
- Infrastructure to address long-term sustainable development through nanotechnology: CO_2 problem, nano-geo-engineering, water filtration and desalinization; this may require establishing a new international organization.
- International mechanism(s) to support nano-ELSI aspects; better coordination of existing national institutes; leverage individual countries' efforts.

9.2.5 Suggested R&D Strategies

- Create an international open source network to promote nanotechnology R&D and applications for sustainable development, other common problems, through precompetitive research (EHS, ELSI; climate change solutions; water filtration; energy and sustainable development technologies) (this might be difficult; it is very competitive now).
- Continue to allocate a portion of R&D projects to EHS and ELSI research and education, and to integrating EHS and ELSI with core R&D.
- Employ standard definitions and research protocols in EHS and ELSI research internationally, i.e., implement ISO, IEC, OECD recommendations.

9.2.6 Several Emerging Issues Have Been Identified

- Labeling is becoming an international issue; there is a contrast between the EU proposed approach aiming to address safety and Taiwan's "nanoMark" approach aimed at addressing authenticity of nanotechnology products.
- Public engagement is now a common interest internationally.

9.3 United States-Australia-China-India-Saudi Arabia-Singapore Workshop (Singapore)

Panel members/discussants

Graeme Hodge (co-chair), Monash University, Australia
Mike Roco (co-chair), U.S. National Science Foundation
Salman Al Rakoyan, King Abdullah Institute for Nanotechnology, Saudi Arabia
Freddy Boey, Nanyang Technical University, Singapore
Craig Johnson, Department of Innovation, Industry, Science and Research, Australia
John Miles, National Measurement Institute, Australia
Murali Sastry, Tata Chemicals Innovation Centre, India
Yuliang Zhao, Institute of High Energy Physics, Chinese Academy of Sciences, China

9.3.1 Key Changes of the Nanotechnology Vision in the Last 10 Years

- Huge progress has been made in putting together building blocks for international governance: International Dialogue on Responsible Development of Nanotechnology (Arlington 2004, Bruxelles 2006, Tokyo 2008), IRGC 10, UNEP, ISO, and OECD.

- International communities and networks of professionals have formed, in nanotechnology and societal implications, with a significant collaborative effort.
- Nanotechnology has moved from being a science and technology dream to a social reality.

9.3.2 Several Major Changes That Are Needed in the Next Decade

- A common international language for nanotechnology and related studies, e.g., ISO standards to be adopted worldwide; characterization [73].
- International joint funding mechanisms to support international standards activities, health and safety testing, other areas of common interest in "precompetitive" research. An alternative would be better coordination, more international co-funding, and leveraging of individual nations' R&D efforts.
- Different countries' interests need to be respected, e.g., developing countries.

9.3.3 Main Scientific/Engineering Advancements and Technological Impacts in the Last 10 Years

- Development of capabilities to do nanoscale science and engineering research around the world.
- Beginning of scaled-up manufacturing capabilities at the nanoscale.
- Evolution from focus on multidisciplinary science and engineering to new multifaceted enabling technologies
- Move from science-only focus in nanotechnology to science and technology for society and the development of beneficial applications.

9.3.4 Key Goals for the Next 5–10 Years

- Scientific communities, industry, and governments should take the lead in undertaking meaningful and proactive public engagement, including better public appreciation/education of the value of the nanotechnology investments and how potential risks are being addressed.
- Open access, collaborative knowledge system(s) for strengthened investment and governance.
- Continued/increased international collaboration in nanotechnology investment; leveraging, sharing of facilities, best use of existing resources.
- Explicit system for incorporating ethical, legal, and other societal issues (ELSI) into nanotechnology governance, such as real-time technology assessment. Although this is not necessarily an issue unique to nanotechnology, and is essentially a broader science issue, nanotechnology could nonetheless set the example.

9.3.5 Needs for Scientific and Technological Infrastructure Include

- Some participants suggested a new international agency for "precompetitive" collaborative R&D.
- An alternative is just better coordination (e.g., following the example of OECD Working Party on Manufactured Nanomaterials [WPMN] in EHS testing).

9.3.6 Emerging Topics and Priorities for Future Nanoscale Science & Engineering Research and Education

- Need for ongoing regulatory review: e.g., incorporation of nanomaterials in existing approved products raises new regulatory issues [74]
- Strengthening of international governance of nanotechnology
- Assessment of societal impacts and regulatory issues for next generations of nanotechnology-enabled products [40]
- International harmonization of patent policies

9.3.7 Several Characteristics of the Implications of Nanotechnology R&D on Society

- Nanotechnology is a lightning rod for debate over the impact of science on society more generally, and to some critics, is a symbol of everything that's wrong in the world
- Potential exists for nanotechnology to further divide the world into haves and have-nots – create a "nano divide" [75]; or, nanotechnology might have the potential to help bridge the divide between north and south [76–78].

References

In addition to the scholarly references listed below [1–78], please refer also to the WTEC "Nano-2" workshop proceedings in Chicago, Hamburg, Tokyo (Tsukuba), and Singapore: Websites for these proceedings are provided in Appendix A. Additional references related to presentations made at those workshops are [79–105].

1. M.C. Roco, R.S. Williams, P. Alivisatos, (eds.), *Nanotechnology Research Directions: Vision for Nanotechnology R&D in the Next Decade* (NSTC, also Springer, 2000, Washington, DC, 1999). Available online: http://www.nano.gov/html/res/pubs.html
2. M.C. Roco, Possibilities for global governance of converging technologies. J. Nanopart. Res. **10**, 11–29 (2008). doi:10.1007/s11051-007-9269-8
3. M.C. Roco, International strategy for nanotechnology research. J. Nanopart. Res. **3**(5–6), 353–360 (2001)
4. National Science Foundation (NSF), *Report: International Dialogue on Responsible Research and Development of Nanotechnology,* (Meridian Institute, Washington, DC, 2004). Available online: http://www.nsf.gov/crssprgm/nano/activities/dialog.jsp

5. D. Guston, (30 Mar). Public engagement with nanotechnology. *2020 Science* (2010), http://2020science.org/2010/03/30/public-engagement-with-nanotechnology
6. M.C. Roco, W.S. Bainbridge (eds.), *Societal Implications of Nanoscience and Nanotechnology* (Kluwer Academic Publishers, Dordrecht, 2001). Available online: http://www.wtec.org/loyola/nano/NSET.Societal.Implications/nanosi.pdf
7. M.C. Roco, W.S. Bainbridge, (eds.), Converging technologies for improving human performance: Nanotechnology, biotechnology, information technology and cognitive science (Springer, Dordrecht, 2003). Available online: http://www.wtec.org/ConvergingTechnologies/Report/NBIC_report.pdf
8. United Nations Educational, Scientific and Cultural Organization (UNESCO), *The Ethics and Politics of Nanotechnology* (UNESCO, Paris, 2006)
9. Center for Nanotechnology in Society at University of California at Santa Barbara (CNS–UCSB), *Emerging Economies/Emerging Technologies: [Nano]Technologies for Equitable Development*. Proceedings of the International Workshop. (Woodrow Wilson Center for International Scholars, Washington, DC, 2009). 4–6 Nov 2009
10. International Risk Governance Council (IRGC), *Appropriate Risk Governance Strategies for Nanotechnology Applications in Food and Cosmetics*, (IRGC, Geneva, 2009). Available online: http://www.irgc.org/IMG/pdf/IRGC_PBnanofood_WEB.pdf
11. Foundation for the Future (FFF) and United Nations Educational, Scientific and Cultural Organization (UNESCO), Humanity and the biosphere. *The Next Thousand Years*. Seminar proceedings, 20–22 Sept 2006, (FFF, Paris/Bellevue, 2007). Available online: http://www.futurefoundation.org/documents/hum_pro_sem7.pdf
12. D. Guston (ed.), *Encyclopedia of Nano-Science and Society* (Sage, Thousand Oaks, 2010)
13. A. Mnyusiwalla, A.S. Daar, P.A. Singer, Mind the gap: Science and ethics in nanotechnology. Nanotechnology **14**(3), R9 (2003). doi:10.1088/0957-4484/14/3/201
14. I. Bennett, D. Sarewitz, Too little, too late? Research policies on the societal implications of nanotechnology in the United States. Sci. Cult. Lond **15**(4), 309–325 (2006)
15. President's Council of Advisors on Science and Technology (PCAST), *Report to the President and Congress on the Third Assessment of the National Nanotechnology Initiative*, (Executive Office of the President, Washington, DC, 2005). Available online: http://www.nano.gov/html/res/otherpubs.html
16. F. Gomez-Baquero, *Measuring the Generality of Nanotechnologies and its Potential Economic Implications*. Paper presented at Atlanta Conference on Science and Innovation Policy, 2009. (IEEE Xplore, 2–3 Oct 2009:1–9, 2009). doi: 10.1109/ACSIP.2009.5367858
17. T. Nikulainen, M. Kulvik, How general are general purpose technologies? *Evidence from Nano-, Bio- and ICT-Technologies in Finland*. Discussion Paper 1208. (The Research Institute of the Finnish Economy, Helsinki, 2009)
18. J. Youtie, M. Iacopetta, S. Graham, Assessing the nature of nanotechnology: Can we uncover an emerging general purpose technology? J. Technol. Transf. **33**(3), 315–329 (2008)
19. P. Shapira, J. Youtie, L. Kay, *National Innovation System Dynamics in the Globalization of Nano-Technology Innovation (Working Paper)* (Georgia Tech Program in Science, Technology and Innovation Policy, Atlanta, 2010)
20. H. Chen, M. Roco, *Mapping Nanotechnology Innovations and Knowledge. Global and Longitudinal Patent and Literature Analysis Series* (Springer, Berlin, 2009)
21. A. Fernandez-Ribas, *Global Patent Strategies of SMEs in Nanotechnology*. Working paper. Science, Technology, and Innovation Policy, (Georgia Institute of Technology, Atlanta, 2009)
22. Z. Huang, H. Chen, L. Yan, M.C. Roco, Longitudinal nanotechnology development (1990–2002): The national science foundation funding and its impact on patents. J. Nanopart. Res. **7**(4–5), 343–376 (2005)
23. J. Wang, P. Shapira, Partnering with universities: A good choice for nanotechnology start-up firms? *Small Business Economics* (Preprint 30 Oct 2009). doi: 10.1007/s11187-009-9248-9
24. D. Hwang, *Ranking the Nations on Nanotech: Hidden Havens and False Threats* (Lux Research, New York, 2010)

25. T. Satterfield, M. Kandlikar, C. Beaudrie, J. Conti, B.H. Harthorn, Anticipating the perceived risk of nanotechnologies. Nat. Nanotechnol. **4**, 752–758 (2009). doi:10.1038/nnano.2009.265
26. N. Pidgeon, B. Harthorn, K. Bryant, T. Rogers-Hayden, Deliberating the risks of nanotechnologies for energy and health applications in the United States and United Kingdom. Nat. Nanotechnol. **4**, 95–98 (2009). doi:10.1038/nnano.2008.362
27. N. Pidgeon, B. Harthorn, T. Satterfield, Nanotech: Good or bad? Chem. Eng. Today **822–823**, 37–39 (2009)
28. P. Hamlett, M.D. Cobb, D.H. Guston, *National Citizens' Technology Forum: Nanotechnologies and Human Enhancement*. CNS Report #R08-0003. (Center for Nanotechnology in Society, Tempe, 2008). Available online: http://www.cspo.org/library/type/?action=getfile&file=88&section=lib
29. C. Bosso (ed.), *Governing Uncertainty: Environmental Regulation in the Age of Nanotechnology* (EarthScan, London, 2010)
30. J. Kuzma, J. Paradise, G. Ramachandran, J. Kim, A. Kokotovich, S. Wolf, An integrated approach to oversight assessment for emerging technologies. Risk Anal. **28**(5), 1197–1219 (2008)
31. S.M. Wolf, G. Ramachandran, J. Kuzma, J. Paradise, (eds.), Symposium: Developing oversight approaches to nanobiotechnology—The lessons of history. J. Law Med. Ethics. **37**(4), 732 (2009)
32. N. Powell, New risk or old risk, high risk or no risk? How scientists' standpoints shape their nanotechnology risk frames. Health Risk Soc. **9**(2), 173–190 (2007)
33. J.A. Conti, K. Killpack, G. Gerritzen, L. Huang, M. Mircheva, M. Delmas, B.H. Harthorn, R.P. Appelbaum, P.A. Holden, Health and safety practices in the nanotechnology workplace: Results from an international survey. Environ. Sci. Technol. **42**(9), 3155–3162 (2008)
34. J.C. Davies, *Oversight of Next Generation Nanotechnology*. Presentation. (Woodrow Wilson Center for Scholars, Apr 2009, Washington, DC, 2009)
35. C. Beaudrie, *Emerging Nanotechnologies and Life Cycle Regulation: An Investigation of Federal Regulatory Oversight from Nanomaterial Production to End-of-Life*. (Chemical Heritage Foundation, Philadelphia, 2010). Available online: http://www.chemheritage.org/Downloads/Publications/White-Papers/Studies-in-Sustainability_Beaudrie.pdf
36. T. Barker, M.L. Lesnick, T. Mealey, R. Raimond, S. Walker, D. Rejeski, L. Timberlake, *Nanotechnology and the Poor: Opportunities and Risks—Closing the Gaps Within and Between Sectors of Society*, (Meridian Institute, Washington, DC, 2005). Available online: http://www.docstoc.com/docs/1047276/NANOTECHNOLOGY-and-the-POOR
37. S. Cozzens, J. Wetmore (eds.), *Yearbook of Nanotechnology in Society, Vol. II: Nanotechnology and the Challenge of Equity and Equality* (Springer, New York, 2010)
38. M.C. Roco, Nanoscale science and engineering: Unifying and transforming tools. AIChE J. **50**(5), 890–897 (2004)
39. V. Subramanian, J. Youtie, A.L. Porter, P. Shapira, Is there a shift to active nanostructures? J. Nanopart. Res. **12**(1), 1–10 (2010). doi:10.1007/s11051-009-9729-4
40. O. Renn, M.C. Roco, White paper on nanotechnology risk governance. (International Risk Governance Council (IRGC), Geneva, 2006). Available online: http://www.irgc.org/Publications
41. M.C. Roco, O. Renn, Nanotechnology risk governance, in *Global Risk Governance: Applying and Testing the IRGC Framework*, ed. by O. Renn, K. Walker (Springer, Berlin, 2008), pp. 301–325
42. S. Murdock, (Nanobusiness Alliance), Personal communication with author Mar 2010
43. President's Council of Advisors on Science and Technology (PCAST), *Report to the President and Congress on the Third Assessment of the National Nanotechnology Initiative* (Executive Office of the President, Washington, DC, 2010). Available online: http://www.nano.gov/html/res/otherpubs.html
44. Lux Research, *The Nanotech Report: Investment Overview and Market Research for Nanotechnology* (Lux Research, New York, 2004)
45. Lux Research, *The Recession's Ripple Effect on Nanotech. State of the Market Report* (Lux Research, New York, 2009)

46. E.A. Corley, D.A. Scheufele, Outreach going wrong? When we talk nano to the public, we are leaving behind key audiences. Scientist **24**(1), 22 (2010)
47. D.A. Scheufele, E.A. Corley, The science and ethics of good communication. Next Gen. Pharm. **4**(1), 66 (2008)
48. D.E. Stokes, *The Pasteur Quadrant* (Brookings Institution Press, Washington, DC, 1997)
49. I. Linkov, F.K. Satterstrom, J. Steevens, E. Ferguson, R.C. Pleus, Multi-criteria decision analysis and environmental risk assessment for nanomaterials. J. Nanopart. Res. **9**(4), 543–554 (2007)
50. T. Tervonnen, I. Linkov, J.R. Figueira, J. Steevens, M. Chappell, M. Merad, Risk-based classification system of nanomaterials. J. Nanopart. Res. **11**, 757–766 (2009)
51. D. Barben, E. Fisher, C. Selin, D.H. Guston, Anticipatory governance of nanotechnology: Foresight, engagement, and integration, in *The New Handbook of Science and Technology Studies*, ed. by E.J. Hackett, O. Amsterdamska, M.E. Lynch, J. Wajcman (MIT Press, Cambridge, 2008), pp. 979–1000
52. R. Sclove, *Reinventing Technology Assessment: A 21st Century Model*, Science and Technology Innovation Program, Woodrow Wilson International Center for Scholars, Washington, DC, 2010. Available online: http://www.wilsoncenter.org/topics/docs/ReinventingTechnologyAssessment1.pdf
53. D.J. Fiorino, *Voluntary Initiatives, Regulation, and Nanotechnology Oversight: Charting a Path*. (Woodrow Wilson Center for Scholars (PEN 19), 2010). Presented 4 Nov 2010. Available online: http://www.nanotechproject.org/process/assets/files/8346/fiorino_presentation.pdf
54. G.A. Hodge, D.M. Bowman, A.D. Maynard, (eds.), *International Handbook on Regulating Nanotechnologies*, (Edward Elgar, Cheltenham, 2010). E-book: 978 1 84844 673 1
55. D. Guston, Innovation policy: Not just a jumbo shrimp. Nature **454**, 940–941 (2008). doi:10.1038/454940a
56. J. Wetmore, E. Fisher, C. Selin (eds.), *Presenting Futures: Yearbook of Nanotechnology in Society* (Springer, New York, 2008)
57. M.N. Nadagouda, A.B. Castle, R.C. Murdock, S.M. Hussain, R.S. Varma, In vitro biocompatibility of nanoscale zerovalent iron particles (NZVI) synthesized using teapolyphenols. Green Chem. **12**(1), 114–122 (2010)
58. B. Harthorn, (4 May). Public participation in nanotechnology—Should we care? *2020 Science* (2010). http://2020science.org/2010/05/04/public-participation-in-nanotechnology-should-we-care
59. D. Berube, C. Cummings, Public perception of risk to nanotechnology in context with other risks. J. Nanopart. Res. (2010). Forthcoming
60. Pew Research Center for the Public and the Press, *Ideological News Sources: Who Watches and Why*, (Pew, Washington, DC, 2010). Available online: http://people-press.org/reports/pdf/652.pdf
61. C. Selin, The sociology of the future: Tracing stories of technology and time. Sociol. Compass **2**(6), 1878–1895 (2008)
62. S.J.H. Graham, M. Iacopetta, Nanotechnology and the emergence of a general purpose technology. Ann. 'Economie Statistique (Ann. Econ. Stat.) **49/50**, 53–55 (2010)
63. X. Li, Y. Lin, H. Chen, M. Roco, Worldwide nanotechnology development: A comparative study of USPTO, EPO, and JPO patents (1976–2004). J. Nanopart. Res. **9**(6), 977–1002 (2007)
64. R.G. Cooper, *Winning at New Products* (Perseus Publishing, Cambridge, 2001)
65. M. Carrier, Two puzzles resolved: Of the Schumpeter-Arrow stalemate and pharmaceutical innovation markets. Iowa Law Rev. **93**(2), 393 (2008)
66. D. Carlton, J. Perloff, *Modern Industrial Organization* (Pearson, London, 2000)
67. U.S. Government Accountability Office (GAO), *Chemical Regulation: Options Exist to Improve EPA's Ability to Assess Health Risks and Manage its Chemical Review Program*, (GAO, Washington, DC, 2005). Report GAO-05-458. Available online: http://www.gao.gov/new.items/d05458.pdf
68. S. Jasanoff, *The Fifth Branch: Science Advisors as Policymakers* (Harvard University Press, Cambridge, 1990)

69. A. Delemarle, B. Kahane, L. Villard, P. Laredo, Geography of knowledge production in nanotechnologies: A flat world with many hills and mountains. Nanotechnol. Law Bus. **6**, 103–122 (2009)
70. Business Insights, *Nanotechnology in Healthcare. Market Outlook for Applications, Tools and Materials, and 40 Company Profiles*. (Business Insights Ltd., London, 2010), Available online: http://www.globalbusinessinsights.com/content/rbdd0035p.htm
71. R. Service, Nanoparticle Trojan horses gallop from the lab into the clinic. Nature **330**, 314–315 (2010)
72. BCC Research, *Nanotechnology in Medical Applications: The Global Market*. (BCC, Wellesley, 2010), Report code: HLC069A
73. E. Richman, J. Hutchison, The nanomaterial characterization bottleneck. ACS Nano **3**(9), 2441–2446 (2009). doi:10.1021/nn901112p
74. L. Breggin, R. Falkner, N. Jaspers, J. Pendergrass, R. Porter, *Securing the Promise of Nanotechnologies: Towards Transatlantic Regulatory Cooperation*. (Chatham House, London, 2009). Available online: http://www.chathamhouse.org.uk/nanotechnology
75. R. Sparrow, Negotiating the nanodivides, in *New global Frontiers in Regulation: The Age of Nanotechnology*, ed. by G.A. Hodge, D. Bowman, K. Ludlow (Edward Elgar, Cheltenham, 2007), pp. 97–109
76. G.A. Hodge, D.M. Bowman, K. Ludlow, Introduction: Big questions for small technologies, in *New Global Frontiers in Regulation: The Age of Nanotechnology*, ed. by G.A. Hodge, D. Bowman, K. Ludlow (Edward Elgar, Cheltenham, 2007), pp. 3–26
77. F. Salamanca-Buentello, D.L. Persad, E.B. Court, D.K. Martin, A.S. Daar, P.A. Singer, Nanotechnology and the developing world. Policy Forum **2**(5), 383–386 (2005)
78. P.A. Singer, F. Salamanca-Buentello, A.S. Daar, Harnessing nanotechnology to improve global equity. Issues Sci. Technol. **21**(4), 57–64 (2005). Available online: http://www.issues.org/21.4/singer.html
79. R.P. Appelbaum, R.A. Parker, China's bid to become a global nanotech leader: Advancing nanotechnology through state-led programs and international collaborations. Sci. Public Policy **35**(5), 319–334 (2008)
80. U. Beck, *Risk Society: Towards a New Modernity* (Sage, London, 1992)
81. L. Bell, Engaging the public in technology policy: a new role for science museums. Sci. Commun. **29**(3), 386–398 (2008)
82. L. Bell, Engaging the public in public policy: How far should museums go? Mus. Soc. Issues **4**(1), 21–36 (2009)
83. R. Berne, *Nanotalk: Conversations with Scientists and Engineers About Ethics, Meaning and Belief in Nanotechnology* (Lawrence Erlbaum Associates, Mahwah, 2005)
84. J. Calvert, P. Martin, The role of social scientists in synthetic biology. EMBO Rep. **10**(3), 201–204 (2009)
85. J. Conti, T. Satterfield, B. Herr Harthorn, Vulnerability and Social Justice as Factors in Emergent U.S. Nanotechnology Risk Perceptions. In review (2010)
86. K. David, P.B. Thompson (eds.), *What Can Nanotechnology Learn from Biotechnology?* (Academic Press (Elsevier), New York, 2008)
87. S. Davies, P. Macnaghten, M. Kearnes, (eds.), Deepening debate on nanotechnology. In: *Reconfiguring Responsibility: Lessons for Public Policy*, (Durham University, Durham, 2009)
88. J.A. Delborne, A.A. Anderson, D.L. Kleinman, M. Colin, M. Powell, Virtual deliberation? Prospects and challenges for integrating the Internet in consensus conferences. *Public Understanding of Science*. (Preprint 9 Oct 2009), (2009). doi: 10.1177/0963662509347138
89. E. Fisher, Ethnographic interventions: Probing the capacity of laboratory decisions. NanoEthics **1**(2), 155–165 (2007)
90. E. Fisher, L.R.L. Mahajan, C. Mitcham, Midstream modulation of technology: Governance from within. Bull. Sci. Technol. Soc. **26**(6), 486–496 (2006)
91. E. Fisher, C. Selin, J. Wetmore (eds.), *Yearbook of Nanotechnology in Society, Vol. I: Presenting Futures* (Springer, New York, 2008)

92. B. Flagg, V. Knight-Williams, *Summative Evaluation of NISE Network's Public Forum: Nanotechnology in Health Care* (Multimedia Research, Bellport, 2008)
93. B.H. Harthorn, K. Bryant, J. Rogers, Gendered risk beliefs about emerging nanotechnologies in the US. In: *Monograph of the 2009 Nanoethics Graduate Education Symposium*. (University of Washington, Seattle, 2009). Available online: http://depts.washington.edu/ntethics/symposium/Nanoethics Special Edition Monograph.pdf
94. B.H. Harthorn, J. Rogers, C. Shearer, *Gender, Application Domain, and Ethical Dilemmas in Nano-Deliberation*. White paper for Nanotech Risk Perception Specialist Meeting, Santa Barbara, 29–30 Jan 2010
95. D.L. Kleinman, J. Delborne, A.A. Anderson, Engaging citizens: The high cost of citizen participation in high technology. *Public Understanding of Science* (Preprint 9 Oct 2009). doi: 10.1177/0963662509347137
96. J. Kuzma, J. Romanchek, A. Kokotovich, Upstream oversight assessment for agrifood nanotechnology: A case study approach. Risk Anal. **28**(4), 1081–1098 (2008)
97. Nanoscale Science, Engineering, and Technology Subcommittee (NSET), Committee on Technology, Office of Science and Technology Policy, *Regional, State, and Local Initiatives in Nanotechnology: Report of the National Nanotechnology Initiative Workshop*, Oklahoma City, 1–3 Apr 2009. (NSET, Washington, DC, 2010). Available online: http://www.nano.gov/html/res/pubs.html
98. Nanoscale Science, Engineering, and Technology Subcommittee (NSET), Committee on Technology, Office of Science and Technology Policy, *Regional, State, and Local Initiatives in Nanotechnology: Report of the National Nanotechnology Initiative Workshop*, Washington, DC, 30 Sept–1 Oct. (NSET, Washington, DC, 2005). Available online: http://www.nano.gov/html/res/pubs.html
99. Nanoscale Science, Engineering, and Technology Subcommittee (NSET), Committee on Technology, Office of Science and Technology Policy. *The National Nanotechnology Initiative Strategic Plan*, (NSET, Washington, DC, 2007). Available online: http://www.nano.gov/html/res/pubs.html
100. National Research Council, Committee on Forecasting Future Disruptive Technologies, *Persistent Forecasting of Disruptive Technologies* (National Academies Press, Washington, DC, 2009)
101. T. Satterfield, *Designing for Upstream Risk Perception Research: Malleability and Asymmetry in Judgments About Nanotechnologies*, White paper for nanotech risk perception specialist meeting, Santa Barbara, 29–30 Jan 2010
102. C. Selin, Expectations and the emergence of nanotechnology. Sci. Technol. Hum. Values **32**(2), 196–220 (2007). doi:10.1177/0162243906296918
103. P. Shapira, J. Wang, From lab to market: Strategies and issues in the commercialization of nanotechnology in China. Asian Bus. Manage. **8**(4), 461–489 (2009)
104. P. Shapira, J. Youtie, A.L. Porter, The emergence of social science research in nanotechnology. Scientometrics (2010). doi:Published online first at 10.1007/s11192-010-0204-x. March 25, 2010
105. C.E. Van Horn, J. Cleary, L. Hubbar, A. Fichtner, *A Profile of Nanotechnology Degree Programs in the United States,* (Center for nanotechnology in society, Tempe, 2009). Available online: http://www.cspo.org/library/reports/?action=getfile&file=186&section=lib

Selected Bibliography (2000–2009)

M. Alexandre, P. Dubois, Polymer-layered silicate nanocomposites: preparation, properties and uses of a new class of materials. Mater. Sci. Eng. R Rep. **28**(1–2), 1–63 (2000)

P. Anger, P. Bharadwaj, L. Novotny, Enhancement and quenching of single-molecule fluorescence. Phys. Rev. Lett. **96**(11), 113002 (2006)

A.S.P. Arico, B. Bruce, J.M. Tarascon Scrosati, W. VanSchalkwijk, Nanostructured materials for advanced energy conversion and storage devices. Nat. Mater. **4**(5), 366–377 (2005)

K. Ariga, J.P. Hilland, Q.M. Ji, Layer-by-layer assembly as a versatile bottom-up nanofabrication technique for exploratory research and realistic application. Phys. Chem. Chem. Phys. **9**(19), 2319–2340 (2007)

M.S. Arnold, A.A. Green, J.F. Hulvat, S.I. Stupp, M.C. Hersam, Sorting carbon nanotubes by electronic structure using density differentiation. Nat. Nanotechnol. **1**(1), 60–65 (2006)

D. Astruc, F. Lu, J.R. Aranzaes, Nanoparticles as recyclable catalysts: the frontier between homogeneous and heterogeneous catalysis. Angew. Chem. Int. Ed. Engl. **44**(48), 7852–7872 (2005)

P. Avouris, Z.H. Chen, V. Perebeinos, Carbon-based electronics. Nat. Nanotechnol. **2**(10), 605–615 (2007)

S.M. Bachilo, M.S. Strano, C. Kittrell, R.H. Hauge, R.E. Smalley, R.B. Weisman, Structure-assigned optical spectra of single-walled carbon nanotubes. Science **298**(5602), 2361–2366 (2002)

A. Bachtold, P. Hadley, T. Nakanishi, C. Dekker, Logic circuits with carbon nanotube transistors. Science **294**(5545), 1317–1320 (2001)

A.A. Balandin, S. Ghosh, W.Z. Bao, I. Calizo, D. Teweldebrhan, F. Miao, C.N. Lau, Superior thermal conductivity of single-layer graphene. Nano Lett. **8**(3), 902–907 (2008)

A.C. Balazs, T. Emrick, T.P. Russell, Nanoparticle polymer composites: where two small worlds meet. Science **314**(5802), 1107–1110 (2006)

F. Baletto, R. Ferrando, Structural properties of nanoclusters: energetic thermodynamic and kinetic effects. Rev. Mod. Phys. **77**(1), 371–423 (2005)

S. Banerjee, T. Hemraj-Benny, S.S. Wong, Covalent surface chemistry of single-walled carbon nanotubes. Adv. Mater. **17**(1), 17–29 (2005)

J.V. Barth, G. Costantini, K. Kern, Engineering atomic and molecular nanostructures at surfaces. Nature **437**(7059), 671–679 (2005)

R.H. Baughman, A.A. Zakhidov, W.A. deHeer, Carbon nanotubes – the route toward applications. Science **297**(5582), 787–792 (2002)

H.A. Becerril, J. Mao, Z. Liu, R.M. Stoltenberg, Z. Bao, Y. Chen, Evaluation of solution-processed reduced graphene oxide films as transparent conductors. ACS Nano **2**(3), 463–470 (2008)

P. Blake, P.D. Brimicombe, R.R. Nair, T.J. Booth, D. Jiang, F. Schedin, L.A. Ponomarenko, S.V. Morozov, H.F. Gleeson, E.W. Hill, A.K. Geim, K.S. Novoselov, Graphene-based liquid crystal device. Nano Lett. **8**(6), 1704–1708 (2008)

L. Bogani, W. Wernsdorfer, Molecular spintronics using single-molecule magnets. Nat. Mater. **7**(3), 179–186 (2008)

A.I. Boukai, Y. Bunimovich, J. Tahir-Kheli, J.-K. Yu, W.A. Goddard, J.R. Heath, Silicon nanowires as efficient thermoelectric materials. Nature **451**(7175), 168–171 (2008)

D. Branton, D.W. Deamer, A. Marziali, H. Bayley, S.A. Benner, T. Butler, M. DiVentra, S. Garaj, A. Hibbs, X.H. Huang, S.B. Jovanovich, P.S. Krstic, S. Lindsay, X.S.S. Ling, C.H. Mastrangelo, A. Meller, J.S. Oliver, Y.V. Pershin, J.M. Ramsey, R. Riehn, G.V. Soni, V. Tabard-Cossa, M. Wanunu, M. Wiggin, J.A. Schloss, The potential and challenges of nanopore sequencing. Nat. Biotechnol. **26**(10), 1146–1153 (2008)

L. Brunsveld, B.J.B. Folmer, E.W. Meijer, R.P. Sijbesma, Supramolecular polymers. Chem. Rev. **101**(12), 4071–4097 (2001)

C. Burda, X.B. Chen, R. Narayanan, M.A. El-Sayed, Chemistry and properties of nanocrystals of different shapes. Chem. Rev. **105**(4), 1025–1102 (2005)

H.J. Butt, B. Cappella, M. Kappl, Force measurements with the atomic force microscope: technique interpretation and applications. Surf. Sci. Rep. **59**(1–6), 1–152 (2005)

Y.W.C. Cao, R.C. Jin, C.A. Mirkin, Nanoparticles with Raman spectroscopic fingerprints for DNA and RNA detection. Science **297**(5586), 1536–1540 (2002)

Q. Cao, H.S. Kim, N. Pimparkar, J.P. Kulkarni, C.J. Wang, M. Shim, K. Roy, M.A. Alam, J.A. Rogers, Medium-scale carbon nanotube thin-film integrated circuits on flexible plastic substrates. Nature **454**(7203), 495–500 (2008)

C.K. Chan, H.L. Peng, G. Liu, K. McIlwrath, X.F. Zhang, R.A. Huggins, Y. Cui, High-performance lithium battery anodes using silicon nanowires. Nat. Nanotechnol. **3**(1), 31–35 (2008)

J.C. Charlier, X. Blase, S. Roche, Electronic and transport properties of nanotubes. Rev. Mod. Phys. **79**(2), 677–732 (2007)

R.J. Chen, Y.G. Zhang, D.W. Wang, H.J. Dai, Noncovalent sidewall functionalization of single-walled carbon nanotubes for protein immobilization. J. Am. Chem. Soc. **123**(16), 3838–3839 (2001)

Z.H. Chen, Y.M. Lin, M.J. Rooks, P. Avouris, Graphene nano-ribbon electronics. Phys. E Low Dimens. Syst. Nanostruct. **40**(2), 228–232 (2007)

J.H. Chen, C. Jang, S.D. Xiao, M. Ishigami, M.S. Fuhrer, Intrinsic and extrinsic performance limits of graphene devices on SiO_2. Nat. Nanotechnol. **3**(4), 206–209 (2008)

B.D. Chithrani, A.A. Ghazani, W.C.W. Chan, Determining the size and shape dependence of gold nanoparticle uptake into mammalian cells. Nano Lett. **6**(4), 662–668 (2006)

H.S. Choi, W. Liu, P. Misra, E. Tanaka, J.P. Zimmer, B.I. Ipe, M.G. Bawendi, J.V. Frangioni, Renal clearance of quantum dots. Nat. Biotechnol. **25**(10), 1165–1170 (2007)

J.N. Coleman, U. Khan, W.J. Blau, Y.K. Gun'ko, Small but strong: a review of the mechanical properties of carbon nanotube-polymer composites. Carbon **44**(9), 1624–1652 (2006a)

J.N. Coleman, U. Khan, Y.K. Gun'ko, Mechanical reinforcement of polymers using carbon nanotubes. Adv. Mater. **18**(6), 689–706 (2006b)

P.G. Collins, K. Bradley, M. Ishigami, A. Zettl, Extreme oxygen sensitivity of electronic properties of carbon nanotubes. Science **287**(5459), 1801–1804 (2000)

P.C. Collins, M.S. Arnold, P. Avouris, Engineering carbon nanotubes and nanotube circuits using electrical breakdown. Science **292**(5517), 706–709 (2001)

Y. Cui, C.M. Lieber, Functional nanoscale electronic devices assembled using silicon nanowire building blocks. Science **291**(5505), 851–853 (2001)

X.D. Cui, A. Primak, X. Zarate, J. Tomfohr, O.F. Sankey, A.L. Moore, T.A. Moore, D. Gust, G. Harris, S.M. Lindsay, Reproducible measurement of single-molecule conductivity. Science **294**(5542), 571–574 (2001a)

Y. Cui, Q.Q. Wei, H.K. Park, C.M. Lieber, Nanowire nanosensors for highly sensitive and selective detection of biological and chemical species. Science **293**(5533), 1289–1292 (2001b)

B.L. Cushing, V.L. Kolesnichenko, C.J. O'Connor, Recent advances in the liquid-phase syntheses of inorganic nanoparticles. Chem. Rev. **104**(9), 3893–3946 (2004)

M.C. Daniel, D. Astruc, Gold nanoparticles: assembly, supramolecular chemistry, quantum-size-related properties, and applications toward biology, catalysis, and nanotechnology. Chem. Rev. **104**(1), 293–346 (2004)

A. Das, S. Pisana, B. Chakraborty, S. Piscanec, S.K. Saha, U.V. Waghmare, K.S. Novoselov, H.R. Krishnamurthy, A.K. Geim, A.C. Ferrari, A.K. Sood, Monitoring dopants by Raman scattering in an electrochemically top-gated graphene transistor. Nat. Nanotechnol. **3**(4), 210–215 (2008)

M.E. Davis, Z. Chen, D.M. Shin, Nanoparticle therapeutics: an emerging treatment modality for cancer. Nat. Rev. Drug Discov. **7**(9), 771–782 (2008)

C. Dekker, Solid-state nanopores. Nat. Nanotechnol. **2**(4), 209–215 (2007)

A.M. Derfus, W.C.W. Chan, S.N. Bhatia, Probing the cytotoxicity of semiconductor quantum dots. Nano Lett. **4**(1), 11–18 (2004)

A.B. Djurisic, Y.H. Leung, Optical properties of ZnO nanostructures. Small **2**, 944–961 (2006)

M.S. Dresselhaus, G. Dresselhaus, R. Saito, A. Jorio, Phys. Rep. **409**(2), 47–99 (2005)

X. Du, I. Skachko, A. Barker, E.Y. Andrei, Approaching ballistic transport in suspended graphene. Nat. Nanotechnol. **3**(8), 491–495 (2008)

X.F. Duan, Y. Huang, Y. Cui, J.F. Wang, C.M. Lieber, Indium phosphide nanowires as building blocks for nanoscale electronic and optoelectronic devices. Nature **409**(6816), 66–69 (2001)

X.F. Duan, Y. Huang, R. Agarwal, C.M. Lieber, Single-nanowire electrically driven lasers. Nature **421**(6920), 241–245 (2003)

B. Dubertret, P. Skourides, D.J. Norris, V. Noireaux, A.H. Brivanlou, A. Libchaber, *In vivo* imaging of quantum dots encapsulated in phospholipid micelles. Science **298**(5599), 1759–1762 (2002)

G. Eda, G. Fanchini, M. Chhowalla, Large-area ultrathin films of reduced graphene oxide as a transparent and flexible electronic material. Nat. Nanotechnol. **3**(5), 270–274 (2008)

J.M. Elzerman, R. Hanson, L.H.W. vanBeveren, B. Witkamp, L.M.K. Vandersypen, L.P. Kouwenhoven, Single-shot read-out of an individual electron spin in a quantum dot. Nature **430**(6998), 431–435 (2004)

X.S. Fang, Y. Bando, U.K. Gautam, C. Ye, D. Golberg, Inorganic semiconductor nanostructures and their field-emission applications. J. Mater. Chem. **18**(5), 509–522 (2008)

R. Ferrando, J. Jellinek, R.L. Johnston, Nanoalloys: from theory to applications of alloy clusters and nanoparticles. Chem. Rev. **108**(3), 845–910 (2008)

M. Ferrari, Cancer nanotechnology: opportunities and challenges. Nat. Rev. Cancer **5**(3), 161–171 (2005)

X.H. Gao, Y.Y. Cui, R.M. Levenson, L.W.K. Chung, S.M. Nie, *In vivo* cancer targeting and imaging with semiconductor quantum dots. Nat. Biotechnol. **22**(8), 969–976 (2004)

P.X. Gao, Y. Ding, W.J. Mai, W.L. Hughes, C.S. Lao, Z.L. Wang, Conversion of zinc oxide nanobelts into superlattice-structured nanohelices. Science **309**(5741), 1700–1704 (2005)

F. Gao, Y. Wang, D. Shi, J. Zhang, M.K. Wang, X.Y. Jing, R. Humphry-Baker, P. Wang, S.M. Zakeeruddin, M. Gratzel, Enhance the optical absorptivity of nanocrystalline TiO_2 film with high molar extinction coefficient ruthenium sensitizers for high performance dye-sensitized solar cells. J. Am. Chem. Soc. **130**(32), 10720–10728 (2008)

B.D. Gates, Q.B. Xu, M. Stewart, D. Ryan, C.G. Willson, G.M. Whitesides, New approaches to nanofabrication: molding, printing, and other techniques. Chem. Rev. **105**(4), 1171–1196 (2005)

S. Gilje, S. Han, M. Wang, K.L. Wang, R.B. Kaner, A chemical route to graphene for device applications. Nano Lett. **7**(11), 3394–3398 (2007)

J. Goldberger, R.R. He, Y.F. Zhang, S.W. Lee, H.Q. Yan, H.J. Choi, P.D. Yang, Single-crystal gallium nitride nanotubes. Nature **422**(6932), 599–602 (2003)

C. Gómez-Navarro, R.T. Weitz, A.M. Bittner, M. Scolari, A. Mews, M. Burghard, K. Kern, Electronic transport properties of individual chemically reduced graphene oxide sheets. Nano Lett. **7**(11), 3499–3503 (2007)

D. Graf, F. Molitor, K. Ensslin, C. Stampfer, A. Jungen, C. Hierold, L. Wirtz, Spatially resolved Raman spectroscopy of single- and few-layer graphene. Nano Lett. **7**(2), 238–242 (2007)

M. Grätzel, Conversion of sunlight to electric power by nanocrystalline dye-sensitized solar cells. J. Photochem. Photobiol. Chem. **164**(1–3), 3–14 (2004)

M.S. Gudiksen, L.J. Lauhon, J. Wang, D.C. Smith, C.M. Lieber, Growth of nanowire superlattice structures for nanoscale photonics and electronics. Nature **415**(6872), 617–620 (2002)

A.K. Gupta, M. Gupta, Synthesis and surface engineering of iron oxide nanoparticles for biomedical applications. Biomaterials **26**(18), 3995–4021 (2005)

M.Y. Han, X.H. Gao, J.Z. Su, S. Nie, Quantum-dot-tagged microbeads for multiplexed optical coding of biomolecules. Nat. Biotechnol. **19**(7), 631–635 (2001)

M.Y. Han, B. Ozyilmaz, Y.B. Zhang, P. Kim, Energy band-gap engineering of graphene nanoribbons. Phys. Rev. Lett. **98**(20), 206805 (2007)

R. Hanson, L.P. Kouwenhoven, J.R. Petta, S. Tarucha, L.M.K. Vandersypen, Spins in few-electron quantum dots. Rev. Mod. Phys. **79**(4), 1217–1265 (2007)

R. Hardman, A toxicologic review of quantum dots: toxicity depends on physicochemical and environmental factors. Environ. Health Perspect. **114**(2), 165–172 (2006)

T.D. Harris, P.R. Buzby, H. Babcock, E. Beer, J. Bowers, I. Braslavsky, M. Causey, J. Colonell, J. Dimeo, J.W. Efcavitch, E. Giladi, J. Gill, J. Healy, M. Jarosz, D. Lapen, K. Moulton, S.R. Quake, K. Steinmann, E. Thayer, A. Tyurina, R. Ward, H. Weiss, Z. Xie, Single-molecule DNA sequencing of a viral genome. Science **320**(5872), 106–109 (2008)

K. Hata, D.N. Futaba, K. Mizuno, T. Namai, M. Yumura, S. Iijima, Water-assisted highly efficient synthesis of impurity-free single-waited carbon nanotubes. Science **306**(5700), 1362–1364 (2004)

J.H. He, An elementary introduction to recently developed asymptotic methods and nanomechanics in textile engineering. Int. J. Mod. Phys. B **22**(21), 3487–3578 (2008)

Y. He, T. Ye, M. Su, C. Zhang, A.E. Ribbe, W. Jiang, C.D. Mao, Hierarchical self-assembly of DNA into symmetric supramolecular polyhedra. Nature **452**(7184), 198–201 (2008)

M.W. Heaven, A. Dass, P.S. White, K.M. Holt, R.W. Murray, Crystal structure of the gold nanoparticle [N(C8H17)(4)][Au-25(SCH2CH2Ph)(18)]. J. Am. Chem. Soc. **130**(12), 3754–3755 (2008)

K. Hennessy, A. Badolato, M. Winger, D. Gerace, M. Atature, S. Gulde, S. Falt, E.L. Hu, A. Imamoglu, Quantum nature of a strongly coupled single quantum dot-cavity system. Nature **445**(7130), 896–899 (2007)

Y. Hernandez, V. Nicolosi, M. Lotya, F.M. Blighe, Z.Y. Sun, S. De, I.T. McGovern, B. Holland, M. Byrne, Y.K. Gun'ko, J.J. Boland, P. Niraj, G. Duesberg, S. Krishnamurthy, R. Goodhue, J. Hutchison, V. Scardaci, A.C. Ferrari, J.N. Coleman, High-yield production of graphene by liquid-phase exfoliation of graphite. Nat. Nanotechnol. **3**(9), 563–568 (2008)

M.C. Hersam, Progress towards monodisperse single-walled carbon nanotubes. Nat. Nanotechnol. **3**(7), 387–394 (2008)

A.A. Herzing, C.J. Kiely, A.F. Carley, P. Landon, G.J. Hutchings, Identification of active gold nanoclusters on iron oxide supports for CO oxidation. Science **321**(5894), 1331–1335 (2008)

A. Hirsch, Functionalization of single-walled carbon nanotubes. Angew. Chem. Int. Ed Engl. **41**(11), 1853–1859 (2002)

A.I. Hochbaum, R.K. Chen, R.D. Delgado, W.J. Liang, E.C. Garnett, M. Najarian, A. Majumdar, P.D. Yang, Enhanced thermoelectric performance of rough silicon nanowires. Nature **451**(7175), 163–167 (2008)

F.J.M. Hoeben, P. Jonkheijm, E.W. Meijer, A.P.H.J. Schenning, About supramolecular assemblies of pi-conjugated systems. Chem. Rev. **105**(4), 1491–1546 (2005)

B.J. Holliday, C.A. Mirkin, Strategies for the construction of supramolecular compounds through coordination chemistry. Angew. Chem. Int. Ed Engl. **40**(11), 2022–2043 (2001)

J.K. Holt, H.G. Park, Y.M. Wang, M. Stadermann, A.B. Artyukhin, C.P. Grigoropoulos, A. Noy, O. Bakajin, Fast mass transport through sub-2-nanometer carbon nanotubes. Science **312**(5776), 1034–1037 (2006)

I. Horcas, R. Fernandez, J.M. Gómez-Rodriguez, J. Colchero, J. Gómez-Herrero, A.M. Baro, WSXM: a software for scanning probe microscopy and a tool for nanotechnology. Rev. Sci. Instrum. **78**(1), 013705 (2007)

M.H. Huang, S. Mao, H. Feick, H.Q. Yan, Y.Y. Wu, H. Kind, E. Weber, R. Russo, P.D. Yang, Room-temperature ultraviolet nanowire nanolasers. Science **292**(5523), 1897–1899 (2001a)

M.H. Huang, Y.Y. Wu, H. Feick, N. Tran, E. Weber, P.D. Yang, Catalytic growth of zinc oxide nanowires by vapor transport. Adv. Mater. **13**(2), 113–116 (2001b)

Y. Huang, X.F. Duan, Y. Cui, L.J. Lauhon, K.H. Kim, C.M. Lieber, Logic gates and computation from assembled nanowire building blocks. Science **294**(5545), 1313–1317 (2001c)

Y. Huang, X.F. Duan, Q.Q. Wei, C.M. Lieber, Directed assembly of one-dimensional nanostructures into functional networks. Science **291**(5504), 630–633 (2001d)

Z.M. Huang, Y.Z. Zhang, M. Kotaki, S. Ramakrishna, A review on polymer nanofibers by electrospinning and their applications in nanocomposites. Compos. Sci. Technol. **63**(15), 2223–2253 (2003)

X.H. Huang, I.H. El-Sayed, W. Qian, M.A. El-Sayed, Cancer cell imaging and photothermal therapy in the near-infrared region by using gold nanorods. J. Am. Chem. Soc. **128**(6), 2115–2120 (2006)

W.U. Huynh, J.J. Dittmer, A.P. Alivisatos, Hybrid nanorod-polymer solar cells. Science **295**(5564), 2425–2427 (2002)

M. Ishigami, J.H. Chen, W.G. Cullen, M.S. Fuhrer, E.D. Williams, Atomic structure of graphene on SiO_2. Nano Lett. **7**(6), 1643–1648 (2007)

P.D. Jadzinsky, G. Calero, C.J. Ackerson, D.A. Bushnell, R.D. Kornberg, Structure of a thiol monolayer-protected gold nanoparticle at 1.1 angstrom resolution. Science **318**(5849), 430–433 (2007)

P.K. Jain, K.S. Lee, I.H. El-Sayed, M.A. El-Sayed, Calculated absorption and scattering properties of gold nanoparticles of different size, shape, and composition: Applications in biological imaging and biomedicine. J. Phys. Chem. B **110**(14), 7238–7248 (2006)

P.K. Jain, X.H. Huang, I.H. El-Sayed, M.A. El-Sayed, Noble metals on the nanoscale: optical and photothermal properties and some applications in imaging sensing biology and medicine. Acc. Chem. Res. **41**(12), 1578–1586 (2008)

J.K. Jaiswal, H. Mattoussi, J.M. Mauro, S.M. Simon, Long-term multiple color imaging of live cells using quantum dot bioconjugates. Nat. Biotechnol. **21**(1), 47–51 (2003)

A. Javey, J. Guo, Q. Wang, M. Lundstrom, H.J. Dai, Ballistic carbon nanotube field-effect transistors. Nature **424**(6949), 654–657 (2003)

W. Jiang, B.Y.S. Kim, J.T. Rutka, W.C.W. Chan, Nanoparticle-mediated cellular response is size-dependent. Nat. Nanotechnol. **3**(3), 145–150 (2008)

R.C. Jin, Y.W. Cao, C.A. Mirkin, K.L. Kelly, G.C. Schatz, J.G. Zheng, Photoinduced conversion of silver nanospheres to nanoprisms. Science **294**(5548), 1901–1903 (2001)

Y.W. Jun, J.S. Choi, J. Cheon, Shape control of semiconductor and metal oxide nanocrystals through nonhydrolytic colloidal routes. Angew. Chem. Int. Ed. Engl. **45**(21), 3414–3439 (2006)

N.W.S. Kam, M. O'Connell, J.A. Wisdom, H.J. Dai, Carbon nanotubes as multifunctional biological transporters and near-infrared agents for selective cancer cell destruction. Proc. Natl Acad. Sci. U.S.A. **102**(33), 11600–11605 (2005)

P.V. Kamat, Meeting the clean energy demand: nanostructure architectures for solar energy conversion. J. Phys. Chem. C **111**(7), 2834–2860 (2007)

P.V. Kamat, Quantum Dot Solar Cells. Semiconductor nanocrystals as light harvesters. J. Phys. Chem. C **112**(48), 18737–18753 (2008)

S.J. Kang, C. Kocabas, T. Ozel, M. Shim, N. Pimparkar, M.A. Alam, S.V. Rotkin, J.A. Rogers, High-performance electronics using dense, perfectly aligned arrays of single-walled carbon nanotubes. Nat. Nanotechnol. **2**(4), 230–236 (2007)

E. Katz, I. Willner, Integrated nanoparticle-biomolecule hybrid systems: synthesis, properties, and applications. Angew. Chem. Int. Ed Engl. **43**(45), 6042–6108 (2004)

E.R. Kay, D.A. Leigh, F. Zerbetto, Synthetic molecular motors and mechanical machines. Angew. Chem. Int. Ed Engl. **46**(1–2), 72–191 (2007)

K.L. Kelly, E. Coronado, L.L. Zhao, G.C. Schatz, The optical properties of metal nanoparticles: the influence of size, shape, and dielectric environment. J. Phys. Chem. B **107**(3), 668–677 (2003)

S. Kim, Y.T. Lim, E.G. Soltesz, A.M. DeGrand, J. Lee, A. Nakayama, J.A. Parker, T. Mihaljevic, R.G. Laurence, D.M. Dor, L.H. Cohn, M.G. Bawendi, J.V. Frangioni, Near-infrared fluorescent type II quantum dots for sentinel lymph node mapping. Nat. Biotechnol. **22**(1), 93–97 (2004)

C. Kirchner, T. Liedl, S. Kudera, T. Pellegrino, A.M. Javier, H.E. Gaub, S. Stolzle, N. Fertig, W.J. Parak, Cytotoxicity of colloidal CdSe and CdSe/ZnS nanoparticles. Nano Lett. **5**(2), 331–338 (2005)

J. Kong, N.R. Franklin, C.W. Zhou, M.G. Chapline, S. Peng, K.J. Cho, H.J. Dai, Nanotube molecular wires as chemical sensors. Science **287**(5453), 622–625 (2000)

X.Y. Kong, Y. Ding, R. Yang, Z.L. Wang, Single-crystal nanorings formed by epitaxial self-coiling of polar nanobelts. Science **303**(5662), 1348–1351 (2004)

A. Kongkanand, K. Tvrdy, K. Takechi, M. Kuno, P.V. Kamat, Quantum dot solar cells. Tuning photoresponse through size and shape control of $CdSe-TiO_2$ architecture. J. Am. Chem. Soc. **130**(12), 4007–4015 (2008)

F.H.L. Koppens, C. Buizert, K.J. Tielrooij, I.T. Vink, K.C. Nowack, T. Meunier, L.P. Kouwenhoven, L.M.K. Vandersypen, Driven coherent oscillations of a single electron spin in a quantum dot. Nature **442**(7104), 766–771 (2006)

R. Krupke, F. Hennrich, H. vonLohneysen, M.M. Kappes, Separation of metallic from semiconducting single-walled carbon nanotubes. Science **301**(5631), 344–347 (2003)

K.N. Kudin, B. Ozbas, H.C. Schniepp, R.K. Prud'homme, I.A. Aksay, R. Car, Raman spectra of graphite oxide and functionalized graphene sheets. Nano Lett. **8**(1), 36–41 (2008)

T. Kuwana, M.R. Mackey, G. Perkins, M.H. Ellisman, M. Latterich, R. Schneiter, D.R. Green, D.D. Newmeyer, Bid, Bax, and lipids cooperate to form supramolecular openings in the outer mitochondrial membrane. Cell **111**(3), 331–342 (2002)

C.W. Lam, J.T. James, R. McCluskey, R.L. Hunter, Pulmonary toxicity of single-wall carbon nanotubes in mice 7 and 90 days after intratracheal instillation. Toxicol. Sci. **77**(1), 126–134 (2004)

D.R. Larson, W.R. Zipfel, R.M. Williams, S.W. Clark, M.P. Bruchez, F.W. Wise, W.W. Webb, Water-soluble quantum dots for multiphoton fluorescence imaging *in vivo*. Science **300**(5624), 1434–1436 (2003)

S. Laurent, D. Forge, M. Port, A. Roch, C. Robic, L.V. Elst, R.N. Muller, Magnetic iron oxide nanoparticles: synthesis, stabilization, vectorization, physicochemical characterizations, and biological applications. Chem. Rev. **108**(6), 2064–2110 (2008)

M. Law, J. Goldberger, P.D. Yang, Semiconductor nanowires and nanotubes. Annu. Rev. Mater. Sci. **34**, 83–122 (2004)

M. Law, L.E. Greene, J.C. Johnson, R. Saykally, P.D. Yang, Nanowire dye-sensitized solar cells. Nat. Mater. **4**(6), 455–459 (2005)

J.H. Lee, Y.M. Huh, Y. Jun, J. Seo, J. Jang, H.T. Song, S. Kim, E.J. Cho, H.G. Yoon, J.S. Suh, J. Cheon, Artificially engineered magnetic nanoparticles for ultra-sensitive molecular imaging. Nat. Med. **13**(1), 95–99 (2007a)

J.S. Lee, M.S. Han, C.A. Mirkin, Colorimetric detection of mercuric ion (Hg2+) in aqueous media using DNA-functionalized gold nanoparticles. Angew. Chem. Int. Ed Engl. **46**(22), 4093–4096 (2007b)

C. Lee, X.D. Wei, J.W. Kysar, J. Hone, Measurement of the elastic properties and intrinsic strength of monolayer graphene. Science **321**(5887), 385–388 (2008)

S. Leininger, B. Olenyuk, P.J. Stang, Self-assembly of discrete cyclic nanostructures mediated by transition metals. Chem. Rev. **100**(3), 853–907 (2000)

N. Lewinski, V. Colvin, R. Drezek, Cytotoxicity of nanoparticles. Small **4**(1), 26–49 (2008)

D. Li, Y.N. Xia, Electrospinning of nanofibers: reinventing the wheel? Adv. Mater. **16**(14), 1151–1170 (2004)

D. Li, M.B. Muller, S. Gilje, R.B. Kaner, G.G. Wallace, Processable aqueous dispersions of graphene nanosheets. Nat. Nanotechnol. **3**(2), 101–105 (2008a)

D. Li, A. Wieckowska, I. Willner, Optical analysis of Hg2+ ions by oligonucleotide-gold-nanoparticle hybrids and DNA-based machines. Angew. Chem. Int. Ed Engl. **47**(21), 3927–3931 (2008b)

X.L. Li, X.R. Wang, L. Zhang, S.W. Lee, H.J. Dai, Chemically derived, ultrasmooth graphene nanoribbon semiconductors. Science **319**(5867), 1229–1232 (2008c)

X.L. Li, G.Y. Zhang, X.D. Bai, X.M. Sun, X.R. Wang, E. Wang, H.J. Dai, Highly conducting graphene sheets and Langmuir-Blodgett films. Nat. Nanotechnol. **3**(9), 538–542 (2008d)

W.J. Liang, M.P. Shores, M. Bockrath, J.R. Long, H. Park, Kondo resonance in a single-molecule transistor. Nature **417**(6890), 725–729 (2002)

C.M. Lieber, Z.L. Wang, Functional nanowires. MRS Bull. **32**(2), 99–108 (2007)

M. Liong, J. Lu, M. Kovochich, T. Xia, S.G. Ruehm, A.E. Nel, F. Tamanoi, J.I. Zink, Multifunctional inorganic nanoparticles for imaging, targeting, and drug delivery. ACS Nano **2**(5), 889–896 (2008)

Z. Liu, W.B. Cai, L.N. He, N. Nakayama, K. Chen, X.M. Sun, X.Y. Chen, H.J. Dai, *In vivo* biodistribution and highly efficient tumour targeting of carbon nanotubes in mice. Nat. Nanotechnol. **2**(1), 47–52 (2007)

W. Liu, M. Howarth, A.B. Greytak, Y. Zheng, D.G. Nocera, A.Y. Ting, M.G. Bawendi, Compact biocompatible quantum dots functionalized for cellular imaging. J. Am. Chem. Soc. **130**(4), 1274–1284 (2008a)

Z. Liu, C. Davis, W.B. Cai, L. He, X.Y. Chen, H.J. Dai, Circulation and long-term fate of functionalized, biocompatible single-walled carbon nanotubes in mice probed by Raman spectroscopy. Proc. Natl Acad. Sci. U.S.A. **105**(5), 1410–1415 (2008b)

L.M. Liz-Marzán, Tailoring surface plasmons through the morphology and assembly of metal nanoparticles. Langmuir **22**(1), 32–41 (2006)

D.L. Long, E. Burkholder, L. Cronin, Polyoxometalate clusters, nanostructures, and materials: from self assembly to designer materials and devices. Chem. Soc. Rev. **36**(1), 105–121 (2007)

C. Loo, A. Lowery, N.J. Halas, J. West, R. Drezek, Immunotargeted nanoshells for integrated cancer imaging and therapy. Nano Lett. **5**(4), 709–711 (2005)

X.W. Lou, L.A. Archer, Z.C. Yang, Hollow micro-/nanostructures: synthesis and applications. Adv. Mater. **20**(21), 3987–4019 (2008)

J.C. Love, L.A. Estroff, J.K. Kriebel, R.G. Nuzzo, G.M. Whitesides, Self-assembled monolayers of thiolates on metals as a form of nanotechnology. Chem. Rev. **105**(4), 1103–1169 (2005)

W. Lu, C.M. Lieber, Nanoelectronics from the bottom up. Nat. Mater. **6**(11), 841–850 (2007)

A.H. Lu, E.L. Salabas, F. Schuth, Magnetic nanoparticles: synthesis, protection, functionalization, and application. Angew. Chem. Int. Ed Engl. **46**(8), 1222–1244 (2007)

W.L. Ma, C.Y. Yang, X. Gong, K. Lee, A.J. Heeger, Thermally stable, efficient polymer solar cells with nanoscale control of the interpenetrating network morphology. Adv. Funct. Mater. **15**(10), 1617–1622 (2005)

J.M. Macak, H. Tsuchiya, P. Schmuki, High-aspect-ratio TiO_2 nanotubes by anodization of titanium. Angew. Chem. Int. Ed Engl. **44**(14), 2100–2102 (2005)

S.A. Maier, P.G. Kik, H.A. Atwater, S. Meltzer, E. Harel, B.E. Koel, A.A.G. Requicha, Local detection of electromagnetic energy transport below the diffraction limit in metal nanoparticle plasmon waveguides. Nat. Mater. **2**(4), 229–232 (2003)

I.L. Medintz, H.T. Uyeda, E.R. Goldman, H. Mattoussi, Quantum dot bioconjugates for imaging, labelling, and sensing. Nat. Mater. **4**(6), 435–446 (2005)

M.A. Meyers, A. Mishra, D.J. Benson, Mechanical properties of nanocrystalline materials. Prog. Mater. Sci. **51**(4), 427–556 (2006)

X. Michalet, F.F. Pinaud, L.A. Bentolila, J.M. Tsay, S. Doose, J.J. Li, G. Sundaresan, A.M. Wu, S.S. Gambhir, S. Weiss, Quantum dots for live cells, *in vivo* imaging, and diagnostics. Science **307**(5709), 538–544 (2005)

M. Moniruzzaman, K.I. Winey, Polymer nanocomposites containing carbon nanotubes. Macromolecules **39**(16), 5194–5205 (2006)

G.K. Mor, K. Shankar, M. Paulose, O.K. Varghese, C.A. Grimes, Use of highly ordered TiO_2 nanotube arrays in dye-sensitized solar cells. Nano Lett. **6**(2), 215–218 (2006a)

G.K. Mor, O.K. Varghese, M. Paulose, K. Shankar, C.A. Grimes, A review on highly ordered, vertically oriented TiO_2 nanotube arrays: fabrication, material properties, and solar energy applications. Sol. Energy Mater. Sol. Cells **90**(14), 2011–2075 (2006b)

B. Moulton, M.J. Zaworotko, From molecules to crystal engineering: supramolecular isomerism and polymorphism in network solids. Chem. Rev. **101**(6), 1629–1658 (2001)

C.J. Murphy, T.K. San, A.M. Gole, C.J. Orendorff, J.X. Gao, L. Gou, S.E. Hunyadi, T. Li, Anisotropic metal nanoparticles: synthesis, assembly, and optical applications. J. Phys. Chem. B **109**(29), 13857–13870 (2005)

C.J. Murphy, A.M. Gole, J.W. Stone, P.N. Sisco, A.M. Alkilany, E.C. Goldsmith, S.C. Baxter, Gold nanoparticles in biology: beyond toxicity to cellular imaging. Acc. Chem. Res. **41**(12), 1721–1730 (2008)

R.W. Murray, Nanoelectrochemistry: metal nanoparticles, nanoelectrodes, and nanopores. Chem. Rev. **108**(7), 2688–2720 (2008)

J.M. Nam, C.S. Thaxton, C.A. Mirkin, Nanoparticle-based bio-bar codes for the ultrasensitive detection of proteins. Science **301**(5641), 1884–1886 (2003)

K.T. Nam, D.W. Kim, P.J. Yoo, C.Y. Chiang, N. Meethong, P.T. Hammond, Y.M. Chiang, A.M. Belcher, Virus-enabled synthesis and assembly of nanowires for lithium ion battery electrodes. Science **312**(5775), 885–888 (2006)

C.M. Niemeyer, Nanoparticles, proteins, and nucleic acids: biotechnology meets materials science. Angew. Chem. Int. Ed Engl. **40**(22), 4128–4158 (2001)

D.J. Norris, A.L. Efros, S.C. Erwin, Doped nanocrystals. Science **319**(5871), 1776–1779 (2008)

K.S. Novoselov, A.K. Geim, S.V. Morozov, D. Jiang, Y. Zhang, S.V. Dubonos, I.V. Grigorieva, A. Firsov, Electric field effect in atomically thin carbon films. Science **306**(5696), 666–669 (2004)

D. Nykypanchuk, M.M. Maye, D. Vander Lelie, D. vander Lelie, O. Gang, DNA-guided crystallization of colloidal nanoparticles. Nature **451**(7178), 549–552 (2008)

M.J. O'Connell, S.M. Bachilo, C.B. Huffman, V.C. Moore, M.S. Strano, E.H. Haroz, K.L. Rialon, P.J. Boul, W.H. Noon, C. Kittrell, J.P. Ma, R.H. Hauge, R.B. Weisman, R.E. Smalley, Band gap fluorescence from individual single-walled carbon nanotubes. Science **297**(5581), 593–596 (2002)

E. Ozbay, Plasmonics: merging photonics and electronics at nanoscale dimensions. Science **311**(5758), 189–193 (2006)

Z.W. Pan, Z.R. Dai, Z.L. Wang, Nanobelts of semiconducting oxides. Science **291**(5510), 1947–1949 (2001)

S.J. Park, T.A. Taton, C.A. Mirkin, Array-based electrical detection of DNA with nanoparticle probes. Science **295**(5559), 1503–1506 (2002)

J. Park, K.J. An, Y.S. Hwang, J.G. Park, H.J. Noh, J.Y. Kim, J.H. Park, N.M. Hwang, T. Hyeon, Ultra-large-scale syntheses of monodisperse nanocrystals. Nat. Mater. **3**(12), 891–895 (2004)

J.H. Park, S. Kim, A.J. Bard, Novel carbon-doped TiO_2 nanotube arrays with high aspect ratios for efficient solar water splitting. Nano Lett. **6**(1), 24–28 (2006)

J. Park, J. Joo, S.G. Kwon, Y. Jang, T. Hyeon, Synthesis of monodisperse spherical nanocrystals. Angew. Chem. Int. Ed Engl. **46**(25), 4630–4660 (2007)

S.Y. Park, A.K.R. Lytton-Jean, B. Lee, S. Weigand, G.C. Schatz, C.A. Mirkin, DNA-programmable nanoparticle crystallization. Nature **451**(7178), 553–556 (2008)

G.R. Patzke, F. Krumeich, R. Nesper, Oxidic nanotubes and nanorods – anisotropic modules for a future nanotechnology. Angew. Chem. Int. Ed Engl. **41**(14), 2446–2461 (2002)

D.R. Paul, L.M. Robeson, Polymer nanotechnology: nanocomposites. Polymer **49**(15), 3187–3204 (2008)

L. Pavesi, L. DalNegro, C. Mazzoleni, G. Franzo, F. Priolo, Optical gain in silicon nanocrystals. Nature **408**(6811), 440–444 (2000)

S. Pavlidou, C.D. Papaspyrides, A review on polymer-layered silicate nanocomposites. Prog. Polym. Sci. **33**(12), 1119–1198 (2008)

D. Peer, J.M. Karp, S. Hong, O.C. Farokhzad, R. Margalit, R. Langer, Nanocarriers as an emerging platform for cancer therapy. Nat. Nanotechnol. **2**(12), 751–760 (2007)

X.G. Peng, L. Manna, W.D. Yang, J. Wickham, E. Scher, A. Kadavanich, A.P. Alivisatos, Shape control of CdSe nanocrystals. Nature **404**(6773), 59–61 (2000)

J.R. Petta, A.C. Johnson, J.M. Taylor, E.A. Laird, A. Yacoby, M.D. Lukin, C.M. Marcus, M.P. Hanson, A.C. Gossard, Coherent manipulation of coupled electron spins in semiconductor quantum dots. Science **309**(5744), 2180–2184 (2005)

J.C. Phillips, R. Braun, W. Wang, J. Gumbart, E. Tajkhorshid, E. Villa, C. Chipot, R.D. Skeel, L. Kale, K. Schulten, Scalable molecular dynamics with NAMD. J. Comput. Chem. **26**(16), 1781–1802 (2005)

C.A. Poland, R. Duffin, I. Kinloch, A. Maynard, W.A.H. Wallace, A. Seaton, V. Stone, S. Brown, W. MacNee, K. Donaldson, Carbon nanotubes introduced into the abdominal cavity of mice show asbestos-like pathogenicity in a pilot study. Nat. Nanotechnol. **3**(7), 423–428 (2008)

L.A. Ponomarenko, F. Schedin, M.I. Katsnelson, R. Yang, E.W. Hill, K.S. Novoselov, A.K. Geim, Chaotic dirac billiard in graphene quantum dots. Science **320**(5874), 356–358 (2008)

H.W.C. Postma, T. Teepen, Z. Yao, M. Grifoni, C. Dekker, Carbon nanotube single-electron transistors at room temperature. Science **293**(5527), 76–79 (2001)

B. Poudel, Q. Hao, Y. Ma, Y.C. Lan, A. Minnich, B. Yu, X.A. Yan, D.Z. Wang, A. Muto, D. Vashaee, X.Y. Chen, J.M. Liu, M.S. Dresselhaus, G. Chen, Z.F. Ren, High-thermoelectric performance of nanostructured bismuth antimony telluride bulk alloys. Science **320**(5876), 634–638 (2008)

M. Prato, K. Kostarelos, A. Bianco, Functionalized carbon nanotubes in drug design and discovery. Acc. Chem. Res. **41**(1), 60–68 (2008)

V.F. Puntes, K.M. Krishnan, A.P. Alivisatos, Colloidal nanocrystal shape and size control: the case of cobalt. Science **291**(5511), 2115–2117 (2001)

X.M. Qian, X.-H. Peng, D.O. Ansari, Q. Yin-Goen, G.Z. Chen, D.M. Shin, L. Yang, A.N. Young, M.D. Wang, S.M. Nie, *In vivo* tumor targeting and spectroscopic detection with surface-enhanced Raman nanoparticle tags. Nat. Biotechnol. **26**(1), 83–90 (2008)

Y. Qin, X.D. Wang, Z.L. Wang, Microfibre-nanowire hybrid structure for energy scavenging. Nature **451**(7180), 809–813 (2008)

T. Ramanathan, A.A. Abdala, S. Stankovich, D.A. Dikin, M. Herrera-Alonso, R.D. Piner, D.H. Adamson, H.C. Schniepp, X. Chen, R.S. Ruoff, S.T. Nguyen, I.A. Aksay, R.K. Prud'homme, L.C. Brinson, Functionalized graphene sheets for polymer nanocomposites. Nat. Nanotechnol. **3**(6), 327–331 (2008)

S.S. Ray, M. Okamoto, Polymer/layered silicate nanocomposites: a review from preparation to processing. Prog. Polym. Sci. **28**(11), 1539–1641 (2003)

U. Resch-Genger, M. Grabolle, S. Cavaliere-Jaricot, R. Nitschke, T. Nann, Quantum dots versus organic dyes as fluorescent labels. Nat. Meth. **5**(9), 763–775 (2008)

M. Rescigno, M. Urbano, B. Valzasina, M. Francolini, G. Rotta, R. Bonasio, F. Granucci, J.-P. Kraehenbuhl, P. Ricciardi-Castagnoli, Dendritic cells express tight junction proteins and penetrate gut epithelial monolayers to sample bacteria. Nat. Immunol. **2**(4), 361–367 (2001)

I. Robel, V. Subramanian, M. Kuno, P.V. Kamat, Quantum dot solar cells. Harvesting light energy with CdSe nanocrystals molecularly linked to mesoscopic TiO_2 films. J. Am. Chem. Soc. **128**(7), 2385–2393 (2006)

N.L. Rosi, C.A. Mirkin, Nanostructures in biodiagnostics. Chem. Rev. **105**(4), 1547–1562 (2005)

N.L. Rosi, D.A. Giljohann, C.S. Thaxton, A.K.R. Lytton-Jean, M.S. Han, C.A. Mirkin, Oligonucleotide-modified gold nanoparticles for intracellular gene regulation. Science **312**(5776), 1027–1030 (2006)

P.W.K. Rothemund, Folding DNA to create nanoscale shapes and patterns. Nature **440**(7082), 297–302 (2006)

C. Sanchez, B. Julian, P. Belleville, M. Popall, Applications of hybrid organic-inorganic nanocomposites. J. Mater. Chem. **15**(35–36), 3559–3592 (2005)

M. Sheng, C. Sala, P.D.Z domains and the organization of supramolecular complexes. Annu. Rev. Neurosci. **24**, 1–29 (2001)

E.V. Shevchenko, D.V. Talapin, N.A. Kotov, S. O'Brien, C.B. Murray, Structural diversity in binary nanoparticle superlattices. Nature **439**(7072), 55–59 (2006)

T. Shimizu, M. Masuda, H. Minamikawa, Supramolecular nanotube architectures based on amphiphilic molecules. Chem. Rev. **105**(4), 1401–1443 (2005)

Y. Si, E.T. Samulski, Synthesis of water soluble graphene. Nano Lett. **8**(6), 1679–1682 (2008)

G.A. Silva, C. Czeisler, K.L. Niece, E. Beniash, D.A. Harrington, J.A. Kessler, S.I. Stupp, Selective differentiation of neural progenitor cells by high-epitope density nanofibers. Science **303**(5662), 1352–1355 (2004)

C. Soci, A. Zhang, B. Xiang, S.A. Dayeh, D.P.R. Aplin, J. Park, X.Y. Bao, Y.H. Lo, D. Wang, ZnO nanowire UV photodetectors with high internal gain. Nano Lett. **7**(4), 1003–1009 (2007)

Y.W. Son, M.L. Cohen, S.G. Louie, Energy gaps in graphene nanoribbons. Phys. Rev. Lett. **97**(21), 216803–216807 (2006a)

Y.W. Son, M.L. Cohen, S.G. Louie, Half-metallic graphene nanoribbons. Nature **444**(7117), 347–349 (2006b)

S. Stankovich, D.A. Dikin, R.D. Piner, K.A. Kohlhaas, A. Kleinhammes, Y. Jia, Y. Wu, S.T. Nguyen, R.S. Ruoff, Synthesis of graphene-based nanosheets via chemical reduction of exfoliated graphite oxide. Carbon **45**(7), 1558–1565 (2007)

E. Stern, J.F. Klemic, D.A. Routenberg, P.N. Wyrembak, D.B. Turner-Evans, A.D. Hamilton, D.A. LaVan, T.M. Fahmy, M.A. Reed, Label-free immunodetection with CMOS-compatible semiconducting nanowires. Nature **445**(7127), 519–522 (2007)

M.E. Stewart, C.R. Anderton, L.B. Thompson, J. Maria, S.K. Gray, J.A. Rogers, R.G. Nuzzo, Nanostructured plasmonic sensors. Chem. Rev. **108**(2), 494–521 (2008)

M.D. Stoller, S.J. Park, Y.W. Zhu, J.H. An, R.S. Ruoff, Graphene-based ultracapacitors. Nano Lett. **8**(10), 3498–3502 (2008)

Y.G. Sun, Y.N. Xia, Shape-controlled synthesis of gold and silver nanoparticles. Science **298**(5601), 2176–2179 (2002)

S.H. Sun, C.B. Murray, D. Weller, L. Folks, A. Moser, Monodisperse FePt nanoparticles and ferromagnetic FePt nanocrystal superlattices. Science **287**(5460), 1989–1992 (2000)

S.H. Sun, H. Zeng, D.B. Robinson, S. Raoux, P.M. Rice, S.X. Wang, G.X. Li, Monodisperse MFe_2O_4 (M = FeCoMn) nanoparticles. J. Am. Chem. Soc. **126**(1), 273–279 (2004)

Z.Y. Tang, N.A. Kotov, M. Giersig, Spontaneous organization of single CdTe nanoparticles into luminescent nanowires. Science **297**(5579), 237–240 (2002)

A.R. Tao, S. Habas, P.D. Yang, Shape control of colloidal metal nanocrystals. Small **4**(3), 310–325 (2008)

D. Tasis, N. Tagmatarchis, A. Bianco, M. Prato, Chemistry of carbon nanotubes. Chem. Rev. **106**(3), 1105–1136 (2006)

A.C. Templeton, M.P. Wuelfing, R.W. Murray, Monolayer protected cluster molecules. Acc. Chem. Res. **33**(1), 27–36 (2000)

E.T. Thostenson, Z.F. Ren, T.W. Chou, Advances in the science and technology of carbon nanotubes and their composites: a review. Compos. Sci. Technol. **61**(13), 1899–1912 (2001)

B.Z. Tian, X.L. Zheng, T.J. Kempa, Y. Fang, N.F. Yu, G.H. Yu, J.L. Huang, C.M. Lieber, Coaxial silicon nanowires as solar cells and nanoelectronic power sources. Nature **449**(7164), 885–889 (2007a)

N. Tian, Z.Y. Zhou, S.G. Sun, Y. Ding, Z.L. Wang, Synthesis of tetrahexahedral platinum nanocrystals with high-index facets and high electro-oxidation activity. Science **316**(5825), 732–735 (2007b)

M. Turner, V.B. Golovko, O.P.H. Vaughan, P. Abdulkin, A. Berenguer-Murcia, M.S. Tikhov, B.F.G. Johnson, R.M. Lambert, Selective oxidation with dioxygen by gold nanoparticle catalysts derived from 55-atom clusters. Nature **454**(7207), 981–983 (2008)

R.Z. Valiev, R.K. Islamgaliev, I.V. Alexandrov, Bulk nanostructured materials from severe plastic deformation. Prog. Mater. Sci. **45**(2), 103–189 (2000)

W.G. Vander Wiel, S. DeFranceschi, J.M. Elzerman, T. Fujisawa, S. Tarucha, L.P. Kouwenhoven, Electron transport through double quantum dots. Rev. Mod. Phys. **75**(1), 1–22 (2003)

L. Vayssieres, Growth of arrayed nanorods and nanowires of ZnO from aqueous solutions. Adv. Mater. **15**(5), 464–466 (2003)

J. Wang, Carbon-nanotube based electrochemical biosensors: a review. Electroanalysis **17**(1), 7–14 (2005)

Z.L. Wang, J.H. Song, Piezoelectric nanogenerators based on zinc oxide nanowire arrays. Science **312**(5771), 242–246 (2006)

J. Wang, M. Musameh, Y.H. Lin, Solubilization of carbon nanotubes by Nafion toward the preparation of amperometric biosensors. J. Am. Chem. Soc. **125**(9), 2408–2409 (2003)

X.D. Wang, C.J. Summers, Z.L. Wang, Large-scale hexagonal-patterned growth of aligned ZnO nanorods for nano-optoelectronics and nanosensor arrays. Nano Lett. **4**(3), 423–426 (2004)

F. Wang, G. Dukovic, L.E. Brus, T.F. Heinz, The optical resonances in carbon nanotubes arise from excitons. Science **308**(5723), 838–841 (2005a)

X. Wang, J. Zhuang, Q. Peng, Y.D. Li, A general strategy for nanocrystal synthesis. Nature **437**(7055), 121–124 (2005b)

X. Wang, L.J. Zhi, K. Mullen, Transparent, conductive graphene electrodes for dye-sensitized solar cells. Nano Lett. **8**(1), 323–327 (2008a)

X.R. Wang, Y.J. Ouyang, X.L. Li, H.L. Wang, J. Guo, H.J. Dai, Room-temperature all-semiconducting sub-10-nm graphene nanoribbon field-effect transistors. Phys. Rev. Lett. **100**(20), 206803–206807 (2008b)

Y. Wang, A.S. Angelatos, F. Caruso, Template synthesis of nanostructured materials via layer-by-layer assembly. Chem. Mater. **20**(3), 848–858 (2008c)

G. Williams, B. Seger, P.V. Kamat, TiO_2-graphene nanocomposites. UV-assisted photocatalytic reduction of graphene oxide. ACS Nano **2**(7), 1487–1491 (2008)

S.A. Wolf, D.D. Awschalom, R.A. Buhrman, J.M. Daughton, S. von Molnár, M.L. Roukes, A.Y. Chtchelkanova, D.M. Treger, Spintronics: a spin-based electronics vision for the future. Science **294**(5546), 1488–1495 (2001)

X.Y. Wu, H.J. Liu, J.Q. Liu, K.N. Haley, J.A. Treadway, J.P. Larson, N.F. Ge, F. Peale, M.P. Bruchez, Immunofluorescent labeling of cancer marker Her2 and other cellular targets with semiconductor quantum dots. Nat. Biotechnol. **21**(1), 41–46 (2003)

Y.N. Xia, P.D. Yang, Y.G. Sun, Y.Y. Wu, B. Mayers, B. Gates, Y.D. Yin, F. Kim, Y.Q. Yan, One-dimensional nanostructures: synthesis, characterization, and applications. Adv. Mater. **15**(5), 353–389 (2003)

Selected Bibliography (2000–2009)   629

J. Xiang, W. Lu, Y.J. Hu, Y. Wu, H. Yan, C.M. Lieber, Ge/Si nanowire heterostructures as high-performance field-effect transistors. Nature **441**(7092), 489–493 (2006)

B.Q. Xu, N.J.J. Tao, Measurement of single-molecule resistance by repeated formation of molecular junctions. Science **301**(5637), 1221–1223 (2003)

H. Yan, S.H. Park, G. Finkelstein, J.H. Reif, T.H. LaBean, DNA-templated self-assembly of protein arrays and highly conductive nanowires. Science **301**(5641), 1882–1884 (2003)

X.N. Yang, J. Loos, S.C. Veenstra, W.J.H. Verhees, M.M. Wienk, J.M. Kroon, M.A.J. Michels, R.A.J. Janssen, Nanoscale morphology of high-performance polymer solar cells. Nano Lett. **5**(4), 579–583 (2005)

J.J. Yang, M.D. Pickett, X.M. Li, D.A.A. Ohlberg, D.R. Stewart, R.S. Williams, Memristive switching mechanism for metal/oxide/metal nanodevices. Nat. Nanotechnol. **3**(7), 429–433 (2008)

Y. Yin, A.P. Alivisatos, Colloidal nanocrystal synthesis and the organic-inorganic interface. Nature **437**(7059), 664–670 (2005)

Y.D. Yin, R.M. Rioux, C.K. Erdonmez, S. Hughes, G.A. Somorjai, A.P. Alivisatos, Formation of hollow nanocrystals through the nanoscale Kirkendall Effect. Science **304**(5671), 711–714 (2004)

T. Yoshie, A. Scherer, J. Hendrickson, G. Khitrova, H.M. Gibbs, G. Rupper, C. Ell, O.B. Shchekin, D.G. Deppe, Vacuum Rabi splitting with a single quantum dot in a photonic crystal nanocavity. Nature **432**(7014), 200–203 (2004)

W.W. Yu, L.H. Qu, W.Z. Guo, X.G. Peng, Experimental determination of the extinction coefficient of CdTe, CdSe, and CdS nanocrystals. Chem. Mater. **15**(14), 2854–2860 (2003)

M. Zheng, A. Jagota, E.D. Semke, B.A. Diner, R.S. Mclean, S.R. Lustig, R.E. Richardson, N.G. Tassi, DNA-assisted dispersion and separation of carbon nanotubes. Nat. Mater. **2**(5), 338–342 (2003)

H. Zheng, J. Wang, S.E. Lofland, Z. Ma, L. Mohaddes-Ardabili, T. Zhao, L. Salamanca-Riba, S.R. Shinde, S.B. Ogale, F. Bai, D. Viehland, Y. Jia, D.G. Schlom, M. Wuttig, A. Roytburd, R. Ramesh, Multiferroic $BaTiO_3$-$CoFe_2O_4$ nanostructures. Science **303**(5658), 661–663 (2004), http://www.sciencemag.org/content/303/5658/661.abstract

G.F. Zheng, F. Patolsky, Y. Cui, W.U. Wang, C.M. Lieber, Multiplexed electrical detection of cancer markers with nanowire sensor arrays. Nat. Biotechnol. **23**(10), 1294–1301 (2005)

K. Zhu, N.R. Neale, A. Miedaner, A.J. Frank, Enhanced charge-collection efficiencies and light scattering in dye-sensitized solar cells using oriented TiO_2 nanotubes arrays. Nano Lett. **7**(1), 69–74 (2007)

Appendices

Appendix I

U.S. and International Workshops

The international study "Nanotechnology Research Directions for Societal Needs in 2020" received input from one national and four international brainstorming meetings titled "Long-term Impacts and Future Opportunities for Nanoscale Science and Engineering" as listed below. Detailed information is available on http://www.wtec.org/nano2/.

Chicago National Workshop
Chicago (Evanston), U.S. March 9–10, 2010
Hosted by WTEC
96 participants
Agenda on http://www.wtec.org/nano2/docs/Chicago/Agenda.html
Sponsored by: NSF

European Union Workshop
Hamburg, Germany. June 23–24, 2010
Hosted by Deutsches Elektronen-Synchrotron (DESY)
60 participants
Agenda on http://www.wtec.org/nano2/docs/Hamburg/Agenda.html
Sponsored by: European Commission (EC), DESY, and NSF

Japan, Korea, and Taiwan Workshop
Tsukuba, Japan (Tokyo region). July 26–27, 2010
Hosted by Japan Science and Technology Agency (JST)
96 participants
Agenda on http://www.wtec.org/nano2/docs/Tokyo/Agenda.html
Sponsored by: JST, MEST, NSC, and NSF

Australia, China, India, Saudi Arabia, and Singapore Workshop
Singapore. July 29–30, 2010
Hosted by Nanyang Technological University (NTU)
61 participants
Agenda on http://www.wtec.org/nano2/docs/Singapore
Sponsored by: Australia, China, India, Saudi Arabia, Singapore, and NSF

Arlington Final Workshop
Arlington, Virginia, United States. September 30, 2010
Hosted by NSF, US
90 participants
Agenda on http://www.wtec.org/nano2/
Webcast on http://www.tvworldwide.com/events/NSFnano2/100930/
Sponsored by: NSF

Public comments: received between September 30 and October 30, 2010

Appendix 2

List of Participants and Contributors

This section includes participants in workshops (see Appendix 1) and individuals who contributed to the present volume. After each person's institutional address is a key to their role. For example, (Arlington) means the person attended the workshop in Arlington; (Contributor) means the person has a contribution in this volume.

Chihaya Adachi
Kyushu University
Yurakuchoi, Chiyoda-ku
Tokyo 100-0006, Japan (Tokyo)

George Adams
Smalley Institute, Rice University
6100 Main Street, Mail Stop 100
Houston, TX 77005 (Contributor)

Wade Adams
6100 Main Street, Mail Stop 100
Houston, TX 77005 (Arlington)

Takuzo Aida
Department of Chemistry
University of Tokyo
7-3-1 Hongo Bunkyo-ku
Tokyo 113-8656, Japan (Tokyo)

Rob Aitken
SAFENANO Director
Research Avenue N
Riccarton, Edinburgh
EH14 4AP, United Kingdom
(Hamburg)

Morris Aizenman
Mathematical and Physical
Sciences, Division of
Astronomical Sciences
National Science Foundation
4201 Wilson Boulevard
Arlington, VA 22230
(Arlington)

Demir Akin
Center for Cancer Nanotechnology
Excellence
School of Medicine, Stanford
University
291 Campus Drive, Room LK3C02,
Dean's Office, Mail Stop 5216
Stanford, CA 94305-5101
(Contributor)

Hiroyuki Akinaga
National Institute of
Advanced Industrial Science
and Technology
1-3-1 Kasumigaseki, Chiyoda-ku
Tokyo 100-8921, Japan (Tokyo)

Aleksei Aksimentiev
Department of Physics, University
of Illinois
263 Loomis Laboratory, 1110 W
Green Street
Urbana, IL 61801-3080 (Contributor)

M. Alam
School of Electrical and Computer
Engineering, Purdue University
Electrical Engineering Building
465 Northwestern Avenue
West Lafayette, IN 47907-2035
(Chicago)

Muhammad Alam
School of Electrical and Computer
Engineering, Purdue University
465 Northwestern Avenue
West Lafayette, IN 47907-2035
(Contributor)

A. Paul Alivisatos
D43 Hildebrand Laboratory
University of California
Berkeley, CA 94720-1460
(Chicago)

Richard Alo
Program Director, Mathematics
National Science Foundation
4201 Wilson Boulevard
Arlington, VA 22230
(Arlington)

Salman Al-Rokayan
King Abdullah Institute for
Nanotechnology
PO Box 2455, Riyadh 11451
Kingdom of Saudi Arabia
(Singapore)

Massimo Altarelli
European XFEL GmbH
Notkestraße 85
22607 Hamburg, Germany
(Hamburg)

Pedro J. Alvarez
Department of Civil and Environmental
Engineering
Rice University
Houston, TX 77251-1892
(Chicago; Contributor)

Rose Amal
School of Chemical Engineering
University of New South Wales
Sydney, New South Wales 2052
Australia (Singapore)

Mostafa Analoui
The Livingston Group
825 3rd Avenue, 2nd floor
New York, NY 10022 (Contributor)

Masakazu Aono
National Institute for Materials Science
1-1 Namiki, Tsukuba
Ibaraki 305-0044, Japan (Tokyo)

Joerg Appenzeller
College of Engineering, Purdue
University
Neil Armstrong Hall of Engineering
Suite 2000
701 W Stadium Avenue
West Lafayette, IN 47907-2045
(Contributor)

Yasuhiko Arakawa
University of Tokyo
7-3-1 Hongo Bunkyo-ku
Tokyo 113-8656, Japan (Tokyo)

Tateo Arimoto
Japan Science and Technology Agency
Kawaguchi Center Building
4-1-8, Honcho, Kawaguchi-shi
Saitama 332-0012, Japan (Tokyo)

Vijay Arora
Fellow, Kraft Foods
3 Lakes Drive
Northfield, IL 60093-2753
(Chicago; contributor)

Appendix 2

Masafumi Ata
National Institute of Advanced
Industrial Science and Technology
1-3-1 Kasumigaseki, Chiyoda-ku
Tokyo 100-8921, Japan
(Tokyo)

Phaedon Avouris
IBM T.J. Watson Center
P.O. Box 218
Yorktown Heights, NY 10598
(Chicago; contributor)

Toshio Baba
Cabinet Office, Government of Japan
1-6-1 Nagata-cho, Chiyoda-ku
Tokyo 100-8914, Japan (Tokyo)

Yoshinobu Baba
Nagoya University
Furo-cho, Chikusa-ku
Nagoya, 464-8601, Japan
(Arlington, Tokyo)

Adra Baca
Corning, Inc.
1 Riverfront Plaza
Corning, NY 14831-0001
(Arlington, Chicago; contributor)

Santokh Badesha
Xerox Corporation
P.O. Box 1000, Mail Stop 7060-583
Wilsonville, OR 97070
(Arlington, Chicago; contributor)

Barbara A. Baird
Chair of Chemical Biology
Baker Laboratory
Cornell University
Ithaca, NY 14853-1301
(Chicago; contributor)

Yoshio Bandou
National Institute for Materials Science
1-2-1 Sengen, Tsukuba-city
Ibaraki 305-0047, Japan
(Tokyo)

Graeme Batley
Commonwealth Scientific
and Industrial Research Organization
P.O. Box 225
Dickson, Australian Capital
Territory 2602, Australia
(Singapore)

Carl A. Batt
Professor of Food Science
Cornell University
Stocking Hall, Room 312
Ithaca, NY 14853-5701
(Chicago; contributor)

Hassan Bekir Ali
World Technology Evaluation Center
4600 N Fairfax Drive, Suite 104
Arlington, VA 22203
(Chicago)

John Belk
Technical Fellow, Boeing Company
100 North Riverside Plaza
Chicago, IL 60606-2016
(Chicago; contributor)

Larry Bell
Senior Vice President for Strategic
Initiatives, Boston Museum of Science
Director, Nanoscale Informal Science
Education Network
1 Science Park
Boston, MA 02114
(Chicago; contributor)

Jayesh R. Bellare
Indian Institute of Technology
Chemical Engineering Bombay
Powai, Mumbai-400 076, India
(Singapore)

Ben Benokraitis
World Technology Evaluation Center
4800 Roland Avenue
Baltimore, MD 21210
(Arlington, Chicago, Hamburg
Singapore, Tokyo)

David Berube
Professor of Communication,
North Carolina State University
201 Winston Hall, Campus Box 8104
Raleigh, NC, 27695-8104
(Arlington, Chicago; contributor)

Jason Blackstock
Fellow, Centre for International
Governance Innovation
57 Erb Street W
Waterloo, ON, N2L 6C2, Canada
(Contributor)

Liam Blunt
University of Huddersfield
Queensgate
Huddersfield ,HD1 3DH, United
Kingdom (Hamburg)

Freddy Boey
Nanyang Technological University
School of Materials Science and
Engineering, Office N4.1-02-05
Singapore 639798 (Singapore)

Dawn A. Bonnell
Department of Materials Science and
Engineering, University of
Pennsylvania
3231 Walnut Street, Room 112-A
Philadelphia, PA 19104 (Arlington,
Chicago, Hamburg, Singapore, Tokyo;
contributor)

Jean-Philippe Bourgoin
Nanoscience Program, Alternative
Energies and Atomic Energy
Commission (CEA)
17, rue des Martyrs
38054 Grenoble cedex 9, France
(Hamburg)

Jurron Bradley
Consulting Director
Lux Research
75 Ninth Avenue, Floor 3R, Suite F
New York, NY 10011
(Chicago; contributor)

C. Jeffrey Brinker
Department of Chemical
and Nuclear Engineering
University of New Mexico
1001 University Boulevard SE
Albuquerque, NM 87131
and
Department 1002
Sandia National Laboratories
Self-Assembled Materials
Albuquerque, NM 87131 (Arlington,
Chicago, Hamburg; contributor)

Fernando Briones Fernández-Pola
CSIC Instituto de Micoelectronica de
Madrid
8 Tres Cantos
E-28760 Madrid, Spain (Hamburg)

Mark Brongersma
Stanford University
Durand Building, 496 Lomita Mall
Stanford, CA 94305-4034
(Chicago; contributor)

Yvan Bruynseraede
Oude Markt 13
Bus 5005 3000 Leuven
Belgium (Hamburg)

Denis Buxton
National Institutes of Health
PO Box 30105
Bethesda, MD 20824-0105
(Contributor)

Christopher Cannizzaro
Physical Science Officer
U.S. Department of State
1900 K Street NW
Washington, DC 20006
(Chicago)

Altaf Carim
U.S. Department of Energy
1000 Independence
Avenue SW, Room 5F-065
Mail Stop EE-2F
Washington, DC 20585 (Contributor)

Appendix 2

Vincent Castranova
Centers for Disease Control
National Institute for Occupational
Safety and Health, Health Effects
Laboratory Division
1095 Willowdale Road
Morgantown, WV 26505-2888
(Chicago; contributor)

Robert Celotta
Center for Nanoscale Science
and Technology, National Institute
of Standards and Technology
100 Bureau Drive ,Mail Stop 6200
Gaithersburg, MD 20899-6200
(Contributor)

Dennis Chamot
National Academies
500 Fifth Street NW, K-951
Washington, DC 2000 (Arlington)

Robert Chang
Department of Materials Science and
Engineering, Northwestern University
2220 Campus Drive
Evanston, IL 60208-3108
(Arlington, Chicago; contributor)

William Chang
National Science Foundation
4201 Wilson Boulevard
Arlington, VA 22230 (Arlington)

Constantinos Charitidis
School of Chemical Engineering
28 Oktovriou, Patision 42
10682 Athens, Greece
(Hamburg)

Hongda Chen
U.S. Department of Agriculture
National Institute of Food and
Agriculture
1400 Independence Avenue SW
Mail Stop 2220
Washington, DC 20250-2220
(Arlington, Chicago)

Hongyuan Chen
Nanjing University
Lab 802/803, Chemistry Building
22 Hankou Road
Nanjing Jiangsu 210093
China (Singapore)

Yet-Ming Chiang
Department of Materials Science
and Engineering ,Massachusetts
Institute of Technology
Room 13-4086, 77 Massachusetts
Avenue
Cambridge, MA 02139 (Chicago)

Eric Chiou
University of California, 37-138
Engineering IV
420 Westwood Plaza
Los Angeles, CA 90095
(Arlington)

Chul-Jin Choi
Korea Institute of Materials Science
797 Changwondaero
Sungsan-gu, Changwon
Gyeongnam 641-831, Korea (Tokyo)

Oscar Custance
National Institute for Materials Science
1-2-1 Sengen
305-0047 Tsukuba, Ibaraki, Japan
(Tokyo)

David Clark
Commonwealth Scientific and
Industrial Research Organization
Materials Science and Engineering
PO Box 218
Lindfield, New South Wales 2070
Australia (Arlington)

Vicki Colvin
Rice University
338 Space Science 201, PO Box 1892
Mail Stop 6
Houston, Texas 77251-1892
(Chicago)

James Cooper
Birck Nanotechnology Center
1205 West State Street
West Lafayette, IN 47907-1257
(Arlington)

Khershed Cooper
Naval Research Laboratory
Code 6354
4555 Overlook Avenue SW
Washington, DC 20375
(Arlington)

Harold Craighead
School of Applied and Engineering
Physics, Cornell University
212 Clark Hall
Ithaca, NY 14853
(Chicago; contributor)

Joanne Culbertson
Office of the Assistant Director
for Engineering, National Science
Foundation
4201 Wilson Boulevard, Room 505
Arlington, VA 22230 (Arlington)

Peter Cummings
Vanderbilt University, 303 Olin Hall
VU Station B 351604
Nashville, TN 37235 (Chicago;
contributor)

Simhan Danthi
National Heart, Lung
and Blood Institute
National Institutes of Health
Building 31, Room 5A52, 31 Center
Drive, Mail Stop 2486
Bethesda, MD 20892 (Arlington)

Dan Dascalu
National Institute for Research and
Development in Microtechnologies
126A, Erou Iancu Nicolae Street
077190
PO Box 38-160, 023573
Bucharest, Romania (Hamburg)

Robert Davis
Department of Chemical Engineering
University of Virginia
102 Engineers' Way, P.O. Box 400741
Charlottesville, VA 22904-4741
(Chicago; contributor)

S. Mark Davis
ExxonMobil Chemical R&D
BTEC-East 2313
5959 Las Colinas Boulevard
Irving, TX 75039-2298
(Chicago; contributor)

Kenneth Dawson
University College Dublin Research
University College Dublin
Belfield, Dublin 4, Ireland
(Hamburg)

Peter Degischer
Institute of Material Science
and Material Technology
Vienna University of Technology
Karlsplatz 13, 1040
Vienna, Austria (Hamburg)

Michael DeHaemer
World Technology Evaluation Center
4800 Roland Avenue
Baltimore, MD 21210 (Arlington)

Daniel DeKee
Program Director, Division of
Engineering, Education, and Centers
Directorate for Engineering 585
National Science Foundation
4201 Wilson Boulevard
Arlington, VA 22230
(Arlington)

Paul Dempsey
Electronics Editor, Engineering
and Technology Magazine
Michael Faraday House
Stevenage
Herts SG1 2AY, United Kingdom
(Arlington)

Appendix 2

Joseph DeSimone
Department of Chemistry
University of North Carolina
Campus Box 3290
Chapel Hill, NC 27599-3290
(Contributor)

Jozef T. Devreese
University of Antwerp, Physics
University Square 1
B2610 Wilrijk, Belgium (Hamburg)

Mamadou Diallo
Materials and Process Simulation
Center, California Institute
of Technology
1200 East California Boulevard
Mail Stop 139-74
Pasadena, CA 91125
and
KAIST (Korea)
291 Daehak-ro
Yuseong-gu, Daejeon 305-701
(Arlington, Chicago,
Hamburg; contributor)

Alain Diebold
College of Nanoscale Science
and Engineering, University at Albany
257 Fuller Road
Albany, NY 12203 (Contributor)

Peter Dobson
Department of Engineering Science
University of Oxford
Parks Road
Oxford OX1 3PJ, United Kingdom
(Hamburg)

Ryoji Doi
Ministry of Economy Trade, and Industry
1-3-1 Kasumigaseki
Chiyoda-ku
Tokyo 100-8901, Japan (Tokyo)

Kazunari Domen
University of Tokyo
7-3-1 Hongo Bunkyo-ku
Tokyo 113-8656, Japan (Tokyo)

Helmut Dosch
Max-Planck-Institut für
Metallforschung
Heisenbergstraße 3
D-70569 Stuttgart, Germany
(Hamburg)

Haris Doumanidis
Directorate of Engineering, Division
of Civil, Mechanical, and
Manufacturing Innovation, National
Science Foundation
4201 Wilson Boulevard
Arlington, VA 22230 (Contributor)

Vinayak P. Dravid
Northwestern University
1131 Cook Hall, 2220 Campus Drive
Evanston, IL 60208-3108
(Arlington, Chicago; contributor)

Calum Drummond
Commonwealth Scientific and
Industrial Research Organization
Material Science
and Engineering-Clayton, Gate 5
Normanby Road
Clayton, Victoria 3168
Australia (Singapore)

Fereshteh Ebrahimi
Department of Materials Science
and Engineering, University of Florida
PO Box 116400
Gainesville, FL 32611
(Chicago; contributor)

Don Eigler
IBM Almaden Research Center
650 Harry Road
San Jose, CA 95120-6099
(Chicago)

Marlowe Epstein
National Nanotechnology
Coordination Office
Stafford II-405, 4201 Wilson
Boulevard
Arlington, VA 22230 (Arlington)

Heather Evans
Office of Science and Technology
Policy, Executive Office of the
President
725 17th Street, Room 5228
Washington, DC 20502 (Arlington)

Bengt Fadeel
Institute of Environmental Medicine
Karolinska Institutet
SE-171 77 Stockholm, Sweden
(Hamburg)

Dorothy Farrell
Office of Cancer Nanotechnology
Research, National Institutes of Health
Building 31, Room 10A52
31 Center Drive, Mail Stop 2580
Bethesda, MD 20892-2580
(Contributor)

Si-Shen Feng
National University of Singapore
Blk E4, 4 Engineering Drive 3, #05-12
Singapore 117576 (Singapore)

Mauro Ferrari
The University of Texas Health
Science Center
1825 Pressler Street, Suite 537D
Houston, TX 77031
(Chicago; contributor)

Bertrand Fillon
Chief Technical Officer, Atomic
Energy Commission, Direction de la
communication Siège
Centre d'études de Saclay
91191 Gif-sur-yvette Cedex
France (Hamburg)

Richard Fisher
Acting Associate Director for Science
Policy and Legislation, National
Institutes of Health
31 Center Drive, Mail Stop 2580
Bethesda, MD 20892-2580
(Contributor)

Patricia Foland
World Technology Evaluation Center
4800 Roland Avenue
Baltimore, MD 21210 (Arlington,
Chicago, Hamburg, Singapore, Tokyo)

Steve Fonash
Nanotechnology Applications and
Career Knowledge Center, 112 Lubert
Building, 101 Innovation Boulevard,
Suite 112, University Park, PA 16802,
USA (Contributor)

Yong Lim Foo
Institute of Materials Research
and Engineering, Materials Analysis
and Characterization
3 Research Link
Singapore 117602 (Singapore)

Lisa Friedersdorf
University of Virginia
395 McCormick Road, PO Box 400745
Charlottesville, VA 22904 (Arlington)

Heico Frima
Directorate-General for Research
European Commission
SDME 2/2
B-1049 Brussels, Belgium (Hamburg)

Martin Fritts
Nanotechnology Characterization Lab
1050 Boyles Street, PO Box B
Building 430
Frederick, MD 21702-1201 (Arlington)

Harald Fuchs
Physikalisches Institut - AG Fuchs
University of Muenster
Wilhelm-Klemm-Str 10
D-48149 Münster, Germany
(Hamburg)

Toshiyuki Fujimoto
National Institute of Advanced
Industrial Science and Technology
1-3-1 Kasumigaseki, Chiyoda-ku
Tokyo 100-8921, Japan (Tokyo)

Appendix 2

Pradeep Fulay
Program Director
Electronics, Photonics
and Device Technology
Division of Electrical
Communications,and Cyber Systems
National Science Foundation
4201 Wilson Boulevard, Room 525
Arlington, Virginia 22230
(Arlington)

Julian Gale
Curtin University of Technology
Nanochemistry Research Institute
GPO Box U1987, Perth
Western Australia 6845
Australia (Singapore)

Sanjiv Sam Gambhir
Director, Molecular Imaging Program
School of Medicine
Stanford University
291 Campus Drive, Room LK3C02
Li Ka Shing Building, 3rd floor
Dean's Office, Mail Stop 5216
Stanford, CA 94305-5101
(Contributor)

Masashi Gamo
National Institute of Advanced
Industrial Science and Technology
1-3-1 Kasumigaseki, Chiyoda-ku
Tokyo 100-8921, Japan
(Tokyo)

Günter Gauglitz
Institut für Physikalische
und Theoretische Chemie
Universität Tübingen
Auf der Morgenstelle 8
D-72076 Tübingen,
Germany (Hamburg)

Inge Genné
Vlaamse Instelling Voor Technologisch
Onderzoek N.V.
Boeretang, Belgium
(Hamburg)

Sarah Gerould
U.S. Geological Survey
12201 Sunrise Valley Drive
Mail Stop 301
Reston, VA 20192 (Arlington)

Louise R. Giam
Department of Chemistry
Northwestern University
2145 Sheridan Road
Evanston, IL 60208-3113 (Contributor)

David Ginger
Department of Chemistry, University
of Washington
Box 351700
Seattle, WA, 98195-1700 (Chicago;
contributor)

Sharon Glotzer
Department of Chemical Engineering
University of Michigan
3074 H.H. Dow Building
2300 Hayward Street
Ann Arbor, MI 48109-2136
(Chicago; contributor)

Bill Goddard
321 Beckman Institute
Mail Stop 139-74
Pasadena, CA 91106 (Chicago)

William A. Goddard
Chemistry 139-74
California Institute of Technology
Pasadena, CA 91125 (Contributor)

Hilary Godwin
Public Health-Environmental Health
Science, University of California
Box 951772
Los Angeles, CA 90095 (Contributor)

Jan-Christoph Goldschmidt
Fraunhofer Institute for Solar Energy
Systems
Heidenhofstraße 2
79110 Freiburg im Breisgau, Germany
(Hamburg)

Larry Goldberg
Senior Engineering Advisor
National Science Foundation
4201 Wilson Boulevard
Arlington, VA 22230 (Arlington)

Justin Gooding
University of New South Wales
School of Chemistry
Sydney 2052, Australia
(Singapore)

David Grainger
College of Pharmacy
University of Utah
30 South 2000 East, Room 301
Salt Lake City, UT 84112-5820
(Chicago; contributor)

Hans Griesser
University of South Australia, Ian
Wark Research Institute
Mawson Lakes Campus
Room IW2-19
South Australia 5095
Australia (Singapore)

Piotr Grodzinski
Office of Technology and Industrial
Relations, National Cancer Institute
Building 31, Room 10, A49
31 Center Drive, Mail Stop 2580
Bethesda, MD 20892-2580
(Chicago; contributor)

David Guston
Consortium for Science, Policy,
and Outcomes, College of Liberal Arts
and Sciences, Arizona State University
P.O. Box 875603
Tempe, AZ 85287-4401
(Chicago; contributor)

Pradeep Haldar
College of Nanoscale Science
and Engineering, University at Albany
257 Fuller Road
Albany, NY 12203 (Chicago)

Jongyoon Han
Research Laboratory of Electronics
Massachusetts Institute of Technology
Room 36-841, 77 Massachusetts
Avenue
Cambridge, MA 02139 (Contributor)

Alex Harris
Chemistry Department
Brookhaven National Laboratory
Building 555, PO Box 5000
Upton, NY 11973-5000 (Contributor)

Barbara Harthorn
Center for Nanotechnology in Society
University of California
Santa Barbara, CA 93106-2150
(Chicago; contributor)

Karl-Heinz Haas
Fraunhofer-Institut fuer
Silicatforschung
Rue du Commerce 31
1000 Brüssel, Belgium
(Hamburg)

Kenji Hata
National Institute of Advanced
Industrial Science and Technology
1-3-1 Kasumigaseki, Chiyoda-ku
Tokyo 100-8921, Japan (Tokyo)

James Heath
Department of Chemistry, California
Institute of Technology
Mail Stop 127-72
Pasadena, CA 91125 (Chicago)

Matt Henderson
World Technology Evaluation Center
1653 Lititz Pike, Suite 417
Lancaster, PA 17601 (Arlington)

Lee Herring
Office of Legislative and Public
Affairs, National Science Foundation
4201 Wilson Boulevard
Arlington, VA 22230 (Arlington)

Mark Hersam
Department of Materials Science and
Engineering, Northwestern University
2220 Campus Drive
Evanston, IL 60208
(Arlington, Chicago, Hamburg
Singapore, Tokyo; contributor)

Masahiro Hirano
Science and Technology Agency
KSP C-1232, 3-2-1 Sakado, Takatsu-ku
Kawasaki 213-0012, Japan (Tokyo)

Huey Hoon Hng
Nanyang Technological University
School of Materials Science and
Engineering, Office N4.1-01-23
Singapore 639798 (Singapore)

Chih-Ming Ho
School of Engineering and Applied
Science, University of California
Engineering IV, Room 38-137
420 Westwood Plaza
Los Angeles, CA 90095-1597
(Chicago)

Michael Hoffmann
Professor of Environmental Science
California Institute of Technology
1200 E California Boulevard
Mail Stop 138-78
105 W Keck Laboratories
Pasadena, CA 91125 (Contributor)

Geoffrey Holdridge
National Nanotechnology
Coordination Office
Stafford II-405, 4201 Wilson Boulevard
Arlington, VA 22230 (Arlington,
Chicago, Singapore, Tokyo)

Nina Horne
Richard and Rhoda Goldman School of
Public Policy, University of California
2607 Hearst Avenue
Berkeley, CA 94720-7320
(Contributor)

Ming-Huei Hong
Center for Nanotechology
Materials Science, and Microsystems
National Tsing Hua University
(NTHU)
No. 101, Section 2, Kuang-Fu Road
Hsinchu, Taiwan 30013
R.O.C. (Tokyo)

Taku Hon-iden
Tsukuba City
Tokoy chiyoda-ku Soto-Kanda
1-18-13 Akihabara Dai Building
8th Floor
Tokyo, Japan (Tokyo)

Hideo Hosono
Tokyo Institute of Technology
4259 Nagatsuta-cho
Midori-ku
Yokohama, 226-8502, Japan (Tokyo)

Wei Huang
Nanjing University of Posts
and Telecommunications
No. 9 Wenyuan Road
Nanjing 210046, China (Singapore)

Evelyn L. Hu
Harvard School of Engineering
and Applied Sciences
29 Oxford Street
Cambridge, MA 02138
(Arlington, Chicago, Hamburg,
Singapore, Tokyo; contributor)

Robert Hwang
Sandia National Laboratories
PO Box 5800
Albuquerque, NM 87185
(Chicago; contributor)

Mark S. Hybertsen
Senior Research Scientist
Brookhaven National Laboratory
530 W 120th Street, Mail Stop 8903
New York, NY 10027 (Contributor)

Yasuo Iida
New Energy and Industrial Technology
Development Organization
MUZA Kawasaki Central Tower
1310 Omiya-cho
Saiwai-ku, Kawasaki
Kanagawa 212-8554, Japan (Tokyo)

Isao Inoue
University of Tsukuba
1-1-1 Tennodai
Tsukuba
Ibaraki 305-8577, Japan (Tokyo)

Eric Isaacs
Argonne National Laboratory
9700 S Cass Avenue
Argonne, IL 60439
(Chicago; contributor)

Satoshi Ishihara
Japan Science and Technology Agency
Kawaguchi Center Building
4-1-8, Honcho, Kawaguchi-shi
Saitama 332-0012, Japan (Tokyo)

Harumi Ito
Japan Science and Technology Agency
Kawaguchi Center Building
4-1-8, Honcho, Kawaguchi-shi
Saitama 332-0012, Japan (Tokyo)

Keith Jackson
National High Magnetic Field
Laboratory
142 Centennial Building, 205 Jones Hall
1530 S Martin Luther King Jr.
Boulevard
Tallahassee, FL32307
(Chicago; contributor)

Chennupati Jagadish
Australian National University
Department of Electronic
Materials Engineering
John Carver 4 22
Canberra ACT 0200
Australia (Singapore)

Patricia Johnson
World Technology Evaluation Center
4800 Roland Avenue
Baltimore, MD 21210
(Chicago, Singapore, Tokyo)

Phil Jones
Director Technical Marketing
and New Ventures, Imerys
100 Mansell Court E
Roswell, GA 30076 (Chicago)

Zakya Kafafi
National Science Foundation
4201 Wilson Boulevard
Suite 1065
Arlington, VA 22230
(Arlington)

David Kahaner
World Technology Evaluation Center
4600 North Fairfax Drive, Suite 104
Arlington, VA 22203 (Tokyo)

Toshihiko Kanayama
National Institute of Advanced
Industrial Science and Technology
1-3-1 Kasumigaseki
Chiyoda-ku
Tokyo 100-8921, Japan (Tokyo)

Naoya Kaneko
Japan Science and Technology Agency
Kawaguchi Center Building
4-1-8, Honcho, Kawaguchi-shi
Saitama 332-0012
Japan (Tokyo)

Barbara Karn
U.S. Environmental Protection Agency
1200 Pennsylvania Avenue NW, 8722F
Washington, DC 20460 (Arlington)

Georgios Katalagarianakis
Directorate-General for Research
European Commission, SDME 2/2
B-1049 Brussels
Belgium (Hamburg)

Kazunori Kataoka
Department of Materials Engineering
University of Tokyo
Hongo 7-3-1
Bunkyo-ku
Tokyo, Japan (Tokyo)

Tomoji Kawai
Sir-Sanken, Osaka University
Mihogaoka 8-1, Ibaraki
Osaka 567-0047, Japan (Tokyo)

Seiichiro Kawamura
Japan Science and Technology Agency
Kawaguchi Center Building
4-1-8, Honcho, Kawaguchi-shi
Saitama 332-0012, Japan (Tokyo)

Masashi Kawasaki
Institute for Materials Research
Tohoku University
2-1-1 Katahira, Aoba
Sendai 980-8577, Japan (Tokyo)

Satoshi Kawata
Department of Chemistry
Osaka University
1-1 Machikaneyama, Toyonaka
Osaka 560-0043, Japan (Tokyo)

Rajinder Khosla
National Science Foundation
4201 Wilson Boulevard, Suite 675
Arlington, VA 22230 (Arlington)

Hak Min Kim
Korea Nanotechnology
Research Society
66 Sangnamdong Changwon
Kyungnam 641-831
Korea (Arlington)

Costas Kiparissides
Center for Research
and Technology Hellas
6th Km Charilaou-Thermi
PO Box 60361
570 01 Thessaloniki
Greece (Hamburg)

Gerhard Klimeck
Purdue University
Birck Nanotechnology Center
Room 1281
West Lafayette, IN 47907 (Contributor)

David Knox
Research Director, MeadWestvaco
1000 Broad Street
Phenix City, AL 42787
(Chicago; contributor)

Wolfgang Kochanek
Metal Part GmbH
Institut für Neue Materialien
Im Stadtwald D2 2
D-66123 Saarbrücken, Germany
(Hamburg)

Astrid Koch
Directorate-General for Research
European Commission, SDME 2/2
B-1049 Brussels
Belgium (Arlington)

Jozef Kokini
University of Illinois
211B Mumford Hall
905 S Goodwin Avenue
Urbana, IL 61801 (Chicago)

Michio Kondo
National Institute of Advanced
Industrial Science and Technology
1-3-1 Kasumigaseki
Chiyoda-ku
Tokyo 100-8921
Japan (Tokyo)

Bruce Kramer
National Science Foundation
4201 Wilson Boulevard
Arlington, VA 22230 (Arlington)

Wolfgang Kreyling
Helmholtz Centre Munich, GmbH
Ingolstädter Landstraße 1
D-85764 Neuherberg
Germany (Hamburg)

Todd Kuiken
Woodrow Wilson International Center
for Scholars
One Woodrow Wilson Plaza
1300 Pennsylvania Avenue NW
Washington, DC 20004 (Arlington)

Ivo Kwon
Ewha Womans University
National Institutes of Health
214 Congressional Lane, Apt 204
Rockville, MD 20852
(Arlington)

Robert Langer
Langer Lab
77 Massachusetts Avenue
Room E25-342
Cambridge, MA 02139-4307
(Chicago)

Richard Leach
National Physical Laboratory
Hampton Road
Teddington
Middlesex TW11 0LW
United Kingdom (Hamburg)

Haiwon Lee
Department of Chemistry
Hanyang University
17 Haengdang-dong
Seoul 133-791, Korea (Arlington)

Jo-Won Lee
Tera Level Nano Devices, KIST 39-1
Hawolgok-dong, Sungbuk-ku
Seoul 136-791, Korea
(Arlington, Tokyo)

Shuit-Tong Lee
Department of Physics and Materials
City University of Hong Kong
Science, Office AC-G6608
Tat Chee Avenue
Kowloon
Hong Kong SAR,
China (Singapore)

Neocles Leontis
Division of Molecular and Cellular
Biosciences, National Science
Foundation
4201 Wilson Boulevard
Arlington, VA 22230 (Arlington)

Grant Lewison
Evaluametrics Ltd.
157 Verulam Road
St Albans AL3 4DW, United Kingdom
(Hamburg)

Michal Lipson
College of Engineering, Cornell
University
Carpenter Hall
Ithaca, NY 14853-2201 (Contributor)

Chien-Wei Li
U.S. Department of Energy
Room 5F-065, Mail Stop EE-2F
1000 Independence Avenue SW
Washington, DC 20585 (Arlington)

Joachim Loo
Nanyang Technological University
School of Materials Science and
Engineering, Office N4.1-01-04a
Singapore 639798 (Singapore)

Mark Lundstrom
School of Electrical and Computer
Engineering, Purdue University
465 Northwestern Avenue
West Lafayette, IN 47907
(Arlington, Chicago, Hamburg,
Singapore, Tokyo; contributor)

Lynnette D. Madsen
National Science Foundation
4201 Wilson Boulevard, Suite 1065
Arlington, VA 22230
(Arlington; contributor)

Lutz Maedler
Universität Bremen
Am Fallturm 1
28359 Bremen, Germany (Hamburg)

Appendix 2

George Maracas
Arizona State University
PO Box 87-5706
Tempe, AZ 85287 (Arlington)

Antonio Marcomini
Università Ca' Foscari Venezia
Dorsoduro 3246-30123
Venice, Italy
(Hamburg)

Jean-Yves Marzin
National Center for Scientific
Research, Campus Gérard-Mégie
3 rue Michel-Ange-F-75794
Paris cedex 16
France
(Hamburg)

Jan Ma
Nanyang Technological University
School of Materials Science and
Engineering, Office N4.1-02-31
Singapore 639798
(Singapore)

Steffen McKernan
RF Nano Corporation
4311 Jamboree Road,
Suite 150
Newport Beach, CA 92660
(Arlington, Chicago)

Scott McNeil
National Cancer Institute
Nanotechnology Characterization
Laboratory
PO Box B, Building 469
1050 Boyles Street
Frederick, MD 21702-1201
(Contributor)

Thomas J. Meade
Department of Chemistry
Northwestern University
Silverman 2504
2145 Sheridan Road
Evanston, IL 60208-3113
(Contributor)

Michael Meador
U.S. National Aeronautics
and Space Administration
Glenn Research Center
21000 Brookpark Road
Cleveland, OH 44135 (Contributor)

Subodh Mhaisalkar
Nanyang Technological University
School of Materials Science and
Engineering, Office N4.1-01-21
Singapore 639798 (Singapore)

John Miles
National Measurement Institute
Unit 1-153 Bertie Street,
Port Melbourne, Victoria 3207
Australia (Singapore)

John A. Milner
Nutritional Science Research Group
Division of Cancer Prevention
National Cancer Institute
National Institutes of Health
6130 Executive Boulevard
Executive Plaza North, Suite 3164
Rockville, MD 20892 (Chicago;
contributor)

Chad A. Mirkin
Department of Chemistry
and Director of the International
Institute for Nanotechnology
Northwestern University,
2145 Sheridan Road, Evanston,
IL 60208 (Arlington, Chicago,
Singapore, Tokyo; contributor)

Yuji Miyahara
National Institute for Materials Science
1-2-1 Sengen
Tsukuba-city
Ibaraki 305-0047, Japan (Tokyo)

Akira Miyamoto
Tohoku University
1-1 Katahira, 2-chome
Aoba-ku
Sendai, 980-8577, Japan (Tokyo)

Nagae Miyashita
Japan Science and Technology Agency
Kawaguchi Center Building
4-1-8, Honcho, Kawaguchi-shi
Saitama 332-0012, Japan (Tokyo)

Dae Won Moon
Korea Research Institute
of Standards and Science
209 Gajeong-ro, Yuseong-gu
Daejeon 305-340, Korea (Tokyo)

Chrit Moonen
National Center for Scientific
Research, Campus Gérard-Mégie
3 rue Michel-Ange - F-75794
Paris Cedex 16, France (Hamburg)

Takao Mori
National Institute for Materials Science
1-2-1 Sengen
Tsukuba-city
Ibaraki 305-0047, Japan (Tokyo)

Seizo Morita
Osaka University
1-1 Machikaneyama, Toyonaka
Osaka 560-0043, Japan (Tokyo)

Jeff Morris
Ronald Reagan Building and
International Trade Center
U.S. Environmental Protection Agency
Room 71184
1300 Pennsylvania Avenue NW
Washington, DC 20004
(Chicago; contributor)

Michael Moseler
Fraunhofer Institute for Mechanics
of Materials
Wöhlerstrasse 11
79108 Freiburg, Germany (Hamburg)

Paul Mulvaney
University of Melbourne
2nd Floor North, Bio21 Institute
Victoria 3010, Australia
(Singapore)

Craig Mundie
Microsoft Corporation
One Microsoft Way
Redmond, WA 98052-6399 (Chicago)

Jose Munoz
Office of Cyberinfrastructure
National Science Foundation
4201 Wilson Boulevard
Arlington, VA 22230 (Arlington)

Shinji Murai
Japan Science and Technology Agency
Kawaguchi Center Building
4-1-8, Honcho, Kawaguchi-shi
Saitama 332-0012, Japan (Tokyo)

James Murday
University of Southern California
Office of Research Advancement
701 Pennsylvania Avenue NW, Suite 540
Washington, DC 20004 (Arlington,
Chicago, Hamburg, Singapore, Tokyo;
contributor)

Sean Murdock
NanoBusiness Alliance
8045 Lamon Avenue
Skokie, Illinois 60077
(Chicago; contributor)

Chris Murray
Department of Chemistry
University of Pennsylvania
231 S 34th Street
Philadelphia, PA 19104-6323
(Chicago; contributor)

Hiroshi Nagano
Japan Science and Technology Agency
Kawaguchi Center Building
4-1-8, Honcho, Kawaguchi-shi
Saitama 332-0012, Japan (Tokyo)

Tomoki Nagano
Japan Science and Technology Agency
Kawaguchi Center Building
4-1-8, Honcho, Kawaguchi-shi
Saitama 332-0012, Japan (Tokyo)

Tomohiro Nakayama
Japan Science and Technology Agency
Kawaguchi Center Building
4-1-8, Honcho, Kawaguchi-shi
Saitama 332-0012, Japan (Tokyo)

Kesh Narayanan
Industrial Innovation and Partnerships
Directorate of Engineering
National Science Foundation
4201 Wilson Boulevard
Arlington, VA 22230 (Arlington)

Jeffrey B. Neaton
Theory of Nanostructured Materials
Molecular Foundry
Lawrence Berkeley National Laboratory
1 Cyclotron Road, Mail Stop 67R3207
Berkeley, CA 94720 (Contributor)

André Nel
Department of Medicine and California
NanoSystems Institute
University of California
10833 Le Conte Avenue, 52-175 CHS
Los Angeles, CA 90095 (Arlington,
Chicago, Hamburg, Singapore, Tokyo;
contributor)

Koon Gee Neoh
National University of Singapore
Blk E5, 4 Engineering Drive 4, #02-34
Singapore 117576 (Singapore)

Elizabeth Nesbitt
International Trade Analyst
for Biotechnology
and Nanotechnology
U.S. International Trade Commission
500 E Street SW
Washington, DC 20436 (Arlington,
Chicago)

Milos Nesladek
University of Hasselt, Campus
Diepenbeek
Agoralaan, Building D
3590 Diepenbeek, Belgium (Hamburg)

Chikashi Nishimura
National Institute for Materials Science
1-2-1 Sengen
Tsukuba city
Ibaraki 305-0047, Japan
(Tokyo)

Susumu Noda
Quantum Optoelectronics Laboratory
Kyoto University
Nishikyo-ku
Kyoto 615-8510, Japan (Tokyo)

Alfred Nordmann
Technical University of Darmstadt
Department of Philosophy
S3 13 364
Royal Palace
64283 Darmstadt, Germany
(Hamburg)

Iwao Ohdomari
Japan Science and Technology Agency
Kawaguchi Center Building
4-1-8, Honcho, Kawaguchi-shi
Saitama 332-0012, Japan
(Tokyo)

Teruo Okano
Institute of Advanced Biomedical
Engineering and Science
Tokyo Women's Medical University
8-1, Kawada-cho
Shinjuku-ku
Tokyo, Japan (Tokyo)

Halyna Paikoush
National Nanotechnology
Coordination Office
Stafford II-405, 4201 Wilson
Boulevard
Arlington, VA 22230
(Arlington, Chicago)

Stuart Parkin
IBM Almaden Research Center
650 Harry Road
San Jose, CA 95120-6099 (Contributor)

Hans Pedersen
Directorate-General for Research
European Commission, SDME 2/2
B-1049 Brussels, Belgium (Hamburg)

Malcolm Penn
Future Horizons Ltd.
44 Bethel Road
Sevenoaks, Kent TN13 3UE
United Kingdom (Hamburg)

Virgil Percec
Department of Chemistry
University of Pennsylvania
231 S 34th Street
Philadelphia, PA 19104-6323
(Contributor)

Thomas Peterson
National Science Foundation
4201 Wilson Boulevard
Room 505
Arlington, VA 22230
(Arlington)

Diana Petreski
National Nanotechnology
Coordination Office
Stafford II-405
4201 Wilson Boulevard
Arlington, VA 22230
(Arlington)

Tom Picraux
Materials Physics and Applications
Mail Stop F612, Los Alamos
National Laboratory
Los Alamos, NM 87545
(Chicago; contributor)

Robert Pinschmidt
Institute for Advanced Materials
Nanoscience, and Technology
University of North Carolina
223 Chapman Hall CB #3216
Chapel Hill, NC 27599-3216
(Arlington)

Michael Plesniak
Mechanical and Aerospace
Engineering, George Washington
University
801 22nd Street NW
Washington, DC 20052 (Arlington)

Ilmari Pyykkö
Department of Otolaryngology
FIN-33014 University of Tampere
Finland (Hamburg)

Dan C. Ralph
Laboratory of Atomic and Solid
State Physics, Cornell University
Clark Hall
Ithaca, NY 14853-2501 (Contributor)

John N. Randall
Vice President, Zyvex Labs
1321 N Plano Road
Richardson, TX 75081 (Contributor)

Mark Ratner
Department of Chemistry
Northwestern University
2145 Sheridan Road
Evanston, IL 60208
(Chicago; contributor)

Lisa Regalla
Science Editor, Twin Cities Public
Television
172 E 4th Street
Saint Paul, MN 55101 (Contributor)

Herbert Richtol
Program Director, Chemistry
National Science Foundation
4201 Wilson Boulevard
Arlington, VA 22230 (Arlington)

Rick Ridgley
National Reconnaissance Office
14675 Lee Road
Chantilly, Virginia 20151-1715
(Contributor)

Thomas Rieker
National Science Foundation
4201 Wilson Boulevard, Suite 1065
Arlington, VA 22230 (Arlington)

Mihail C. Roco
Senior Advisor for Nanotechnology
National Science Foundation
4201 Wilson Boulevard
Arlington, VA 22230
(Arlington, Chicago, Hamburg
Singapore, Tokyo; contributor)

Juan Rogers
Georgia Institute of Technology
School of Public Policy
685 Cherry Street NW
Atlanta, GA 30332-0345
(Arlington)

John Rogers
Chair in Engineering Innovation
3355 Beckman Institute
Department of Materials Science and
Engineering, University of Illinois
1304 W Green Street
Urbana, IL 61801 (Contributor)

Sven Rogge
Professor of Physics, Delft University
of Technology
Lorentzweg 1
2628 CJ Delft, The Netherlands
(Contributor)

Celeste Rohlfing
National Science Foundation
4201 Wilson Boulevard, Suite 1005
Arlington, VA 22230 (Arlington)

Gregory Rorrer
Directorate for Engineering, Chemical
Bioengineering, Environmental, and
Transport Systems, National Science
Foundation
4201 Wilson Boulevard
Arlington, VA 22230 (Arlington)

Zeev Rosenzweig
University of New Orleans
CSB 238, 2000 Lakeshore Drive
New Orleans, LA 70148 (Arlington)

Skip Rung
Oregon State University
308 Education Hall
Corvallis, OR 97331-3502
(Arlington, Chicago; contributor)

James A. Ruud
GE Global Research
One Research Circle
Niskayuna, NY 12309
(Chicago; contributor)

Sayeef Salahuddin
Electrical Engineering Division
University of California
253 Cory Hall
Berkeley, CA 94720-1770
(Contributor)

Yves Samson
Atomic Energy Commission
Grenoble-Institute for Nanoscience
and Cryogenics
17 rue des Martyrs
38054 Grenoble, Cedex 09, France
(Arlington)

Ashok Sangani
Program Director, Particulate
and Multiphase Processes
Directorate for Engineering
Chemical, Bioengineering
Environmental, and Transport Systems
National Science Foundation
4201 Wilson Boulevard
Arlington, VA 22230 (Arlington)

Nobuyuki Sano
University of Tsukuba
1-1-1 Tennodai
Tsukuba
Ibaraki 305-8577, Japan (Tokyo)

Linda Sapochak
Solid State and Materials Chemistry
National Science Foundation
4201 Wilson Boulevard
Arlington, VA 22230 (Arlington)

John Sargent
Science and Technology Policy
Resources, Science and Industry
Division
Congressional Research Service
Library of Congress
101 Independence Avenue SE
Room 423
Washington, DC 20540 (Arlington)

Takayoshi Sasaki
National Institute for Materials Science
1-2-1 Sengen
Tsukuba City
Ibaraki 305-0047, Japan (Tokyo)

Murali Sastry
Tata Chemicals Innovation Centre
S no 1139/1, Ghotavde Phata
Urawde Road
Pirangut Industrial Area
Mulshi, Pune , 412 108
India (Singapore)

Tatsuo Sato
Asian Technology Information
Program
MBE 225, Tokyo Toranomon Building
1-1-18 Toranomon
Minato-ku, Tokyo 105-0001
Japan (Tokyo)

Nora Savage
Nano Team Leader, U.S.
Environmental Protection Agency
Office of Research and Development
National Center for Environmental
Research
1200 Pennsylvania Avenue NW
Mail Stop 8722F
Washington, DC 20460
(Arlington, Chicago; contributor)

George Schatz
Department of Chemistry
Northwestern University
Ryan Room 4018, 2145 Sheridan Road
Evanston, IL 60208-3113
(Chicago; contributor)

Erik Scher
Siluria
2625 Hanover Street
Palo Alto, CA 94304 (Chicago;
contributor)

Jeff Schloss
National Institutes of Health
9000 Rockville Pike
Bethesda, Maryland 20892 (Contributor)

John Schmitz
NXP Semiconductors
P.O. Box 80073
5600 KA Eindhoven
Netherlands (Hamburg)

Jean-Christophe Schrotter
Chemin de la Digue
78603 Maisons Laffitte Cedex
France (Hamburg)

Norman Scott
Biological and Chemical Engineering
Cornell University
216 Riley Hall
Ithaca, NY 14853-5701
(Chicago; contributor)

Mizuki Sekiya
National Institute of Advanced
Industrial Science and Technology
1-3-1 Kasumigaseki
Chiyoda-ku
Tokyo 100-8921, Japan (Tokyo)

Cynthia Selin
Center for Nanotechnology in Society
Arizona State University
PO Box 875603
Tempe, AZ 85287-5603
(Chicago; contributor)

Mark Shannon
2132 Mechanical Engineering
Laboratory, University of Illinois
1206 W Green Street, Mail Stop 244
Urbana, IL 61801
(Chicago; contributor)

Philip Shapira
Georgia Institute of Technology
D. M. Smith Building, Room 107
685 Cherry Street
Atlanta, GA 30332-0345
(Chicago; contributor)

Robert Shelton
World Technology Evaluation Center
1653 Lititz Pike, Suite 417
Lancaster, PA 17601
(Arlington, Chicago)

Tadashi Shibata
University of Tokyo
Hongo 7-3-1, Bunkyo-ku
Tokyo, Japan (Tokyo; contributor)

Hiromoto Shimazu
Japan Science and Technology Agency
Kawaguchi Center Building
4-1-8, Honcho, Kawaguchi-shi
Saitama 332-0012, Japan (Tokyo)

Nira Shimony-Eyal
Hebrew University of Jerusalem
Jerusalem 91120
Israel (Hamburg)

Takahiro Shinada
Japan Science and Technology Agency
Kawaguchi Center Building
4-1-8, Honcho, Kawaguchi-shi
Saitama 332-0012
Japan (Tokyo)

Richard Siegel
217 Materials Research Center
Rensselaer Polytechnic Institute
Troy, NY 12180 (Arlington
Chicago; contributor)

Michelle Simmons
University of New South Wales
Atomic Fabrication Facility
Centre for Quantum Computer
Technology
Newton Building, Room W103
Sydney, New South Wales 2052
Australia (Singapore)

Alex Simonian
National Science Foundation
Room 565
4201 Wilson Boulevard
Arlington, VA 22230 (Arlington)

Lewis Sloter II
Associate Director, Materials
and Structures Research/Technology/
Weapons Systems Office
Department of Defense
1777 N Kent Street 9030
Arlington, VA 22209-2110 (Arlington)

Jun-ichi Sone
National Institute for Materials Science
1-2-1 Sengen
Tsukuba city
Ibaraki 305-0047, Japan (Tokyo)

Kim Sooho
Korea Institute of Materials Science
797 Changwondaero
Sungsan-gu, Changwon
Gyeongnam 641-831, Korea (Tokyo)

Clivia Sotomayor Torres
Centre d'Investigació en Nanociencia I
Nanotecnologia
Campus UAB, Building Q-2nd Floor
08193 Bellaterra, Spain (Hamburg)

Francesco Stellaci
Professor of Materials Science and
Engineering, Massachusetts Institute
of Technology
Room 13-4053, 77 Massachusetts
Avenue
Cambridge, MA 02139 (Contributor)

J. Fraser Stoddart
Department of Chemistry
Northwestern University
2145 Sheridan Road
Evanston, IL 60208-3113 (Contributor)

Michael Stopa
Center for Nanoscale Systems
Harvard University
11 Oxford Street, LISE 306
Cambridge, MA 02138
(Chicago; contributor)

Anita Street
U.S. Department of Energy, Office of
Intelligence and Counterintelligence
1100 Independence Avenue SW
Washington, DC 20585 (Arlington)

Shigeru Suehara
Cabinet Office, Government of Japan
1-6-1 Nagata-cho, Chiyoda-ku
Tokyo 100-8914, Japan (Tokyo)

Ming-Huei Suh
Nanyang Technological University
50 Nanyang Avenue
Singapore 639798 (Tokyo)

Sang-Hee Suh
Korea Institute of Science and
Technology
335 Gwahangno, Yuseong-gu
Daejeon 305-806, Korea (Tokyo)

Jyrki Suominen
Directorate-General for Research
European Commission, SDME 2/2
B-1049 Brussels, Belgium
(Arlington, Hamburg)

Tsung-Tsan Su
Industrial Technology Research
Institute
Building 67, Room 211-1
195 Sec. 4 Chung Hsing Road
Chutung Hsinchu
Taiwan 31040
ROC (Arlington, Tokyo)

Wei-Fang Su
Nanyang Technological University
50 Nanyang Avenue
Singapore 639798 (Tokyo)

Tatsujiro Suzuki
University of Tokyo
Hongo 7-3-1, Bunkyo-ku
Tokyo, Japan (Tokyo)

Yoshishige Suzuki
Osaka University
Mihogaoka 8-1, Ibaraki
Osaka 567-0047, Japan (Tokyo)

Hidenori Takagi
University of Tokyo
Hongo 7-3-1, Bunkyo-ku
Tokyo, Japan (Tokyo)

Akira Takamatsu
Japan Science and Technology Agency
Kawaguchi Center Building
4-1-8, Honcho, Kawaguchi-shi
Saitama 332-0012, Japan (Tokyo)

Hidetaka Takasugi
Japan Science and Technology Agency
Kawaguchi Center Building
4-1-8, Honcho, Kawaguchi-shi
Saitama 332-0012, Japan (Tokyo)

Kunio Takayanagi
Tokyo Institute of Technology
2-12-1 Ookayama
Meguro-ku
Tokyo 152-8550, Japan (Tokyo)

Masahiro Takemura
National Institute for Materials Science
1-2-1 Sengen
Tsukuba city
Ibaraki 305-0047, Japan (Tokyo)

Suk-Wah Tam-Chang
Macromolecular, Supramolecular, and
Nanochemistry Program, National
Science Foundation
4201 Wilson Boulevard
Arlington, VA 22230 (Arlington)

Appendix 2

Kazunobu Tanaka
Japan Science and Technology Agency
Kawaguchi Center Building
4-1-8, Honcho, Kawaguchi-shi
Saitama 332-0012
Japan (Tokyo)

Syuji Tanaka
Japan Science and Technology Agency
Kawaguchi Center Building
4-1-8, Honcho, Kawaguchi-shi
Saitama 332-0012
Japan (Tokyo)

Naoko Tani
Semiconductor Portal, Inc.
East 17F Akasaka Twin Tower
2-17-22, Akasaka, Minato-ku
Tokyo 107-0052
Japan (Tokyo)

E. Clayton Teague
National Nanotechnology Coordination Office
Stafford II-405, 4201 Wilson Boulevard
Arlington, VA 22230
(Arlington)

Vasco Teixeira
University of Minho
Largo do Paço
4704-553 Braga, Portugal
(Hamburg; contributor)

Alan Tessier
Environmental Biology
National Science Foundation
4201 Wilson Boulevard
Arlington, VA 22230
(Arlington)

C. Shad Thaxton
Institute for Bionanotechnology in Medicine, Northwestern University
Robert H. Lurie Building
303 E Superior Street, Room 10-250
Chicago, IL60611 (Contributor)

Tom Theis
IBM
1101 Kitchawan Road
Yorktown Heights, NY 10598
(Chicago; contributor)

Treye Thomas
Consumer Product Safety Commission
4330 E West Highway, Suite 600
Bethesda, MD 20814 (Arlington)

Harry F. Tibbals
University of Texas Southwestern Medical Center
5323 Harry Hines Boulevard
Dallas, TX 75390-9004
(Contributor)

Gregory Timp
Department of Electrical Engineering
Notre Dame University
275 Fitzpatrick Hall
Notre Dame, IN 46556 (Contributor)

Sally Tinkle
National Nanotechnology Coordination Office
Stafford II-405, 4201 Wilson Boulevard
Arlington, VA 22230 (Arlington)

Sandip Tiwari
School of Electrical and Computer Engineering, Cornell University
410 Phillips Hall
Ithaca, NY 14853-2501
(Chicago; contributor)

Christos Tokamanis
Directorate-General for Research
European Commission, SDME 2/2
B-1049 Brussels, Belgium (Hamburg)

Donald A. Tomalia
Dendrimer and Nanotechnology Center, Central Michigan University
2625 Denison Drive
Mt. Pleasant, Michigan 48858
(Contributor)

Mark Tuominen
Department of Physics and Co-director
of the Center for Hierarchical
Manufacturing and MassNanoTech
University of Massachusetts, Amherst
411 Hasbrouck Laboratory
Amherst, MA 01003
(Arlington, Chicago,
Hamburg; contributor)

Kohei Uosaki
National Institute for Materials Science
1-2-1 Sengen
Tsukuba City
Ibaraki 305-0047, Japan (Tokyo)

Brian Valentine
U.S. Department of Energy
1000 Independence Avenue SW
Washington, DC 20585
(Arlington)

Rick Van Duyne
Chemistry Department, Northwestern
University
2145 Sheridan Road
Evanston, IL 60208-3113 (Chicago)

Usha Varshney
Directorate for Engineering, Electrical
Communications, and Cyber Systems
National Science Foundation
4201 Wilson Boulevard
Arlington, VA 22230
(Arlington)

Latha Venkataraman
Nanoscale Science and Engineering
Center, Columbia University
530 W 120th Street
New York, NY 10027-8903
(Contributor)

Lijun Wan
Institute of Chemistry
Chinese Academy of Sciences
Zhongguancun North First Street 2
100190 Beijing, China
(Singapore)

Chen Wang
National Center for Nanoscience
and Technology
No.11 ZhongGuanCun BeiYiTiao
100190 Beijing,
China (Singapore)

Kang Wang
California NanoSystems Institute
Department of Electrical Engineering
University of California
Box 951594
Los Angeles,
CA 90095 (Contributor)

Masahiro Watanabe
Japan Science and Technology Agency
Kawaguchi Center Building
4-1-8, Honcho, Kawaguchi-shi
Saitama 332-0012,
Japan (Tokyo)

Satoshi Watanabe
University of Tokyo
Hongo 7-3-1, Bunkyo-ku
Tokyo, Japan (Tokyo)

Scott Watkins
Commonwealth Scientific
and Industrial Research
Organization Materials Science
and Engineering
Office: Bayview Avenue
Post: Private Bag 10
Clayton, Victoria 3168
Australia (Singapore)

Andrew Wee
National University of Singapore
2 Science Drive 3,
Office: S13-03-12
Singapore 117542
(Singapore)

Udo Weimar
Institute of Physical Chemistry
Auf der Morgenstelle 8
D-72076 Tübingen
Germany (Hamburg)

Paul S. Weiss
California NanoSystems Institute,
University of California
570 Westwood Plaza, Building 114
Los Angeles, CA 90095
(Chicago; contributor)

Jeffrey Welser
Semiconductor Research Corporation
1101 Slater Road, Suite 120
Durham, NC 27703
and
IBM Almaden Research Center
650 Harry Road
San Jose, CA 95120 (Arlington,
Chicago, Hamburg; contributor)

Wolfgang Wenzel
Karlsruhe Institute of Technology
PO Box 3640
76021 Karlsruhe
Germany (Hamburg)

Rosemarie Wesson
Engineering Division, National Science Foundation
4201 Wilson Boulevard
Arlington, VA 22230
(Arlington)

Paul Westerhoff
Professor and Interim Director
School of Sustainable Engineering
and the Built Environment, Civil
Environmental, and Sustainable
Engineering Programs
Del E. Webb School of Construction
Programs, Arizona State University
Engineering Center
G-Wing Room 252
Tempe, AZ 85287-5306
(Contributor)

Mark Wiesner
Duke University
Box 90287, 120 Hudson Hall
Durham, NC 27708-0287
(Chicago; contributor)

Ellen Williams
Department of Physics
University of Maryland
College Park
MD 20742-4111 (Chicago)

R. Stanley Williams
HP Senior Fellow, Information
and Quantum Systems Lab
Hewlett-Packard Laboratories
1501 Page Mill Road
Palo Alto, CA 94304 (Contributor)

Grant Wilson
Department of Chemical Engineering
The University of Texas at Austin
1 University Station C0400
Austin, TX 78712-0231 (Chicago)

Adam Winkleman
National Academies
500 Fifth Street NW, K-951
Washington, DC 2000 (Arlington)

Stuart A. Wolf
NanoStar, University of Virginia
395 McCormick Road
Charlottesville, VA 22904
(Arlington, Chicago, Singapore
Tokyo; contributor)

Maw-Kuen Wu
Academia Sinica
128 Academia Road, Section 2
Nankang
Taipei 115, Taiwan (Tokyo)

Omar Yaghi
Professor, Department of Chemistry
and Biochemistry, University of
California
Box 951594
Los Angeles, CA 90095 (Contributor)

Yiyan Yang
Institute of Bioengineering and
Nanotechnology
31 Biopolis Way, The Nanos, #04-01
Singapore 138669 (Singapore)

Tetsuji Yasuda
National Institute of Advanced
Industrial Science and Technology
1-3-1 Kasumigaseki
Chiyoda-ku
Tokyo 100-8921
Japan (Tokyo)

Jackie Ying
Institute of Bioengineering and
Nanotechnology
31 Biopolis Way, The Nanos, #04-01
Singapore 138669
(Singapore)

Naoki Yokoyama
National Institute of Advanced
Industrial Science and Technology
1-3-1 Kasumigaseki
Chiyoda-ku
Tokyo 100-8921
Japan (Tokyo)

Shinji Yuasa
National Institute of Advanced
Industrial Science and Technology
1-3-1 Kasumigaseki
Chiyoda-ku
Tokyo 100-8921
Japan (Tokyo)

Hua Zhang
School of Materials Science and
Engineering, Office N4.1-02-25
Nanyang Technological University
Singapore 639798
(Singapore)

Yong Zhang
Faculty of Engineering, National
University of Singapore
E1-05-18, Division of Bioengineering
117576 Singapore (Singapore)

Xiang Zhang
University of California
5130 Etcheverry Hall, Mail Stop 1740
Berkeley, CA 94720-1740
(Chicago; contributor)

Yuliang Zhao
Chinese Academy of Sciences, Institute
of High Energy Physics
19B YuquanLu, Shijingshan District
100049 Beijing, China (Singapore)

Liu Zhongfan
Institute of Chemistry and Molecular
Engineering, Peking University
Cheng Rd, Haidian District
Beijing 202, 100871, China
(Singapore; contributor)

Haoshen Zhou
National Institute of Advanced
Industrial Science and Technology
1-3-1 Kasumigaseki
Chiyoda-ku
Tokyo 100-8921, Japan (Tokyo)

Otto Zhou
Department of Physics, University of
North Carolina
Phillips Hall, CB #3255
Chapel Hill, NC 27599-3255
(Contributor)

Appendix 3

NNI Timeline in Selected Publications, 1999–2010

1999

R.W. Siegel, E. Hu, M.C. Roco, Nanostructure science and technology. Adopted as official document of U.S. National Science and Technology Council (NSTC) (Springer (previously Kluwer), 1999), Available online: http://www.wtec.org/loyola/pdf/nano.pdf

M.C. Roco, R.S. Williams, P. Alivisatos (eds.), *Nanotechnology Research Directions: Vision for Nanotechnology R&D in the Next Decade*. Adopted as official document of NSTC (Springer (previously Kluwer) 2000, Washington, DC, 1999), Available online: http://www.nano.gov/html/res/pubs.html

S. Smalley, Testimony at the first congressional hearing to the U.S. House of Representatives Committee on Science, subcommittee on basic research, Available online: http://www.merkle.com/papers/nanohearing1999.html. 22 June 1999

National Science and Technology Council (NSTC) and Interagency Working Group on Nanoscience, Engineering, and Technology (IWGN), Nanotechnology—shaping the world atom by atom (Document for public outreach, Washington, DC, 1999), Available online: http://www.wtec.org/loyola/nano/IWGN.Public.Brochure/

2000

National Science and Technology Council (NSTC) and Nanoscale Science, Engineering and Technology Subcommittee (NSET) report, National Nanotechnology Initiative (NSTC/NSET, Washington, DC, July 2000), Available online: http://www.nano.gov/html/res/nni2.pdf

M.C. Roco, W.S. Bainbridge (eds.), Societal implications of nanoscience and nanotechnology. NSF and DOC report (Springer (previously Kluwer), 2001), Available online: http://www.wtec.org/loyola/nano/NSET.Societal.Implications/nanosi.pdf

2001

National Nanotechnology Coordination Office (NNCO) was established through a memorandum of understanding signed by NSF, DOD, DOE, NIH, NIST, NASA, EPA, and DOT

M.C. Roco, International strategy for nanotechnology research. J. Nanopart. Res. **3**(5–6), 353–360 (2001)

2002

National Research Council (NRC), *Small Wonders, Endless Frontiers. Review of the National Nanotechnology Initiative* (National Academy Press, Washington, DC, 2002), Available online: http://www.nano.gov/html/res/small_wonders_pdf/smallwonder.pdf

M.C. Roco, Nanotechnology – a frontier in engineering education. Int. J. Eng. Educ. **18**(5), 488–497 (Aug 2002)

2003

National Science and Technology Council (NSTC), Report of the national nanotechnology initiative workshop. Materials by design (NSTC, Arlington, June 2003), Available online: http://www.nano.gov/NNI_Materials_by_Design.pdf

M.C. Roco, W.S. Bainbridge, *Converging Technologies for Improving Human Performance: Nanotechnology, Biotechnology, Information Technology and Cognitive science* (Springer (previously Kluwer), Dordrecht, 2003), Available online: http://www.wtec.org/ConvergingTechnologies/Report/NBIC_report.pdf

National Science and Technology Council (NSTC) and Nanoscale Science, Engineering and Technology subcommittee (NSET), Regional, tate, and local initiatives in nanotechology (NSTC, Washington, DC, 30 Sept–1 Oct 2003), Available online: http://www.nano.gov/041805Initiatives.pdf

Congress, The 21st century nanotechnology research and development act. 108th congress. Public law 108–153 (2003), Available online: http://frwebgate.access.gpo.gov/cgi-bin/getdoc.cgi?dbname=108_cong_public_laws&docid=f:publ153.108.pdf

2004

National Science and Technology Council (NSTC) and Nanoscale Science, Engineering and Technology subcommittee (NSET), National nanotechnology initiative (NSTC, Washington, DC, Dec 2004), Available online: http://www.nano.gov/NNI_Strategic_Plan_2004.pdf

National Science Foundation (NSF), Report: international dialogue on responsible research and development of nanotechnology (Meridian Institute, Arlington/Washington, DC, 2004), Available online: http://www.nsf.gov/crssprgm/nano/activities/dialog.jsp

2005

President's Council of Advisors on Science and Technology (PCAST), Report to the president and congress on the third assessment of the national nanotechnology initiative (Executive Office of the President, Washington, DC, 2005), Available online: http://www.nano.gov/html/res/otherpubs.html

2006

National Research Council (NRC), *A Matter of Size: Triennial Review of the National Nanotechnology Initiative* (National Academies Press, Washington, DC, 2006), Available online: http://www.nap.edu/catalog.php?record_id=11752

2007

National Science and Technology Council (NSTC) and Nanoscale Science, Engineering and Technology subcommittee (NSET), *The National Nanotechnology Initiative Strategic Plan* (NSTC/NSET, Washington, DC, 2007), Available online: http://www.nano.gov/html/res/pubs.html

M.C. Roco, W.S. Bainbridge, *Nanotechnology: Societal Implications,* vol 1, vol 2 (Springer, Boston, 2007)

2008

President's Council of Advisors on Science and Technology (PCAST), *National Nanotechnology Initiative (NNAP)* (PCAST, Washington, DC April 2008), Available online: http://www.nanowerk.com/nanotechnology/reports/reportpdf/report118.pdf

National Science and Technology Council (NSTC) and Nanoscale Science, Engineering and Technology subcommittee (NSET), Strategy for nanotechnology-related environmental, health, and safety (Feb 2008)

2009

National Science and Technology Council (NSTC) and Nanoscale Science, Engineering and Technology subcommittee (NSET), Nanotechnology-enabled sensing. Report of the National Nanotechnology Initiative workshop (Arlington May 2009), Available online: http://www.nano.gov/html/res/NNI_Nanosensing.pdf

National Science and Technology Council (NSTC) and Nanoscale Science, Engineering and Technology subcommittee (NSET), Regional, state, and local initiatives in nanotechnology. Report of the National Nanotechnology Initiative workshop (Oklahoma City/Washington, DC, April 2009), Available online: http://www.nano.gov/html/res/pubs.html

2010

President's Council of Advisors on Science and Technology (PCAST), Report to the president and congress on the third assessment of the national nanotechnology initiative (Executive Office of the President, Washington, DC, 2010), Available online: http://www.nano.gov/html/res/otherpubs.html

M.C. Roco, C. Mirkin, M. Hersam (eds.), Nanotechnology research directions for societal needs in 2020. World Technology Evaluation Center (WTEC) report (Springer 2010)

National Science and Technology Council (NSTC) and Nanoscale Science, Engineering and Technology subcommittee (NSET), Regional, state, and local initiatives in nanotechology, 1–3 April 2009 (Oklahoma City, 2010), Available online: http://www.nano.gov/html/meetings/nanoregional-update/index.html (To be published in Dec. 2010)

National Science and Technology Council (NSTC) and Nanoscale Science, Engineering and Technology subcommittee (NSET), *The National Nanotechnology Initiative Strategic Plan* (NSTC/NSET, Washington, DC, 2010), Available online: http://www.nano.gov/html/res/pubs.html (Note: to be published in Dec. 2010)

Appendix 4

NNI Centers, Networks, and Facilities

Centers and Networks of Excellence

Centers and networks provide opportunities and support for multidisciplinary research among investigators from a variety of disciplines and from different research sectors, including academia, industry and government laboratories. Such multidisciplinary research not only leads to advances in knowledge, but also fosters relationships that enhance the transition of basic research results to devices and other applications. The multidisciplinary centers are listed below, organized by funding agency.

National Science Foundation

Engineering Research Center

Center for Extreme Ultraviolet Science and Technology, University of Colorado–Boulder (http://euverc.colostate.edu/)
Center for Sinthetic Biology, University of California–Berkeley (http://www.synBERC.org/)

Science and Technology Center

Nanobiotechnology Center, Cornell University (http://www.nbtc.cornell.edu/)
Emergent Behaviors of Integrated Cellular System, Center at MIT (http://ebics.net)
Center for Energy Efficient Electronics Science, University of California–Berkeley (http://www.e3s-center.org/)

Nanoscale Science and Engineering Centers

Center for Hierarchical Manufacturing, University of Massachusetts–Amherst (http://www.umass.edu/chm/)

Center for Nanoscale Systems (NSEC), Cornell University (http://www.cns.cornell.edu/)

Science of Nanoscale Systems and their Device Applications (NSEC), Harvard University (http://www.nsec.harvard.edu/)

Center for Biological and Environmental Nanotechnology, Rice University (http://www.ruf.rice.edu/~cben/)

Center for Integrated Nanopatterning and Detection (NSEC), Northwestern University (http://www.nsec.northwestern.edu/)

Center for Electronic Transport in Molecular Nanostructures (NSEC), Columbia University (http://www.cise.columbia.edu/nsec/)

Center for Directed Assembly of Nanostructures (NSEC), Rensselaer Polytechnic Institute (http://www.rpi.edu/dept/nsec/)

Center for Scalable and Integrated Nano-Manufacturing (NSEC), University of California–Los Angeles (http://newsroom.ucla.edu/page.asp?id=4601)

Center for Chemical-Electrical-Mechanical Manufacturing Systems (NSEC), University of Illinois, Urbana–Champaign (http://www.nano-cemms.uiuc.edu/)

Center on Templated Synthesis and Assembly at the Nanoscale, University of Wisconsin (http://www.nsec.wisc.edu/)

Center for Probing the Nanoscale, Stanford University (http://www.stanford.edu/group/cpn/)

Center for Affordable Nanoengineering of Polymeric Biomedical Devices, Ohio State University (http://www.nsec.ohio-state.edu/)

Center of Integrated Nanomechanical Systems, University of California–Berkeley (http://nano.berkeley.edu/coins/)

Nano-Bio Interface Center, University of Pennsylvania (http://www.nanotech.upenn.edu/)

Center for High Rate Nanomanufacturing, Northeastern University (http://www.nano.neu.edu/)

Materials Research Science and Engineering Centers

These four MRSECs are fully dedicated to nanotechnology research:

Center for Nanoscale Science (MRSEC), Pennsylvania State University (http://www.mrsec.psu.edu/)

Center for Quantum and Spin Phenomena in Nanomagnetic Structures (MRSEC), University of Nebraska, Lincoln (http://www.mrsec.unl.edu/)

Center for Research on Interface Structure and Phenomena (MRSEC), Yale University (http://www.crisp.yale.edu/index.php/Main_Page)

Genetically Engineered Materials (MRSEC), University of Washington (http://www.mrsec.org/centers/university-washington)

Appendix 4

In addition, many other MRSECs have one or more interdisciplinary research group(s) focused on nanoscale science and engineering topics. See http://www.mrsec.org/ for more information.

NSF Nanoscale Science and Engineering Networks

Network for Computational Nanotechnology (http://www.ncn.purdue.edu/)
National Nanotechnology Infrastructure Network (http://www.nnin.org/)
Oklahoma Network for Nanostructured Materials (http://www.okepscor.org/default.asp)
Nanoscale Informal Science Education Network (http://www.nisenet.org/)
Network for Nanotechnology in Society, Arizona State University–University of California–Santa Barbara (http://cns.asu.edu/resources/nsf.htm)
Experimental Program to Stimulate Competitive Research, University of New Mexico (http://www.nmepscor.org/)

Centers for Learning and Teaching

National Center for Learning & Teaching in Nanoscale Science & Engineering (http://www.nclt.us/)
NSF's directory of R&D Centers (http://www.nsf.gov/about/partners/centers.jsp)

National Science Foundation with the Environmental Protection Agency

Centers for Environmental Implications of Nanotechnology

University of California–Los Angeles (http://cein.cnsi.ucla.edu/pages/)
Duke University (http://www.ceint.duke.edu/)

Department of Energy

Nanoscale Science Research Centers

Center for Nanoscale Materials, Argonne National Laboratory (http://nano.anl.gov/)
Molecular Foundry, Lawrence Berkeley National Laboratory (http://foundry.lbl.gov/)
Center for Integrated Nanotechnologies, Sandia and Los Alamos National Laboratory (http://cint.lanl.gov/)
Center for Functional Nanomaterials, Brookhaven National Laboratory (http://www.bnl.gov/cfn/)
Center for Nanophase Materials Science, Oak Ridge National Laboratory (http://www.cnms.ornl.gov/)

Department of Defense

Institute for Soldier Nanotechnologies, Massachusetts Institute of Technology (http://web.mit.edu/isn/)
Center for Nanoscience Innovation for Defense, University of California–Santa Barbara, Riverside and Los Angeles (http://www.instadv.ucsb.edu/93106/2003/January21/new.html)
Institute for Nanoscience, Naval Research Laboratory (http://www.nrl.navy.mil/content.php?P=MULTIDISCIPLINE)

National Aeronautics and Space Administration (NASA)

University Research, Engineering, and Technology Institutes
(NASA funding now discontinued)

Institute for Intelligent Bio-Nanomaterials & Structures for Aerospace Vehicles, Texas A&M University (http://tiims.tamu.edu/)
Biologically Inspired Materials Institute (BIMat), Princeton University (http://bimat.princeton.edu/html/overview.html)

National Institute for Occupational Safety and Health

Nanotechnology Research Center, Robert A. Taft Lab (http://www.cdc.gov/niosh/topics/nanotech/)

National Institute of Standards and Technology

Center for Nanoscale Science and Technology, NIST Gaithersburg (http://www.cnst.nist.gov/)

National Institutes of Health

Nanotechnology Characterization Laboratory, NCI-Frederick (http://ncl.cancer.gov/)

NHLBI Program of Excellence in Nanotechnology

Integrated Nanosystems for Diagnosis and Therapy, Washington University (http://www.nhlbi-pen.info/)

Nanotechnology: Detection & Analysis of Plaque Formation, Emory University Georgia Tech (http://pen.bme.gatech.edu/)
Nanotherapy for Vulnerable Plaque, Burnham Institute (http://www.pennvp.org/)
Translational Program of Excellence in Nanotechnology, Massachusetts General Hospital (http://cmir.mgh.harvard.edu/nano/tpen)

Nanomedicine Development Centers

Center for the Optical Control of Biological Functions, University of California–Berkeley (http://nihroadmap.nih.gov/nanomedicine/fundedresearch.asp)
Center for Cell Control, University of California–Los Angeles (http://ccc.seas.ucla.edu/)
Phi2 DNA, Purdue University (http://www.vet.purdue.edu/PeixuanGuo/NDC/)
Nanomedicine Center for Nucleoprotein Machines, Georgia Institute of Technology (http://www.nucleoproteinmachines.org/)
National Center for Design of Biomimetic Nanoconductors, University of Illinois–Urbana–Champaign (http://www.nanoconductor.org/)
Center for Protein Folding Machinery, Baylor University (http://proteinfoldingcenter.org/)
Nanomedicine Center for Mechanobiology, Columbia University (http://www.mechanicalbiology.org/)
Engineering Cellular Control:Synthetic Signaling and Motility Systems, University of California–San Francisco (http://qb3.org/cpl/)

Centers of Cancer Nanotechnology Excellence

The Siteman Center of Cancer Nanotechnology Excellence, Washington University (http://www.siteman.wustl.edu/)
Center of Nanotechnology for Treatment, Understanding, and Monitoring of Cancer (NANOTUMOR), University of California, San Diego (http://ntc-ccne.org/)
Carolina Center of Cancer Nanotechnology Excellence, University of North Carolina (http://cancer.med.unc.edu/ccne/)
Center for Cancer Nanotechnology Excellence Focused on Therapy Response, Stanford University (http://mips.stanford.edu/public/grants/ccne/)
MIT-Harvard Center of Cancer Nanotechnology Excellence, Massachusetts Institute of Technology (http://nano.cancer.gov/action/programs/mit/)
Nanotechnology Center for Personalized and Predictive Oncology, Emory University Georgia Institute of Technology (http://www.wcigtccne.org/)
Center for Cancer Nanotechnology Excellence, Northwestern University (http://www.ccne.northwestern.edu/)
Nanosystems Biology Cancer Center, California Institute of Technology (http://www.caltechcancer.org/)

Appendix 5

Glossary

AAAS	American Association for the Advancement of Science
AAO	Anodic aluminum oxide
AC	Alternating current
AD	Alzheimer's disease
ADF	Annular dark field
ADME/Tox	Absorption, distribution, metabolism, excretion, and toxicity
AD-PEG	Adamantane-polyethylene glycol
AD-PEG-Tf	Adamantane-polyethylene glycol-tranferrin
AF&PA	American Forest & Paper Association
AFM	Atomic force microscope/y
AIMD	*ab initio* molecular dynamics
AIST	Advanced Industrial Science and Technology
ALE	Atomic layer epitaxy
ALS	Amyotrophic lateral sclerosis
AMD	Age-related macular degeneration
ANC	Alliance for Nanotechnology in Cancer (NCI)
ANF	Asia Nanotechnology Forum
ANSI	American National Standards Institute
APM	Atomically precise manufacturing
ASEE	American Society for Engineering Education
ASTC	Association of Science Technology Centers
ATE	Advanced Technical Education (NSF)
AuNP	Gold nanoparticle
BAL	Bronchoalveolar lavage
bcc	Body-centered cubic
BNC	Boron-nitrogen-carbon
BEM	Boundary element methods
BES	Office Basic Energy Sciences (DOE)
BHJ	Bulk heterojunction
BiSFET	Bilayer pseudospin field-effect transistor

BNL	Brookhaven National Laboratory
BNNT	Boron nitride nanotube
BOSE	Board on Science Education (NRC)
BP	Base pair
BPL	Beam pen lithography
BTU	British thermal unit
CAREER	Faculty Early Career Development Award (NSF)
CAS	Chinese Academy of Sciences
CBEN	Center for Biological and Environmental Nanotechnology
CC	Computational complexity
CCD	Charge-coupled device
CCNE	Center for Cancer Nanotechnology Excellence
CCNE-TR	Center for Cancer Nanotechnology Excellence Focused on Therapy Response
CCS	Carbon capture and storage
CCSSO	Council of Chief State School Officers
CDC	Carbide-derived carbon
CDP	Cyclodextrin-containing polymer
CDR	Carbon dioxide reforming
CEA	Alternative Energies and Atomic Energy Commission (France)
CEIN	Center for the Environmental Implications of Nanotechnology (NSF)
CEPA	Canadian Environmental Protection Act (1999)
C-FDA	5-carboxyfluorescein diacetate
CIGS	Copper indium gallium selenide
CINN	Research Center on Nanomaterials and Nanotechnology
CM	Carrier multiplication or chemical manufacturing plants
CMD	Classical molecular dynamics
CMOS	Complementary metal-oxide-semiconductor
CMP	Chip multiprocessors
CNDP	Critical nanoscale design parameter
CNMS	Center for Nanophase Materials Sciences
CNPP	Cancer Nanotechnology Platform Partnerships
CNRS	French National Center for Scientific Research
CNS-ASU	Center for Nanotechnology in Society at the Arizona State University
CNSE	College for Nanoscale Science and Engineering (SUNY)
CNS-UCSB	Center for Nanotechnology in Society at the University of California, Santa Barbara
CNT	Carbon nanotube
CNTC	Cancer Nanotechnology Training Centers (NCI)
CNTFET	Carbon nanotube field-effect transistor
CPSC	Consumer Product Safety Commission
CPU	Central processing unit
CRADA	Cooperative research and development agreement
CSIC	Spanish National Research Council
CSO	Civil society organization

Appendix 5

CSTP	Committee for Scientific and Technological Policy
CT	Computed tomography
CTE	Constant of thermal expansion
CTR	Crystal truncation rod
CVD	Chemical vapor deposition
DAPI	4',6-diamidino-2-phenylindole
DDA	Discrete dipole approximation
DEF	Dendrimer-enhanced ultrafiltration
DFT	Density functional theory
DG	Conductivity (cantilever sensors)
DGU	Density gradient ultracentrifugation
DME	Diabetic macular edema
DNA	Deoxyribonucleic acid
DOC	Department of Commerce
DOD	Department of Defense
DOE	Department of Energy
DoEd	Department of Education
DPN	Dip-pen nanolithography
dsDNA	Double-stranded DNA
dw	Resonant-frequency (cantilever sensors)
Dx	Diagnostics
EC	European Commission/Community
ECE	Electrocaloric effect
ECM	Extracellular matrix
EDX	Energy-dispersive x-ray (spectroscopy)
EELS	Electron energy loss spectroscopy
EERE	Office of Energy Efficiency and Renewable Energy (DOE)
EFM	Electric force microscopy
EFRC	Energy Frontier Research Center (DOE)
EHS	Environmental health and safety
EIT	European Institute for Innovative Technology
ELISA	Enzyme-linked immunosorbent assay
ELSI	Ethical, legal, and social implications
EM	Electron microscopy
EMI	Ethyl-methylimmidazolium
ENM	Engineered nanomaterial
EOF	Electroosmotic flow
EPA	U.S. Environmental Protection Agency
EPR	Electron paramagnetic resonance (spectrometry, spectra)
ERC	Engineering Research Centers (NSF)
ESEM	Environmental scanning electron microscopy
EU	European Union
EUV	Extreme ultraviolet
EUVL	Extreme ultraviolet lithography
EWS-FL11	Ewing's sarcoma fusion gene

EXAFS	Extended x-ray absorption fine-structure spectroscopy	
fcc	Face-centered cubic	
FDA	Food and Drug Administration	
FDFD	Finite-difference frequency domain	
FDTD	Finite difference time domain	
FEG	Field-emission gun	
FET	Field-effect transistor	
FFF	Foundation for the Future	
FFT	Fast Fourier transform	
FIB	Focused ion beam	
FIFRA	Federal Insecticide, Fungicide, and Rodenticide Act	
FP	Food preservatives	
fs	Femptosecond	
f_T	Cutoff frequency	
FTIR	Fourier transform infrared spectroscopy	
GB	Grain boundaries	
GDP	Gross domestic product	
GFP	Green fluorescent protein	
GGA	General gradient approximation	
GMP	Good manufacturing practice	
GOALI	Grant Opportunities for Academic Liaison with Industry (NSF)	
GPCR	G-protein-coupled receptor	
GPU	Graphical processing unit; general-purpose unit	
GUI	Graphical user interface; government-university-industry	
GW	Green's function G and screened interaction W (GW approximation)	
HAADF	High-angle annular dark field	
HDL	High-density lipoprotein	
HDS	Hydrodesulfurization	
HeIM	Helium ion microscopy	
HELA	Squamous carcinoma cell line (named for Henrietta Lacks)	
HEPA	High-efficiency particulate air (filter)	
HF	Hartree-Fock	
HIV	Human immunodeficiency virus	
HOMO	Highest occupied molecular orbital	
HPC	High-performance computing	
HPV	Human papillomavirus	
HTE	High-throughput experimentation	
HTS	High-throughput screening	
IANH	International Association of Nano Harmonization	
ICON	International Council on Nanotechnology	
IDE	Investigational device exemption	
IEC TC113	IEC Technical Committee on Nanotechnology standardization for electrical and electronic products and systems	
IEC	International Electrotechnical Commission	
IgE	Immunoglobulin E	

IGU	Insulating glass unit
IGERT	Integrative Graduate Education and Research Traineeship program (NSF)
IMEC	Belgian nanoelectronics research organization
IMRE	Institute of Materials Research & Engineering (Singapore)
IND	Investigational new drug
INS	Inelastic neutron scattering
IP	Intellectual property
IPCC	Intergovernmental Panel on Climate Change
IPS	Instructions per second
IQE	Internal quantum efficiency
IR	Infrared
iREU	International Research Experience for Undergraduates
IRGC	International Risk Governance Council
ISE	Informal Science Education program(s) (NSF); Fraunhofer Institute for Solar Energy Systems
ISL	*in situ* leaching
ISM	*in situ* mining
ISO	International Organization for Standardization
ISTEC-CNR	Institute of Science and Technology for Ceramics
IT	Information technology
ITRI	Industrial Technology Research Institute
ITRS	International Technology Roadmap for Semiconductors
I/UCRC	Industry & University Cooperative Research Centers (NSF)
IUPAC	International Union of Physical and Applied Chemistry
IWGN	Interagency Working Group on Nanoscience, Engineering and Technology
IWM	Fraunhofer Institute for Mechanics of Materials
iWSG	International Winter School for Graduate Students
JST	Japan Science and Technology Agency
KIT	Karlsruhe Institute of Technology
KRISS	Korea Research Institute of Standards and Science
LAUSD	Los Angeles Unified School District
LBNL	Lawrence Berkeley National Laboratory
LCA	Life-cycle analysis/assessment
LCAT	Lecithin:cholesterol acyl transferase
LC-APPI-MS	Liquid chromatography-atmospheric pressure photoionization-mass spectrometry
LCO	Lithium cobalt oxide
LDA	Local density approximation
LDH	Lactate dehydrogenase; layered double hydroxide
LDL	Low density lipoprotein
LED	Light-emitting diode
LEF	Laboratory Experience for Faculty
LFP	Lithium iron phosphate

LG	Liquid natural gas
LIB	Lithium ion battery
LITEN	Laboratory for Innovation in New Energy Technologies and Nanomaterials
LME	Large multinational enterprises
LMS	Lithium manganese spinel
LNCO	Lithium nickel cobalt oxide
LOD	Limit of detection
LP	Lipoprotein
LTO	Lithium titanate
LUMO	Lowest unoccupied molecular orbital
M&A	Merger and acquisition
MBE	Molecular beam epitaxy
MC	Monte Carlo
MD	Molecular dynamics
MDR	Multidrug-resistance
MFC	Micro fuel cell
MEA	Membrane electrolyte assembly
MEG	Magnetoencephalography
MEMS	Microelectromechanical systems
MF	Microfiltration
MFM	Magnetic force microscopy
MIP	Macrophage inflammatory protein
MMP	Micron-sized magnetic particle
MOF	Metal organic framework
MONA	Merging Optics and Nanotechnologies
MOS	Metal oxide semiconductor
MOSFET	Metal oxide semiconductor field-effect transistor
MP	Magnetic particle
MR	Magnetic resonance
MRAM	Magnetic random access memory
MRFM	Magnetic resonance force microscopy
MRI	Magnetic resonance imaging
mRNA	Messenger RNA
MRSEC	Materials Research Science and Engineering Centers (NSF)
MS	Mass spectrometry
MSDS	Material Safety Data Sheets
MSNP	Mesoporous silica nanoparticle
MTJ	Magnetic tunnel junction
MTT	Metabolic and cytotoxic activity (assay)
MTV	Multivariate
MTV-MOF	Multivariate metal organic framework
MWCNT	Multiwalled carbon nanotubes
NA	Numerical aperture
NACK	National Applications and Career Knowledge Center

Appendix 5

nano-SIMS	Nano-secondary ion mass spectrometry
NAS	National Academy of Sciences
NASA	National Aeronautics and Space Administration
NBD	7-nitrobenz-2-oxa-1,3-diazol-4-yl
NBIC	Nanotechnology, biotechnology, information technology, and cognition
NC	Nanocrystalline
nc-AFM	Non-contact atomic force microscopy
NCAM	Nanocapillary array membrane
NCI	National Cancer Institute
NCL	National/Nanotechnology Characterization Laboratory
NCLT	National Center for Learning and Teaching
NCN	Network for Computational Nanotechnology
NCTF	National Citizens' Technology Forum (Arizona State University CNS)
NCTM	National Council of Teachers of Mathematics
NCTR	National Center for Toxicological Research (FDA)
NDC	Nanodisk-code
NEGF	Non-equilibrium Green's function (theory)
NEHI	Nanotechnology Environmental Health Implications (EHS working group)
NEI	Neuroscience Education Institute
NEMS	Nanoelectromechanical systems
NGA	National Governors Association
NGO	Nongovernmental organization
NHLBI	National Heart, Lung, and Blood Institute Programs of Excellence in Nanotechnology
NHGRI	National Human Genome Research Institute
NIAID	National Institute of Allergy and Infectious Disease (FDA)
NIEHS	National Institute of Environmental and Health Sciences (FDA)
NIH	National Institutes of Health
NIL	Nanoimprint lithography
NIMS	National Institute for Materials Science
NIOSH	National Institute for Occupational Safety and Health
NIRT	Nanoscale Interdisciplinary Research Team (NSF)
NIST	National Institute of Standards and Technology
NISE Net	National Informal Science Education Network
NM	Nanomaterials
NMC	Nickel magnesium cobalt
NMR	Nuclear magnetic resonance (spectrometry)
NMT	Nanofabrication Manufacturing Technology (Pennsylvania State University)
N&N	Nanoscience and nanotechnology
NNCO	National Nanotechnology Coordination Office
NNI	National Nanotechnology Initiative
NNIN	National Nanotechnology Infrastructure Network
NNUN	National Nanofabrication User Network

NOD/SCID	Non-obese diabetic severe combined immunodeficiency (mice)
NP	Nanoparticle
NR	Neutron reflectivity
NRC	National Research Council
NRI	Nanotechnology Research Initiative
nRTD	Nano resonant tunneling diode
NSE	Nanoscale science and engineering
NSEE	Nanoscale Science and Engineering Education (NSF program)
NSEC	Nanoscale Science and Engineering Center (NSF)
NSES	Neutron spin echo spectroscopy or National Science Education Standards
NSET	Nanoscale Science, Engineering and Technology subcommittee of the Committee on Technology of the National Science and Technology Council
NSF	National Science Foundation
NSOM	Near-field scanning optical microscopy
NSRC	National Science Resources Center (DOE)
NSTA	National Science Teachers Association
NT	Nanotube or nanotechnology
NTU	Nanyang Technological University
NUE	Nanotechnology Undergraduate Education in Engineering (NSF program)
NV	Nitrogen-vacancy
NW	Nanowire
NZVI	Nanoscale zero-valent iron
OA/RA	Osteoarthritis/rheumatoid arthritis
OCM	Oxidative coupling of methane
OECD	Organization for Economic Co-operation and Development
OECD CSTP	Committee for Scientific and Technological Policy
OECD WPMN	OECD Working Party on Manufactured Nanomaterials
OECD WPN	OECD Working Party on Nanotechnology
OJL	Optoelectronics Joint Research Laboratory
OLED	Organic light emitting diode
ONAMI	Oregon Nanoscience and Microtechnologies Institute
OPE	Oligo(phenylene ethynylene)
OSHA	Occupational Health and Safety Administration
OSTP	Office of Science and Technology Policy
OWL	On-wire lithography
PA	Peptide amphiphiles
PALM	Photo-activated localization microscopy
PC	Polycrystalline
PCAST	President's Council of Advisors on Science and Technology
PCM	Phase change memory
PCR	Polymerase chain reaction
PEFC	Polymer electrolyte fuel cell

PEG	Polyethylene glycol
PEI	Polyethyleneimine
PEM	Polymer electrolyte membrane
PEN	Project for Emerging Nanotechnologies
PF	Power factor
PFM	Piezoelectric force microscopy (or piezoresponse force microscopy)
Pgp	Permeability glycoprotein
PGM	Platinum group metals
PHEV	Plug-in hybrid electric vehicle
PhoReMoST	Nano<u>Pho</u>tonics to <u>Re</u>alize <u>Mo</u>lecular <u>S</u>cale <u>T</u>echnologies (EU Center of Excellence)
PI	Principle investigator
PM	Particulate matter
PMN	Premanufacture notice; polymorphonuclear
PNA	Peptide nucleic acid
PNC	Polymer nanocomposite
POC	Point-of-care
PPL	Polymer pen lithography
PPP	Private–public partnership
PPV	Poly(phenylene vinylene)
PRINT	Pattern replication in non-wetting templates technology (UNC)
PSA	Prostate specific antigen
PTB	Benzodithiophene polymers
PV	Photovoltaic(s)
PVDF	Polyvinylidene fluoride
Q	Ultrahigh optical quality factor (i.e., low optical loss)
QCL	Quantum cascade laser
QD	Quantum dot
QENS	Quasi-elastic neutron scattering
QM	Quantum mechanical/molecular
qPCR	Quantitative PCR
QSAR	Quantitative structure–activity relationship (models)
RAFT	Reversible Addition-Fragmentation chain Transfer
RAM	Random-access memory
R&D	Research and development
RC	Resistor-capacitor
RCT	Reverse cholesterol transport
REACH	Registration, Evaluation and Authorization CHemicals Regulation of the European Union (enacted 18 December 2006)
ReaxFF	Reactive force field
REBO	Reactive empirical bond-order (model)
RES	Reticulo-endothelial system
RET	Research Experience for Teachers (NSF)
REU	Research Experience for Undergraduates (NSF program)
RF	Radio frequency

RF-FET	Radio-frequency field-effect transistor
RFID	Radio frequency identification
RNA	Ribonucleic acid
RNase	Ribonuclease
RO	Reverse osmosis
ROS	Reactive oxygen species
RRM2	Ribonucleotide reductase subunit 2
RSA	Random sequential adsorption
RTD	Resonant tunneling diode
RTTA	Real-time technology assessment
SAD	Surface area dose
SA/V	Surface-area-to-volume
SAXS	Small-angle x-ray scattering
SBIR	Small Business Innovation Research programs (various U.S. government agencies)
SEI	Social and ethical issues
SEM	Scanning electron microscopy
SEMATECH	SEmiconductor MAnufacturing TECHnology, a non-profit consortium guiding basic research in semiconductor manufacturing
SERS	Surface-enhanced Raman scattering / spectroscopy
SFG	Sum frequency generation
SGER	Small Grant for Exploratory Research (NSF)
SHG	Second harmonic generation
SHIM	Scanning helium ion microscopy
SI	Systeme Internationale (international system of units)
SIA	Semiconductor Industry Association
SIMS	Secondary ion mass spectrometry
Si-NW	Silicon nanowires
siRNA	Small interfering RNA (or short interfering RNA or silencing RNA)
SKBR3	Human breast cancer cell line
SLC	Small Learning Community
SME	Small and medium-sized enterprise
SNOM	Scanning near-field optical microscopy
sNSOM	Scattering-type near-field scanning optical microscopy
SNUR	Significant New Use Rule
SOFC	Solid electrolyte fuel cell
SPM	Scanning probe microscopy
SQUID	Superconducting quantum interference device
SRAM	Static random access memory
SRC	Semiconductor Research Corporation
SRM	Solar radiation management
S&T	Science and technology
STC	Science and Technology Centers (NSF)
STED	Stimulated emission depletion

STEM	Scanning transmission electron microscopy or Science, Technology, Engineering, and Mathematics
STM	Scanning tunneling microscopy
STORM	Stochastic optical reconstruction microscopy
STT	Spin-transfer torque
STT-RAM	Spin torque transfer random access memory
STTR	Small Business Technology Transfer programs (various U.S. Government agencies)
SUNY	State University of New York
SVHC	Substance(s) of very high concern
SWCNT	Single-walled carbon nanotubes
SWNT	Single-walled nanotubes or single-walled carbon nanotubes
SWIR	Shortwave infrared
SWOT	Strengths, weaknesses, opportunities, and threats (analytical/strategic planning methodology)
TC229	Nanotechnologies Technical Committee of the ISO
TCE	Trichloroethylene
TEAM	Transmission electron aberration-corrected microscopy
TEM	Transmission electron microscopy
Tf	Transferring
T-FET	Tunneling field-effect transistor
TFFC	Thin-film flip-chip
TFSI	bis(trifluoro-methane-sulfonyl)imide
TFT	Thin film transistor
THz	Terahertz
TMP	Technology Maturation Program (BNL)
TMR	Tunneling magnetoresistive radio
TM&S	Theory, modeling, and simulation
TNF	Tumor necrosis factor
TSCA	Toxic Substances Control Act
Tx	Therapeutics
UA	United atom
UC CEIN	University of California Center for the Environmental Impact of Nanotechnology
UCSB	University of California, Santa Barbara
UCLA	University of California, Los Angeles
UF	Ultrafiltration
UFP	Ultrafine particles
UM	University of Melbourne
UNC	University of North Carolina
UNEP	United Nations Environment Program
UNESCO	United Nations Educational, Scientific and Cultural Organization
UNIDO	United Nations Industrial Development Organization
USDA	United States Department of Agriculture

UV	Ultraviolet
VC	Venture capital
VLSI	Very-large-scale integration
Voc	Open-circuit voltage
WH	White House
WPMN	Working Party on Manufactured Nanomaterials (OECD)
WPN	Working Party on Nanotechnology (OECD)
WTEC	World Technology Evaluation Center
WWTP	Wastewater treatment plant
XAFS	X-ray absorption fine structure
XANES	X-ray absorption near-edge spectroscopy
XAS	X-ray absorption spectroscopy
XFEL	European X-ray Free-Electron Laser
XPS	X-ray photoelectron spectroscopy
XR	X-ray reflectivity
XSW	X-ray standing wave
ZEB	Zero energy buildings
ZIF	Zeolitic imidazolate frameworks
zL	Ultrasmall-volume pores
ZMW	zero-mode waveguide
ZT	Thermoelectric figure of merit

Author Index

A
Adams, G., 539, 540
Akin, D., 55
Aksimentiev, A., 50–52
Alam, M.A., 29, 31–33, 55, 56
Alvarez, P.J.J., 159, 188, 191, 192
Analoui, M., 601
Appenzeller, J., 31, 379, 399–401
Arora, V., 561
Avouris, P., 31, 375, 379, 381–383, 399–403, 480, 483

B
Baca, A., 417
Badesha, S., 159, 202
Baird, B.A., 339
Batt, C., 349
Belk, J., 109, 467
Bell, L., 501, 535, 536
Berube, D., 508, 517, 595
Blackstock, J., 250
Bonnell, D.A., 71, 74–76, 78, 84, 85, 90, 91, 97, 99, 100
Bradley, J., 603
Brinker, C.J., 221, 251, 261, 271, 290
Brongersma, M.L., 417, 418, 422, 426
Buxton, D., 548

C
Carim, A., 542
Castranova, V., 159, 161, 163, 167, 169, 170, 180, 190–192, 196–198, 205, 306, 334
Celotta, R., 542
Chang, R.P.H., 501, 510, 534
Craighead, H., 71
Cummings, P.T., 29, 34, 36, 51, 56–58, 63

D
Davis, R.R., 445, 446, 450–452, 456, 458, 465
Davis, S.M., 445–447, 461, 462
De Simone, J., 332
Diallo, M., 221, 227, 228, 232, 234, 235, 237, 246, 247, 251
Diebold, A., 540, 541
Doumanidis, H., 143
Dravid, V.P., 71, 93, 96

E
Ebrahimi, F., 470–472
Eigler, D.M., 71, 76, 84, 86

F
Farrell, D., 305
Ferrari, M., 159, 177, 178
Fisher, R., 548
Fonash, S.J., 501, 537

G
Gambhir, S.S., 181, 310, 321, 322
Giam, L.R., 116–118
Ginger, D.S., 71, 76, 78, 261, 281
Glotzer, S.C., 35
Goddard, W.A., 32, 63, 228, 234, 235, 246
Godwin, H.A., 159, 177, 185, 197, 199
Grainger, D.W., 159, 167, 177, 202
Grodzinski, P., 159, 308, 346, 546
Guston, D.H., 23, 250, 561, 562, 564, 571, 581, 585, 590, 592, 595, 596

H
Han, J.Y., 228, 245
Harris, A., 463

Harthorn, B.H., 177, 185, 194, 561, 571, 581, 585, 593, 595
Hersam, M., xiii, 467, 480–483, 490, 493, 494, 501, 555, 557
Hoffmann, M.R., 246
Horne, N., 597
Hu, E.L., 6, 22, 250, 417, 420, 434, 437, 445, 467
Hwang, R., 109
Hybertsen, M.S., 31–33, 47–49

I
Isaacs, E.D., 71, 81, 82

J
Jackson, K., 71
Jones, M.R., 123, 124
Jones, P., 467, 473

K
Klimeck, G., 57–59, 540
Knox, D., 469

L
Lipson, M., 424, 432, 433
Lundstrom, M.S., 29, 31, 60, 63, 399, 400, 405, 440

M
Madsen, L.D., 137
McNeil, S.E., 178, 545
Meador, M.A., 484–486
Milner, J., 193, 228
Mirkin, C.A., xiii, 53, 56, 109, 110, 114, 116–124, 150, 152, 305, 310–317, 320, 321, 325–328, 361, 362, 366
Morris, J.W., 159, 471, 594, 599
Murday, J.S., 11, 253, 254, 292, 295, 483, 484, 501, 502, 505, 506, 509, 520, 522, 532, 550–552
Murdock, S., 578
Murray, C.B., 121, 475

N
Neaton, J.B., 31–33, 47–49, 380
Nel, A.E., 159, 161–165, 167–169, 174–177, 180, 185, 190–192, 195–199, 202, 206–208, 210, 305, 307, 310, 329–332, 334, 360–362

P
Parkin, S.S.P., 52, 384
Percec, V., 134, 136
Picraux, T., 403

R
Ralph, D.C., 31, 52, 379, 380, 384, 396–398
Randall, J.N., 131, 132
Ratner, M.A., 121
Regalla, L., 549
Ridgley, R., 19, 140
Roco, M.C., xiii, 1–6, 9, 10, 13, 16, 17, 20–22, 30, 72, 162, 223, 418, 445, 467, 468, 502, 506, 524, 561, 562, 564, 565, 567–569, 573, 575–579, 585, 590, 598, 605, 608, 610, 612
Rogers, J.A., 31–33, 486, 487
Rogge, S., 57–59, 540
Rung, S., 588
Ruud, J., 109, 467

S
Salahuddin, S., 52, 53
Savage, N., 159, 227, 237
Schatz, G.C., 53, 54, 114, 120, 122
Scher, E., 445
Schloss, J.A., 51, 546
Scott, N.R., 159, 193, 229
Selin, C., 585, 590, 596
Shannon, M.A., 188, 226, 227, 237, 238, 290
Shapira, P., 11, 561, 568–570, 575
Shibata, T., 150
Siegel, R.W., 6, 22, 467, 479, 489, 505
Stellaci, F., 247, 248
Stoddart, J.F., 310, 329–332
Stopa, M., 49

T
Teixeira, V., 147, 291
Thaxton, C.S., 56, 305, 311, 313, 315, 326–329, 361
Theis, T., 375
Tibbals, H.F., 352, 360
Timp, G., 50–52
Tiwari, S., 379, 544
Tomalia, D.A., 133–135, 232, 235
Tuominen, M.T., 109, 113, 114, 122, 130, 145, 147

V
Venkataraman, L., 31–33, 47–49

W
Wang, K.L., 115, 270, 404
Weiss, P.S., 71, 76, 77, 385, 467, 475
Welser, J.J., 375, 380, 405
Westerhoff, P., 161, 200–202
Wiesner, M.R., 159, 169, 170, 185
Williams, R.S., 3, 17, 30, 31, 55, 56, 72, 223, 418, 445, 467, 502, 561
Wolf, S.A., 61, 375, 384, 395, 396, 408, 409, 571

Y
Yaghi, O.M., 235, 237, 249, 250

Z
Zhang, X., 109
Zhongfan, L., 152, 495
Zhou, O., 318

Subject Index

A

Advances in nanotechnology, 31, 71, 100, 211, 228, 235, 279, 325, 341
Aerosols, 110, 169, 223, 225, 250, 576
Aerospace, xviii, 140, 398, 484–486, 493, 579, 650, 666
AFM. *See* Atomic force microscope
Assembly (directed, guided), 76, 110, 124–126, 176, 475, 543, 664
Atomic force microscope (AFM), 73–75, 99–101, 130, 419, 452, 550, 551
Australia, vi, xiii, 17, 63–65, 100–102, 134, 152, 210–212, 254–255, 295–299, 366–368, 409–411, 440–441, 494–495, 507, 557–558, 570, 610–612, 632, 634, 635, 637, 639, 641, 642, 644, 647, 648, 653, 656

B

Batteries, xxv, xxxiii, 41, 61, 62, 76, 145, 233, 263, 264, 267–269, 274, 277, 279, 283, 290–293, 297, 299, 378, 398, 408, 476, 485, 489, 604
Bioengineering, 210, 308, 366, 394, 557, 651, 657, 658
Biomarkers, 308, 310–312, 322, 337, 341, 345–348, 352, 354, 357, 358, 361, 364, 367, 602
Biomedicine, ix, 86, 178, 310, 315, 340, 343–345, 360–362, 438
Biomimetics, 11, 126, 127, 145, 148, 152, 273, 327–328, 494, 667
Biosensors, 31, 55–56, 188, 193, 239, 360, 440, 475, 483, 547

Biotechnology, xiv, xxxv, xxxvii, 126, 238–240, 344, 349–351, 394, 495, 502, 503, 571, 572, 649, 660
Bottom-up (nanotechnology), 10, 31, 38, 41, 189, 190, 292, 293, 327, 338, 351, 364, 388, 403, 404, 543, 562–563

C

Carbon nanotubes, xxx, xxxvi, xxxvii, 19, 32, 33, 55, 74, 75, 113, 131, 140–143, 150, 152, 163–165, 171, 181, 192, 193, 228, 231, 233, 234, 268–269, 273, 289, 291, 317–318, 343, 379, 380, 399–402, 480–483, 485–486, 604
Catalysis (at nanoscale), 445, 465
Catalyst (nanostructured), xxxvii, 7, 394, 446–449, 453, 455–458
Cell biology, 307, 337–340
Ceramic, 24, 83, 110, 137, 232, 357, 360, 468, 491, 576
Chemical industry, 8, 26, 191, 203, 523, 586
China, viii, xiii, 3, 17, 61, 63–65, 100–102, 152, 210–212, 233, 240, 254–255, 272–273, 294–299, 366–368, 409–411, 440–441, 451, 494–495, 557–558, 566, 569, 570, 575, 610–612, 632, 637, 643, 646, 656, 658
Chip (on a), 57, 148, 358, 365, 377, 423, 432–433
Climate, xiv, xxxii, 20, 23, 25, 193, 221–255, 261–263, 278–280, 295, 297, 368, 394, 553, 572, 580, 596, 610
CMOS. *See* Complementary metal oxide semiconductor
Coatings (nanostructured), 109, 148
Colloids, 199–201, 281, 576
Combinatorial approaches, 125, 476, 494

685

Commercial applications, xxvi, 80, 135, 170, 382, 451, 456, 489
Community (nanotechnology, professional, multidisciplinary), ix, xx, xxxviii, 2, 7–8, 13–14, 25, 365, 502, 543, 562, 587
Competitiveness, xxiv, 18, 21, 391, 492, 545, 563
Complementary metal oxide semiconductor (CMOS), xxxvi, 86, 131, 376, 380, 382, 390, 391, 401, 404, 406, 408, 410, 422, 423, 427, 483, 484, 505, 540, 541, 543
Composites (nanostructured), 179
Cyberlearning, 521–523

D
Definition (of nanotechnology), xiii, ix, xiv, 2–4, 12, 18, 211, 592
Dendrimers, 37, 133–136, 180, 188, 228, 246–247, 334, 343, 493, 655
Dendrimersomes, 136–137
Department of Commerce (including NIST), 13, 19, 21, 202, 388, 389, 391, 502, 504, 506, 521, 523, 526, 527, 533, 542–543, 545, 547, 586, 589, 659, 660, 666
Department of Defense (DOD), 13, 15, 19, 146, 429, 527, 565, 653, 660, 666
Department of Energy, 13, 19, 21, 35, 44, 89, 146, 229, 230, 267–269, 271, 272, 278, 388, 389, 391, 429, 504–506, 512, 521, 526, 527, 533, 542, 566, 636, 646, 654, 656, 660, 665
Design of nanomaterials, xxxi, xxxii, 9, 35–37, 39, 176–179, 194, 211
Dispersions, 40, 140, 178, 191, 192, 230–231, 250, 433, 468, 491
DNA, 50–55, 94, 111, 120–124, 133, 134, 176, 193, 229, 306, 312, 314–317, 320, 321, 323, 325–327, 337, 338, 349–352, 358, 360, 368, 394, 488, 667
DOD. *See* Department of Defense

E
EC. *See* European Commission
Economic impact, ix, xviii, xxxv, 343, 391, 451–452, 599
Education (training), xxiv, 43–45, 145, 183–185, 206, 253, 299, 502, 521, 527, 554, 555, 565
EHS. *See* Environmental, health and safety

Electrical energy storage, 138, 267–268, 274
Electron microscope (microscopy), 72, 77–80, 87–89, 93, 94, 127, 180–182, 200, 340, 383, 402, 419, 428, 448–449, 452, 455, 462, 487, 538, 552
ELSI. *See* Ethical, legal and other societal issues
Emerging areas, ix, xxvii, 7, 20, 23, 25, 229, 385, 467–495, 573
Energy conservation (efficiency), xx, 20, 25, 125, 189, 190, 224, 229, 232, 236, 239–240, 244, 261–299, 394, 451, 456, 458, 475, 478, 505, 525, 572
Energy conversion, xvi, xviii, xx, xxii, xxiv, xxxiii, 14, 20, 25, 125, 126, 135, 149, 150, 223, 229, 232, 236, 239, 240, 244, 261–299, 417, 484, 494, 572
Energy storage, xvi, xxiv, 137, 138, 152, 162, 263, 266–269, 274, 278, 283, 291, 293, 543
Environment (nanoscale processes in), 161
Environmental, health and safety (EHS), ix, x, xiv, xvi, xxiv, xxvi, xxvii, xxxi, xxxviii, 3, 10, 15, 20, 24, 25, 128, 147, 159–212, 492, 506, 522, 532, 546, 562–568, 571, 573, 578–580, 584–587, 595, 607, 608, 610, 612
Environmental technologies, 541
Ethical, legal and other societal issues (ELSI), ix, x, xvi, xx, xxiv, xxvi, xxxviii, 20, 24, 562, 563, 565–568, 573–574, 578, 580, 583–585, 587, 606–611
EU. *See* European Union
European Commission (EC), 171, 207, 405, 435, 608, 631, 640, 644, 645, 650, 654, 655
European Union (EU), xiii, 16, 17, 21, 60–61, 97–99, 147–150, 170, 171, 173, 202, 203, 207–208, 251–253, 290–292, 295, 360–362, 405–407, 434–437, 452, 490–492, 507, 552–555, 564, 566, 567, 569, 571, 573, 574, 605–608, 610, 631
Experimental methods, 36, 73
Exposure (environmental), 163

F
Films (nanostructured, nanocomposite), 229, 234, 265, 266, 273, 288, 296
Fuel cells, 86, 145, 290–293, 299, 394, 451, 463, 485, 493
Fullerene, 141, 189, 201, 281, 475

G

Gene (delivery), 136
Geoengineering, xviii, xxv, xxxii, 242, 250–251, 563, 609
Germany, viii, xiii, 3, 60–61, 97–99, 134, 135, 207–208, 251–253, 290–292, 360–362, 405–407, 434–437, 449, 490–492, 507, 552–555, 569, 570, 605–608, 631, 634, 639–641, 645, 646, 648, 649, 656, 657
Global partnerships, 100, 521–523, 526, 530
Governance of nanotechnology, xiv, xv, xxvii, xxxviii, 4, 7, 16–25, 130, 561–612
Government (R&D, funding), 18
Graphene, xvii, xviii, xxx, xxxv–xxxvii, 24, 31, 80, 95, 110, 113, 127, 130–132, 141, 148, 150–152, 233, 273, 297, 380–383, 385, 392, 402–403, 409, 493, 543

H

Health (care), xx, 8, 14, 25, 26, 178, 222, 244, 333, 340, 343, 360–362, 365, 376, 394, 424, 565, 582, 604
High-performance computing, 32, 34, 39–40, 42, 44, 394
High-performance materials, xiv, 20, 25, 229, 467–495, 573
Hydrogen storage, xvi, xxv, xxxii, xxxiii, 137, 145, 268–269, 291, 495

I

Imaging, xxix, xxxiv–xxxvi, 73, 74, 76–79, 81, 86–88, 92, 93, 95–97, 99–101, 135, 136, 165, 169, 177, 178, 180–182, 192, 200, 273, 306, 307, 317–322, 324, 329, 330, 332, 341, 345–348, 352, 356, 357, 361, 363, 364, 367, 368, 382, 402, 418, 419, 430, 431, 434, 435, 438, 483–484, 547, 548, 551–552, 602, 641
India, vi, viii, xiii, 63–65, 100–102, 152, 210–212, 254–255, 295–299, 366–368, 409–411, 440–441, 494–495, 557–558, 569, 570, 575, 610–612, 632, 635, 652
Industry (collaboration, partnership), 128, 255, 457, 523, 526
Information technology (IT), xiv, xxiv, 11, 22, 93, 366, 380, 387, 393, 423, 502, 505, 522, 541, 593, 604, 652, 660
Infrastructure (R&D), viii, ix, xx, 14, 15, 565
Innovation (and commercialization), xx, 172, 451, 562, 563, 568–571, 579, 597

Institutes, 12, 13, 19, 43, 45, 46, 89, 102, 128, 135, 146, 184, 208, 255, 346, 429, 504, 509, 524, 546–549, 553, 554, 557, 582, 609, 636, 638, 640, 646, 647, 652, 666
Instrumentation, ix, xv, xxix, xxxi, xxxviii, 14, 15, 19, 42, 44, 72, 77, 86, 89, 93, 94, 96–100, 102, 127, 169, 180–183, 185–186, 206, 212, 277, 295, 317–318, 394, 417, 438, 448, 455–457, 477, 486, 495, 504, 515, 523, 525, 526, 530, 550–552, 555
International (interactions, collaborations, partnerships), ix, x, xx, xxvii, 26, 46, 60, 209, 210, 555, 573–575, 586, 588, 609, 611
International perspectives, viii–xx, 21, 59–65, 97–102, 147–152, 207–212, 251–255, 290–299, 306, 360–368, 405–411, 434–441, 464–465, 490–495, 530, 552–558, 574, 587, 605–612
Investigative tools, 20, 25, 29–65, 71–102, 515, 567
In vitro diagnostics, 308, 310–317, 341, 344, 345, 361
In vivo imaging, 310, 317–322, 347, 361, 367, 547
IT. *See* Information technology

J

Japan, viii, xiii, xxxv, 3, 6, 16, 17, 61–63, 97, 99–100, 134, 150–152, 208–210, 253–254, 272, 292–295, 362–366, 408–409, 437–440, 464, 493–494, 555–557, 564, 566, 569–571, 573, 574, 605, 608–610, 631, 633–635, 637, 639–645, 647–649, 651–656, 658

K

Korea (South), viii, xiii, 3, 17, 61–63, 99–100, 150–152, 208–210, 253–254, 292–295, 362–366, 408, 437–440, 464, 493, 555–557, 566, 569, 570, 574, 608–610, 631, 637, 639, 645, 646, 648, 653, 654

L

Labeling, 120, 152, 171, 187, 192, 195, 208, 571, 610
Legislation, 171, 193, 204, 571–572, 576, 577, 640

Lighting, xxxiii, 229, 232, 264, 269, 272, 274–276, 279, 283–285, 296, 298, 417, 433, 435, 483

Lithography, xviii, xxx, xxxvi, 73, 83–84, 97, 110, 111, 114–118, 128, 130, 132, 133, 149, 151, 334, 378, 385, 389, 392, 406, 419, 428, 541, 543, 551, 552

M

Manufacturing (nanoscale), 3, 30, 73, 109, 162, 221, 271, 307, 388, 435, 446, 468, 506, 562

Memristors, 379

Metamaterials, ix, xvii, xxx, xxxvi, 24, 60, 110, 112, 113, 125, 152, 418, 421–422, 424–425, 441, 474–475, 478

Metrology, 14, 15, 20, 25, 30, 35, 39–40, 57–59, 71–103, 112, 125, 128, 133, 146, 273, 378–379, 385, 389, 441, 553, 567, 577, 580, 582

Molecular electronics, 32, 49, 111, 120–121, 131, 409

Molecular motor, xxix

Molecular recognition, 235, 488

Molecular simulation (dynamics), xxvii, 30, 34, 36, 37, 40, 46, 51, 54, 55, 57, 62, 65, 340

Monolayers, xxxvi, 77, 91, 133, 234, 290, 378, 381, 382, 385, 402

Moore's Law, xxxvi, 24, 57, 350, 358, 376–377, 505

N

Nanobiotechnology, 10, 25, 127, 129, 162, 229, 337–340, 346, 349–352, 363–364, 366, 601

Nanocrystal(s) (metals, ceramics), xxxvii, 24, 110, 121, 137–140, 144, 276, 283, 285, 287, 289, 330, 379, 455, 468, 470–475, 484, 576, 594, 602

Nanoelectronics, xviii, xxii, xxviii, xxix, xxxv, xxxvi, 7, 14, 15, 20–22, 25, 26, 31, 76, 86, 98, 114, 130–131, 280, 375–411, 426, 434, 484, 505, 509, 540, 541, 553, 565, 579, 586, 588, 589

Nanofibers, xvi, xxxii, 143, 144, 179, 298, 335, 336, 469–470

Nanofluidics, xv, xxxvii, 228, 229, 245–246, 253, 486–489

Nanoimprint, xviii, xxx, 83–84, 97–98, 110, 119, 130, 133, 147, 149–151, 378, 389, 428, 543, 552

Nanoinformatics (nanobioinformatics), xxvi, xxxix, 38, 130, 161–162, 176, 182, 183, 194, 212, 565, 584

Nanomagnetics, xxii, xxxv, xxxvi, 20, 25, 83, 130–131, 375–411

Nanomanufacturing, ix, x, xxvi, xxix, xxx, 10, 14, 15, 19, 29, 30, 38, 39, 42, 46, 72, 83–84, 88–89, 112, 125–130, 133, 143–147, 150, 152, 194, 294, 388, 475, 478, 543, 565, 584, 586

Nanomedicine, xviii, xxv, xxxiv, xxxv, 26, 127, 129, 135, 148–150, 177, 178, 208, 280, 305–310, 337, 343, 344, 362, 363, 366, 541, 548–549, 601

Nanoparticle(s), 2, 31, 73, 111, 165, 228, 265, 306, 425, 445, 467, 517, 564

Nanopatterning tools, 73, 84, 97–98, 124–126, 151, 542

Nanophotonics, xxxvi, 20, 25, 113, 417–441, 457, 464, 543

Nanostructure(s) on surfaces, xviii, xxv, 126, 488

Nanosystem (nanoscale system, nanobiosystem), ix, xviii, xx, xxii, xxvii, xxix, xxx, xxxiv, 6, 8–11, 20, 25, 26, 30, 37, 38, 41, 46, 47, 60, 62, 72, 83, 98, 100, 101, 125, 130, 149, 152, 163, 175, 177, 208, 267, 305–368, 376, 404–405, 468, 562–563, 573, 575–578, 588, 593, 601, 604

Nanotube, 19, 32, 74, 113, 162, 228, 268, 317, 379, 447, 468, 604

Nanowire devices, 76, 403–404

National Nanofabrication Infrastructure Network (NNIN), 388–389, 504, 507, 509, 521, 522, 526, 533, 544–545, 574, 586, 591, 593

National Nanotechnology Initiative (NNI), viii, ix, xviii, 1–26, 41, 42, 47, 73–74, 87, 162, 170, 222–223, 230, 391, 429, 452, 477, 501, 504–507, 515, 520, 522, 524, 528, 529, 532, 539, 564–568, 574, 576, 579, 581, 584, 586, 588, 589, 595, 603, 659–662

Near-field scanning optical microscopy (NSOM), 87, 90–91, 118, 419, 422

Network for Computational Nanotechnology (NCN, nanoHub), 44, 504, 539–540, 665

NNI. *See* National Nanotechnology Initiative

NSOM. *See* Near-field scanning optical microscopy

Nuclear reactors, 289

Numerical methods, 29, 38–39, 41–43, 45–47

Subject Index 689

O

Optical fiber, 423
Optics, xxxvi, 79, 93–97, 422, 424, 425, 428, 429, 435–436, 551–552

P

Photonics, xxxvi, 14, 43, 111, 112, 126, 130, 148, 151, 269, 285, 334, 384, 464, 480
Photosynthesis, xxviii, 254, 263, 266–267, 273, 291–293, 441
Photovoltaics (PVs), xxv, xxxiii, xxxvii, 64, 76, 86, 139–141, 240, 246, 263–266, 272, 273, 279–283, 291–293, 296, 403, 408, 434, 435, 441, 468, 474, 475, 479, 484–485
Physical infrastructure, xiv, xiv, xxiv, xxv, 25, 100, 195, 501–558, 573, 574, 579
Plasmonics, ix, xiv, xv, xxviii, xxxiii, xxxvi, 20, 24, 25, 73, 84, 86, 97, 112, 120, 125, 130, 152, 273, 403, 417–441, 457, 464
Platform (R&D, precompetitive, nanotechnology), xxvi, 577–578, 583, 610
PNCs. *See* Polymer nanocomposites
Polymer nanocomposites (PNCs), 231, 489–490
Public perceptions, 23, 186, 571–573, 577, 595
PVs. *See* Photovoltaics

Q

QDs. *See* Quantum dots
QMs. *See* Quantum mechanics
Quantum behavior, x, 50, 427
Quantum computing, xviii, xxviii, xxxvi, 49, 50, 58, 61, 133
Quantum confinement, xv, 281, 284
Quantum dots (QDs), xvii, xxviii, xxxiii, 39, 49–50, 65, 119, 148, 152, 162, 273, 284, 322, 394, 420, 438, 474–475, 483–485, 601–602
Quantum mechanics (QMs), 32, 36, 62–65, 420, 427, 429, 431

R

R&D investment and implementation strategies, 46, 89, 185–187, 243–244, 255, 277–278, 344–345, 366, 390–392, 429–430, 456, 477–478, 494, 526–530, 582–583

Regulatory policies (regulations), xxvii, 161, 163, 170–174, 365, 583
Responsible (governance, development), xv, xxvii, 4, 5, 7, 18, 20, 22, 24, 25, 130, 561–612
Ribonucleic acid (RNA), 133, 326, 331, 347, 352, 368

S

Safe design of nanomaterials, xxxii, 176–179, 194
Saudi Arabia, xiii, 63–65, 100–102, 254–255, 295–299, 366–368, 409–411, 440–441, 494–495, 557–558, 610–612, 632
Scanning probe microscopes (SPM), 74, 76, 84, 85, 88, 100, 101, 538, 550
Scanning tunneling microscope (microscopy, STM), 73–77, 84, 90, 99, 100, 127, 132, 133, 428, 550, 551
Self-assembling (self-assembled), x, xxi, xxviii, xxix, xxx, xxxvi, 30, 37, 38, 42, 61, 62, 77, 83, 100, 109–111, 113–115, 121, 124, 130, 134–136, 144, 147, 148, 151, 189, 234, 247, 264, 272–274, 281, 282, 290–293, 323–324, 334–336, 343, 366, 378, 380, 385, 388, 389, 392, 396, 406, 428, 446–447, 451, 455, 473–475, 491, 492, 512
Sensors (nanoscale), 284, 389, xvi, xxv, xxx, xxxi, 229, 241–242, 308, 321, 348, 352, 354–356, 359, 360, 365, 392
Simulation (numerical, computer, modeling and), x, xxviii, xxix, 23, 29–65, 129, 209, 367, 410–411, 492, 504, 515, 539, 544
Singapore, xiii, 17, 59, 63–65, 97, 100–102, 152, 210–212, 254–255, 295–299, 366–368, 409–411, 440–441, 464, 494–495, 557–558, 605, 610–612, 632
Small Business Innovative Research (SBIR), 506, 544, 589
Small Business Technology Transfer (STTR), 506, 589
Social impact, 18
Social (societal) impact, 10, 14, 47–48, 62, 112, 368, 456, 478, 506, 567–568, 575, 612
Societal development, xv, 4, 572–573, 577
Societal implications, xxxix, 14, 16, 20–23, 25, 162, 376, 517, 518, 524, 549, 562, 567, 568, 582, 590, 591, 593, 594, 596, 608, 609, 611, 659, 661

Spintronics, ix, 24, 131, 376, 379, 381, 383–384, 386, 396, 408, 410, 493, 543
SPM. *See* Scanning probe microscopes
Standard(s), 10, 49, 72, 111, 163, 226, 261, 307, 418, 458, 483, 504, 563
Stem cells, xxx, xxxv, 309, 333, 334, 338, 342, 344–346, 355, 359–361, 365, 517
STTR. *See* Small Business Technology Transfer
Superlattices, xvii, 121–124, 139, 270, 276
Supramolecular, 76, 111, 134, 135, 151, 232, 406, 493
Sustainability, xx, xxv, xxvi, xxxii, xxxiii, 10, 20, 25, 94, 162, 174, 185, 187–190, 193, 194, 221–255, 261–299, 368, 451–452, 456–458, 462, 474, 475, 478, 492, 562, 566, 572–573, 580, 584–585, 589
Sustainability (sustainable development), x, xiv, xvi, xxii, xxiv, xxvii, xxxii, 23, 26, 207, 222, 223, 243, 244, 253, 279, 564, 566, 582, 593, 609, 610
Synthesis (nanostructures, material), 109, 111, 114, 243, 284, 346, 429–430, 435, 447–448

T

Taiwan, viii, xiii, 3, 17, 61–63, 99–100, 150–152, 208–210, 253–254, 292–295, 362–366, 408–409, 437–440, 464, 493–494, 507, 555–557, 569–571, 608–610, 631
Templating (template), 110, 111, 113, 119, 123, 133, 138, 328, 332, 334, 351, 365, 385, 409, 447, 555
Theory, modeling and simulation (TM&S), xxvii, xxviii, xxix, 20, 23, 29–65, 139, 492, 504, 515, 539
Theranostics, 177–178, 307, 309, 310, 319, 320, 322, 330–331, 345, 346, 356, 364, 366, 604

Therapeutics (Tx), xvii, xviii, xxv, xxxvi, xxxiv, 7, 14, 21, 53, 73, 98, 126, 152, 162, 177–179, 305–310, 315, 322–334, 337, 338, 342, 344–348, 352, 356, 358, 361, 363, 364, 366–368, 476, 601–604
Thermal insulation, 271
Thermoelectrics, xxxvii, 41, 64, 76, 81, 263, 270–272, 276, 279, 285–288, 291, 292, 294, 297, 468, 475, 476, 479, 493
Tissue engineering, xxxv, 112, 127, 145, 148, 306, 307, 337, 342–343, 346
TM&S. *See* Theory, modeling and simulation
Toxicity, xxvi, xxxi, 163, 167, 168, 171, 176–178, 185–186, 191, 192, 196, 197, 199–200, 208, 211, 212, 253, 322, 325, 326, 346, 362, 563, 584, 594, 602, 609
Tx. *See* Therapeutics

V

Vision (looking into the future), 24–26, 30–31, 310, 433, 446

W

Waste (remediation), xvi, xxxii, 96, 129, 135, 166, 188–190, 195, 234, 235, 242, 252, 417, 431–432, 506
Water (filtration, purification), ix, xvi, xviii, xxv, xxxii, 23, 37, 129, 139–140, 150, 188, 239, 246, 255, 263, 495, 609, 610
Wiring, xxv, xxix, xxxvii, 48, 111, 119, 121, 129, 140–143, 230, 247, 253–254, 283–284, 379, 395, 485–486, 547

Z

Zeolites (zeolitic materials), xvi, xxxii, 138, 188, 227–228, 242, 250, 447–448, 451, 461, 462